Lecture Notes in Networks and Systems

Volume 524

The series "Lecture Notes in Networks and Systems" publishes the latest developments in Networks and Systems—quickly, informally and with high quality. Original research reported in proceedings and post-proceedings represents the core of LNNS.

Volumes published in LNNS embrace all aspects and subfields of, as well as new challenges in, Networks and Systems.

The series contains proceedings and edited volumes in systems and networks, spanning the areas of Cyber-Physical Systems, Autonomous Systems, Sensor Networks, Control Systems, Energy Systems, Automotive Systems, Biological Systems, Vehicular Networking and Connected Vehicles, Aerospace Systems, Automation, Manufacturing, Smart Grids, Nonlinear Systems, Power Systems, Robotics, Social Systems, Economic Systems and other. Of particular value to both the contributors and the readership are the short publication timeframe and the world-wide distribution and exposure which enable both a wide and rapid dissemination of research output.

The series covers the theory, applications, and perspectives on the state of the art and future developments relevant to systems and networks, decision making, control, complex processes and related areas, as embedded in the fields of interdisciplinary and applied sciences, engineering, computer science, physics, economics, social, and life sciences, as well as the paradigms and methodologies behind them.

Indexed by SCOPUS, INSPEC, WTI Frankfurt eG, zbMATH, SCImago.

All books published in the series are submitted for consideration in Web of Science.

For proposals from Asia please contact Aninda Bose (aninda.bose@springer.com).

More information about this series at https://link.springer.com/bookseries/15179

Michael E. Auer · Samir A. El-Seoud ·
Omar H. Karam
Editors

Artificial Intelligence and Online Engineering

Proceedings of the 19th International Conference on Remote Engineering and Virtual Instrumentation

Springer

Editors
Michael E. Auer
CTI Global
Frankfurt, Germany

Samir A. El-Seoud
British University in Egypt
El Shorouk, Egypt

Omar H. Karam
British University in Egypt
El Shorouk, Egypt

ISSN 2367-3370 ISSN 2367-3389 (electronic)
Lecture Notes in Networks and Systems
ISBN 978-3-031-17090-4 ISBN 978-3-031-17091-1 (eBook)
https://doi.org/10.1007/978-3-031-17091-1

This Springer imprint is published by the registered company Springer Nature Switzerland AG
The registered company address is: Gewerbestrasse 11, 6330 Cham, Switzerland

Preface

It is a great privilege for us to present the proceedings of the 19th "International Conference on Remote Engineering and Virtual Instrumentation: Artificial Intelligence and Online Engineering" (REV2022) to the authors and delegates of this event and to the wider, interested audience. Initiated in 2004, REV has been held in Villach (Austria), Brasov (Romania), Maribor (Slovenia), Porto (Portugal), Duesseldorf (Germany), Bridgeport (USA), Stockholm (Sweden), Brasov (Romania), Bilbao (Spain), Sydney (Australia), Porto (Portugal), Bangkok (Thailand), Madrid (Spain), New York (USA), Duesseldorf (Germany), Bengaluru (India), Georgia (USA), and Hong Kong.

This year, REV2022 has been organized and hosted at The British University in Egypt (BUE), Cairo, Egypt (https://www.bue.edu.eg/) from 28 February to 02 March 2022. REV2022 has been held as a hybrid conference and attracted 121 scientists and industrial leaders (with 34 onsite and 87 online participants) from more than 50 countries.

The REV conferences are the annual conferences of the International Association of Online Engineering (IAOE, www.online-engineering.org) and the Global Online Laboratory Consortium (GOLC, http://online-engineering.org/GOLC_about.php).

REV is a series of annual events concerning the area of remote engineering and virtual instrumentation.

The REV2022 conference takes up the following topics in its variety and discusses the state of the art and future trends under the global theme "Artificial Intelligence and Online Engineering":

- Artificial Intelligence and Online Engineering
- Cross Reality
- Big Data
- Data Science and Online Engineering
- Biomedical Engineering
- Collaborative Work in Virtual Environments
- Online Labs in Industry and Engineering Education

- Human–Machine Interaction & Usability
- Internet of Things & Industrial IoT
- Industry 4.0
- Cyber Physical System
- Cyber Security
- M2M Concepts
- Networking, Edge & Cloud Technology
- Remote & Crowd Sensing
- Smart Objects
- Smart Environments (Smart Home, Smart Building, Smart City)
- Standards & Standardization Proposals for Online Labs
- Remote Technologies and Health Care
- Accessibility for People with Disabilities

A focus during REV2022 was put on digitalization and global changes in the society, special trends in engineering education and biomedical engineering. Also, the conference presentations focused on the impact of artificial intelligence and data-driven algorithms on engineering solutions in automation, mobility, and smart cities.

The conference was opened by the CEO and founding president of IAOE, Michael E. Auer, who underlined the importance to have this conference with long lasting traditions for first time on the African continent. In his greeting, the First Vice-Chancellor of BUE, Prof. Yehia Bahei-El-Din, stated, that BUE has been investing for several years now in modern instructional infrastructure such as several online labs just as a remotely accessible wind channel for experiments in aerodynamics.

REV2022 offered an exciting technical program as well as networking opportunities concerning the fundamentals, applications, and experiences in the field of Online Engineering and related new technologies, including

- Internet of Things & Industrial Internet of Things (Industry 4.0),
- Online & Biomedical Engineering,
- Data Science, Machine learning, and Artificial Intelligence,
- Cross & Mixed Reality, and
- Remote Working Environments.

In total, four special sessions have been organized at REV2022, namely

1. The Impact of Digitization in Engineering Training
2. Flavour and Mood - Multimodal Data Sensing and Analysis
3. Advances in the Applications of Informatics and Computer Science
4. Smart Education for Sustainable Digital Era

Five outstanding scientists and industry leaders accepted the invitation for keynote speeches:

1. **Christian Gütl**, Graz University of Technology, Austria
2. **Naoko Fukami,** Japan Society for the Promotion of Science Promotion Society (JSPS) Cairo, Egypt
3. **Hesham H. Ali**, University of Nebraska at Omaha, NE, USA
4. **Tarek M. Sobh**, Lawrence Technological University, Southfield, MI, USA
5. **Frederic Andres**, National Institute of Informatics, Tokyo, Japan

The conference was organized by the Faculty of Informatics and Computer Science (ICS) of BUE with support from the Faculty of Engineering of BUE. Prof. Samir Abou El-Seoud and Prof. Omar Karam served as local chair and local co-chair.

Submissions of Full Papers, Short Papers, Work in Progress, Poster, Special Sessions, Workshops, and Tutorials have been accepted.

All contributions were subject to a double-blind review. The review process was extremely competitive. We had to review about to 170 submissions. A team of over 90 program committee members and reviewers did this terrific job. Our special thanks goes to all of them.

Due to the time and conference schedule restrictions, we could finally accept only the best 68 submissions for presentation or demonstration.

The conference was supported by Phoenix Contact as Gold Sponsor and Japan Society for Promotion of Science (JSPS).

As always Sebastian Schreiter did an excellent job to edit this book.

The 20th REV conference will be held at Aristotle University of Thessaloniki in Greece in March 2023.

Michael E. Auer
Samir A. El-Seoud
Dominik May
Omar Karam

Committees

General Chair

Michael E. Auer	Founding President and CEO of the IAOE, CTI Frankfurt/Main New York, Vienna, Bangalore

REV2022 Co-chairs

Dominik May	IAOE President, USA
Omar Karam	BUE, Egypt
Samir A. El-Seoud	BUE, Egypt

Program Co-chairs

Doru Ursutiu	IAOE Honorary President, Romania
Andreas Pester	The British University in Egypt (BUE), Egypt
Hani Ghali	BUE, Egypt

International Advisory Board

Abul Azad	President Global Online Laboratory Consortium, USA
Alberto Cardoso	University Coimbra, Portugal
Amr El Abbadi	University of California Santa Barbara, USA
Ashraf Saad	King Salman International University, Egypt
Bert Hesselink	Stanford University, USA
Claudius Terkowsky	TU Dortmund University, Germany
Cornel Samoila	University of Brasov, Romania Denis Gillet, EPFL Lausanne, Switzerland
Doru Ursutiu	University of Brasov, Romania
Franz Schauer	Tomas Bata University, Czech Republic

Hamadou Saliah-Hassane	Université TÉLUQ, Montréal, Canada
Krishna Vedula	IUCEE, India
Reem Bahgat	Cairo University, Cairo, Egypt
Teresa Restivo	University of Porto, Portugal
Tarek Sobh	University of Bridgeport, USA
Vasant Honavar	Penn. State University, USA
Veteran Sunil	Salford University, UK
Yehia Bahei-El-Din (Chair)	The British University in Egypt
Zorica Nedic	University of South Australia

Technical Program Chair

Sebastian Schreiter	IAOE, France

IEEE Liaison

Manuel Castro	UNED Madrid, Spain

Workshop and Tutorial Chairs

Sherif Aly	American University in Cairo (AUC), Cairo, Egypt
Valerie Varney	RWTH Aachen, Germany

Special Session Chairs

Alexander Kist	University of Southern Queensland, Australia
Ihab Adly	The British University in Egypt (BUE), Egypt

Publication Chair and Web Master

Sebastian Schreiter	IAOE, France

International Program Committee

Abeer Hamdy	The British University in Egypt (BUE), Egypt
Akram Abu-Aisheh	Hartford University, USA
Ann Nosseir	The British University in Egypt (BUE), Egypt
Andreas Pester	The British University in Egypt (BUE), Egypt
Ananda Maiti	University of Southern Queensland, Australia
Ahmad Mostafa	The British University in Egypt (BUE), Egypt
Alexander Kist	University of Southern Queensland, Australia

Contents

Data Science and Online Engineering

Collaborative Work in Virtual Environments

The Impact of Digitization in Engineering Education and Training

A Novel Dynamometer: The Gripwise

Maria Teresa Restivo[1](✉), Diana Urbano[2], Tiago Andrade[3], Paulo Abreu[2], and Fátima Chouzal[2]

[1] LAETA, University of Porto, A3ES, Lisbon, Portugal
trestivo@fe.up.pt
[2] LAETA, Faculty of Engineering, University of Porto, Porto, Portugal
{urbano,pabreu,fchouzal}@fe.up.pt
[3] GripWisetech.com, Porto, Portugal
tiago.andrade@wisifytech.com

Abstract. Hand grip strength is the most used muscle function indicator for clinical purposes. The BodyGrip system, trademark Gripwise, is a novel patented cloud-based embedded system and a promising tool for assessment/prevention, rehabilitation, screening and occupational therapy, opening interesting new possibilities.

This work pretends to shortly describe a novel embedded system for hand grip evaluation, offering many new capabilities, good characteristics, the potentiality of contributing for sharing data storage in the cloud, patented, under technology transfer and validated by the new findings accepted in the community.

The Gripwise digital dynamometer system is a Class I Medical Device with a resolution of 1N capable to assess the strength profile with an acquisition rate of 100 samples/second. Several pilot studies were carried on in order to validate the device. Currently, other are on track. Some insights on the novel system together with a set of studies' syntheses may show its potential.

The Gripwise system has already different accessories making it a multifunction solution for different muscle groups assessment.

Keywords: Smart devices · Hand grip descriptors · IoT system

1 Introduction

Many studies have shown that the maximum value of the hand grip strength (HGS) has been found as a relevant indicator for the individuals general muscle condition and, therefore, of their general health status, thus being the most used muscle function indicator for clinical purposes.

In fact, studies have found that HGS is associated with indicators of morbidity, short- and long-term mortality, dependence, disability, among others [1]. On the other hand, low HGS values are associated with frailty indicators [2], related to age, sex, length of hospital stays, low vitamin D level and increased risk of mortality, among others.

Since the decrease in muscle strength appears before the muscle structure and composition changes are detected [3], isometric HGS has been widely used to track, diagnose and monitor health aspects related to frailty and functional and nutritional status. However, few studies are focus on the association of these health aspects with muscular endurance, that is, the ability of the muscle to maintain a high level of strength

M. E. Auer et al. (Eds.): REV 2022, LNNS 524, pp. 3–11, 2023.
https://doi.org/10.1007/978-3-031-17091-1_1

(in this case, its maximum strength) when asked to repeatedly develop it. Endurance, which depends on the ability of mitochondrial energy production, is vital for many every day repetition tasks or sustained support [4]. But, usually, the devices only provide the maximum of the hand grip strength value (HGS).

2 The Novelty of the System

The prototype BodyGrip system, consisting of a novel dynamometer and a dedicated software application, allows the assessment of isometric evaluation of the hand grip strength and its evolution in defined time intervals (HGSt). The dynamometer was conceived to make possible to measure applied compression and traction forces and to estimate the energy expended simultaneously. Thus, in addition to measuring the grip strength of the hand, it also allows measuring strength of different muscle groups, using specific accessories for each group. The measurements are record in a local or remote database and also allows to digitally monitor the isometric profile force during the test period.

The system presents multifunctionality, measuring the force strength of different muscle groups (by the use of different accessories) and allowing to follow its evolution over selectable time intervals. It estimates the energy expended and the average and instantaneous associated power. This is possible with the novel design of its load cell that considerably decreases the sensitivity to hand positioning and handling torsion effects, Fig. 1 (a). The BodyGrip, Fig. 1(b), offers automatic scaling and operating ranges adaptable to the user, and offers wirelessly communication with personal computer applications or mobile devices and/or human-machine interface.

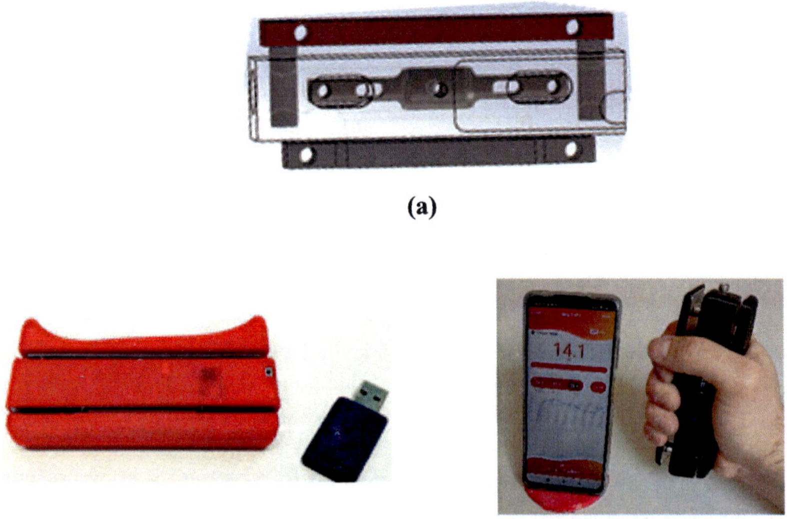

(b) (c)

Fig. 1. **a)** Novel Load Cell: low sensitivity to hand positioning and handling torsion effects. **b)** BodyGrip system. **c)** Gripwise system

The BodyGrip allows to record user data, to define the test time interval and to monitor the respective force time profile in the graphical application window, in real time (sampling of 10 ms). Any additional post-processing algorithms can be programmed and incorporated into the system. Lastly, Fig. 1 (c) presents the commercial solution Gripwise.

The prototype offers a light solution (2.30 N), compact dimensions (0.114 m × 0.022 m × 0.045 m), easy portability, with a range of ±1000 N, with good sensitivity, resolution and precision.

In its basic function, in measuring HGS, it has already been validated against the reference standard, the Jamar dynamometer [5].

The US patent was granted in 2020, and the European has been recently granted in 2021 [6].

In 2020, the technology was licensed to Gripwise Tech, a University of Porto spin-off and in 2021, the first production phase had been focused on the frailty phenotype of the elderly [7, 8].

The commercial version of BodyGrip, Gripwise, integrates with iOS and Android applications, a WebAPP and a cloud platform where distinct types of data are split into two different European servers. Gripwise solution is GDPR (General Data Protection Regulation) compliant and is prepared to integrate medical software through the HL7/FHIR standard. FHIR - Fast Healthcare Interoperability Resources for Healthcare Interoperability, and HL7for the exchange, integration, sharing, and retrieval of electronic health information.

The manufactured Gripwise systems are provided to the users after a controlled process against a working standard, called the Griptester, which certifies the system quality.

The Griptester comprises a high-quality customized load cell and its signal conditioning that connects to the internet, and mobile device with a dedicated APP that communicates together with the internet and with the Gripwise via Bluetooth.

During the Gripwise system verification process, the mobile application collects Griptester load data via network, Gripwise output data via Bluetooth and the system serial number by scanning the identifier QR Code.

The application offers three calibration test modes: fully automatic, semi-automatic and manual, providing a final result of successful or failed test. In the first case, the results are within the defined error limits and the Gripwise system calibration data is immediately remotely registered on a server. The costumer, using his/her system serial number, can consult the online calibration recorded values on the dealer's website: https://gripwisetech.com/onboarding/.

3 Studies

3.1 Former Results Based on BodyGrip

Two studies conducted with the BodyGrip prototype had already shown the potentiality that can be achieved with such a system. The first study [7], focused on identifying force *vs.* time parameters which could contribute to better describe individual strength. Five parameters related to the capability to develop and to maintain strength in percentage of maximum hand grip strength, were extracted from the force *vs* time profile. They were obtained during a hand grip strength test of 15 s in a sample group of 94 university students. Contrary to maximum HGS, such parameters were found to be independent of the participant's anthropometric characteristics. The findings point to the possibility to characterize the force development during a test into three stages.

Next, we will summarize the proposed descriptors as well as the most relevant findings within reference [7] since the new insights revealed in the present work are based on those studies.

A typical curve describing hand grip strength during the time interval of 10 s duration test, is depicted in Fig. 2, where the descriptors proposed in [7] are also indicated.

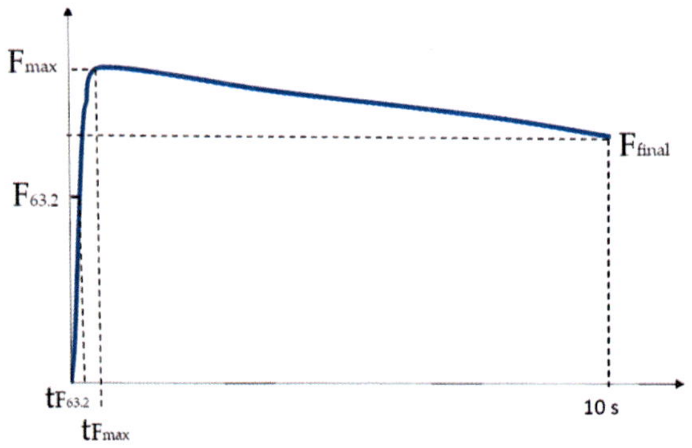

Fig. 2. Hand grip strength profile for the time interval of 10 s.

The profile shown in Fig. 2 is very similar to the first-order system response to a step function. Due to the meaningful quantities found for that type of physical behavior, the value F_63.2, corresponding to (1–1/e) F_{max}, was chosen as one of the parameters to be analyzed. The HGSt curve can therefore be divided in three stages: from $t = 0$ s to tF_63.2, from tF_63.2 to tF_{max} and from tF_{max} until the end of the test, in this case 10 s.

The two first stages of the HGSt profile are related to the ability to develop strength and the third stage is an indication of the ability to sustain maximum force.

With the five parameters shown in Fig. 2, the following variables, normalized to F_{max} were defined:

- pF_{av}—average force during 10 s test expressed in percentage of F_{max};
- $pRate_1 = (F_{63.2}/tF_{63.2})/F_{max}$—rate of change of force up to $tF_{63.2}$ expressed in percentage of F_{max};
- $pRate_2 = (F_{max} - F_{63.2})/(F_{max} * (tF_{max} - tF_{63.2}))$—the rate of change of force from $tF_{63.2}$ to tF_{max}, expressed in percentage of F_{max};
- $pRate_3 = (F_{final} - F_{max})/(F_{max} * (10 - tF_{max}))$—the rate of change of force from the time of maximum force until the end of the test, in percentage of F_{max}.

Therefore, the outcomes of reference [7] can be summarized as:

- There are no gender differences in the time intervals up to $tF_{63.2}$ and to tF_{max};
- The normalized quantities are independent of the anthropometric data. Therefore, the anthropometric are not confounding variables;
- There is a strong relationship between normalized average HGS, pFav and the normalized rate, pRate3, confirming that pFav is a measure of the ability to sustain maximum hand grip strength [9];
- There are no gender differences in the normalized descriptors. This conclusion agrees to one of the outcomes of reference [7] where it was found that the variable pFav, is similar for young adult females and males;
- Considering two different categories of pFav, it is possible to distinguish between subjects that show different rates of force development in the first and second stages of the HGSt profile.

In the second study [8], the HGSt data was obtained among an elderly group that performed a 10 s duration hand grip strength test. Along with this measurement, a Tilburg Frailty Indicator (TFI) test was conducted. This goal was to investigate the relationship between the HGSt parameters, and the frailty condition indicated by the TFI measure. An artificial neural network model was constructed using as inputs the HGSt parameters, F_{max}, tF_{max}, $F_{63.2}$, $tF_{63.2}$ and F_{final} and as output the outcome of the TFI test. The approach looked to the idea of identifying frailty-related conditions from the HGSt defined parameters, using artificial neural networks. Results obtained were able to identify frailty matching of TFI scores reaching a minimum accuracy of 75%. Despite the small sample size used in this study, this system appears to be a promising tool for exploring new frailty-related features based on HGSt tests.

The procedure for testing HGSt with BodyGrip is the same in both studies and is described in any of the works.

The fact that the HGSt test can be performed in just ten seconds and the acquired data processed immediately, providing quick results, opens the possibility to configure the new commercialized Gripwise with the capability to be used to identify and alert for frailty risk, apart from its use as an evaluation instrument of HGSt.

3.2 New Insights

The present work aims at giving new insights concerning the behavior of HGSt as a function of time, extending the conclusions obtained in reference [7].

The main question to be tackled is "Are there any differences in the HGSt profile descriptors across the two very distinct age groups?".

Table 1 shows the distribution of cases per gender and age group. The "Young" sample is from reference [7] and the elderly correspond to the sample of reference [8].

The convenience sample consists of 187 subjects, with a female percentage of 76.4%, of which 48,9% are elderly, and a male percentage of 23.6%, of which 52.3% are elderly. The sample is unbalanced towards women in both age groups.

Table 1. Sample characterization

	Young		Elderly	
	Female N = 73	Male N = 21	Female N = 70	Male N = 23
Median Age (Min – Max)	20 (19–37)	22 (19–38)	76 (66–91)	77 (67 – 90)

Tables 2 and 3 show the averages of the HGSt descriptors defined in the previous section, respectively for the young and elderly groups. It also shows the significance of the non-parametric Mann Whitney-U test to assess gender differences in the parameters within each age group.

Table 2. Elderly Group: Means, standard deviations and gender descriptors comparison.

Descriptor (Unit)	Gender		Significance
Mean ± SD	Female N = 70 (75.3%)	Male N = 23 (24.7%)	p
F_{max} [N]	155 ± 39	229 ± 66	<0.001
tF_{max} [s]	2.41 ± 1.01	2.34 ± 1.10	0.765
$F_{63.2}$ [N]	97 ± 24	142 ± 39	<0.001
$tF_{63.2}$ [s]	0.60 ± 0.29	0.61 ± 0.32	0.452
F_{final} [N]	117 ± 40	180 ± 64	<0.001
F_{av} [N]	128 ± 38	190 ± 57	<0.001
pF_{av}	82 ± 7.6	83 ± 7.7	0.582
$pRate_1$ [1/s]	131 ± 65	134 ± 76	0.789
$pRate_2$ [1/s]	28 ± 23	32 ± 26	0.236
$pRate_3$ [1/s]	−3.22 ± 1.5	−2.7 ± 1.1	0.976

Table 3. Young Group: Means, standard deviations and gender descriptors comparison.

Descriptor (Unit)	Gender		Significance
Mean ± SD	Female N = 73 (76.7%)	Male N = 21 (22.3%)	P
F_{max} [N]	201 ± 61	368 ± 120	<0.001
tF_{max} [s]	1.89 ± 0.84	1.96 ± 1.08	0.765
$F_{63.2}$ [N]	125 ± 38	227 ± 75	<0.001

(*continued*)

Table 3. *(continued)*

Descriptor (Unit)	Gender		Significance
Mean ± SD	Female N = 73 (76.7%)	Male N = 21 (22.3%)	P
$tF_{63.2}$ [s]	0.67 ± 0.40	0.75 ± 0.42	0.452
F_{final} [N]	149 ± 54	273 ± 107	<0.001
F_{av} [N]	165 ± 52	293 ± 102	<0.001
pF_{av}	82 ± 4.3	81 ± 4.9	0.582
$pRate_1$ [1/s]	115 ± 59	111 ± 63	0.789
$pRate_2$ [1/s]	42 ± 25	51 ± 42	0.236
$pRate_3$ [1/s]	−3.18 ± 1.12	−3.18 ± 1.09	0.976

As one can see, the conclusions of reference [7] can be extended to the elderly group. The normalized quantities are, for this age group, also independent of gender.

Therefore, we can compare the averaged quantities between young and elderly, without distinguishing males and females.

Table 4 shows the averages standard deviations for the parameters, besides the force variables, that were found to be statistically significantly different across age groups.

The significance of the Mann-Whitney-U test to compare the means as well as the respective effective sizes [10] are also displayed. The comparisons show that both groups have the same normalized average HGS, pFav revealing similar ability to sustain maximum force. Differences in the time to achieve maximum force are statistically significant but with a small effect size. The normalized rates of development from the beginning of the test until 63.2% of maximum HGS, $F_{63.2}$, are similar in both groups, but the rate from that $F_{63.2}$ to F_{max} is significantly bigger for the elderly, with a medium effect size.

Table 4. Averages, standard deviations, comparison test results and respective effect sizes.

Descriptor (Unit)	Age Group		Significance	Effect size
Mean ± SD	Elderly N = 93	Young N = 94	P	R
tF_{max}	2.39 ± 1.03	1.89 ± 0.89	<0.001	0.28
$pRate_2$ [1/s]	29 ± 23	45 ± 29	<0.001	0.41

Figure 3 depicts a sketch of the different stages of HGSt for the two age groups.

These new insights provide interesting information concerning the descriptors of the HGS time profile.

It takes elderly, in average, more time to achieve maximum force, although the time it takes to reach 63.2% of maximum force is the same time for young and elderly. Also,

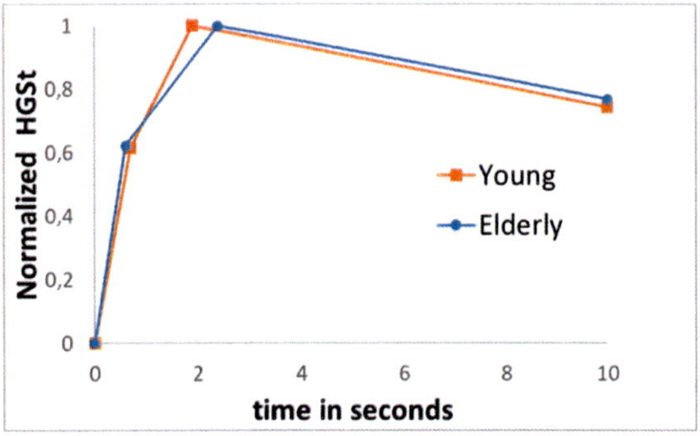

Fig. 3. Stages of HGSt for young and elderly groups

during a 10 s test, the average force expressed as function of maximum HGS is the same for elderly and young. The differences in the way HGSt develops with time will be explored in future works described in the next section.

4 Current and Future Studies Based on Gripwise

Due to the pandemic situation of the past two years, it has been very difficult to collect data to proceed with the investigation concerning HGSt. Very recently, (with a rate of vaccination of almost 90% in the country), there is the opportunity to integrate a study in collaboration with a research group in physical activity, health, and leisure area, where several physical condition tests are being conducted, together with 10 s hand grip strength tests already using the GripWise system. This will enable analysing the relationship between the HGSt parameters and individual performance in physical functionality tests. The data will be collected before and after a program of six months where community-dwelling elderly will be performing multicomponent exercise for physical fitness [11]. The relationships between the HGSt parameters and the results of the tests assessing physical functionality pre and post program will be investigated.

Another perspective of future study is focused in using the GripWise system to explores rehabilitation of the hand with musculoskeletal pathologies, to investigate whether and how the parameters HGSt vary over the rehabilitation time [12]. This study will involve physical therapists and psychiatrists.

5 Food for Thought

The analyses performed along the different studies already conducted based on HGSt show that there are many possibilities that can be developed bringing relevant contributions to broadening the knowledge on the hand grip strength as an important biomarker. New possibilities will be investigated aiming at establishing HGSt testing as a quick, easy and non-invasive tool to screen health related issues and physical functional ability decline, thus permitting timely interventions. Additionally, such a tool can be of great help in monitoring the hand rehabilitation process with quantitative measures, allowing, among other things, better prediction of rehabilitation time.

The characteristics of the GripWise system are unique to achieve the project goals. Moreover, embedded with AI, such a low-cost new generation device could be a powerful prediction tool of physical and health related conditions, as well as of evolution of upper limb traumatic rehabilitation.

Acknowledgments. Researchers are grateful to the support of the Laboratory for Energy, Transports and Aeronautics (LAETA), under funding of project REF: UIDB/50022/2020 from Portuguese Foundation for Science and Technology.

References

1. Rantanen T, Volpato S, Ferrucci L, Heikkinen E, Fried LP, Guralnik JM (2003) Handgrip strength and cause-specific and total mortality in older disabled women: exploring the mechanism. J. Am. Geriatr. Soci. 51(5):636–641. https://doi.org/10.1034/j.1600-0579.2003.00207.x
2. Reeve TE et al (2018) Grip strength measurement for frailty assessment in patients with vascular disease and associations with comorbidity, cardiac risk, and sarcopenia. J. Vasc. Surg. 67:1512–1520
3. Lopes J, Russell DM, Whitwell J, Jeejeebhoy KN (1982) Skeletal muscle function in malnutrition. Am. J. Clin. Nutr. 36:602–610
4. Aagaard P, Simonsen EB, Andersen JL, Magnusson P, Dyhre-Poulsen P (2002) Increased rate of force development and neural drive of human skeletal muscle following resistance training. J. Appl. Physiol. 93(4):1318–1326
5. Guerra RS, Amaral TF, Sousa AS, Fonseca I, Pichel F, Restivo MT (2017) Comparison of Jamar and Bodygrip dynamometers for handgrip strength measurement. J. Strength Cond. Res. 31:1931–1940
6. Restivo, M.T., Quintas, M.R., Silva, C.M., Andrade, T.F., Santos, B.F.: Device for measuring muscle strength and energy. US20180249940, (2020-12-08), EP16791083.5, (2021-12-01), WO2017046715 (2017-03-23)
7. Urbano D, Restivo MT, Amaral TF, Abreu P, Chousal M (2020) An attempt to identify meaningful descriptors of handgrip strength using a novel prototype: preliminary study. Information 11(12):546. https://doi.org/10.3390/info11120546
8. Urbano D et al (2021) Handgrip strength time profile and frailty: an exploratory study. Appl. Sci. 11(11):5134. https://doi.org/10.3390/app11115134
9. Leyk D et al (2007) Hand-grip strength of young men, women, and highly trained female athletes. Eur. J. Appl. Physiol. 99:415–421
10. Fritz CO, Morris PE, Richler JJ (2011) Effect sizes estimates: current use, calculations, and interpretation. J. Exp. Psychol. Gen. 141(1):2–18
11. Kang S, Hwang S, Klein AB, Kim SH (2015) Multicomponent exercise for physical fitness of community-dwelling elderly women. J. Phys. Ther. Sci. 27(3):911–915
12. Shiratori AP, Iop Rda R, Borges Júnior NG, Domenech SC, Gevaerd Mda S (2014) Evaluation protocols of hand grip strength in individuals with rheumatoid arthritis: a systematic review. Rev. Bras. Reumatol. 54:140–147

Software Configurable Hardware-Based Remote Laboratory System

Tinashe Chamunorwa[1](✉), Doru Ursuţiu[1,2], Cornel Samoilă[1,3], Horia Hedesiu[4], and Horia Alexandru Modran[1]

[1] Transilvania University of Brasov, Brasov, Romania
{chamunorwa.tinashe,udoru,csam,horia.modran}@unitbv.ro
[2] Academy of Romanian Scientists, Bucharest, Romania
[3] Romanian Academy of Technical Sciences, Bucharest, Romania
[4] Technical University Cluj-Napoca, Cluj-Napoca, Romania
horia.hedesiu@emd.utcluj.ro

Abstract. Working and learning remotely has been a rising and steadily increasing phenomenon. Of particular interest is the conducting of laboratory experiments over remote facilities. The ever-evolving electronic systems application requirements and the dynamic operating environments have necessitated the need for configurable systems. Internet of Things (IoT) is now driving the industrial revolution 4.0 and hence learning remotely is inevitable. In the same vein, configurable remote laboratories are an important aspect of the electronic engineering revolution. Configurability enables laboratory platforms to run with multiple sets of hardware and to design systems capable of handling future developments and plugins, bringing more support to students performing laboratory work during the Covid-19 era. This is a study to develop a relatively inexpensive, configurable, and cloud-based remote laboratory platform for electronic engineering students. The laboratory platform provides real-time interaction with two myRIO devices. The devices are used as the instructor's hardware experimental setup and are in the faculty laboratory and functioning as IoT nodes. LabVIEW myRIO toolkit was used to design the finite state machines which contain the lab applications. Each device consists of a tri-state finite state machine with each state being a laboratory session. Six basic electronic engineering experiments were used, and students can access them from any internet-connected device through common browsers. The system can be configured to have more myRIO devices and more experiments per device.

Keywords: HTTP · IoT · MyRIO · State machine · SystemLink cloud

1 Introduction

The abrupt emergence of the Covid-19 pandemic caused a sudden turnaround in general life activities. Students in universities and industrial training institutes around the world were greatly affected by the pandemic. The use of laboratories was disallowed by many

M. E. Auer et al. (Eds.): REV 2022, LNNS 524, pp. 12–22, 2023.
https://doi.org/10.1007/978-3-031-17091-1_2

local authorities [1]. The pandemic prompted the surge of remote laboratories in electronic engineering and a lot of solutions were proffered. The surge in remote laboratories has been encouraged by IoT which in turn supports industrial revolution 4.0. The IoT concept helps in learning electronic engineering courses. A basic IoT ecosystem consists of sensors, a processor unit, actuators, networking, and a cloud platform [2].

A lot of work has been done in electronic engineering remote laboratories. This study seeks to add to the pool of ideas in this important field. Developments in practical education continued during the pandemic with an interesting robotic laboratory-based system suggested in [1] for electronic engineering experiments. PocketLab, is a hands-on remote-ready system bringing together software, sensors, lessons, and a community [3]. In [4] a remote lab is described; the system is prototyped on a NI ELVIS board. NI myDAQ devices are then used for final implementation. The setup uses one PC with multiple myDAQ devices. Presented in [5] is an introductory practical lab on PSoC based system. It has a PCB board with a buck-booster converter half- bridge-based topology. The system is restricted to power electronics and a multi-concept will be more useful for an educational system.

This paper presents the design and implementation of a configurable remote laboratory training platform. A learning platform is presented consisting of six experiments of selected electronic engineering concepts which can be accessed and performed on any internet-connected device through usual browsers. The system consists of remote control, simulated experiments, and IoT-based experiments. The laboratory platform offers interaction with hardware in real-time. The system provides support for learning and training in the field of electronic engineering to students and industrial trainees.

2 Laboratory Platform Development

Figure 1 shows the block diagram which represents the developed remote laboratory ecosystem. For this design, two myRIO devices were connected to the host computer although more can be added. One myRIO device was connected through Wi-Fi broadcast from the device and the other device was connected by USB cable. The devices were identified using unique IP addresses. Each myRIO device contains three laboratories giving a total of six experiments. Additional experiments can be added according to requirements. For this design, two experiments can run concurrently one from each myRIO device. The LabVIEW software in the computer connects to the internet over TCP/IP. The SystemLink Cloud platform connects to browsers using the HTTP protocol. The two myRIO devices function as the instructor experimental set-up and are situated in the faculty laboratory. This setup allows the student to select and run different experiments from the cloud.

LabVIEW and MyRIO toolkit were central to the design of the laboratory system. Figure 2 shows the LabVIEW project hierarchy [7].

- The project hierarchy shows two myRIO devices in the project which are identified by IP addresses.
- The applications (VIs) below each myRIO device in the hierarchy are deployed and executed from that device.

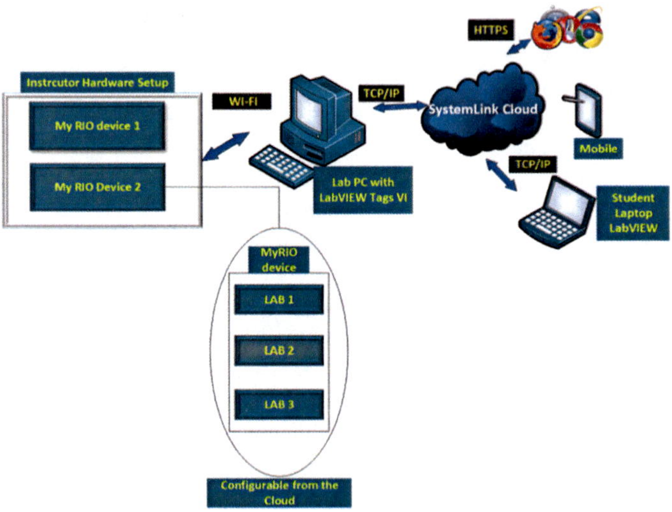

Fig. 1. Block diagram of the developed laboratory system.

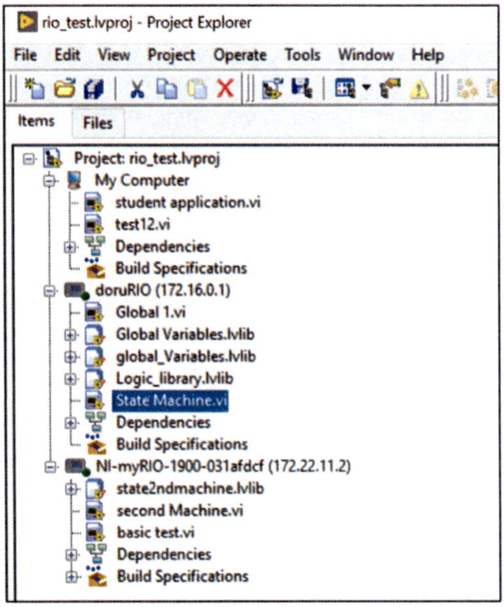

Fig. 2. The LabVIEW project hierarchy.

- The VIs under My Computer and above the NI myRIO devices are executed in the computer and used to connect the LabVIEW and the cloud platform.
- Each myRIO device runs a finite state machine that contains three states.
- Each state is a laboratory application (see Fig. 3).

- The Student Application VI connects the two myRIO devices applications with the SystemLink Cloud.
- The transfer of data from State Machine VI running in the myRIO device to VI running in the computer is enabled by global variables.

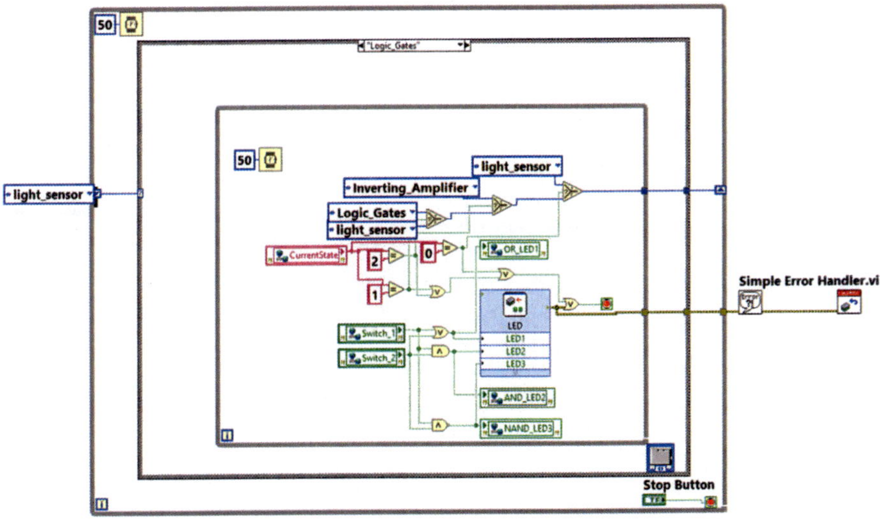

Fig. 3. One of the state machines running on a MyRIO device.

- The finite state machine contains three states.
- Each state has a design for a single lab.
- The transition from one to another is done from the cloud platform through the Enum function.

Communication is also established between the LabVIEW software running on the computer and the SystemLink Cloud. Figure 4 shows part of the VI which achieves that connection.

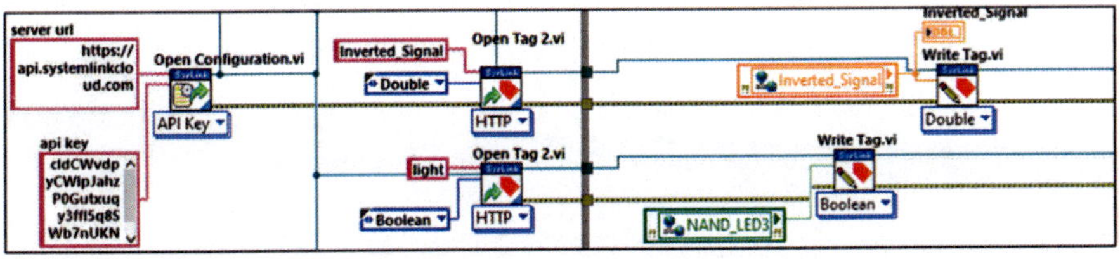

Fig. 4. Configuration for LabVIEW to SystemLink connection.

- The transfer of data between the VI running in the computer and the cloud is enabled by the server URL and the tag path.
- The API key ensures security between LabVIEW and SystemLink Cloud.
- Data exchanged between the cloud and LabVIEW has a unique pathname corresponding to the tag name in the cloud.
- Each pathname has a unique name and Open Tag VIs are configured to HTTP protocol.
- A path passing Boolean data type will have a tag of Boolean data type.

The tags were configured and managed in the data section of the SystemLink Cloud interface. The section gives an option to create tags as shown in Fig. 5.

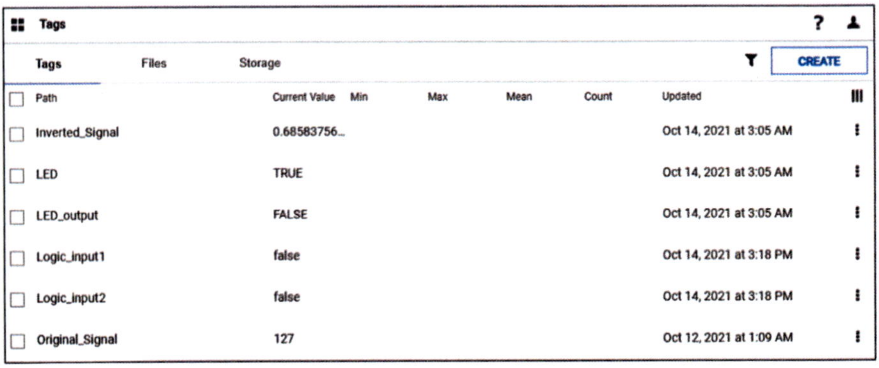

Fig. 5. Tag creation and configuration interface.

The Web interface was created using the SystemLink Cloud Graphical User Interface. The interface provides widgets for inputs, containers, decorations, and outputs as shown in Fig. 6. A typical widget can be customized for its color, shape, orientation, and tag value.

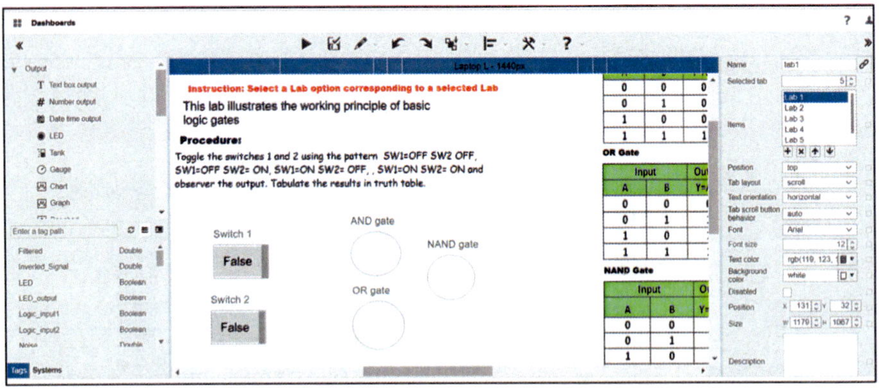

Fig. 6. SystemLink Cloud editor interface [7].

3 Selected Results

This section presents selected results from the experiments performed. Two myRIO devices were used to contain experimental applications. Six experiments were de-signed, and the main achievements of the study are as follows.

- Use of multiple myRIO devices.
- Two experiments running concurrently.
- Loading of multiple applications in MyRIO device
- The System is configurable from a cloud platform.
- Interactive SystemLink Cloud interface.

The six experiments are:

- Signal to Noise ratio (graphical visualization)
- Low pass filtering.
- Digital Data Multiplexing.
- Ambient Sensor Data Monitoring
- Inverting Amplifier.
- Basic Logic gates.

Every lab Web interface shows:

- A brief background theory on the experiment.
- Procedure for the experiment.
- An interactive interface for the experiment.

These laboratories interactive interfaces can be accessed at the link: https://www.systemlinkcloud.com/dashboard/freeform/player?name=Remote%20Lab&skylineId=cf27eb1d-4424-44e1-9734-a4f049151984#Remote%20Lab

The laboratory interfaces as seen from the internet browser are shown. A Student selects a lab option corresponding to a selected lab by selecting on the tab and the drop-down, menu labeled 'lab Option'. The experiment interface shown in Fig. 7 illustrates the working principle of a 2-1 multiplexer. A multiplexer is a device that allows one of several analog or digital input signals which are to be selected and transmits the input that is selected into a single medium. The inputs to this multiplexer are a flashing signal and a switch. The output is shown both on the LabVIEW front panel and a remotely connected myRIO device's LED. This experiment illustrates the basic concept of remote control.

In this experiment the students:

- Toggle the selector switch, ON and OFF.
- Observe which input is transferred to the output and compares with the truth table.
- The output is also shown on a remote myRIO device connected via the internet.

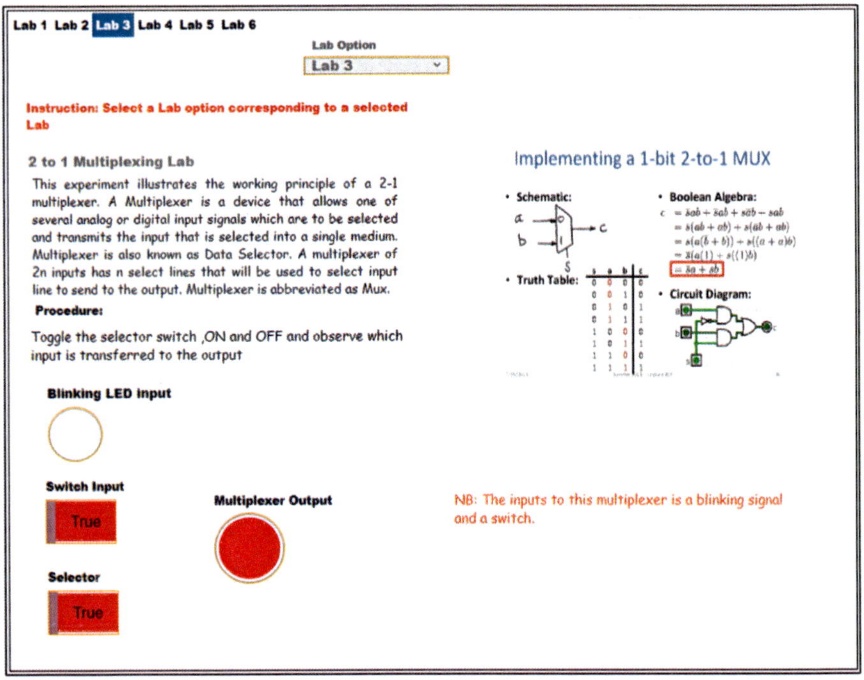

Fig. 7. Web interface for multiplexer experiment.

The Lab interface shown in Fig. 8 illustrates the working principle of an inverting amplifier. This is a simulated experiment. In this lab students do the following procedure:

- Changes the values of Resistor 1 from 10 to 100 while keeping Resistor 2 constant.
- Changes the values of Resistor 2 from 10 to 100 while keeping Resistor 1 constant. In all the cases the students observe the changes in the waveforms.

The lab interface in Fig. 9 illustrates the working principle of basic logic gates. This interface is connected to a remote myRIO device which also replicates the LabVIEW front panel logic gates results.

The Procedure followed by students is as follows is:

- The student toggles the switches 1 and 2 using the pattern SW1 = OFF SW2 OFF, SW1 = OFF SW2 = ON, SW1 = ON SW2 = OFF, SW1 = ON SW2 = ON and observes the output and compares the result with provided truth tables.

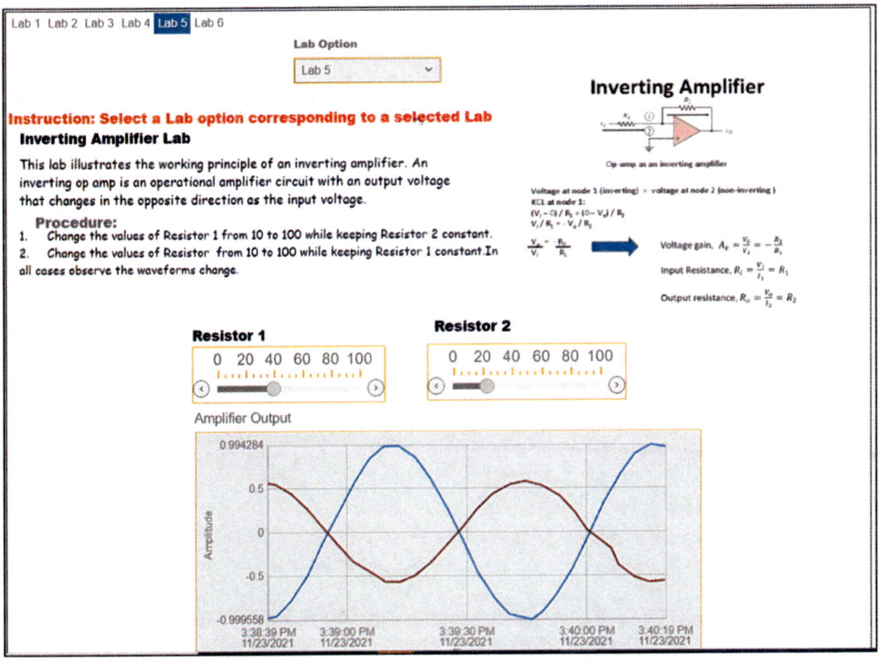

Fig. 8. Web interface for inverting amplifier.

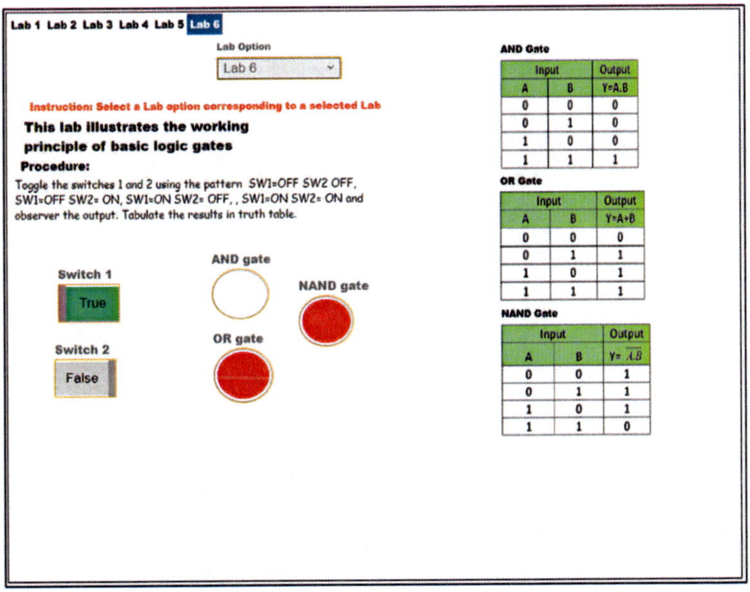

Input		Output
A	B	Y=A.B
0	0	0
0	1	0
1	0	0
1	1	1

Input		Output
A	B	Y=A+B
0	0	0
0	1	1
1	0	1
1	1	1

Input		Output
A	B	$Y=\overline{AB}$
0	0	1
0	1	1
1	0	1
1	1	0

Fig. 9. Web interface for basic logic gates

The experiment interface shown in Fig. 10 is the output of an ambient light monitoring process. The sensor is connected to a remote myRIO device which functions as an IoT node. Students can design and develop a LabVIEW VI which can be used to select set points and trigger alarms when thresholds are exceeded. This experiment illustrates the basis of remote system monitoring and control in IoT.

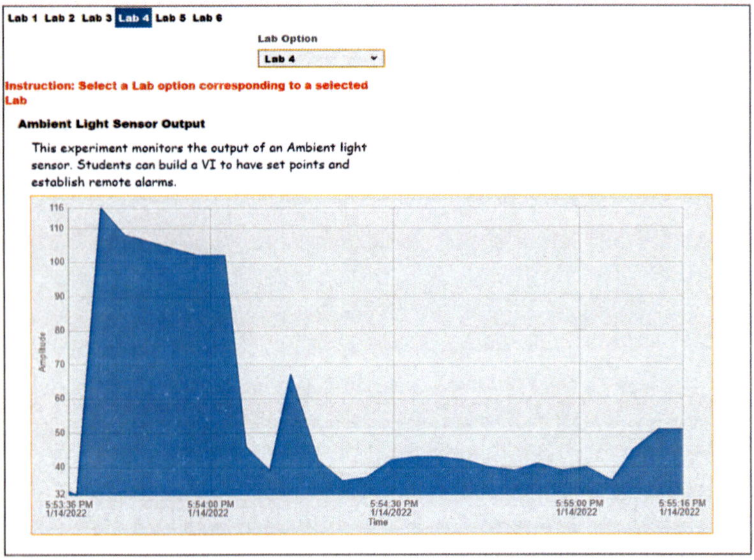

Fig. 10. Ambient light sensor output interface

The experiment in Fig. 11 illustrates the concept of signal-to-noise ratio (SNR) simulation from LabVIEW. SNR is the ratio between the desired information or the power of a signal and the undesired signal or the power of the background noise. It compares the level of the desired signal to the level of background noise and its unit of expression is typically decibels (dB). Also, a ratio greater than 0 dB or higher than 1:1, signifies more signal than noise.

In this experiment, the students initialize both the signal and noise sliders at value 20. They observe the graphs and move the signal slider to 40 and 80 observing the graphs. In Fig. 11 the brown graph shows the original clean signal, and the blue graph shows the result of adding uniform noise.

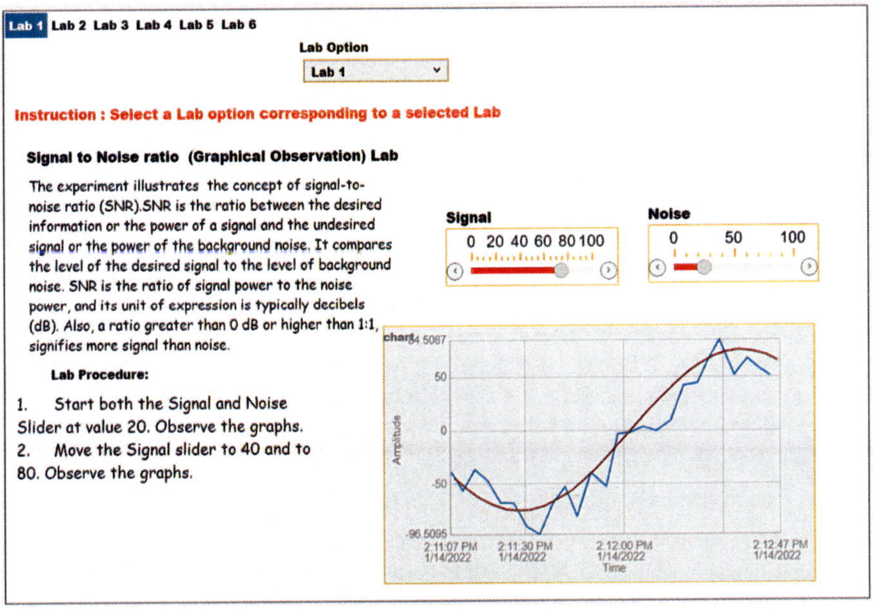

Fig. 11. Signal to noise ratio lab interface

4 Conclusions

The study successfully demonstrated the proof of concept to develop a remote laboratory platform. Electronics engineering basic concepts were used to demonstrate the system's functionality. The interface faces some delays in showing results depending on the network latency. The experiment interface can be accessed and manipulated from any internet-connected device's screen. Although the two myRIO devices and three applications per device were used in this design, the number of the devices and the experiments can be increased to suit requirements. This design has illustrated IoT, simulated, and remote device control experiments. The study is a continuous process and efforts are underway to further develop the remote laboratory platform. This study will lead to the development of a new device for educational and research purposes. The versatility and intuitive nature of LabVIEW help in the rapid concept-to-development process. More sensors and peripherals devices like Arduino will be added so that the platform can implement many applications. The system will be helpful for faculties that run on low budgets, especially the developing countries.

5 Discussions and Recommendations

Trials were done on the system using a limited number of students and colleagues in the last semester. Suggestions have been made for the experiments to show interaction with the remote hardware. The Laboratory platform needs integration into a learning management system (LMS). An LMS system will help manage students' laboratory booking systems [8].

Acknowledgments. My heartfelt gratitude goes to NI Romania for their deep support in this research. I would also like to thank Prof.Dr. Doru URSUTIU, my supervisor, for all his efforts to secure the kits and for being helpful in the writing of this report.

References

1. Zamora-Hernández I, Rodriguez-Paz MX, González-Mendivil JA, Zarate-Garcia JA (2021) A remote robot based lab to develop competencies in engineering students during Covidl9 pandemic. In: 2021 IEEE Global Engineering Education Conference (EDUCON), pp 624–630. https://doi.org/10.1109/EDUCON46332.2021.9453895
2. Rahman NA, Idris MR, Baharudin KS (2020) Development of educational kit for IoT online learning. Int J Technol Innov Hum 1(1):26–32. https://doi.org/10.29210/881001
3. Wood T (2020) Effective PocketLab sensor use in an engineering dynamics course. In: ASEE virtual annual conference content access. ASEE, Virtual Online. https://doi.org/10.18260/1-2--34500
4. Klinger T, Kreiter C, Pester A, Madritsch C (2020) Low-cost remote laboratory concept based on NI myDAQ and NI ELVIS for electronic engineering education. In: IEEE global engineering education conference (EDUCON). IEEE, Porto. https://doi.org/10.1109/EDUCON45650.2020.9125086
5. Ulrich B (2017) A modern, versatile, and low-cost educational system for teaching DC/DC converter control with analog, digital and mixed-signal methods. In: International conference on research and education in mechatronics (REM). IEEE, Wolfenbuettel, pp 1–8. https://doi.org/10.1109/REM.2017.8075234
6. SystemLink Cloud homepage (2021) https://www.systemlinkcloud.com/dashboard/freeform/editor?name=Remote%20Lab&skylineId=cf27eb1d-4424-44e1-9734-a4f049151984#Remote%20Lab. Accessed 28 Nov 2021
7. NI Homepage (2021) https://www.ni.com/ro-ro/support/documentation/supplemental/08/best-practices-for-managing-ni-labview-applications-using-the-pr.html. Accessed 30 Nov 2021
8. Mateos V, Gallardo A, Richter T, Bellido L, Debicki P, Villagra VA (2011) LiLa booking system: architecture and conceptual model of a rig booking system for on-line laboratories. Int J Online Biomed Eng 7(4):26–35. https://doi.org/10.3991/ijoe.v7i4.1837

Work in Progress: Noise Reduction Through Artificial Intelligence Techniques: An Introductory Study

Horia Alexandru Modran[1](✉), Doru Ursuțiu[1,2], Cornel Samoilă[1,3], and Tinashe Chamunorwa[1]

[1] Transilvania University of Brașov, Brașov, Romania
{horia.modran,udoru,chamunorwa.tinashe}@unitbv.ro
[2] Romanian Academy of Scientists, Bucharest, Romania
[3] Romanian Academy of Technical Sciences, Bucharest, Romania

Abstract. Signals from the real world, resulting from various data acquisition methods, have many components which can influence the quality of the data. One of the most important factors is the noise. It can cause different kind of problems in Artificial Intelligence algorithms, as the learning model can interpret the noise as a data pattern. Because of this, reducing or eliminating the noise from a signal is a crucial process.

This paper is focusing on treating the problem of signal denoising. A method to eliminate the noise from signals by using of a Convolutional denoising auto-encoder was developed. The developed autoencoder has six layers: an input layer, two convolutional layers serving as encoder, two convolutional layers as decoder and an output layer. One of the main advantages of this approach is that, compared to other methods used for denoising signals, autoencoders can be trained with the downstream classifier, improving the accuracy of the system.

This introductory study will be extended and improved so that the autoencoder presented in this paper will to be able to eliminate thermal noise from a signal using a similar machine learning model.

Keywords: Signal denoising · Artificial intelligence · Deep learning · Neural networks · Autoencoder

1 Introduction

Real-world signals, which are the result of data acquisition techniques, contain many components that can influence the quality of the data. Among these problems, the noise is an important factor.

There is a high probability of human error during data collection, but, at the same time, the devices that collect data may be inaccurate, resulting in different kind of noise in the data set. Without proper treatment, the noise can create problems in different Artificial Intelligence algorithms, because the model can treat the noise as a pattern and

M. E. Auer et al. (Eds.): REV 2022, LNNS 524, pp. 23–32, 2023.
https://doi.org/10.1007/978-3-031-17091-1_3

can try to generalize from it. Therefore, addressing and eliminating the noise from the data set is a very important step.

Denoising signals is a known issue and has a long history. Therefore, several techniques exist, like finite and infinite impulse response filters, wavelet-based methods, filtered reside technique and empirical mode decomposition. In recent years, multiple novel approaches to signal filtering, using artificial intelligence algorithms and deep neural networks appeared [1].

This paper is focused on the problem of denoising a signal - specifically, a method to denoise signals by using of a Convolutional Denoising Autoencoder (CDAE – detailed in Sect. 4). An advantage of using this kind of approach is that, compared to other traditional methods used for denoising, autoencoders can be trained with the downstream classifier as well, which can improve the overall accuracy [2].

Among the deep learning architectures are the auto-encoders and Convolutional Neural Networks (CNN), which can be used together to create Convolutional Autoencoders. [2].

Autoencoders are multilayered artificial neural networks that have a training method and a symmetric input-output architecture. They are using an unsupervised machine learning technique, by trying to replicate the exact input example.

Several machine learning models have been applied for denoising different types of signals. Feng et Al. [3] used a SVM for estimating the noise level of a White Gaussian Noise channel. In other papers that were analyzed [4, 5], the modulation classification in presence of different values of Signal-to-Noise Ratios and spectral identification are performed using a CNN classifier.

The Denoising autoencoders are used to reconstruct clean data from a corrupted (noised) input [6]. In contrast to standard auto-encoders, in this case, noisy samples are given as input and the model learns to reconstruct a denoised sample of them. In order to perform this operation, both the clean and the noisy version of each sample has to be provided to the deep neural network model. In general, convolutional autoencoders perform better than classical encoders for processing images or other types of signals [7].

2 Denoising Techniques

There are several common methods for denoising a signal:

- Deep deactivation automatic coding method - autoencoders are useful for eliminating noise. Because it might be the case that distinctive noise can be learned from the signal or data, noisy data can be provided as an input and obtaining denoised data as output. Every autoencoder comprises two main elements: an encoder that encodes the input data, the second one being the decoder. The basic principle of an automatic noise-canceling encoder is to force the hidden layers of the neural network to learn to generalize, in such a way that it could recreate the original input data.
- Principal Component Analysis (PCA) is the technique of computing the principal components and using them for performing a change of basis on the dataset, selecting only the important principal components, and removing the others [8]. It can be used

as well to eliminate corrupted data from a signal, while still retaining the important characteristics of the data. Mohan Babu et. Al. [9] used this algorithm to eliminate noise from the data in two steps.

- Fourier transform – some recent research proved that noise can be easily eliminated from the signal if it contains a pattern. In this technique, the signal is converted from the time domain into the frequency domain by calculating the Fourier Transform. The theory involves that most of the noisy can be eliminated from the signal by maintaining the frequencies that contain the essential features and removing all other frequencies [10].

Because autoencoders can be feed with noisy data and can serve as denoisers, a deep autoencoder was chosen to be used in this experiment. The main structure of the deep convolutional autoencoder is illustrated in Fig. 1.

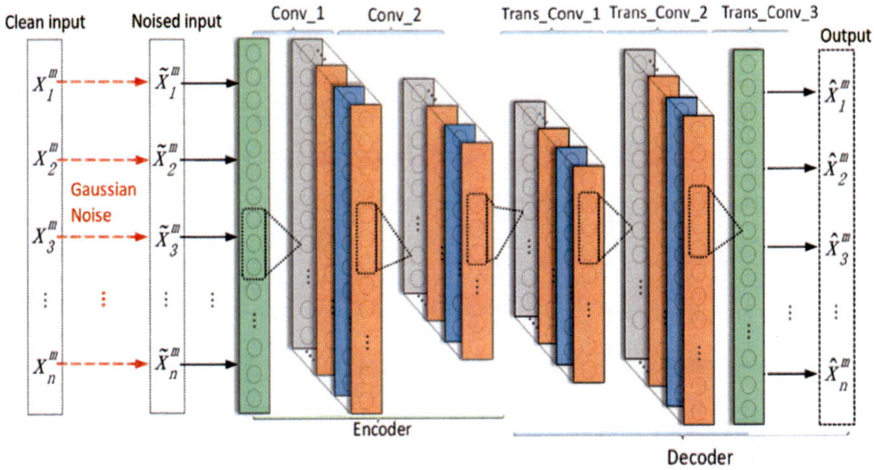

Fig. 1. Autoencoder

3 Data Synthesis

In this experiment a deep convolutional denoising autoencoder was implemented, through a three-step process:

- generating a large data set of sine wave signal samples
- creating a large set of noised sinewave samples by adding random Gaussian to the clean signals
- developing an autoencoder that can denoise the noisy sine wave inputs, thus eliminating noise for new data (signals that the machine learning model never saw)

For this process, the Tensorflow and Keras deep learning libraries were used to create an autoencoder for denoising the signal. Two aspects were analyzed first - the data that was used and a high-level view of the model.

Sine wave samples were used as pure signals and as automatic coding targets Fig. 2 shows the shape of the signal, while Fig. 3 illustrates a noisy signal sample.

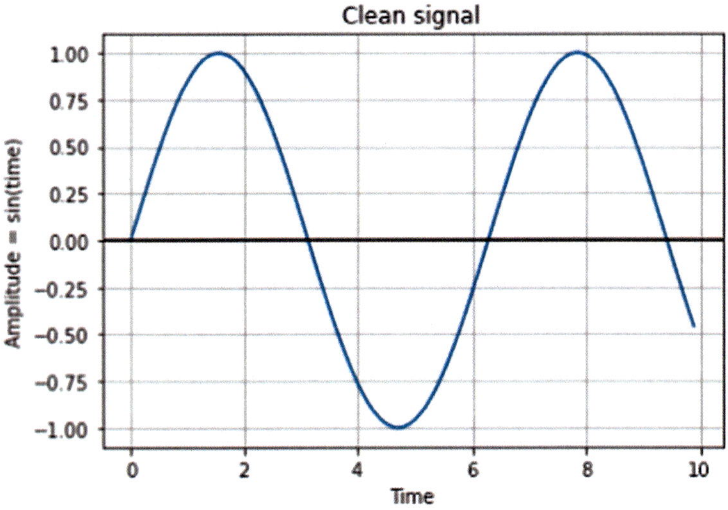

Fig. 2. Original signal shape

For the model around 100.000 samples were used. For each sample, a random value of Gaussian noise was added. As can be seen in Fig. 3, the original signal becomes noisy.

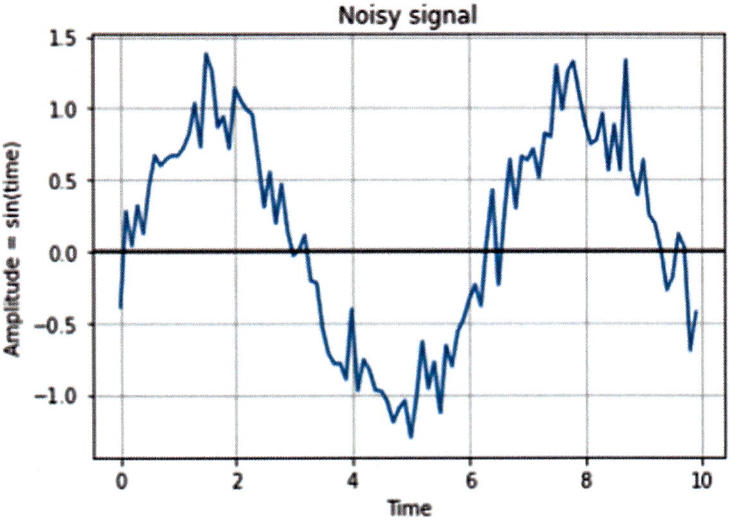

Fig. 3. Noisy signal shape

4 Artificial Intelligence Model

The architecture of the machine learning model that was developed is illustrated in Fig. 4 and comprises the following six layers:

- an input layer
- 2 conv1D convolutional layers (encoder)
- 2 Conv1D transposition layers (decoder)
- a Conv1D output layer, with a sigmoid activation function

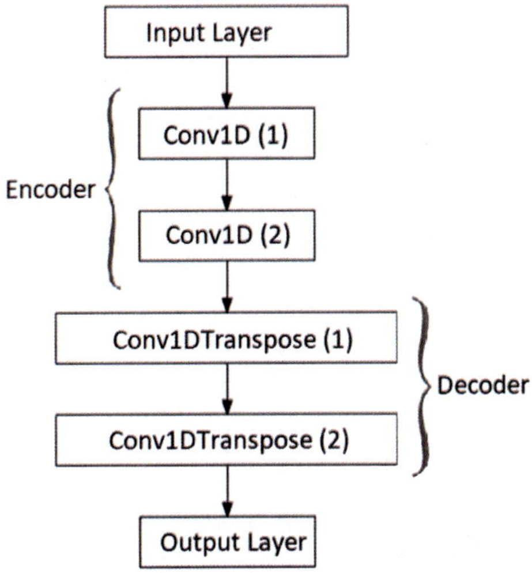

Fig. 4. Neural network architecture

The generation of sine wave signals contains three steps, to generate samples such as the signal shown in Fig. 3:

- defining the configuration settings for the signal generator
- data generation (pure waveforms)
- saving waveforms in files and previewing a subset of them

The following code snippet was used to generate the 100.000 samples:

```
#Number of samples (100.000)
num_samples = 100000

#Configuration
num_elements = 1
interval_per_element = 0.01
total_num_elements = int(num_elements / interval_per_element)
starting_point = int(0 - 0.5*total_num_elements)
num_samples_visualize = 1

# Containers for samples and subsamples
samples = []
xs = []
ys = []
```

The second step consists of adding noise to the samples that were generated in the previous step. The Python code used for this operation is shown below:

```
# Add noise to data
noisy_samples = []
for i in range(0, len(x_val)):
  if i % 100 == 0:
    print(i)
  pure = np.array(y_val[i])
  noise = np.random.normal(0, 1, pure.shape)
  signal = pure + noise_factor * noise
  noisy_samples.append([x_val[i], signal])
```

After generating the data, the auto-encoder creation step follows, which involves the following operations:

- set configuration parameter of the model
- importing and preparing the data
- training the model
- visualization of the waveforms in the test set for which the noise was eliminated (to visually validate if the model works properly)

The neural network model was configured with the following parameters:

```
# Model configuration
input_shape = (150, 1)
batch_size = 150
no_epochs = 5
train_test_split = 0.3
validation_split = 0.2
verbosity = 1
max_norm_value = 2.0
```

The value of the parameters used have the following meanings:

- input_shape must have the same dimensions as the Conv1D input layer, is thus (150,1).
- the batch size was set to 150 - it proved to be the optimal value, both in terms of performance and prediction time
- a small number of epochs was used, as it was noticed that the model hasn't improved with a higher value
- about 30% of the dataset was used as test data, while 20% of it was retained for validation purposes.

The neural network developed for the autoencoder has the following characteristics:

- the Conv1D layers of the encoder contains 128 and 32 filters, respectively, and they use a Rectified Linear Unit (ReLU) activation function.
- the Conv1DTranspose layers of the decoder have 32 and 128 filters
- the final Conv layer of the output doesn't change the shape of the data, but it changes the number of channels back to 1.

The Python code for creating the autoencoder is shown below:

```
# Create the model
model = Sequential()
model.add(Conv1D(128, kernel_size=3, activation='relu',
                 kernel_constraint=max_norm(max_norm_value),
                 kernel_initializer='he_uniform', input_shape=input_shape))
model.add(Conv1D(32, kernel_size=3, activation='relu',
                 kernel_constraint=max_norm(max_norm_value),
                 kernel_initializer='he_uniform'))
model.add(Conv1DTranspose(32, kernel_size=3, activation='relu',
                          kernel_constraint=max_norm(max_norm_value),
                          kernel_initializer='he_uniform'))
model.add(Conv1DTranspose(128, kernel_size=3, activation='relu',
                          kernel_constraint=max_norm(max_norm_value),
                          kernel_initializer='he_uniform'))
model.add(Conv1D(1, kernel_size=3, activation='sigmoid',
                 kernel_constraint=max_norm(max_norm_value),
                 padding='same'))

model.summary()
```

After completing the training process, it is necessary to test whether the model works properly. To do this, some reconstructions were generated: a noisy sample from the test set was added (data that the model has not seen before) and it was tested whether it looks like the pure waveform signal. Figure 5 illustrates the test result performed on a sample.

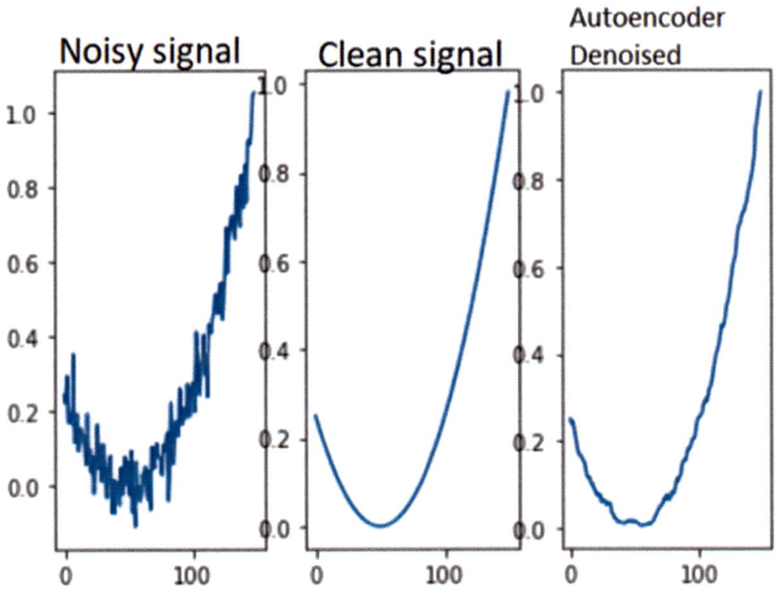

Fig. 5. Testing the autoencoder

As can be seen in Fig. 5, the encoder has learned to eliminate much of the noise. The noisy samples are not yet completely noiseless, but we obtained a Signal-to-Noise Ratio (SNR) of about 23–24 dB.

5 Further Developments

The autoencoder presented in this paper will be improved in the future research, to be able to eliminate thermal noise from a signal. For this purpose, data will be acquired using the BioRadio device from Great Lakes Neurotechnologies and Elvis II from National Instruments. A channel for data acquisition was configured on this device with a sampling rate of 2 kHz and a resolution of 1 μV. The data can then be imported and processed in the LabVIEW programming environment. The encoder needs to be feed with about 100.000 samples of pure signal and as many thermal noise signal samples. The architecture of the machine learning model will be similar to that of the autoencoder presented in Sect. 4.

The aim is to perform the data processing using edge AI techniques, so that the Artificial Intelligence algorithm to run directly on the PSoC6 microcontroller. Thanks to the launch of the ModusToolbox solution from Cypress/Infineon in May 2021 [11], the process of developing and running pre-trained machine learning models directly on the PSoC6 microcontroller was greatly simplified. The model can be developed in Python using libraries such as TensorFlow and Keras and then imported into ModusToolbox [12]. As Fig. 6 illustrates, the imported model can be subsequently optimized for embedded systems and its performance can be re-validated before being deployed on the PSoC6 device.

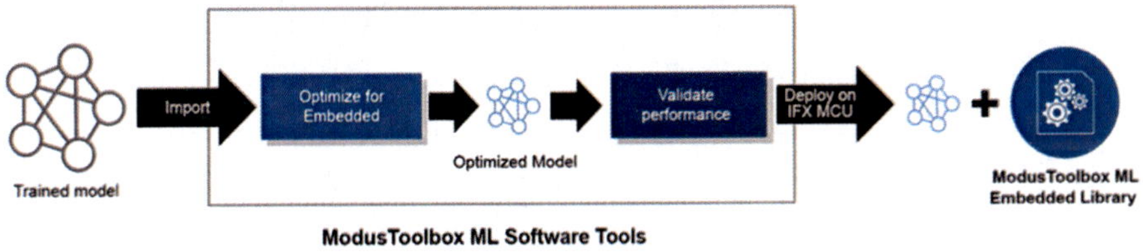

Fig. 6. Deploy machine learning model on PSoC6 [13]

6 Conclusion

Because the noise can cause different kind of problems in Artificial Intelligence algorithms, as the learning model can treat the noise as a data pattern, eliminating the noise from the signal is a crucial step.

This paper proposed an Artificial Intelligence solution for the problem of signal denoising. The method presented is using a convolutional denoising autoencoder to eliminate the noise from signals. It contains six layers: input layer, encoder with two 1D convolutional layers, decoder with two convolutional 1D transposed layers, and output layer. Compared to the traditional methods used for eliminating noise from signals, autoencoders can be trained with the downstream classifier, which is improving the accuracy of the model.

As it was presented in Sect. 4, the encoder has learned to eliminate much of the noise, but the noisy samples are not yet completely noiseless. Therefore, the current study will be improved both in term of performance and extended, to be able to eliminate thermal noise from a signal using an autoencoder having similar architecture as the one used in this experiment.

References

1. Antczak, K.: Deep Recurrent Neural Networks for ECG Signal Denoising, Computer Science series, Neural and Evolutionary Computing arXiv:1807.11551 (2018)
2. Almazrouei, E., Gianini, G., Mio, C., Almoosa, N. Damiani, E.: Using AutoEncoders for radio signal denoising. In: Proceedings of the 15th ACM International Symposium on QoS and Security for Wireless and Mobile Networks, pp. 11–17 (Nov 2019). https://doi.org/10.1145/3345837.3355949
3. Feng V, Yu Chang S (2012) Determination of wireless networks parameters through parallel hierarchical support vector machines. IEEE Trans Parallel Distrib Syst 23(3):505–512. https://doi.org/10.1109/TPDS.2011.156
4. O'Shea TJ, Corgan J, Clancy TC (2016) Convolutional Radio Modulation Recognition Networks. In: Jayne C, Iliadis L (eds) EANN 2016, vol 629. CCIS. Springer, Cham, pp 213–226. https://doi.org/10.1007/978-3-319-44188-7_16
5. O'Shea, T., Roy, T., Erpek, T.: Spectral detection and localization of radio events with learned convolutional neural features. In: Proceedings of the Signal Processing Conference (EUSIPCO), pp. 331–335. IEEE (2017). https://doi.org/10.23919/EUSIPCO.2017.8081223
6. Vincent, P., Larochelle, H., Lajoie, I., Bengio, Y., Manzagol, P.A.: Stacked denoising autoencoders: Learning useful representations in a deep network with a local denoising criterion. Journal of Machine Learning Research **11**, 3371–3408 (2010). https://doi.org/10.5555/1756006.1953039

7. Lee, D., Choi, S., Kim, H.: Performance evaluation of image denoising developed using convolutional denoising autoencoders in chest radiography, Nuclear Instruments and Methods in Physics Research, Section A: Accelerators, Spectrometers, Detectors and Associated Equipment **884**, 97–104 (2018). https://doi.org/10.1016/j.nima.2017.12.050
8. Jolliffe, I., Cadima, J.: Principal component analysis: a review and recent developments. Phil. Trans. R. Soc. A. 3742015020220150202 (2016). https://doi.org/10.1098/rsta.2015.0202
9. Mohan Babu, Y., Subramanyam, M.V., Giri Prasad, M.N.: PCA based image denoising, Signal & Image Processing: An International Journal (SIPIJ) **3**(2) (April 2012). https://doi.org/10.5121/sipij.2012.3218
10. Bi, G., Zheng, S.: A Novel Method of Signal Denoising, Physics Procedia **33**, 858–865 (2012). ISSN 1875-3892, https://doi.org/10.1016/j.phpro.2012.05.146
11. Infineon's new ModusToolbox™ Machine Learning enables TinyML for secure AIoT. [Online] at https://www.infineon.com/cms/en/about-infineon/press/market-news/2021/INFCSS202105-071.html. Last accessed 30 September 2021
12. Machine Learning on the Edge: [Online] at https://www.cypress.com/solutions/machine-learning-solutions. Last accessed 3 November 2021
13. ModusToolbox Machine Learning: [Online] at https://www.cypress.com/products/modustoolbox/machine-learning Last accessed 19 November 2021

Printed Wearable Accessories for Electrochemical Applications

Petru Epure[1(✉)], Doru Ursutiu[2,3], Cornel Samoila[2,4], and Petru P. Epure[1]

[1] EPI-SISTEM SRL, str. Livezii nr. 17A, Sacele, Romania
petru.epure@epi.ro
[2] Universitatea Transilvania din Brasov, Blv. Eroilor nr. 29, Brasov, Romania
[3] Academy of Romanian Scientists, Bucharest, Romania
[4] Technical Sciences Academy of Romania, Bucharest, Romania

Abstract. This paper presents a method to provide rapid prototyping for electrochemical measurement configurations useful for wearable applications. The first stage of development of the electrodes is to prepare a 3D model and then to transform the model in a real electrochemical interface using hybrid inkjet technologies. The final version quality of the electrode depends of several factors related to ink and substrate physics. Measuring electrochemical response of the body sweat is a major way of investigation for medical engineering. The results are important for remote health evaluation.

Keywords: Prototyping · Inkjet · Electrochemistry · Wearable · Remote measurement

1 Introduction

The biosensors market was valued at $17,500.0 million in 2018 and is expected to reach $38,600.2 million by 2026, registering a CAGR (Compound Annual Growth Rate) of 10.4% from 2019 to 2026 [1]. Biosensors are analytical devices that can convert a biological response into an electrical signal. Biosensors majorly comprise three parts, including a component that identifies an analyte and produces signal, a signal converter combined with a digitalization device. Biosensors find their applications in food & beverages, environmental monitoring, home care diagnostics. And various nano mechanical biosensors. In addition, biosensor applications are also widely dominant in the field of plant biology to study metabolic processes [2].

The wearable market was in 2019 about 32.63 billion USD and will evolve with a CAGR of 15.9% between 2020 and 2027. Just for the smartwatch devices the sales in 2019 were at 148 million units having a grow of about 30–35%. The evolution will continue for the next 5–10 years in the same pace. The IoT (Internet of Things) market will reach 1.1 trillion USD between 2018 and 2023 giving a significant impact on the global internet network. It is estimated that by 2025 over the world will exist more than 21 billion connections dedicated to IoT [3]. The evolution of this huge infrastructure involves big efforts to be made by companies from various sectors providing hardware and software parts or modules.

M. E. Auer et al. (Eds.): REV 2022, LNNS 524, pp. 33–40, 2023.
https://doi.org/10.1007/978-3-031-17091-1_4

2 System Description

Prototyping of wearable electrochemical sensors is difficult for inkjet methods. The traditional way is to perform a serigraphic method meaning that a textile mask has to be prepared for each material or each thin film that the user intents to apply. Between the thin film application is it mandatory to perform a thermic treatment in order to obtain a solid thin film surface. The inkjet process involves different types of inks. Silver ink is usually flexible and inert in unstable or extreme conditions. It can be laid on plastics, paper, PET, Kapton, PEN, ABS and cured between 70 and 123 °C for 20 to 40 min. Carbon ink is highly flexible and stable which enables high speed printing and results in resistant layers. It can be laid on plastics, paper, PET, PEN, Polyester and cured between 107 to 120 °C. Carbon ink is widely used in low voltage electronics, resistors, etc. Dielectric ink has the role to protect the contacts and the inactive surface of the substrate being resistant to extreme temperature and humidity conditions. It cures under UV light for 10 to 15 min. An important aspect of inkjet printing is the viscosity, density and granulation. Viscosity ensures a controlled flux when applying a thin layer on the substrate. Different tips are used for a wide range of characteristics. Higher viscosity means larger tips of 0.4 mm or 0.5 mm but this comes at the expense of the resolution when printing. Granulation is important and reflects the quality and resolution if the process uses tips of 0.2 mm interior diameter. Tests were conducted on polyamide and phenolic sheets with textile inserts to determine how usable this technique for rigid and flexible substrates. Related to biosensors it is wellknowned that screen printed electrodes can be functionalized to detect heavy metals or contaminants [4] and to be used for a limited number of times.

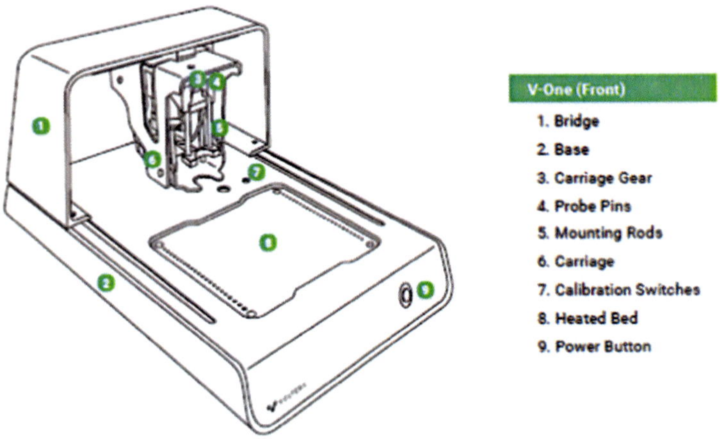

Fig. 1. Architecture of the inkjet device

In Fig. 1 is presented the architecture of V-ONE device used for initial prototyping of wearable electrodes. V-ONE is a 3D inkjet printing device that uses Loctite 479 for the silver traces and Loctite 6017 for the carbon traces.

Testing was done using silver and carbon ink. The substrate must be fixed well in place while printing to prevent errors resulted from positioning. At the start of the printing process a "zero" level is to be determined which is dependent on the thickness of the substrate. Marking are to be made so that all of the layers line up on the substrate in the case of multilayer prints. An example of testing is presented in Fig. 2 at low resolution.

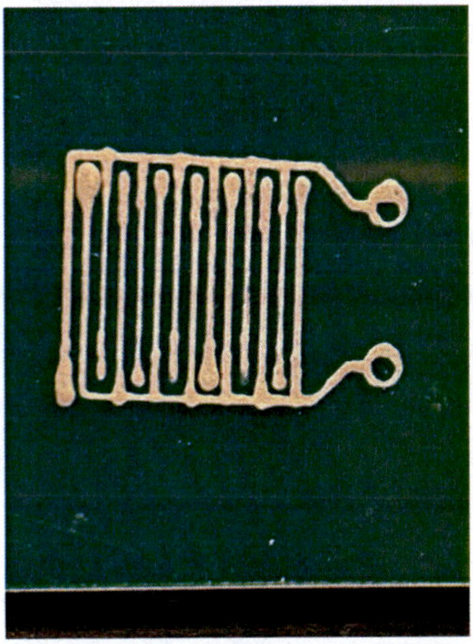

Fig. 2. Testing on rigid substrate, printing errors due to flux overload of silver ink

The presented architecture can be the frame for any measurement application based on biosensors or electrochemical sensors. In this case the configuration use screen printed electrodes manufactured on alumina substrates and containing WE (working electrode), RE (reference electrode) and CE (counter electrode). There are studies regarding the way of preparation for metalic electrodes [5] but also how to use carbon electrodes for wearable sensors [6]. An example of such electrode is presented in Fig. 3 at high resolution.

By using 4 different diameters for the printing tip (0.2 mm, 0.25 mm, 0.33 mm, 0.41 mm) the method is capable of laying multiple layers of different thicknesses. The advantage comes from economy of electroactive substances, with printing waste being minimal and in the range of microliters. Diminishing waste of expensive inks leads to a smother electrode development process.

Electrode testing can be made using impedance measurements like in Fig. 4 where the left picture is Nyquist representation and the right picture is Bode plot. The shape of the electrodes can be functionalized using different substances for different applications. Electrochemical tests were conducted using PBS, citric acid solutions, etc. The electrochemical measurements are compensated by temperature and humidity factors [7]. In this situation the calibration curve is changed mainly because of pH changing due to temperature and humidity [8]. Calibration curve is affected mainly in a slope changing, the increase of acidity is giving a bigger slope of the calibration curve. In this case a

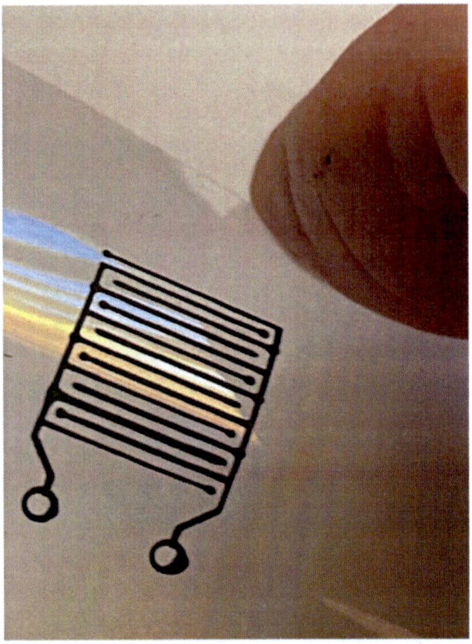

Fig. 3. Testing on flexible substrate, slight overprinting on the electrodes contacts

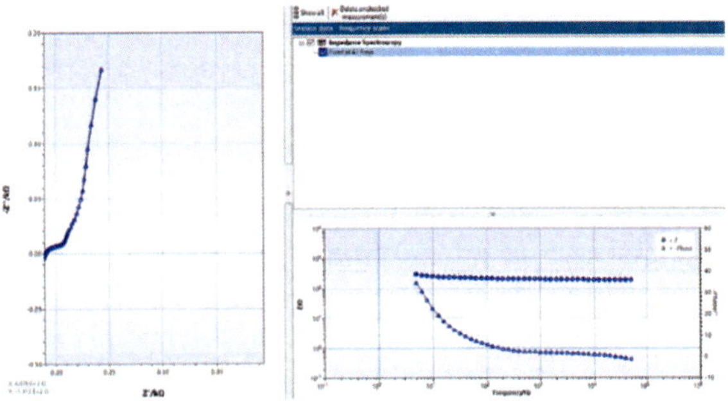

Fig. 4. Impedance measurement performed on a interdigitated electrode

current variation is giving a lower variation of concentration when the acidity is higher. All this characteristics are available for the linear range of the sensor.

3 CAD Software Description

CAD models can include different layers for each material used in the print. An electrode model can be described like in Fig. 5 where the right picture presents 3 materials on a white substrate and the left picture presents 2 materials on a blue substrate and no protective ink. In Fig. 6 there are 2 different layers with the carbon ink being drawn in purple and the silver ink in red. These layers are printed one after the other and the

Fig. 5. CAD Model description of an electrode

then cured. The thermic curing process has impact on the texture of the electrode and so different curing regimes result in slight variations in appearance.

There is a specific frame box made using the 3D-DESIGN technologies and then 3D-PRINT prototyping facilities. The box volume is adjustable and fits the dimensions of the SmartScreen.

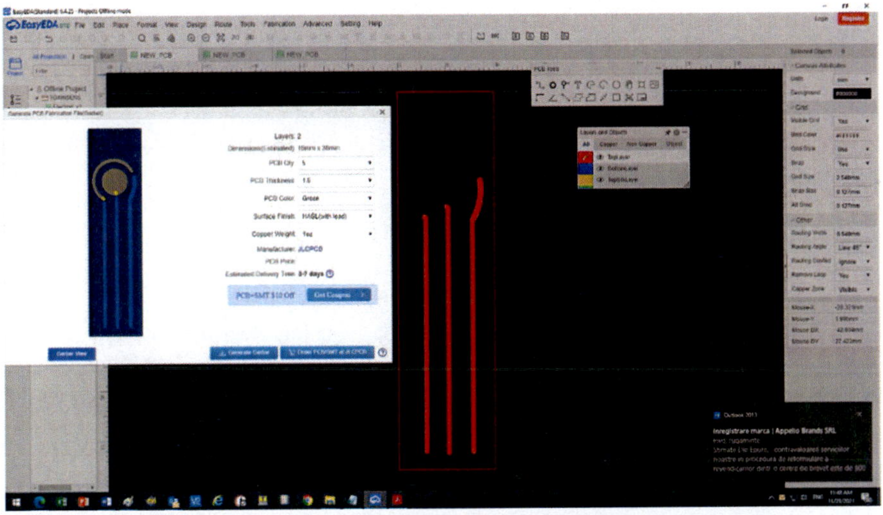

Fig. 6. Gerber electrode model

The 3D printing is done following a 3D model made using CAD software on a Gerber format. Voltera V-ONE comes with specific software [9] that enables quick prototyping using multiple layers for the electrode. In Fig. 7 is presented a model on a single material representing an interdigitated electrode as a base for a future sensor.

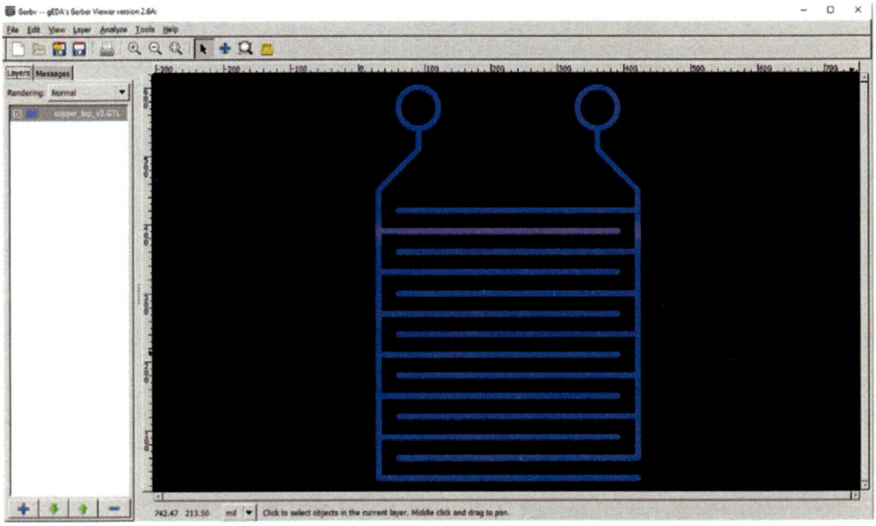

Fig. 7. Gerber interdigitated electrode model

4 Parameters for Inkjet Models

Voltera is capable to print and cure multiple types of inks on different surfaces. The surface must be able to withstand 120 °C. The equipment has the possibility of dispersing small quantities of ink similar to titration methods. The ink layer on the substrate is cured with a heating pad that reaches 240 °C [10]. In Fig. 8 is presented how the temperature control is achieved using a software where PID elements were integrated with high precision.

A practical application is to functionalize electrodes capable to detect some bacterial toxins. The electrode looks like in Fig. 9 and the working electrode is made using a carbon ink mixed with specific nanomaterials. The functionalization is useful in order to provide a better selectivity and a very good sensitivity. The special ink is applied on the working electrode and then a curing session is mandatory to ensure the proper electrical conductivity. In this mode, the electrodes will provide the possibility to detect toxins in hospitals or restaurants.

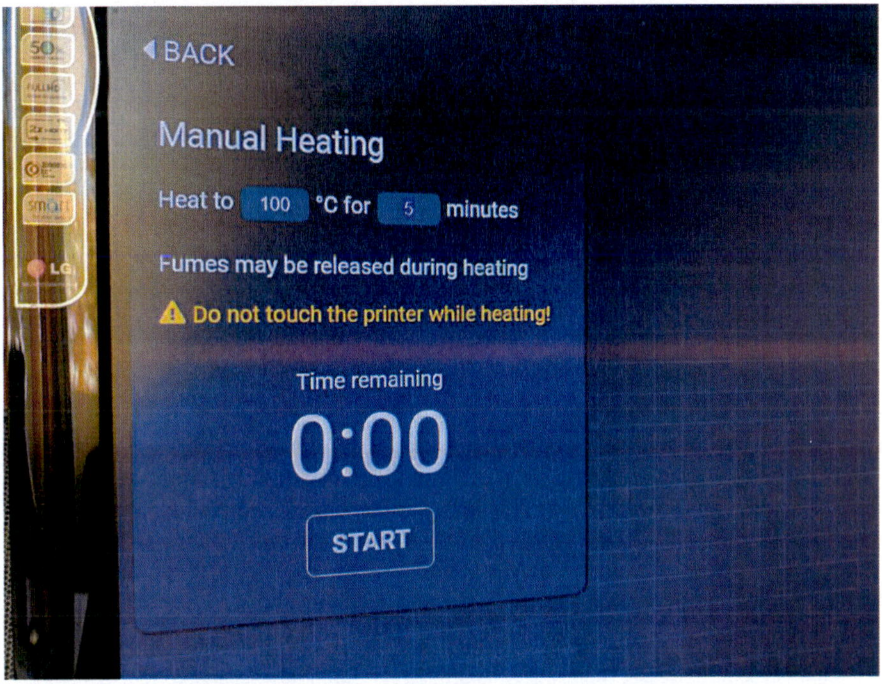

Fig. 8. Thermal treatment provide by inkjet printer

Fig. 9. Practical application for contaminants detection

5 Conclusion

The use of 3D modeling combined with the printing capabilities of an equipment such as Voltera V-ONE enables quick prototyping for electrodes on multiple substrates. Using flexible inks and flexible substrates leads to a biosensor that, with the help of the necessary electronics can be transformed into a wearable device which serves electrochemical detection in the medical field.

Acknowledgement. This work was supported by a grant of the **Ministry of Research, Innovation and Digitization, CNCS/CCCDI – UEFISCDI**, project number **PN-III-P2–2.1-PED-2019–2443** acronym **Toxinsens**. The project team will continue to collaborate in order to achieve new methods, electrodes and sensors.

References

1. Grandview webpage: https://www.grandviewresearch.com/press-release/global-biosensors-market
2. Statista webpage: https://www.statista.com/statistics/1101442/iot-number-of-connected-devices-worldwide/ Last accessed 25 November 2021
3. Stoical BE et al (2021) Uncovering the behavior of screen-printed carbon electrodes modified with polymers molecularly imprinted with lipopolysaccharide. Electrochemistry Communication 124
4. Saenchoopa A et al (2021) A disposable electrochemical biosensor based on screen-printed carbon electrodes modified with silver Nanowires/HPMC/Chitosan/Urease for the detection of mercury (II) in water. Biosensors 11:351
5. Ravariu C, Manea E, Babarada F (2017) Masks and metallic electrodes compounds for silicon biosensor integration. J Alloy Compd 697:72–79 March
6. Serbanescu M et al (2017) Smart, Low Power, Wearable Multi-Sensor Data Acquisition System for Environmental Monitoring, 10th International Symposium on Advanced Topics in Electrical Engineering (ATEE). Bucharest, ROMANIA, MAR 23–25:118–123
7. Titoiu AM, Porumb R, Fanjul-Bolado P, Epure P, Zamfir M, Vasilescu A (2019) Detection of allergenic lysozyme during winemaking with an electrochemical aptasensor. Electroanalysis 31:2262–2273
8. Vasilescu A, Fanjul-Bolado P, Titoiu AM, Porumb R, Epure P (2019) Progress in electrochemical (Bio) sensors for monitoring wine production. Chemosensors 7:66
9. Voltera pages: https://www.voltera.io/solutions/education Last accessed 25 November 2021
10. Henkel pages: https://www.henkel-adhesives.com/us/en/products/industrial-coatings/inks-coatings.html Last accessed 25 November 2021

AES Hardware Implementation Based on FPGA with Improved Throughput

Andreea Cristina Suiu Cristea and Balan Alexandra(✉)

Transilvania University of Brasov, Brașov, Romania
alexandra.balan@unitbv.ro

Abstract. AES (Advanced Encryption Standard) is one of secured symmetric key cryptographic algorithms. AES can be implemented software or hardware. Compared to software implementation, the hardware implementation is much faster, being used in applications which require higher data rate, such as routers. This paper presents the hardware implementation of AES by using Altera SoCKit board. This hardware implementation is based on efficient methods to implement internal operations, Key generation and AES controller. The paper presents a comparison of two implementations that may be used by students during the cryptographic laboratory in order to explore the operation of the AES encryption algorithm.

Keywords: AES-256 · High throughput · FPGA · Key expansion · AES controller

1 Introduction

Functionally, AES is a block cipher, with 128-bits dimension, symmetric, able to encrypt or decrypt information by using encryption keys with different length: 128-bits, 192-bits or 256-bits. It is symmetric because it uses the same key for both encryption and decryption processes [1].

In this paper, the implementation of the algorithm was optimized by using combinational logic to implement internal stages and by using efficient methods to implement the AES Controller and the Key Expansion. The present implementation aims to reduce the critical path delay and to increase working frequency. A comparison between the classical implementation of the AES algorithm and the optimized implementation is Results and Discussion section. Both implementations are made by the authors.

The implementation could be used on training the engineers in hardware security mechanisms studied in the courses of the cybersecurity master programs. Section 3 describes in detail all the mathematical operations used in the AES algorithm. The following Section provides a practical example of AESimplementation using hardware description languages using all the formulas presented in Sect. 3.

M. E. Auer et al. (Eds.): REV 2022, LNNS 524, pp. 41–50, 2023.
https://doi.org/10.1007/978-3-031-17091-1_5

2 State of the Art

Over time, different researchers tried to implement AES on FPGA, using a smaller area and reaching the best possible speed. One of this implementations is authored by Farooq and Faisal Aslam who suggested an AES implementation on spartan-6 FPGA, using five different methods for design to achieve the high performance [2]. This implementation uses 161 CLB Slices and works at 886,64 MHz. The Varhade and Kasat implementing AES on cyclone-II Altera, using 1746 CLB Slices and 32678-bits memory [3]. One of the most interesting AES implementation is suggested by Wadi and Zainal. They proposed a series of modifications, such as decrease number of rounds and redesign of S-box for increase the frequency [4]. Another implementation is approached by Wang to the purpose of speed improvement. He used pipelining and parallel processing [5]. Kaur and Vig have implemented in an effective manner the AES algorithm by simultaneous processing of several rounds [6]. This fact led to increase speed, but, also, to increase area.

3 AES Algorithm

The AES algorithm is symmetric, block cipher and iterative type in nature [2]. AES-256 allows a higher encryption rate which makes it more secure than AES-128 and AES-192. Generally, the algorithm can be divided in two parts. The first part consists of the encryption and decryption operations. It involves the all-necessary stages for performing encryption or decryption. The second part consists of the key generation for each round.

All operations in AES (i.e. addition, multiplication and division) are performed on 8-bits over the finite field GF (28) with the irreducible polynomial represented in Eq. 1 [1]:

$$m(x) = x^8 + x^4 + x^3 + x + 1 \tag{1}$$

Thus, each processed block is 16 bytes which are arranged as a 4 x 4 square matrix named state array. This state array is modified at each stage of encryption or decryption. After the final stage, the resulting state array is copied into the output matrix. The encryption or decryption process consists of four transformations in every round, except the final round where one of transformations is ignored: AddRoundKey, SubByts, ShiftRows and MixColumns.

The round key array is added to the state array over GF (2^8). This operation "translates" as a XOR operation in the form state matrix 4x4.

The Sub Bytes is a non-linear operation. This involves replacing each byte in the array with a corresponding byte value in the look-up table with a fixed size of 256 bytes. It is based on two operations: Multiplicative Inverse in Galois field GF (2^8) and Affine Transformation.

The Sub Bytes is a non-linear operation. This involves replacing each byte in the array with a corresponding byte value in the look-up table with a fixed size of 256 bytes. It is based on two operations: Multiplicative Inverse in Galois field GF (2^8) and Affine Transformation.

$$S(X) = AffineTransformation\left(X^{-1}\right) \tag{2}$$

$$AT = \begin{pmatrix} 1\,1\,1\,1\,1\,0\,0\,0 \\ 0\,1\,1\,1\,1\,1\,0\,0 \\ 0\,0\,1\,1\,1\,1\,1\,0 \\ 0\,0\,0\,1\,1\,1\,1\,1 \\ 1\,0\,0\,0\,1\,1\,1\,1 \\ 1\,1\,0\,0\,0\,1\,1\,1 \\ 1\,1\,1\,0\,0\,0\,1\,1 \\ 1\,1\,1\,1\,0\,0\,0\,1 \end{pmatrix} \times \begin{pmatrix} i_7 \\ i_6 \\ i_5 \\ i_4 \\ i_3 \\ i_2 \\ i_1 \\ i_0 \end{pmatrix} \times \begin{pmatrix} 0 \\ 1 \\ 1 \\ 0 \\ 0 \\ 0 \\ 1 \\ 1 \end{pmatrix} \tag{3}$$

where, AT means Affine Transformation and i $= X^{-1}$.

The Inverse Sub Bytes operation is also a non-linear transformation. The only difference between Sub Bytes and Inverse Sub Bytes is that use different look-up tables. For encryption is named S-box and for decryption is named Inverse S-box. It based, also, on two operations: Multiplicative Inverse in Galois field GF (2^8) and Inverse Affine Transformation (Fig. 1).

$$AT^{-1} = \begin{pmatrix} 0\,0\,1\,0\,0\,1\,0\,1 \\ 1\,0\,0\,1\,0\,0\,1\,0 \\ 0\,1\,0\,0\,1\,0\,0\,1 \\ 1\,0\,1\,0\,0\,1\,0\,0 \\ 0\,1\,0\,1\,0\,0\,1\,0 \\ 0\,0\,1\,0\,1\,0\,0\,1 \\ 1\,0\,0\,1\,0\,1\,0\,0 \\ 0\,1\,0\,0\,1\,0\,1\,0 \end{pmatrix} \times \begin{pmatrix} i_7 \\ i_6 \\ i_5 \\ i_4 \\ i_3 \\ i_2 \\ i_1 \\ i_0 \end{pmatrix} \times \begin{pmatrix} 1 \\ 1 \\ 1 \\ 0 \\ 0 \\ 0 \\ 0 \\ 0 \end{pmatrix} \tag{4}$$

The Shift Rows is an operation which consists of cyclically shifted of each row of state with a certain number of steps. Thus, the first row of the state remains unchanged, the second row is shifted by one byte, the third is shifted by two bytes and the fourth row is shifted by three bytes. At encryption the cyclically shifted is at left and at decryption the cyclically shifted is at right.

The Mix Columns is a linear transformation. It is performed over columns of the state which are treating as the coefficients of polynomial over GF (2^8). Each column of the state is multiplied modulo $(X^4 + 1)$ with a fixed polynomial a(x) given in Eq. 5 in case of encryption [4].

$$a(x) = \{03\}x^3 + \{01\}x^2 + \{01\}x + \{02\} \tag{5}$$

The computation for the Mix Columns transformation in matrix multiplication is presented in Eq. 6.

$$\begin{pmatrix} s'_0 \\ s'_1 \\ s'_2 \\ s'_3 \end{pmatrix} = \begin{pmatrix} 02\ 03\ 01\ 01 \\ 01\ 02\ 03\ 01 \\ 01\ 01\ 02\ 03 \\ 03\ 01\ 01\ 02 \end{pmatrix} \times \begin{pmatrix} s_0 \\ s_1 \\ s_2 \\ s_3 \end{pmatrix} \tag{6}$$

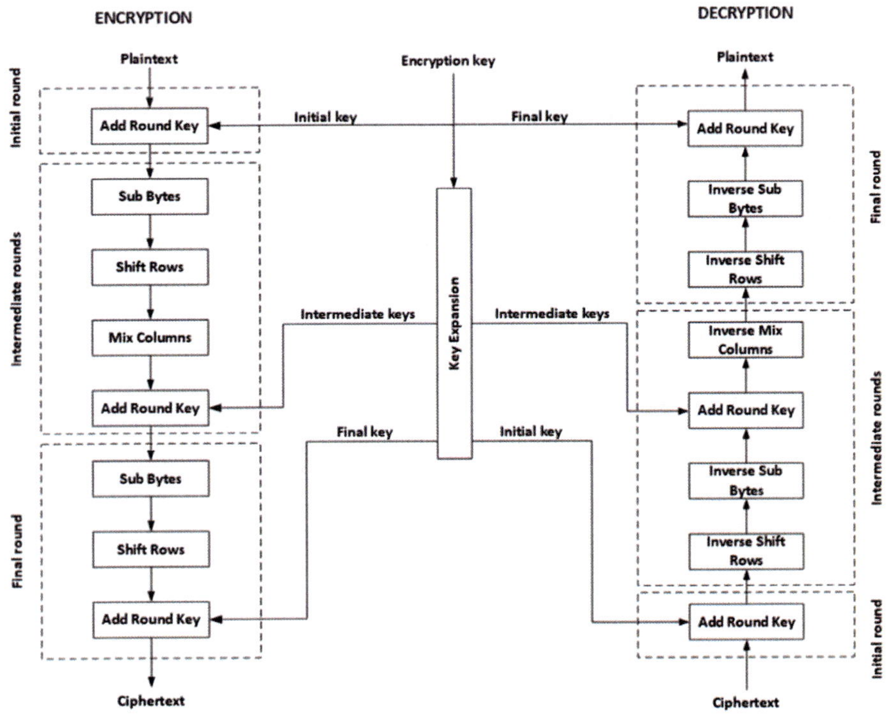

Fig. 1. Block diagram of AES encryption and decryption.

The Inverse Mix Columns multiplies the polynomial which is formed by each column of the state with $a^{-1}(x)$ modulo $(X^4 + 1)$, where $a^{-1}(x)$ is given by Eq. 7 [4].

$$a^{-1}(x) = \{0B\}x^3 + \{0D\}x^2 + \{09\}x + \{0E\} \tag{7}$$

The computation for the Inverse Mix Columns transformation in matrix multiplication is presented in Eq. 8.

$$\begin{pmatrix} s'_0 \\ s'_1 \\ s'_2 \\ s'_3 \end{pmatrix} = \begin{pmatrix} 0E & 0B & 0D & 09 \\ 09 & 0E & 0B & 0D \\ 0D & 09 & 0E & 0B \\ 0B & 0D & 09 & 0E \end{pmatrix} \times \begin{pmatrix} s_0 \\ s_1 \\ s_2 \\ s_3 \end{pmatrix} \tag{8}$$

The Key Expansion is used to generate a new key for each round of the algorithm, based on initial encryption key. For AES-256 are necessary 14 keys, one being the initial key and other 13 keys being generated. The Key Expansion consists of three operations: SubWord(), RotWord() and Rcon[i] which are performed at state level. SubWord() operation involves replacing the first four byte in the key array with a corresponding byte value in the look-up table (S-box). RotWord() is an operation that performs a cyclic permutation by one byte. Rcon[i] is the round constant word array.

4 Proposed Implementation for AES 256 Algorithm

The present implementation aims to reduce the critical path delay and to increase working frequency by using efficient methods to implement the AES Controller and the Key Expansion, which are described below.

The classical implementation of internal stages is based on combinational logic. Each stage is based on calculation of the arithmetic operation using the combinational elements such as logic gates and multiplexers. This method is used to reduce the latency of the entire design and to have a reusable architecture, each stage being a stand-alone module that contains an independent logic from other elements of the design.

The Add Round Key stage is implemented with an XOR gate, as it is in theory, and the Shift Rows stage is implemented by inverting bits in the 128-bit data block. The Sub Bytes stage is based on the S-box implemented with look-up table because it is faster in this design than the S-box implemented with logic gates. Note that Inverse Shift Rows and Inverse Sub Bytes stages are implemented in the same way, with the difference that another look-up table is used, called Inverse S-box.

From all the internal stages, Mix Columns is implemented differently from the theory, using a minimized calculation of the functions involved, because it contains the most complex operations, which would lead to a low frequency. The proposed Mix Columns implementation is based on calculation the multiplication by 01, 02, 03, 09, 11, 13 and 14 using combinational logic. This method reduces the critical path of this transformation which would have been generated if it had used look-up tables. For multiplication with 01 is no processing time, but for others is.

According to Eqs. (5–6), multiplication with 02 can be written as follows:

$$2 \times a = (a_7 a_6 a_5 a_4 a_3 a_2 a_1 a_0) \times (00000010) mod\left(x^8 + x^4 + x^3 + x + 1\right) \quad (9)$$

Knowing this, the multiplications with the other coefficients can be written according to the multiplication with 02, as follows:

$$3 \times a = 2 \times a + a \quad (10)$$

$$9 \times a = (((2 \times a) \times 2) \times 2) + a \quad (11)$$

$$11 \times a = ((((2 \times a) \times 2) \times a) \times 2) + a \quad (12)$$

$$13 \times a = ((((2 \times a) \times a) \times 2) \times 2) + a \quad (13)$$

$$14 \times a = ((((2 \times a) \times a) \times 2) \times a) \times 2 \quad (14)$$

This method provides an efficient implementation of Mix Columns and Inverse Mix Columns transformation, reducing complexity and delays.

5 Results and Discussion

The design is simulated by using ModelSim Student Edition simulator. After verification that the simulation works, the design is synthesized and implemented on Cyclone-V Altera.

The AES Controller is implemented using finite-state machine and internal registers, according to the block diagram of the algorithm. The finite-state machine is very simple, with few states and decision blocks. It has four major parts through which the data blocks flow: the initialization part in which the necessary encryption keys are calculated, the first round in which a single stage takes place, respectively Add Round Key, the intermediate rounds in which several stages take place in the same order, such as Add Round Key, Sub Bytes, Shift Rows and Mix Columns and the final round in which the final result is produced. In order to avoid critical paths, registers were used at the outputs of some states from the finite-state machine.

The proposed implementation of Key Expansion is based on using sequential circuits. Thus, Key Expansion is implemented using finite-state machine and combinational logic for the intern operations mentioned in previous section.

The block diagram for the finite-state machine used to control the Key Expansion is shown in Fig. 2. This method reduces the delay paths which are generated by combinational logic used in previous work.

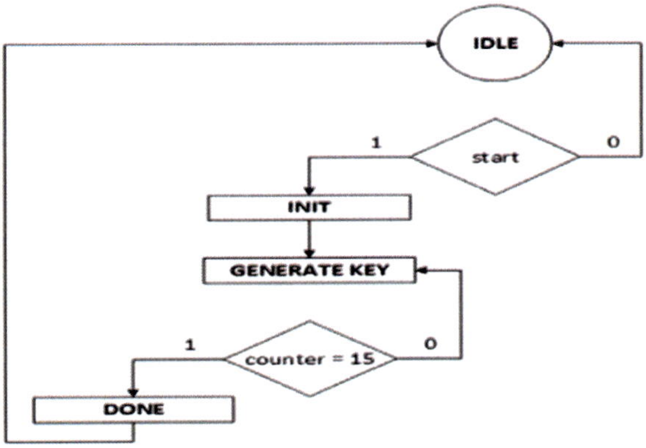

Fig. 2. Finite-state machine for key expansion.

FPGA implementation of AES-256 is carried out using Altera SoCKit with Quartus Prime Lite Edition as a synthesis tool. The results are expressed in term of area and delay. The total logic utilization for the proposed design is 2963 ALM Slices and the maximum clock frequency is 107.43 MHz. The implementation results are shown in Table 1.

Hardware implementation of AES algorithm can be characterized by multiple performance parameters such as latency and throughput [1]. Throughput is computed manually by using Eq. 15. A typical unit of throughput is Mbit/s or Gbit/s. Encryption or decryption latency is described as the time consumed to encrypt or decrypt a single block of

Table 1. Implementation results

Resource Utilization	Used	Available	Utilization
Logic Utilization (ALMs)	2963	41910	7%
Total Registers	4801	-	-
Total Pins	157	499	31%
Total Block Memory Bits	65536	5662720	1%
Maximum Clock Frequency	107.43 MHz	-	-

plaintext, respectively a single block of ciphertext.

$$Throughput = \frac{128 - bits \times ClockFrequency}{Latency} \tag{15}$$

Comparison between present implementation and previous implementation, the classical one, in terms of throughput and maximum frequency is given at Table 2.

Table 2. Throughput and maximum frequency comparison with previous work

Implementation	Device	Operation	Throughput [Mbit/s]	Maximum Frequency [MHz]
Proposed Implementation	5CSXFC6D6F31C6	Encryption	50.555	107.43
Previous Implementation	5CSXFC6D6F31C6	Encryption	27.066	48
Proposed Implementation	5CSXFC6D6F31C6	Decryption	48.249	107.43
Previous Implementation	5CSXFC6D6F31C6	Decryption	25.18	48

Comparison between present implementation and previous implementation in terms of area occupied is given at Table 3.

In the following, the data from the tables will be presented in the form of bar graphs. Thus, the differences between the two implementations can be better observed.

The comparison of clock frequency achieved by this implementation and previous implementation is shown in Fig. 3. In this figure, it is observed that present implementation has high frequency, while for previous design it is relatively less. This difference appears because of elimination of long combinational paths and, also, because of elimination of look-up tables in the Mix Columns stage where the time was wasted to read from ROM.

The comparison of throughput achieved by this implementation and previous implementation is shown in Fig. 4. In this figure, it is observed that present implementation

Table 3. Logic utilization and number of registers comparison with previous work

Implementation	Device	Logic Utilization [ALMs]	Number of Registers
Proposed Implementation	5CSXFC6D6F31C6	3938	4909
Previous Implementation	5CSXFC6D6F31C6	5248	2598

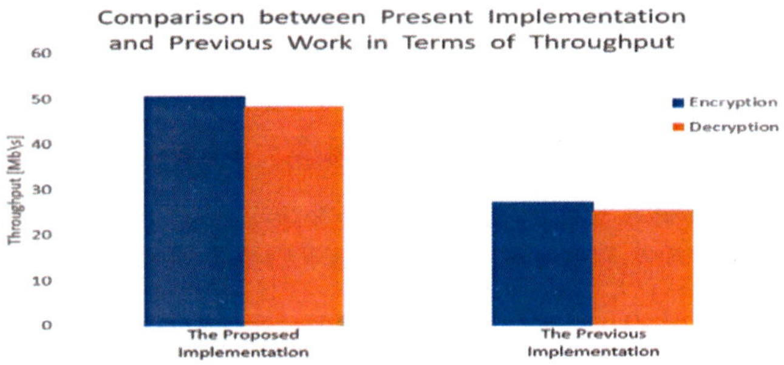

Fig. 3. Comparison between present and classical implementation in terms of throughput.

has high encryption and decryption throughput, while for previous design it is relatively less. From Eq. 15, it can be seen that throughput is a function that depends on the clock frequency and latency. The clock frequency in previous work is much smaller than the clock frequency in this work (it was explained before why), hence this difference. Also from this figure, it is observed that throughput of encryption is higher than throughput of decryption. The clock frequency is almost the same for both encryption and decryption processes, but latency is increasing at decryption. As a result, due to this increase in latency, there is a corresponding decrease in the achieved throughput.

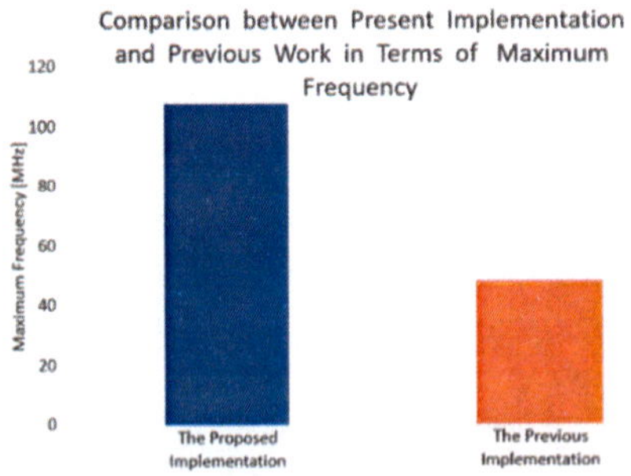

Fig. 4. Comparison between present implementation and previous work in terms of maximum frequency.

From the Fig. 5, it is observed that the present implementation of AES-256 occupies less area, while the previous implementation of AES-256 occupies high area in terms of logic utilization. This is because the key expansion was implemented sequentially, not combinational. Also from this figure, it is observed that the current implementation has a higher number of used registers than the initial one. This increase is due to the use of registers in long combinational paths to avoid critical delays.

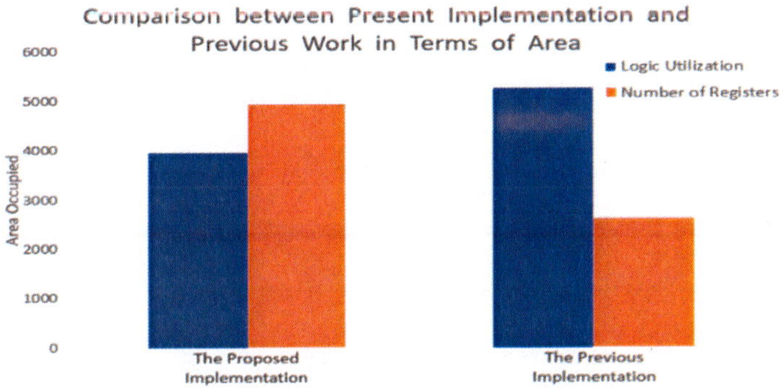

Fig. 5. Comparison between present implementation and previous work in terms of area.

6 Conclusions

In this paper, it is presented an FPGA implementation for AES-256 using look-up tables for S-box and other efficient methods for the most complex parts of AES-256 such as Mix Columns, Key Expansion and AES Controller. This implementation of AES-256 using Altera SoCKit with Cyclone V FPGA integrated and achieves high throughput of 50.55 Mbps and maximum operating frequency of 107.43 MHz. This implementation aimed at increasing the frequency and throughput, a result which was achieved. Even if the design wasn't made from scratch, but based on a previous implementation, it was demonstrated that changing a few important elements, it is reaches an almost double frequency.

References

1. Srinivas NS (2016) Akramuddin: FPGA based hardware implementation of AES Rijndael algorithm for encryption and decryption. In: International conference on electrical, electronics, and optimization techniques (ICEEOT). IEEE, pp 1769–1776
2. Farooq U, Aslam MF (2017) Comparative analysis of different AES implementation techniques for efficient resource usage and better performance of an FPGA. J King Saud Univ Comput Inf Sci. 295–302
3. Varhade SA (2015) Kasat: implementation of AES algorithm using FPGA & its performance analysis. Int J Sci Res. 4(5)

4. Wadi SM (2013) Zainal: rapid encryption method based on AES algorithm for grey scale HD image encryption. Proc Technol 11(3):51–56
5. Wang W, Chen J, Xu (2012) An implementation of AES algorithm Based on FPGA. In: International conference on fuzzy systems and knowledge discovery. IEEE, pp 1615–1617
6. Kaur S (2007) Vig: efficient implementation of AES algorithm in FPGA device. In: International conference on computational intelligence and multimedia applications (ICCIMA), vol 2. IEEE, pp 179–187

The Impact of Digitization on the Romanian Music Therapists

Fulvia Anca Constantin(✉)

Transilvania University of Brasov, Brasov, Romania
fulvia.constantin@gmail.com

Abstract. The aim of this paper is to show the benefits of digitalization in music therapy training and measurements. It also takes into consideration whether the use of music technology helps music therapists and their clients to develop certain abilities for successfully healing. Moreover, we investigate the way the modern, digitalized instruments help in diagnosing and improving the health of people with either simple or complex needs. A total of 44 music therapists were interviewed and their answers were carefully discussed. The findings suggest that digitalized music technology (i) allows music therapists to diagnose and conduct valuable therapy sessions within therapeutic contexts which has not previously been identified and (ii) offers clients a sense of accomplishment, success and independence in spite of existing health problems. We reveal that digital transformation and technology use easier the therapist's way and provides numerous opportunities to permit access to music-therapy for people with problems/needs.

Keywords: Music therapy · Digitalization · Measurement · Training · Technology

1 Introduction

1.1 Overview

One of the most benefic developments of the present is digitalization. From training to practice, from engineering and health to arts the use of digital technology is a must, a necessity.

Digital technologies, and by those we understand any electronic tools or systems, electronic devices and resources that generate, transmit, store or process data, are used in a creative or non-creative-based approach depending on the field of work and the knowledge of the user.

At present, the educational process gradually removes everything that becomes outdated or dysfunctional. The benefit of digitalization is the key to innovation, creativity and competitiveness, offering students numerous abilities and opportunities. They become more specialized and capable, their work being more dynamic and productive.

Specifically, music therapy training uses digital technology not only in the teaching area, but also in playing music, in working with music and measuring the effects of it.

M. E. Auer et al. (Eds.): REV 2022, LNNS 524, pp. 51–57, 2023.
https://doi.org/10.1007/978-3-031-17091-1_6

From mobile technology used in school to sophisticated devices used in the therapeutic process there is a wide spectrum of digital equipment ready to be used with great results. To only bring an example: it is proven that music-based digital therapeutics has the potential to facilitate rehabilitation so that sensor-automated, progressive, and individualized rhythmic loco-motor training program can be implemented successfully to train walking speed [1].

The technical (e.g. engineering and IT) training is a must in this case. Music therapists look for technology to support a client's active participation to therapy sessions or to enlarge his musical expression. Using technology in music therapy has benefits not only for the client, but also for the therapist and the interdisciplinary treatment team. However, there are clinical contra-indicators that indicate when technology might be avoided [2].

Moreover, we note the emerging technology - new discovered sensors that acquire data types from real-world and then transfer it to the cyber world processed and analyzed. An example is medical cyber-physical systems (MCPS) used in hospitals for high-quality and secure healthcare [3]. These sensors are able to collect patient information containing health data. Collected data are sent via the wireless communication medium. Wired sensors can also be used; however, wireless sensors provide more flexibility and comfort to both the caregiver and the patients. The data collected by the sensors can be stored in a server and made accessible to clinicians. While designing CPS architecture for healthcare applications, special attentions need to be paid to ensure data security [4].

Another example of sensors of last generation used in medical field is the biosensors coupled to Field Effect Transistors (BioFET.) We talk about the electron device co-integrated with the bio-receptor elements, demand that involves compatibility among the technologies and materials (e.g. enzymatic transistor designed to be sensitive to glucose.) [5] Also the integrating technology (such as computers, mobile devices like smartphones and tablets, digital cameras, social media platforms and networks, software applications, so on) is used not only in practice but also in management and collecting data from and about the client.

If for providing music there are used certain devices and/ or software, for measuring the effects of music the technology develops continuously, a whole bunch of non-invasive devices being available. We only give the examples of (i) the MindWave Mobile that measures and outputs the EEG power spectrums (waves), (ii) the EEG cap, (iii) the Wearable EEG Meditation Headset, (iv) Muse – the Brain Sensing Headband, (v) Flowtime Biosensing Meditation Headband, (vi) The Versus Headset, so on.[1]

It is proven that technology facilitates the bonding and understanding between music therapist and his patient that is transforming the patient's therapeutic process and provides an effective methodology for session analysis [6]. No doubt, using technology means to combine elements of human computer interaction (HCI) with music psychology/therapy, so on, that can be used to increase access to active music-making opportunities.

[1] See https://www.diygenius.com/hacking-your-brain-waves/ for a description of the models.

1.2 Literature Review

Why is Music Working no Matter the Technology Used?

Either for relaxing, preventing or healing, music therapy is one of the non-invasive methods that have as central element: neuro-psychology. The way it acts is explained by the vibrational therapy given by the sound. The Romanian inventor Gogu Constantinescu placed his life work and applications on the theory of *sonicity*. Basically, he proved that energy can be transmitted by vibration. Through a fluid pipe (e.g. water), mechanical energy can be transferred from a source to a consumer[2] [7].

Years later, Olivia Dewhurst Maddock stated that sound is the result of a vibratory movement of particles and objects. Even if our ears can only perceive a small fraction of the vast vibratory spectrum that exists everywhere, both created by nature and man-made, the sound and vibration produced by it has the ability to deeply penetrate the physical, emotional and spiritual level and make changes. Properly applied, sound can act therapeutically, restoring the musical integrity of body and soul [8].

Since the nineteenth century, the development of scientific research has made possible to analyze and measure the physiological effects of music on breathing, heart rate, circulation, blood pressure, which has gradually proven the value of sound and therapeutic music. Scientific studies have confirmed the idea that resonance is the basis of healing through sound and music. When the sound waves penetrate into the body, they transmit live vibrations to cells that help restore and consolidate a healthy form [9].

Therefore, the human being continuously transmits and receives the frequencies corresponding to the physical, emotional, mental and spiritual life. These frequencies could be natural or created by technology.

However, there are powerful arguments on the resonance phenomenon, not only theoretical but also practical (e.g. quantum physics). When we begin to look at everything that exists from this perspective, the world's reality appear to be composed of specific energies that have a certain frequency of vibration. From this point of view, our perspective changes because we discover and have the proof that energy fields exist. The conclusion is that everything vibrates at certain rhythms depending on specific frequencies. So, from the most subtle aspects to the coarsest forms of manifestation, everything is ordered in certain vibration frequencies [10].

Edgar Cayce who said that regardless of the illness, the patient is always a person who suffers from a lack of vibrational coordination between his three bodies (physical, mental and spiritual). This non-coordination can be corrected through musical vibrations: „Tones and sounds will be the way through which the body's energies can be coordinated; is the basis that will allow them to function perfectly." [11] The examples cold go further as the latest studies on neuro-science with the newest technology available reveal more proofs of how music works on humans.

About Digitalization and Technology

The 21-century indicated an increased technology usage since it became accessible. No matter the size of the devices - smartphones, tablets or laptops, they all became necessary at school, at work or home. This burst of technology has also led to changes in how we

[2] In 1922, the treaty was published in Romanian as *The Sound Therapy*.

interact with music as pre-installed software has become more accessible. Developments in hardware and software, and creation of new instruments that utilize technology, have limited the creation and production of music, but also the application of music therapy [12].

A comprehensive literature review of technology and its implication in music therapy education, theory and practice counted seven types of technologies: "(i) adapted musical instruments, (ii) recording technology (iii) electric/electronic musical instruments, (iv) computer applications, (v) medical technology, (vi) assistive technology for the disabled and (vii) technology-based music/sound healing practices" [13] (p. 291). Three other types of technologies involving sound waves are: (viii) analysis and logging of data about client progress, (ix) creating and hearing listening material, and (x) medical technology that involves sound waves. There are examples of using all these technology but currently, in Romania very few types are to be found in the cabinets.

The modern clinical and therapeutic music therapy uses digital handheld music devices (multitasking ones) aiding in music-making, basic in the practice of music therapy with one of the four methods of music therapy (recreating, improvising, listening, and composing) [14]. The creation and development of applications allow music therapists to use musical methods in parallel with practices using old-fashioned instruments.

In Romania, music therapy is still at beginning, there is more to learn. The application of music is mostly passive, receptive so the therapist makes sure that the client can hear the music in very good conditions with no distortions. He uses techniques to measure the effect of sound on devices that allow biomedical measurements such as EEG, EEG, ECG, HR, BP, or temperature that also provides the level of relaxation or awareness during listening music. Sometimes, clients are encouraged to play music but are used mostly regular instruments not electronic ones. The experience will change.

2 Short Research

2.1 Purpose of the Research

The purpose of this research was to determine if the Romanian music therapists use digital music technology in their work, and in what conditions. There is only one Master Program in Music Therapy - that of Transilvania University of Brașov. Out of 83 Music therapy students that graduated until now, a number of 47 people were invited to complete a 6 - question rapid survey. We validated 44 participants. Their age is between 27 and 59 years old. We wanted to know (i) what is their main line of work – music therapy, psychology, medicine, education or any other, (ii) if they use digital technology in their line of work, (iii) if their clients are disturbed by the use of digitalization during therapy session, (iv) if therapists have an easy access to measurement technology, (v) how music therapists acquired knowledge about using the music and music therapy technology, (vi) if they adapted their choices of technology on gender, age, and/or geographical location, and if they could give a positive example of using digitalization.

The completion rate was of 100%. All forty-four music therapists were involved in the experiment. The survey was e-mailed alongside the instruction and a protection of data document. The answers were received quickly, some of them in minutes or hours. There was no need to reduce the variables that could alter the results of the study.

2.2 Discussions of Answers

Regarding the questionnaire, we had the next answers:

(i) **What is your main line of work – music therapy, psychology, medicine, education or any other?** Out of the 44 respondents 13 declared being music therapists / psychotherapists, 7 respondents declared using music therapy in medicine, 14 use music therapy adjacent education, 6 are social workers and 4 are combined arts therapists.

(ii) **Do you use digital technology in your line of work? If yes, please enlarge the answer.** 72,7% (n = 32) of respondents gave a positive answer adding that they use the technology in regular basis. Their answer included computer software, recording technology (more or less sophisticated), electric/electronic musical instruments (e.g. digital piano - electronic keyboard), medical technology (wireless blood pressure, temperature and pulse measurement devices), and frequency measurements with sensors (Mindwave type). The rest of 27,3% of respondents said that they sometimes use technology. None of them use new medical devices. It is the case of several teachers, social workers or art-therapists.

(iii) **Do you have an easy access to technology/measurement technology?** Although all music therapists declare that they have access to technology, most of the respondents (78%) gave a negative answer regarding measurement or sophisticated technology. The reasons varied between the costs and the lack of it on the market (e.g. Mindwave, Muse, etc.) Other difficulties involved the lack of knowledge and information about the technology.

(iv) **Are your clients disturbed by the use of the existent technology during therapy sessions?** The answer indicated that 60% of the adult clients were not disturbed by the electronic devices/ technology. They gladly accepted to see the effect brought by music. The other answers were: so-so (20%) and rejection (20%).

(v) **How did you acquire the necessary knowledge about using the digital music and music therapy technology?** Answering this question 52,3% of respondents (n = 23) declared that they self-taught the basics of digital music technology. Regarding the music therapy technology they responded that apart from using the instructions the devices came with, it was of help the internet courses and journal articles.

(vi) **Did you adapt your choices of technology on gender, age, and/or geographical location?** The answer was *no* for the gender and location, but *yes* for the age variable. Sometimes is hard to impose some kind of technology on children, mostly on children with disabilities. For example, the use of an electric piano create no problems at all, thus the use of sensor mobile technology is not accepted by the children. It is essential for therapist to show patience, creativity and adaptability.

Please Give Me a Positive Example of Using Digitalization

There were only few examples given and all of those indicated that more training in music technology related to clinical practice is needed, with attention given to ways to make more technology accessible to a variety of learners. The most complete answer came from a clinical neuro – psychologist that was trained in the field of music therapy. Apart from computer software, electronic instruments, sensor based measurement devices he has access to imaging technology (X-rays, CT scan, MRI and fMRI MRI ultrasound and other nuclear medicine imaging, including positron-emission tomography (PET) so his diagnosis and therapy session is complete and the results likely positive.

3 Conclusion

It is clear that incorporating music technology into clinical practice has numerous potential benefits, even though it can create gaps between the developer, clinician, and client, mostly for music therapists who are not immersed in technology. The music therapy practitioners may find it difficult to keep up with changes in technology, and figure out how to combine and integrate them into their practice. However, the utilization of technology provides exceptional access to music-making for those that could not access traditional musical instruments. The possibility of using technology is evident and there is still a growing need for digitalization, for using technology in music therapy which would lead to a more prepared new generation of music therapists.

Nowadays, as the worldwide Coronavirus (COVID-19) pandemic alters our society, music therapists adapt their therapy delivery to ensure client safety. One solution is telehealth, but for this music therapist's needs practice-based knowledge, understanding of clients, available and proper technology and a virtual service that permits technical capabilities. In Romania music-therapists alongside psycho-therapists accepted the challenge. The services meet the proposed goal, are accessible, appropriate, and develop the therapists' abilities and knowledge about technology. Although the technology is not the newest one, depending on the clients' acceptance and abilities, a new step toward digitalization was made. Music therapists successfully embraced telehealth as necessity and continue healing.

References

1. Hutchinson K et al (2020) A music-based digital therapeutic: proof-of-concept automation of a progressive and individualized rhythm-based walking training program after stroke. Neurorehabilitation and Neural Repair
2. Magee W, Burland K (2008) Using electronic music technologies in music therapy: opportunities, limitations and clinical indicators in British. J. Music Ther. 22(1):3–15
3. Dey N, Ashour AS, Shi F, Fong SJ, Tavares JMRS (2018) Medical cyber-physical systems: a survey. J. Med. Syst. 42(4):1–13. https://doi.org/10.1007/s10916-018-0921-x
4. Haque SA (2014) Review of cyber-physical system in healthcare. Int. J. Distrib. Sens. Netw.
5. Ravariu C, Manea E, Babarada F (2017) Masks and metallic electrodes compounds for silicon biosensor integration. J. Alloys Compd. 697:72–79
6. Magee W, Burland K (2008) Exploratory study of the use of electronic music technologies in clinical music therapy. Nord. J. Music Ther. 17(2):124–141

7. Cezar C (1984) Introdution to Sonology. Musical Publishing, Bucharest
8. Dewhurst Maddock O (2003) Sound Therapy. Teora Publishing, Bucharest
9. Pop I et al (2007) Sonicity applications – Experiments. Perfomantica Publishing, Iaşi
10. Misje R (2013) Music Technology in Music Therapy. Unpublished manuscript, University of Bergen, Bergen, Norway
11. Magee WL (2012) Music and medicine: applications of electronic music technologies in addressing psychic pain [video]. https://www.youtube.com/watch?v=u9-OJs93naA. Accessed 11 Dec 2021
12. Crowe BJ, Rio R (2004) Implications of technology in music therapy practice and research for music therapy education: a review of literature. J. Music Ther. 41(4):282–320
13. Nagler JC (2011) Music therapy methods with hand-held music devices in contemporary clinical practice: a commentary. Music Med. 3(3):196–199
14. Knight A (2013) Uses of iPad® applications in music therapy. Music Ther. Perspect. 31(2):189–196

New Laccase Portable Monitor

Petru Epure(✉), Andra Perju, and Petru P. Epure

EPI-SISTEM SRL, str. Livezii nr. 17A, Sacele, Romania
petru.epure@epi.ro

Abstract. This paper presents an embedded system that measures the laccase activity in vineyard. The second stage of development of the system integrates a potentiostat for electrochemical measurements performed on screen printed electrodes upgraded with GPS coordinates to localize the danger. The infection risk is favored according to several climatic factors. Monitoring the vineyard every week become important according to humidity, temperature, sun exposure and many other parameters. In this way we can predict the possible appearances of the disease in the incipient phase.

Keywords: Winemaking · Disease · Electrochemistry · Biosensors · Measurement & control · Embedded C · IoT

1 Introduction

The biosensors market was valued at $17,500.0 million in 2018 and is expected to reach $41.8 billion by 2028, registering a CAGR (Compound Annual Growth Rate) of 7.9% from 2021 to 2028 [1]. Biosensors are analytical devices that can convert a biological response into an electrical signal. Biosensors majorly comprise three parts, including a component that identifies an analyte and produces signal, a signal converter combined with a digitalization device. Biosensors find their applications in food & beverages, environmental monitoring, home care diagnostics. And various nano mechanical biosensors. In addition, biosensor applications are also widely dominant in the field of plant biology to study metabolic processes.

The wearable market was in 2020 about 40.65 billion USD and will evolve with a CAGR of 15.9% between 2020 and 2027[2]. Just for the smartwatch devices the sales in 2019 were at 148 million units having a grow of about 30–35%. The evolution will continue for the next 5–10 years in the same pace. The IoT (Internet of Things) market will reach 1.1 trillion USD between 2018 and 2023 giving a significant impact on the global internet network. It is estimated that by 2025 over the world will exist more than 30.9 billion connections dedicated to IoT [3]. The evolution of this huge infrastructure involves big efforts to be made by companies from various sectors providing hardware and software parts or modules. Electrochemical detection of laccase activity is not very spread because most of determinations are made using spectrometry. Recent portable potentiostat offers the opportunity the evaluate the laccase activity on field in short time.

M. E. Auer et al. (Eds.): REV 2022, LNNS 524, pp. 58–66, 2023.
https://doi.org/10.1007/978-3-031-17091-1_7

2 Device Description

The main challenge in the wine-making industry is vineyard infection caused by laccase, which affects 15–20% of global production. The measuring system is designed to be an useful instrument for wine makers to respond to this type of threat or similar [4]. The platform is made up of a detection system combined with an evaluation and decision software. On site evaluation for laccase during the production stages of wine-making results in better decisions towards stopping the fungi attack in a timely manner.

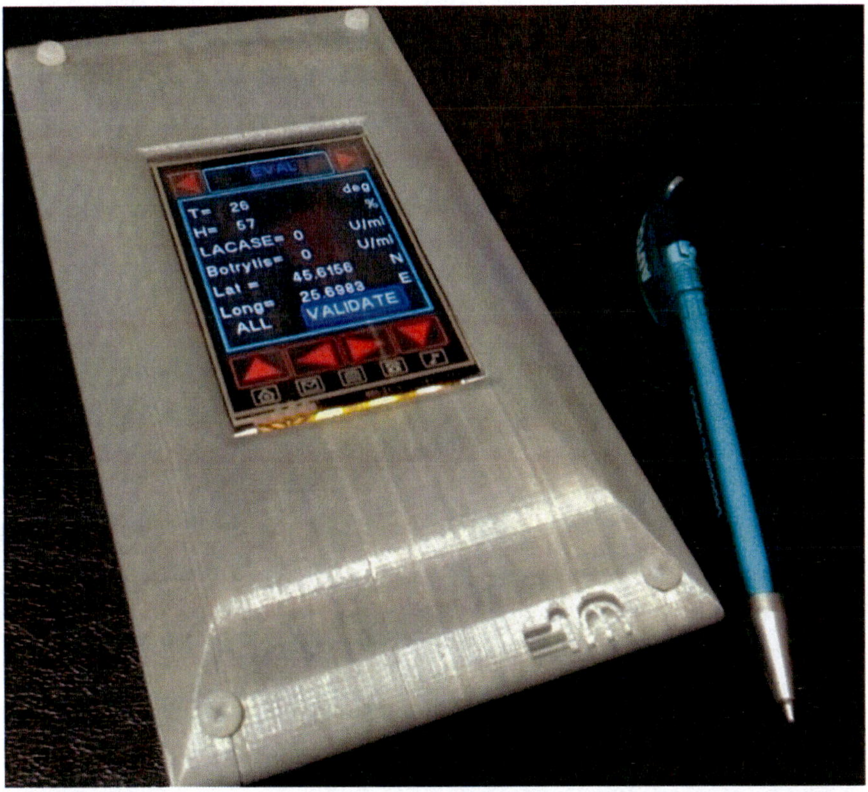

Fig. 1. Portable device showing data evaluation after an electrochemical evaluation

The equipment includes gold electrochemical sensors, GPS sensor and a GPRS module to send data about the fungi infection in the vineyard. Technic specifications were fixed in place so that the portable device is able to fully implement the necessary methods of electrochemistry [5], localization and data processing with the results being available in screen and also online via Cloud.

The device is made up of multiple modules. The power supply module facilitates the portability of the device with a 3.7 V 10000 mAh battery, enough for all power requirements. The battery is assisted by auxiliary circuits for charging and monitoring. The electrochemical module is represented by a specialized ADuCM355 chip with PICO firmware like in Fig. 2. For electrochemical procedures the method parameters [6] must be first loaded onto the chip. This is done with an UART connection with the main microcontroller. The GSM and GPRS modules made by Spanish partner are designed to give localization capabilities to the device while also facilitating data transfer to the Cloud. The GPRS works with 2G communication and requires 500 mA with the GPS requiring 1A. Communication between this module in Fig. 2 and the main controller is done using in private I2C protocols. The control module is build around a ARM Cortex-M3 from Cypress Semiconductors (Infineon group). The PSOC 5 is reconfigurable and allows for digital routing. Firmware is developed in-house at EPI-SISTEM and controls the ADuCM355 electrochemical chip, the VIZIC interface and the GSM GPRS module. The graphic interface is displayed on a SmartGPU2 screen from Vizic Technology and allows for touch-screen applications and commands.

Fig. 2. Detailed modules around the PSOC controller

The working methods are implemented within the device and so the end user can start from the GUI an amperometric measurement as well as measurements for temperature and humidity. Data sets are processed using the assigned calibration curve. After data validation the data package is sent to the Cloud via the GPS/GPRS module. If the transmission is hindered then the data package is saved in the module and is to be sent when conditions permit. The transmission protocol follows several proprietary directives put in place by the Spanish partner.

A working protocol was put in place for this type of determinations using optimized electrodes. Preliminary tests were made in the partners laboratory using the portable

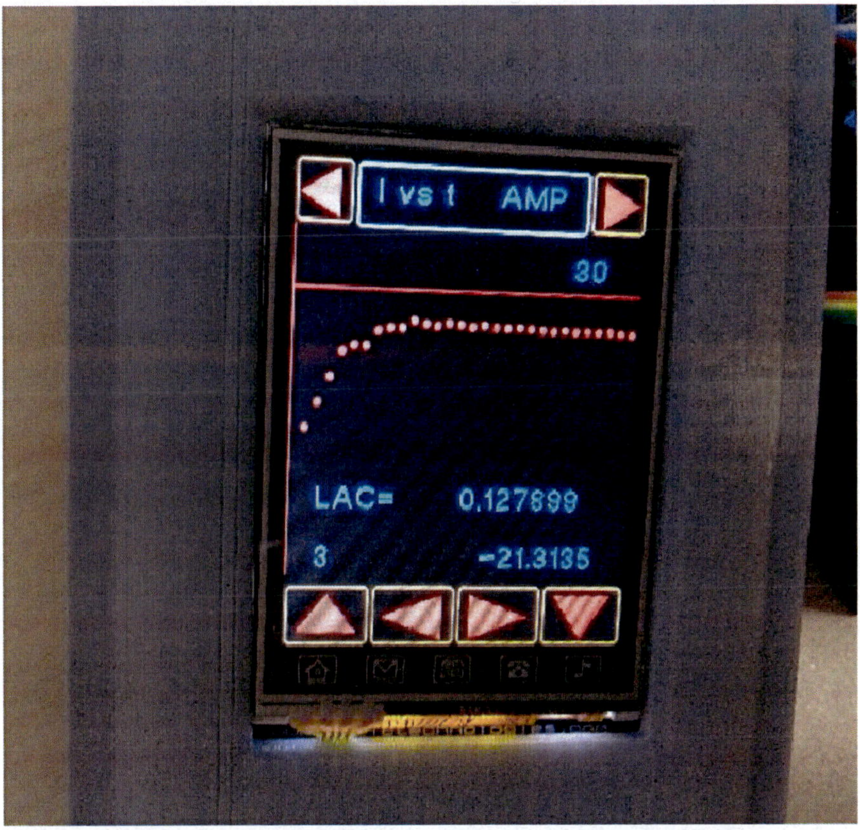

Fig. 3. Graphical representation based on amperometric response

device in parallel with specialized laboratory equipment. This process was important in developing the device and eliminating potential faults of the electrochemical interface. Tests were conducted with standards with known concentrations using the protocol created by electrochemists. Multiple tests like in Fig. 3 were run to determine statistical errors and calibrate the device providing calibration curves function of ambient temperature.

3 Software Description

The software development includes proprietary protocols used for communication with the electrochemical module the GPS/GPRS module and data processing. Software was developed using PSOC Creator and Kiel C. The GPS/GPRS module ensure a connection via a private APN protected by password. Data are sent using a TCP protocol to the cloud and the module will receive back a confirmation sequence. If the confirmation is ACK the data had been recorded in the database if the structure is according to the syntax. If not, data are stored on a table for further examination. If the confirmation is

NACK then the module stores unsent data until the next connection to APN. Normal transmission data is managed by a directive and the previous unsent data are managed with a special directive. Some technical incidents caused the revision of the firmware and a new version was install in order to fix some functional errors. Multiple datasets were transmitted during the summer of 2021 resulting in identifying and fixing issues on both sides of the transmission.

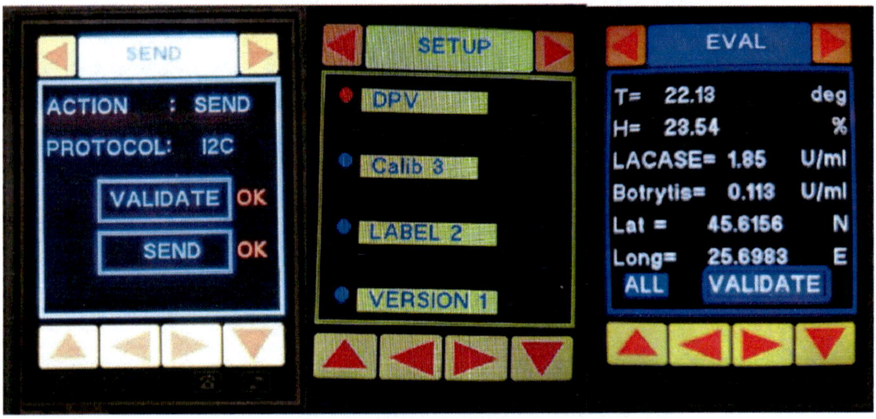

Fig. 4. Graphical user interface menus

Data can be reviewed after the electrochemical measurement is complete and before the data set is send to the Cloud like in the Fig. 4 - right. The data transfer initiated like in Fig. 4 left takes about 3–4 s or around 13 s if the network is busy. Practical tests have showed that the stability of the screen-printed gold electrodes is good and temperature measurement made by the specific sensor is used to compensate and modify the result based on calibration curves Fig. 4 - middle.

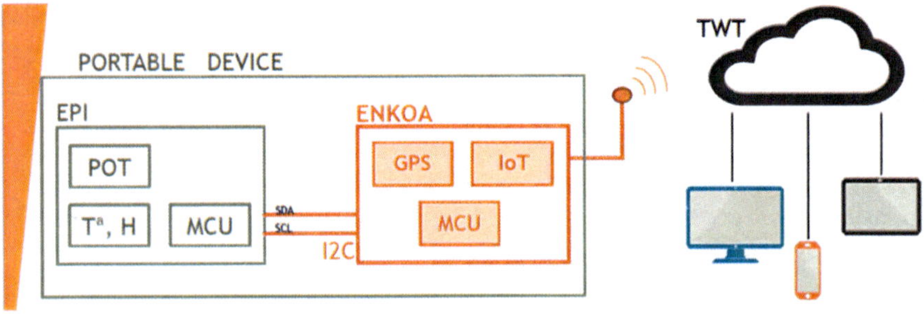

Fig. 5. Scheme of the system showing how the tasks are divided between partners

The device uses a special labeling model in 6 characters with the first character being the grape variety, the next 2 being the number of the sample, next one the level of grape infection and the last 2 characters being used for localization. This label is used in the

Cloud for database management and the dataflow is build according to Fig. 5. Access to the database is granted by user and password to perform security verifications. Multiple tables are contained in the database with each one keeping track of different parameters recorded by the device. The tables include information regarding the time of the data packages, localization, temperature, humidity, laccase and botrytis concentrations. The tables can be downloaded and analyzed with specific software.

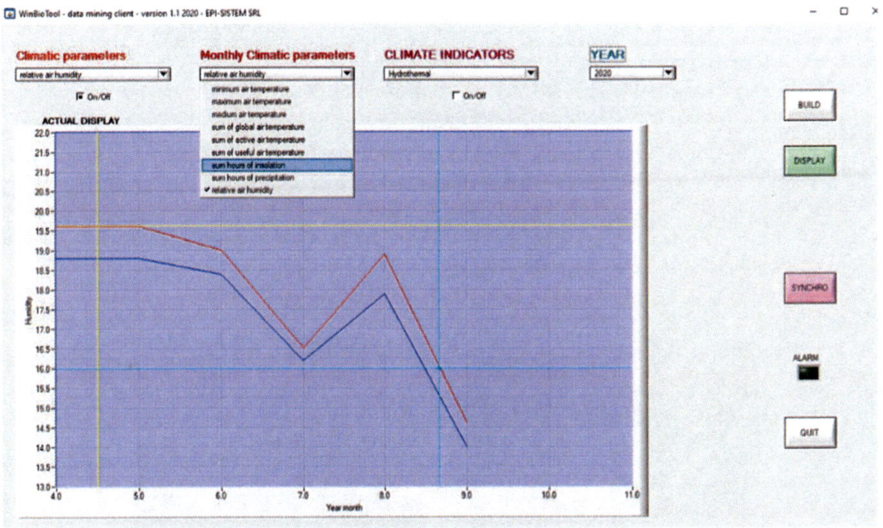

Fig. 6. Offline software application

The specific software can be a cloud version accessible like a web application for a number of credentials. Another version of specific software can be used by local users via an offline database consolidation by importing latest measurements tables and applying filtering criteria like in Fig. 6.

4 IoT Mapping

The validation of the system took place at the headquarters and in the laboratory of a vineyard in Spain where the method developed by Romanian specialists was explained to the local vineyard specialists from Eltciego. The cloud representation of the laccase activity is showed in a physical map like in Fig. 7. Although the weather was rather unfavorable, the mixed team was properly organized and performed measurements in the vineyard. It was found that in that area the GPS signal was not available and GPRS coverage did not allow immediate data transmission. The device has possiblity to resend data when the GPRS coverage is available. Storage is keeping data for 2 days.

The measurements were performed using a methodology similar to that explained in Chapter 2 adapted to the existing conditions in the lab and vineyard [7]. No traces of fungal infection have been found, a fact confirmed by laboratory tests. Participants

Fig. 7. Vineyard areas evaluation on the map (Rioja Area, Spain)

understood the proposed methodology and how the device can be exploited and in this sense several improvement ideas have been formulated for a future version.

Fig. 8. Device testing in the vineyard

In the Fig. 8 there are some aspects from the validation process in the field/vineyard. With the occasion of the validation a new data communication to Cloud was tested in several locations. To be noted that the GSM coverage is not very extensive and that in some areas the GPS/GPRS module retains the measured data until it reconnects again to the private network access using GPRS new connection. This incident was also been reported in both countries involved (in the vineyard areas of all project partners).

All these measurements were possible using several versions of PCB's (Fig. 9) adapted to PSOC5 microcontroller, AduCM335 electrochemical microcontroller, DHT11 sensor and powered by a LiPo battery at 3.7 V. Control of the configuration was design based on SmartGPU2 from Vizic Technologies, part of the menus are presented in Fig. 4. The reliability of the device is very good for this stage of development. We anticipate a new version of the device with advanced functionalities already discussed.

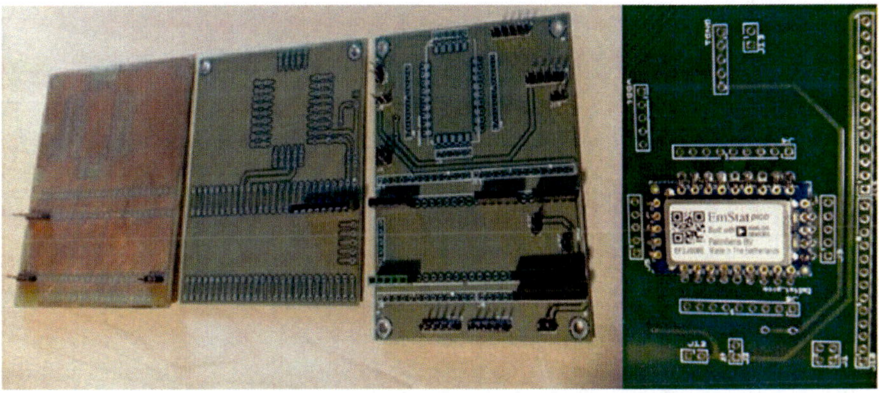

Fig. 9. Development boards for PSOC and electrochemical chip

5 Conclusion

This paper presents a series of activities related to the development of a new portable device using embedded programming and 3D technologies. At this stage it was implemented the measurement protocol, it was calibrated the electrochemical response according to TH measurements. The device showed in Fig. 1 implemented amperometry and voltammetry determination for characterization electrochemical level of laccase infection in the vineyard or winery. Several variants of housing for the device have been made through 3D design technology and 3D print in order to assimilate the module developed by project partners. Functional specifications have been drafted and can now be used for portable device optimization. A second improved version of the interface has been made using graphic controls in embedded technology by using a SmartGpu2 device with touch capabilities. The software application has good stability features and is available to the user data and graphs related to data measured during live measurements.

This portable device is a result of collaboration between private SME's engineers and specialists from universities and research institutes. The portable solution presented in this paper was validated in vineyard by Spanish specialists and approved by comparison with lab determinations. The previous experience of all partners [6–8] was conjugated to create a new measuring method and a new portable measuring device. Using this electrochemical device the laccase infection risk is assessed in line with the calibration curve and the user can map out the infection spread with the help of multiple measured points. Based on this information the wine making process gets a head start in the fight with fungi attacks. Despite the fact that several solutions were proposed in the past [9] and the electrochemical determination is available since 80's [10] the portable measurement solutions are not spread on the market yet.

Acknowledgement. This work was supported by a grant of the **Ministry of Research, Innovation and Digitization, CNCS/CCCDI – UEFISCDI**, project **ERANET-MANUNET-III-WINBIOTOOL-1**, contract no.150/2020 within PNCDI III. Our team will continue developments in order to provide an enhanced version of the portable device.

References

1. Grandview webpage (2019). https://www.grandviewresearch.com/press-release/global-biosensors-market
2. Wearable Technology Market Size Analysis Report 2028 (grandviewresearch.com) (2018)
3. Statista webpage (2020). https://www.statista.com/statistics/1101442/iot-number-of-connected-devices-worldwide/. Last accessed 25 Nov 2021
4. Titoiu AM, Porumb R, Fanjul-Bolado P, Epure P, Zamfir M, Vasilescu A (2019) Detection of allergenic lysozyme during winemaking with an electrochemical aptasensor. Electroanalysis 31:2262–2273
5. Vasilescu A, Fanjul-Bolado P, Titoiu AM, Porumb R, Epure P (2019) Progress in electrochemical (bio)sensors for monitoring wine production. Chemosensors 7:66
6. Gaspar S, Brinduse E, Vasilescu A (2020) Electrochemical evaluation of laccase activity in must. Chemosensors 8:126
7. Quintela S, Villaran M, Elejalde E (2011) Occurrence of ochratoxin a in rioja alavesa wines. Food Chem 126:302–305
8. Epure, P., Ursutiu, D., Samoila, C.: Remote Monitoring and Control of Winemaking Parameters. In: 2019 5th Experiment International Conference (exp.at'19), pp. 214–219 (2019). https://doi.org/10.1109/EXPAT.2019.8876474
9. Cuadrado MU, Pérez-Juan PM, De Castro ML, Gómez-Nieto MÁ (2005) A fully automated method for in real time determination of laccase activity in wines. Anal Chim Acta 553:99–104
10. Zouari N, Romette J-L, Thomas D (2007) Continuous-flow estimation of laccase activity in rotten grape juice by a computerized electrode. J. Chem. Technol. Biotechnol. 40:195–201

Online Labs in Industry and Engineering Education

Development of a Simulator Tool for Teaching the Autonomous Vehicles Behavior

Ioana-Diana Buzdugan(✉), Ioana-Alexandra Roșu, and Csaba Antonya

Transilvania University of Brasov, Blvd. Eroilor 29, 500036 Brasov, Romania
ioana.buzdugan@unitbv.ro

Abstract. The paper is proposing a high-level simulation tool for teaching and learning autonomous vehicle behavior. Simulation tools offer numerous points of interest compared to conventional teaching strategies. Different concepts are presented, some of them allowing interactions between the user and simulated equipment (vehicles, sensors). A simulation tool has been developed based on Kolb's Experiential Learning Theory (ELT) in a pedagogical manner. The teaching-learning methodology presented in this study is a reflection on interactive, collaborative, and experiential learning environment for autonomous vehicles curriculum, using Kolb's theory as an educational approach to promote student learning and development, preparing them for real-world situations, and an interactive driving simulator as a teaching tool.

Keywords: Autonomous vehicle · Driving simulator · Kolb's theory

1 Introduction

The development of autonomous vehicles in various operating environments (urban, industrial, etc.) is a subject that concerns today all major vehicles manufacturers on the one hand, but on the other hand, it is also a topic for industrial and university research centers. Autonomous vehicles are currently at the top of university and industrial research because of their numerous benefits like improved safety, decreased congestion, increased mobility, and so on [1].

The effectiveness of automotive courses encourages students to be curious, productive, innovative, and enterprising, thus involving the generation of ideas and taking action. The quality of the academic educational environment is decisively determined by the blended learning methods of teaching and the students' opinion [2].

Compared to conventional teaching methods, simulation tools have several advantages, they can demonstrate different perspectives, allow users to interact with simulated equipment (vehicles, sensors), and provide them feedback that allows them to develop their knowledge and abilities [3].

This paper presents the use of Kolb's theory which includes four stages [4, 5], presented in Sect. 3.1–3.4, to develop the base of a series of simulation applications. The structure of this paper is presented in three sections. Section 2 presents the simulator and evaluates its efficiency, as a teaching and learning tool. Section 3 describes the suggested teaching-learning methodology which is based on Kolb's Experiential Learning Theory. In the last section we presented some general conclusions as well as our future work.

M. E. Auer et al. (Eds.): REV 2022, LNNS 524, pp. 69–80, 2023.
https://doi.org/10.1007/978-3-031-17091-1_8

2 Driving Simulator for Experiential Learning

In the academic courses, more precisely in the engineering area, is well known that equations, text, diagrams, tables, and graphs represent ways of presenting the results obtained following the performed calculations in the research [6]. According to [3], one of the most challenging aspects of academic activities is trying to make a connection between theoretical and analytical results to real-world situations. Thus, learning how to integrate these aspects into knowledge contributes to the students' development in the implementation and design of real-world systems. One approach to providing all this knowledge is based on using a driving simulator, where the students not only study but also have the possibility to practice the application of their understanding.

The proposed simulator used for studying the autonomous vehicles' behavior in different situations offers the possibility of interaction with vehicles through the steering wheel and pedals, and provides the user a significant knowledge of the system behavior in a multimodal driving environment. This increases the efficiency of teaching courses by creating an interactive course for students.

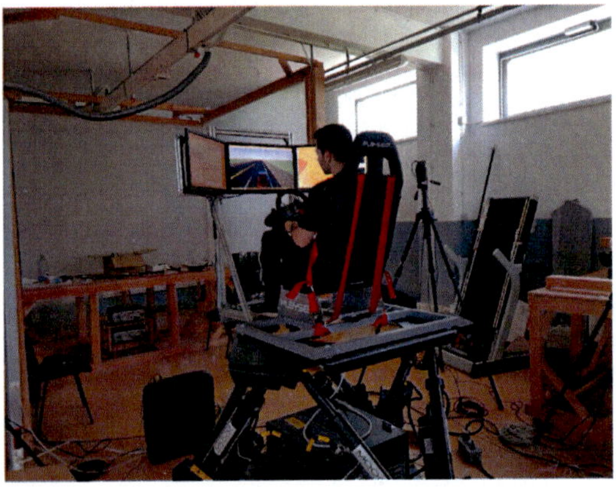

Fig. 1. The vehicle simulator

2.1 The Simulator Environment

According to [7], driving simulators may be used to experience the presence of autonomous vehicles and visualize their behavior according to different driving scenarios.

Our vehicle simulator consists of a driving seat mounted on top of a Moog hexapod system[1] (Fig. 1). This system provides the driver with inertial feedback into six degrees of freedom (i.e., roll, pitch, yaw, heave, sway, and surge) as well as visualization on a

[1] Moog motion system, https://www.moog.com/products/motion-systems.html, last accessed 2021/12/21.

multi-screen display. The interaction of the user with the vehicle is performed directly through the steering wheel and pedals.

The driving simulation visualization is based on the Epic Games Unreal simulation environment (Fig. 2) and is displayed on three screens in front of the user. The Epic Games Unreal engine[2] is a graphics game engine that includes visual development tools for a high level of the environment details for a real driving experience. In addition, the user can choose from a variety of driving scenarios and the vehicle dynamics are synchronized with the Moog hexapod motion system, thus it can create a reactive environment and can influence the decisions that autonomous vehicles have to make in different situations. The selected driving scenarios are Free Driving, Vehicle Following, Lane Changing, and Overtaking Maneuver.

Fig. 2. The driving simulation visualization in Unreal Engine - lane changing maneuver

2.2 Simulation Framework

A simulation framework is a flexible tool that can be modified according to the user's preferences and needed. The architecture of a high-level simulation framework are presented in [8]. The proposed autonomous vehicle simulation framework is composed of modules that contain vehicle dynamics – with speed control, vision-based perception, decision-making algorithms, and 3D – 2D visualization. This provides the ability for students to change then integrate those modules to visualize their effect on the vehicle performance in simulation [9].

The vehicle dynamics model, as well as the filtering and motion-cueing algorithms, are developed in MATLAB-Simulink software. Those algorithms convert the driver's

[2] Epic Game Unreal Engine, https://www.unrealengine.com/en-US/, last accessed 2021/12/21.

calculated inertial forces into displacement commands of the hexapod system in three directions (surge, sway and heave) and three rotations (roll, pitch, and yaw) [10].

A finite decision set (free drive, vehicle following, changing lane, overtaking, and emergency brake) provides a starting point for modeling traffic participants (actors). The bicycle kinematic model is used to represent the vehicle dynamics in the longitudinal and lateral directions – see Fig. 3 [11]. This model offers the fundamental characteristics of a vehicle through simplifying assumptions in normal driving conditions. The bicycle kinematic model consists on longitudinal velocity, lateral velocity, and yaw angle – ψ (with which the front wheel can be turned), and motion parameters (longitudinal position - x, lateral position - y, and heading angle - θ). The model represents a vehicle that contains two axles and the length between them is called wheelbase – l.

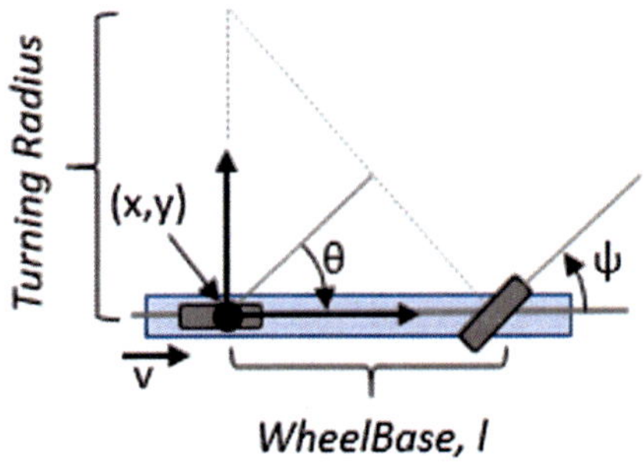

Fig. 3. The bicycle kinematic model

The velocity of the actors is computed using a Stateflow chart[3] with a deterministic finite-state automaton (Fig. 4). According to [12] a finite state-machine is defined as a computational mathematical model that describes the system as being in one of a finite number of states at any time in response to some inputs. The states' change is referred to as a transition. The states are accelerating/decelerating, cruising with constant velocity, the stop state, and lane changing maneuver. The transition between the states is computed according to the relative distance in the x and y directions of the actors. A transfer function approximates the dynamics of vehicles, the connection between the throttle angle and vehicle acceleration [13]. The relevant subsystem of the longitudinal speed control adjusts the acceleration commands based on the driving state in order to create the required longitudinal speed. Stay-in-lane acceleration, acceleration during lane changing, and target gap acceleration during target gap selection are all examples of acceleration states [14].

As shown in the Stateflow chart of Fig. 4, the model is built upon the fact that a driver can be in the following driving states:

- Free-driving – the ahead vehicle is too far away to have any influence on the ego vehicle. In this state the driver just tries to reach a certain desired speed;

[3] Stateflow chart, https://www.mathworks.com/help/stateflow/gs/finite-state-machines.html, last accessed 2021/12/21.

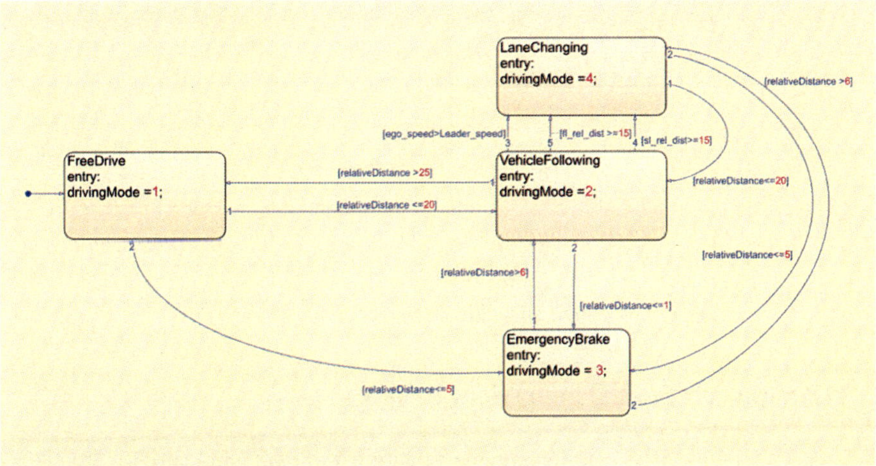

Fig. 4. The Stateflow chart between states

- Approaching and Vehicle-Following – the ego vehicle is approaching the ahead vehicle in an effort to achieve the desired safe-gap. In this state, the ego vehicle (which is currently moving at a higher speed in comparison to the ahead vehicle) decelerates so that it matches the speed of the ahead vehicle when the desired safe-gap is achieved. In the vehicle-following case, the driver is closely following the ahead vehicle while adhering to the safe-gap threshold and with almost zero speed difference;
- Emergency braking – if the distance between the ego vehicle and the ahead vehicle falls below the required safe-gap, the driver reacts to it by rapidly decelerating. Depending on the current gap and relative speed of the vehicles, the reaction could be aggressive like the application of emergency brakes. It can happen if the ahead vehicle decelerates abruptly;
- Lane Changing - the ego vehicle moves in its lane (right lane) and the ahead vehicle is moving too slowly as shown in the above figure (Fig. 5). The ego vehicle is constantly looking for the opportunity to overtake the ahead vehicle. On finding an overtaking opportunity (if the distances between the ego vehicle and ahead, left-ahead, left-rear are in the safe-gap area), it moves to the adjacent lane, overtakes the ahead vehicle by moving parallel to it and then moves back to its precedent lane. If the traffic conditions allow, the ego vehicle can even overtake more than one slow vehicle in a single overtake maneuver.

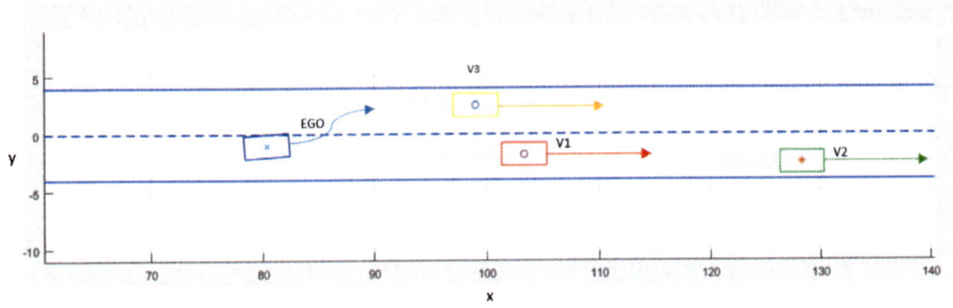

Fig. 5. The lane changing maneuver – 2D visualization

3 Teaching Method. Theoretical Framework

The autonomous vehicle course includes a simulation framework and a driving simulator. This course focuses on 'learning by doing' as recommended by Kolb's experiential learning theory (ELT). This course improves teaching-learning with the following major advantages:

- It stimulates students' curiosity and commitment to the course. All students have a purpose and must fulfill a certain objective that is essential to the accomplishment of the overall goal in various ways;
- Students have to study, analyze and simulate the given scenarios, which always requires additional questions, readings, and source investigations;
- It encourages and develops students' abilities to observe and understand the aspects they have simulated and come up with new ideas. It also allows them to modify the simulation framework and create their scenario.

According to [4, 5] learning is a complex process in which knowledge is generated as a result of the transformation of experience. Kolb's Theory identifies two experiences: Concrete Experience (CE) and Abstract Conceptualization (AC) as well as two types of experience transformation: Reflective Observation (RO), Active Experimentation (AE) [2]. There are two components to Kolb's Theory. One of the highest results relies on experiential learning when the students learning process is based on a recurring series of events of the four stages presented in the bellow sections: (Sect. 3.1) which is based on the existing experience followed by (Sect. 3.2) where the experience is observed, which leads to (Sect. 3.3) generations of new ideas or modifications of the existing experience which are then (Sect. 3.4) used to help solve further similar situations based on the acquired knowledge [15–17]. The secondary component of Kolb's Theory focused on four distinct learning styles: diverging (learn by imagining and understanding – CE and RO), assimilating (observation and thinking – AC and RO), converging (doing and thinking – AC and AE) and accommodating (doing and feeling – CE and AE) [3]. Therefore, it is obvious that different learning styles are involved in every stage of the learning cycle.

In the following sections are presented the teaching method within the topic of "Autonomous vehicles behavior".

3.1 Kolb's Concrete Experience

In the first stage, Kolb's Concrete Experience integration, the professor presents a general autonomous vehicle theory as well as a comprehensive overview of vehicle dynamics and driving rules in terms of addressing the subject to be created. Also, the professor provides an introduction of the simulation framework (Fig. 6) in MATLAB-Simulink software, presents the components of the vehicle simulator (Fig. 1) and how they can experiment and visualize different traffic situations. The driving state of the simulator user (student) influences the autonomous vehicle behavior determining their critical decisions.

After this introduction, the students are prepared to modify the presented characteristics and also to simulate the proposed driving scenario by integrating different modules, trying to understand the decision-making algorithm behind them. The decision-making algorithm is created by using a Stateflow chart to describe how this model reacts to input signals, events, and time-based conditions (Fig. 4). Students may observe details about autonomous vehicle behavior by using the Unreal Engine graphics gaming engine interface (Fig. 2).

This activity create a better understanding of the close connection between theoretical and analytical aspects.

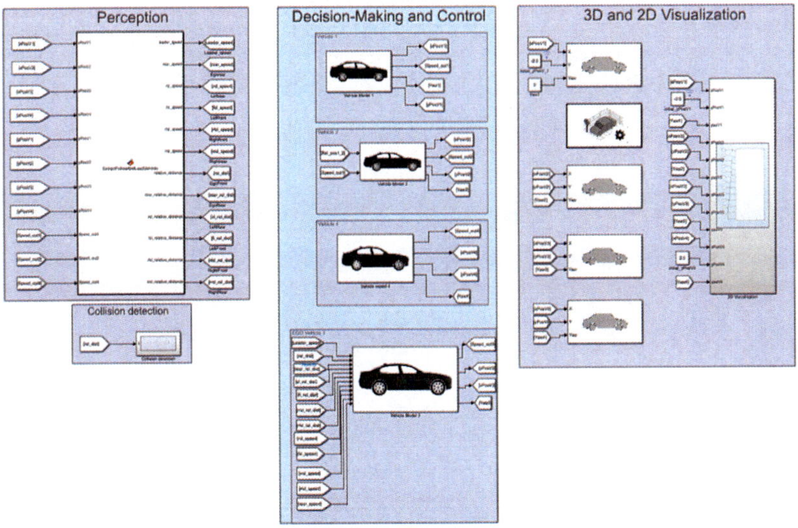

Fig. 6. The simulation framework

3.2 Kolb's Reflective Observation

The second stage is Kolb's Reflective Observation, where students should discuss autonomous vehicle behavior in uncertain situations, observed in the previous activity. Driving scenarios contain a series of uncertain situations, such as collision, aggressive driving, reaction time, safe/dangerous operation, and the student must observe each autonomous vehicle reaction. In the simulation framework, each vehicle has a different velocity profile thus having a specific behavior imposed by the value of accelerations and velocity, distance to the ahead vehicle, distance to the left/right-front/rear vehicle.

In Fig. 7 and Fig. 8 are presented several vehicle speeds and positions with different colors: vehicle 1 – yellow, vehicle 2 – blue, vehicle 3 – red, and vehicle 4 – green. The variation of the vehicles' speed is influenced by the proposed decision-making process. Through these variations, the behavior (passive, normal, aggressive) of the vehicle can be observed. In Fig. 8 it is showed the position of the vehicles from where the distances between the vehicles can be observed, following the decision of behavior. Students can plot the vehicle's speed to identify how the vehicle reacts based on the imposed conditions.

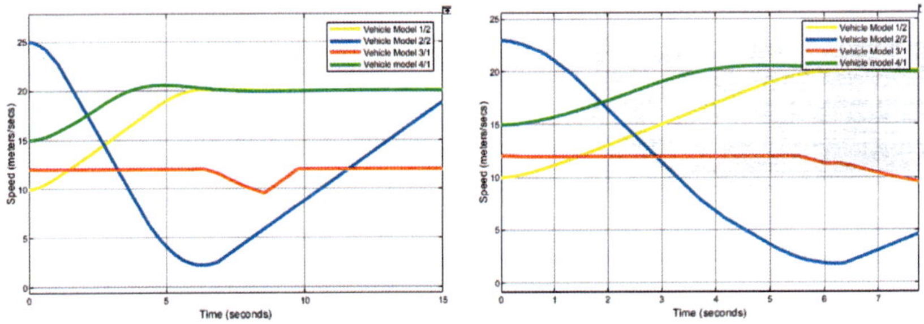

a. Lane changing b. Collision between vehicle 3 and vehicle 2

Fig. 7. Vehicles' speed variations

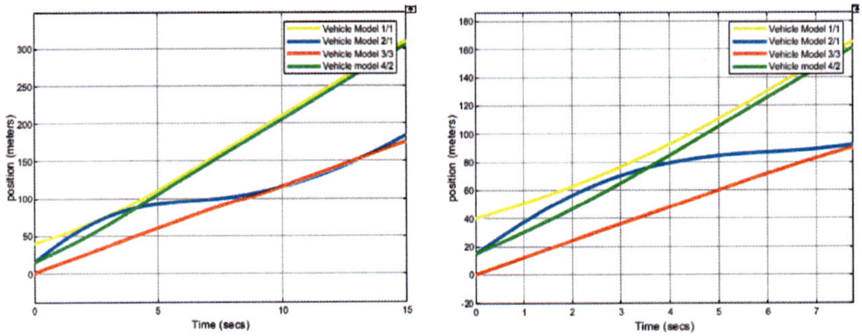

a. Lane changing b. Collision between vehicle 3 and vehicle 2

Fig. 8. Vehicles' position

The students are encouraged to debate the autonomous vehicle performance based on key questions such that: "How to avoid an accident? Has the relative distance between vehicles an important influence on the autonomous vehicles' behavior in uncertain situations?", "What influences an accident and how quickly an autonomous vehicle can make decisions in different driving scenarios?" and so on.

This activity will help students to analyze the discrepancies between their understanding and experience.

3.3 Kolb's Abstract Conceptualization

The professor emphasizes the theoretical part in the third stage of Kolb's theory, and the students are engaged in analyzing and creating their model containing the vehicle dynamics, sensors, and decision-making algorithm of the autonomous vehicles with the use of the simulation framework. Due to the simplicity and the modular structure of the framework, the student has the freedom to modify any parameter (Table 1).

Table 1. List of parameters that can be changed

Parameters	Results
Relative distance between two vehicles	Influence the logic behind the driving state of the vehicles
Minimum brake distance	Depending on the imposed conditions, an attempt is made to avoid a collision
Vehicle speed, maximum speed	Respecting the road rules – the ego vehicle's desired speed should be greater than the overtaken vehicle speed
Safety margins	Respecting safety requirements for overtaking maneuver
Lateral acceleration	Evaluate the safety and comfort (normal or aggressive behavior) by the distribution of the lateral acceleration with which the overtaking maneuver is performed
Gap acceptance	Determine the minimum gap (critical gap) for a safe maneuvering

Students can experiment with changing some of the controlling physical parameters of the vehicle dynamics and Stateflow chart and evaluate the outcomes. If during the simulation, the collision takes place, the simulation stops, and the student must restore the previous conditions and observe what influences the parameters have on the model.

This approach during the autonomous vehicles' course offers a better understanding of the behavior of the autonomous vehicles and also provides answers to previously asked questions.

3.4 Kolb's Active Experimentation

The fourth stage, also known as Kolb's Active Experimentation integration – testing stage, is when the simulator tries to increase students' curiosity about autonomous vehicles' behavior and sensors with the purpose of connecting the virtual information with real-world issues.

At this stage, the student has the opportunity to gain new experiences by creating and simulating different driving scenarios developed in the previous stage of the cycle, such as intersections or roundabouts (Fig. 9), also under the teacher's guidance. Thus

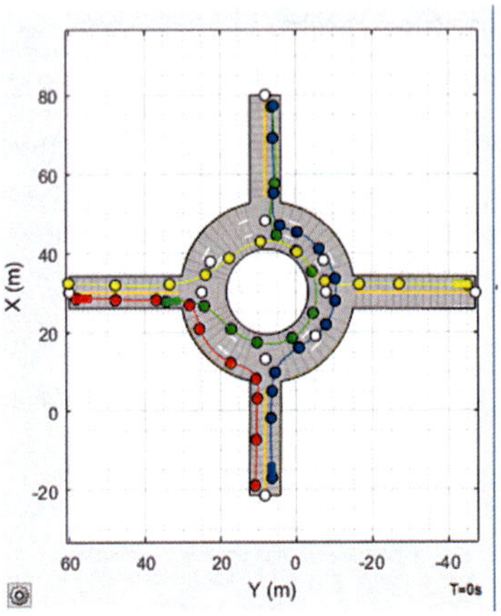

Fig. 9. The roundabout scenario

improving the simulation framework and obtaining a better understanding of autonomous vehicles' behavior.

During the lectures period, it is recommended to support students to fill a questionnaire that provides the quality and the relevance of the proposed ideas in the teaching-learning process.

4 Conclusion and Future Work

Students have an essential role in the evolution of the academic curriculum and can also contribute to the development of autonomous vehicles in a realistic way knowing how to deal with real-world situations. The teaching-learning methodology presented in this paper supports the involvement of a simulator tool as a teaching tool by using Kolb's theory. Simulation-based learning offers a wide range of opportunities to practice challenging skills in academic education.

While traditional learning is focused on the teacher and the aim is to explain knowledge and skills through the transferring of information, experiential learning is focused on students and the aim is to develop knowledge and skills through experience and to become more flexible in different learning situations. Since students learning styles are different (diverging, assimilating, converging, accommodating), this pedagogical methodology can also be seen as a teamwork approach, supporting the interaction with the professor and simulating new ideas which contribute to the development of each learning style. Furthermore, the ability to modify the dominant physical parameters of the ego vehicle proves to be an educational tool, as students rapidly understand how changing a vehicle parameter resulted in a critical change in vehicle performance. Following the simulation, the student is able to obtain a solid understanding of autonomous behavior.

Seeing that students' motivation can play a major role in the automotive domain and also in any learning areas leads to a desire to increase student motivation even more which could follow in successful learning outcomes and this is a significant reason for additional study. In order to evaluate the ideal and realistic process, the teaching method of this research should be examined and expanded in a succession of realistic case studies, and also the evolution of the students will be followed during two years. This will be done in future work.

Acknowledgements. This work was supported by a grant of the Romanian Ministry of Research, Innovation and Digitization, CNCS/CCCDI – UEFISCDI, project number PN-III-P2–2.1-PED-2019–4366 within PNCDI III (431PED).

References

1. Rodrigues M, McGordon A, Gest G, Marco J (2018) Autonomous navigation in interaction-based environments – a case of non-signalized roundabouts. IEE Trans. Intell. Veh. 3(4):425–438
2. Threeton MD, Walter RA (2009) Automotive technology student learning styles and their implications for faculty. The Pennsylvania State University, pp 11–15
3. Hulme K, Kasprzak E, Enghlish K, Moore-Russo D, Lewis K (2009) Experiential learning in vehicle dynamics education via motion simulation and interactive gaming. International Journal of Computer Games Technology, vol 2009. Hindwai Publishing Corporation, pp 2–4
4. Kolb DA (1976) Management and the learning process. Calif Manage Rev 18(3):21–31
5. Kolb DA (1984) Experiential learning: experience as the source of learning and development. Prentice Hall, Englewood Cliffs, NJ, pp 20–38
6. Ahmad A, Latib NA (2014) Teaching in automotive practical task: practice in vocational colleges. In: 4th world congress on technical and vocational education and training (WoCTVET). Elsevier Ltd., Malaysia, pp 291–293
7. Antonya C, Buzdugan ID (2021) Multimodal environment for studying the behavior of autonomous vehicles in traffic situations. In: Auer ME, Centea D (eds) ICBL 2020, vol 1314. AISC. Springer, Cham, pp 347–355. https://doi.org/10.1007/978-3-030-67209-6_37
8. Saraoglu M, Morozov A, Janschek K (2019) MOBATSim: model-based autonomous traffic simulation framework for fault-error-failure Chan analysis. In: Intelligent autonomous vehicles – 10th IAV 2019, Poland, pp 240–244
9. O'Brien M, Gupta K, Emran B, Najjaran H (2018) Autonomous vehicle control: teaching tool and simulation. In: IEEE frontiers in education conference (FIE), San Jose, CA, USA, pp 1–8
10. Antonya C, Carabulea L, Pauna C (2018) Predictive actuation of a driving simulator. In: Burnete N, Varga B (eds) International congress of automotive and transport engineering 2018. Springer, Cham, pp 128–135
11. Matlab, Simulink, Mathworks Inc Homepage (2021). https://www.mathworks.com/help/robotics/ref/bicyclekinematicmodel.html. Last accessed 15 Nov 2021
12. Xing X, Zhou T, Chen J, Yu, Z (2021) A hazard analysis approach based on STPA and finite state machine for autonomous vehicles. In: IEEE Intelligent Vehicles Symposium (IV), pp 150–155
13. Matlab, Simulink, Mathworks Inc Homepage (2021). https://www.mathworks.com/help/releases/R2017b/mpc/examples/design-an-adaptive-cruise-control-system-using-model-predictive-control.html?searchHighlight=lane%20following%20control&s_tid=doc_srchtitle. Last accessed 15 Nov 2021
14. Toledo T, Koutsopoulos HN, Ben-Akiva M (2007) Integrated driving behavior modeling. Transp. Res. Part C: Emerg. Technol. 15(2):96–112

15. Merhtash M, Yuen T, Balan L (2019) Implementation of experiential learning for vehicle dynamic in automotive engineering: roll-over and fishhook test. In: The 12th international conference interdisciplinary in engineering, pp 769–773
16. Kolb AY, Kolb DA (2005) Learning style and learning spaces: enhancing experiential learning spaces: enhancing experiential learning in higher education. Acad. Manag. Learn. Educ. 4(2):192–212
17. Kolb DA (2014) Experiential learning: experience as the source of learning and development. FT press

Production Cell Structure Constructor for Remote Laboratory Experiments

Karsten Henke[1], Mykhailo Poliakov[2(✉)], Heinz-Dietrich Wuttke[1], Johannes Nau[1], and Oleksii Poliakov[3]

[1] Ilmenau University of Technology, 98684 Ilmenau, Germany
[2] Zaporizhzhia Polytechnic National University, Zhukovsky street, 64, Zaporizhzhia 69063, Ukraine
polyakov@zntu.edu.ua
[3] National Technical University of Ukraine "Igor Sikorsky Kyiv Polytechnic Institute", pr-t Peremohy, 37, Kyiv 03056, Ukraine

Abstract. The paper presents a method for extending the diversity of virtual experiments in remotely controlled laboratories. Using the example of experiments controlling the model of a manufacturing cell, it is shown how a variety of different experiments can be achieved by synthesizing new variants of the structure of this manufacturing cell. The manufacturing cell is used to process workpieces and consists of elements such as conveyors, turntables, processing devices, and others. Each of these devices is represented by a variety of discrete sensors and actuators, where the sensors are inputs and the actuators are outputs to the control system, which must be designed by the student as a finite state machine (FSM). The devices with their sensors and actuators, as well as visual and virtual device models, are grouped together as a structure constructor. Here, the visual model of a device is a set of graphic objects whose visual properties depend on the values of the tags of the virtual model. The virtual model of the device specifies the graphical (tag values) and functional (values of the inputs of the FSM) response of the device model to the control signals of the FSM. The requirements for the structure of the manufacturing cell are determined by the sequence of operations for processing the workpiece. As a result of the synthesis of the manufacturing cell structure, the following are determined: the composition and location of the devices on the working field of the system; their visual reactions to temporal changes in the values of the outputs of the FSM; the values of the inputs in response to the control signals (=outputs of the FSM); and the values of the variables of the relationship with neighboring devices. The proposed method can be useful in organizing a laboratory workshop on the design of parallel and hierarchical control automata.

Keywords: Remote laboratory · Structure constructor · Production cell

1 Introduction

In the context of the coronavirus pandemic, engineering education is undergoing a phase of accelerated adoption of distance learning methods, including the field of training in

M. E. Auer et al. (Eds.): REV 2022, LNNS 524, pp. 81–91, 2023.
https://doi.org/10.1007/978-3-031-17091-1_9

building digital systems. An important place in distance learning technologies is occupied by methods of remote engineering [1], in particular, methods of performing remote experiments with physical and virtual models of control objects [2, 3]. As noted in [4], the tasks of transition from conducting separate remote experiments in certain disciplines by individual students to the full coverage of laboratory courses in all disciplines of the specialty by all students are relevant for remote laboratories. Within the framework of this global task, there are such particular tasks as increasing the variety of experiments and reducing the time for preparing an experiment. A sufficient variety of experiments is achieved by providing each student with at least one individual design problem. This prevents the projects from being copied. However, this task must be solved with financial constraints on the purchase of physical models. One of the ways to reduce the preparation of an experiment is to use design tools as part of a remote laboratory. In this paper, a method for solving the set tasks is considered on the example of the Production Cell model of the remote laboratory Grid of Online Laboratory Devices Ilmenau (GOLDi) [5].

2 State of Art

Models of elements and systems of production equipment [6] are an important object of study for future specialists in digital control systems for the following reasons:

1. Competence in the design and maintenance of production systems enhances their employability.
2. Experiments with these models in a visual form reflect the theoretical methods of software implementation of complex automatic behavior of production equipment, coordination of the work of individual elements of the system and building a hierarchy of control devices.
3. Students know the logic of the functioning of physical models of industrial devices from everyday life.

It is known software for 3D modeling of production systems using visual components, for example the product Visual Components [7]. The advantages of this tool include high quality visualization and a large base of components for designing a production system. However, these tools are complex enough to be used in courses on designing the behavior of technological components at the FSM level.

The paper [8] describes approaches to the design of production systems aimed at increasing their robustness. These approaches are focused on increasing the stability of the system already at the stage of conceptual design, but they also do not allow designing the automatic behavior of its elements.

In [9], an example of using virtual reality tools for designing a production system is given, but these tools do not allow generating variants of the structure of the automaton behavior of the system.

The work [10] describes the concept and examples of building digital twins for virtual factories. However, this concept is not detailed to the level of twins of the behavior of elements of production cells.

The physical model of Production Cell is a set of equipment [5, 11, 12] consisting of belt conveyors, rotary tables, a processing device and a transport carriage (see Fig. 1). This equipment is a control object in a system that simulates the operation of an automatic line for the machining of parts.

Fig. 1. Production cell physical model

The student's task is to develop a control device for this equipment at the logical level. At this level, each sensor in the equipment corresponds to a bit in the register of inputs, and to each actuator, a bit in the register of outputs of the control device. The control logic takes into account the levels at the inputs and the state of the device when forming the levels at the outputs. The model of such control is a state machine. An example of a finite automaton for controlling a belt conveyor is shown in Fig. 2 [5].

The educational topic, the development of which is fixed in the course of this remote experiment, is the design of a system of parallel finite automata. Various technologies for the implementation of control devices are available on the GOLDi remote laboratory website. The construction of control devices based on the formalism of finite state machines (FSMs) is one of such technologies. When designing control automata in the GOLDi laboratory, the Graphical interactive FSM Tool (GIFT) [5] is used. An example of designing a control machine using the GIFT tool is shown in Fig. 3.

The GIFT tool allows you to specify all the elements of a control machine tuple: inputs, outputs, states, and transition and output functions. At the same time, the limited functional capabilities of the existing elements of the designed system (for example, unidirectional movement of the conveyor belt, rotation of the table platform only by 90°, and others) do not allow modifying the structure of the system of the physical model of Production Cell with the existing number of devices included in it. This fixed structure corresponds to the visual and virtual model.

That is, the physical, visual and virtual models of the system object remain unchanged. Accordingly, the possibilities of modifying control automata and the variety of experiments with the Production Cell model are limited.

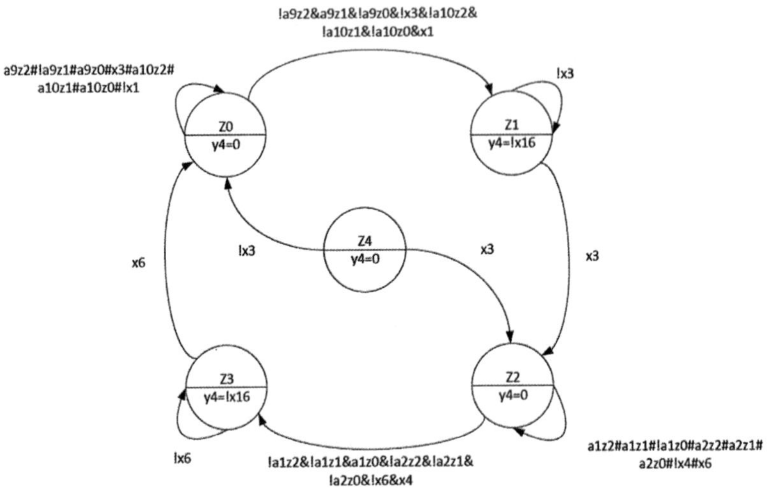

Fig. 2. Conveyor line control machine example

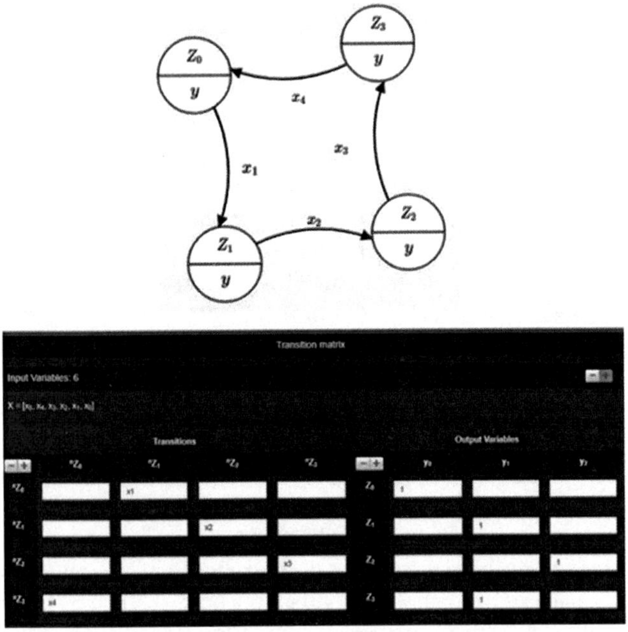

Fig. 3. An example of designing a control machine using the GIFT tool

3 Structure Constructor

The limitations on the number of devices included in the physical model of Production Cell are not essential for the corresponding virtual experiment. In this experiment, a different Production Cell structure can be modeled, including one with multiple processing devices.

The initial data for such a design is the route of the workpiece passing through the Production Cell processing devices. After determining the composition of the machines, the route includes elements for transporting the workpiece – belt conveyors and rotary tables. Moreover, some processing devices, conveyors, rotary tables can be used many times at various stages of the route of the workpiece. This data is used to build a higher-level Moore control automaton. The states of this automaton correspond to the stages of the route of passage of the workpiece. Outputs in the active state enable the operation of the equipment control machine at this stage. The route of the part starts from the device "Source of blanks" and ends with the device "Warehouse of parts".

The process of preparing/designing an experiment includes the following stages:

1. Formation of a model of the structure of a production cell.
2. Display of the procurement route on the generated model of the structure of the production cell.
3. Formation of visual and virtual models of production cell devices.
4. Determination of inputs and outputs of state machines that control the operation of individual devices in the system.
5. Designing control machines for each device in the system.

The stage of forming a model of the structure of a production cell is performed in the environment of the "Structure Designer" software tool. It includes the placement of processing devices, devices for loading, moving and unloading the workpiece on the working field of the system.

An example of this tool's screen is shown in Fig. 4. The left column of the screen contains graphic images of the devices, which are "dragged" into the working area of the system. These are such devices as: PW – parts warehouse; SB – source of blanks; PM – processing device (side view); PM – processing device (front view); CB – conveyor belt (horizontal); RT – Rotary table; CB – conveyor belt (vertical).

The system working field is located in the central part of the designer screen. This field is divided into cells that have a row-column address. Moving the workpiece along the lines will be called horizontal, and along the columns – vertical. In this case, the horizontal movement can be left-to-right (from column 1 to column 6) and right- to-left (from column 6 to column 1). Accordingly, vertical movement can be bottom- up (from line A to line G) and top-down (from line G to line A).

The cell address of the working field is associated with the device that occupies this cell. Each cell of the working area can be occupied by one device from the left column of the screen, subject to the following restrictions:

- There can be only one SW device and located at address A1.
- There can be only one PW device and located at address F6.

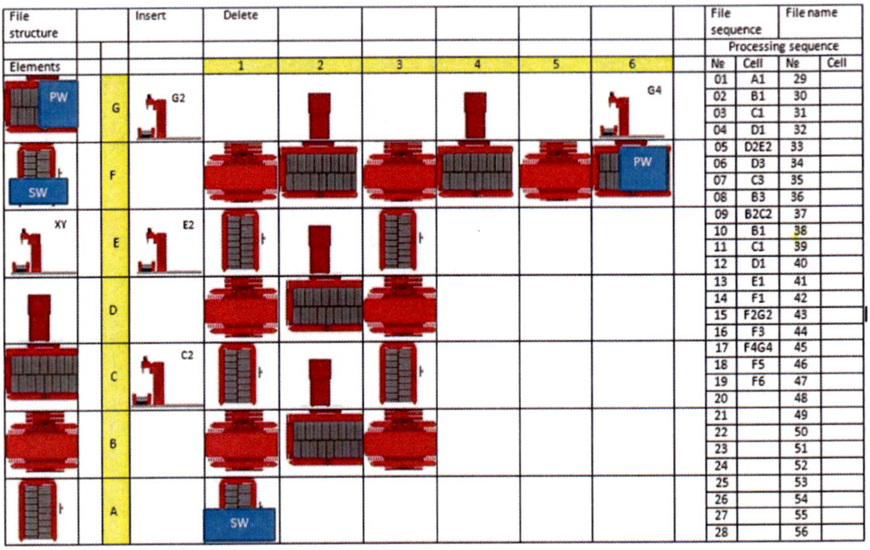

Fig. 4. Example of the structure builder tool screen

- There can be several CB devices in the system. Table 1 shows the options for installing these devices, taking into account the fact that the conveyor belt moves in one direction.
- There can be several RT devices in the system. Table 2 shows the options for installing these devices, taking into account that the movement of the conveyor belt RT is carried out from the entrance (the point of entry of the workpiece onto the table) to the exit (the point of departure of the workpiece from the table).

Table 1. Conveyor belt installation options.

Device type	Direction of movement	Designation	Example in Fig. 4
Conveyor belt (vertical)	Upwards	CBvb	C1
Conveyor belt (vertical)	Top down	CBvt	C3
Conveyor belt (horizontal)	From left to right	CBgl	F2
Conveyor belt (horizontal)	From right to left	CBgr	B2

- There can be several PM devices in the system. Each PM device is shown in two views (front and side view). These devices also differ in type, which determines the algorithm of operation. For example, a drilling device does not move the workpiece horizontally during vertical movement of the tool, unlike milling machines.

- PM devices – processing machine (front view) are installed "above" CB devices – conveyor belt (horizontal). For example, a PM (front view) machine is installed in cell G4 above the conveyor in cell F4.

Table 2. Installation options for the turntable

Entry position	Descent direction 1	Descent direction 2	Designation	Example in Fig. 4
From below	Up	To the right	1RT	D1
Left	To the right	Way down	2RT	F3
Above	Way down	To the left	3RT	B3
On right	To the left	Up	4RT	B1

- PM – processing machine (side view) devices are installed in the immediate vicinity of the corresponding PM (front view). For example, PM (front view) is set in cell G4.

On the right side of the designer screen, there is a window for the route of moving the blank from the source of blanks to the parts warehouse. The route is described through a sequence of device addresses that the workpiece passes through the working field of the system. Thus, on one structure, different routes of the workpiece can be realized, and the selected sequence of workpiece processing operations can be implemented using systems of different structures. This increases the number of options for tasks for designing an experiment, but at the same time creates additional problems in the construction of visual and virtual models of the system, which are discussed below.

The visual model of the system consists of visual models of devices.

The visual model of the device is a simplified top view of the device at the current time with a workpiece moving on it or without a workpiece. The workpiece is depicted in a simplified way – with a rectangle. The movement of the workpiece through the device can be displayed continuously or discretely, for example, in three positions: at the beginning, in the middle and at the end of the movement. Images of a device, its elements, and a blank are managed graphic objects. Graphics object tags change their visual properties, such as coordinates, rotation angle, color of the object on the screen, and others. The meaning of the tags is determined in the virtual model during the experiment.

The virtual model of the device manages the graphic properties of the images of devices and the workpieces on them at the current moment in time. In this case, the logic of the type is used: "if the workpiece at time t1 entered the conveyor, the conveyor operates normally; there are no restrictions on moving the workpiece from the side of the devices connected to the conveyor, then in n seconds the workpiece will be in the middle of the conveyor." This event should be reflected visually on the screen of the virtual experiment and by generating the values of the inputs of the control machine that emulate the operation of the device's sensors. The parameter n sets the speed of movement of the workpiece along the device. The block diagram of the interconnection between the operating and control automata, virtual and visual models of the device of the production cell is shown in Fig. 5.

At the next stage, the inputs and outputs of the state machines that control the operation of individual devices in the system are determined. It does this by linking the base I/O sets of each device to its location in the production cell system, and by generating signals of the readiness of this device to receive the workpiece from the previous device and the readiness to transfer the workpiece to the next device along the route of the workpiece movement. This should also taking into account the readiness of the previous device along the route of the workpiece movement to transfer and the next device – to accept the workpiece.

At the last stage of preparing a virtual experiment, the design of control automata for each device in the system is carried out. In this case, the state machines of various devices interact with each other. The logic of interaction is determined by the composition of their sensors and actuators, the parameters of the flow of workpieces and the sequence of their processing. Let us consider the simplest case: there is only one workpiece in the system at a time and the route of its movement does not change.

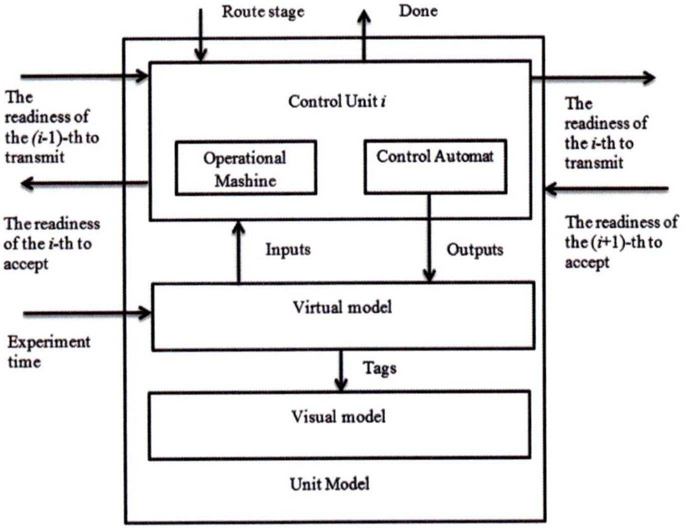

Fig. 5. Structural diagram of the interconnection of the models of the device of the production cell

Let the route of movement be described in the structure of a finite state machine with a fixed sequence of state changes. Each of these states is tied to the activity of a device located in a cell of the working field of the system. In this case, for the i-th state of this finite automaton, we will consider neighboring states corresponding to the cells of the working field of the system, which have a common side and participate in the process of moving the workpiece. An example of displaying the working field of the system and the route described in Fig. 4 is shown in Fig. 6. The numbers in the rounded rectangles represent the number of the stage of the workpiece path, the onesided arrows represent the direction of movement, and the two-sided arrows represent the workpiece machining operation on the device.

Looking at Fig. 6 it follows that the rotary table D1 at operation 4, when receiving the workpiece, interacts with the conveyor C1, and when dispensing – with the conveyor D2. In addition, at operation 12, the same table interacts with the E1 conveyor to dispense the workpiece.

The problem of describing the automatic behavior of a set of devices on the route of movement of the workpiece can be solved by a student in several ways, which are the subject of study in a remote laboratory.

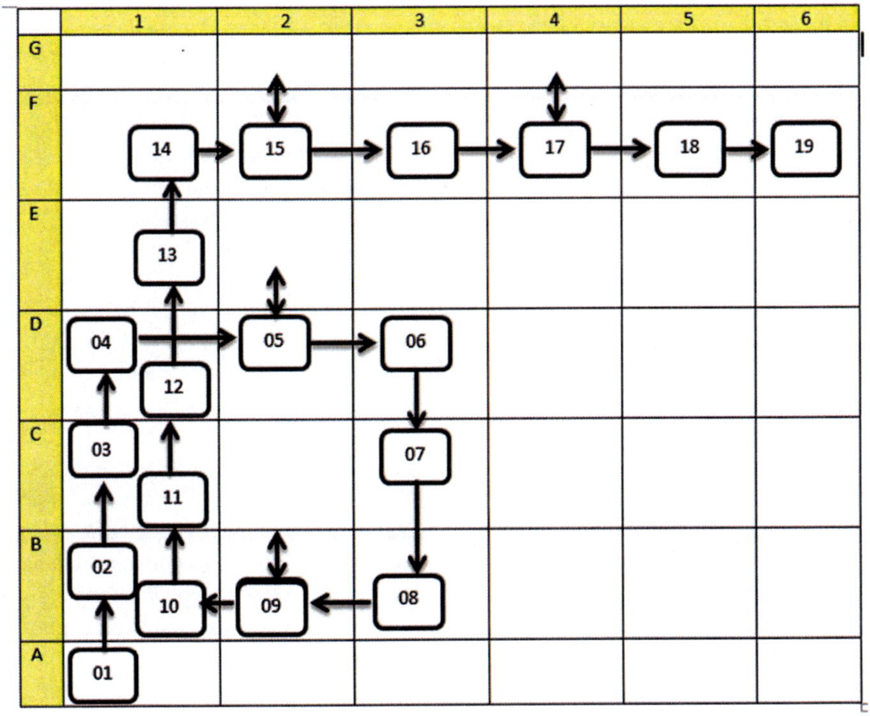

Fig. 6. Display of the workpiece route by device of the production cell

For example, by building a system of parallel operating finite automata. The corresponding set of automata is given in the section "Example: Production Cell parallel" on the website of the GOLDi laboratory [4]. Secondly, by building a hierarchy of control automata, with the interaction of the levels of the hierarchy through a typical element [13]. In this case, the task of the upper-level automaton in each state is to activate the corresponding controlled automaton at the lower level. Third, by using non-binary automata that allows parallel activity in several adjacent states at each cycle of the system operation [14]. For the three states, this is the activity of post-activity, activity and pre-activity. For example, consider three adjacent devices along the route shown in Fig. 4: RT (cell D3), CB (cell C3), and RT (cell B3). In step 7, the activity of the conveyor C3 is to move the conveyor belt from top to bottom. The post-activity of the D3 turntable is to turn it back to its original state - in line with the D2 conveyor. The pre-activity of the B3 rotary table is the movement of the belt to receive the workpiece from the C3 conveyor.

Thus, to conduct a remote experiment, a visual model of the working field of the system, a model of the route of the workpiece, and virtual models of the devices of the production cell should be prepared. These models are created using the constructor considered in this work. In addition, one of the methods described above prepares the control automata of the devices.

4 Conclusion

Increasing the range of virtual experiments with control automata in a remote laboratory is an urgent task at the stage of mass transition to distance learning for students of technical specialties. This problem is solved by combining various typical models of the system devices in the experiment. In this work, the models of the Production Cell are selected as such devices [5]. Each version of the structure of a virtual experiment differs in the nomenclature and location of typical devices of a production cell and requires the creation of an original set of control machines, a visual and virtual model of the control object.

The structure of a production cell system for conducting a virtual experiment is designed by a remote student using the Structure Constructor tool. The initial data for the design is the sequence of workpiece processing operations. Based on these data, the student determines the composition and location of devices on the working field of the system and the structure of the top-level control machine. The functions of the upper-level automaton include the activation of the control automaton of the device on which the current operation is performed. Device control automata represent the lowest level of the system automata hierarchy. The Structure Builder also allows you to build a visual and virtual model of a remote experiment. In the course of this experiment, a student on a remote computer screen observes the process of moving the workpiece along a given route, changes in the position of visual models and, based on this, makes a conclusion about the quality of his project for controlling a production cell.

The proposed Structure Constructor is planned to be used in the remote laboratory GOLDi as a tool for constructing visual and virtual models of a production cell for a given route of movement of workpieces.

References

1. Remote and virtual tools in engineering: monograph/general editorship Karsten Henke, Dike Pole, Zaporizhzhia, Ukraine, 250 p. (2015)
2. Azad AKM et al (eds.) (2012) Internet accessible remote laboratories: scalable e-learning tools for engineering and science disciplines. IGI Global. https://doi.org/10.4018/978-1-61350-186-3
3. Gravier C, Fayolle J, Bayard B, Ates M, Lardon J (2008) State of the art about remote laboratories paradigms - foundations of ongoing mutations. Int. J. Online Biomed. Eng. (iJOE) 4(1):1–9. https://doi.org/10.3991/ijoe.v4i1.480

4. Poliakov M, Wuttke H-D, Henke K (2020) Quality and efficiency indicators of remote laboratories. In: Auer ME, Ram B. K (eds) REV2019 2019, vol 80. LNNS. Springer, Cham, pp 143–154. https://doi.org/10.1007/978-3-030-23162-0_14
5. GOLDi-labs cloud Website (2022). http://goldi-labs.net. Accessed 04 Jan 2022
6. Abele E et al (2017) Learning factories for future oriented research and education in manufacturing. CIRP Annals 66(2):803–826. https://doi.org/10.1016/j.cirp.2017.05.005
7. Visual Components company Website (2022). www.visualcomponents.com. Accessed 04 Jan 2022
8. Bellgran M, Säfsten K. Production system design and evaluation for increased system robustness. In: Proceedings of Second World Conference on POM and 15th Annual POM Conference, Cancun, Mexico, April 30–May 3 2004. ABSTRACT 002-0536
9. Yang X et al (2015) Manufacturing system design with virtual factory tools. Int. J. Comput. Integr. Manuf. 28:25–40. https://doi.org/10.1080/0951192X.2013.800948
10. Yildiz E, Møller C, Bilberg A (2021) Demonstration and evaluation of a digital twin-based virtual factory. Int. J. Adv. Manuf. Technol. 114(1–2):185–203. https://doi.org/10.1007/s00170-021-06825-w
11. Catalog – Feltron Zeissler. https://www.feltron-zeissler.de/shop/318-simulationsmodelle
12. Product Overview Simulation Models. Catalog – Feltron Zeissler (2009). https://www.feltron-zeissler.de/shop/img/cms/PDF-Kataloge/Pruefungsmaterial/06_Simulationsmodelle/Modelluebersicht_Simulationsmodelle.pdf. Accessed 04 Jan 2022
13. Poliakov M, Subbotin S, Poliakov O (2021) Interoperability of integrated hierarchical systems. Syst. Technol. 2(133):68–78. https://doi.org/10.34185/1562-9945-2-133-2021-08
14. Poliakov M, Subbotin S, Poliakov O (2021) Performance indicators of models of non-binary control automates. In: 2021 IEEE 16th international conference on the experience of designing and application of CAD systems (CADSM), pp 38–42. https://doi.org/10.1109/CADSM52681.2021.9385220

Embedded Systems Engineering Education Using Cost Effective Pocket Labs Platform

Christian Madritsch(✉)

Carinthia University of Applied Sciences, CUAS, Villach, Austria
c.madritsch@fh-kaernten.at

Abstract. This paper describes the cost-effective use of Pocket Labs in Embedded Systems Engineering education. First, the rationale behind this approach is explained by comparing it to the traditional method of using laboratory equipment. Next, the goals and contents of existing Embedded Systems Lectures are explained. The focus is put on the lectures Microcontroller Programming (Bachelor Level) as well as Digital Signal Processing Applications (Master Level). The following section explains two Lab Exercises and derives requirements for the equipment used from them. The section on Pocket Labs Platform illustrates our current solution and its main properties. The last chapter compares the traditional with our Pocket Labs infrastructure and gives conclusions and an outlook.

Keywords: Embedded systems · Pocket labs · Microcontroller · Digital signal processing · Lab equipment

1 Introduction

Microcontroller Programming, part of our bachelor program, as well as the course Digital Signal Processing Applications, part of our master program, require specialized laboratory equipment like oscilloscopes, function generators, dynamic spectrum analyzer, etc. for the effective realization of various programming exercises. This expensive specialized equipment is only available in very few and dedicated labs at our university. This is restricting the flexibility of lecture schedule planning as well as the possibility for students to work on their free time on those exercises. Furthermore, this kind of equipment represents a significant financial investment.

Multi-purpose laboratories, like the Real-Time and Image-Processing (RTISP) lab at our university, are not equipped with specialized measurement or signal generation hardware. Those labs are a combination between lecture room and workspace for students.

Nevertheless, some exercises require the use of e.g. oscilloscopes in order to capture waveforms, or e.g. function generators to produce proper input signals for e.g. digital-signal-processing applications implemented on microcontroller platforms.

Using NI myDAQ in combination with off-the-shelf and custom miniSystems creates a cost-effective Pocket Labs platform [1], which can partly replace expensive Lab

M. E. Auer et al. (Eds.): REV 2022, LNNS 524, pp. 92–100, 2023.
https://doi.org/10.1007/978-3-031-17091-1_10

Equipment. The first part of this paper gives a brief overview about the goals and the contents of our Embedded Systems education. Next, the main requirements for the practical realization of programming exercises are shown using examples from both courses. The following chapter presents the Pocket-Labs platform and shows how it is used during lectures and lab exercises. Finally, a comparison between traditional lab equipment and our Pocket Labs platform will depict advantages and disadvantages and discuss the main benefits.

The lecture Microcontroller Programming using Pocket Labs is already in its second run and improvements have been identified and are being applied. The lecture Digital Signal Processing Applications is currently ongoing, the experiences of previous courses helped to shape its outline.

Using Pocket-Labs instead of regular lab equipment can be an effective way to reduce the overall costs of infrastructure, it enables a more flexible room planning, and it allows students to work at home on their tasks and assignments. Although this approach cannot be applied at every level of engineering education, the Embedded Systems courses were a perfect fit.

2 Embedded Systems Education

At CUAS, the faculty of Engineering and IT (EngIT) is dedicated to Engineering Education. The Engineering branch of our faculty is specializing on system level education, the IT branch on user and application level. Several bachelor and master study programs in different fields, like mechanical engineering, systems engineering, or medical IT have been established since many years. The embedded systems education is mainly concentrated at the engineering branch. During the first year of related bachelor programs, the computer science and programming basics are focusing on the facilitation of skills like:

- C programming (unlike C#, C++, or Java at the IT branch),
- hardware basics (computer architectures, CPU, system bus, memory map, IO) and
- Embedded Systems Design Patterns (state machine, polled loop systems).

The second year contains lectures like Microcontroller Programming (1 and 2), and Digital Signal Processing.

2.1 Microcontroller Programming 1 and 2 (Bachelor Level)

In Microcontroller Programming, the first step is to differentiate between different kinds of Embedded Systems (ASIC, discrete Logic, Embedded PC, Microcontroller based, PLC, ...).

Next, specific Microcontroller architectures are discussed in detail (in our case ARM Cortex M4 and M7) as well as their implementations using physical IC components (ST Microelectronics STM32F4xx and STM32F7xx) [2]. During this section, the individual on-chip peripheral modules like GPIO, ADC, DAC, PWM, Timers, UART, USART, SPI, I2C, I2S, Interrupt Controller, and DMA are covered (see Fig. 1).

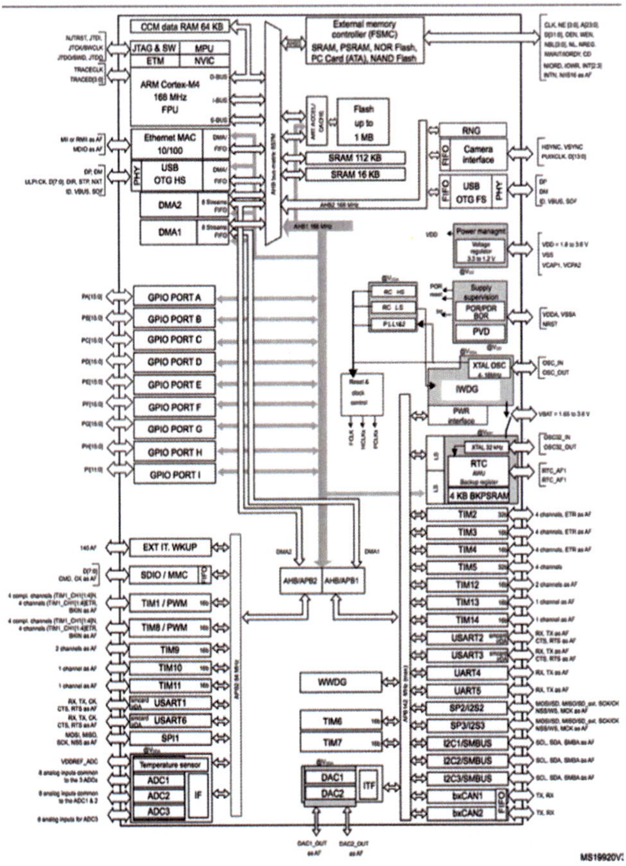

Fig. 1. STM32F407VG device overview [3]

Furthermore, the development process using an Integrated Development Environment – IDE (in our case Keil uVision with Arm MDK) is explained [4]. Project creation, source code editing, building (compile and link) and debugging are also part of this step.

Since software development using STM Microcontrollers is supported by a semi-automatic configuration tool (STM32 CubeMX, see Fig. 2), the individual steps in order to automatically generate the initialization code for the peripheral modules is studied.

After the Microcontroller details are clear, the next step is to move of-chip to the development board used (STM F4 Discovery). On-board LEDs, User Buttons, MEMS Microphone, and a Digital Amplifier are just a few components we are using here.

Covering the software architecture aspect, the simplest variant used is a polled loop system – in combination with one or more state machines. After the discussion of advantages and disadvantages, the Interrupt based architecture is introduced and combined with both, polled loop systems and state machines.

Finally, a Real-Time Operation System (RTOS) is used to combine preemptive priority multitasking/multithreading with interrupt-based systems (FreeRTOS or RTX).

Fig. 2. STM32 CubeMX configuration tool [5]

All previously mentioned contents are extensivly combined with hands-on laboratory exercises, homework assignments, and at the end of the second course with a final project.

2.2 Digital Signal Processing Applications (Master Level)

In Digital Signal Processing Applications, the first step is to distinguish model-based design using Matlab and Simulink from physical implementations using Microcontrollers and DSPs. Here, system aspects like latency, run-time, response-time, and code size play an important role.

After this is established, one DSP platform is used to implement the signal-processing chain (see Fig. 3). It consists of the capture of external signals (e.g. audio using a microphone or Line-In), the data transport into the computational system (DSP), the application of DSP algorithms (e.g. FIR filter, spectral analysis, …) and finally the output of the manipulated signal to external outputs (using DAC).

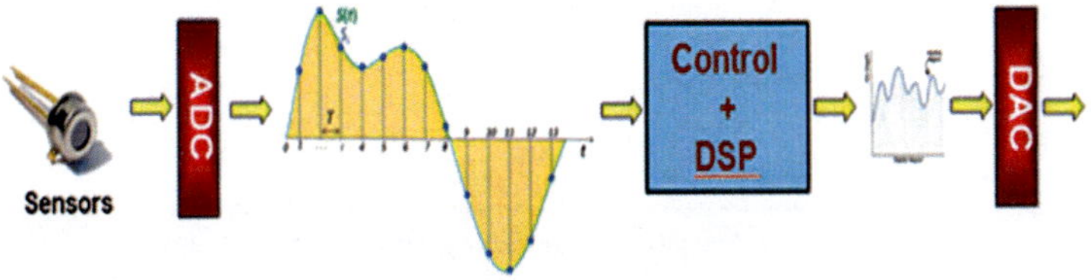

Fig. 3. DSP signal processing chain

In DSP Applications, the lecture is accompanied with practical elements like laboratory exercises, homework assignments, and the final project.

3 Main Requirements

In order to understand how to derive main requirements for Pocket Labs, two example Laboratory exercises are explained with an emphasis on what kind of Measurement or Signal generation equipment is needed to perform these tasks.

3.1 Debouncing of the User Button (Microcontroller 1)

Mechanical Switches or Buttons experience the phenomenon, that after the user activates the device, the output voltage does not immediately change from e.g. GND to VCC but instead bounces several times between these values until a steady state is reached (see Fig. 4). This is caused by the mechanical properties of the device.

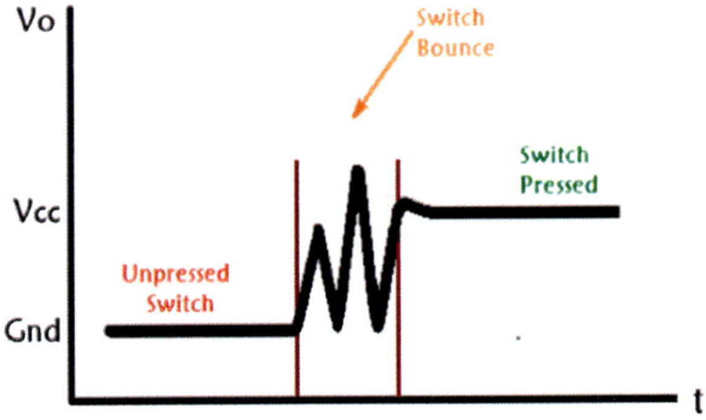

Fig. 4. Bouncing of a switch

Since the bouncing effect may cause several interrupts on the Microcontroller, this effect needs to be eliminated. First, students have to observe this behavior using an Oscilloscope. The bouncing happens usually 10 to 100 times over the period of 1 ms, which requires an Oscilloscope with a sampling rate of 200 kS/s [6].

Next, two debouncing strategies are applied: physical and in software. Physical debouncing is adding a Low-Pass Filter to the signal line of the switch. Here, students need to use a Multimeter to measure electrical characteristics like Voltage and Resistance. Debouncing in software means that the Interrupt will be disabled during the bouncing period of the signal line.

3.2 Finite Impulse Response Filter (DSP Applications)

The application of Finite Impulse Response (FIR) filters is a very basic task in DSP. At the beginning, students need to compute the filter characteristics using Matlab [7]. Next, the signal processing chain needs to be setup and programmed. The input signal will be white noise in the frequency range of HiFi-audio (up to 22 kHz) generated by a Function Generator.

The signal will be captured by the on-chip ADC and transported via DMA into the main memory of the DSP processor. There, the FIR filter is applied. Two different implementation styles are compared: manual implementation using nested for-loops and the build-in filter provided by ARM for the DSP processor. Students need to compare the efficiency of both approaches by measuring the runtime using test points (GPIO) with an Oscilloscope [8].

Finally, the resulting waveform will be output using the on-chip DAC and the on-board Power Amplifier. Students need to measure the spectral information of the input signal and the filtered signal using a Dynamic Signal Analyzer (see Fig. 5) and compare the results with the computed filter characteristics from Matlab.

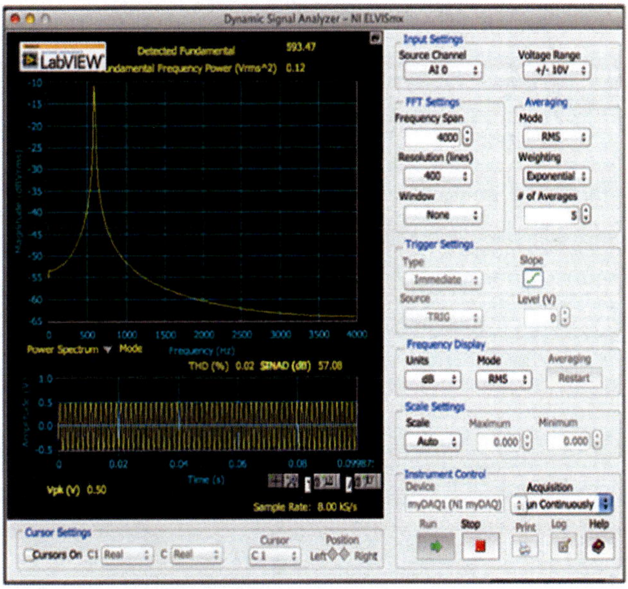

Fig. 5. Dynamic signal analyzer

Summary of Instruments. From those two example laboratory exercises, the following requirements can be derived:

- Oscilloscope: 2 Channels, 200 kS/s
- Multimeter: Voltage (20 V), Resistance
- Function Generator: White Noise, 22 kHz
- Dynamic Signal Analyzer: 2 Channels, 44 kS/s

4 Pocket Labs Platform Used

CUAS has many years of experience in the use of Remote- and Pocket Labs. Since Remote Labs obviously do not match the basic requirements, Pocket Labs were considered. NI myDAQ [9] in combination with an extension miniSystem was chosen is the ideal way to go.

4.1 NI myDAQ

The NI myDAQ base system consists of the myDAQ device, cables, and connectors as well as software to configure the system operated by a Windows PC. NI myDAQ can be used for Data Acquisition and Data Generation:

- Digital Multimeter (DC−, AC-Voltage, DC−, AC-Current, Resistance, Diode);
- Two Analog Inputs (16 bits, 200 kS/s, +−10 V);
- Eight Digital Inputs (5 V);
- Audio Input (+−2 V, 1.5 Hz to 400 kHz);
- Two Analog Outputs (16 bits, 200 kS/s, +−10 V);
- Eight Digital Outputs (3.3 V, 4 mA);
- Audio Output (+−2 V).

The myDAQ system can be extended by using miniSystems. In Fig. 6, myDAQ in combination with an extension miniSystem is depicted. This miniSystem allows students to use myDAQ as an Oscilloscope, Function Generator, or Dynamic Spectrum Analyzer.

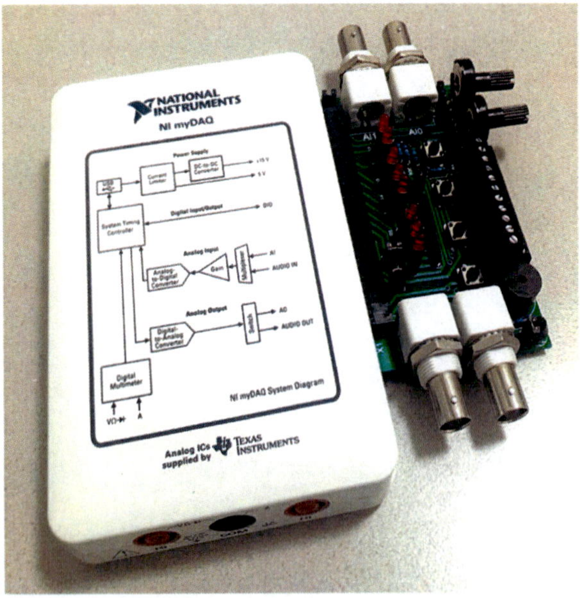

Fig. 6. myDAQ with miniSystem

NI myDAQ can be operated by using ready-made Virtual Instruments as depicted in Fig. 7.

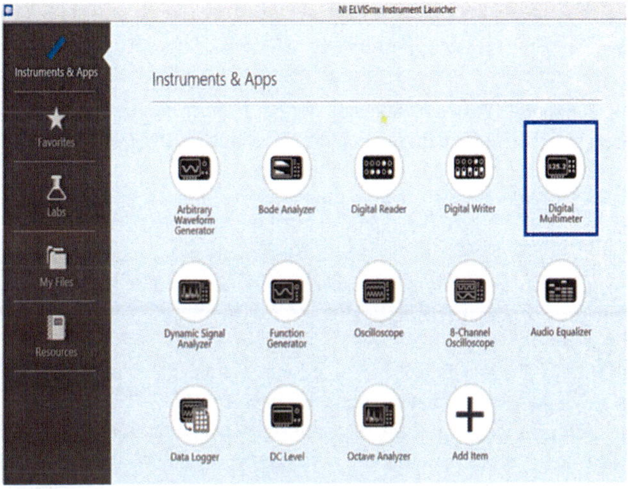

Fig. 7. NI ELVIS instrument launcher

5 Comparison, Conclusions, and Outlook

The increase of the number of students in engineering related studies leads to the problem of the availability of suitable rooms. Regular lecture rooms are disposable in a larger number, but specialized Labs are limited. Also, the investment in specialized Labs is a significant financial burden, since a fully equipped Lab exceeds 100 k€ by far. In comparison, Pocket Labs in combination with miniSystems and probes cost less than 300 € per unit. Although not all lab exercises are suitable, e.g. communication technologies related lectures have a greater demand on sampling rates, resolution, and precision of the measurement devices, many other lectures can profit by using Pocket Labs (e.g. Microcontroller Programming, Digital Signal Processing Applications, Control Systems, Computer Science, etc.).

Another aspect is the ever-changing mode in which study programs are designed [10]. At our university, we have full-time, part-time, distance, extended modes which require a highly flexible way of planning. Students want to work at any time and at any place on their assignments. The use of Pocket Labs introduces a high level of flexibility.

At the end of the mentioned lectures, students have been asked using a questionnaire whether they prefer to work with traditional lab equipment or with Pocket Labs [11] instead. The result of this questionnaire was that there is a very high acceptance of using Pocket Labs, mainly due to the fact that they are used to working with them from the first semester on and that the uniform and concise user-interface of the Virtual Instruments used allows for a rapid learning curve.

In future, additional lectures at CUAS will adopt the Pocket Labs concept for suitable lectures and thus allowing us a high degree of flexibility together with a cost-effective solution in engineering education.

References

1. Madritsch C, Klinger T, Pester A, Schwab W (2016) Work in progress: using pocket labs in master degree programs. International conference on interactive collaborative learning, Great Britain
2. Martin T (2016) The designer's guide to the cortex-M processor family, 2nd edn. Newnes
3. ST Homepage: www.st.com/en/evaluation-tools/repo/stm32f4discovery.html. Last Accessed 06 2021
4. Keil Homepage: www.keil.com/mdk5/uvision. Last Accessed 06 2021
5. ST Homepage: www.st.com/en/evaluation-tools/uintest/32f746gdiscovery.html. Last Accessed 05 2021
6. Horowitz P, Hill W The art of electronics, 3rd edn. Cambridge University Press
7. Mathworks Homepage: www.mathworks.com. Last Accessed 12 2020
8. Madritsch C, Klinger T, Pester A (2019) Heterogeneous computing exercises in digital signal processing applications. International conference on interactive collaborative learning, Bangkok, Thailand
9. NI Homepage: www.ni.com. Last Accessed 12 2020
10. Werth W, Madritsch C (2019) Innovative elements for a balanced and successful study and work program in part-time engineering education. International conference on interactive collaborative learning, Dubai, UAE
11. Madritsch C, Klinger T (2017) Work in progress: pocket labs in IoT education. International conference on interactive collaborative learning, Budapest, Hungary

Development of a Remote Experiment for Practical Work in Physics at the University: The Case of Free Fall

Zineb Laouina[1(✉)], Lynda Ouchaouka[1], Mohamed Moussetad[1], and Mohamed Radid[2]

[1] Physics Department LIMAT Laboratory, Hassan II University - Casablanca, Casablanca, Morocco
Laouina.zineb1@gmail.com

[2] Chemistry Department, Hassan II University - Casablanca, Casablanca, Morocco

Abstract. Laboratory experimentation plays a crucial role in scientific education and engineering in higher education. The traditional hands-on labs are the natural scenarios where practical skills can be improved but thanks to Information and Communication Technologies (ICT), virtual and remote labs can provide a framework where Science, Technology, Engineering and Mathematics (STEM) disciplines can also be developed. This paper deals particularly with remote access to a real laboratory equipment using contemporary computer and network technology for creating the environment that will enable a remote user to perform the required practical experiments in physics.

The main goal of this paper is to build an interactive system to control and manipulate distance practical experiments in physics by the students enrolled in the first year of undergraduate program of Physics and Chemistry at the University.

In this paper, we have focused on the design and creation of a prototype to make the equipment of the practical experiments of laboratories accessible to students via the Internet network. In order to eliminate time and space constraints through affording flexibility, protect students while conducting dangerous experiments and reduce the cost and the maintenance of laboratory equipment.

Keywords: Remote lab · Free fall · E-lab

1 Introduction

Elabs are computer environments designed to allow learners to conduct remote experiments through the Internet network [1]. They offer the opportunity to bypass the difficulties imposed by the massification of higher education, so that students can combine learning through experimentation and trial and error to conceptualize what they have learned in experimental subjects.

More and more universities are committed to the development of distance learning experiments as an integral part of their academic offer, in order to strengthen teaching/learning among students. This is achieved through the combined implementation of multi-user interfaces and virtual classrooms that allow the development of

M. E. Auer et al. (Eds.): REV 2022, LNNS 524, pp. 101–111, 2023.
https://doi.org/10.1007/978-3-031-17091-1_11

synchronous collaborative work, as students and teachers can control the laboratory equipment together and interact simultaneously.

This innovation leads to the emergence of new uses for students and new practices for teachers, such as:

- The appropriation of ICT, interactive boards, application servers, remote control
- Simulation/driving software
- Remote control of electronic devices with interfaces
- Remote control of scientific equipment
- Synchronous or asynchronous guidance
- Option of recording practical work sessions

In the present work, we intend to design and realize a remote manipulation: the free fall, which is part of the experiments planned in the Mechanics module for the students of the first year of university enrolled in the Physics Chemistry course. This work follows a first work that we realized for the manipulation "simple pendulum" [2]. These two manipulations are part of the Elab of the Faculty of Science Ben m'sik, which offers other manipulations such as:

- Measuring and Monitoring of Temperature Changes
- Velocity of sound in air
- Simple pendulum
- Virtual lab
- 12 simulated physics experiments
- Maghreb Network of Remote Laboratories

2 Background

Within our research team, several thesis works [3, 4] have been involved in solving the problems of simulated, virtual and distance experiences, in order to strengthen the faculty's distance learning dipositive. Our Elab, includes several online manipulations, simulated, serious games, which focus on: "development of a distance practice for laboratory experiments on measuring instruments" [5], "the creation of advanced online university laboratories" [6] in addition to participation in European projects of capacity building to improve the quality of higher education Tempus Project Escience [7], and Erasmus Project Express [8] (Fig. 1).

The architecture of our Elab is presented below:

Fig. 1. E-lab of the Faculty of sciences Ben M'sik

Objectives of the Remote Experience

The objective of our work is to facilitate access to the experience through an automated solution and connected to the network.

Constraints

- Handling must be done from a distance
- The height configuration must be variable

To carry out this experiment, we followed the steps below.

3 Experiment

This paper is organized as follows: a first section describes the experiment, contains the mechanical design, the electronic and the IT parts.

3.1 Description of the Experiment

Principle of Galileo's law: "Free fall":

A body that is dropped without initial velocity follows an accelerated rectilinear motion downward. As long as air friction is negligible, all bodies fall in the same way regardless of their mass [9].

A ball of mass is in free fall if it is subjected only to the action of its weight $P = m * g$, where g is the acceleration of gravity.

Considering the free fall of a body released without initial velocity from a height h above the ground. The object begins to move under the action of its weight.

Initial conditions: $V(0) = 0$ and $z(0) = h$.

Hourly equations of motion:

$$z(t) = \frac{1}{2}gt^2 \quad et\ V(t) = gt$$

where $V(t)$ is being the speed of the ball at time (t) (Fig. 2).

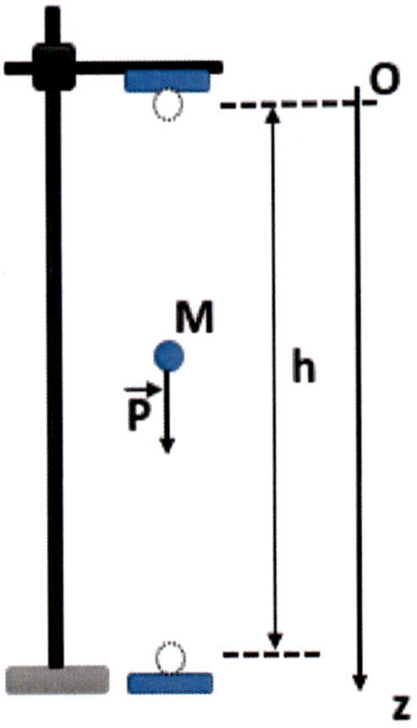

Fig. 2. Representation of the principle of Galileo's law

Pedagogical Objectives

- To study the movement of a body in line
- Determine the relationship between height and time of free fall

Principle of Operation

- A ball held by a magnet is released at time **t = 0** without initial velocity.
- An optical sensor measures the travel time **t** of the ball from its launching point for different heights **h**.

The Work Required From Learners

1. For different drop heights **h**, measure the travel time of the ball and calculate the quotients $\mathbf{h/t^2}$.

h (m)	0	25	50	75
t (s)				
h/t^2				

2. Graph the height of fall **h(m)** as a function of time **t(s).**
3. Calculate the average value of the quotient $\mathbf{h/t^2}$, then compare this value obtained with **g/2** with ($\mathbf{g = 9, 8\ m/s^2}$)
4. Express the drop height h of the ball as a function of **g** and $\mathbf{t^2}$.

3.2 Mechanical Part

Our approach was to start with the design of the prototype of the practical experiment, and then we developed our prototype. All the pieces of the experiment were designed in the SolidWorks Software as shown in Fig. 3.

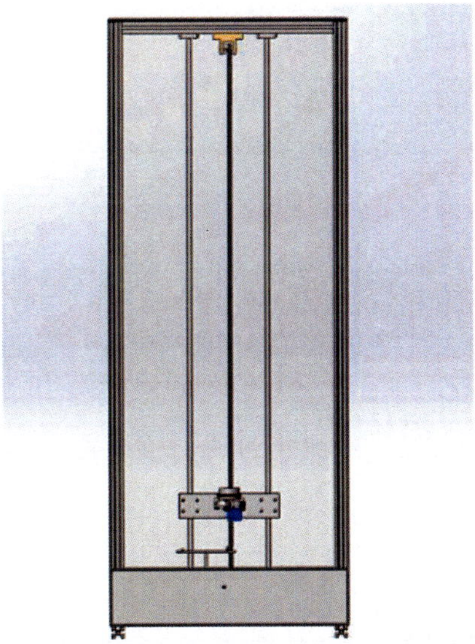

Fig. 3. The design of the prototype for the remote experiment of Free Fall

After the design, all the elements were made by the 3D printer, the laser cutting and the vinyl cutting in order to produce the following prototype, Fig. 4.

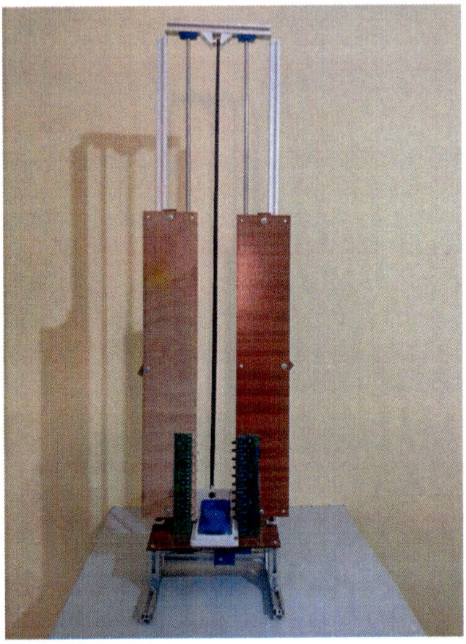

Fig. 4. Prototype developed for the practical experiment of the Free Fall

3.3 Electronical Part

3.3.1 Implementation

The Controller

The control board is the unit responsible for monitoring the experimentation bench directly. The following figure shows the connection with the external elements (Fig. 5):

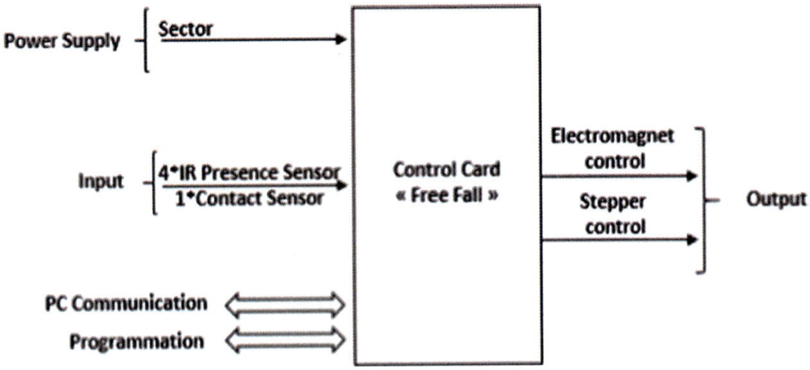

Fig. 5. Synoptic of the control card

The Microcontroller
It is an 8-bit microcontroller of AVR type, we will choose the Arduino board. It is interfaced with sensors and actuators to control the experimentation bench.

The Controller
The controller is an 8-bit AVR microcontroller.

The Sensors
In order to take the values of the chronometer during the descent of the metal ball, four sensors of presence of infrared type are arranged vertically along the trajectory of the fall. An additional sensor is added to detect the advent of the contact of the ball on the ground.

The Actuators
A Nema 17 stepper motor is used to move the ball launch platform vertically. This allows to modify the height of the fall during the experiments.

On the other hand an electromagnet is used to hold/release the metal ball to trigger the experiment.

Communication
The controller needs to retrieve the control commands from the user are USART serial port in ASCII format.

3.3.2 System Architecture:

Hardware System Architecture
The system consists of the following elements:

The Experimental Bench
It is an electromechanical system able to restore the experience of the free fall and to carry out the measurements. This bench is controlled through a controller which receives the orders of the outside and translates them into an electric command.

The Server
It is the element that exposes the user interface and the web services in order to make possible the remote control-command. It is directly connected to the controller of the experimentation bench as well as to a camera to be able to broadcast the progress of the experiment (Fig. 6).

3.4 IT Part

3.4.1 Uploading of the Handling

The board's manual control software is based on the LabVIEW graphical programming environment. It is ergonomically designed to guide the user through the experience. The user can visualize in real time thanks to an IP camera. A router switch ensures the

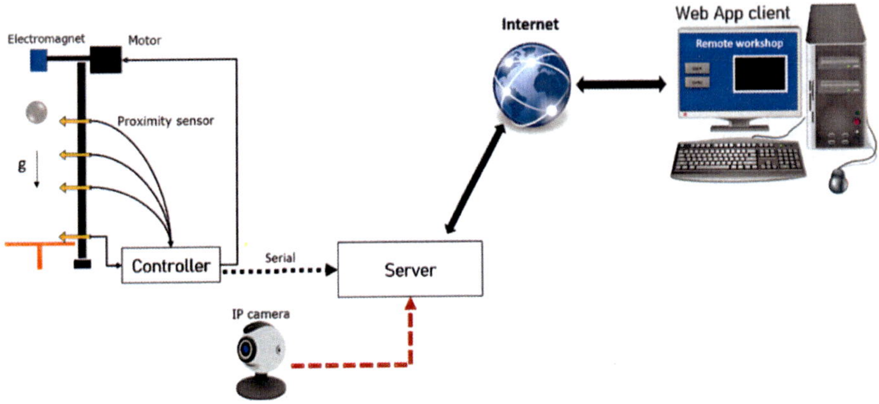

Fig. 6. Block system diagram

connection of the server and the camera to the Internet. The user has the possibility to choose a target position and send it to the control board.

The user interface on the PC is an HTML web page provided by a LabVIEW software server. The access to the page is done in the server (Fig. 7).

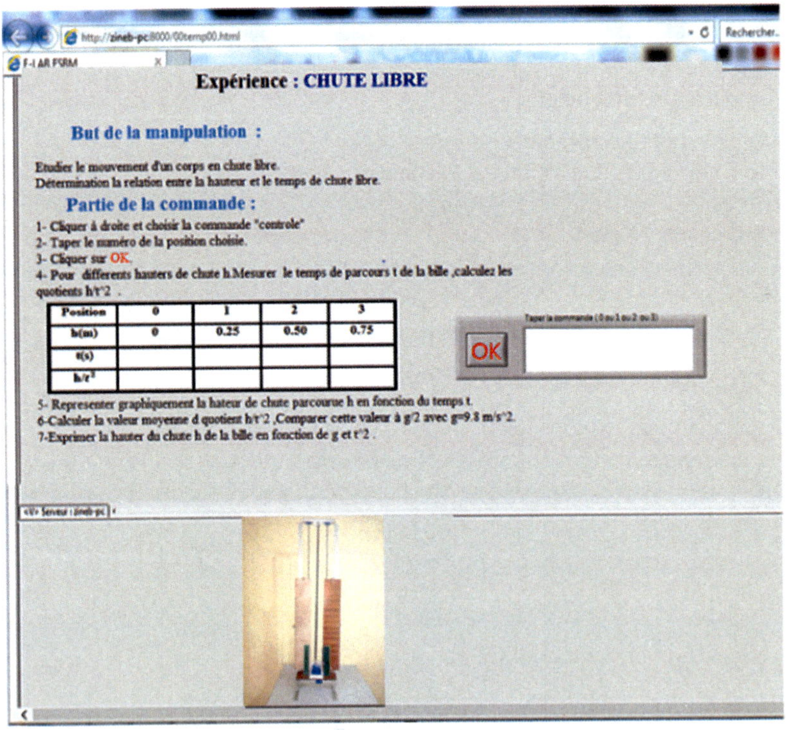

Fig. 7. User interface

The user interface shown below will be available to users who connect remotely to the web server. The web server is a service offered by the LabView software which is hosted on a computer (or physical server). The web page (user interface) is designed in an easy-to-use way to guide the students in the practical work, in our case in French language which is the language of learning in the current context.

The positions are provided in the table; then the ball is released, and the user is led to time the height h by viewing on the camera interface to deduce the time t (Fig. 8).

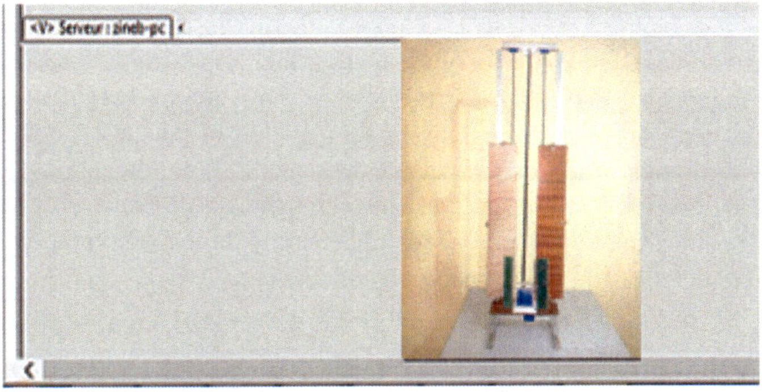

Fig. 8. Camera view user interface

This example shows the operating mode of the free fall experiment:

The student will choose position 1 for example (Fig. 9):

1. He types the command then ok

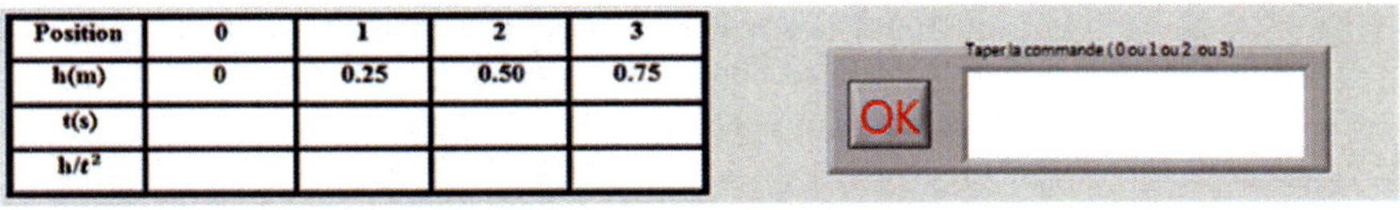

Position	0	1	2	3
h(m)	0	0.25	0.50	0.75
t(s)				
h/t^2				

Fig. 9. Positions of the height h

2. The motor that manages the ball mechanism will be activated until position 1
3. The ball will be released by the electromagnet.
4. The student will be asked to record on a stopwatch the time needed for the ball to complete its trajectory from the corresponding height h in our case h = 0.25 m). The choice of the stopwatch is dictated to involve the student and create interactivity.
5. The student will transfer the result obtained by the stopwatch on the table provided.
6. He/She will have to make the graph that represents the height of h travelled as a function of time t

7. Then he/she will calculate the average value of the quotient h/t^2, then compare this value to g/2.
8. Then express the value of fall h of the ball according to g and t^2.

4 Conclusion

In this paper, we have worked on the design and creation of a prototype to make the equipment of the practical experiments of laboratories accessible to students via a server.

Our work is in line with the work of our team, which has carried out several research projects on practical work at distance. In this work we have presented a global architecture (software and hardware) for the remote implementation of practical laboratory works in the field of physics. The progress of our work will allow students not only to do simulations and virtual instrumentation, but they can also do real-world experiments linked with the physics of the phenomenon under study through observation or control. This is done remotely with TCP/IP network connection.

5 Perspectives

The perspectives of this work is to generalize the use of online experiments to other scientific disciplines, in order to have an Elab that will support students.

In order to achieve the expected learning objectives, and to increase their chances of success.

References

1. Ainsworth JD, Buchan IE (2009): e-labs and work objects: towards digital health economies. In: Mehmood R, Cerqueira E, Piesiewicz R, Chlamtac I (eds) EuropeComm 2009. LNICST, social informatics and telecommunications engineering, vol 16. Springer, Heidelberg, pp 205–216. https://doi.org/10.1007/978-3-642-11284-3_22
2. Laouina Z et al (2021) Manufacturing and developing remote labs in physics for practical experiments in the university. In: Auer ME, May D (eds) REV 2020, vol 1231. AISC. Springer, Cham, pp 193–204. https://doi.org/10.1007/978-3-030-52575-0_16
3. Khazri Y (2016) Gestion dev la file d'attente dans une plateforme des travaux pratiques en physique à distance, research thesis, the Hassan II University of morocco
4. Al Sabri A (2021) Mise en oeuvre d'une téléscience de travaux pratiques
5. Kbiach M, et al (2017) EXPERES, un projet innovant pour répondre aux besoins des universités marocaine en matière d'intégration des TICs pour la modernisation des travaux pratiques. https://doi.org/10.13140/RG.2.2.22786.09921
6. Al Sabri A, et al (2020) Remote lab experiments: measuring and monitoring of temperature changes. Int J Online Biomed Eng (iJOE) 16(4). https://doi.org/10.3991/ijoe.v16i02.12219
7. Khazri Y, Al Sabri A, Sabir B, Toumi H, Moussetad M, Fahli A (March 2017) Development and management of a remote laboratory in physics for engineering education (E-LAB FSBM). In: Proceedings of the 2nd international conference on big data, cloud and applications, pp 1–6

8. Khazri Y, Al Sabri A, Sabir B, Toumi H, Moussetad M, Fahli A (2017) Development and management of a remote laboratory in physics for engineering education (E-LAB FSBM), pp 1–6. https://doi.org/10.1145/3090354.3090460
9. Sabir B, Khazri Y, Moussetad M, Touri B (2017) Hardware and software co-design of Arabic alphabets recognition platform for blind and visually impaired persons. Open Electr Electron Eng J 11:193–200. https://doi.org/10.2174/1874129001711010193
10. Ouchaouka L, Laouina Z, Moussetad M, Talbi M, Elkouali N (2021) The effectiveness of a learner-centered pedagogical approach with flipped pedagogy and digital learning environment in higher education feedback on a cell biology course. Int J Emerg Technol Learn (iJET) 16:4–15. https://doi.org/10.3991/ijet.v16i12.19125

Latency Resistant Safety in Distributed Remote Laboratories

Johannes Nau(✉), Pierre Helbing, Karsten Henke, and Detlef Streitferdt

Technische Universität Ilmenau, Ilmenau, Germany
johannes.nau@tu-ilmenau.de

Abstract. This work will focus on the problems of creating a safe distributed laboratory. We explicitly will not discuss how to make individual elements of an experiment safe, as this is highly application-dependent. Instead, the goal is to find and evaluate different methods to detect and respond to fault conditions that an individual laboratory device might detect. Specifically, the methods should differentiate between user-based faults and those introduced through network communications. We develop a mathematical model to simulate distributed laboratories. We will introduce (time-dependent) network latency and jitter between all elements. Based on the model, a discrete event simulation is created. This simulation environment simulates three different fault detecting methods: the token method, the timestamp method, and the full-state-transfer method. We will compare detection ratios, bandwidth usage, and memory usage between the three methods based on the simulation.

Keywords: Virtual and remote laboratories · Distributed laboratories · Control engineering education · Operational safty

1 Introduction

Online Laboratories usually consist of several experiments. These are either completely simulated (virtual laboratories) or consist of different hardware setups (remote laboratories). In most cases, these experiments are premade and offer no possibility of changing the hardware setup. Notably, the remote lab platforms VISIR [1] and our current remote lab GOLDi [2] offer some flexibility in the hardware setup. Recently there has been the idea of decomposing the monolithic hardware setup of most experiments into smaller *laboratory devices* that can be recombined into new experiments and shared across institutional borders. Besides the benefit of the added flexibility that can lead to a richer learning environment, this concept allows for a new type of online laboratories that combine the approach of take-home labs [3] with remote laboratories. This new type of online Laboratories is called hybrid take-home laboratories. It allows the student to interact with a remote part of an experiment through some local take-home equipment. The new project CrossLab [4] among other topics, is researching these new types of distributed online laboratories.

M. E. Auer et al. (Eds.): REV 2022, LNNS 524, pp. 112–124, 2023.
https://doi.org/10.1007/978-3-031-17091-1_12

One aspect of remote labs is operational safety. Even for monolithic experiments, this challenge is common. As the topic of safety is one to consider whenever systems interact with the real world, the industry has standardized how to approach the problem [5]. Safety is hard to solve in distributed systems because the internet connecting different parts of the system or the experiment does not give any timing guarantees. One way to not rely on these guarantees is to make each part inherently safe (Internal Safety), so no matter how other parts interact with a laboratory device, it will not reach an unsafe state. Our research group already studied operational safety with our existing remote lab GOLDi [2], which allows interconnecting two laboratory devices with each other.

In this context, it is important to differentiate between fault conditions that occur because of the network latency and those that occur because of the faulty programming or configuration of the user. The first type of fault (e.g., when a limit switch is hit, but due to network latency, the corresponding motor is not immediately stopped) is called infrastructure fault [6] and should be masked for the user. The latter type of fault (e.g., when a limit switch is hit, but due to programming errors, the corresponding motor is not immediately stopped) is called user fault [6] and should be reported to the user.

This paper focuses on different methods to differentiate between the above-mentioned fault conditions. The novelty is that we model our methods with unlimited interconnected laboratory devices in mind instead of just two, as we did in our previous research. We will first describe a mathematical framework to embed our research question in (Sect. 2). Based on that framework, we define three different algorithms (Sect. 3) and test them in a simulated environment (Sect. 4). Finally, we discuss the results of the tests (Sect. 5).

2 Mathematical Problem Statement

To understand the problem of latency-resistant safety in distributed remote laboratories, we first need to define what precisely a distributed experiment is and when an experiment is in a faulty state.

2.1 Stationary Experiment

We begin by describing a stationary (i.e., non-distributed) experiment. Since an experiment is essentially a system, we start with the formal definition of a system, where a system S is defined by a 4-tuple:

$$S = (X, T, R, x_0) \tag{1}$$

where X is the set of all possible states of the system, T is the set of all possible times of the system, $R : X \times T \to X$ is the rule function of the system, and $x_0 \in X$ is the initial state of the system.

We now introduce a subspace $F \subseteq X$ called the fault state space of the system. The following 5-tuple then defines a stationary experiment:

$$\overline{E} = (X, F, T, R, x_0) \tag{2}$$

As a stationary experiment can not have infrastructure faults, naturally, every fault that occurs, i.e., a time $t \in T$ is reached, such that $R(x_0, t) \in F$, is caused by a programming error of the user and therefore a user fault.

Before we move on to the mathematical definition of a distributed experiment, we assume that each state in the state space is a tuple of n variables $v_1, v_2, \ldots, v_n$.

2.2 Distributed Experiment

A distributed experiment will now split the experiment over m different devices. We make the assumption, that each variables v_i can only be modified by one device, which means that we can simply split the rule function R into a tuple of rule functions $\hat{R} = \left(R^1, R^2, \ldots, R^m\right)$. Every $R^i : X \times T \rightarrow X^i$ will output a state $x^i \in X^i$ such that the concatenation of all x^i is the state of the system. Note, that this assumes without loss of generality that node 1 will be controlling the variables $v_1, v_2, \ldots, v_{j-1}$, node 2 will be controlling the variables $v_j, v_{j+1}, \ldots, v_k$, and so on, until node m is controlling remaining variables up until v_n.

We can model the network and processing latencies by defining a propagation function $D(i, j, t_2) \mapsto t_1 : \{1, \ldots, n\} \times \{1, \ldots, m\} \times T \rightarrow T$, that outputs the time t_1 which is the latest time that the value of v_i is known to node j at time t_2. In other words: $t_2 - t_1$ is the time it takes to propagate the value of v_i from the originating node to node j, i.e. the latency.

The following 6-tuple now defines the distributed experiment:

$$\hat{E} = \left(X, F, T, \hat{R}, D, x_0\right) \tag{3}$$

In most remote experiments, only the latest value of a variable is considered for any calculation. It is reasonable to assume the same for our rule functions R^i in the distributed experiment model. Therefore the time behavior of the state can be described by the following equation:

$$x(t) = \begin{pmatrix} v_1(t) \\ v_2(t) \\ \vdots \\ v_n(t) \end{pmatrix} = \text{concatenate}_{i=1}^{m} R^i\left(x_l^i(t), t\right) \tag{4}$$

where $x_l^i(t)$ is the local state of the i-th node:

$$x_l^i(t) = \begin{pmatrix} v_1(D(1, i, t)) \\ v_2(D(2, i, t)) \\ \vdots \\ v_n(D(n, i, t) \end{pmatrix} \tag{5}$$

It is important to note that the fault space will also decompose into local fault spaces F^i for each node i. That is because each node—or laboratory device—will bring its own safety implementation, and therefore fault space. The overall fault space will just be the union of all local fault spaces: $F = \bigcup_{i=1}^{m} F^i$.

Now, whenever a fault occurs, i.e., $x(t) \in F$ for some time $t \in T$, it can either be an infrastructure fault or a user fault. Our formal definition is that whenever an equivalent stationary model is fault-free, every fault in the distributed experiment is an infrastructure fault. The definition of an equivalent stationary model is complicated or, in some cases, even impossible in the general case. However, when we make the assumption, that the time is discrete, i.e. $T \subseteq \mathbb{N}$ and each rule function R^i will only be dependent on the variables in the previous time step, we can define the equivalent stationary model as trivial by defining $D(i, j, t) = t - 1$, which will allow to collapse the distributed experiment $\hat{E}$ from Eq. (3) into a stationary experiment $\overline{E}$ from Eq. (2).

The goal is to find a way to detect user faults in distributed experiments without detecting any infrastructure faults, without knowing the rule function (as the rules can and will be modified by the user), and the propagation function (as the propagation function represents the unpredictable network latency).

3 Detection Algorithms

We came up with three different methods to possibly detect user faults, that we call the *token method*, the *timestamp method* and the *full-state-transfer method.*

3.1 Token Method

When a node detects a faulty local state $x_l^i(t) \in F^i$, it will send a token to all other nodes. When a node receives a token, it will return it to the originating node. The original node will determine the faulty state as a user fault as soon as it receives a token from all other nodes and the fault condition has not been cleared, i.e., the local state has been faulty for the whole time.

Mathematically this is described by the following fault detection function:

$$f_{\text{token}}(t) \equiv \text{true} \Leftrightarrow \exists i : \forall \min_{j=1}^{m} D^t\Big(i, j, D^t(j, i, t)\Big) \leq t' \leq t : x_l^i(t') \in F^i \qquad (6)$$

where $D^t(i, j, t)$ is the token propagation function for a token from node i to node j, that is definied analogously to the propagation function D.

Discussion. This method ensures that each node has enough time to resolve the fault condition when we assume that the token propagation will be at least as slow as the variable propagation. However, it is quite likely that this method fails in complex experiment scenarios that involve multiple hops between nodes to resolve a fault condition. It should be possible to develop more sophisticated fault detection algorithms on this principle that take these scenarios into account. However, this is outside of the scope of this paper.

3.2 Timestamp Method

We assume that all nodes are perfectly synchronized and syntonized for the timestamp method. That means that each event can contain a timestamp. Based on this, the delay for each event can be calculated, and the propagation function can be observed up until the running time. A timestamp is attached to every variable, which is updated on every change of its associated variable by its controlling device. These timestamps will be propagated to all other devices as part of the variable propagation. The other devices can now use the timestamps to calculate the delays between themselves and the sending device. Furthermore, we introduce new parameters for these delays. These delays can be described as a 3-tuple:

$$d = (s, r, v) \tag{7}$$

The above tuple d is the delay from the sending device s to the receiving device r and has a value of v. These delays are also part of the state, and each delay is controlled by the receiving device r. Every time a device r receives a message from another device s the value v of the corresponding delay $d = (s, r, v)$ is updated. A device can use the newest value or the minimal, maximal, or average value. This change is then propagated to all other devices. If a faulty local state ${x^i}_l \in F$ is detected by a device, it can now use its local information about the delays to calculate an estimate for how long it should take to resolve the faulty state. If the device remains in a faulty state even after the estimated time has passed, it will determine the faulty state as a user fault. Should the device resume to a non-faulty state during the estimated time, the node will determine the faulty state as an infrastructure fault.

The set of saved delays for a device i at time t will be called $D_{i,t}$ and $dv_{i,t} : D_{i,t} \to \mathbb{N}$ will return the value for a given saved delay of device i at time t. So for $d = (s, r, v) \in D_i(t) : dv_{i,t}(d) = v$. For this paper the following fault detection function was defined:

$$f_{\text{timestamp}}(t) \equiv \text{true} \Leftrightarrow \exists i, t_f \in T :$$
$$t_f = t - \sum_{d \in D_{i,t_f}} \left(\frac{dv_{i,t_f}(d)}{|D_{i,t_f}|} \right) \cdot m \wedge \left\{ x_l^i(t') \mid t_f \leq t' \leq t \right\} \subseteq F^i \tag{8}$$

The estimated delay can be described informally as the average of the saved delays in a device times the number of devices in the experiment. Furthermore, since the delays to and from other devices are not known at the beginning of an experiment, an estimate of 1 is used.

3.3 Full-State-Transfer Method

The idea of the full-state-transfer method is that a node is not allowed to introduce a variable change that will result in a faulty state. However, each node has a different local state, so it is not trivial to make this decision. However, this should become trivial when each variable change also transmits the local state,

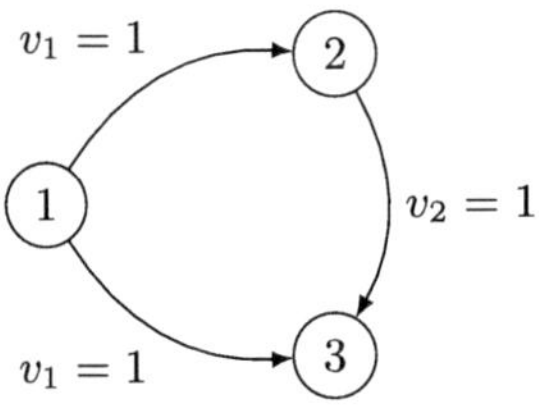

Event	full-state space at node 3
v_1 to node 2	$\left\{\begin{pmatrix}0\\0\\0\end{pmatrix}\right\}$
v_1 to node 3	$\left\{\begin{pmatrix}0\\0\\0\end{pmatrix}, \begin{pmatrix}1\\0\\0\end{pmatrix}\right\}$
v_2 to node 3	$\left\{\begin{pmatrix}1\\0\\0\end{pmatrix}\right\}$

Fig. 1. Full-State-transfer method

which was the basis for the variable calculation. Nevertheless, we also need to consider that even the local state is based on the assumption of the local state of other nodes. Take a look at Fig. 1: When node 1 transmits a variable change of v_1 to node 2 and 3. That will result in a faulty local state in node 3 because only a change of v_2 would resolve the fault. When node 3 now transmits a variable change of v_3 to node 1 together with its local state, node 1 would detect a user fault, even though node 2 had not the time to react according to the initial update of v_1. The resolution for this problem is that we also need to include the local state of node 2 in what we call *full-state space* that we transmit from node 3 to node 1.

As the same argument can be made recursively to all preceding nodes, we conclude that we need to transmit a full-state as a set that is the union of all full-states transmitted for each variable. We can—and indeed must—restrict the full-state space to a subspace that ensures that the variable each node controls are set to the exact value for each node. The full-state space for a variable i is therefore defined as follows:

$$\widetilde{\mathcal{X}}^i(t) = \{x_l^i(t)\} \bigcup_{j=1}^{n} \mathcal{X}^{\mathcal{N}(j)}(D(i,j,t)) \tag{9}$$

$$\mathcal{X}^i(t) = \left\{ \begin{pmatrix} v_1 \text{ if } \mathcal{N}(1) \equiv i \text{ else } \widetilde{v_1} \\ v_2 \text{ if } \mathcal{N}(2) \equiv i \text{ else } \widetilde{v_2} \\ \vdots \\ v_n \text{ if } \mathcal{N}(n) \equiv i \text{ else } \widetilde{v_n} \end{pmatrix} \forall \begin{pmatrix} \widetilde{v_1} \\ \widetilde{v_2} \\ \vdots \\ \widetilde{v_n} \end{pmatrix} \in \widetilde{\mathcal{X}}^i(t) \right\} \tag{10}$$

where $\mathcal{N}(j)$ is the controling node for variable j.

The detection of a user fault is then simply done with the following fault detection function:

$$f_{\text{full-state}}(t) \equiv \text{true} \Leftrightarrow \exists i, j : \mathcal{X}^{\mathcal{N}(j)}\Big(D(i,j,t)\Big) \subseteq F^i \tag{11}$$

Discussion. In theory, this method should be quite robust against detecting infrastructure faults as user faults because the full-state space essentially encodes every possible variable composition responsible for a node not responding with a resolving variable change. However, as the full-state space can be a very large set, this method is not very bandwidth efficient compared to other methods.

4 Simulation

For the simulation, we used a discreet event simulation written in python. The complete source for the simulation we ran is available under [7]. To allow for an easy generation of simulation scenarios, we made the additional assumption that all variables are boolean. In that way, we can easily generate rule functions by creating a lookup table for every variable combination, which is reasonable when the rule function is not dependent on too many variables. We run an initial simulation with a broad range of parameters to analyze the influence of the parameters on the simulation runtime. Based on that simulation, we parameterized the simulation to get a simulation time of under 60 seconds for a single experiment run to simulate 2000 timesteps[1]. The parameters chosen for the simulation are:

- Number of nodes: between 2 and 10
- Number of variables per node: between 1 and 5
- Number of variables accounted for in each lookup table: between 1 and 5
- Maximum number of variables in the whole experiment: 13

In addition to the parameters above, every simulation will get a random propagation function. The propagation function is combined as a weighted average of the base functions shown in Fig. 2 The propagation functions are in detail defined as follows:

- *Uniform*: A time independent function drawing a number from a uniform distribution between 20 and 100
- *Exponential*: A time independent function drawing a number from an exponential distribution with he probability density of $f(t) = \frac{1}{10}\exp\left(-\frac{t}{10}\right)$. Additional a fixed delay of 20 is added to the drawn delay.
- *Normal* A time independent function drawing a number from a normal distribution with mean 60 and a standard deviation of 12 truncated between 20 and 100.

For the simulation we choose to simulate the extrem cases, where one base function is weighted at 100 percent. This simplifies the analises of the simulation results. In other words—we choose one of the following weight vectors:

[1] It is noteworthy that even at this stage, we can see a considerable difference between the methods resource usage. The parameters were chosen because of the full-state-transfer method. The Timestamp method could handle simulations with at least 20 variables in the same runtime, and the token method could easily handle well over 40 variables.

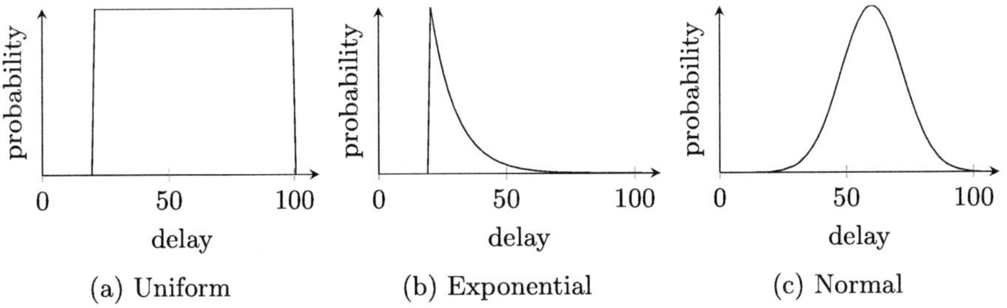

Fig. 2. Propagation functions

$$\begin{pmatrix}1\\0\\0\end{pmatrix}, \begin{pmatrix}0\\1\\0\end{pmatrix}, \begin{pmatrix}0\\0\\1\end{pmatrix}$$

Note that the time is unitless in the simulation and represents time steps. It can be multiplied by any time delta between steps to get an actual time. The different propagation functions are also shown in Fig. 2.

We first run a latency-free model for each scenario and collect all states that the model traverses. After this, we simulate with the selected random propagation function. Again we collect all states that the model traverses. We then compare the two sets of states and determine the states that only the distributed model traversed. We then select two fault spaces: One that only includes states that the distributed model traversed, which by our definition above should only produce infrastructure faults. The other fault space is the set of states that only both models traversed. With this fault space, a user fault should appear. We then run a final simulation with each method for both fault spaces.

5 Results

We ran about $2.4 \cdot 10^5$ scenarios, which included six individual simulations (for each of the three methods and the two fault spaces). So in total, there are $1.4 \cdot 10^6$ individual simulations, which took about 40 h to run in parallel on our hardware.

During the simulation, we collected the following metrics for every method:

- if and when a fault is detected with the infrastructure fault space
- if and when a fault is detected with the user fault space
- the needed bandwidth for each method
- the memory usage per node
- the longest duration that a fault condition is met in a node

To simplify the analysis, we only ever consider the data in the first node. Because the scenario generation is symmetric, this can be done without loss of any generality.

	Prediction	
Actual	Positive	Negative
Positive	121, 260	107, 435
Negative	26, 541	154, 190

(a) full-state-transfer

	Prediction	
Actual	Positive	Negative
Positive	155, 178	84, 860
Negative	71, 574	120, 514

(b) timestamp method

	Prediction	
Actual	Positive	Negative
Positive	41, 556	198, 483
Negative	15, 546	176, 543

(c) token method

Fig. 3. Confusion matrices for the different methods

From those, $2.3 \cdot 10^4$ simulations timed out, and $1.4 \cdot 10^5$ simulation did not produce any fault by chance, which leaves us at $1.3 \cdot 10^6$ simulations that we used to analyze the results.

The separation of our fault space allows us to determine the false negative and false positive rates for all of our approaches. A false negative occurs if no error is detected in the user fault space, while a false positive occurs if an error is detected in the infrastructure fault space. First let us have a look at the confusion matrix (Fig. 3) and statistical classification metrics (Fig. 4) gathered from them. The results show that there is no clear winner here. While the full-sate-transfer method is the most accurate, the timestamp method has almost the same accuracy with better sensitivity. The token method has a very low accuracy (remember that a random process would reach a 50% accuracy) and therefore is not usable in a real-world application. The metrics are not or only very little dependent on the propagation function, as Fig. 5 shows.

We analyzed the above metrics concerning how long a fault condition is present in the simulation. The results are shown in Fig. 7. For the analysis, we binned all the results into bins with at least 1000 samples in them. That explains why the graphs start at different values. Also note, that there is not a lot of

	Sensitivity	Specifity	Accuracy
Full-State-Transfer	53.02%	85.31%	**67.28%**
Timestamp	**64.65%**	62.74%	66.80%
Token	17.31%	**91.91%**	50.47%

Fig. 4. Classification metrics

	Sensitivity			Specifity			Accuracy		
	(a)	(b)	(c)	(a)	(b)	(c)	(a)	(b)	(c)
Full-State-Transfer	54%	52%	54%	85%	85%	86%	**68%**	**67%**	**68%**
Timestamp	**65%**	**64%**	**65%**	57%	66%	65%	61%	65%	65%
Token	19%	16%	18%	**92%**	**94%**	**90%**	51%	50%	50%

Fig. 5. Classification metrics dependent on the propagation functions uniform (a), normal (b), and exponential (c)

	Detection Delay		
	Control Fault	Infrastructure Fault	Average
Full-State-Transfer	1.2	18.5	3.4
Timestamp	8.5	38.3	13.7
Token	138.9	226.2	261.4

Fig. 6. Detection delay for the different methods

data available for fault durations above 100 timesteps of length. Lastly, when reaching values around 2000, we see artifacts from the fact that at this time, the simulation stopped. The graph shows that the token method only detects errors at a minimum fault length of 40 timesteps, which equals exactly one round trip time with the minimum latency of 20 timesteps.

The detection delay is also an interesting metric to analyze. Because in our definition of control faults, there is no way of knowing when it occurs, we choose

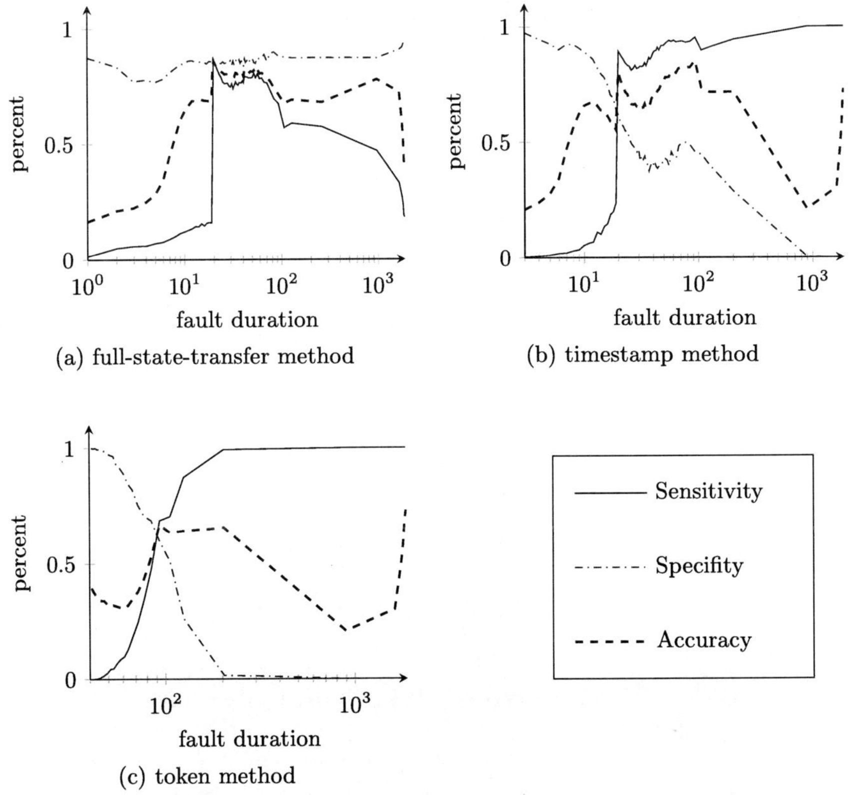

Fig. 7. Classification metrics dependent on the duration of the fault

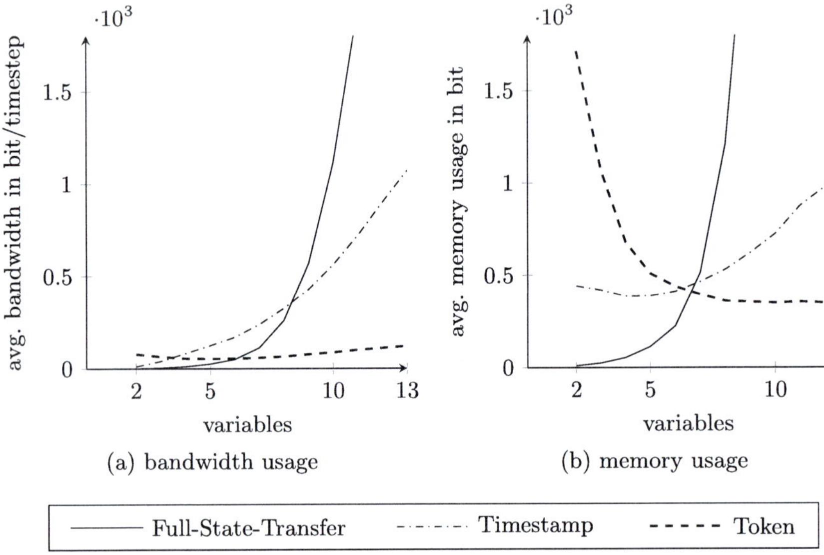

(a) bandwidth usage (b) memory usage

Fig. 8. Resource usage for different methods

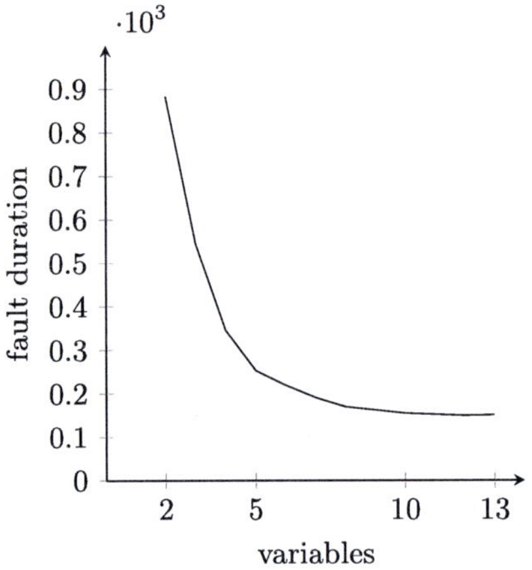

Fig. 9. Fault duration over variable count

a differential approach here. That means that we look at the times that the different methods detect a fault in any given situation. We take the minimal detection time as a reference and use the difference of the detection time of the remaining methods as detection delay. Figure 6 shows the averaged results in response to both the control faults and infrastructure faults, even though only the detection delay in the control fault scenarios is relevant. However, it is interesting to note that the detection delay is more prominent for infrastructure faults across all three methods. Unsurprisingly, the detection delay for the full-state-transfer method is the lowest, as it is explicitly designed to transmit all information necessary for fault detection delay during the experiment. The timestamp method uses a similar approach but needs a detection delay margin to safeguard against any jitter in the network latency. Therefore, it follows the detection delay of the full-state-transfer method closely.

Lastly, we can look at the memory and bandwidth usage of all described approaches. Both are depicted for differing numbers of simulated variables in Fig. 8. The full-state approach uses the most bandwidth and memory by a wide margin. The least demanding algorithm in terms of bandwidth and memory usage is the token-based algorithm. The timestamp-based approach is still considerably less resource-intensive than the full-state approach and more expensive than the token-based approach. In general, the number of variables increases bandwidth usage for all approaches. However, the increase factor of the full-state-transfer algorithm is considerably higher than that of the other approaches, with the token-based having the smallest. Also interesting is the decrease in the average memory usage for the token-based approach as the number of variables increase (which also is seen slightly for the timestamp method). An explanation for this could be that simulations with fewer variables tend to produce more faults that last longer, causing the nodes to save much more token references than in larger simulations. This explanation is supported by Fig. 9.

6 Conclusion and Further Work

This paper presented a novel approach to formally model distributed remote laboratories. The approach is based on the general definition of a system that we extended by a fault space and a propagation function. For simplicity reasons, we made some assumptions that made it easier to reason about two different faults: user faults and infrastructure faults. Nonetheless, the formal description can also be used to reason about such systems more generally.

We also introduced new methods to detect user faults. The token-based and timestamp-based methods use principles in distributed systems that we have already seen in industrial applications. The method of full-state-transfer is novel and, based on our measures, delivers comparable or even better results, although the resource usage is comparatively high. None of the methods discussed here are suitable for real-world applications, as the accuracy is far too low for such usage.

With our simulation based on the mathematical framework introduced in this paper, we were able to establish a tool that can be used to compare the performance of the different methods without building a real-world remote laboratory. Our results show that while performing well in detection accuracy and specificity, the full-state-transfer approach may still be better suited for detecting faults at the beginning of an experiment. Also, it is only suitable for use in experiments with a low amount of variables considering the fast increasing bandwidth and memory requirements. Furthermore, the timestamp-based approach described in this paper has proved to be an effective method for detecting faults with the highest amount of faults detected overall, an acceptable detection delay, and moderate bandwidth and memory requirements. As described in this paper, the token-based method performed worst in many cases, with the notable exception being bandwidth and memory usage. Maybe combining the approaches could yield even better properties than any of the three discussed approaches. Another interesting metric could be the number of distinct errors detected during a simulation. A distinct error could be defined as a faulty state f that lasts from time t_1 to time t_2, meaning not every instance of f between t_1 and t_2 is seen as a different error but rather the same.

Furthermore, we believe that each of the discussed detection methods can be further refined and yield better results in the future. We hope that this work is helpful as a baseline for further research.

Acknowledgements. This work is funded by the Stiftung Innovation in der Hochschullehre, Treuhandstiftung in Trägerschaft der Toepfer Stiftung gGmbH in the project CrossLab [4] (project number FBM2020-VA-182-3-00590).

References

1. Tawfik M et al (2013) Virtual instrument systems in reality (VISIR) for remote wiring and measurement of electronic circuits on breadboard. IEEE Trans Learn Technol 6(1):60–72. https://doi.org/10.1109/TLT.2012.20
2. Henke K, Vietzke T, Wuttke H-D, Ostendorff S (2016) GOLDi - grid of online lab devices ilmenau. Int J Onl Eng 12(04):11. https://doi.org/10.3991/ijoe.v12i04.5005
3. Durfee W, Li P, Waletzko D (2004) Take-home lab kits for system dynamics and controls courses. In: Proceedings of the 2004 American control conference, 2004, vol 2, pp 1319–1322. https://doi.org/10.23919/ACC.2004.1386757
4. CrossLab project website. https://cross-lab.org
5. Functional safety of electrical/electronic/programmable electronic safety-related systems - Part 1: General requirements. International Standard IEC 61508-1
6. Henke K, Vietzke T, Wuttke H-D, Ostendorff S (2015) Safety in interactive hybrid online labs. Int J Onl Eng 11(03):56. https://doi.org/10.3991/ijoe.v11i3.4557
7. Nau J, Helbing P: REV2022 Simulation. GitHub repository. https://github.com/jonau/REV2022-Simulation

Remote Laboratory to Develop Hardware for IoT Applications

Kyathsandra Chandrashekar Narasimhamurthy(✉), Vankalakunti Suchitra, Tumakuru Prasanna Sushma, Satish Kumar Kavana, Ramesh Kumar Deepashree, and Krishnagiri Sanna siddappa Harshitha

Siddaganga Institute of Technology, Tumkur, Karnataka, India
{kcnmurthy,suchitrav}@sit.ac.in

Abstract. Nowadays monitoring of most of the appliances are being done using IoT (Internet of Things) applications. The challenge in developing these IoT applications is to measure the real time natural physical parameters like temperature, light, wind etc. using sensors and how to interface actuators to meet the requirements. This paper outlines the development of an innovative Remote access system that provides a unique platform for the user to interface the sensors and actuators. There is a provision for developing the code to verify the customized test cases. An important feature of this system is user can visualize the real time implementation of the project which will make the IoT application development faster.

Keywords: Sensors · Actuators · IoT · Interface · Arduino

1 Introduction

Learning is effective when the concepts are understood and those concepts are visualized by doing relevant experiments. Creating the facility for experiential learning is the key to make learning effective. Simulation tools and virtual laboratories are the available platforms for experiential learning. Even though these will help to understand the concepts far better than the conventional black board teaching, however they won't give the real time actual results. On the contrary the Remote laboratory systems the user get the access of modules and able perform the desired experiments live and even measure various parameters. However, remote laboratory systems have their own challenges, limitations in implementation and usage. There are many remote laboratory systems available in various universities in different domains. Most of the remote laboratory facilities have limitations in terms of number of experiments available, number of variations possible in a given experiment and number of measurable parameters [1–7]. An ideal remote laboratory system is one where user can perform experiments of their own choice, vary all parameters and measure all possible readings for an effective analysis.

The challenges of the remote laboratory system can be listed as follows.

1) Number of users can use the system at a time.

M. E. Auer et al. (Eds.): REV 2022, LNNS 524, pp. 125–132, 2023.
https://doi.org/10.1007/978-3-031-17091-1_13

As the remote lab systems give the real time access of the module for user, unless there multiple modules are available it becomes challenging to provide the same module to many users as each one wants to perform the experiments for their own set of values at their own pace. If only a single module is there, then we need to create a queuing system.

2) Number of variable parameters available for user control.
It's essential to provide the access to the users all possible parameters that affect performance of an experiment. Creating real time remote interface of these parameters to the user is a challenging task.
3) Visualization of the response.
Response of the experiments is essential in understanding the concepts. Depending on the type of the experiment the responses will be in the form of waveforms, meter readings, displacements etc. Making these responses to reach user is a challenge.
4) Maintenance of the remote laboratory system.
As the remote lab system is available for 24×7, it's challenging to keep the system in working condition for all the time. It is possible to automate the maintenance of the remote lab systems.

This paper Sensors and Actuators section gives the details of the conceptual design of the Remote lab system for IoT applications. The design of complete system is explained in section design and development. The real time working of the remote lab system is discussed in results and discussion section. Finally the summary of the IoT remote laboratory system is done in conclusion section.

2 Sensors and Actuators

To develop any IoT applications it's necessary to identify the physical parameter that is measurable and alterable. Meaning, the IoT applications are feasible where it's possible to sense the physical parameters and actuate according to the need. As all IoT applications need to be implemented for real time scenario, therefore sensing and actuating will be the first and final blocks of the complete system along with connectivity and security issues of the design. Hence the sensors and actuators plays a major role in development of any IoT application.

So even before start developing the prototype of hardware for an IoT application if it's possible to get the real time test results of sensors and actuators it will helpful in the design of robust system. The proposed remote laboratory system is designed to create a platform to get the variations of the desired real time physical parameter along with flexibility to choose relevant sensors and actuators is the biggest challenge. The Remote laboratory will provide the remote user an easy access to vary the physical parameters to mimic the real time scenario for the appropriate sensors. The remote lab will also give the choice of pairing the sensors and actuators to user so that the test results can be obtained as per requirements.

The Remote laboratory system for development of hardware for IoT applications is implemented using the three sensors and three actuators. The sensors used are Light Dependent Resistor (LDR), Temperature sensor and Opto-coupler. The actuators are bulb, DC motor and AC motor. The combination of LEDs and LDR is used to Mimic the variations in intensity of natural light as shown in Fig. 1a. The temperature variation and sensing is realized using a heater controlled by AC mains and temperature sensor respectively as shown in Fig. 1b. Stepper Motors are used to mimic the blocking of Opto-couplers to realize the motion of objects as shown in Fig. 1c. The actuators used in the remote laboratory are AC bulb, DC and AC motor as shown in the Fig. 1d–f respectively.

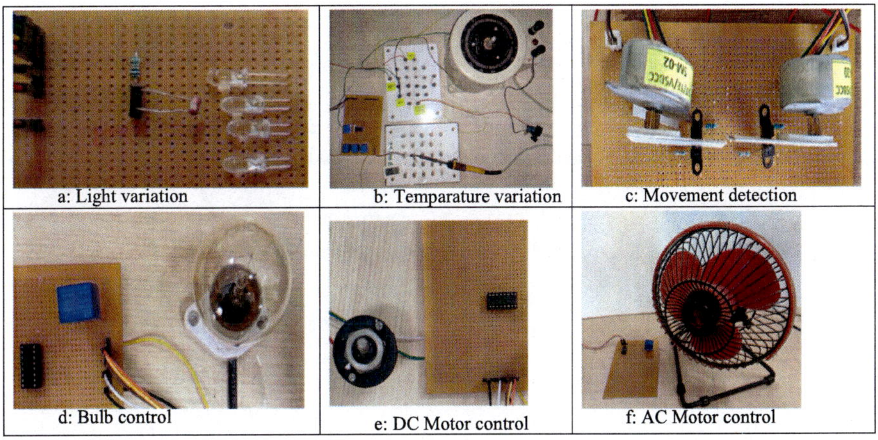

Fig. 1. Sensors and actuators used in implementation of remote laboratory.

Using the LED-LDR combination it's possible to test the applications based on light intensity ranging from dark, dim and bright light conditions by appropriately switching the LEDs. Using this type of sensors it's possible to implement Light automation at home and also in security related applications to trigger a siren or turn on a bulb if shadow is detected in a protected area.

Heater and transformer combination feature of the remote lab will help the user to implement any application where the operating temperatures range from 23 °C to 140 °C. Test readings to develop an IoT application for designing a Air conditioning system can be easily obtained. Such type of test results help to finalize the system design sooner.

The Opto-coupler based sensor can interface bidirectional DC motor; this will help to develop smart and secured doors in private and public places. User has the freedom to combine multiple sensor inputs to drive one or multiple actuators also this will help to design complex IoT applications.

3 Design and Development

The Remote lab system to develop the hardware for IoT applications has many challenges. First one is to provide the online access to the system. Creating a circuitry to realize variable physical parameters like temperature, light and motion, then providing the choice of pairing of sensors-actuators. Then provide a platform for testing customized code. Finally, video steaming of the real-time execution of the application will be done.

Online access to the Remote laboratory system is conceptually shown in Fig. 2. It consists of server and a subsystem (in future more subsystems will be added). The sensors, actuators and other interface circuitry are interfaced to the Arduino controller. Two cameras are used to provide the video live steaming.

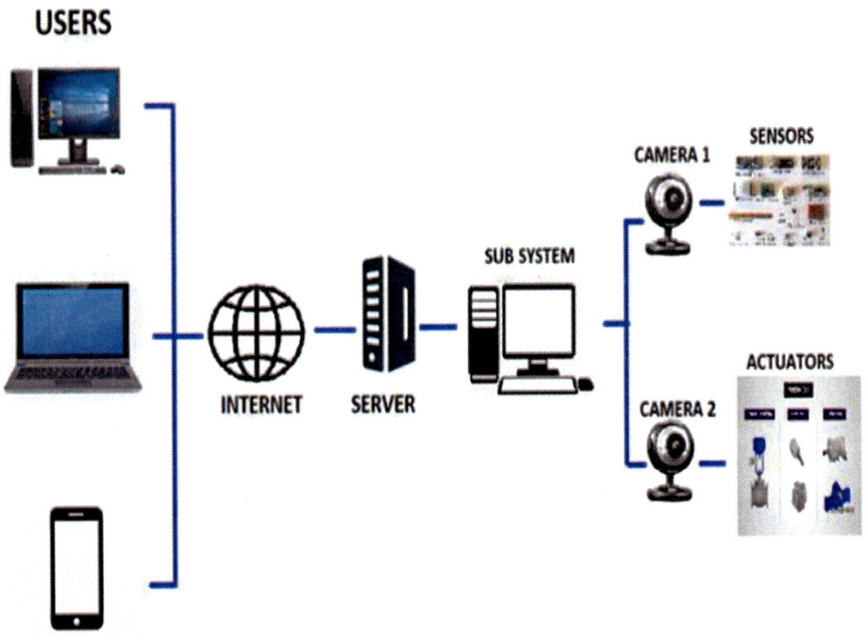

Fig. 2. Conceptual block diagram of remote laboratory for IoT applications.

In the phase-I of the proposed Remote lab for IoT applications, as discussed in the previous section, three sensors and actuators are used. To pair the sensors and actuators as per the users wish a controller named as "Master Arduino" is used as shown in Fig. 3. Once the paring is established, user can write the customized code to implement the desired application on to the "User Arduino". The script running on the User Arduino will trigger the actuator paired to the sensor as shown in Fig. 3. The port information of sensors and actuators will be provided to the user for scripting the code. The equivalent voltage levels of the physical scenarios like various lighting condition for LDR that is dark, dim light and bright light and various temperature levels for interfacing temperature sensors and opto coupler outputs will be provided to the user to write the effective coding.

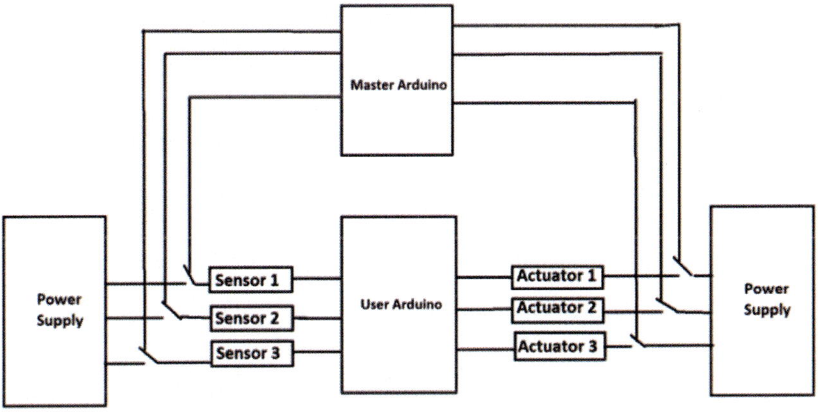

Fig. 3. Pairing of sensors and actuators and their interface

The complete flow of the proposed remote lab system for one application is shown in Fig. 4 and 5. Python code is used to provide the GUI to the user. The first page of the proposed system is as shown in Fig. 4a. Authenticated user can select Light Dependent Resistor (Light sensor) one of the available sensor from the dropdown option as shown in Fig. 4b. Later user bulb as one of the actuator to be operated as per the need again from the dropdown option as shown in Fig. 4c. Later user can set the desired physical condition for testing purpose. In case of LDR sensor the physical quantity to be set is the intensity of the light. The light intensity is mimicked in this project by tunning ON/OFF many LEDs. As shown in Fig. 4d its possible to mimic 4 different physical light intensity conditions like; Too dark, Dark, Moderate and Bright.

The selection of sensor, actuator and desired physical conditions are activated by the Arduino. Now user has established the required hardware interface. Now, it's time to write a code to test the desired application. The user can write the code by clicking on "Open Code Editor" option as shown in Fig. 5 (a) & (b) and finally can edit the code in the console as shown in the Fig. 5(c). The required port information of various devices can also be seen in Fig. 5(c). This code will be uploaded onto User Arduino.

As the conduct experiment button is clicked the entire system will be activated and as per the desired condition the physical conditions will be created, sensor output is signal conditioned and is processed in the Arduino and the actuator will be operated accordingly. The entire phenomenon will be shown live to the remote user using two cameras one at the sensor and the other at the actuator side. The user can repeat the entire process for any sensor, actuator and available physical conditions. Results of this entire process will be discussed in next section.

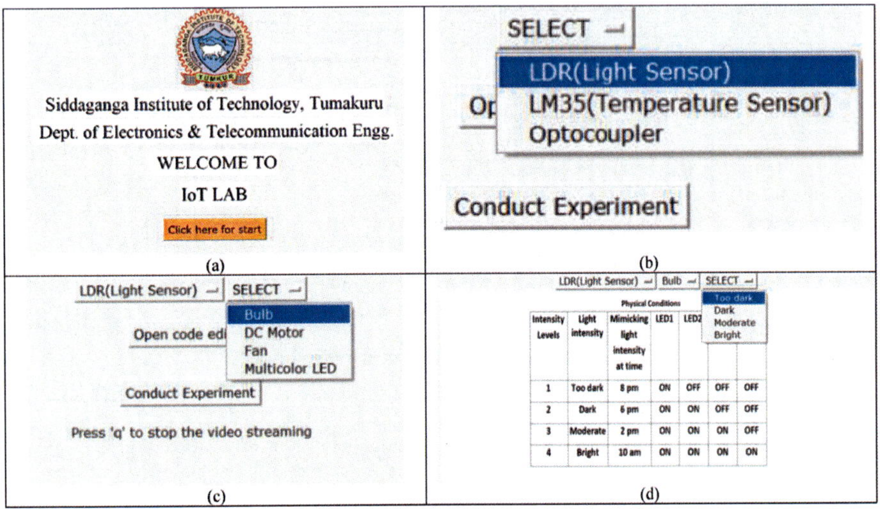

Fig. 4. Process steps in selection of sensor and physical condition. (a) First page, (b) Selection of sensor, (c) Selection of actuator (d) selection of physical condition

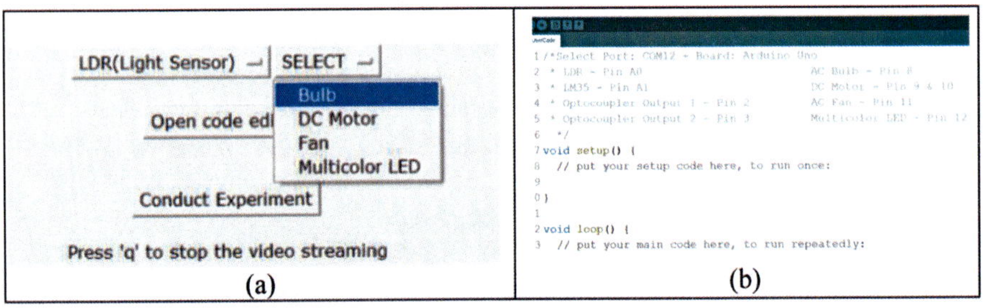

Fig. 5. (a) Selection of user editor console, (b) Editor console template

4 Results and Discussions

In this section let's see how the selected sensors read the different physical conditions that are created. Table 1 give the details of the voltages at the LDR sensor for different intensity level of the light (derived from LEDs). The actuator that needs to be controlled is a Bulb. The sensor and actuator used for this particular example are shown in Fig. 1 (a) and 1(d) The task assumed for this application is, if the intensity of light becomes bright the bulb should be tuned OFF. The table shows the voltage read by the Arduino in decimal value, using this user can decide the condition for Bulb. The last column shows the Bulb turning OFF as the intensity of the light becomes bright. This type of application can be used in the automatic street light control. This can also be a part of GREEN HOUSE projects.

Table 2 lists the results obtained while interfacing temperature sensor to AC motor (fan). The necessary variations in the temperature is realized using the multi-tapped transformer to feed different voltages to the heater to realize different temperature. The

Table 1. Results of LDR interfaced with LEDs light intensity

Level	Light intensity	Voltage (mV)	Decimal value	Bulb condition
1	Too dark	870	179	ON
2	Dark	800	165	ON
3	Moderate	760	157	ON
4	Bright	620	127	OFF

temperature is sensed using a temperature sensor LM35. Table 2 give the details of the ac voltages obtained at various tapings of the transformer, decimal equivalent of the temperature sensor output read by the Arduino and the condition of the fan (it's designed to turn-on fan if the temperature exceeds 50 °C). The sensor and actuator used for this particular example are shown in Fig. 1 (b) and 1(f). This type of application can be used in the any application where its used to monitor the temperature which ranges from maintenance of industry Boilers to Air-conditioned home.

Table 2. Results of temperature sensor and fan

Level	Transformer voltage, V	Temperature °C	Decimal value	Fan condition
1	40	29.3	14	OFF
2	120	47.6	24	OFF
3	180	85	65	ON
4	230	98	72	ON

Table 3 shows the condition of DC motor (which used to mimic the door) against condition of the opto couplers (which indicates the presence of persons near the door). The movement of the person near the door is mimicked by interrupting the opto couplers by the moving the obstacle in the Opto coupler slot. The obstacles are. The Table 3 gives 4 possible outputs from the pair of opto sensors. It shows the condition of the DC motor like clock-wise, anti clock-wise and halts conditions.

The above three combinations of sensors and actuators are just an example how the remote lab system can be used for developing the IoT applications. Since user can use their own code and able to vary the physical conditions it becomes very handy to develop the applications by mix matching the sensor and actuator.

Table 3. Results of Opto sensors and DC motor for various conditions

Level	Real time scenario	Opto 1	Opto 2	Motor rotation	Door condition
1	No one on either door	LOW	LOW	Stop	Closed
2	Entering	HIGH	LOW	Clockwise	Opens
3	Exiting	LOW	HIGH	Anti-clock	Closes
4	Enter and exit	HIGH	HIGH	Stop	Open

5 Conclusion

The proposed Remote lab system for developing the IoT application is an unique innovative approach to provide a platform with a set of sensors and actuators. The system will provide a unique opportunity to select the sensors and actuators for implementing any complex real time application. The most interesting feature of the system is realization of various physical conditions, and their live streaming and user can test their own code to fine tune their design. This remote lab will provide the remote user to verify their code with an opportunity to visualize the live steaming of the real time implementation of the task. This Remote lab system is aimed at reducing the development time to provide feasible solution for complex real time issues.

References

1. GO-LAB. https://www.golabz.eu/
2. LabsLand. https://labsland.com/en/labs
3. TU Berlin: Remote Farm. https://remote.physik.tuberlin.de/experimente/
4. iLab. https://icampus.mit.edu/projects/ilabs/
5. Ponce OA, Pagán Maldonado N (2017) Educational research in the 21st century: challenges and opportunities for scientific effectiveness. Int J Educ Res Innov 8:24–37
6. Hsiao KF, Chen NS, Huang SY (2012) Learning while exercising for science education in augmented reality among adolescents. Interact Learn Environ 20(4):331–349
7. Narasimhamurthy KC, Orduna P, Rodríguez-Gil L, Bharath GC, Susheen Srivatsa CN, Mulamuttal K (2021) Analog electronic experiments in ultra-concurrent laboratory. In: Auer M, May D (eds) REV 2020. AISC, vol 1231. Springer, Cham, pp 37–45. https://doi.org/10.1007/978-3-030-52575-0_3

Promotion of Remote Experimentation in Three Latin American Countries

Carlos Arguedas-Matarrita[1](✉), Eric Montero-Miranda[1], Fiorella Lizano-Sánchez[1], Gabriela Varela[2], Jorge Esteban Maeyoshimoto[3], Gabriel Medina[3], and Ignacio Idoyaga[3]

[1] Universidad Estatal a Distancia, San José, Costa Rica
carguedas@uned.ac.cr
[2] Consejo de Formación en Educación-(CFE)-ANEP, Montevideo, Uruguay
[3] Universidad de Buenos Aires, Buenos Aires, Argentina

Abstract. This paper presents the first results of the diagnosis of a project that seeks to promote the development and use of Remote Laboratories in Argentina, Costa Rica and Uruguay. In the current context, educational institutions must integrate resources that allow remote experimental work to achieve the knowledge and skills necessary for learning science at both the secondary and University levels. The project is developed under the modality of research based on design, in this diagnostic stage a total of 448 teachers completed a specially designed instrument, among the main results it was obtained that the participants consider experimental work very important in the teaching and learning of science, however, the frequency with which experimental activities are carried out is low, and the need for focused resources to carry out experimental work in science is also pointed out. From this starting point, the project proposes the design of new remote access laboratories and teaching sequences to overcome the difficulties encountered.

Keywords: Remote Laboratories · Online education · Experimental sciences

1 Introduction

The emergency remote teaching (ERT) [1], articulated since 2020 at all levels of the Latin American educational system in response to the health crisis of COVID-19, reveals that technological and teacher training needs persist in educational institutions. Consequently, there are many organizational and methodological tensions that managers and teachers face when education is articulated in digital environments or in hybrid settings.

The teaching of natural sciences faces particular challenges when education incorporates digital mediations [2]. The experimental character of these disciplines must be especially considered by teachers in their planning. Particularly, the experimental activities (EA) didactically designed, which involve the manipulation of variables and empirical work, are a privileged way of knowing in the area. Furthermore, these types of activities are the strategy of choice when seeking to promote learning of intellectual and sensorimotor procedures [3] typical of experimental work and professional practice in science and technology degrees.

M. E. Auer et al. (Eds.): REV 2022, LNNS 524, pp. 133–142, 2023.
https://doi.org/10.1007/978-3-031-17091-1_14

Working in virtual teaching and learning environments, on different platforms with different applications, allows to sustain teacher-student communication and carry out multiple activities. However, there are still few strategies to support EA in digital environments. Metaphorically, the virtual Classroom was extended, but the challenge of extending the laboratory remains. Along these lines, in 2020 the extended laboratory model (EL) [4] is proposed as a theorical approach to this problem.

The conformation of the SciLabs program, which integrates the efforts of research teams based at the Universidad de Buenos Aires, in Argentina, at the Council of Education Training, in Uruguay, and at the Universidad Estatal a Distancia of Costa Rica, responds to the need to find alternatives to sustain experimental teaching in digital environments in Latin America. This research and development program, which proposes work within the framework of the EL model, seeks to design Remote Laboratories and teaching and learning sequences (TLS) that include them. Currently 14 trained researchers participate, 10 researchers in training, most of them scholarships holders and master's and doctoral thesis students, and dozens of mid- and higher-level teachers in practice. In this way, an interdisciplinary group is formed where natural sciences teachers, physics, chemistry and biology teachers, designers, sociologists and technologists work hand-in-hand. Funding for the program comes from competitive funds from the Universidad de Buenos Aires, the National Research and Innovation Agency of Uruguay, and the Organization of American States.

The objective of this work is to discuss the characteristics and present the first results of the diagnostic stage of the SciLabs program.

1.1 The Extended Laboratory Model

The model proposes the systematic and synergistic use of different strategies and devices to carry out experimental activities. It is a set that seek to establish an experimental hybrid to recover EA in education that incorporates digital measurements. In a non-exclusive way, work on extended laboratory logic can include the performance of simple experimental activities, the use of smartphones as measurement instruments, simulations, virtual laboratories and remote laboratories (RL). The latter are central and structuring of the teaching proposals in this model.

A RL is a set of software and hardware technologies that allows students and teachers to carry out real experimental activities of a high level of sophistication at a distance. The instruments are operated from anywhere and at any time through a graphical user interface [5] allowing the analytical treatment of empirical data [6]. In this way, the educational use of remotes laboratories holds the promise of increasing the autonomy and degrees of freedom of students in relation to laboratory practices, allowing self-regulation of learning. In addition, it enables teachers to personalize EA to meet the educational needs of different student profiles.

RLs can be classified in real-time remote laboratories (RTL) and deferred remote laboratories (DL). In an RTL the manipulation of the instruments, the observation and the real experience occur synchronously. These types of laboratories are limited in the number of students who can use it at the same time and involve higher maintenance costs. They are frequent in the teaching of physics. In a DL the EA is real but the access to it is deferred. These laboratories are based on a set of recordings and they are powerful alternatives for teaching chemistry and biology.

The use of RL in mid- and high- education, particularly in physics and engineering, reports encouraging results [7]. They favor the creativity and initiative of the students [8] and many competences that are the object of education [9]. However, their diffusion in Latin America and the training of teachers for their use in teaching proposals is still scarce. Consequently, it is necessary to advance in the concretion of programs and projects that allow promoting the remote experimentation in the region as a recovery strategy for EA in the face of future hybrid educational scenarios [10].

2 Methodology

The way of working, oriented to the development of Remote Laboratories in real time and deferred, their installation in Latin America and the generation of TLS appropriate to the context and published as Open Educational Resources, corresponds to a research based on design (see Fig. 1).

The methodology, as in all design-based research [11], includes cooperation between teachers responsible for courses and researchers to identify relevant educational needs in the context of the research and propose original solutions based on research results. The work proposal is based on the realization of iterative cycles of implementation, retrospective analysis and refinement of the RLs and TLS, in addition to the pragmatic and rigorous use of methodologies of the qualitative and quantitative approaches to analyze data from different sources (see Fig. 2). Thus, the research which aims to design Technology and teaching devices, includes strategies such as conducting interviews, questionnaires, observations, documentary analysis and tasks of various kinds.

The actions planed in the program included as part of the initial diagnosis, an exploratory study based on the performance of EA before and during the ERT. First, six interviews were conducted with key informants: physics, chemistry and biology teachers who worked at the intermediate and higher level in Argentina, Costa Rica and Uruguay. Content analysis [12] and grounded theory [13] were used to generate different study categories. Based on the latter, multiple-choice questions and Likert-type statements were generated for the design, secondly, of a data collection instrument, a questionnaire was generated that could be applied to a large number of teachers. The validation of this instrument was carried out by means of a round of experts. Then, the questionnaire was applied remotely to the teachers who participated in the presentation day of the SciLabs program.

The questionnaire used for data collection was structured in three sections. The first, collects the affiliation data (subject, seniority, educational level, country, others) of the participating teachers. The remaining sections inquire about the performance of EA: before and during the pandemic.

The study of the EA carried out before and during the pandemic focused on recognizing the frequency (never, rarely, often, always) that participants proposed the EA in their educational practices and the importance they attached to it (not important, little importance, something of importance, a lot of importance). Moreover, teachers were consulted about the nature of EA proposed during the pandemic (home activities, use of

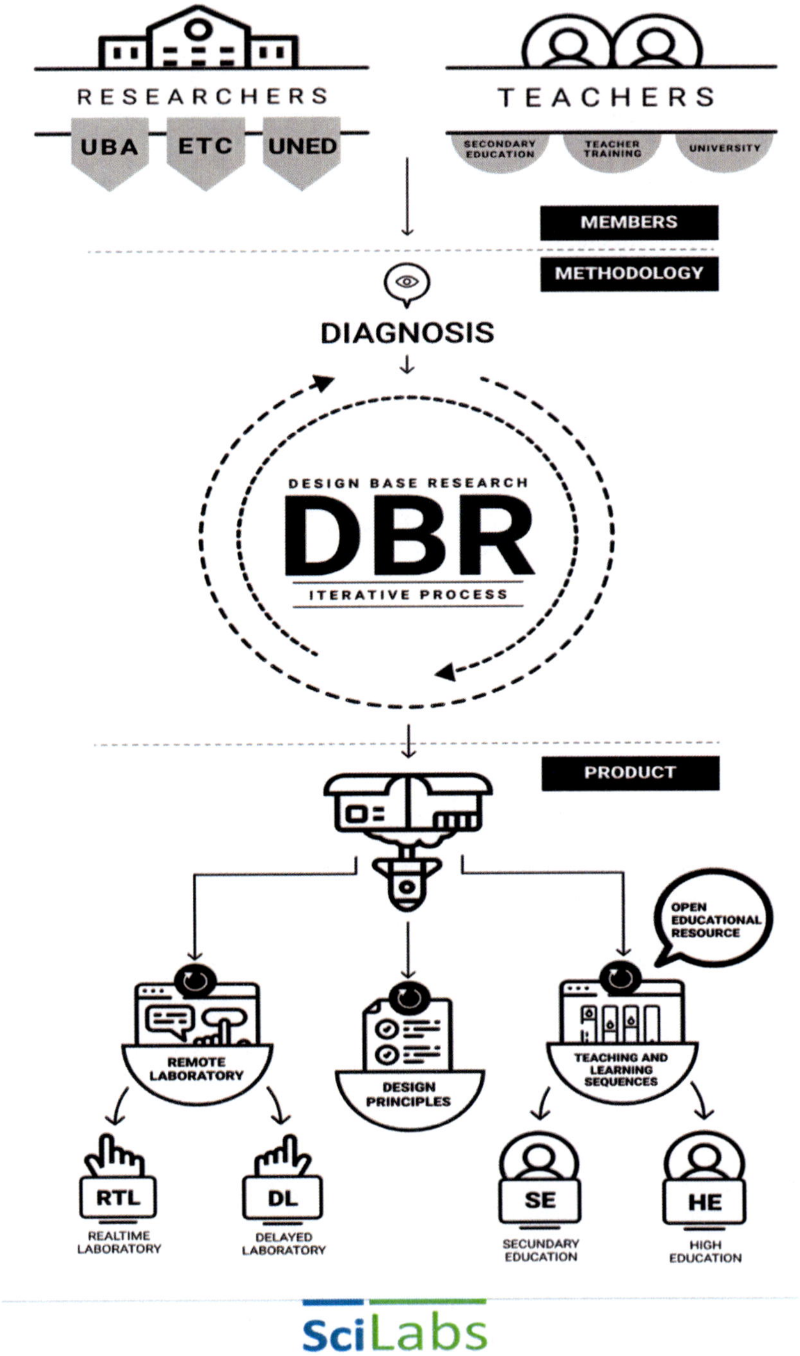

Fig. 1. SciLabs project work methodology.

smartphones, simulators, etc.) and about other areas of interest for the development of RL.

Data analysis resorted to descriptive statistics and inferences. Absolute and relative frequencies were calculated, and contingency tables were constructed. In all cases, the SPSS-26 software was used.

3 Results

A total of 448 participants completed the instrument, registering a distribution by country of 58.7% (n = 263) for Argentina, 35.3% (n = 158) for Uruguay, 4.5% (n = 20) for countries who do not participate in the project and 1.6% for Costa Rica (n = 7). In relation to the disciplinary area in which they work, it was found that around 80% of the participants work in biology and chemistry (see Fig. 1). Regarding the educational level, it was found that 51.8% (n = 232) work at the intermediate level, 27.7% (n = 124) at the university level and 20.5% (n = 92) in non-university teacher training institutes.

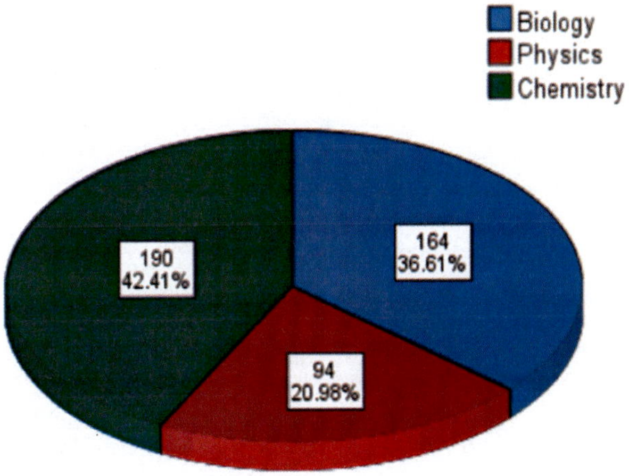

Fig. 2. Disciplinary area in which the participants work.

The participation of chemistry and biology teachers in this study is especially relevant, since they are vacancy areas in which deferred or real-time laboratory developments are very scarce. Consequently, this high participation of teachers of these subjects in the presentation of the SciLabs program could be realizing the need for developments and resources to carry out digital EA in these disciplines.

Respect to the age ranges to the age ranges of the participants, it was found that experiences teachers with more than 15 years of experience and new teachers with less than 5 years of experience are well represented in the sample (see Table 1). Therefore, it can be thought that teachers from different backgrounds approach the SciLabs program since they share the need to seek alternatives to carry out EA in digital environments.

Regarding the realization of EA before and during the pandemic it was found that these activities are carried out with low frequency during the pandemic (see Table 2),

Table 1. Results for the age ranges of the participants.

Age range	Frequency	Percent
Between 10 and 15 years	75	16.7
Between 5 and 10 years	80	17.9
More than 15 years	152	33.9
Less than 5 years	141	31.5
Total	448	100.0

despite the fact that in the natural sciences experimental work plays a central role in the teaching and learning processes, access to laboratories was limited due to social distancing measures. The decrease in the frequency of EA in the pandemic highlights the need for programs and projects that provide alternatives for teachers to sustain teaching with EA in digital or hybrid environments.

Table 2. Contingency table on the performance of experimental activities before and during the pandemic.

		How often were experimental activities carried out in your classes before the pandemic?				Total
		O	N	R	A	
How often are experimental activities carried out in your classes during the COVID-19 pandemic?	O	54	0	8	22	84
	N	23	8	30	17	78
	R	137	2	81	47	267
	A	3	0	0	16	19
Total		217	10	119	102	448

O: Often; N: Never; R: Rarely; A: Always.

Another relevant aspect in this diagnosis is to try to establish the importance that the participants attribute to the experimental activities, it was found that although they are carried out less frequently during the pandemic than before the pandemic, they are always activities considered of great importance in science teaching. This highlights the inability of teachers to sustain these types of activities in remote learning and the need to work to find creative solutions (Table 3).

In relation to the experimental activities carried out by the students in the courses during the pandemic, the carrying out of experimental activities was obtained more frequently:

- Through the use of instruments and household inputs (f = 239),
- Using recorded videos on in a synchronous video conference (f = 228),

Table 3. Contingency table on the importance of experimental activities before and during the Pandemic.

		What importance/value do you assign to the experimental activities that you carried out in your course before the pandemic?				Total
		SI	AI	NI	LI	
What importance/value do you assign to the experimental activities that you carry out in your course during the COVID-19 pandemic?	SI	30	80	1	1	112
	AI	14	286	1	2	303
	NI	1	4	3	0	8
	LI	9	11	1	4	25
Total		54	381	6	7	448

SI: Something of importance; AI: A lot of importance; NI: Not important; LI: Little importance.

- Using a simulator (f = 187),
- Through video analysis programs (f = 153),
- Using a virtual laboratory (f = 102),
- Using a mobile phone (f = 95).

Table 4 shows the frequencies of coincidence in topics, whose perception of the participants represents an opportunity to approach from developments in RL. The total values do not represent all the responses, since those that have viability in the development of RL resources were selected.

Table 4. Frequency of disciplinary contents according to the perception of the participants.

Natural Science					
Chemistry	Frequency	Physics	Frequency	Biology	Frequency
Separation of mixtures	5	Radioactivity	7	DNA sequencing	8
Chemical reactions	32	Electronics /Arduino	4	Biological fluids	4
Kinetics	4	Balance/Rotations	2	Microscope	28
Thermodynamics	9	Magnetic field	9	Cultures/microbiology	11
Chemical balance	8	Mechanics	5	Photosynthesis	7
Volumetry (A-B)	7	Óptics	6	Cell diffusion	7
Biochemistry	5	URM y UVRM (Kinematics	15		

(continued)

Table 4. (*continued*)

Natural Science					
Chemistry	Frequency	Physics	Frequency	Biology	Frequency
Solutions	6	Electric field	9		
Acidity level (pH)	3	Electrical circuits	5		
Biomolecules*	15	Radiation from a black body	2		
Food analysis	2	Sound/waves	5		
Gas law	4	Energy conservation	5		
Instrumental analysis	10	Dynamic	12		
Chemical bond	2	Fluids	1		
Electrochemistry	6				

In the discipline of Chemistry, the content of chemical reactions stands out, where several of the participants agreed on the need for RL oriented to organic synthesis and traditional reactions of general chemistry. Moreover, a coincidence is established in laboratories related to the study of biomolecules, both in Chemistry and Biology. Responses are highlighted where developments are requested with equipment for high-precision instrumental analysis (spectroscopy and high-resolution chromatography). Finally, topics such as acid-base volumetry and gas laws are highlighted, where developments are already available [14, 15].

In the area of Physics, the need for RL in kinematics, dynamics and electric and magnetic fields stands out. In this sense, one of the groups participating in the project, as well as other groups, have practices focused on these topics.

Finally, in Biology there is a demand in RL microscopy and microbiology, with some other responses with similar frequency in DNA sequencing, photosynthesis and cell diffusion. In the case of the microscope, the coincidences point to experiments with prokaryotic and eukaryotic animal and plan cells and processes of cell division.

4 Conclusions

The first results of the diagnostic stage make clear the needs of teachers and educational system in the region to have viable alternatives to carry out EA in digital environments or in hybrid educational settings. This becomes an urgent and strategic schedule given the definition of education formats in the post-pandemic.

The notion held by teachers of the need to incorporate experimental activities is limited by the low availability of tools and resources appropriate to the context. In this sense, a program like SciLabs, which proposes the design of RL and TLS based on educational research, provides an opportunity to overcome the difficulties encountered.

The diagnosis made will allow to orient the work to the real needs of science education in the Latin American region and forms a first product that allows RL developers to identify vacancy areas.

The choice of a design-based research model will allow the rapid transfer of the results to the classrooms and the participation of the teaching staff in the project. On the one hand, highly contextualized concrete products such as RL, TLS and open educational resources will be generated and al the teaching material necessary for the implementations. On the other hand, original knowledge will be generated as design principles that have a high potential for generalization.

Acknowledgements. This work was carried out within the framework of the FSED-2-2020-1-163647 project of the National Research and Innovation Agency of Uruguay.

References

1. Hodges C, Moore S, Lockee B, Trust T, Bond A (2020) The difference between emergency remote teaching and online learning. https://er.educause.edu
2. Byers W (2002) Promoting active learning through small group laboratory classes. Univ Chem Educ 6:28–34
3. Lorenzo M (2020) Revisando los trabajos prácticos experimentales en la enseñanza universitaria. Aula Universitaria (21), e0004. https://doi.org/10.14409/au.2020.21.e0004
4. Idoyaga IJ, Vargas-Badilla L, Nahuel Moya C, Montero-Miranda E, Garro-Mora AL (2020). El Laboratorio Remoto: una alternativa para extender la actividad experimental. Campo Universitario 1(2):4–26. https://campouniversitario.aduba.org.ar/ojs/index.php/cu/article/view/17
5. Royo J (2018) Diseño digital. Ediciones Paidós Ibérica
6. Arguedas-Matarrita C et al (2019) Remote experimentation in the teaching of physics in Costa Rica: first steps. In: Proceedings of the 2019 5th experimental international conference, Madeira, Portugal. https://doi.org/10.1109/EXPAT.2019.8876553
7. Arias E, Arguedas-Matarrita C (2020) El trabajo experimental en la enseñanza de la Física en tiempos de pandemia mediante el uso de la aplicación II Ley de Newton en la UNED de Costa Rica. Revista Innovaciones Educativas 22(Especial):7–18. https://doi.org/10.22458/ie.v22iEspecial.3204
8. Aramburu Mayoz C et al (2021) FPGA remote laboratory: experience in UPNA and UNIFESP. In: Auer ME, May D (eds) REV 2020, vol 1231. AISC. Springer, Cham, pp 112–127. https://doi.org/10.1007/978-3-030-52575-0_9
9. Moussa M, Benachenhou A, Belghit S, Adda Benattia A, Boumehdi A (2021) An implementation of microservices based architecture for remote laboratories. In: Auer ME, May D (eds) REV 2020, vol 1231. AISC. Springer, Cham, pp 154–161. https://doi.org/10.1007/978-3-030-52575-0_12
10. Lugo MT, Loiácono F (2020) Planificar la educación en la pospandemia: de la educación remota de emergencia a los modelos híbridos. Educación Y Tecnología 3(1). Recuperado a partir de. https://publicaciones.flacso.edu.uy/index.php/edutic/article/view/2
11. de Benito Crosetti B, Salinas Ibáñez JM (2016) La Investigación Basada en Diseño en Tecnología Educativa. Revista Interuniversitaria de Investigación en Tecnología Educativa. https://doi.org/10.6018/riite2016/260631
12. Laurence Bardin Ediciones AKAL, 3 abr 1991–192
13. Glaser BG (2004) Remodeling grounded theory. Forum qualitative social research 5(2)

14. Idoyaga IJ et al (2021, octubre-diciembre) Conocimientos del profesorado universitario sobre la enseñanza de la química con laboratorios remotos. Educación Química 32(4). https://doi.org/10.22201/fq.18708404e.2021.5.79189
15. Arguedas-Matarrita C et al (2022) Design and development of an ultra-concurrent laboratory for the study of an acid–base titration (ABT) at the universidad estatal a distancia (UNED), Costa Rica. In: Auer ME, Bhimavaram KR, Yue X-G (eds) REV 2021, vol 298. LNNS. Springer, Cham, pp 122–130. https://doi.org/10.1007/978-3-030-82529-4_13

Rural, Remote and Real: Democratizing the Access to Science in Rural Schools Using Remote Experiments

Javier García-Zubía[1](✉), Verónica Canivell[1], Diego Casado[1], Ignacio Angulo[1], Unai Hernandez-Jayo[1], Mariluz Guenaga[1], Cristina Gimenez[2], and Luis Rodríguez-Gil[2]

[1] University of Deusto, Avda. Universidades 24, 48007 Bilbao, Spain
zubia@deusto.es
[2] LabsLand, Bilbao, Spain

Abstract. Rural schools are not, in general, well equipped for the STEM disciplines, but the students in rural areas should have the same opportunities that the students in the urban areas. The use of remote experiments could mitigate this situation. Remote experimentation democratizes the access to science and technology. This papers describes the R3 Project: Rural, Remote & Real.

Keywords: Remote experiments · Remote laboratories · Rural schools

1 Introduction and Scope

1.1 FECYT, Spanish Foundation for Science and Technology

FECYT (www.fecyt.es/en) is a Spanish governmental office that depends on the Ministry of Science and Innovation. Its main objective is "FECYT works to strengthen the link between science and society through actions that promote open and inclusive science, scientific culture and education", and one of the main activities is develop STEM disciplines mainly among the young people.

Every year there is a call for project proposals. In 2020 more 400 proposals were submitted, the threshold was 85/100.

R3 Project was submitted to the 2020 call for proposals and it was approved. The R3 Project started in September 2021.

Rural Education in Spain

The population of Spain is around 50 million people. The 90% of Spain is considered rural and only the 20% of the population live there. The population of the rural areas is elderly. Figure 1 clearly shows that the population distribution in Europe is not homogeneous [1].

Spain is the fourth European country in terms of depopulation. The most risky countries are Estonia, Finland and Latvia.

M. E. Auer et al. (Eds.): REV 2022, LNNS 524, pp. 143–150, 2023.
https://doi.org/10.1007/978-3-031-17091-1_15

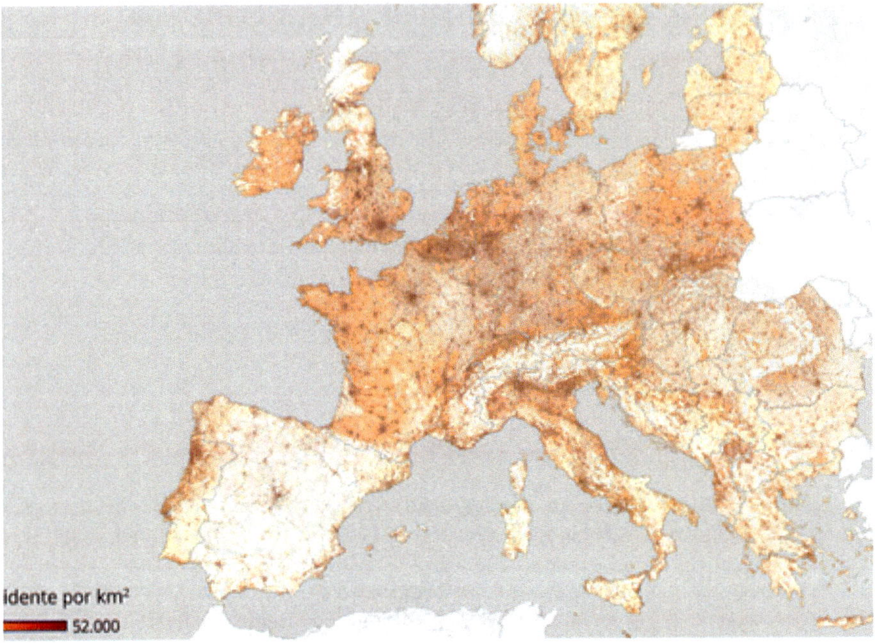

Fig. 1. Distribution of population in Europe

A way of fixing rural population in Spain is to have schools, at least primary schools. In this scenario, there are some CRAs (Centro Rural Agrupado, Grouped Rural School) since 1980, when the Spanish Government planned them. A CRA is a group of small schools disseminate in a rural area.

The main characteristic of the CRAs and its schools is that they are usually small (from 5 to 20 students) and they grouped the students of different ages (from 3 to 12). Using this approach, the rural areas have schools. For instance, the students can be grouped in two classrooms: 6–9 years old, and 9–12 years old, even all the students can be in a unique classroom. Around 75.000 students are enrolled in these schools.

In this situation the CRAs do not have laboratories, or if they have laboratories they are basic.

The Spanish Ministry of Education has designed a plan [2] for the development of these CRAs with 14 actions that will power the rural schools. This plan is aligned with the SDG Goal 4: "**ensure inclusive and equitable quality education and promote lifelong learning opportunities for all**" and it is part of the 2030 Agenda [3].

CRAs are small, with not so many students, neither teachers, but in general they have good Internet connection, that in many cases is with optical fiber. The 2015 PISA report [4] remarks that students in rural areas obtain better results compared to students in urban areas. And this result is even more clear for Spain, Finland, Estonia and Latvia, that are the countries with more depopulation.

Therefore, Spain needs the rural schools, they have good potential and good results, but in some aspects, they have problems. For instance, it is hard to install a STEM laboratory in a CRA. The R3 Project -Rural, Remote and Real- is focused in giving the students access to real experiments without having them, through the remote experimentation.

2 R3 Project General Description

R3 Project is supported by FECYT, University of Deusto and LabsLand. FECYT supports and disseminates the project, Deusto researchs and LabsLand provides the remote experiments. Its main objective is to promote STEM disciplines among primary school students in Spanish rural areas. R3 Project tries to mitigate the lack of experimental resources in rural schools.

Project R3 is a one-year project and it has four main tasks:

- Explaining the Project R3 and selecting rural schools and teachers.
- Training teachers in the remote experimentation and the remote labs offered by LabsLand.
- Deploying in the classroom the remote experiments selected by the teachers.
- Evaluate the Project R3 with teachers, students and stakeholders.

The previous four steps are repeated two times in a year, from September to January and from January to June. After the first iteration, the results analysis with the involved teachers will drive the second iteration.

In total twenty teachers should be involved in the project: ten teachers in each phase. It is expected that all the teachers and students to be from rural areas, and this is easy to determine for primary school students, but not for secondary and high school students. These students are concentrated in the main village of each area with, maybe, thousands of citizens, and because of this teachers can think that they are not rural teachers, but for the R3 Project, they are. Anyway, the project expects that the majority of the involved teachers are primary school teachers.

The teachers need support in three directions:

- STEM (Science, Technology, Engineering and Mathematics)
- Remote experiments
- Remote experiment platform

They need support in STEM because the primary school teachers are not, in general, experts in science, and neither in technology. The teachers need to be confident with the experiments that they will use. All projects are driven by the teachers, they are the STEM providers of the rural students.

They also need to know what a remote experiment is, and what is not. And finally they need to know how to use the remote experiments in the classroom because they are accessed through internet.

3 R3 Project Remote Experiments

Remote experimentation is part of Technology Enhanced Learning. Using this tool, a student can make a real experiment without having being in front of the physical devices. That is, the student will control and observe the experiment through the Internet. He or she will access the experiment using a web browser, after their hands will be the mouse, their eyes will be the webcam, and everything will be in the web interface.

There are many advantages associated to remote experimentation, and some disadvantages (LIBRO). Regarding to the rural schools, the main advantage is that as they do not have hands-on laboratories, they can make experiments using remote laboratories, and they only need an Internet connection. Using the words of Carisa Bohus pioneer in 1996 "Second Best To Being There", SBTBT [5].

During the R3 Project, a key aspect is to explain to the involved teachers that a remote experiment is not a virtual experiment, and neither a simulation. Sometimes this is not easy because they tend to associate computers with virtuality. It was established by [6] that the sense of immersion in a remote experiment affects to the learning process. It is important to remark that before teaching the students, the teachers have to learn with the same remote experiments.

Finally the teachers need to know if the use of remote experiments has a positive effect in the students learning process, that is, is it a good idea to use remote experiments with students? The answer is positive [7].

After this introduction, the teachers must try and play with the remote experiments provided by LabsLand (www.labsland.com). In https://proyecto-r3.labsland.com/ the teachers can access and test 28 remote experiments (see Fig. 2).

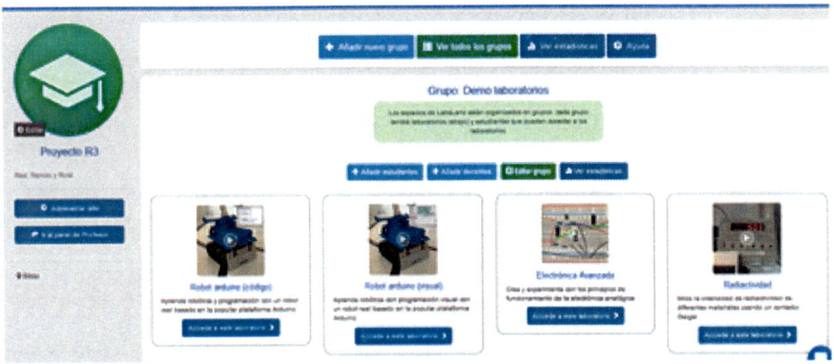

Fig. 2. Set of remote experiments in R3 Project

The remote experiments provided by LabsLand for R3 Project can be grouped in four areas:

- Physics. Experiments like Floatability, Electricity, Pendulum, etc.
- Chemistry. Experiments like Radioactivity, Boyle's Law, Gay-Lussac's Law, etc.
- Biology. Experiments like Planarian, Microscope, etc.

- Technology. Experiments like Robotics, Arduino, 3D Printer

Each teacher will choose a remote experiment, or a set of them, according to the students' curriculum, their interest or other aspects.

The remote experiments offered by LabsLand are agnostic, that is, they can be use, in principle, in any classroom, by any teacher, for any student, only attending to the scientific level. In a LabsLand remote experiment, the user can only see the experiment itself and some measuring devices like clock, counters, data, etc. There is not any result or analysis, all this info should be obtained by the teacher and the students. Figure shows the Floatability Remote Experiment, in which we can only see if the object floats or sinks (see Fig. 3).

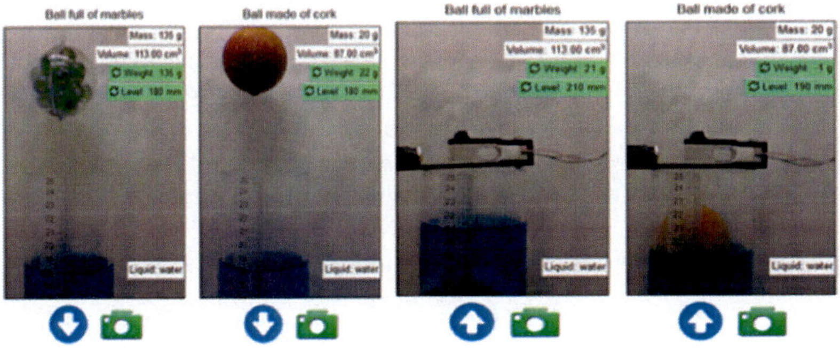

Fig. 3. Archimedes/floatability remote experiment

In addition to the set of remote experiments, R3 Project provides the teachers a didactical material: files and videos. These files –lessons– are short and they describe how a teacher could use the remote experiment in his classroom, that is, how they "should" do it. It is only a reference for him. In general, the lessons have 4–5 pages, but there is one exception: the robotics and the Arduino. In this case, the activity in the classroom will be longer than with the other experiments. For instance, the teacher will need around ten sessions to teach the student how he or she can program and control the movement of the robot (see Fig. 4), whilst for Archimedes experiment only one session will be needed.

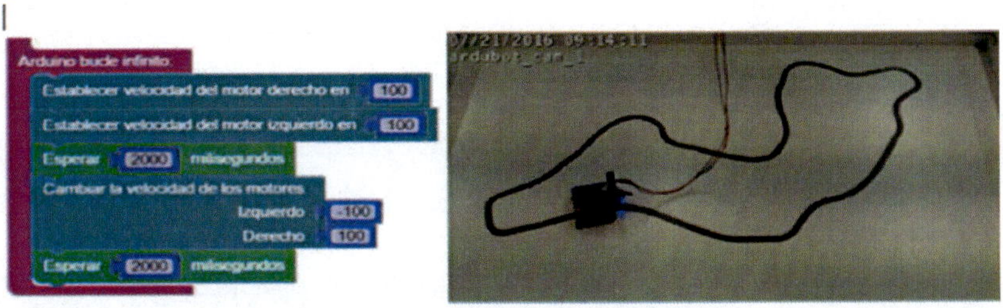

Fig. 4. Robotics remote experiment

All materials produced can be found in [8] (in Spanish).

But teachers may not only need didactic support in the classroom, but also know how students will access the remote experiment, where will they click to access the robot. LabsLand provides three different solutions for this:

- Direct Link. In the R3 Space of the school, the teacher will follow a two steps process, and after a link will be send to the teacher. Finally the teacher will share this link with the students through a LMS, the blackboard, etc.
- LMS Link. In the R3 Space of the school, the teacher will follow a process (more than two steps) to obtain a link that will be included in the LMS: Moodle, Sakai, Google Classroom, etc.
- Manual. The teacher will manually add all the students, one by one. This is not recommended.

In the second approach, as the student is in the school LMS, he or she only needs to click on the activity. The access will be direct. But in the first approach, as the school does not have a LMS (or the teacher does not want to use a LMS), first the student will create an account, and after he or she will access the remote experiment using user/password. In this case, GDPR will be respected. The Direct Link is the easiest method for the teacher, but it is not the easiest accessing method for the students, and vice versa for the LMS Link: harder (just a bit) for the teacher and easier for the student.

Figure 5 shows a R3 Space of a school with the three accessing methods for students.

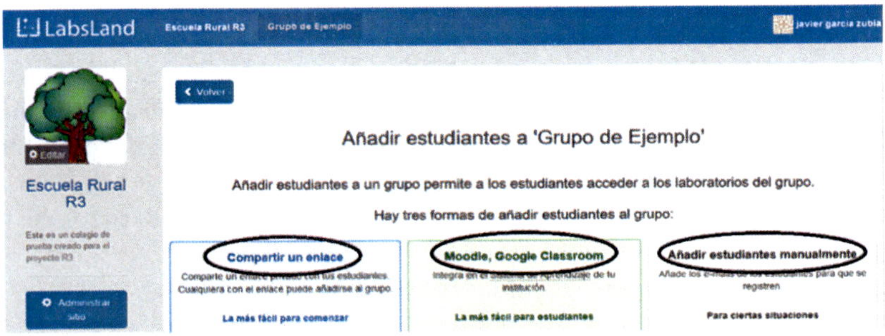

Fig. 5. Example of R3 project space

At this moment, the student could access the remote experiment using any web browser under any OS without plug-ins or problems with firewalls.

4 R3 Project Results

R3 Project is running now, it is just in the middle of the first phase. The project was introduced to the community, teachers and schools were invited to join the project, the set of remote experiments was established, the materials were produced and now the teachers should deploy the remote experiments in the classroom. The initial goal of R3 Project is to support 20 teachers in total, ten teachers in each phase.

R3 Project organized two informative sessions with the collaboration of an interest group in rural education (www.escuelarural.net), also with the collaboration of the Office for Innovative Education of the Regional Basque Government (http://www.berritzeguneak.net/). Sixty-four people attended these two sessions: teachers, school directors, teachers' trainers, and other interested people.

Now, in the first phase, there are 35 people of 20 rural schools involved in the project. At this moment, they organize the use of the selected remote experiments in their classrooms. It is important to remark that at this moment, the R3 Project has reached the goal for all the project. It seems that at this moment the R3 Project is successful.

The people involved in the R3 Project filled a 5-point Likert scale questionnaire [9].

Now, in R3 Project there are 24 teachers and other 12 persons are directors, teachers' trainers, etc. These 24 teachers teach in more than 70 classrooms, the majority of them (66%) are primary school classrooms, 24% are secondary school classrooms, and the other 10% are high school classrooms.

Around 90% of the teachers say that they like to innovate in the classroom, and they are open to introduce new topics even if they are not in the curricula and they are a bit risky. Close to the 50% of the teachers say that they are not comfortable when teaching science, and more of the 50% of the teachers think that the R3 Project will help them to teach science.

About the students, the teachers say that the 80% of them (3–5 in the Likert scale) use computers and web without problems and more than 90% of the students enjoy the innovation in teaching.

About the schools, the teachers say that 50% of the schools have enough computers/tablets (4–5 in the Likert scale), but 25% of the schools are not well equipped (1–2 in the Likert scale). Close to the 40% of the schools have a good Internet connection (4–5 in the Likert scale), but the 30% (1–2 in the Likert scale) does not have it. Fifty % of the schools are well equipped in science.

The most liked experiments are Robotics, Archimedes, Planarians and 3D Printer, but all experiments are interesting for at least one teacher. 80% of the teachers say that they like to have educational materials related with the experiments, but at the same time 85% of them say that they will write their own materials. All of them would like to share their materials and comment the materials produced by the other teachers.

More than 85% of the involved teachers are ready to inform to R3 Project about the results in the classrooms with photos, videos and reports. In general the schools know and manage the GDPR correctly.

5 Conclusions ad Future Work

Rural schools are not, in general, well equipped for the STEM disciplines, but the students in rural areas should have the same opportunities that the students in the urban areas. The use of remote experiments could mitigate this situation. Remote experimentation democratizes the access to science and technology.

The R3 Project provides remote experiments to rural schools to give them access to real experiments using internet. The R3 Project is still running but the available results are in the correct way.

The future work should be to develop new remote experiments more focused in primary school, because the current remote experiments are more suitable for secondary and high schools. In addition, it should be interesting to measure the impact in the learning process using a pre-post questionnaire.

Acknowledgements. The R3 Project (FCT 20-15490) is supported by FECYT: Fundación Española de Ciencia y Tecnología.

References

1. Epdata. https://www.epdata.es/datos/despoblacion-espana-datos-estadisticas/282. Accessed 14 Nov 2021
2. Ministry of Education (2018), p 548. http://www.educacionyfp.gob.es/dam/jcr:40be89f4-2ee9-44fc-8a61-5b927f804531/i18cee08-f-propuestas-de-mejora.pdf
3. SDG Objectives. https://sdgs.un.org/goals. Accessed 14 Nov 2021
4. PISA report 2015. http://www.oecd.org/publications/pisa-2015-results-volume-ii-9789264267510-en.htm. Accessed 14 Nov 2021
5. Shor MH, Bohus C, Atkan B (2011) Second best to being there. An historical perspective. In using remote labs in education. Universidad de Deusto, pp 27–52. ISBN 978-84-9830-335-3
6. Corter JE, Nickerson JV, Esche SK, Chassapis C (2007) Constructing reality: a study of remote, hands-on, and simulated laboratories. ACM Trans Comput Hum Interact 14(2), Article 7
7. García-Zubía J (2021) Remote laboratories. Empowering STEM education with technology World Scientific Publishing, pp 107–115
8. R3 Project Repository. https://drive.google.com/drive/folders/1xFnwl2XAJIaB2juPIb3JWStfF1STllGF?usp=sharing. Accessed 14 Nov 2021
9. R3 Project Questionnaire. https://docs.google.com/forms/d/1-gcHGJQ3I0oBdoEo3UCVm82fRCZu8UDX_dsNVHPGLqc/edit#responses. Accessed Nov 2021

Interactive Online Aerodynamics Experiment: Lift and Drag on an Aerofoil Wing

Ahmed Aboelezz[1], Peter Makeen[1,2](✉), and Hani Ghali[1,2]

[1] Centre for Emerging Learning Technologies (CELT), The British University in Egypt (BUE), Cairo, Egypt
Peter.Makeen@bue.edu.eg

[2] Electrical Engineering Department, Faculty of Engineering, The British University in Egypt (BUE), Cairo, Egypt

Abstract. Following the Covid_19 pandemic situation, the development of future online teaching and learning methodologies should focus on providing remote (distant) students with the same required skills as in a typical standard operation. In engineering education, experimentation has a critical role for students to acquire key physical concepts. Wing lift and drag measurements are the most important experiments in aerodynamics courses, where the design of any proposed online experimentation method should ensures that the student will gain the same practical skills as in a normal standard on-campus situation. Consequently, this design must ensure that the required theoretical, analytical concepts, and experimental skills are fully covered in any proposed online learning mechanism for the current and post Covid_19 situation. This paper presents an effort towards the development of new teaching and learning mechanism for aerodynamics experiments along with the corresponding assessment for online implementation in post Covid_19 situation. The experiment has been developed using an online mixed-reality (MR) approach, which will provide the students the ability to operate the experiments from anywhere at any time through access to an online platform that hosts the mixed-reality (MR) experiment.

Keywords: Distance learning · Online labs · Experimental aerodynamics · Force measurements

1 Introduction

In general, wind tunnel experiments are often used to test everything from buildings, aircraft, and automotive designs, and they serve a wide range of specializations ranging from architecture, civil, mechanical, and electrical. During the test, the used model is placed inside the wind tunnel's test section, where air is pushed to flow around and through it. Wind tunnel experiments are classified into four types: aerodynamic forces on the model, pressure distribution of the wind force on the object, spatial change of wind speed around the model, and visualization of air flow around the model to offer relevant aerodynamic information about the model.

Usually, students perform these experiments in a typical on-campus environment to get the required measurable data and to confirm their understanding of the theoretical

M. E. Auer et al. (Eds.): REV 2022, LNNS 524, pp. 151–157, 2023.
https://doi.org/10.1007/978-3-031-17091-1_16

background behind the acquired results. However, learning basic aerodynamics concepts remains a difficult effort in a contingency situation where on-campus amenities are severely limited, if not prohibited altogether.

Virtual experiments, or simulation-based experiments, have been presented as a ready replacement for real-world situations. Students' involvement and connection with the software tool, on the other hand, remains limited, with little understanding of relevant essential topics. [1] proposes a virtual aerospace laboratory in which real-life examples such as movies of low-speed and supersonic flows, lift generation, and various flying maneuvers are employed. The lack of contact between students and the experiment is the method's primary flaw.

Remote experimentation appears as an appealing alternative, allowing distant students to control and observe a real-time physical experiment setup and related equipment and sensors. Various technological possibilities have aided the execution of such a scenario. The use of the NI LabView software tool to build the experiment's user interface (UI), which includes control knobs, sliders, indicators, and charting/graphing options, integrated with a data acquisition system to perform electronic control on the real-time experiment's setup, is one of the most widely used techniques.

A web-based, interactive laboratory experiment in turbomachine aerodynamics is proposed in [2, 3]. Control units, motors, and two video cameras have been added to an existing laboratory, and NI LabView has been utilized to construct a user interface so that students can run the experiment completely remotely. A survey of students' opinions on both on-campus and remote labs reveals that remote labs have considerable advantages.

In [4] a full scenario of an online aerodynamics experiment has been developed to operate fully remotely along with corresponding real-time visualization at the British University in Egypt (BUE). In addition, two extra online tasks have been applied within the experiment pre-lab activities and post-lab assessment to ensure the coverage of the corresponding experiment's ILOs.

[5] proposes a complete remote online experiment for a phasor measuring unit (PMU) system. [6] proposes a 3D scene combustion control system for a 1000 MW thermal powerplant operation. LabVIEW was used to simulate three DC motor control system experiments (system modelling, speed control, and position control studies) in [7]. Those tests were done with the intention of contributing to the online teaching labs. A tank level system with remote control Proportional-Integral-Derivative (PID) controllers has been implemented in [8]. A web browser with a graphical user interface could be used to control the system across a local or internet network.

The Mixed-reality (MR) approach used in this paper uses real measured data which have been obtained through normal on-campus experiment operation, where obtained data are stored for use in the online experiment. On the other hand, to enhance student's engagement, videos for the real experiment operation are also included along with step-by-step instructions of the experiment's procedure.

The student will be able to change different experimental parameters and examine their effect on the overall output. This is easily achieved due to the amount of stored data that have been obtained from the operation of the experiment under different conditions.

The MR experiment integrates both the concept of online experiment associated with real data. In other words, students will operate the experiment and get real output data identical to the case of a normal on-campus operation. The MR experiment is accessible online from anywhere using any browser, where step-by-step instructions of the experiment operation as a normal real on-campus lab are included.

One of the key benefits of the MR technique is that both on-campus and distance (remote) students will receive the same results from the experiment, ensuring that the ILOs are covered in the same way in both circumstances. In other words, distant students will run the experiment and receive real-world results, much like a typical on-campus operation.

2 Experiment Setup

In the proposed experiment, the student will understand the change in wing aerodynamics forces due to angle of attack and wind speed change. In addition, the student will learn about new concepts such as stall and separation and how the geometry of the aerofoil, in general, could change the aerodynamics performance. The starting point in conducting the current experiment is to design and manufacture the force measurements system shown in Fig. 1. Two 5 kg load cells one for lift and one for drag were connected with an L shape connector. The force system was placed on the wind tunnel using an L shape connector also and was fixed to a rotating plate to enable change the wing angle of attack. The wing was directly fixed to the force system through the L shape connector. The rotating plate was connected to a servo motor shown in Fig. 2 to enable automatic change in angle of attack. The load cells and the servo motor are connected with the ARDUINO. A wood wing with NACA 0018 aerofoil was manufactured and used in the current research which is shown in Fig. 3. Furthermore, the wind speed inside the wind tunnel was measured by hot wire anemometer, these values was used to calibrate the relation between the wind speed and the motor control input.

Fig. 1. The force measurement system

Fig. 2. Servomotor system

Fig. 3. Wing installed on the force measurement system.

3 Experiment Outcomes

Lift and drag measurements for NACA 0018 wing shown in Fig. 4 have been obtained in this experiment. The student will chose wind speed for the operation conditions of the experiment, following that the student will chose the starting angle of attack and measure the lift and drag forces, then change the angle of attack and measure the lift and drag forces again This operation is repeated till the student take all the required measurements to setup the lift and drag curves, and the overall operation is repeated for different wind speed. The lift is increasing with increasing the angle of attack till the angle of 20° where the stall has appeared, then the lift is decreasing as shown in Fig. 5. The drag is increasing with increasing the angle of attack till the stall angle but the drag keeps increasing this time as shown in Fig. 6.

Fig. 4. NACA 0018 aerofoil

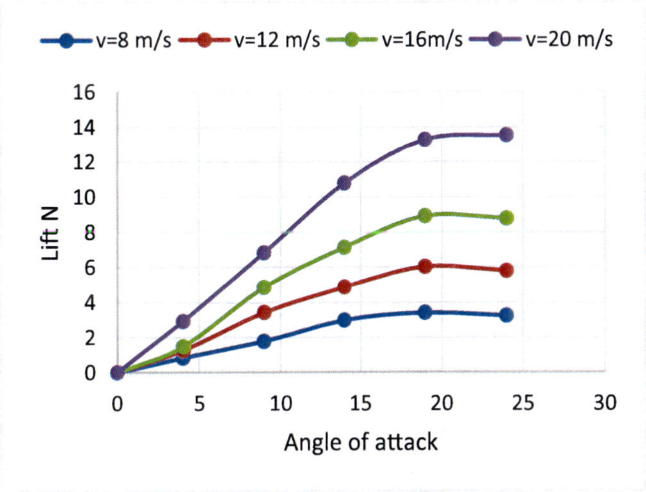

Fig. 5. Lift measurements

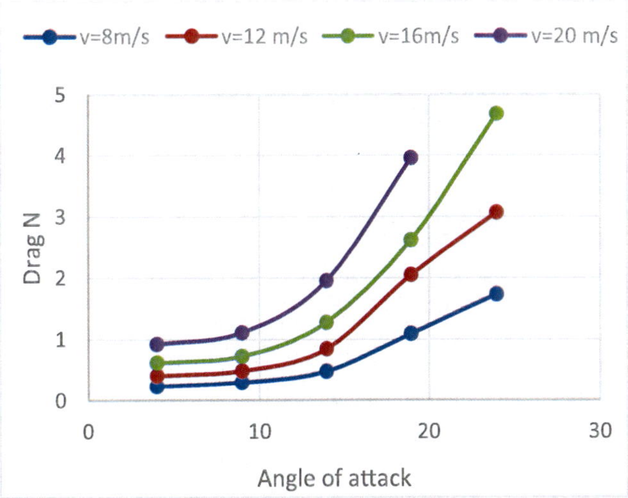

Fig. 6. Drag measurements

4 Mixed-Reality (MR) User Interface

As previously stated, this paper proposes a novel method that combines real data measurement with an online web interface to allow the student to run the experiment at any time and from any location utilizing the experimental data which have been collected previously. Figure 7 shows the user interface (UI) that allows the student to adjust the experiment wind speed and angle of attack. For each experiment condition, the user can see instant data on the observed lift and drag force, which is displayed in the table shown in Fig. 7. In the shown example, the user has data for a specific wind speed at different angles of attack, and the online tool allows the user to visualize the results shown in Fig. 8. The MR online experiment is available at https://celt-bue.com/#!/apps/liftAndDragVirtualLab).

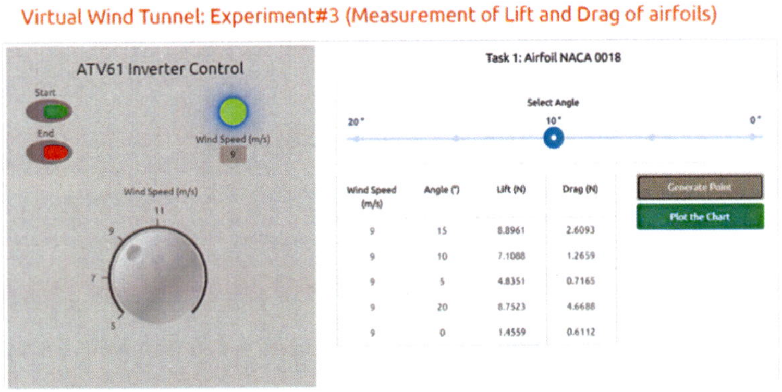

Fig. 7. User interface (UI) for the online experiment

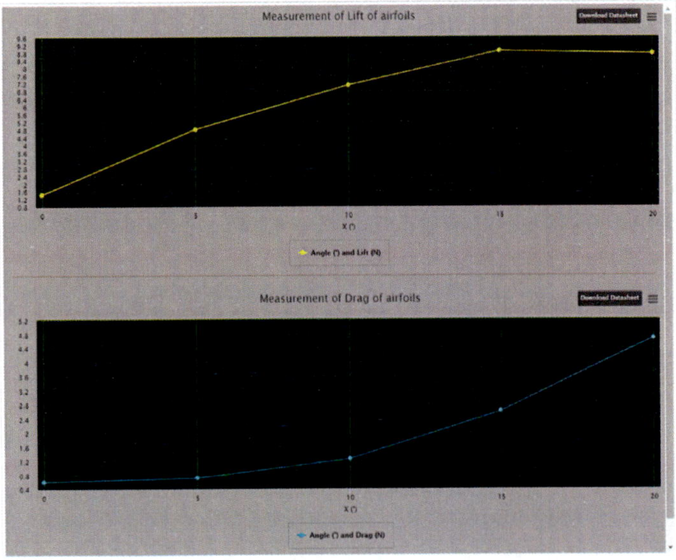

Fig. 8. User interface (UI) for the outcomes

5 Conclusions

The Lift and Drag forces on an aerofoil wing have been measured, remotely, using an aerodynamics experiment based on the MR technique. Students will learn about the aerodynamic forces acting on a wing as the wind speed changes. The lift and drag measurements for the NACA 0018 wing were successful conducted, and the patterns are consistent with theoretical as well as published data. This is achieved based on the proposer design and calibration of the developed force system. The student can use the online tool given in the current study to test any form of aerofoil and evaluate the change in aerodynamic behaviour. This experiment can be carried out online at any time and from any location, using the illustrative instructions and the required videos and photos obtained during the actual experiment's operation. It can also be applied beyond the COVID-19 era, with blended learning becoming more widespread in higher education institutions. Artificial intelligent could be used in the future work to predict the values for the point which was not measured in the experiment and this will enable a more freedom for the user. Also, more aerofoil types should be tested and added to the tool.

References

1. Higuchi H, Henning GA (1996) Development of a virtual aerospace laboratory for undergraduate education. Comput Appl Eng Educ 4(1):19–26
2. Cranston G, Lock G (2012) Techniques to encourage interactive student learning in a laboratory setting. Eng Educ 7(1):2–10
3. Navarathna N et al (2010) Web-based, interactive laboratory experiment in turbomachine aerodynamics. J Turbomach 132(1)
4. Aboelezz A, Makeen P, Ghali H (2021) Integrated online wind tunnel experiments and assessment for contingency scenario: case study at the british university in egypt (BUE). In: Auer ME, Centea D (eds) ICBL 2020, vol 1314. AISC. Springer, Cham, pp 441–446. https://doi.org/10.1007/978-3-030-67209-6_47
5. Mohammed AK, El Zoghby HM, Elmesalawy MM (2020) Remote controlled laboratory experiments for engineering education in the post-COVID-19 era: concept and example. In: 2020 2nd Novel intelligent and leading emerging sciences conference (NILES). IEEE, pp 629–634
6. Guan S, Hu W, Zhou H, Liu G-P (2020) A method of remote experiment for complex energy system: an example for process control experiment of thermal power plants. In: 2020 IEEE 4th Conference on Energy Internet and Energy System Integration (EI2). IEEE, pp 1501–1506
7. Alli KS (2019) A LabVIEW-based online DC servomechanism control experiments incorporating PID controller for students' laboratory. Int J Electr Eng Educ, 0020720919868142
8. Ikhlef A, Kihel M, Boukhezzar B, Mansouri N, Hobar F (2016) Online PID control of tank level system. In: 2016 IEEE global engineering education conference (EDUCON). IEEE, pp 281–284

Teaching Control System Using Online Mixed-Reality Experiment During and After COVID-19 Era

Peter Makeen[1,2](✉), Ahmed Aboelezz[1], and Hani Ghali[1,2]

[1] Centre for Emerging Learning Technologies (CELT), The British University in Egypt (BUE), Cairo, Egypt
Peter.Makeen@bue.edu.eg

[2] Electrical Engineering Department, Faculty of Engineering, The British University in Egypt (BUE), Cairo, Egypt

Abstract. Recently, a huge number of universities worldwide are having their education interrupted, partially or fully by the COVID-19 global pandemic especially teaching laboratories for engineering students. In this paper, an online Mixed Reality (MR) experiment is initiated to serve the electrical and mechanical engineering students during and after the COVID-19 era. The MR approach is based on the use of real measured data which are obtained through normal on-campus experiment operation, where the obtained results are stored for use in online experiments. Hence, both on-campus students and distance (remote students) will get the same outcomes from the experiment. In this paper, a linear positioning control system MR experiment is implemented based on the Proportional-Integral-Derivative (PID) controllers. The user "student" can control the speed of an aerofoil wing fixed on a controllable positioning carriage, understand analytically and experimentally the open-loop control system, compare the results with the closed-loop control system using the P, PI, PID classical controllers and is evaluated throughout a self-assessment. The experiment is fully designed and implemented using NI LabView software using NI myRIO-1900 embedded controller device to control the aerofoil horizontal movement and manipulate the PID gain parameters.

Keywords: Distance learning · Online labs · Control systems

1 Introduction

The Covid-19 pandemic has imposed great challenges on teaching and learning requiring new mechanisms and methodologies to cope with the required social distancing. Universities have switched from standard face-to-face operation to different online teaching and learning scenarios. This situation has created a direct need for an effective learning mechanism to ensure the efficient delivery of the intended learning outcomes (ILOs), especially for the practical component in engineering education. Any proposed scenario should ensure that the student will gain the same practical skills as in a normal standard on-campus situation [1]. Consequently, this should ensure that both the theoretical,

M. E. Auer et al. (Eds.): REV 2022, LNNS 524, pp. 158–165, 2023.
https://doi.org/10.1007/978-3-031-17091-1_17

analytical, and experimental skills are fully covered in any proposed online learning mechanism for the current and post Covid-19 situation.

In [2] a case study of online learning in the department of electrical engineering, Universitas Negeri Semarang during the Covid-19 pandemic has been investigated. It is concluded that online learning ensures a reliable information technology facility, innovation in delivering the teaching material which attracts the students, the possibility of lectures to do a practicum with simulation, and an increase in student participation during the online sessions. Consequently, researchers were developing online experiments to enhance engineering education and ensure interdisciplinarity between various departments.

In [1] a full scenario of an online aerodynamics experiment has been developed to operate fully remotely along with corresponding real-time visualization at the British University in Egypt (BUE). In addition, two extra online tasks have been applied within the experiment pre-lab activities and post-lab assessment to ensure the coverage of the ILOs. A complete remote online experiment for a phasor measurement unit (PMU) system has been proposed in [3]. In [4] a 3D scene combustion control system of 1000 MW thermal powerplant process has been proposed. In [5] three DC motor control system experiments (system modelling, speed control, and position control experiments) have been simulated using LabVIEW. Those experiments were a view of contributing to the online teaching laboratories. In [6] remote control Proportional-Integral-Derivative (PID) controllers of a tank level system have been implemented. The user could control the system through a web browser via a graphical user interface using a local or internet network.

This paper aims to propose a new teaching and learning mechanism for the control system laboratory based on the P, PI, PID controllers of linear positioning system throughout an online mixed-reality (MR) approach. It will provide the students with the ability to operate the experiments from anywhere at any time through access to an online platform that hosts the MR experiment. Unlike virtual experiments, which are mainly based on simulation models [7], the MR approach is based on the use of real measured data which have been obtained through normal on-campus experiment operation by the instructor, where the obtained results are stored for use in online experiments. In addition, to enhance students' learning experience, videos for the real experiment operation are also included along with step-by-step instructions of the experiment's procedure.

One of the main advantages of the MR approach is that both on-campus students and online students will get the same outcomes from the experiment, ensuring that the ILOs are covered identically for both cases. The MR experiment integrates the concept of virtual experiment with the concept of real data. In other words, students will operate the experiment and get real output data identical to the case of a normal on-campus operation.

2 Experiment Setup and Methodology

The student "user" in this experiment controls the linear speed movement of an aerofoil wing and studies the difference between the open-loop and closed-loop control systems. In addition, the student can manipulate the PID gain parameters to observe the dynamic behaviour of the system.

The proposed experiment set up is consisting of an aerofoil wing fixed on a horizontal linear traverse system of a ball screw with Nut SFU1605 (16mm Diameter – 5 mm Pitch – 1000 mm Length), limit switch which is responsible for the instantaneous stop of the carriage while moving in a wind tunnel test section hosted by the Centre for Emerging Learning Technologies (CELT) at the British University in Egypt (BUE), Nema 23 Stepper Motor 24Vdc, 3A, 1.3 Nm, and 700 rpm, and Rotary encoder which used to measure the distance of the aerofoil precisely. All the mentioned components are represented below in Fig. 1 .

Fig. 1. The proposed electro-mechanical experimental setup.

2.1 Experiment's PID Controller

Proportional-Integral-Derivative (PID) controllers are common feedback and widely utilised in almost all research fields because of the simplicity of use and good output impact and behaviour of the systems [8]. The student will take the theoretical background of this controller at the corresponding lectures and tutorials then will use the MR laboratory to integrate the theoretical background and practical implementation of such controllers. The main equation of the PID controllers can be described as

$$C(s)_{PID} = K_c\left(e(t) + \frac{1}{T_i}\int_0^t e(t).dt + T_d\frac{de}{dt}\right) \tag{1}$$

where, K_c is the controller gain, T_i is the integral time in minutes which is called the reset time, and T_d is the derivative time in minutes which is called the rate time.

In this experiment, the user "student" will gain the comparison experience of the open-loop and closed-loop control systems in the presence of PID controllers as shown in Fig. 2 . In addition, the student will be able to observe the behaviour of the system output at different conditions as will be described in the following subsections.

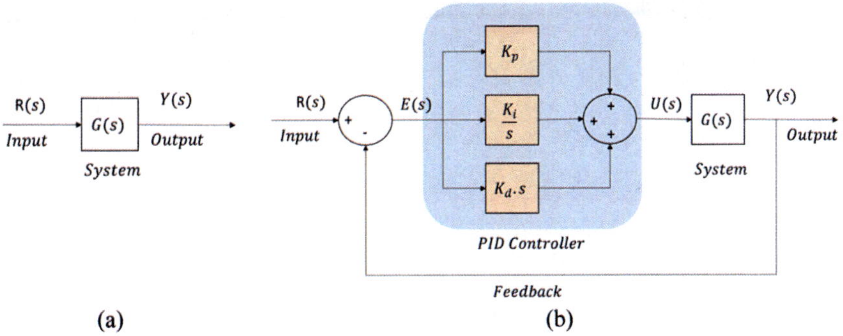

Fig. 2. Schematic diagram of the open-loop and closed-loop based on PID controller control systems respectively.

2.2 Experiment's User Interface

LabView software has been used to design the user's interface (UI) control panel as shown in Fig. 3 using NI myRIO-1900 embedded controller. The user "student" has full access to control the speed of the carriage in the open-loop system and control the closed-loop system by entering the required set point and manipulating the gain parameters of the PID controllers as shown in Fig. 3. The user will observe the effect of each parameter on the output of the system, compare between open-loop and closed-loop control systems and will compare between the P, PI, and PID controllers.

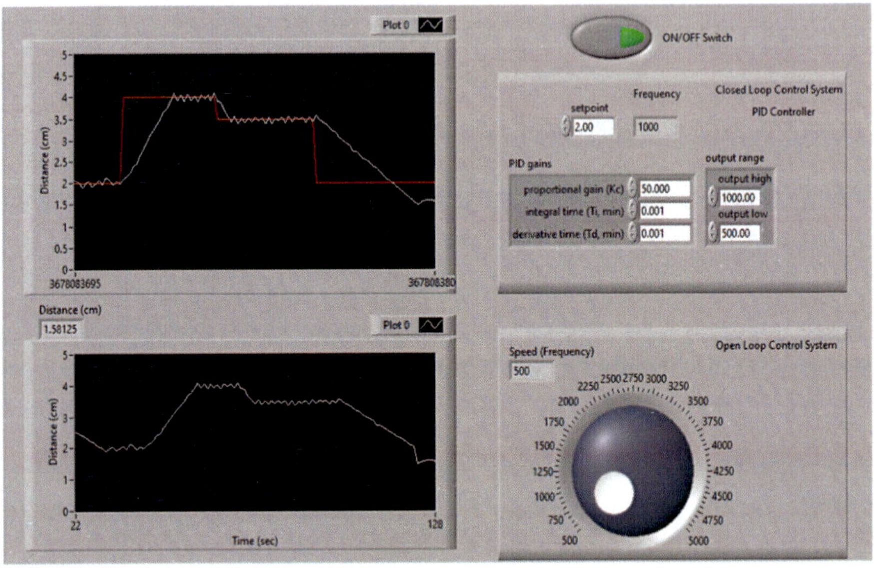

Fig. 3. Student's interface and controlled parameters using LabVIEW.

2.3 Experiment's User Outcomes

Whenever the user "student" is satisfied from performing the experimental lab and getting all the required results, the student should investigate the results and observe the dynamic behaviour of the system while varying the PID gain parameters. As shown in Fig. 4 set of graphs will be obtained from the experiment. The graphs describe the behaviour of the carriage's distance and the motor regulation ratio while reaching the pre-set desired distance value.

Firstly, the user "student" should run the experiment while varying only the P parameter and observe the total time to reach the pre-set goal and the motor regulation ratio as shown in Figs. 4-a and 4-b. It is observed that whenever the P parameter increased the carriage reached the desired distance faster in almost 27 s however in the lower P values the time reached 56 s. Also, the motor regulation reached 100% and then slow down whenever getting closer to the target (4.5 cm).

Secondly, the user should run the PI controller by fixing the P parameter at the fastest behaviour obtained from the P controller test then start to vary the I parameter and observe the behaviour as shown in Figs. 4-c and 4-d. The carriage reached the targeted value in 13 s which is almost 50% from the P controller and the motor regulation ratio during the movement.

Thirdly, the effect of varying the D parameter to the system is concluded from Figs. 4-e and 4-f. The carriage reached the destination in 13 s the same as the PI controller but there is different behaviour from the motor to reach the targeted value

Finally, A clear comparison between the PID controllers will appear to the student to gain the required experience concern these controllers as shown in Figs. 4-g and 4-h.

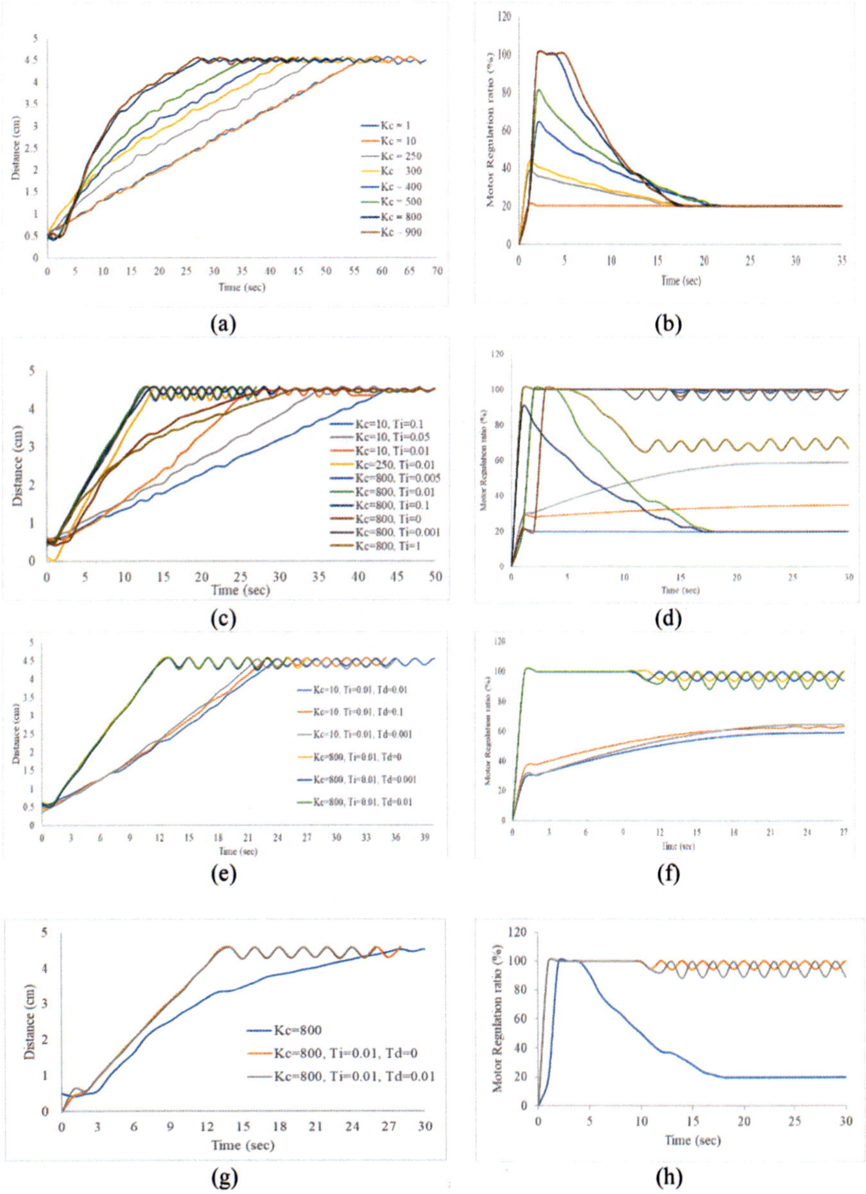

Fig. 4. The outcome graphs from the experimental setup while implementing the P, PI, and PID controllers.

2.4 Experiment's User Self-assessment

At the end of each stage, the user "student" will be self-assessed based on the conceptual understanding of the P, PI, and PID controllers as shown in Fig. 5. After each part of the experiment, the student has to select his answers, and cannot proceed to the next stage unless choosing the correct answers as shown in the P, and PID controllers in Figs. 5-a and 5-b respectively.

Finally, a summary of the different PID controllers concerning the peak time, maximum percentage overshoot, and settling time will be obtained as in Table 1. and also this conclusion has been stated in other applications in [5, 9].

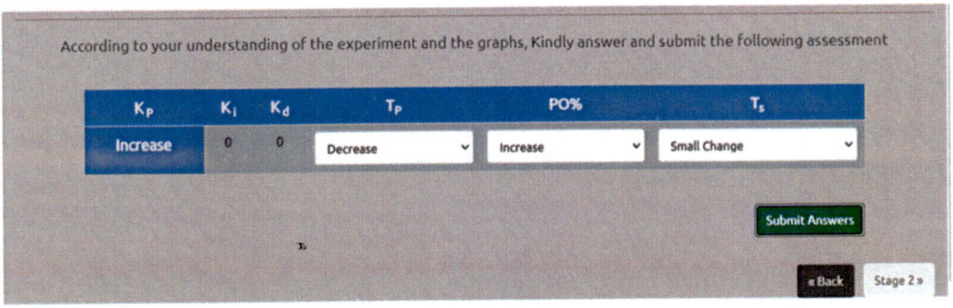

(a)

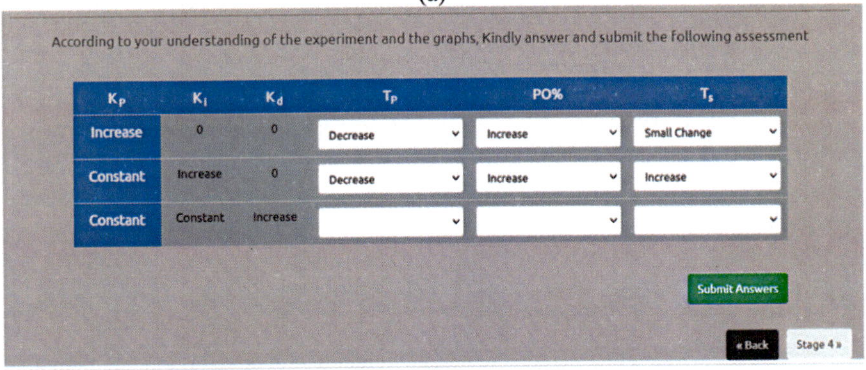

(b)

Fig. 5. Stages of the self-assessments for P, and PID controllers in the experiment.

Table 1. Comparison between P, PI, and PID controllers of the proposed system.

K_p	K_i	K_d	T_p	PO%	T_s
Increase	0	0	Decrease	Increase	Small change
Constant	Increase	0	Decrease	Increase	Increase
Constant	Constant	Increase	Minor decrease	Minor decrease	Minor decrease

where K_p, K_i, K_d are the proportional, integral, and derivative gains, respectively. PO% represents percentage overshoot. Also, T_p, and T_s are the peak time and settling time respectively

3 Conclusion

In this paper, An electro-mechanical experiment based on the MR approach is proposed to control the horizontal linear movement of an aerofoil wing, and control the accuracy of reaching a predetermined set distance point while studying and implementing the open-loop and the closed-loop control systems using the traditional P, PI, PID controllers. This experiment can be performed online from anywhere and at any time, with the illustrative instructions and the required videos and images obtained during the real operation of the experiment to be used in the virtual experiment's interface. Also, it can be used beyond the COVID-19 era where blended learning may be generalized in higher educational institutions. The proposed control system mixed-reality experiment makes the ability of students to access the on-campus lab and perform the experiment with the same instructions, procedures, and results of the online teaching and gain the same ILOs.

An improvement in the controller could be investigated in future work where the PID controllers shall be compared with advanced fuzzy logic and artificial intelligence (AI) controllers. Also, online assessments would be enhanced to evaluate the theoretical and experimental understanding of students and to emphasize their weaknesses and strengths.

References

1. Aboelezz A, Makeen P, Ghali H (2021) Integrated Online Wind Tunnel Experiments and Assessment for Contingency Scenario: case study at the British University in Egypt (BUE). In: Auer ME, Centea D (eds) Visions and Concepts for Education 4.0, vol 1314. Advances in Intelligent Systems and Computing. Springer, Cham, pp 441–446. https://doi.org/10.1007/978-3-030-67209-6_47
2. Sunardiyo S (2021) Online learning in higher education during the Covid-19 pandemic: a case study in the department of electrical engineering. Universitas Negeri Semarang. IOP Conference Series: Earth and Environmental Science: IOP Publishing. p 012020
3. Mohammed AK, El Zoghby HM, Elmesalawy MM (2020) Remote controlled laboratory experiments for engineering education in the post-COVID-19 era: concept and example. In: 2nd novel intelligent and leading emerging sciences conference (NILES). IEEE, pp 629–634
4. Guan S, Hu W, Zhou H, Liu G-P. A method of remote experiment for complex energy system. An example for process control experiment of thermal power plants. In: IEEE 4th conference on energy internet and energy system integration (EI2). IEEE, pp 1501–1506
5. Alli KS (2019) A LabVIEW-based online DC servomechanism control experiments incorporating PID controller for students' laboratory. Int J Elect Eng Educ. 0020720919868142
6. Ikhlef A, Kihel M, Boukhezzar B, Mansouri N, Hobar F (2016) Online PID control of tank level system. In: IEEE global engineering education conference (EDUCON). IEEE, pp 281–284
7. Reeves SM, Crippen KJ (2021) Virtual laboratories in undergraduate science and engineering courses: a systematic review, 2009–2019. J Sci Educ Technol 30(1):16–30
8. Sekarsari K, Tata T (2021) Performance analysis of PID control in DC Brushless motor using trial and error method. IOP Conf Ser Mater Sci Eng: IOP Publishing 1098(4):042027
9. Bhilai S (2014) Comparative study of P, PI and PID controller for speed control of VSI-fed induction motor. Int J Eng Dev Res 2:2740–2744

Using Mobile-Learning at Transilvania University in the Pandemic and Post-pandemic

Fulvia Anca Constantin(✉)

Transilvania University of Brasov, Brasov, Romania
fulvia.constantin@gmail.com

Abstract. Speeded by the pandemic virus conditions, universities and colleges adapted their education to new distance forms. Therefore, higher education faculties were able to consider the opportunity to expand their learning possibilities with mobile learning. They encouraged students to learn, to research, collaborate, and share ideas among them with the aid of internet and technology development. The aim of this paper is to present students and educators' attitudes and efforts towards the use of M-learning in diverse domains of higher education, pointing out the established benefits and problems. As M-learning implies server updates, new contracts for speed internet and hiring more IT specialists in order to identify the existing strengths and weaknesses and to facilitate the development of the technology infrastructure, we present the case of online teaching in the departments of Transilvania University. We want to have an overview of the two years of teaching online during pandemics and the impact it had on the students.

Keywords: Mobile learning · Pandemic · Education · Online teaching · Transilvania University

1 Introduction

Due to the COVID-19 pandemic the worldwide educational institutions looked for new forms of distance teaching. If on one hand, the universities invested in technology and speed internet, on the other, the students looked for devices to allow them to get distance–education.

Transilvania University provided M-learning devices, quality access to the Internet and protection of personal data/privacy, all of extremely importance, given the fact that a large part of work was transferred to the online environment so students and employers could handle it from home.

From a psychological point of view the government restrictions were very hard to be accepted and numerous cases of depression and anxiety were reported. Being hard to maintain social contacts, people looked to find other ways to connect. However, according to Nguyen et al., women are more talkative and usually tend to communicate about any subject with family and friends through phone, platforms and social media, while men use concentration (e.g. computer games), technical focus and sports reasons to start socialize and communicate with each other [1].

M. E. Auer et al. (Eds.): REV 2022, LNNS 524, pp. 166–173, 2023.
https://doi.org/10.1007/978-3-031-17091-1_18

When, in 13 of March 2020, the Romanian government anounced the lockdown and emergency state people were in panic. The issued decree no. 195 of March 16, 2020 regarding the establishment of the state of emergency on the Romanian territory, announced by the president of Romania and published in the Official Gazette no. 212 of March 16, 2020, came on top of the evolution of the international epidemiological situation caused by the spread of SARS-CoV-2 coronavirus in more than 150 countries [2].

At that moment, approximately 160,000 people were infected and more than 5,800 died. The declaration of "pandemic" done by the World Health Organization (11 of March 2020), was taking into account the experience of countries severely affected by the evolution of the virus and the measures that had a positive impact in limiting its spread. There were targeted actions regarding public health concomitant with limiting or interrupting non-essential socio-economic activities. In particular, these actions were seen as restricting the exercise of fundamental rights and freedoms, not necessary establishing exceptional states in order to prevent the community spread of the infection, seeking the evolution of the epidemiological situation in Romania and the public health risk assessment which indicates a massive increase in the number of people infected with SARS-CoV-2 coronavirus. Anyway, taking into consideration that the exceptionally social and economic measures to limit SARS-CoV-2 coronavirus infection among the population would have a particularly serious impact, (i) mainly on the right to life, and (ii) on the right to health of individuals, it was decided to establish a state of emergency. It was an urge of need for it. The declared reason was to reduce the negative effects on the health of people and on the economy caused by measures taken at national and international level to combat the spread of the SARS-CoV-2 coronavirus [3].

The companies' admin and IT specialists worked hard on remoting the work, therefore in two-day time most of the people worked from home. This is also the case of Transilvania University. Starting with 14th of March, the jobs were remoted and preparation for on-line teaching was ready. Luckily, Transilvania University was ready to support such an action. And we talk here about the technical equipment (platform, server, devices for professors) and qualitative speed internet connection. [4] Therefore, the education of students moved entirely to the Internet. It was need for performant M-learning technology.

At the beginning of the pandemic, it was expected that this situation would only last for a while; however, it lasted several months. So, for the second semester 2019–2020 all the university departments offered teaching online with the use of various platforms. In a short time, profesors, students and administrative staff had to learn how to work with the e-learning platform. The process was for some rather difficile, the differences between devices and of digital competencies became evident. Thus, it was notable the willingness and flexibility to switch to distance learning [5].

Although there were plenty of arguments supporting online teaching to only remind continuing education, the comfort offered by the home environment and time flexibility, a serie of problems were evident: home internet connection, modern devi-ces, lack of computer abilities, lack of motivation and socialization. By the end of the university year students were divided in two: those for which distance learning was beneficial, an opportunity to achieve their objectives and those who consider it not a proper education, but a weak, risky one, stating that "the dynamic development of digital technologies and still more available connection to the Internet gives aggressors more space to attack potential victims." [6].

With some exceptions regarding the vocational study programs such as Music and Physical Education, Transilvania University of Brasov remained closed from March until July 2020 and then again partially from October 2020 until February 2021. Thus, the 2020–2021 school year brought at Transilvania University a new implemented program named the *Hybrid Programme* which included online teaching for the courses and physical ('face to face') attendance for seminaries and laboratories. The first semester was acceptable in this form but then, by the second semester (March 2021) it was everything back to normal.

1.1 Describing the Research

As a methodological approach we included a systematic review based on selecting, identifying, and synthesizing the observations, questionnaires, and interviews providing the methods used at Transilvania University in all domains of study.

We evaluated the used methods to point out the challenges and to describe an empirical assessment of the effectiveness of M-learning. Therefore, we focussed on findings, analyzing and interpreting those in order to justify the already made changes in teaching practice and to propose others. We were gathering data from all Transilvania University programs. Based on it we pointed out the psychological, educational and technological changes that have been created, the needed conditions for the use of mobile technologies in a pedagogical way that further more encouraged students to have an authentic learning. Also, we showed that the availability and uptake of devices have led to new manner and means of communication, apart from socializing and entertainment.

We anticipated gathering data from all Transilvania University programs which we did it. Based on it we point out the psychological, educational and technological changes that have been created, the needed conditions for the use of mobile technologies in a pedagogical way that further more encourage students to have an authentic learning.

2 Methodology

As stated before, the purpose of the research was to have an overview on the application of planned distance learning and the use of mobile technology.

2.1 The Participants

Transilvania University of Brasov has 18 specializations and 30 departments. A number of 20255 students were enrolled in the academic year 2020–2021. Out of these, 16263 were undergraduate students, 3529 were master's degree students, 463 PhD students and 476 were resident doctors of UnitBv. With regard to the teaching staff, we count 721 professors and assistents and 289 associates, mostly specialists in different areas of activity.

The number of participants in the research was of 256 people. Out of these, 214 were students and 42 were professors or associates. For the sake of the research there was no distinction made between full time professors or assistants and associates and, also, no distinction between undergraduate and graduate students.

2.2 The Research Questions

No doubt, the Covid-19 pandemic represented a new line in teaching at university and school's level by moving teaching to the online environment. We conducted the research in the same way meaning that interviews were done over the phone and a written questionnaire was sent via e-mail to each one of the participants. Also, a large part of research was done through observation.

Talking about their experience in the interview, the students brought in discussion things as the quality of their devices, the speed of the internet, the home environment, the interest shown in the subject, and professor's personal charisma. Some of the respondents asked about personal laptops or tablets from the university, others already had the needed devices. We must say that, through the *Man and Computer Programme,* all the teaching staff, professors and associates including PhD students, received a personal tablet or laptop to be used until the last day of their studies.

The main question of the research was, though, about mobile–learning. Adiacent questions referred to the type of technology used (last generations of computer, tablet, smartphone and their marks), the difficulties they had during classes, how did they learn to use the platform, how was the connection to the platform, what type of internet they had, if they encountered some form of risk by using or not the Internet either during or after the Covid-19 pandemic. Another research question was how the ratio of time spent on the Internet in connection with university classes or homework preparation and extracurricular activities changed.

Despite the fact that the distance learning was planned fastlly, it was updated carefully bringing up an e-learning platform quite logical and easy to work with. Also, we have to say that the learning courses were not novelties at Transilvania University, most of PhD and master students using it before. However, significant complications were reported to appear in the educational process referring to videoconferencing tools, the speed of internet, interrupts of connection given by the overusage, and sometimes the software used. For example, some of the older devices didn't support the download of certain materials. Some of professors or students found this situation becoming stressful so it was completed a list of extensive range of possible challenges and proposals of solutioning. Among them there were technological investments (qualitative modern brands of laptops and tablets, more IT specialists being hired, working on socio-economic factors, finding solutions for intrusion and disturbance situations, eliminating the lack of digital competence by organizing trainings for all, problems with assessments, the workload, and the question of compatibility of online learning with subjects that needs physical presence such as choir, or vocal or physical training. [7] Although there were pointed out challenges, both, professors and students recognized the opportunities the online teaching education brought for professional development and growth.

2.3 The Research Method

As the main research method, we specified the use of the questionnaire, and for clarification a short interview. The research was completed step by step.

We started with an initial research structured in the form of interview applied to 6 participants. Based on results, we were adjusting the questions and conducted the actual research. As we mentioned before, the questionnaire was applied focusing on the period during and post Covid-19 pandemic crisis.

Professors and students were chosen randomly making sure to have a whole university departments' representation. There were a total of 256 participants of which 197 were women (76,9%) and 59 were men (23,1%). It was a full-time and part-time professors and students' participation in the age range of 19 to 67 years old. Respondents were divided into two categories (professors and students) for comparison regarding the ability to use M-learning. The results were not statistically significant.

With regard to the distribution of participants we present the number in the table below. (Fig. 1) Moreover, it was considered the percentage of respondents using M–learning, not taking into account the type of device used if it was modern and good enough to be used.

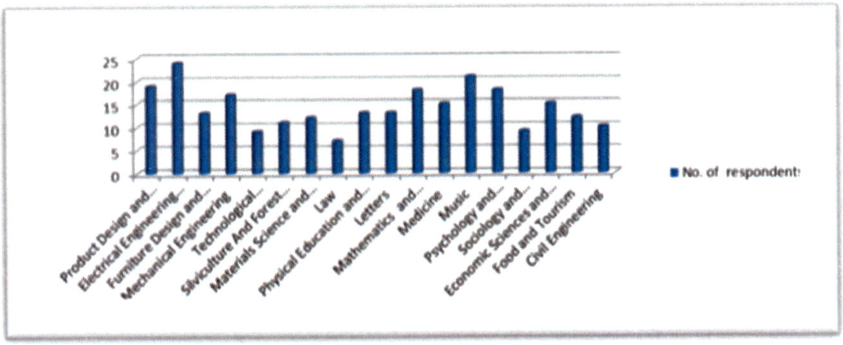

Fig. 1. Distribution of the respondents per specialization

Regarding the Mobile learning usage, it was calculated that 98,1% of students and 100% of professors used Mobile-learning devices. Different percentages were with regard to the ability of using modern devices for learning. As professors were trained and mostly self–trained, 72,9% of the students declare having difficulties in accessing and working on the platform in the first month. The number raised to 89,7% on the second month of usage. Apart of the lack of knowledge in handling the devices and the e-learning platform, students complained about the weak connection evident when too many attendants were on-line.

In order to analyze the data obtained from the questionnaire statistical calculations were performed with the SPSS software platform.

3 Results and Discussion

The main research questions were: (i) how many students used modern mobile-learning devices, (ii) what difficulties they had using those and, (iii) what technology they used during and after the Covid-19 pandemic. We were also interested if the ratio of time spent on the Internet in connection to the school activities changed comparing to the time before the pandemic.

With no exception all respondents (N = 256) said they learned from home during the pandemic although some of them suggested difficulties or declared not being happy about the situation. As level of satisfaction, 72% of respondents said that they were satisfied, 12% were indifferent and 16% were not satisfied with the situation. On the comments they establish as pro mobile–learning motivations as: the imposed government's policy (63,3%), health safety (62,8%), comfort of working from home (37,5%), better working environment at home (28,2%), less stressful than working from the university (11%). The negative comments included the inability of keeping a student-life balance (43%), not having a reliable internet connection at home (18%) losing the learning environment including studying in the resourcesful library (44%). Between the things students missed the most were the social relations with their coleagues (68,5%) while professors missed the most the technology and the resources. (42,8%, N = 18).

While everybody was free to choose its own mobile-learning device (it was noted technology produced by Apple, Samsung, Dell, HP, Asus, Lenovo, and others) the same e-learning platform ensured uniformity in the access methos. Sometimes there was no compatibility between the type of device and the provided platform. Though, while respondents were satisfied with the level of technological support offered by the university, their personal devices sometimes made hard the possibility of attending the classes. Also, both males and females agreed that they had some basic knowledge of using a computer, but they needed training that was provided in order to learn how to use the e-learning application. One responded mentioned: "we are grateful that Transilvania University's reaction to the pandemic situation was so effective and fast, being prepared and of help in troubled situations, solving problems, offering training when necessary and being so open-minded." Regarding the courses, a professor of psychology mentioned that "the results expressed that changes to the courses' structure were not as intense as was expected maybe because students' self-study more seriously." [8] As a fair number of students worked harder, others complained that the number of assignments increased during the on-line period which took away their free time. This idea could be found also in a 2021 study, the authors reporting a heavier workload during pandemic [8].

With regard to the ratio of time spent on the Internet in connection with university classes or homework preparation, all respondents declared spending more time doing homeworks, research or attending classes than navigated on the internet for entertainment and other extracurricular activities.

4 Conclusions

We anticipated gathering data from all Transilvania University programs. Based on it we pointed out the psychological, educational and technological changes that have been created, the needed conditions for the use of mobile technologies in a pedagogical way fact that encouraged students to have an authentic learning. Also, we showed that the availability and uptake of devices have led to new manners and means of communication (e.g. education), apart from socializing and entertainment.

As seen, the participants didn't emphasized differences between men and women, professors and students. There were slight differences between the groups of respondents but ther were not statistically significant. We didn't take into consideration other psychological factors such as shyness in responding or the desire to show more than the reality. However, the support provided by Transilvania University indicated in a high percentage promising results meaning that the respondents agreed that their departments provided them materials, information and support.

We must say that the date of research/data collection was in the middle of pandemic not in the beginning of it which gave the respondents time to adjust to the situation. Moreover, as some of the students stated some of the existing difficulties were solved and forgotten.

No doubt, at university level, the activities that expand the learning possibilities with the aid of internet and technology development is helpful and encouraging mainly in the pandemic time. It sees new approaches to studying from the point of view of methodologies, location, subject matter domains, context or device types, and educational level. It helps students to gain a holistic understanding through the use of mobile learning, to have global access to information, and to experiment and simulate the practice in diverse domains. This was also the case of Transilvania University of Brasov.

However, the study is not complete. It has several limitations. One of them is the low number of respondents, less than 1% of the total of university's members. Another problematic aspect is using just members of Transilvania University not having the possibility to compare the findings with other universities so we can not claim the results as being general. Also, we didn't talk about the medical consequences that experienced after using the computer daily for long hours, to only count: vision problems, weight gain, and computer dependence. We just consider this research as basis for further investigation.

References

1. Nguyen, M., Hargittai, E., Marler, W.: Digital inequality in communication during a time of physical distancing: the case of COVID-19. Comput. Hum. Behav. **120**(5), 106717 (2021)
2. http://legislatie.just.ro/Public/DetaliiDocumentAfis/223831. Accessed 15 Nov 2021
3. http://legislatie.just.ro/Public/DetaliiDocumentAfis/223831. Accessed 18 Nov 2021
4. Priyadarshani, H., Jesuiya, D.: Teacher's perception on online teaching method during COVID19: with reference to school level teachers at faculty of education, the open University of Sri Lanka. Shanlax Int. J. Educ. **9**(2), 132–140 (2021)
5. Frolova, E., Rogach, O., Tyurikov, A., Razov, P.: Online student education in a pandemic: new challenges and risks. Eur. J. Contemp. Educ. **10**(1), 43–52 (2021)
6. Miková, J., Chráska, M.: University Students' Risky Behavior in Online Environment during the Covid-19 Pandemic, ICEEPSY 2021, European Publisher (2021)

7. Makori, A., Agufana, P.: Cyber bulling among learners in higher educational institutions in sub-saharan africa: examining challenges and possible mitigations. High. Educ. Stud. **10**(2), 53–65 (2020)
8. Niemi, H.M., Kousa, P.A.: Case study of students' and teachers' perceptions in a finnish high school during the COVID pandemic. Int. J. Technol. Educ. Sci. (IJTES) **4**(4), 352–369 (2021)

Online Sucker-Rod Pumping with Photovoltaic Driven System Sizing Tool for Oil and Gas Industrial Sector

Mohamed Elyamany[1], Sameh O. Abdellatif[2(✉)], and Hani Ghali[2]

[1] Petroliium Engineering and Gas Technology Deprartment, Faculty of Energy and Environmental Engineering, The British University in Egypt, Cairo, Egypt

[2] The Electrical Engineering Department, and FabLab, at the Centre of Emerging Learning Technologies CELT, British University in Egypt (BUE), Cairo 11387, Egypt

Sameh.osama@bue.edu.eg

Abstract. The current study demonstrate an attempt to implement an online tool capable of sizing sucker-rod pumps (SRPs) used in oil and gas industry. This tool aims mainly to investigating the utility of integrating renewable energy resources in the oil and gas industry. As well as managing the energy consumption estimated from SRPs with respect to actual infield measurements. Along with optimizing the production cost and reducing it with maintaining the stability of the oil production rate. In addition, identifying the economic feasibility of the enrolment of renewable energy (PV) in the oil and gas industry. Finally, the tool explore the levelized cost of energy (LCOE) variation against the field location while considering various filed locations in Egypt.

Keywords: Photovoltaic online tools · Sucker-rod pumping · Oil and gas industry · Optimization

1 Introduction

In line with the weak economic growth and a partial breakdown in some of the specific factors that stimulated energy demand in 2018, growth in energy markets decreased in 2019 [1]. All (other than nuclear) fuels grew slower than their aperture of 10 years and, for the fourth time in six years, coal consumption was decreasing. Nonetheless, renewable power has continued to rise record-high and has contributed largely (41%) to primary energy growth, exceeding nuclear power for the first time. The decline in energy demand, coupled with the transition to natural gas and renewables in the fuel mix, resulted in a substantial slow-up in carbon growth, but partially only reversed the unusually strong rise observed in 2018 [1]. Natural gas and renewable energy consumption are increasing over time while oil, coal, and nuclear are facing consumption reduction, only hydroelectricity is almost having the same level of consumption over the past 25 years [1].

Sucker rod pumping is considered as the conventional technique for artificial lifting for crude oil [1]. There is a period in the life of an oil well when external energy needs to be supplied to help or reach the desired production. Around 85% of the wells requiring

M. E. Auer et al. (Eds.): REV 2022, LNNS 524, pp. 174–185, 2023.
https://doi.org/10.1007/978-3-031-17091-1_19

the use of an Artificial lift for onshore wells use sucker rod or beam pumps because most operators know that they are simple, reliable, and familiar [2–4]. Since average oil outputs for a well take place every day during 12 consecutive months with a maximum value of 15 barrels or when, in a period of twelve consecutive months, a daily average gas output is considered to be of a peak value not exceeding 90,000 cubic feet, a well is considered a stripper for a tax reason. On the ground, a well is generally known as a stripper well if a well becomes marginal mainly due to low output. It should be observed that each oil field or a developed well has a specific price, which defines its critical point of benefit, below which a field or well is to become unprofitable [5]. If the well no longer generates sufficient oil volumes to produce sufficient profits from oil sales and thus is suspended, closed, or abandoned either because of high costs of operation or due to limited volumes of output or because of both [6]. Such wells need a higher price or a decreased operating/production price [per barrel of oil produced or thousand cubic feet (Mcf) of natural gas] for continuous activity. Conditions such as site access (remote ability), fluid viscosity, well depth and energy requirement (artificial lifting required), fluid composition and separating, processing, and disposal is a must [5]. The power requirement for an artificially lifted petroleum production system is also desirable for various production conditions and at different phases in the lifetime of an oil well. In order to drive and maintain output at the desired rate or level, the rating of the primary mover is required [6]. When designing the artificial lift, it is usually hard to find a standard engine (prime mover) to match the estimated cyclic load after estimating the necessary fluid load for the well. The lowest engine that fulfills the load requirement is a standard procedure. Oversize the engines eventually leads to wells being over-pumped, pump efficiency decreases, and consequently lifetime for the pump. It is common to update the pumping mode during the design process (stroke length, pump displacement, strokes per minute) [7]. This improvement changes the speed and torque specifications and creates a pumping system that can gradually pump wells over time [8].

In Egypt, several oil and gas companies that use sucker rod pumps have encountered many problems during the lifting operations especially those problems related to electrical power supplying as described in [9]. Furthermore, the electrical power consumed has been found to be different from one well to another depending on several factors which are flow rate, specific gravity of fluid, fluid level, weight of sucker rod, stroke length and pump speed [7]. As a result, prime mover should be adequately sized to provide sufficient power to lift the production fluid and to overcome friction loss in the rod string and polished rod. Consequently, several researchers have proposed numerous solutions to enhance the power supply reliability and sustainability in remote areas [9–13]. Studies have shown that its appeal is synonymous with low investment and maintenance costs, adjustable depths and outflows, high energy performance and the likelihood of working on a variety of temperatures using varied fluid compositions and viscosities [14].

Based on the previous introduction, the main issue that makes the oil and gas industry profitable is the production cost. The aging of the well leads to an increase in the production cost of the oil until a point that there is no profit from producing the oil from it, even if the well still has a reserve. One of the factors contributing to the production cost is the conventional power sources used to electrically supply artificial lifting systems such as diesel [2]. Conventional power sources have many issues regarding price fluctuation, transportations, and even harmful environmental gas emissions [3]. Therefore, controlling the production cost and reducing it with maintaining the stability of the oil production rate is the main challenge of this research.

The research starts with data collection, followed by load requirement calculations (see Fig. 1). Then setting the proposed system alternatives and the techno-economic analytical model must be developed. The sizing validation of the proposed system must be made, then the parametric sweep optimization and the water cycle optimization models must be developed. After calculating the Levelized Cost of Energy (LCOE) and the optimum system must be chosen based on it. A comparison between the existing system and the proposed optimum system must be done. Applying the settings of the optimum system on the whole field is shown in, followed by the optimum location.

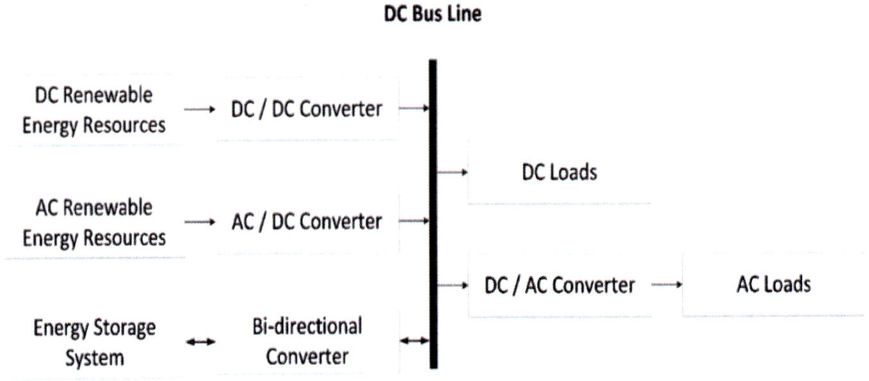

Fig. 1. HRES DC-coupling settings.

2 Mathematical Model

This section is focused on determining the economic feasibility and viability of implementing PV solar energy in the oil & gas sector for a small company. Meeting the electrical demand in a reliable and sustainable manner for an off-grid remote oil system is the main requirement of the company. The overall flowchart of the research is shown in Fig. 2. It starts with the data collection followed by the load requirement calculations. Then setting the proposed system alternatives and the techno-economic analytical model. After calculating the LCOE and the optimum system have to be chosen based on it. A comparison between the existing system and the proposed optimum system must be done.

The PV system sizing using the intuitive method is going through some steps to approach the final system value; thus, it is designed to cover the load needs. The main factors that have a significant effect on the system size are the load demand and the PSH.

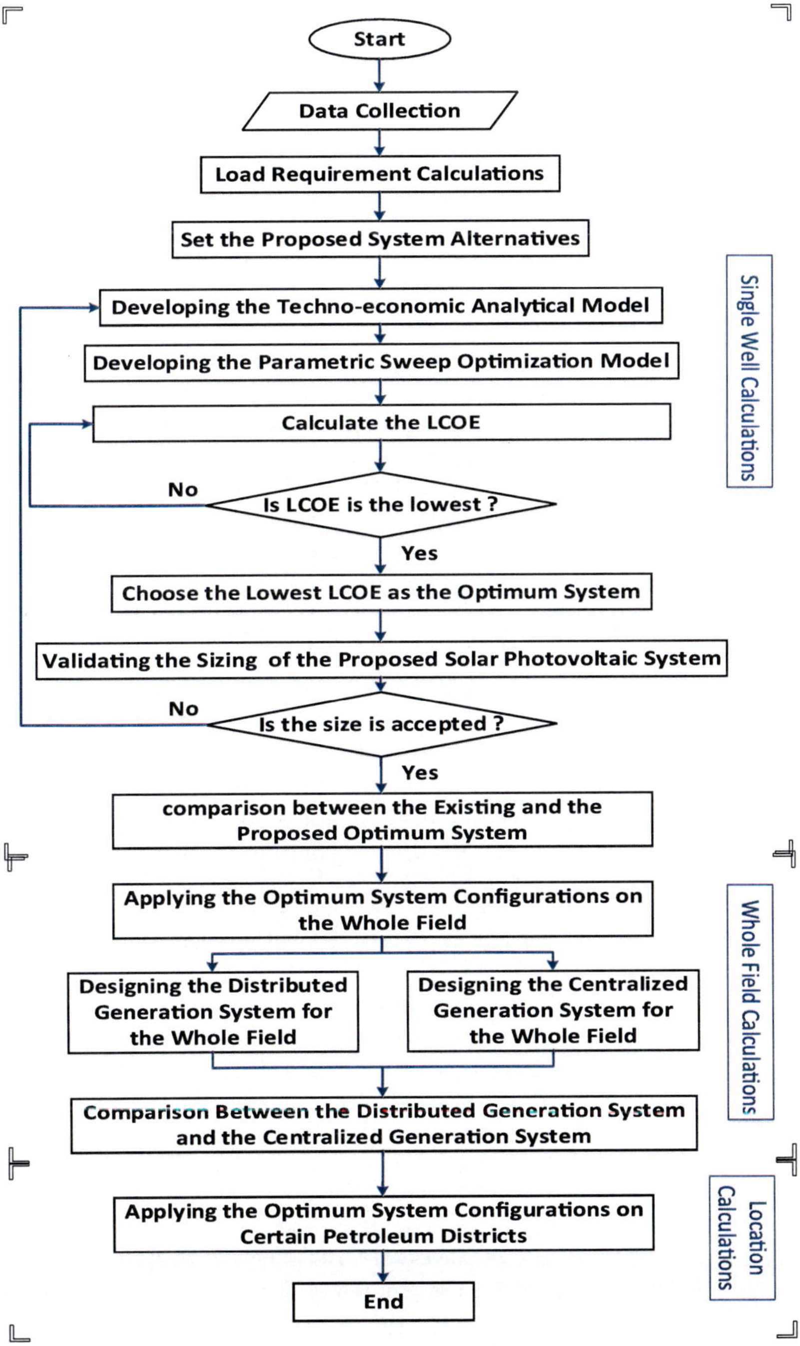

Fig. 2. Methodology flowchart

The author of [102] used the following equations to calculate the total PV parameters, and to get the size of the PV module in the system (3.7) is being used.

$$P_{PV} = E_L S_f / \eta_{inv} \eta_R PSH \tag{1}$$

where E_L is the total energy demand per day, S_f is safety factor that represents the losses of the dust, cable joints and cloudy times, η_{inv} is inverter efficiency, η_R is voltage regulator efficiency, and PSH is peak sun hours in the proposed area. PSH of the case study location can be calculated based on the solar irradiation of its location. If there is any storage system connected with the power system, the size of this system could be calculated by:

To calculate the ampere hour of the storage system, the following equation is being used.

$$E_B = \frac{S_f * E_L}{\eta_B DOD} \tag{2}$$

$$I_B = E_B / V_B \tag{3}$$

where E_B is the battery storage capacity in Wh, η_B: battery efficiency, and all other general losses in the system, DOD: allowable value to be discharged from the battery, I_B is the battery storage capacity in Ah, and V_B is the battery voltage.

The inverter is a very important part of the system as it is the main connection between the PV system and the load of the grid. It is different from the on-grid system than the off-grid system as in the on-grid system it has an extra task which is to synchronize the PV system with other power sources or with the large power grid. Based on the ABB inverter specifications, the ratio between the DC input power and the AC output power is 1.2. But in this thesis and based on the condition of this case study, the oversized inverter has to be used and the ratio between the DC and the AC is 1:1 as a protection and to increase the lifetime of the inverter. The size of the chosen inverter size based on [87] is the same with the PV system size and this research is using the same design concept, therefore, the size of the inverter could be calculated by using (1 to 3). The selected system parameters that has been used in this research are presented in Table 1..

To set the cost of the system parameters the specifications, the data sheet, and the manual of the system components must be considered. In [87] the Annualized Total Life Cycle Cost (ATLCC) of the system could be calculated by;

$$ATLCC = \sum\nolimits_{Device} C_{cap,a} + C_{o\&m,a} + C_{rep,a} + C_{s,a} \tag{4}$$

where, $C_{cap,a}$: the capital costs, $C_{o\&m,a}$: operation and maintenance costs, $C_{rep,a}$: the replacement costs, and $C_{s,a}$: salvage value cost. In this study the $C_{rep,a}$ is counted inside the $C_{o\&m,a}$ while $C_{s,a}$ is considered to be zero. Therefore, ATLCC has been formulated as follows:

$$C_{cap,a} = \frac{C_{cap}}{\frac{(1+ndr)^{L_s}-1}{ndr(1+ndr)^{L_s}}} \tag{5}$$

Table 1. Parameters of the system calculations

System parameters	Settings
Safety factor (SF)	1.25
Inverter efficiency (ηinv)	0.9
Voltage regulator efficiency (ηR)	0.9
Peak sun hour (h/day) PSH	5
Depth of discharge (DOD)	0.5
SPV efficiency	0.16
Battery efficiency (ηB)	0.8
Solar radiation at STC (W/m2) Gref	1000
Room temperature at STC (°C) Tref	25
Total lifetime of the system (Year) LS	20

$$C_{o\&m,a} = \frac{C_{o\&m}}{\frac{(1+ndr)^{L_s}-1}{ndr(1+ndr)^{L_s}}} \tag{6}$$

$$ndr = \left[\left(\frac{1 + interest\%}{1 + inflation\%}\right) - 1\right] \tag{7}$$

where C_{cap} and $C_{o\&m}$ are the total capital cost, operating and maintenance expenses during the expected lifespan of the system, respectively. L_S: is the total lifetime of the according to the study, and ndr is the net of inflation rate discount.

The total capital costs of the system could be calculated as follows:

$$C_{cap} = C_{cap,PV} + C_{cap,Inv} + C_{cap,OC} + C_{cap,Bat} + C_{cap,C} + C_{cap,DG} \tag{8}$$

where $C_{cap,PV}$, $C_{cap,Inv}$, $C_{cap,OC}$, $C_{cap,Bat}$, $C_{cap,C}$ and $C_{cap,DG}$ are the total capital cost for PV array, inverter, other costs, battery, the controller and diesel generator respectively. The operation and the maintenance costs of the system could be calculated by following equation:

$$C_{o\&m} = C_{o\&m,PV} + C_{o\&m,Inv} + C_{o\&m,OC} + C_{o\&m,Bat} + C_{o\&m,C} + C_{o\&m,DG} \tag{9}$$

where $C_{o\&m,PV}$, $C_{o\&m,Inv}$, $C_{o\&m,OC}$, $C_{o\&m,Bat}$, $C_{o\&m,C}$ and $C_{o\&m,DG}$ are the total operating and maintenance cost for PV array, inverter, other costs, battery, the controller and diesel generator respectively. The maintenance cost ($C_{o\&m}$) term contains some other costs ($C_{o\&m,O}$) inside it such as the dust cleaning cost, and some other costs. These costs have a direct effect on the LCOE of the system, therefore, it must be considered.

As the system will be installed in the desert, the dust will be deposited on the panels which can easily affect the PV panel performance and efficiency. Therefore, a dust cleaner is imperative to clean the panels on a regular basis. This cleaner has 5 years lifetime and needs replacement every five years. There will be some other costs involved during the lifespan of the system. These costs will include the labor cost, cleaning cost, … etc. The LCOE (EGP/kWh) for the system of this study over 20 years, could be resulted from the following equation:

$$LCOE = \frac{\mathrm{ATLCC}}{\mathrm{E_L}} \tag{10}$$

To calculate the ATLCC of any system, the inflation and interest rates of the project location must be counted. Based on [103, 104], the average inflation rate for Egypt from 1984 till 2021 is 11.45% while the interest rate in 2019 is 5% for the renewable energy projects based on the Central Bank of Egypt (CBE). For the cost equations, the inflation, and the interest rates of Egypt need to be provided. The inflation rate has been taken as an average for the 40years as it will affect the income of the project. Meanwhile, the interest is using the rate of 2019 only as it is the year of the project loan.

3 System Modelling

In a remote place in the western desert of Egypt, there is an oil field represented by multiple sucker rod pumping systems (SRPs). These SRPs which are also called "beam pumping", provide mechanical energy to lift the oil from the bottom hole of the well to the surface. Due to its simplicity, efficiency, and easiness, it is the most common way to lift the oil in the last stage of the primary recovery phase of the well such as this research case study. This stage has a very low bottom-hole pressure; therefore, the SRPs is used to pump the oil up and maximize the oil production rates. It is also suitable for slim holes, multiple completions, high-temperature, and viscous oils. SRPs design varies from one well to another, therefore, there is also a difference in their power needs. The case study field consists of 10 different wells with different specifications, and each well has its own SRPs. There is a separate diesel generator to run each SRPs individually in each well. The first well is taken as an example for the whole field to apply the proposed hypothesis. This well have SRP unit type C-640D-305–144 from LUFKIN, which means it can handle a maximum polished rod limit of 30500 lb. And can operate up to 640 000 lb torques with the maximum stroke length 144 inches. This research is investigating all the power scenarios that can supply the SRPs with the needed energy by taking the technical requirements of the load and the economic factors into consideration. Therefore, the mechanical and electrical specifications of the SRPs that is used in this case study are listed in Table 2. and Table 3.. Meanwhile, Fig. 3 shows the SRPs and its main parameters where Pe is the input electrical power to the system's prime mover, PRHP is the polish rod horsepower and Phydr is the output mechanical power. Sucker-rod strings transfer the motion from the driving machine on the wellhead to the downhole oil pump. The electrical pumping unit is used to operate a down-hole pump to push underground emulsion (oil and water emulsion) into a pump- barrel. The weight of the sucker rod and pump-barrel are lifted by the ram until it touches the top of the pump stroke then a fluid reservoir is released from the emulsion and the pump is ready for the downstroke [100]. The sucker rod load is unsteady due to the distinction of the amperes consumed in the upstroke and the down stroke. During the movement of the pump through upstroke the maximum load will be reached, and a maximum ampere is consumed by the motor. On the other hand, a minimum ampere is consumed during the downstroke as shown in Table 3.. Hence, the actual load is calculated using two ranges of amperes (one is considered for the upstroke and the other one for the down stroke).

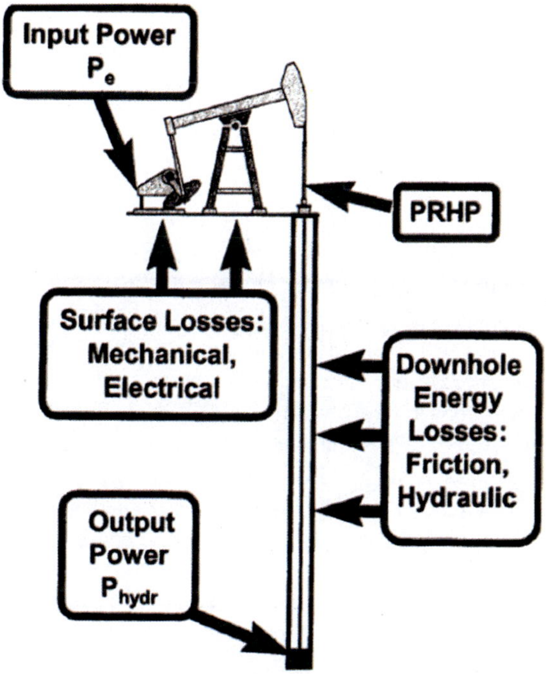

Fig. 3. Energy requirement of sucker-rod pump [100]

Table 2. The technical specifications of well no. one

Flow rate (bbl./D)	400
Specific gravity	0.845
Pump depth (ft)	3950
Weight of sucker rod string (lb)	8000
Stroke length (in)	144
Pump speed (spm)	6.2

Table 3. The electrical specifications of the SRPs' prime mover

Volts (V)	380
Current (Amp) rated	86
Current (Amp) actual	30 for down stroke – 67 for up-stroke
Power factor	0.86
Rate power (KW)	48.6

4 Results and Discussion

The parametric sweep model that is used in this research has generated the following figure. As shown in Fig. 4, the stand-alone diesel generator has the lowest LCOE till the PV contribution reaches 30% then the hybrid system takes the lead in the lowest LCOE value. This shows that the hybrid system before 30% of the PV system has no economic benefit to this research case study. Furthermore, there is an overlap at the point 54% (PV system contribution), where the LCOE reaches the lowest level. This proves that the selection of the PV system contribution at 54% was the correct selection at techno-economic section as the best hybrid system configuration.

Usually, when a small system, or any system are designed, some basic components are existing whatever the system size is. Therefore, at the beginning of the graph, these components have a huge effect on the system cost. Meanwhile at the middle of the graph, the cost of these components already distributed in a larger system size. This will cause a lower LCOE as much we proceed in the track of the curve, till the point where the battery system started to join the design.

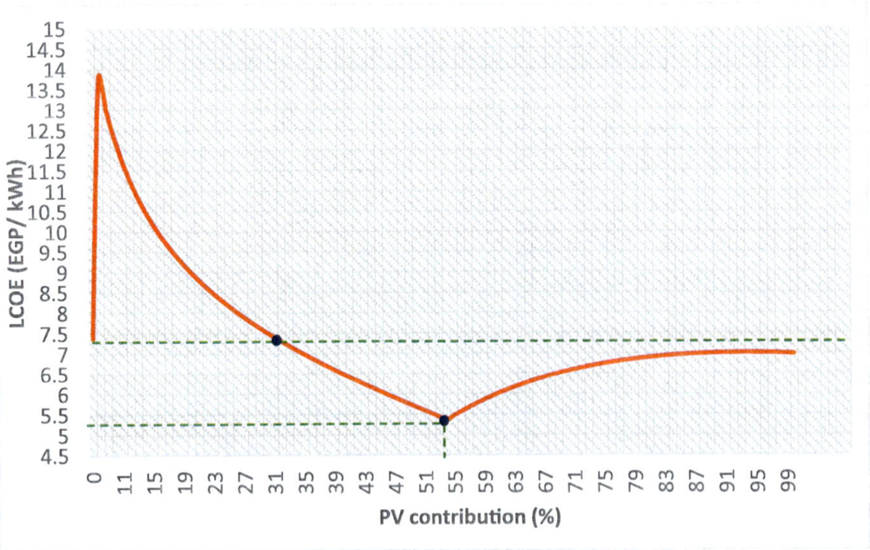

Fig. 4. Parametric sweep curve

As per the discussion in the techno-economic analysis section, the optimum hybrid system is distributed between 54% energy share from the photovoltaic system and 46% from the diesel generator system. The total energy required for the sucker rod pumping system per day is 658.8 kWh. The share of the PVS is 355.78 kWh while the diesel share is 303.075 kWh. To generate the 355.78 kWh, it needs an 89.2 kWp solar system. As shown in Table 4, and by using the sizing tool that is used, the total photovoltaic system size that is used to generate the same amount of energy is 91.5 kWp. The error between the sizing tool result and the research result is calculated by (3.18) and it is 0.025 (2.5%) which within the allowable limit. Therefore, it could be concluded that the proposed size of the photovoltaic system is capable to generate the required energy that is needed to run the SRPs.

As a direct result of the increase in the PVS size that is generated from the sizing tool, the number of PV panels increased also from 262 panels to 269 panels. Meanwhile, the area of the system increased from 895 m^2 to 1095 m^2. Based on the PVS size for both models, the total required power rating for the DC/AC inverter in the sizing tool is 76.25 kW, while the required power rating for the proposed model is 89.2 kW. As the rating of the chosen DC/AC inverter in both models is 27 kW, the required inverter number for the proposed model is 4 while the sizing tool needs only 3 inverters. This difference in the number of the inverters is due to a ratio that is used in the sizing tool between the AC power and the DC power. In the proposed system case, the worst-case scenario has been used therefore, the number of inverters is more than the sizing tool. This change also has a direct relation with the increase in the capital cost of the proposed model than the sizing tool.

Table 4. Comparison between the optimum system and the sizing tool system.

Parameters	Optimum system specifications	PV-ON sizing tool system specifications
Solar PV (kWp)	89.2	91.5
Diesel generator (kW)	120	120
PV panels (unit)	262	269
Area (m2)	895	1095
27 KW DC/AC inverters (unit)	4	3
Total capital cost (EGP)	871177	850925
Diesel consumption (L/y)	61898	61898
Generator renting cost (EGP/y)	421575	421575
Total fuel cost (EGP/y)	618980	618980
Total electrical production (kW/y)	273057	273888
LCOE (EGP/kW)	5.35	5.3

It worth mentioning that in Table 4, the sizing tool is dealing with the PV system only and the total power system is consisting of both PV and diesel generator systems. Therefore, the diesel generator data including the diesel consumption, rental rate, and the total fuel cost are included in the table. The real comparison in Table 4 is between the PV system specifications, meanwhile the other data are to complete the specifications of the total power system. The capital cost of the proposed model 871177 EGP, while 850925EGP for the sizing tool with 20252 EGP difference. This change is due to different assumptions that have been used in both models such as safety factors, error ratio, losses assumptions, … etc. The change in these assumptions doesn't make any problem for the proposed system size design as they variate from location to location and from customers to another. The total electrical production per year for the proposed system is 273057 kWh/y, while it is 273888 kWh/y. We can use (4.18 (RMSE)) to calculate the total error between the proposed system and the sizing tool result, and it shows that the error is 0.3% which is very small to be considered. The LCOE of the sizing tool is 5.30 EGP per kWh while it is 5.35 EGP per kWh for the proposed system. It shows that the error between both LCOE is 0.9% which is even less than 1%. This percentage is very small, and it is in the acceptable error range. Concludes that the proposed system size is very much

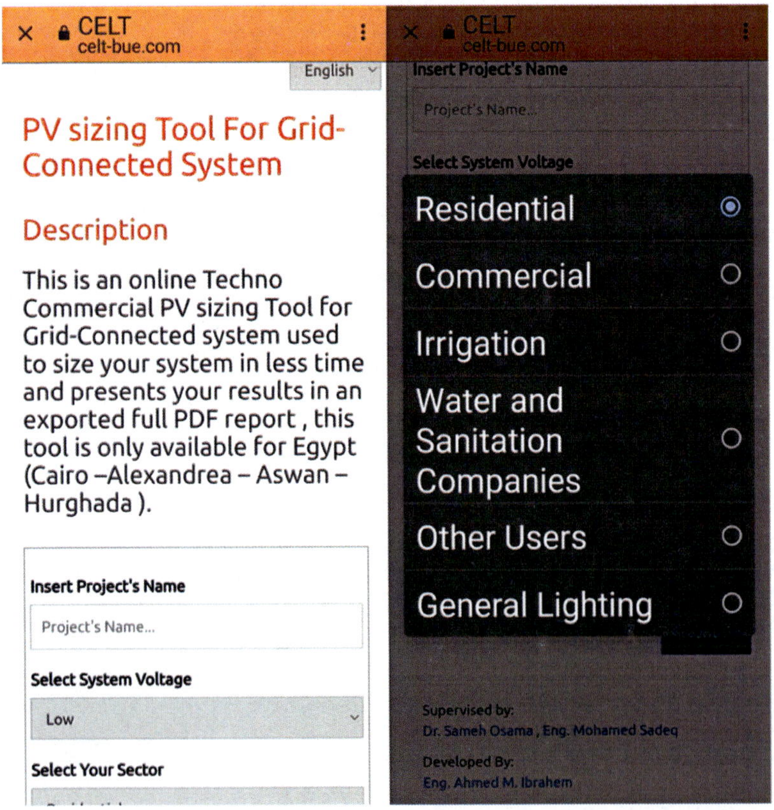

Fig. 5. PV-ON online tool

acceptable to supply the load with all the required energy. It must be highlighted that the cheapest system does not mean it is more feasible or more applicable in practical life. In this type of study that is concerned about a true case study, as much as the proposed system is practical, as much it became optimum in the eyes of the investors.

5 Conclusion

In every district of the Egyptian land, there are different meteorological specifications that leads to a change in the production rate of the PV system. Consequently, the proposed hybrid system of 54% PV, and 46% DG is used again over five different locations distributed over the petroleum locations of Egypt. The Specifications of well no. 1 are used as a reference system for this application. The PV system size was 89.2 kWp to generate 54% of the total energy required for the load. The losses percentage considered as 25% for the total production.

References

1. Mantey EK, Wolodko BB (1998) Method and apparatus for controlling downhole rotary pump used in production of oil wells, ed: Google Patents
2. Dave MK, Ghareeb Mustafa M (2017) Performance evaluations of the different sucker rod artificial lift systems. In: SPE Symposium: Production Enhancement and Cost Optimisation
3. Feng Z-M, Tan J-J, Li Q, Fang X (2017) A review of beam pumping energy-saving technologies. J Petroleum Explor Prod Technol 8(1):299–311. https://doi.org/10.1007/s13202-017-0383-6
4. Di Tullio MT, Marfella F (2018) Enhanced sucker rod pumping model: a powerful tool for optimizing production, efficiency and reliability. In: SPE Middle East Artificial Lift Conference and Exhibition
5. Osaretin CA, Iqbal T, Butt S (2020) Optimal sizing and techno-economic analysis of a renewable power system for a remote oil well. AIMS Electron. Electric. Eng. 4:132
6. Hicks A (1986) Using fiberglass sucker rods in deep wells. In: SPE Deep Drilling and Production Symposium
7. Takacs G (2015) Sucker-rod pumping handbook: production engineering fundamentals and long-stroke rod pumping: Gulf Professional Publishing
8. Skinner DR (1984) Efficient use of electric power in production operations. J Petroleum Technol 36:1326–1334
9. Khatita MA, Ahmed TS, Ashour FH, Ismail IM (2014) Power generation using waste heat recovery by organic Rankine cycle in oil and gas sector in Egypt: a case study. Energy 64:462–472
10. Elhousieny M, Khalil A (2017) Photovoltaic/gen-set hybrid system at Egyptian western desert. In: Offshore Mediterranean Conference and Exhibition
11. Castaldo F, Brocco V, Cimino R, De Ghetto G, Rossini, S (2015) Development of innovative solar-fossil hybrid plant in a remote O&G area. In: Offshore Mediterranean Conference and Exhibition. Offshore Mediterranean Conference
12. Said M, El-Shimy M, Abdelraheem M (2015) Photovoltaics energy: improved modeling and analysis of the levelized cost of energy (LCOE) and grid parity–Egypt case study. Sustain Energy Technol Assess 9:37–48
13. Diab F, Lan H, Ali S (2016) Novel comparison study between the hybrid renewable energy systems on land and on ship. Renew Sustain Energy Rev 63:452–463
14. Torres LH, Schntman L (2013) Sucker-rod pumping system of oil wells: Modelling, Identification and Process Control. IFAC Proc. 46(24):260–265

Online Photovoltaic Pump System Sizing Tool for Educational Purposes

Abdullah Hatem[1], Ahmed Ashraf[1], I. M. Mahmoud[1], Sameh O. Abdellatif[2](✉), and Hani Ghali[2]

[1] Renewable Energy Department, Faculty of Energy and Environmental Engineering, The British University in Egypt, Cairo, Egypt

[2] The Electrical Engineering Department, and FabLab, at the Centre of Emerging Learning Technologies CELT, British University in Egypt (BUE), Cairo 11387, Egypt

Sameh.osama@bue.edu.eg

Abstract. In this paper, a full sizing of on-grid Photovoltaic water pumping system (PVWPS) for Agricultural Irrigation purposes is carried out. The proposed system is identified as solar water pump to storage tank system in addition to the method of irrigation being drip irrigation. Sizing of such system is considered as a validation for the application of such systems which will ultimately encourage the use of such systems. The method of irrigation is explained. Moreover, the required pressure for this type of application is investigated. Results are verified and shows full photovoltaic sizing to the system with economic data. The implemented code is considered as a seed for an online tool to be used either in industrial sector or educational purposes.

Keywords: Photovoltaic water pumping system (PVWPS) · Centrifugal pump · Submersible pump · Irrigation · Drip irrigation

Table 1. Nomenclature

Symbol	Name	Unit
AMEU	Average monthly electricity usage	___
TDH	Total dynamic head	m
Hs	Static head suction	m
PVD	Vessel pressure at discharge	___
PVS	Vessel pressure at suction	m
N	Number of stages	__
ρ	Density	Kg/m^3
VS	Fluid velocity suction	m/s
VD	Fluid velocity discharge	m/s
Dd	Pipe internal diameter at discharge	mm

(*continued*)

M. E. Auer et al. (Eds.): REV 2022, LNNS 524, pp. 186–197, 2023.
https://doi.org/10.1007/978-3-031-17091-1_20

Table 1. (*continued*)

Symbol	Name	Unit
Ds	Pipe internal diameter at suction	mm
RE	Reynolds number	—
V	Fluid velocity	m/s
D	Pipe internal diameter	mm
$L_{e/D}$	Length to diameter ratio	—
M	Dynamic viscosity	cP
P	Power	Kw
Q	Volumetric flow rate	m^3/hr.
u	Specific volume	m^3/kg
P	Absolute pressure	Pa
Z	Elevation of the liquid	m

1 Introduction

Following the approach for using clean energy, renewable energy sources began to rise. Sources such as wind, photovoltaic and hydropower are now dominating the renewable energy market as they proved their capability to cover the increasing demand for electricity in societies. Photovoltaic technologies are the most common type of the renewable energy sources this is due to the fact that it has a wider range of applications either domestic or commercial in addition to having multiple installation techniques such as ground or roof mounted it also comes in last place when talking about maintenance costs. This quickly led to the approach of powering almost any electrical load using photovoltaic technologies. Such an example is water pumping. Photovoltaic water pumping systems (PVWPS) were introduced as the perfect alternative for diesel powered systems [1]. Avoiding the environmental impact whilst providing the same application with significantly lower running costs. Photovoltaic water pumping systems operate on both on-grid and off-grid systems. Photovoltaic water pumping system has a variety of configurations, which are fully dependent on the application of the system [2]. Figure 1 shows the introduced system, which is best, described as a pump to storage system [3].

The proposed system uses a submersible pump and operates to cover agricultural irrigation process whilst covering the domestic water needs for a residential building. The system includes:

- PV modules
- Inverter
- Submersible pump
- Storage tank sizes
- Pipes
- Valves and fittings for the pipes

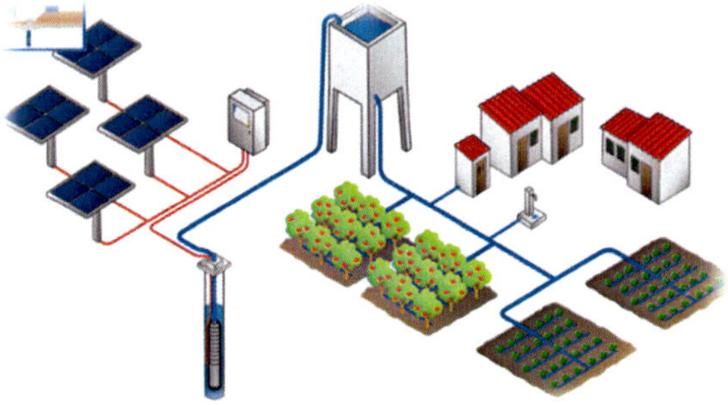

Fig. 1. Schematic for a photovoltaic water pumping system

PV modules are installed on the ground and with suitable sizing with direct relation to the pump size and the working hours required by the system [4]. Furthermore, a suitable power for the inverters is calculated to withstand the amount of power extracted from the PV modules and to ensure that the AC current will be efficiently transformed into Direct current to fully function the water pump. A submersible pump is used in such systems due to the height of the liquid underground in wells [5]. Using wells as a water source usually require a pump to extract water from elevations up to 100 m [6]. Submersible pumps offer multiple stages giving the pump more power to pump fluids to these elevations from the point of suction to the point of discharge or delivery [7]. Using storage tanks is considered a variable factor in the system. This is mainly due to the type of crops that need the water for example: a well-known method of irrigation is drip irrigation. This method requires water to be pumped at a slow rate into multiple crops at the same time not to over flood the roots of the crop especially in the early stages of development which decrease the amount of water used [8]. As a well-studied irrigation could have led to a successful project that could increase the productivity in agriculture [9]. Such application will not be a possibility when using a powerful multi-stage pump which why the use of storage tanks is introduced [10]. These tanks are often installed at a relevant elevation to help provide enough pressure for the liquid to reach the crops. In addition to sizing the photovoltaic system and the size of the pump the pipes, valves and fittings are some important factors to be taken into consideration in this matter. There are numerous types of pipes ranging from steel, plastic and PVC pipes. Each type has a pressure limit for different applications. Besides purchasing the types of valves to control the irrigation network and achieve total control over the vessel pressure in the pipes at the delivery.

The purpose of modelling a PVWPS is to predict the performance of the system and to prove that the design of this system is efficient. In this paper sizing of a PVWPS using either centrifugal or submersible pumps is carried out using MATLAB software. This includes full PV sizing for either on-grid or off-grid system. An accurate representation for the flow of orders of the system is represented in a simplified flowchart showing the sequence of logical operations calculated to output the full sizing data. The constructed code demonstrate the initial step toward having an online accessible tool to be used in sizing water pumping systems. The targeted tool will be evaluated by both industrial customers and operators as well as research-based users. The targeted tool will follow the same procedures as in our previous investigation in [11].

2 Mathematical Model

To size a photovoltaic water pump, one must study the principles of this application. These principles are translated into a set of equations as follows:

One of the main considerations when sizing a photovoltaic system is the install type and location which is ultimately used in the calculation of the maximum power of a PV system [11]. The maximum power in photovoltaic systems highly depends on the available area to install the system and the mounting type either ground, roof or industrial shed mounting. Each type requires a certain amount of area rooftop mounting needing the most space while ground mounting has the least required area as follows:

$$\text{Maximum power} = \text{Available area}/\text{Mounting system area (KWP)}$$

The average monthly electricity usage is the addition of the usage throughout the year which is subcategorized as summer, winter, spring and autumn. The average will be calculated by dividing the total electricity over 12 months as follows:

$$\text{AMEU} = (\text{winter, summer, spring and autumn})/12\,\text{months (KWh/month)}$$

Bernoulli set of equations, which produces electrical power as follows:

$$P_h = \frac{\rho * g * H_t * Q}{3.6 * 10^6 * \mu_m} \tag{1}$$

The Energy balance equation for any fluid is written as:

$$uP + gZ + \frac{V^2}{2} = K(constant) \tag{2}$$

To compare the flow between any two points on a streamline utilizing Bernoulli equations where 1 and 2 represent two different points of a fluid on the same streamline is shown as:

$$u(1)P(1) + gZ(1) + \frac{V1^2}{2} = u(2)P(2) + gZ(2) + \frac{V2^2}{2} \tag{3}$$

Reynolds number is often described as a ration between forces (motion and its resistance) which sets limits to the types of flow such as laminar or turbulent as shown below [12]:

$$R_e = \frac{\rho * V * D}{M} \tag{4}$$

The Energy content per unit mass of the liquid at the reservoir tank, which is the one of the indications of the distance a fluid could flow, is calculated as follows:

$$E = \frac{P1}{\rho} + Z1g + \frac{V1^2}{2} \tag{5}$$

Fluid velocities vary between the two points inside of pump the suction and discharge points. This difference is present due to the variation in the flowrate and the internal diameter at the two ends as shown below:

$$V_D = \frac{354 * Q}{D_d^2} \tag{6}$$

Logically the speed on the discharge must be significantly higher than the suction to provide enough pressure for the application as follows:

$$V_S = \frac{354 * Q}{D_s^2} \tag{7}$$

Vessel pressures in suction side of the system is normally considered as an input. However, this modification is present to be able to use this value with the appropriate SI units as follows:

$$P_{Vd} = P_{VD} * \frac{10200}{\rho} \tag{8}$$

The vessel pressure is more sensitive to the suction side due to the overflowing of water in a limited space while in the discharge side is purely dependent on the pressure output of the fluid as shown below:

$$P_{VS} = P_{VS} * \frac{10200}{\rho} \tag{9}$$

The number of stages in a submersible pump is reliant on the discharge pressure or speed required which is calculated from the total dynamic head. The larger the required output the more powerful the pump must be. Consequently, more than one impeller shall be used which is the idea of multi-stage pumps. Given that in this equation the result must be approximated to the next big significant figure no matter what the decimal is as shown below [12]:

$$N = \frac{TDH}{H_s} \tag{10}$$

The Length of pipes required could be mathematically calculated for accurate representation using the following method where Le/D is the length to diameter ratio:

$$Ls = L(suction\,pipe) + \sum\left(\frac{Le}{D}\right) * Ds \tag{11}$$

The pump performance curves are considered as an essential measurement to the effect of the pump on the overall system. This graph enterprise the idea that when the required flowrate increases the pump can no longer pump the fluid at the same elevation therefore informing the user for the need to purchase a more powerful pump [13]. Users could avoid this potential by exploring pump performance curves, which are available at any dealer when purchasing the water pump in addition to setting the elevation limit for the certain pump as shown in Fig. 2. Accurate sizing significantly reduces initial costs and also helps in decreasing maintenance costs.

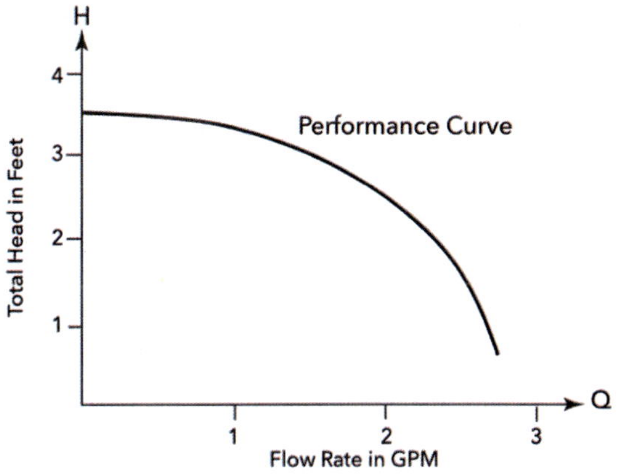

Fig. 2. Pump performance curve standard

3 System Modelling

A simple configuration of a PVWPS consists of: PV module (cells and arrangement), inverter, battery (in case of off-grid systems), storage tank (if needed), a suitable sized water pump and pipes with their valves and fittings. Water pumps are complicated in design yet easy in use. It operates on one of two principles which are dynamic or positive displacement. The two most commonly used pumps both operate on the dynamic principle which are centrifugal and submersible pumps [14]. Centrifugal pump is often installed on the surface of any water source. As the fluid enters the cavity of the pump the impeller begins to rotate making the fluid gain mechanical energy which therefore pumps the liquid with higher pressure. This defines the two sides of the pumping system, which are suction and discharge suction being at the water source and discharge meaning the water outlet [15]. On the other hand, the submersible pump is a type of pump installed in deep wells. It is designed to be submerged as form the name submersible. It is designed to be vertically installed, overflowing the cavity; rotating the impeller to push the fluid out [16].

Figure 3 shows the flow chart of designing PVWPS that includes the required stages for design the system. However, this process often requires multiple stages because it is designed to pump water from very deep wells. In a scenario where the user needs to pump water to a high elevation a submersible pump is required as it can produce a high pressure in several stages, which helps in increasing the water elevation, and if the user needed to pump water in to a small elevation a centrifugal pump will be used. In case the user used submersible pump, a simple variable is needed to be provided such as density, elevation, and losses. Therefore, the result could be got for the suction and discharge calculations and centrifugal pump case another variable is needed to be provided such as flowrate, viscosity, pipe diameter so the result could be obtained for the suction and discharge. After the user gets the result for the suction and discharge, the number of working hours has been decided for the system to get the system power calculations. If the user's system contains an inverter a calculation for inverter will be done to calculate its size, then it shows the full final sizing data output. In case the user's system does not require inverter, the full final sizing data output is including number of PV cells, size of pump would show up directly.

4 Results

After careful considerations and applying a certain load using photovoltaic water pumping system accurate results are obtained and explained in the below. This output data provides the user with sufficient data to turn this output into a real-life system avoiding problems such as oversizing the system or under sizing the system. Oversizing such a system significantly increase the total cost of the project and causes an increase in the dissipated energy which potentially drives the customers away from using this photovoltaic technology. Under sizing is also a very important issue as the customer will invest a great amount of money and they won't get the potential return to the system which fatherly drives users away from photovoltaic technologies.

Case Study 1: Submersible Pump. A small farm owner in Ismailia (30.5965° N, 32.2715° E) has no water source and seeks to be in the area of agricultural irrigation. This owner considered the option of a photovoltaic water pumping system. As the farmer did not have a water source, a well was constructed. This means that the owner will need a pump that deals with high elevations to pump fluids which is the submersible pump. The well did not have to be deep as the water depth is very narrow which is why a two-stage submersible pump is suggested. The pump suggested is 1.5 KW pump with only 48% efficiency as shown in Table 2. This is because the owner designed a drip irrigation system as the flowrate required was estimated to be 1.5 m^3/h. The drip irrigation method does not require high pressure when pumping the fluids and is an ideal method of irrigation for limited time crops such as tomatoes.

Calculating the costs of such a system presents in two sides, which are the mechanical side and the electrical side. Calculating the electrical aspect of the system, which includes the type and number of PV cells and the inverter sizing. The mechanical aspect is present in selecting the water pump and the size of the storage tanks. The system might need storage tanks, which ranges about 1000 EGP per 1000 L. A water pumps this small only needed two monocrystalline PV cells and an appropriate inverter has been sized (Table 3).

Case Study 2: Centrifugal Pump. Another farm owner in Ismailia (30.5965° N, 32.2715° E) grows trees. Trees need a significantly more amount of water than seasonal crops, which means that the owner will need to choose an alternative method for irrigation, which will be surface irrigation. This irrigation method requires significantly more amount of water per unit time. The flowrate of the system was estimated to be 120 m^3/h that required a much more powerful pump. The owner had a viable water source nearby (the source being a lake) which did not require the water pump-to-pump water from a great height which is why a centrifugal pump is suggested. The centrifugal pump suggested is a 23.5 KW pump with a high efficiency of 72% as shown in Table 4.

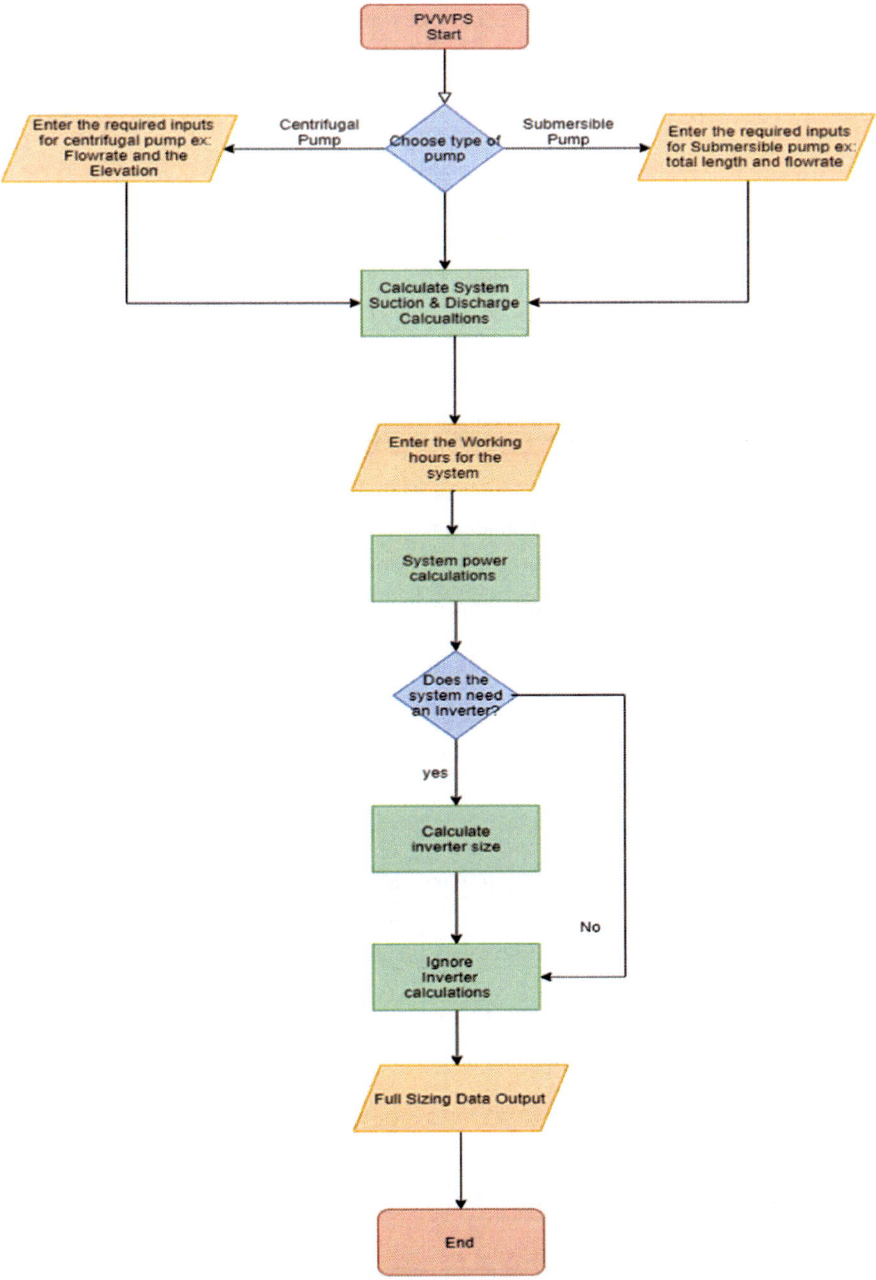

Fig. 3. Flowchart for PV water pumping system.

Table 2. Submersible pump results

Name	Value
Number of stages	2
Efficiency	48%
Pump Power	1.443
Electrical power/month	96 KW
Number of PV modules	2
Type of PV modules	Jinko 335-W mono
Inverter power	534 W AC
Annual produced energy (ABE)	$1.15*10^3$ KWH
Lifetime produced energy	$2.67*10^4$ KWH

Table 3. Submersible pump cost results

Type	Model	Unit price
PV cells	Polycrystalline	2250 EGP
Inverter	5 KW	12600 EGP
Pump (Submersible)	1.5 KW	4500 EGP
Tanks (1000 L)	Plastic	1000 EGP

Table 5 show the cost study of the system which including Monocrystalline photovoltaic cells usually cost about 2,250 EGP per cell. This system requires 30 PV module to provide a steady supply to the pump not to affect the efficiency that means that the approximated cost will be 67,500 EGP.

Table 4. Centrifugal pumps results

Name	Value
Efficiency	71.80%
Pump power	23.31 KW
Electrical power/month	1482 KW/month
Number of PV modules	30
Type of PV modules	Jinko 335-W mono
Inverter power	8.2 KW AC
Annual produced energy (ABE)	17770 KWh
Lifetime produced energy	410500 KWh

Table 5. Centrifugal pump cost results

Type	Model	Unit price
PV cells	Polycrystalline	2250 EGP
Inverter	5 KW	12600 EGP
Pump (Centrifugal)	25 KW	20000 EGP
Tanks (1000 L)	Plastic	1000 EGP

One of the most important factors of pump selection is considering the height needed to pump the water (elevation of the liquid). This has a direct relation with the number of stages of a submersible pump as shown in Fig. 4. Number of stages is directly proportional with the dynamic head. Another significant factor when sizing a water pump is the head vs flowrate ratio. As the elevation increases (the height required to pump the fluid) increases the lower the flowrate output by the pump as shown in Fig. 5. This phenomenon occurs due to several factors such as the specific drawdown, losses due to friction either inside the pump or the pipes and vessel pressure (pressure inside the pipes). Such relation could be concluded in the following figure.

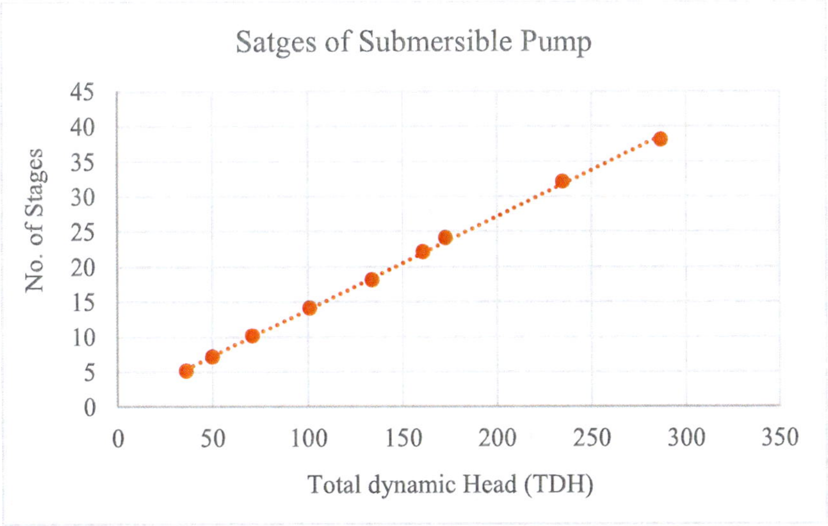

Fig. 4. Stages of a submersible pump

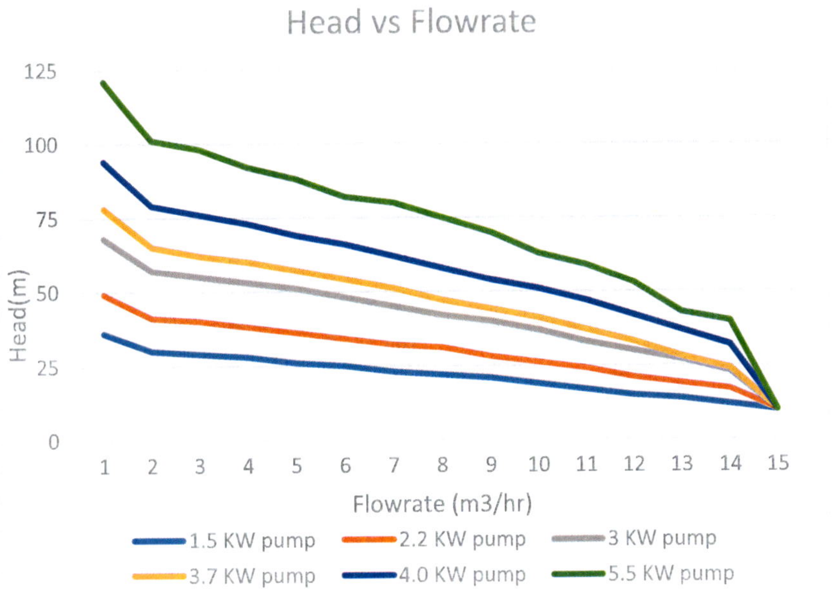

Fig. 5. Results of head vs. flow rate graph

5 Conclusion

Photovoltaic water pumping systems (PVWPS) were initially introduced as an alternative to diesel powered pump, which had a significantly high running and maintenance costs. These diesel-powered pumps were a threat to the atmosphere due to the amount of pollution they cause. PVWPS eliminates such factors and has proven quality and applicability over the period of its existence eliminating the environmental factor and significantly lowering the running and maintenance costs. Full sizing of on-grid PVWPS was done based on two different scenarios with different load profiles and different types of irrigation methods. The proposed systems could easily be transferred to a real-life system with the amount of data output from the sizing of the system. Another system could easily be sized using the flowchart in Fig. 3 following the steps to approach such a conclusion for different load profiles of other systems. An individual farmer can afford such low price pumping system in a developing country like Egypt. The designed system is economically feasible to meet the irrigation challenges faced during dry season.

References

1. Narale P, Rathore NS, Kothari S (2013) Study of solar PV water pumping system for irrigation of horticulture crops. Int J Eng Sci Invention 2(12):54–60
2. Muralidhar K, Rajasekar N (2021) A review of various components of solar water-pumping system: configuration, characteristics, and performance. Int Trans Electr Energy Syst Wiley Online Library 31(9):e13002

3. Bode C, Schmitz G (2018) Dynamic simulation and comparison of different configurations for a coupled energy system with 100% renewables. In: 12th international renewable energy storage conference, IRES 2018, vol 155, pp 412–430
4. Zarie MM, Makeen P, Abdellatif SO, Mahmoud IM, Abdel-Salam TS (2019) Techno-economic feasibility of photovoltaic system for an educational building in Egypt (Case Study). In: 2019 international conference on innovative trends in computer engineering (ITCE), Aswan, pp 521–525
5. Belgacem BG (2012) Performance of submersible PV water pumping systems in Tunisia. Energy Sustain Dev 16(4):415–420
6. Bakelli Y, Kaabeche A (2019) Optimal size of photovoltaic pumping system using nature-inspired algorithms. Int Trans Electr Energy Syst Wiley Online Library 29(9):e12045
7. Yang Y, Zhou L, Shi W, Zhaoming H, Yong H, Xiao, Y (2021) Interstage difference of pressure pulsation in a three-stage electrical submersible pump. J Petroleum Sci Eng 196:107653
8. Yang W, et al (2021) Drip irrigation incorporating water conservation measures: effects on soil water–nitrogen utilization, root traits and grain production of spring maize in semi-arid areas. J Integrative Agric 20(12):3127–3142
9. Eltawil M, Alhashem H, Alghannam O (2021) Design of a solar PV powered variable frequency drive for a bubbler irrigation system in palm trees fields. Process Saf Environ Prot 152:140–153
10. Zhu J, et al (2018) Surfactant effect on air/water flow in a multistage electrical submersible pump (ESP). Exp Thermal Fluid Sci 98:95–111
11. Sadeq M, Abdellatif S (2021) PV-ON: an online/bilingual PV sizing tool for grid-connected system, case studies in Egypt. Int Trans Electr Energy Syst Wiley Online Library 31(7):e12910
12. Jack TK (2014) Excel^{TM} in centrifugal pump selection. Int J Eng Res Dev 10(7):28–50
13. Zhou W, Qiu N, Wang L, Gao B, Liu, D (2018) Dynamic analysis of a planar multi-stage centrifugal pump rotor system based on a novel coupled model. J Sound Vib 434:237–260
14. Lyden A, Tuohy P (2022) Planning level sizing of heat pumps and hot water tanks incorporating model predictive control and future electricity tariffs. Energy 238(A):121731
15. Raj PP, Gupta P, Kumar P, Paramesh G, Rohit K (2020) Model and analysis of rotor and impeller of eight stage centrifugal pump. Mater Today Proc 21(1):175–183
16. Haque M, Islam M, Haniu H, Akhter M (2019) Energy efficiency improvement of submersible pumps using in barind area of Bangladesh. Energy Procedia 160:123–130

A Low-Cost Spectrometry Remote Laboratory

Alex Villazon(✉), Omar Ormachea, Angel Zenteno, and Adriana Orellana

Universidad Privada Boliviana (UPB), Cochabamba, Bolivia
{avillazon,oormachea,angelzenteno1,adrianaorellana1}@upb.edu

Abstract. The Covid-19 pandemic has precipitated the digital transformation in education worldwide and has exposed weaknesses and limitations in laboratory and experimental activities, mainly in the field of engineering. This forced us to provide rapid answers through a change in practice in our renewable energy program, from a conventional hands-on classroom experiment to a remote spectrometry laboratory. In this context, using costly spectrometry equipment that was not adapted to be operated remotely, was not an option.

In this paper we describe how we adapted our low-cost spectrometry technology (which is based on a 3D-printed mini-spectrometer and a smartphone) to deploy a remote laboratory as a rapid solution, due to the impossibility of using conventional and costly spectrometers, which work only for on-campus learning. This adaptation was helpful, not only to have several spectrometers available for a higher number of students, but also to allow teachers to prepare asynchronous activities that can be realized without their presence. We applied Internet of Things (IoT) technology for remotely controlling the experiments and used Machine Learning to automatically calibrate our low-cost smartphone spectrometer. We believe that such a low-cost spectrometry remote laboratory can benefit developing countries and enable the development of MOOC and MOOL type courses.

Keywords: Educational innovation · Internet of Things · Machine Learning · Distance education · Low-cost smartphone spectrometer

1 Introduction

Since their early development in the 90's, remote laboratories where mainly driven by the high cost of laboratory equipment, the need of trained personnel to operate them, and have become a viable solution to provide remote access to share expensive equipment [1]. There is a large body of remote laboratories in different areas (for academic, research and industrial applications) [2–4], which have evolved with the latest development of computer and communication technologies [5], and in some cases, have become commercial products (e.g.,

M. E. Auer et al. (Eds.): REV 2022, LNNS 524, pp. 198–209, 2023.
https://doi.org/10.1007/978-3-031-17091-1_21

the LabsLand[1] and Labwork[2] remote labs spin-off companies from University of Deusto and the Hong Kong Polytechnic University, respectively).

Recently, the global Covid-19 pandemic has precipitated the digital transformation in education worldwide and has exposed weaknesses and limitations in laboratory and experimental activities, mainly in the field of engineering. This situation forced us to provide rapid answers through a change in practice in our renewable energy program, from conventional hands-on classroom experiments to a remote spectrometry laboratory. In this context, using costly spectrometry equipment that was not adapted to be operated remotely, was not an option.

Even though there are some spectrometry remote laboratories [6–8], they have several drawbacks, thus limiting their applicability to our problem: Firstly, all the aforementioned remote spectrometry laboratories use costly spectrometers, either commercially available (e.g., Ocean Optics USB650 in [6] or the AvaSpec-2048 multi-channel spectrometer in [8]) or non-commercial 'in-house' specialized spectrometer (e.g., a Mössbauer spectrometer developed in [7])). Secondly, they rely on commercial or closed software, and therefore are not simple to integrate on a web platform for remote control of the experiments. Finally, they rely on a calibration procedure, which is typically manual or semi-automatic, that can be tedious and not adapted for a remote laboratory (e.g., to restart the equipment after each usage, to have a clean setup for the next user).

To cope with the aforementioned drawbacks, we leveraged and adapted our low-cost smartphone spectrometry technology [9] to rapidly and successfully deploy a low-cost spectrometry remote laboratory. The contributions of this paper for the remote laboratory community can be summarized as follows:

- The use of low-cost spectrometry technology, which uses a 3D-printed mini-spectrometer and a smartphone, thus avoiding the use of costly spectrometry equipment.
- The development of low-cost spectrometry remote lab kits that can be easily replicated, so that several spectroscopy experiments can be deployed and shared through a web application.
- The use of Internet of Things (IoT) technology allowing the user to remotely control the smartphone spectrometer and the spectrometry experiments.
- The use of a Machine Learning algorithm to automatically calibrate the spectrometer, so that it can be ready to a next experiment.

The rest of the paper is structured as follows: Sect. 2 describes the basic spectrometry technology (hardware and software), on which our remote lab is based on; Sect. 3 depicts the hardware and software architecture of our remote laboratory, whereas Sect. 4 describes the development of the remote lab. Finally, Sect. 5 concludes the paper.

2 Low-Cost Smartphone Spectrometer

We have developed a low-cost spectrometer technology that is based on a 3D-printed mini-spectrometer and a smartphone with an App to process the images

[1] https://labsland.com/en/about.

[2] https://www.labwork.com.hk/about-us.

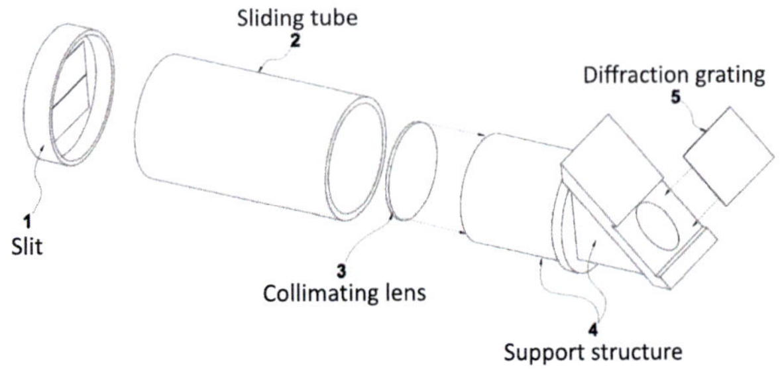

Fig. 1. Basic configuration of the transmission mini-spectrometer.

captured through the smartphone's camera [9]. Originally, we applied our smartphone spectrometer to the biomedical and industrial areas. However, the pandemic brought us the opportunity to apply it in engineering education through the adaptation to build our low-cost spectrometry remote lab.

2.1 3D-printed Mini-Spectrometer

Our low-cost spectrometry technology uses a 3D-printed mini-spectrometer which is attached to the camera of a smartphone. As can be seen in Fig. 1 the mini-spectrometer follows the basic configuration of a transmission spectrometer. It contains a 100 µm slit (1), a 4.5 cm sliding tube (2), a focusing collimating lens (3) which addresses the parallel beam to a 1000 lines/mm diffraction grating (4), all assembled on a support structure (5). The camera of the smartphone is positioned in front of the hole where the diffraction grating is placed, whereas the light signal crosses straight from the slit to the camera's sensor.

For processing the spectral images obtained by the mini-spectrometer, we developed a multi-platform smartphone App (Android and iOS) which can process, visualize, store and transmit spectral data via an Internet connection. The resulting mini-spectrometer was 3D printed with a precision of 0.1 mm and the use of a PLA filament (with high density black plastic to avoid ambient light to pass through). As can be seen in Fig. 2 the device includes a support for optical components and a clip that couples the spectrometer to any smartphone independently of the camera position, thus easing the mounting procedure.

Concerning the spectral resolution, our low-cost smartphone spectrometer is able to resolve spectral lines with values less than 2 nm, which is comparable to several costly commercial spectrometers[3]. Furthermore, the spectral resolution of our smartphone spectrometer can be improved by using a slit with a smaller width, a diffraction grating with more than 1000 lines/mm and a smartphone with a higher resolution camera. The approximate cost of the 3D-printed spectrometer is of ≈30 USD (i.e., 2 orders of magnitude less than a commercial spectrometer with similar characteristics).

[3] See several compact spectrometers at https://www.oceaninsight.com/products/spectrometers/microspectrometer/.

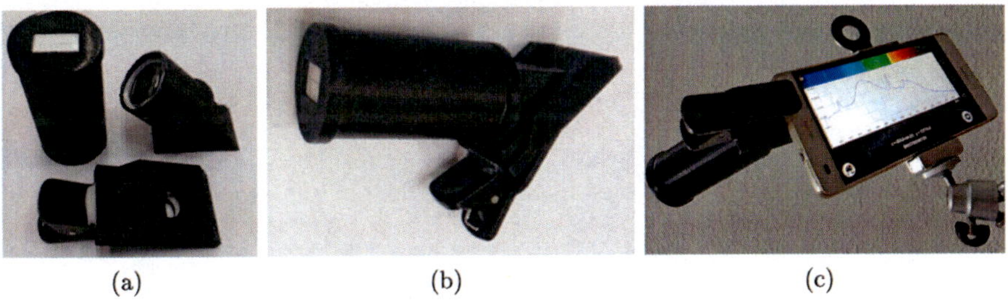

Fig. 2. (a) 3D printed components of the mini-spectrometer; (b) the assembled mini-spectrometer; and (c) the mini-spectrometer attached to the smartphone.

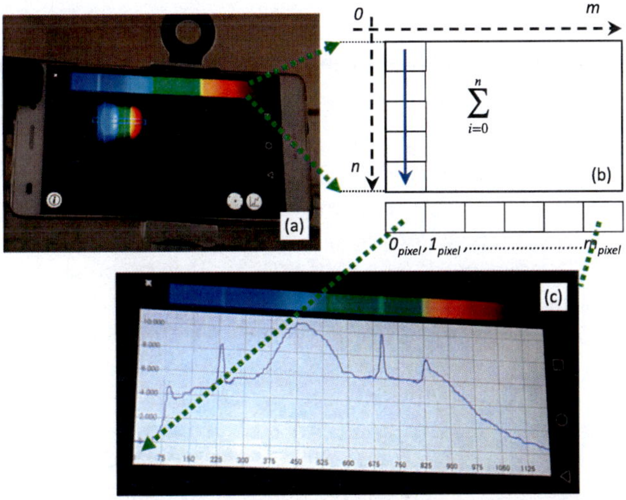

Fig. 3. The process to obtain the spectral curves with: (a) the spectrum image with a user-selected rectangle shown on top; (b) the vertical sum of the selected rectangle (as a per-pixel array) and; (c) the spectral curve obtained from the array (in arbitrary units).

2.2 Smartphone Spectrometry App

In general, to obtain the spectrum, spectrometers require processing the data generated when light reaches the sensor. With conventional spectrometers, this is done through special hardware and software that is often closed and proprietary and requires a license. In our approach use the smartphone's camera as the sensor to obtain digital images that are processed by our smartphone spectrometry App available for free[4].

[4] https://upb-edu.gitlab.io/spectraupb/.

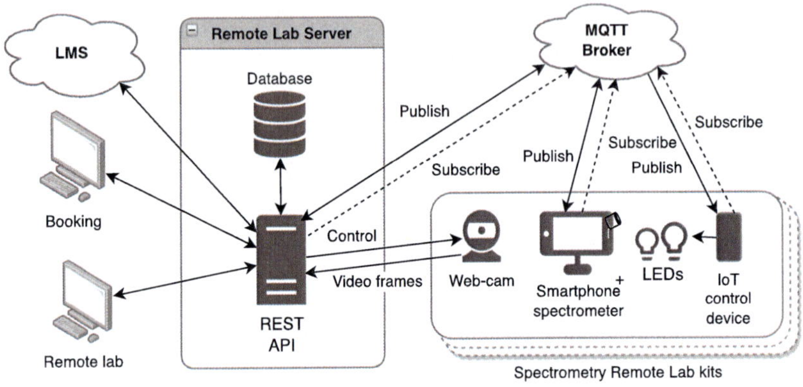

Fig. 4. Architecture of the low-cost spectrometry remote laboratory.

In short, our spectrometry App works as follows: It receives the image from the mini-spectrometer (through the smartphone's camera) which is then processed as a matrix of pixels (i.e., the Red, Green and Blue - RGB components). The user can select a rectangular region of the area of interest in the image (see Fig. 3(a)) where the spectrum is clearly visible. This rectangular area defines the new coordinate's origin of the corresponding pixel matrix of size mxn. The selected region is then converted to gray-scale and the values for each column of the matrix are added vertically to obtain an array of size m with values in arbitrary units (see Fig. 3(b)) corresponding to each pixel index in the X axis. The array is used to generate a line chart representing the spectral curve (see Fig. 3(c)), which is however not yet calibrated as a function of the wavelength. The Y axis of the spectral curve corresponds to the intensity of the spectral line in arbitrary units. Finally, the spectrometer requires to be calibrated as a function of the wavelength, using known spectral lines as references, e.g., Mercury or Neon spectral lines from conventional fluorescent lamps, for manual or automatic calibration as will be described in Sect. 4.

3 Architecture of the Remote Lab

The architecture of the remote lab consists of three main parts: The client-side web app, the server-side remote lab web app, and the spectrometry remote lab kits as shown in Fig. 4.

The client-side web app provides a *booking* interface to reserve a time slot to use one of the available spectrometry remote lab kits, and a *remote lab* interface to remotely control the kits and perform the spectrometry experiments. Both client-side web apps, interact with the server-side remote lab web app through a REST API (which handles authentication, control and communication). The server has a local database, where the users, booking information, data of the experiments are stored. The server has also interfaces to communicate with external Learning Management Systems (LMS), such as Moodle, through the Learning Tools Interoperability (LTI)[5]. The spectrometry remote lab kits are composed by the mini-spectrometer attached to a smartphone, a set of LED lights, a webcam to visualize the whole experimental setup, and an electronic control system based on a wirelessly connected IoT device.

The communication infrastructure is based on MQTT[6] which is a popular and standard IoT communication framework. The communication between the server and the smartphone is bidirectional, i.e., the smartphone sends spectral information to the server for visualization, whereas the user can control the spectrometer directly from the web client, which triggers control messages that are sent to the smartphone through a MQTT Broker. For this, both the server and the smartphone spectrometer subscribes to MQTT topics (to receive messages) and publish to specific MQTT topics (to send messages). The main advantage of using MQTT is because the communication is fully asynchronous and indirect, so that we can add as many spectrometry remote lab kits as needed (no pre-registration required).

To control the LED lights that are used for the experiments, we include an electronic IoT control device (see Fig. 5). The IoT device is connected to the internet through Wifi and to the LEDs through a custom electronic circuit (via analog connectors) to send or receive control signals and data. The IoT device is subscribed to the MQTT 'control' topic to receive commands asynchronously, to turn on/off or dimmer the LEDs.

To have a global overview of the experiment, a webcam is connected directly to the server which re-distributes the video stream to all the clients that are either controlling the remote lab or observing the experiment. The webcam is also controlled through its own API to move, focus or zoom the video stream.

4 Development of the Remote Lab

To implement the spectrometry remote laboratory, we had to adapt our low-cost spectrometry technology and solve two main issues.

- We needed to remotely control both the spectrometer and the devices used for the experiment. For this, we had to extend our smartphone spectrometry software and had to develop a web application that allows students to remotely control the parameters of the spectrometer, observe the spectra in real-time, measure the wavelengths of the spectra, store the data generated in the web platform, and carry out experiments and practical exercises. Our solution uses IoT devices to remotely control the parameters of intensity, and turn-on, turn-off or dimmer the LEDs that are used as light sources with different spectral characteristics.

[5] http://www.imsglobal.org/activity/learning-tools-interoperability.

[6] MQTT (Message Queue Telemetry Transport) is a lightweight, publish-subscribe network protocol that transports messages between devices.

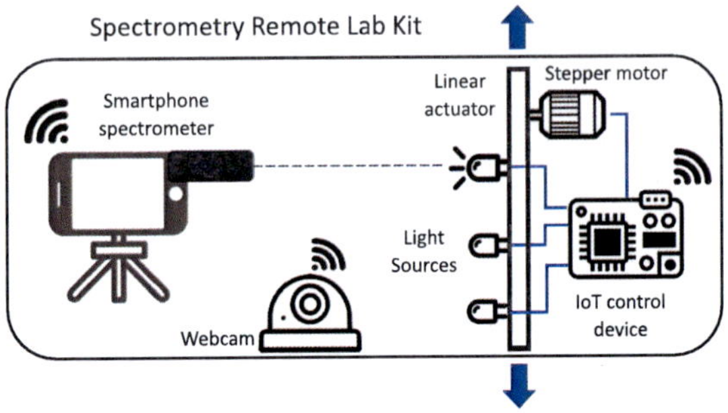

Fig. 5. Overview of the spectrometry remote lab kit.

- The baseline smartphone spectrometer required manual wavelength calibration, i.e., each time that the device was restarted, the user had to visually identify and select reference spectral lines (e.g., using a light source with Mercury). This was highly impractical for a remote laboratory that should operate 24/7. Therefore, we had to improve our smartphone spectrometry software for automatic wavelength calibration using a Machine Learning algorithm, which is able to identify the reference lines, and can be ready without user intervention.

In the following we describe the development of our low-cost spectrometry remote lab through: the IoT control system, the automatic calibration using Machine Learning techniques, and the modifications that were required on the spectrometry App.

4.1 IoT-Based Control System

To simplify the deployment of the remote lab, we developed an IoT control device, which is based on a low-cost embedded ESP32[7] micro-controller. The ESP32 was programmed using MicroPython[8], which is a small subset of the Python standard library, optimized for micro-controllers and constrained environments. The use of the ESP32 micro-controller simplifies implementation of the control system, since it includes a full TCP/IP stack for Wifi communication.

The ESP32 micro-controller also includes analog and digital general-purpose input/output (GPIO) pins that can be programmed to read and write data signals, and control actuators. For our remote labs, a simple electronic circuit was deployed to connect the ESP32 to the LED lights and a stepper motor. The ESP32 was also programmed to asynchronously receive control commands from the server, through a subscription to the corresponding MQTT 'control' topic.

To correctly position and point the spectrometer to the different light sources (i.e., LEDs for the experiment and other light sources for the calibration), the

[7] https://www.espressif.com/en/products/socs/esp32.
[8] https://micropython.org/.

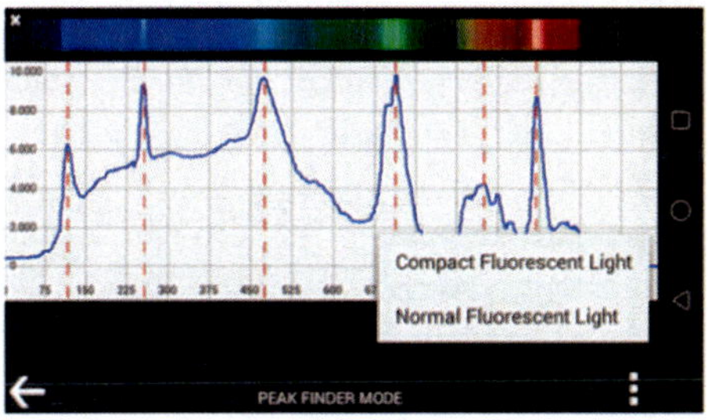

Fig. 6. Peak detection on the spectral curves for manual wavelength calibration.

IoT control device also drives a stepper motor that moves a belt driven linear actuator, on which the light sources are mounted (see Fig. 5).

The implementation of the ESP32-based control system was straightforward thanks to the use of IoT technology ($\approx$250 lines of MicroPython code).

4.2 Automatic Calibration Using Machine Learning

The calibration of the spectrometer as a function of the wavelength requires the use of known spectral lines as references. In our baseline approach (before the remote lab), we used conventional fluorescent lamps containing Mercury with well-known spectrum lines. The calibration required two principal reference lines in the visible region. We chose a peak in the blue and in the green part of the Mercury spectrum[9], at 435.83 nm and at 546.07 nm, respectively. To select the wavelength references, we applied a peak detection algorithm to the spectral curves (see Fig. 6). The user could then visualize and select in the touch screen the peaks corresponding to the related references. Once selected, the App extrapolated the values of the wavelengths for each pixel, to have the accurate wavelength in the X axis. The calibrated values were displayed and the user could select any place in the spectrum to obtain the corresponding wavelength values in real-time.

For a remote lab setting, the manual calibration was highly impractical and tedious. We therefore improved the calibration using a Machine Learning algorithm to identify the shape and peaks of the reference spectrum lines.

Overview of the automatic calibration: First, we changed the reference source light from a conventional fluorescent Mercury lamp, to a tiny Neon bulb, due to space constraints of spectrometry remote lab kit. When the smartphone spectrometer is restarted or the user pushes a 'Calibrate' button on the web app, the IoT control device receives an MQTT message to move the stepper motor to the initial position of the Neon bulb, which is also turned on. The spectrum image is then automatically detected, the rectangle is also automatically selected and

[9] http://umop.net/spectra/spectrum.php?elem=Hg.

the auto-focus of the camera is enabled. If no satisfactory spectrum is detected, the stepper motor slightly moves the linear actuator to scan the surroundings area and find the spectrum. Then, the App's mode is changed to show the spectra lines, and the Machine Learning algorithm is executed to automatically identify the correct wavelength references. Once the mini-spectrometer is calibrated, it sends a message to the IoT control device, to turn off the Neon bulb, and to move the linear actuator to the position where the LEDs are located. The spectrometry remote lab kit is then ready to be used.

The Machine Learning algorithm: For the automatic calibration, we applied supervised Machine Learning, which requires labelled dataset to train the algorithm so as to classify data or predict outcomes accurately. In our case, the training dataset corresponds to the images of the spectrum of the reference element (i.e., Neon) captured with our mini-spectrometer and labelled with their intensity peaks lines[10] (i.e., at 585.25 nm and at 640.22 nm, in the 'yellow' and 'red' part of the visible spectrum, respectively). To avoid manually collecting the training set, we applied *data augmentation* [10] (size change, left or right shift, noise addition, and brightness variations) to a smaller set of 30 spectra images, obtaining 300 different images to pre-process and label the training set.

The pre-processing consists of detecting peaks from each intensity plot of the spectral curves, through a program written in Python. Then, we labelled that data, marking where our 'yellow' and 'red' reference peaks are and their respective RGB pixel color values. The dataset *features* for each peak detected during pre-processing includes the position, intensity and color (RGB values). The target variables are the 'yellow' and 'red' Neon references, encoded as ordinal variables (with 0 as the left-most peak and 14 as the last peak). Thus, our model aims to predict which detected peaks correspond to the known references. We used k-fold cross-validation [11] with $k = 5$ to choose the best suited Machine Learning algorithm, i.e., train the model with 4/5 of available data and measure accuracy in the never seen 1/5, repeat the process 5 times and calculate the mean accuracy. Using this approach we selected the Support Vector Machine (SVM) algorithm [11] with a linear kernel that achieved the highest accuracy. With the reference peaks detected, we can proceed to calibration. Figure 7 depicts the overall preparation process for the Machine Learning-based automatic calibration. Once the model was trained, we integrated a SVM classifier library in our spectrometry smartphone App to execute the automatic calibration algorithm in real-time.

4.3 Adaptation of the App for the Remote Lab

To better cope with the requirements for the remote laboratory, we adapted our spectrometry App as follow:

- We integrated an MQTT client directly in the App, to send and receive control and data messages to the web app through the server.

[10] http://umop.net/spectra/spectrum.php?elem=Ne.

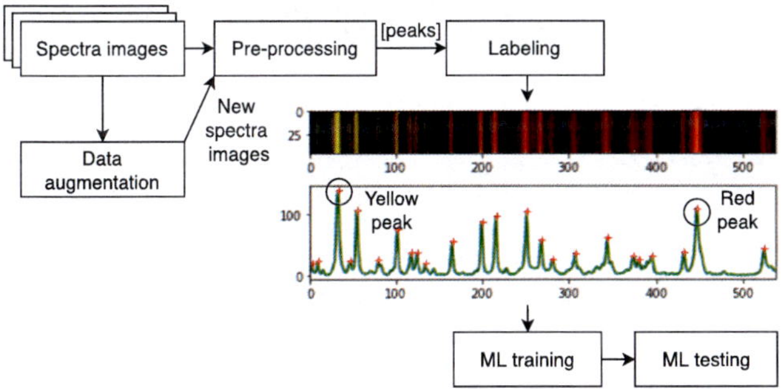

Fig. 7. Preparation steps for the machine learning-based automatic calibration.

- Instead of streaming the screen of the smartphone App (which may consume too much bandwidth), we send an array with the spectrum and the wavelength values directly through MQTT in real-time.
- To control the spectrometer, we added several buttons in the client-side web app that trigger remote actions on the smartphone spectrometer through MQTT messages:
  - The 'Save' button stores a snapshot of the data on the server,
  - The 'Calibrate' button sends a message to both the smartphone spectrometer and the IoT control device to run the automatic wavelength calibration, and
  - The 'Exposure' slider triggers a modification of the exposure time on the camera of the smartphone.
- We included the selection of the wavelength in the spectral curve, to obtain the intensity value directly on the client-side web app. All the stored spectral data is listed and available to the user, after the end of the remote lab session.

The implementation of all the components of the remote lab was done using MicroPython for the IoT control device, Angular[11] for the client-side web apps (booking and remote lab interfaces), the REST API with Django[12], the server with Python, the database is MySQL[13], and the spectrometry smartphone software is multi-platform (Android and iOS) implemented in Dart with the Flutter[14] UI development framework. Note that even though Flutter provides a single code base for Android and iOS, we had to implement some native libraries (in C++) for fast and efficient processing of the images and part of the Machine Learning algorithm.

The final spectrometry remote lab web interface is shown in Fig. 8 where the user can perform the experiments, control the spectrometer, visualize the spectral data, store the experimental data and observe and control the spectrometry remote lab kit through its webcam.

---

[11] https://angular.io/.
[12] https://www.django-rest-framework.org/.
[13] https://www.mysql.com/.
[14] https://flutter.dev/.

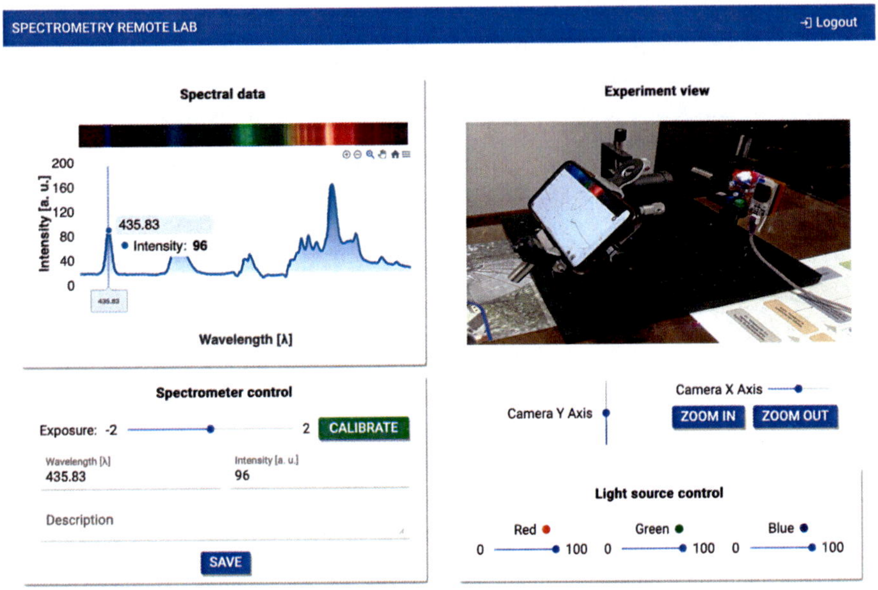

Fig. 8. The Spectrometry remote lab client-side web interface.

We applied our low-cost spectrometry remote laboratory to the area of solar energy (renewable energy program), allowing students to remotely operate the spectrometer and analyze spectral phenomena individually (e.g., black body and colour temperature spectra from different sources). This is because, in many educational programs, solar spectral phenomena are usually presented rather superficially; there are no specific labs for students to learn how to read spectra, understand solar spectra, colour temperature, and use spectrometers.

5 Conclusion

This work shows the potential use of low-cost 3D-printed spectrometry technology to rapidly deploy remote laboratories. Thanks to the communication capabilities of IoT devices and the use of Machine Learning for automatic calibration, we were able to integrate our spectrometry software with a web application for remote operation, control, real-time visualization and spectral data collection from our low-cost smartphone spectrometer. The resulting remote laboratory allows students to perform asynchronous and autonomous spectrometry activities, and improve their conceptual learning gains. We believe that such a low-cost spectrometry remote laboratory can benefit the academic and research communities, in particular those in developing countries and enable the development of MOOC and MOOL type courses.

Acknowledgement. This work was partially funded by the ERASMUS+ Project "EUBBC-Digital" (No. 618925-EPP-1-2020-1-BR-EPPKA2-CBHE-JP).

References

1. Gravier C, Fayolle J, Bayard B, Ates M, Lardon J (2008) State of the art about remote laboratories paradigms - foundations of ongoing mutations. Int J Online Biomed Eng 4(1):1
2. Gustavsson I et al (2016) Lab sessions in VISIR laboratories. In: 2016 13th International conference on remote engineering and virtual instrumentation (REV), February 2016, pp 350–352
3. Santana I, Ferre M, Izaguirre E, Aracil R, Hernandez L (2013) Remote laboratories for education and research purposes in automatic control systems. IEEE Trans Indust Inf 9(1):547–556. https://doi.org/10.1109/TII.2011.2182518
4. Vagaš M, Sukop M, Varga J (2017) Design and implementation of remote lab with industrial robot accessible through the web. Appl Mech Mater 859:67–73
5. Garcia-Zubia J, Orduna P, Lopez-de Ipina D, Alves GR (2009) Addressing software impact in the design of remote laboratories. IEEE Trans Indust Electron 56(12):4757–4767
6. Thoms LJ, Girwidz R (2016) Training and assessment of experimental competencies from a distance: optical spectrometry via the Internet. Il Nuovo Cimento C 38(3):1–10. https://doi.org/10.1393/ncc/i2015-15113-3
7. Silaev Jr AA, Godovikov SK, Postnikov EB, Radchenko VV, Silaev Sr AA (2013) Remote access mössbauer spectrometry. Bull Russ Acad Sci Phys 77(6):790–794. https://doi.org/10.3103/S1062873813060324
8. Zimin A, Shumov A, Troynov V, Zemtsov I (2020) Online plasma diagnostics in the remote spectroscopy laboratory for practical training and experimental research. Information 11(1):9
9. Ormachea O, Villazón A, Escalera R (2017) A spectrometer based on smartphones and a low-cost kit for transmittance and absorbance measurements in real-time. Opt Pura Apl 50(3):239–249. https://doi.org/10.7149/OPA.50.3.49053
10. Andriyanov NA, Andriyanov DA (2020) The using of data augmentation in machine learning in image processing tasks in the face of data scarcity. J Phys Conf Ser 1661:012018. https://doi.org/10.1088/1742-6596/1661/1/012018
11. Anguita D, Ghio A, Ridella S, Sterpi D (2009) K-fold cross validation for error rate estimate in support vector machines. In: Proceedings of the 2009 International Conference on Data Mining, DMIN 2009, 13–16 July 2009, Las Vegas. CSREA Press, pp 291–297

Efficient Collaborative and Cooperative Laboratory Experimentation System for Online Engineering, Science and Technology Education

Ahmed M. Abd El-Haleem[1,2](✉), Noha A. Elmosilhy[3], Abdullah I. Salama[1], Serag El Din Eman[4], and Mahmoud M. Elmesalawy[1]

[1] Helwan University, Cairo, Egypt
{asalama,melmesalawy}@h-eng.helwan.edu.eg
[2] British University in Egypt (BUE), Cairo, Egypt
ahmed.abdelhaleem@bue.edu.eg
[3] Canadian International College (CIC), Cairo, Egypt
noha_elmosilhy@cic-cairo.com
[4] National Telecommunication Institute (NTI), Cairo, Egypt
eman.serag@nti.sci.eg

Abstract. The presence of the COVID-19 pandemic forced the educational systems all over the world to shift their activities to be hold remotely using online learning systems. Creating an efficient remote learning system that facilitate the transition to e-learning and distance education has become a must, especially in practical sectors such as Engineering, Science and Technology that require laboratory-demanded courses. Focusing only on the individual-based experiment where a single user can access and conduct the experiment, dismissing the structure of group-based laboratory experiment, can't reflect comprehension construction as in the real on-site laboratories. In this paper, a group-based online learning system is proposed to provide a collaborative and cooperative virtual learning environment for laboratory experimentation taking into consideration different aspects that may impact the interactions between students. We divided the whole group-based laboratory experimentation platform process into four main parts: experiment creation using integrated authoring tool, experiment configuration and scheduling, monitored run-time process, and pre/post session configuration. We also proposed a runtime experiment student's web-based graphical user interface that represents developed features that successfully achieve flexible, scalable and reusable system with the aim of maintaining satisfactory and effective user experience.

Keywords: Group-based · Online laboratory experiment · Collaboration · Cooperation · Laboratory learning system

1 Introduction

The presence of COVID-19 pandemic, forced all the educational systems all over the world to shift their activities to be hold remotely using online learning

M. E. Auer et al. (Eds.): REV 2022, LNNS 524, pp. 210–221, 2023.
https://doi.org/10.1007/978-3-031-17091-1_22

systems. Focusing on how to create an efficient remote learning system to fulfil the different educational system requirements and ease the transition to e-learning and distance education is still a challenge, especially in practical sectors such as Engineering, Science and Technology that require laboratory-demanded courses.

Many researches investigate how to hold an online laboratory experiment as in [1], in which the authors review the student's feedback and the gain from implementing remote lab using the Virtual Instrument Systems in Reality (VISIR) which allows the student to work in a real electronic lab environment. In [2], project called Forging Online Education through FIRE (FORGE) is introduced. This project is concerned with complementing the online courses with effective online laboratory experiment, while using the facilities of Future Internet Research and Experimentation (FIRE) in an effective way. The FORGE architecture shows smooth integration with the existing learning Management systems (LMSs). In [3], the authors investigate the development of remote laboratory architecture focused on the embedded systems based on Redis. In [4], the authors design Smart Laboratory Learning Objects (SLLOs) which help the developers to manage the same online laboratory that connects many user interfaces with different types. The online laboratory management can include lab stations, lab activities and connection information. In [5], an online virtual lab tool is developed called WebLab-Deusto, which allows students to access real lab equipment remotely.

All of these previous researches considered only the individual based experiment in which single user can access and conduct the experiment, dismissing the structure of group-based laboratory experiment. Many practical laboratory experiments in the Engineering, Science and Technology sectors, require forming a group consisting of a number of students who can collaborate, simultaneously interact and sharing knowledge while performing the laboratory experiment [6]. Providing a collaborative laboratory experimentation environment which allow group of students to synchronously collaborate, perform group discussion, and support fairness contributions from all group members to reflect the intended learning outcomes of the student's collaboration as in physical laboratories is still a challenge.

In [7], the authors redesign the DIESEL architecture in order to support synchronous remote collaborative environment to enable multiple students from different areas to collaborate in performing a laboratory experiment. The proposed architecture also supports a scheduling system in which a user can reserve a session and invite his colleagues to start working on a remote experiment. However, the proposed architecture requires installing an application on each student's workstation and instructor in order to be able to perform remote lab experiment.

In [8], the authors focus on providing a collaborative environment using MicroNet software that integrated with LMS, the MicroNet component is the Collaboration Server which responsible for the audio/video stream management in the CRL session. Although the authors aim to provide a collaborative integrated platform, the provided design is limited for sharing video and audio, dismissing features and resources that allow students to engage simultaneously in collaborative activities. In [9], the authors aim to offer a collaborative virtual learning environment by using three students' productive interactions, as elaborated explanations, argumentation and mutual regulation of cognitive processes. Mutual regulation of cognitive processes, defines how students in each group can interact in shared modes of cognitive regulations. For example, a number of students forming a group to perform a laboratory experiment in which the input of the experiment is distributed among these students and the students negotiate the values and share the activities to coordinate their experimental design. Although the authors in [9] investigate properly different collaborations aspects and their effects on learning objects, however, they do not consider a how students can access and interact simultaneously while performing the same laboratory experiment and how students can share their outputs.

In [10], the authors investigate three learning paradigms that emulate a traditional physical laboratory, however, this investigation was limited to capture collaboration between students as only invitation in LMS, where the director user can invite multiple users and they can share director activity. The authors considered only the mutual engagement (synchronous collaboration), which does not reflect the real group-based experiment structure that is divided into two types, mutual engagement and cooperative work. In cooperative work, the laboratory experiment is divided into many parts, in which each student responsible for a part and then their share their outputs for final experiment results. In [11], the authors proposed the requirements and architecture design of efficient LLS which can be used by Engineering, Science and Technology sectors that have laboratory experimentation activities.

Based on above literature and extending what proposed in [11], the main contribution of this paper can be summarized as follows:

- Design an efficient group-based laboratory experimentation system that supports collaborative and cooperative online laboratory learning to reflect the intended learning outcomes of the student's active learning and comprehension construction as in the real on-site laboratories.
- Present two group-based interaction models, sequential and concurrent, considering in each interaction model, session time-slot reservation, group management, way of communication and sharing aspects.
- Design group-based experiment policies assurance system to reflect mutual regulation of cognitive processes to guarantee each student's interactions that guide interpretation and construction of concepts.
- Provide network cloud storage that ease the collaborative and cooperative interaction process between group members.
- Provide an integration between the proposed collaborative and cooperative group-based online laboratory experimentation platform and domain network directory management service.
- Provide general and reusable virtual laboratory resources (VLR) management for laboratory software and hardware resources.
- Design runtime experiment student's web-based graphical user interface that represents developed system features

The paper is organized as follows: Sect. 2 presents detailed description of the high-level architecture of the proposed system, Sect. 3 describes the whole system layered architecture, followed by the group-based interaction types designs and student's web-based graphical user interface in Sect. 4, while Sect. 5 states the conclusions.

2 High-Level Architecture of the Proposed Collaborative and Cooperative Laboratory Learning Management System

Our proposed group-based laboratory collaborative and cooperative experimentation platform is detailed designed considering all required components with defined interfaces for successfully achieving flexible, scalable and reusable system with the aim of maintaining satisfactory and effective user experience. The system high-level architecture, shown in Fig. 1, consists of the group-based experiment service which is responsible for experiment configuration, experiment parts and students' distribution, Network cloud storage management, as well as experiment sharing and control run time engine. The architecture also includes the domain network directory interface and remote access host management for the lab machines pool and instruments, while considering authoring tool, course experiment manager and notification manager which responsible for notifying students with their experiments related aspects.

Authoring Tool: Is a customized software tool embedded in LLS that allows the teacher/instructor to create a new experiment in the form of laboratory learning object (LLO) that fit the course learning objectives and outcomes. The authoring tool system main function is content organization and delivery, and also provides different types of assessments.

Experiment Database: A database that contains all laboratory experiments information and laboratory learning objects created by a teacher/instructor using the authoring tool.

Course Experiment Database: A database that contains all courses information and their copies of added laboratory experiments LLOs to the course, which is created by a teacher/instructor.

Course Experiment Manager: Is responsible for course experiments management, in which the teacher can add a pre-created experiment to the course and manage its schedule for the students.

Students Profile Shared Storage: Is the high-capacity network-attached storage (NAS) that allows saving and retrieval of data from a centralized location for authorized users, as the students shall remotely access a different lab machine during each experiment, hence users' data should be available by means of roaming user profiles.

Notification Manager: Which is responsible for notifying students with their new scheduled experiments, along with their left steps that may include choosing their teammates and assigning each group member his part in the experiment.

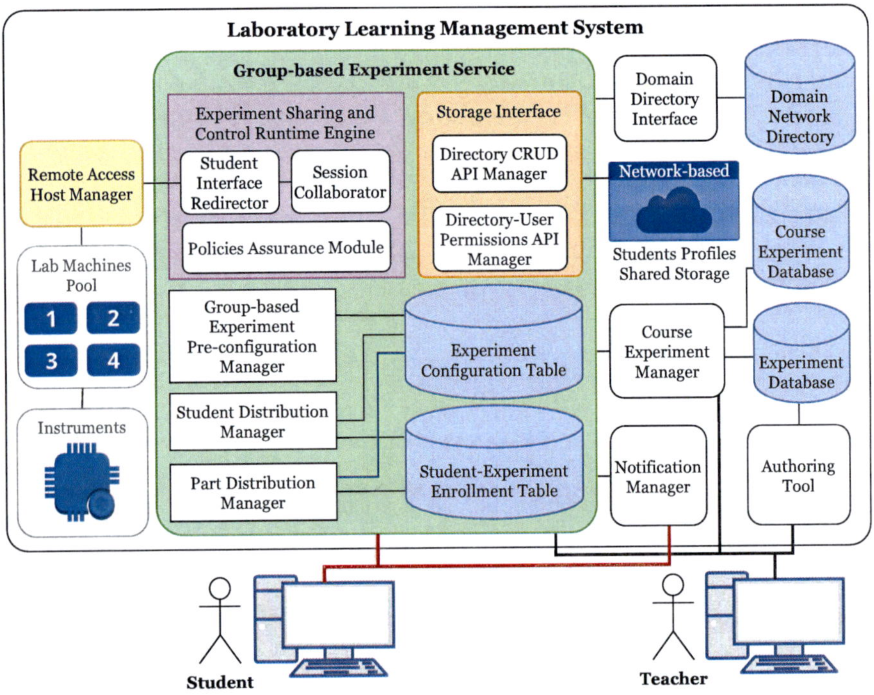

Fig. 1. High-level architecture of the proposed collaborative and cooperative laboratory learning management system

Remote Access Host Manager: Is the application that hosts the lab machine remote access connection and provides two types of connection, one with granted control, and the other without, for only screen sharing ability.

Domain Directory Interface: Is the application programming interface (API) for LLS to interface with the domain network directory, in order to obtain users' credentials and discover all lab machines available to be used by students to complete their experiments remotely.

Group-Based Experiment Service: Consists of the following five modules to achieve the required proposed group-based collaboration and cooperation system:

1. **Group-Based Experiment Pre-configuration Manager**: Manages the configuration data which is needed before starting and during the experiment

into experiment configuration table, such as time-slot reservation, students' group construction, set experiment-running policies and the way of interaction with the experiment, which is classified into sequential interaction and parallel interaction.

2. **Student Distribution Manager**: Manages students group construction as number of students in each group and data of each student enrolled in experiment, and is responsible for distribution of students to the groups.
3. **Part Distribution Manager**: Manages the distribution of all experiment parts on each student in the same group. A part can be a set of steps assigned to each student, with defining type of the set either consecutive or distributed steps.
4. **Experiment Sharing and Control Module (Runtime Engine)**: Is the runtime engine during the experiment and includes:
   - **Student Interface Redirector**: Responsible for switching student's interface to instantly access and control lab machine to perform assigned part within experiment, or to only see the screen-share of the lab machine. This switching mode used only in group-based laboratory experiment that requires students' sequential interaction. The difference between the sequential interaction and parallel interaction is described in details in the next section.
   - **Session Collaborator**: Starts a session sharing between students within experiment. Session sharing guarantee collaboration and cooperation between student's group, where students can simultaneously interact and share knowledge while performing the laboratory experiment.
   - **Policies Assurance Module**: Is responsible for students' progress control measures, where a set of policies is configured, as choose the maximum and minimum number of students involved in each group, set a minimum contribution progress percentage that assure each student even participation within experiment.
5. **Storage Interface**: Consists of two API-related modules responsible for the students' profile shared storage folders management:
   - **Directory CRUD API Manager**: Responsible for the create, read, update and delete (CRUD) of directories that will be accessible by students accessing the lab machine remotely.
   - **Directory-User Permissions API Manager**: Is responsible for changing access permission for the created directories. A student can have no/limited granted access for a directory. Directory granted access can be read-only or read/write permission, which can change before starting, during, and after finishing the experiment.

3 Layered Architecture of the Proposed Collaborative and Cooperative Laboratory Learning Management System

The system implements a five-tier business model: the system users' layer, the experiment sharing and control runtime engine layer, the group-based experiment service layer, the laboratory learning management system (LLMS) layer, and the infrastructure layer, as shown in Fig. 2.

The system users layer represents the student and the teacher roles. The student receives audio, video, and text from his group colleagues, and is able to remotely access the lab machine from the infrastructure layer via the run-time engine layer.

The experiment sharing and control run-time engine layer consists of the applications responsible for managing and maintaining the running concurrent sessions and control the remote lab machines which are accessed through a web browser. The run-time engine provides remote access host manager which hosts the lab machine remote access and have two types of shares. The lab machine granted control share type is sent to the master student who is in control, and the lab machine screen-only share type is sent to the students not in control. These two share types are made possible via the student interface redirector, which is responsible for the machine screen area shown in Fig. 5. A recommended implementation for the remote access host manager is Myrtille [12] built using .NET framework via C# language, where the average session uses about 200 KB/s of bandwidth.

This layer is also responsible for the audio, video and text chat between the group student members, which is done by means of the session collaborator. A good implementation that supports multi-participants, video stream, audio stream, and text chat is Conference Call [13] built using PHP Web socket (Ratchet). Another implementation for it is built using socket.io and NodeJS, but is ideal for not more than 4 students in one group.

The running experiment have certain policies that ensures evenly contribution between the same group members within the experiment. This contribution can be calculated by means of experiment parts finished or time. Policies assurance module is responsible for that via direct control on the student interface redirector if needed.

The group-based experiment service layer consists of the applications responsible for configuring experiments, students' distribution and student-to-part distribution managers. The process of assigning students to the same group is the responsibility of the student distribution manager, whether this distribution is manual by means of the teacher choice or by means of the students' desires, or automatic by means of specific criteria determined by the teacher. Each experiment is fragmented into multiple parts, these parts can be assigned to specific students via the part distribution manager. The student distribution and part distribution managers results are saved in student-experiment enrollment database table. The experiment is configured via the group-based experiment pre-configuration manager, and saved in the experiment configuration database table.

The LLMS layer consists of the core modules, managers and interfaces with external systems. The teacher is able to create new experiments via the customized authoring tool, and the LLO is saved in the experiment database. Adding a created experiment to course, teacher is responsible of, is possible via

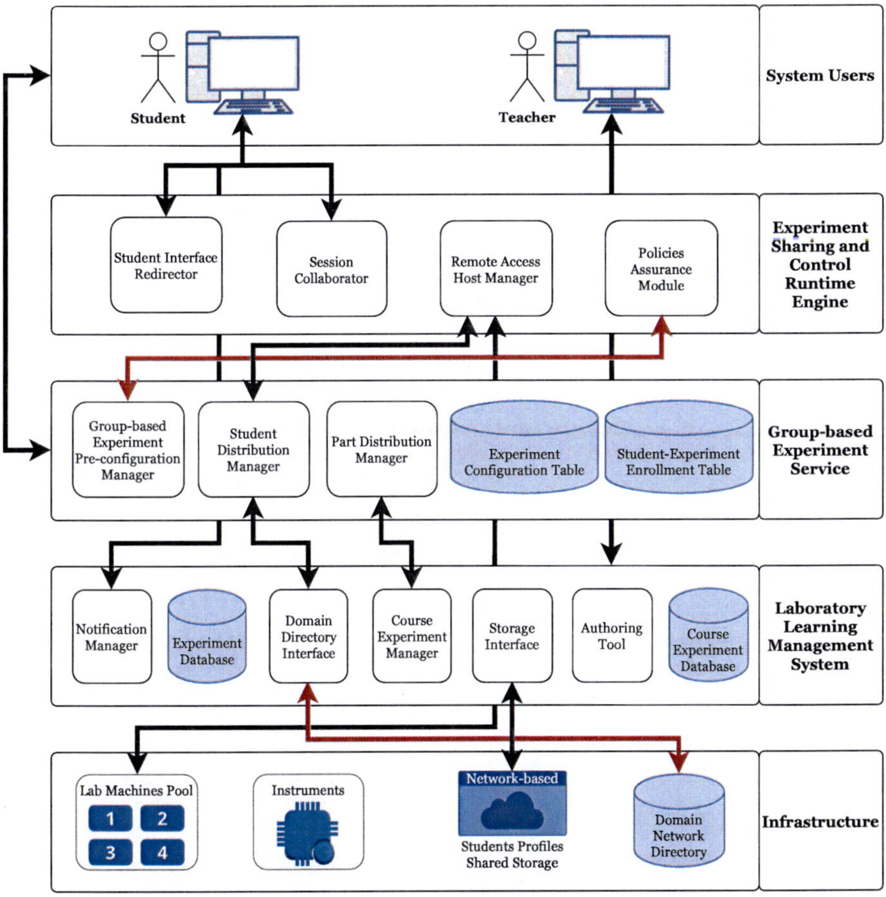

Fig. 2. Layered architecture of the proposed Collaborative and Cooperative laboratory learning management system

the course experiment manager, where a copy of the experiment LLO is added to the course and saved in the course experiment database. New added experiments to courses are notified to the students via the notification manager. The system is able to access the domain network directory via the domain directory interface, and access the network-based students' profiles shared storage via the storage interface.

The infrastructure layer consists of all of the hardware resources such as physical and virtual lab machines, instruments such as experimental boards and devices, the students' profiles shared storage, and the domain network directory. A web service is used as a gateway between all layers to allow the system users modules and system modules and services to access the database. This approach was preferred as it allows the separation of the system users and system modules and services from the data storage process.

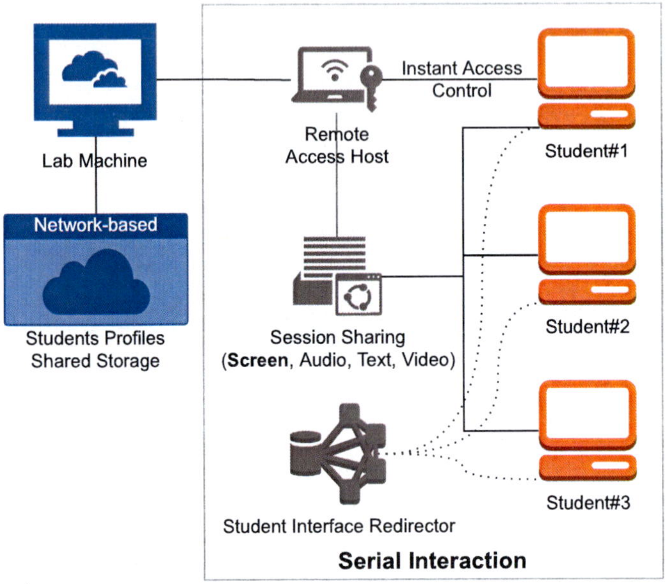

Fig. 3. Group-based laboratory sequential interaction

4 Group-Based Laboratory Experiment Interaction Types and Student User Interface

Group-based laboratory experiment defines two ways of student's interaction types based on the experiment requirements, the sequential interaction between the same group students, and the concurrent interaction, as well as the student user interface while performing the experiment, as defined in the following subsections. For each interaction type, each group should reserve a time-slot to perform experiment to guarantee available software/hardware resources within the selected time-slot.

4.1 Sequential Interaction

In sequential interaction, known as cooperative interaction, as shown in Fig. 3, only one student (master) can instantly remotely access and control lab machine to perform his assigned part using remote access host module, While other students in group can co-operate and share the master student's activities using session sharing module. Session sharing module allow screen, audio, text and video sharing. Students in turns can switch interface and control the lab machine to finish assigned parts using students interface switch module. When a student finish his assigned part, the results will be saved in lab machine under student profile in a group shared folder.

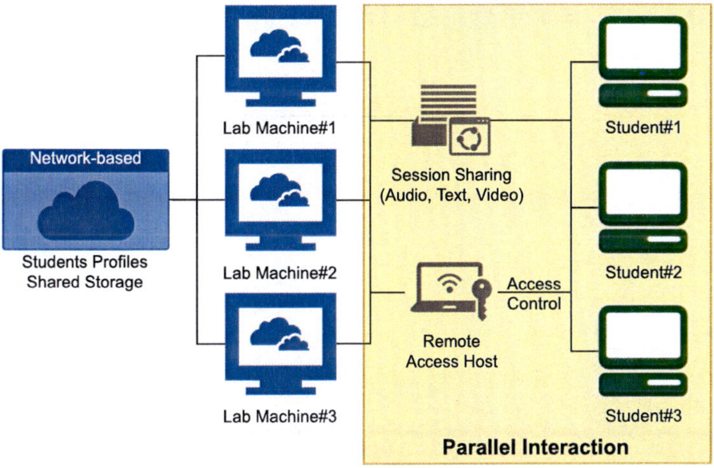

Fig. 4. Group-based laboratory concurrent interaction

4.2 Concurrent Interaction

In the concurrent interaction, known as collaborative interaction, as shown in Fig. 4, group of students can mutually engage in a coordination way to perform experiment objectives efficiently. In concurrent interaction, each student can access his own lab machine using remote access host module, and use session sharing module to collaborate with his colleagues within the group by means of audio, video, and text. Each student can share his activity using the shared storage module.

4.3 Student Graphical User Interface

Figure 5 shows run-time experiment student's web-based graphical user interface design. The design is divided into six parts. Left-hand side shows the experiment content and assessments, which added by the instructor through the authoring tool. The experiment content may consist of text, images, tables, etc. to reflect the experiment objectives, procedures, expected results, etc., while the assessments include number of different types of questions, such as multiple-choice, true/false, matching, etc., to measure the student progress and evaluate his performance.

The center of the student's graphical user interface shows the remotely connected lab machine screen whether controlled by the student (master) or it's just a shared-screen of the master's screen in case of sequential interaction. The right-hand side of the screen shows the info of the student's group connected colleagues. The progress of each student is shown in a bar located at the top of the web screen.

The bottom of the web screen contains the chat area, the student's audio settings to mute/unmute and change volume, and video settings to switch camera

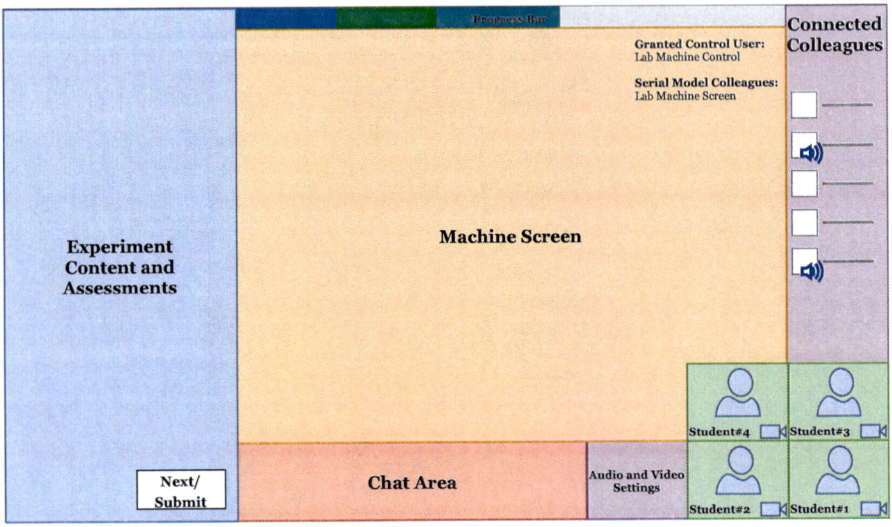

Fig. 5. Runtime experiment student's web-based graphical user interface

on/off, and the multi-video conference area showing the video coming from each participant student.

5 Conclusion

In this paper, we proposed a group-based laboratory experimentation platform that supports collaborative and cooperative virtual learning environment that reflect the intended learning outcomes of the student's active learning and comprehension construction as in physical laboratories. The proposed design supports individual-based and group-based types of experiments that can be configured within the experiment creation process. The group-based experiment includes sequential and concurrent student interactions that reflect cooperative and collaborative interactions respectively, taking into consideration session time-slot reservation, group management, way of communication and sharing aspects. The proposed design also guarantees each student interaction using policies assurance system that guides interpretation and construction of concepts.

Acknowledgment. This research is a part of research project supported by the Academy of Scientific Research and Technology (ASRT) through Egypt RDI/P4/1/20-21 program (2021–2023).

References

1. May D, Reeves B, Trudgen M, Alweshah A (2020) The remote laboratory VISIR-Introducing online laboratory equipment in electrical engineering classes. In: 2020 IEEE frontiers in education conference (FIE). IEEE, pp 1–9
2. Mikroyannidis A, et al (2017) Forge: an elearning framework for remote laboratory experimentation on fire testbed infrastructure
3. Zapata-Rivera LF, Larrondo-Petrie MM, Weinthal CP (2019) Generation of multiple interfaces for hybrid online laboratory experiments based on smart laboratory learning objects. In: 2019 IEEE frontiers in education conference (FIE). IEEE, pp. 1–8
4. Villar-Martinez A, Rodriguez-Gil L, Angulo I, Orduna P, Garcia-Zubia J, López-De-Ipiña D (2019) Improving the scalability and replicability of embedded systems remote laboratories through a cost-effective architecture. IEEE Access 7:164164–164185
5. García-Zubia J, et al (2010) Secondlab: a remote laboratory under second life. In: IEEE EDUCON 2010 conference. IEEE, pp 351–356
6. Herrera OA, Fuller DA (2011) Collaborative model for remote experimentation laboratories used by non-hierarchical distributed groups of engineering students. Austral J Educ Technol 27(3)
7. Callaghan MJ, Harkin J, McGinnity TM, Maguire LP (2007) Paradigms in remote experimentation. Int J Onl Eng 3(4)
8. Bochicchio MA, Longo A (2010) Extending LMS with collaborative remote lab features. In: 2010 10th IEEE international conference on advanced learning technologies. IEEE, pp 310–314
9. Lima EF, Siebra CA (2021) Design of learning objects for collaboration promotion and their effects on students' behaviour. Informat Educ 20(1):85–106
10. de la Torre L, et al (2013) Providing collaborative support to virtual and remote laboratories. IEEE Trans Learn Technol 6(4):312–323
11. Elmesalawy MM, et al (2021) AI-based flexible online laboratory learning system for post-COVID-19 era: requirements and design. In: 2021 international mobile, intelligent, and ubiquitous computing conference (MIUCC). IEEE, pp 1–7
12. Myrtille (2021) A native HTML5 Remote Desktop Protocol. https://github.com/cedrozor/myrtille
13. Conference Call (2021) A conference call implementation using WebRTC and Ratchet Web Socket. https://github.com/amirsanni/conference-call-ratchet

Mixed Reality and Gamification in Distance Learning Education: The Virtual Enterprise Planning Simulator to Learn ERP Strategies

Carlos Alberto González Almaguer[1,2](✉), Verónica Saavedra Gastélum[1,2], Ángeles Carolina Aguirre Acosta[2,3], Claudia Zubieta Ramírez[2,4], Alejandro Acuña López[2], Pedro Oscar Pérez Murueta[1,2], and Luis Javier Morales Rivas[1,2]

[1] Tecnologico de Monterrey, School of Engineering and Sciences, Ave. Eugenio Garza Sada 2501, 64849 Monterrey, NL, Mexico
cgonzalz@tec.mx

[2] Tecnologico de Monterrey, Architecture, Art and Design School, Institute for the Future of Education, Ave. Eugenio Garza 2501, 64849 Monterrey, NL, Mexico

[3] Tecnologico de Monterrey, Vice Rectory of Educational Innovation and Academic Regulation, Ave. Eugenio Garza Sada 2501, 64849 Monterrey, NL, Mexico

[4] Tecnologico de Monterrey, Vice Rectory of Academic Affairs, Ave. Eugenio Garza Sada 2501, 64849 Monterrey, NL, Mexico

Abstract. The worldwide confinement due to Covid19 has boosted the development of creative learning activities in academia, including role-playing games and multidisciplinary workshops brought to students' homes. Tecnologico de Monterrey, like most universities in the world, is delivering classes remotely using internally developed and external platforms designed to continue the practices that were done every day in laboratories, facilities, and partner companies and institutions.

One of the biggest challenges of distance education is creating fun and practical learning. Integrating innovative technologies into the learning process is necessary to achieve this challenge. One of these technologies is Augmented Reality (AR), which allows virtual objects to be superimposed in the real world to "coexist" with the objects in real-world surroundings.

In this research, we generated an ERP (Enterprise Resource Planning) simulator based on the educational platform "Tec21 Car Assembler" . Using Virtual Reality (VR), the students applied the virtual classroom's theoretical learning to practice, guided by an academician collaborating with training partners from companies and institutions partnering with Tecnologico de Monterrey to transfer knowledge to students.

Keywords: Augmented reality · Virtual reality · Gamification · Educational innovation · Higher education

M. E. Auer et al. (Eds.): REV 2022, LNNS 524, pp. 222–230, 2023.
https://doi.org/10.1007/978-3-031-17091-1_23

1 Introduction

1.1 The ERP Systems

In the '70s, the material requirements planning systems (MRPs) appeared due to the need for manufacturing plants to plan the acquisition of the raw materials required to satisfy their customers' orders. By the 90s, enterprise requirement planning (ERP) systems were created, helped by production planning software. These systems sought to integrate all the companies' operating entities and evolved to become "systems applications and products" (SAP) [1].

The continuous lags between delivery times promised in the clients' orders and the actual deliveries led to analyses of the root causes. Issues include the need to integrate demand, the lack of materials tickets, the weak analyses of line capacity, and the lack of suppliers and integration of their information. The problem generated an exponential development of methodologies for production planning, purchase of materials, and materials handling. The main problem was the isolated work in companies' departments and entities. The concept of systems, seeing everything as a set of interrelated elements to achieve a common goal, led to the systematization of tools that allowed production planning to become efficient and meet the needs of their customers.

Distance learning of these concepts is not easy. Simulators and computer platforms can be used to understand the concepts and interactions of an MRP, but these focus on a single area or function, not systemically. Thus, we were motivated to use ERP with a car assembly simulator incorporating augmented and virtual reality. The students could see and manipulate the product parts and understand the system interactions of all the components until the finished product is delivered to the customer.

Virtual and augmented reality are catalysts in best distance learning practices for students to acquire and develop problem-solving and systemic thinking skills.

2 Development

Since 2003, the Tecnologico de Monterrey School of Engineering has sought to design academic activities where knowledge of production planning, inventory control, experiment designs, quality, administration, and evaluation would introduce or reinforce projects such as the "Tec21 Car Assembler". This STEM practice involved assembling scale cars using Meccano models.

In March 2021, the distance learning activity had to be redesigned. We used the automotive EON platform with augmented reality lessons similar to the Meccano subassemblies that we use in laboratory practices. The aim was that the students experience the university activity physically similar to those in automotive manufacturing plants.

Simulating an ERP system [2] (nowadays best known as SAP) is to apply systemic thinking to integrate the different assembly departments so students can understand the importance of seeing the company as a system, the flow of information, and various interfaces that must work to reduce delivery times, missing materials, and increase the company's productivity, among other things.

To teach our students these concepts, we employ SAP modules that can only be used within the Technological facilities in computer labs due to software license restrictions. The Covid-19 health confinements challenged us to design a simulator that replicates the manufacturing processes of a car assembly plant based on Meccano [4] and Lego models.

This activity design emulates an Enterprise Resource Planning (ERP) system. It makes it easier for students to design a company with fully linked activities and departments, this is necessary because if the flow of information is lost, they will not be able to meet the assembly demand to meet customers' requests on Customer Orders (CO). The ERP facilitates the construction of working models and mechanical devices.

We designed the simulation with a challenging role play to make it more attractive. During the August-December 2020 semester, there was a negative experience in applying the virtual plant. The students did not question the learning or the advice of the teaching team, but they commented about the large amount of time invested in doing the deliverables and the time that the students had to wait for feedback.

When a person participates in a board game, the motivation is to win. Developing competencies in students makes it attractive to spend hours and hours transforming methodologies and problem situations into fun challenges to engage them in the learning process. The advantages of gamification [5] are:

- Making learning fun and interactive.
- Addicting students to learning.
- Providing students with the opportunity to experience real utility.
- Providing immediate feedback.
- Enhancing the learning experience.

The most representative processes of a car-assembler simulation start with marketing and sales, which generates the customers' sales orders that lead to the purchase order, which advances to the materials purchaser. The latter should consult the productivity index. For example, if the purchase order is for 40 cars, it would be a mistake to buy the spare kits for that amount of product if the productivity index suggests buying more to counteract the design and quality problems that might arise (see Fig. 1).

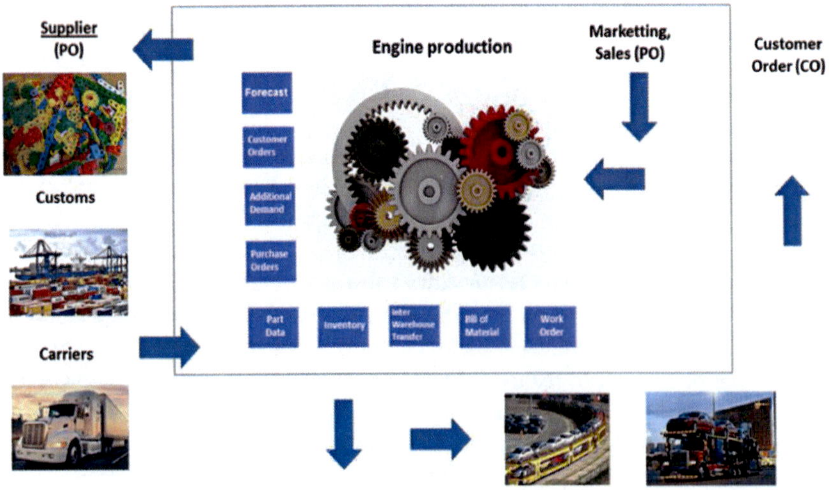

Fig. 1. VEP EPR architecture

Each family of parts has historical data parameterized with the manufacturing times. The delivery times are calculated with the VEP [7]. At this point in the simulation, it is crucial to calculate forecasts to anticipate the purchase orders made by the customers and parts orders to suppliers so that parts arrive on time to reduce product delivery times.

The simulation programming must clearly define the processes and map the information flow to assemble five different Meccano models. Errors are simulated from probability distributions; the dice, hourglasses, and other playful tools are substitutes in our activity.

The objective of the simulation is to deliver to the client their orders in the requested time, with the lowest cost, the highest quality, and the highest profit margin for the company. In our gamification, the students always have access to audiovisual material with advice in each module and instant feedback on the mistakes made during the process.

The simulator runs on an online platform so that the teacher can parameterize or emphasize the academic area of interest. In this way, many students can play simultaneously in different courses.

What is the advantage of this simulator compared to those already offered by universities and system developers? The answer is that the simulator is not limited to a single aspect of ERP but, as seen in Fig. 2, it analyzes modules of purchases, customers, suppliers, demand, materials purchases, inventory management, product engineering, process, quality, and logistics. The simulator belongs to Deterding's [6] spectrum of the "Gamification of learning," which refers to "the use of game elements in contexts that are not games" or "the process of making activities more like a game" Buckley and Doyle [7]. Thus, the simulator is an alternative to game-based learning. When implemented in education, gamification has multiple benefits: improving student participation, increasing their motivation and encouraging curiosity, and generating attractive and enjoyable learning environments.

Our investigation found other options on the market with concepts similar to the VEP, but they are limited to a single module and do not include augmented reality. Universities like Harvard and MIT have solutions such as the "Beer Game" developed by MIT to instruct industrial engineers, Flexsim to simulate discrete events, and "Fishbank" [9] to teach about management resources, to mention a few examples.

Harvard has an operations management simulation called "Balancing Process Capacity and Global Supply Chain Management Simulation". However, having only one aspect, the ERP is not complete. On the other hand, it can be parameterized by the teacher and the game. Its graphical interface is very attractive to the students (See Fig. 2).

Virtual reality has been an excellent complement to the simulator, as it focuses on the design, engineering, and research related to immersive interactive systems. These virtual reality systems synthesize environments, or worlds, simulating reality by rendering three-dimensional images, sounds, and forcing feedback [10]. This technology allows students to immerse themselves in environments and processes altered to simulate their reality. It allows educators to intensify students' experiences in research and learning, stimulate their imagination, and motivate them to commit to the intended research and learning processes.

In the transition stage of the simulator, we used libraries from the EON XR Reality platform that bears a strong resemblance to the Meccano devices that we use for physical

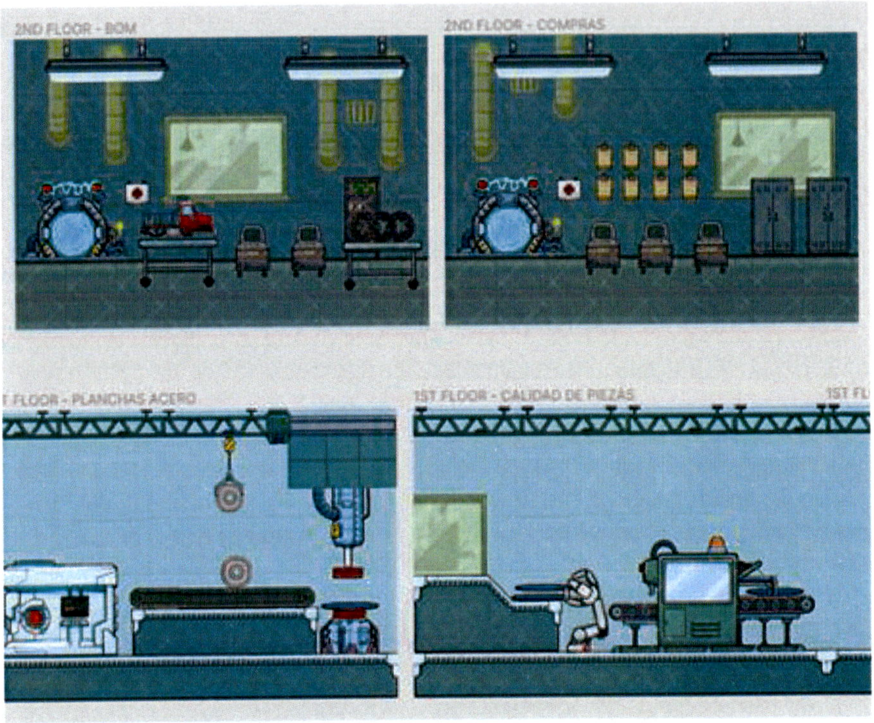

Fig. 2. Initial design of the student interface.

Fig. 3. Automotive lesson teaching the Bill of Material concepts.

simulations. Through collaborative work with the Tecnologico de Monterrey School of Design, we managed to model all the cars and digitize them in virtual and augmented reality. The students learned to manipulate and know all the parts found in the simulator's bill of materials (BOM) (See Fig. 3).

Virtual and Augmented Reality allowed the environment to become interactive, colorful, and three-dimensional. It provided us with an immersive experience, where the user can witness experiences and events that appear to be real.

Virtual and augmented reality experiences have been shown to increase student motivation and facilitate better research, critical thinking, problem-solving, and communication skills through collaborative exercises. The advantages of VR and AR technology are reflected in the results for our students in learning achievement, motivation, and attitude that comprise the learning outcomes. Previous studies have reported that this type of technology improves learning performance and a positive attitude toward activities, where students show emotions of happiness as they "learn through playing" [12].

According to Panneta (2020) [13], various essential trends in emerging technologies will significantly affect companies, society, and people in the coming years. They will allow companies and society to regain confidence in technology. The growth of AR technology is exponential. Innovative educational applications in the classroom are increasingly frequent, allowing students and teachers to "interact" more thoroughly with their activities and in the teaching-learning process.

3 Results

Before using the virtual enterprise planning (VEP) simulator, a knowledge test on automotive engineering was applied to the students taking the subject of Project Evaluation IN2025. The basic statistics are in Table 1.

Table 1. Summary statistics for before and after VEP.

Sample	Sample size	Median	Std. deviation
Before VEP	26	74.2	13.2
After VEP	26	92.69	5.33

With a sample size of 26 students and a mean of 74.23, we can appreciate that the learning obtained with the traditional classroom method was good. From the variability of the grades in the graph, we can infer that the group was non-homogeneous and that the learning was not standardized. The lowest grade was 50 (See Fig. 4).

With the beta version of the VEP simulator using the EON platform, the mean increased to 92.69, and the standard deviation decreased to 13.17 after using the simulator. We can observe that knowledge was homogeneous; the minimum grade was 80, and the maximum was 100.

Using the student's t-test with two samples, we statistically validated a significant increase in student learning using the simulator; the research protocol was followed.

Thus, two knowledge tests were applied focusing on the practice (topic) whose achievement we wanted to measure. The first test was applied before using the simulator. After they carried out the practice, the second was given to students to measure statistically significant learning (Fig. 5).

The hypothesis was tested through the mean of the grades in the exams (M1 and M2). Our study defined "knowledge gain" as the difference between M1 before the PEV

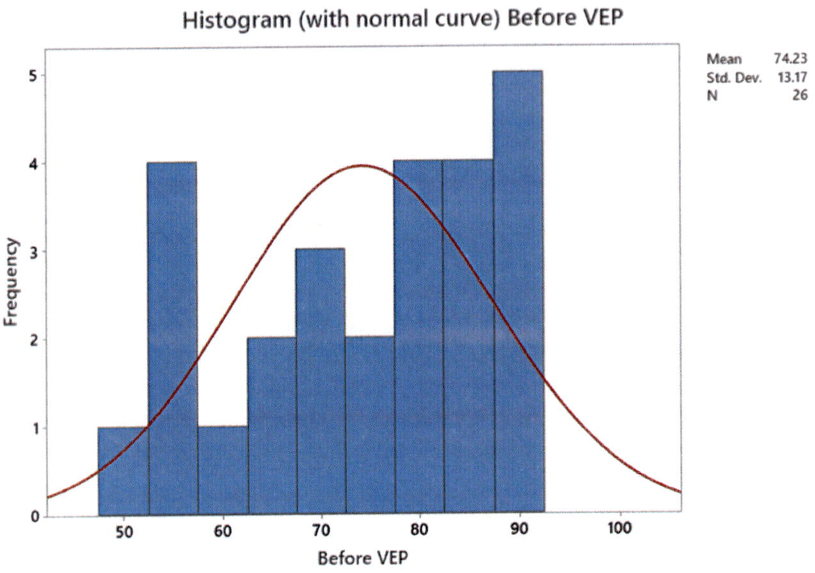

Fig. 4. Histogram of the test results before the VEP lesson.

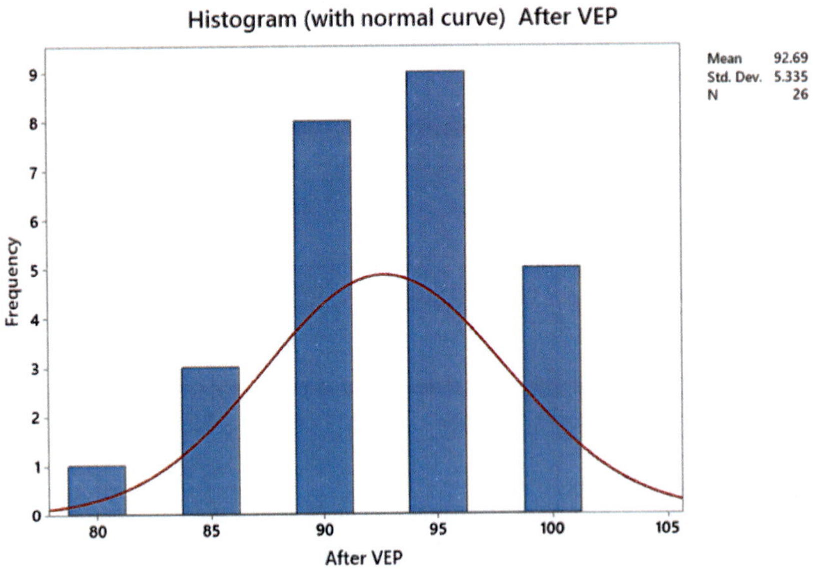

Fig. 5. Histogram of the test results after the VEP lesson.

and M2 after applying the PEV. When analyzing the individual M1 scores vs. M2, we used a two-sample student t-test to validate that they were statistically different and that the mean of M2 was significantly higher.

H_0: $\mu_1 = \mu_2$: The average of the grades obtained by the students is the same regardless of whether they use the simulator.
H_1: $\mu_1 \neq \mu_2$: The average of the grades obtained by the students is different.

We sought to reject the null hypothesis H_0, which would demonstrate the veracity of H_1t and validate the assumption that students improved their academic performance using the simulator in industrial engineering topics.

For each subject, specific practices of the virtual plant will be designed in the future, where the elements of the test (exam) are linked so that the student has a more detailed knowledge of each and can, through experimentation, link the theory learned in class with the practice activities of the virtual plant.

With the use of Minitab software, we calculated each exam results for:

- Mean
- Standard deviation
- Variance
- Median
- Quartiles

Comparing the means before using the simulator vs. after shows the increase in learning for students who used the PEV simulator.

Applying the two-means test gives results that prove that the variable under study fulfills the hypothesis.

4 Conclusions

Virtual and augmented reality are invaluable tools for distance education, growing exponentially due to Covid-19 confinements. These tools have an educational component, but they also help students' emotional stability as they experience total immersion using the different platforms. Learning ERP systems is not easy. However, creating the PEV simulator has motivated students to learn, have fun, and, most importantly, increase their knowledge.

The construction of the PEV represented many challenges, but in these confining circumstances, adding the online game component let us leverage natural human competitiveness. We guided students to learn better strategies to obtain more points in the simulator, which reflected the learning of the complex topics about quality, and production planning, which are not easy to learn in a conventional classroom. The results obtained were excellent and showed the effectiveness of gamification and mixed reality in distance learning.

Acknowledgement. The authors wish to acknowledge the technical support of Writing Lab, Institute for the Future of Education, Tecnologico de Monterrey, Mexico, in the production of this work.

The authors would like to acknowledge the financial support of NOVUS 2020, ID 199 an initiative of Tecnologico de Monterrey, Mexico, in the production of this work.

References

1. https://www.igi-global.com/chapter/evolution-erp-systems/18445
2. https://observatory.tec.mx/edu-news/world-bank-tec21-tec-de-monterrey-educational-model
3. Stretch-Stephenson SM, Rika Houston H, Germano MA (2015) Integrating technology into marketing courses via SAP enterprise resource planning (ERP) system. J Bus Econ 6(3):424–430. ISSN 2155-7950
4. Marriot R (2012) Meccano. Shire Publications. ISBN 13: 9780747810568
5. Gonzalez Almaguer CA, Maya López M, Acuña López A, Caballero Montes E, Zubieta Ramirez C, et al (2020) Stem competency-based learning for engineering and design students of the educational model Tec21. In: DS 104: proceedings of the 22nd international conference on engineering and product design education (E&PDE 2020), VIA design, VIA University in Herning, Denmark, 10–11 September 2020
6. Holloway S (May 2018) Gamification in education: 4 ways to bring games to your classroom. https://tophat.com/blog/gamification-education-class/
7. Deterding S, O'Hara K, Sicart M, Dixon D, Nacke L (2011) Gamification: using game design elements in non-gaming contexts. In: Proceedings of the conference on human factors in computing systems, pp 2425–2428
8. Buckley P, Doyle E (2016) Gamification and student motivation. Interact Learn Environ 24(6):1162–1175
9. https://forio.com/simulate/mit/fishbanks-spanish/simulation/login.html
10. https://hbsp.harvard.edu/search?type=Simulation&discipline=Operations+Management
11. Meinhold R (2018) Virtual reality. Salem Press Encyclopedia of Science. https://0-search-ebscohost-com.biblioteca-ils.tec.mx/login.aspx?direct=true&db=ers&AN=89250621&lang=es&site=eds-live&scope=site
12. Akçayır M, Akçayır G (2017) Advantages and challenges associated with augmented reality for education: a systematic review of the literature. Educ Res Rev 20:1–11
13. Panetta K (2020) 5 trends drive the gartner hype cycle of emerging technologies. Obtenido de. https://www.gartner.com/smarterwithgartner/5-trends-drive-the-gartner-hype-cycle-for-emerging-technologies-2020/
14. Gonzalez Almaguer CA, Saavedra V, Acuña López A, Caballero Montes E, Aguirre A, Zubieta Ramírez C (February 2021) Distance learning through simulators and virtual platforms for the teaching of industrial engineering within the Tec 21 educational model. In: DSDE 2021: 2021 4th international conference on data storage and data engineering, Barcelona, Spain, pp 93–99

Internet of Things and Industry 4.0

Utilization of a Digital Twin for an Electric Vehicles Smart Charging Station for Future Use with Engineering Students

Sergiy Korotunov[1], Galyna Tabunshchyk[2](✉), and Peter Arras[3]

[1] National University Zaporizhzhia Polytechnic, Zaporizhzhia, Ukraine
[2] Software Tools Department, National University "Zaporizhzhia Polytechnic", Zaporizhzhia, Ukraine
galina.tabunshchik@gmail.com
[3] KU Leuven, Campus De Nayer, Sint-Katelijne-Waver, Belgium
peter.arras@kuleuven.be

Abstract. Studying digital twin technology in engineering education is a challenge of implementing multidisciplinary knowledge in a comprehensive manner. to scale up competences of the students. In the paper authors consider technologies and approaches which are used for digital twins and propose a digital twin for the electric charging station. The goal of the digital twin is to optimize the grid side of the system. In this paper authors look at the possibilities in the use of the digital for predictive maintenance of the system. The interconnection of building blocks of the digital twin and connection to the physical system from the primary electric energy source over the charging system to the customer side (user who will charge her/his car) is taken into account.

Keywords: Digital twin · Electric vehicle · Smart grid

1 Introduction

Smart grid [1] is a concept that is used to describe the next generation electricity network. This network has following characteristics that makes it better than traditional one – high configurability, reactivity and self-control capability. Electric vehicles (EV) are important targets for the smart grid to manage. They are always equipped with batteries having high energy-storage capacity, so their charging creates large electric load on the power system. Electric vehicles need to be charged more often than vehicles with internal combustion engine and it takes from 30 min up to 12 h. That makes it necessary to build a large and scalable charging infrastructure which embraces fast charging stations, battery swapping stations, and individual charging points for slower charging.

The simulation with real-time feedback of the smart grid will help to optimize the network and can help to make predictions for the future use of the grid so that investments for expansion of the network, maintenance and error estimates become available. This type of simulation technology, known as a digital twin (DT), is a key emerging technology

M. E. Auer et al. (Eds.): REV 2022, LNNS 524, pp. 233–240, 2023.
https://doi.org/10.1007/978-3-031-17091-1_24

in Industry 4.0 [2]. As such, introducing and working with digital twins in engineering education is necessary to prepare the engineers of tomorrow for their future jobs.

The use of a DT can save a lot of time and make education more cost-efficient and diverse. In this paper a DT of a smart grid in relation to the charging problematics of electric vehicles is described. The creation of the DT for a smart grid can help students get better understanding of the concept and allow them to use different optimization techniques and methods to achieve better overall performance.

2 Literature Review

For Digital Twin there are many definitions.

A Digital Twin – is a virtual representation of the physical objects and processes that is used as a digital counterpart of a physical system [3, 4]. It can be used for educational purposes and help students fully understand the subject in consideration without a need of having its physical representation.

A DT is a digital functional representation of a living or non-living physical entity [5]. DT refers to a digital representation of potential and actual physical assets (physical twin), processes, people, places, systems and devices that can be used for various purposes [6]. The digital replica provides both the elements and the dynamics of how an Internet of things (IoT) device operates and lives throughout its life cycle [7].

Digital twin (DT) consists of three parts: physical product, virtual product and the linkage between physical an virtual product [8].

Summarized, definitions of DT technology all contain three important characteristics. Firstly, it is a functional digital representation of a product or process which behavior is similar to the real system. Secondly there a digital connection between the physical model and the corresponding virtual model or virtual prototype (VP). Finally, the digital connection between the digital twin and the physical model is used to collect real-time data from the physical model using sensors. In driven models the digital connection also serves as the control channel for the physical system.

DT integrate a number of different technologies: IoT to set the digital thread between the physical model and the virtual model, artificial intelligence, machine learning and software analytics to create dynamic digital simulation models that update and change similar to their physical counterparts.

A digital twin continuously updates itself from multiple (sensor or data) sources to represent the real-time status, working condition or position of a process or product. The DT also integrates historical measured and stored data to update the digital model in order to be the best possible replica of reality.

The monitoring and updating of the digital model is called the digital thread of the DT. The digital thread is the whole of IT-related objects and processes that connect the digital replica to the physical model or process. It consists in the sensors for the monitoring and measuring of the physical model, controlling software and actuators/controlling hardware that can change the state of the physical model.

DT technology is said to be one of the key technologies for Industry 4.0 [9].

DT are used for different tasks, such as fault diagnosis in photo voltaic installations such as rooftop and building integrated photo voltaic systems [10], simulation of the smart grid [11] or autonomous electrical charging stations [12].

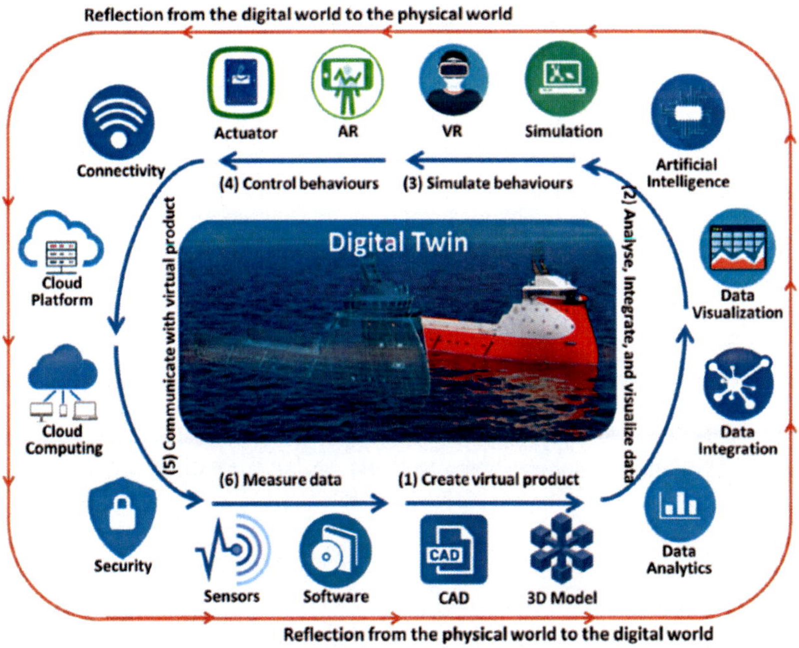

Fig. 1. Technologies used in digital twins [5]

DT are becoming an integral part of the digital transformation which impacts all aspects of life and as such has a great impact on the educational system as well. Students are the future users and designers/implementers of digitalization. So they need to be trained to conceive, build and maintain digital twins too [13].

3 Basic Design of the Developed Digital Twin

The main goal of teaching approaches on how to work with Industry 4.0 to students is to provide them with ability to design, build, test and optimize a virtual prototype and use it along with its own DT. The virtual prototype is the digital replica of reality and is built in the design stage to optimize the design. The combination with DT-technology to integrate real-time measurements and data acquisition, combining with sensors and actuators to provide feedback allows to test and optimize the VP depending on real and not only assumed data and parameters.

VP and DT can be created for basically any cyber-physical system, that combines and integrates both cyber and physical level in its work. The proposed case is to consider one of the most actual of such systems – smart grids, specifically for to the purpose of creating and simulating efficient EV-charging systems. This DT should integrate infrastructure and process of charging EVs. EVs are becoming extremely popular nowadays due to variety of reasons starting from environment concerns and finishing with strive to solve fossil fuel and their large carbon-footprint issues This setup can be used by engineering students to study complex systems and experiment with Industry 4.0 solutions in a real-time way.

EV-charging system is a complex system that scales well from lab-size to the full life-size. It also can incorporate renewable energy sources for sustainability courses. The first step is the creation of the VP. It was decided to use electronic circuits to charge 1.25 V Li-ion-batteries to simplify to lab scale the hardware part of the system and fully emulate charging of the EVs. Charging can be done from net-powered, wind-powered and solar energy. It is possible to charge from a specific source independently or combine multiple sources in demand peaks. In a smart grid the charging system can be not only consumer on the net, but also an energy provider, therefore it is possible to "sell" extra energy to the net.

The DT should be created in order to implement:

- safety control and identification of the customer who wants to charge his vehicle; The created VP should be controlled with an RFID-card or from the Internet.
- check the charge level of the connected batteries;
- charging time monitoring;
- setting prices per kWh of energy;
- communication with the user;
- ending of charging when the battery is fully charged;
- optimization of the multiple EVs handling;
- manage load by enabling different energy sources based on the demand;
- "selling" extra energy when demand is low;
- autonomous deciding what energy source will be used for charging;
- prioritizing renewable energy sources.

Scheme of the developed VP (**Fig. 2**) combines cyber and physical levels. On the physical level it contains from:

A – energy sources (net – 1, solar – 2 and wind – 3);

B – charging of the EV;

C – indicator representing process of "selling" extra energy to the grid;

D – battery and charge controller.

On the cyber level it uses data from sensors and actuators to execute monitoring, testing and optimization, presenting to the end user graphical representation of the collected data and possible actions.

The DT allows engineering students to perform all tasks which will be available in commercial setups, and they can experiment with different approaches and different human-machine interfaces. DT has the equipment needed to charge the batteries, connected customer identification and payment system, which is equipped with various energy sources. It allows one to switch power modes, offers different payment methods and different user interfaces.

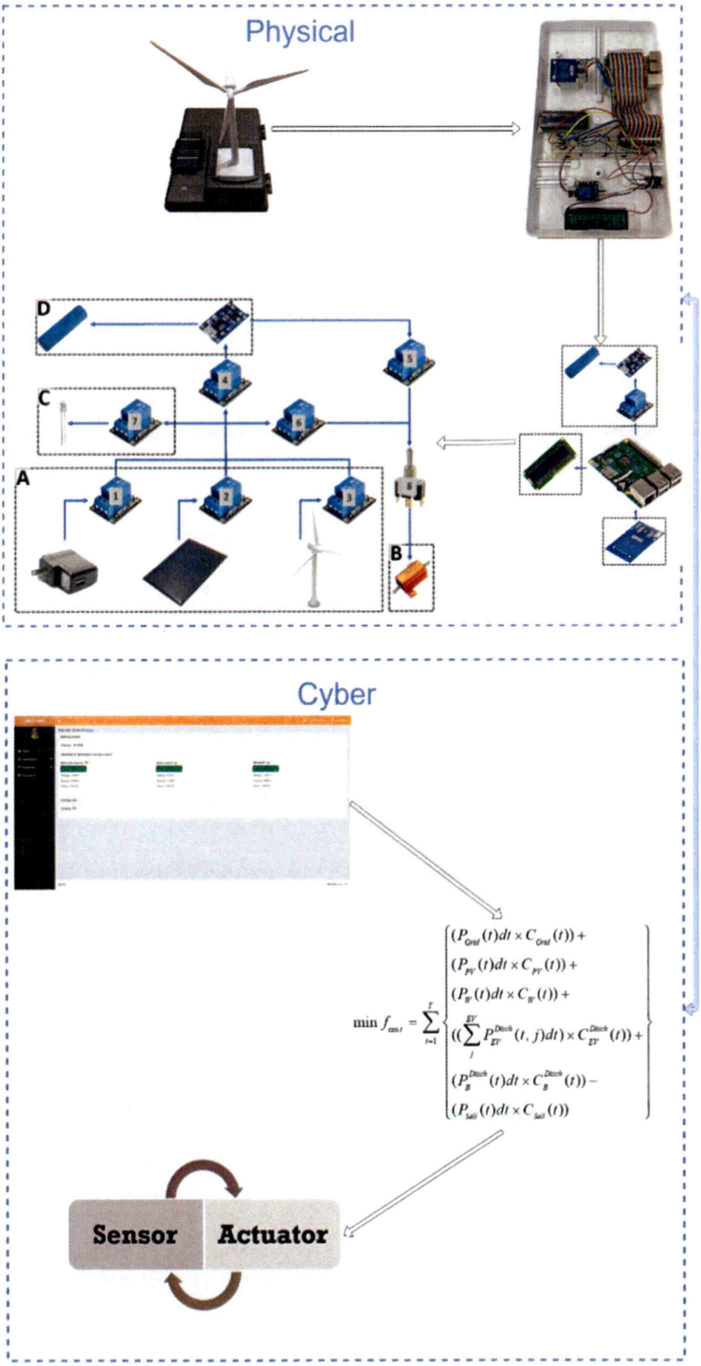

Fig. 2. Scheme of the virtual prototype of the electric vehicles charging system

4 Predictive Maintenance Analytics Implementation in Digital Twin

Predictive maintenance (PDM) is playing a key role in the Industry 4.0 [14, 15]. To receive high precision PDM there are implemented a variety of methods.

Typical maintenance tasks for EV charging stations include: regular inspection of connector contacts; inspection of the charging cable to look for signs of wear; inspection of the connector to look for cracks, breaks or exposed metal; inspection of the holster to ensure the connector rests firmly in the holster; replacement of filters in DC charging stations [16] and others.

In some cases, the equipment maintenance or repair can be done remotely by software. Other situations require a technician to be dispatched, and if so, they can arrive armed with advance information about the problem and the right parts and tools to fix it.

When there are developing cyber-physical systems additional attention is payed to the self-aware and self-maintenance systems.

To implement predictive maintenance to the DT the software architecture should be changed to the three level architecture, with would contain cloud database, which contains key performance indicators, which are collected from the physical systems. Usage of the predictive analytics would help to prioritize the next best actions to troubleshoot the problem, diagnose the underlying issues, and identify the most likely solution. One of the advantages of such three level architecture is flexibility of the solutions at each level.

5 Conclusions

A digital twin was developed in order to model the control system of an electric vehicles charging station, which in its turn is a smart grid that provides ability to manage incoming power from different energy sources depending on demand and availability. Developing of the digital twin in this case allows to avoid creating a physical model saving costs and time but enables all necessary calculations to be executed. The setup can be used by engineering students to learn, experiment, and study complex cyber physical systems in a real-time fashion.

On the digital twin of the EV charging system a predictive maintenance analytics scheme is being implemented. The transformation of the controlling software to a three level architecture will allow much more flexibility to enable predictive and preventive maintenance.

Acknowledgement. This work is carried out partly with the support of Erasmus + KA2 project 619034-EPP-1-2020-1-UA-EPPKA2-CBHE-JP "Cross-domain competences for healthy and safe work in the 21st century" (WORK4CE).

References

1. Tabunshchyk G, Korotunov S (2020) Analysis of approaches to the simulation and verification of cyber-physical systems. Radio Electronics Computer Science Control. 57–68. https://doi.org/10.15588/1607-3274-2020-3-5
2. Cinar ZM, Nuhu AA, Zeeshan Q, Korhan O (2020) Digital twins for industry 4.0: a review. In: Calisir F, Korhan O (eds) Industrial Engineering in the Digital Disruption Era. Lecture Notes in Management and Industrial Engineering. Springer, Cham, pp 193–203. https://doi.org/10.1007/978-3-030-42416-9_18
3. Negri E, Fumagalli L, Macchi M (2017) A review of the roles of digital twin in CPS-based production systems. Procedia Manufacturing. 11:939–948. https://doi.org/10.1016/j.promfg.2017.07.198
4. Arras P, Tabunshchyk G (2022) Implementation of digital twins for industry 4.0 in the engineering study program. In: Tonkonogyi V, Ivanov V, Trojanowska J, Oborskyi G, Pavlenko I (eds) Advanced Manufacturing Processes III. Lecture Notes in Mechanical Engineering. Springer, Cham, pp 677–686. https://doi.org/10.1007/978-3-030-91327-4_65
5. Toa F, et al (2018) Digital twin-driven product design. Int J Prod Res 57(12):3935–3953. https://doi.org/10.1080/00207543.2018.1443229
6. Assawaarayakul C, Srisawat W, Ayuthaya SDN, Wattanasirichaigoon S (2019) Integrate digital twin to exist production system for industry 4.0. In: 2019 4th technology innovation management and engineering science international conference (TIMES-iCON), Bangkok, Thailand, pp 1–5 (2019). https://doi.org/10.1109/TIMES-iCON47539.2019.9024430
7. Pires F, Cachada A, Barbosa J, Moreira AP, Leitão P (2019): Digital twin in industry 4.0: technologies, applications and challenges. In: 2019 IEEE 17th international conference on industrial informatics (INDIN), Helsinki, Finland, pp 721–726, . https://doi.org/10.1109/INDIN41052.2019.8972134
8. Glaessgen EH, Stargel D (2012) The digital twin paradigm for future NASA and US air force vehicles. In: 53rd structures, structural dynamics, and materials conference: special session on the Digital Twin, Honolulu, US (2012)
9. Parrott A, Warshaw L (2017) Industry 4.0 and the digital twin Manufacturing meets its match. Deloitte Development LLC, p 20
10. Jain P, Poon J, Singh JP, Spanos C, Sanders SR, Panda SK (2020) A digital twin approach for fault diagnosis in distributed photovoltaic systems. IEEE Trans Power Electron 35(1):940–956. https://doi.org/10.1109/TPEL.2019.2911594(2020)
11. Arras P, Tabunshchyk G, Korotunov S, Okhmak V (2019) Modeling and simulation of the services for vehicle charging infrastructure interaction. In: 10th IEEE international conference on intelligent data acquisition and advanced computing systems: technology and applications (IDAACS), Metz, France, pp 330–333. https://doi.org/10.1109/IDAACS.2019.8924449
12. Arras P, Tabunshchyk G, Korotunov S, Okhmak V (2020) Cost optimization simulation for electric vehicle charging infrastructure. In: IEEE European technology and engineering management summit (E-TEMS), Dortmund, Germany, pp 1–4. https://doi.org/10.1109/E-TEMS46250.2020.9111715
13. Saracco R (2019) Digital twins: bridging physical space and cyberspace,". Computer 52(12):58–64. https://doi.org/10.1109/MC.2019.2942803
14. Lee J, Kao H, Yang S (2014) Service innovation and smart analytics for industry 4.0 and big data environment. Procedia CIRP [s.l.] 16:3–8 (2014)
15. Liu C, Tang D, Zhu H, Nie Q (2021) A Novel predictive maintenance method based on deep adversarial learning in the intelligent manufacturing system. IEEE Access 9:49557–49575. https://doi.org/10.1109/ACCESS.2021.3069256

16. Adopting Field Service Software to Improve Maintenance for EV Charging Station Praxedo Software Inc. North America, © Praxedo (2021)
17. Barinov A: EV charging company case study: data analysis for predictive maintenance solution https://www.ai.intelliarts.com/post/ev-charging-company-case-study-data-analysis-for-predictive-maintenance-solution

HYPERCam v1: Hyperspectral Camera Based on a Low-Cost Hardware and Software Architecture for Food Analysis Through Image Processing

Unai Hernandez-Jayo[1](✉), Haizea Martin[1], Jose Manuel Amigo[2,3], and Javier Garcia-Zubia[1]

[1] University of Deusto, 48007 Bilbao, Spain
unai.hernandez@deusto.es
[2] IKERBASQUE. Basque Foundation for Science, 48011 Bilbao, Spain
[3] Department of Analytical Chemistry, University of the Basque Country UPV/EHU, 48080 Bilbao, Spain

Abstract. A hyperspectral image makes it possible to collect and process information in the electromagnetic spectrum. Through the analysis of these images, information on the chemical composition of the sample under study can be obtained. The field of application of this type of technology is very extensive: the food industry, pharmaceuticals, forensic analysis or the field of restoration are just some of the most widespread examples. This article presents the work-in-progress to obtain a low-cost and portable hyperspectral imaging camera, so that it can be used in the fields of application mentioned above, as well as in certain subjects such as computer image analysis, machine learning, etc. The camera designed and tested in the first prototype is based on the use of several low-cost embedded systems and simple electronics and mechanics, so that it is an easily replicable DIY solution.

Keywords: Hyperspectral image · Embedded systems

1 Introduction

It is a fact that today's society continually demands information about the products it consumes, and food is perhaps one of the areas in which it is most concerned. The current trends on environmentalism, natural products, traceability, local consumption make it necessary for products to have the most complete labeling possible so that the final consumer has the maximum data on their life cycle. The previously mentioned parameters form part of what is understood about quality, and quality is an ideal multiparametric concept strongly linked to the human appeal for food and safety concerns. Assessing these parameters in the food we consume guarantees that the consumer will have more confidence about its food and, moreover, that less food waste will be generated [1].

Encouraged by this demand (and obligation, depending on the country), industrial processes must have mechanisms that guarantee the composition of the products they

M. E. Auer et al. (Eds.): REV 2022, LNNS 524, pp. 241–251, 2023.
https://doi.org/10.1007/978-3-031-17091-1_25

offer to the market. For this reason, food industries must have mechanisms in their production lines that help them obtain information about each of the product units they are manufacturing so that those that do not meet minimum quality or quality standards can be eliminated on time [1].

RGB Vision systems have been used for many years to separate food with defects, with a different shape or colour than the consumer's preference [2]. Nevertheless, the consumers' continuous demands in quality make that new quality parameters must be controlled that cannot be registered by using RGB. Here, vis-NIR-Hyperspectral and multispectral devices are becoming a plausible choice to determine parameters like the amount of protein, fat, sugar, salt in different products that otherwise could not be quantified [3].

Hyperspectral imaging (HSI), imaging spectrometry [4] is a spectral sensing technique in which an object is photographed using several well-defined optical bands in a broad spectral range [5]. This technology has been used in various fields of application since the 1960s [6]. The key difference with color vision systems is that a spectral signature is recorded for each pixel containing from dozen (multispectral), several hundred or even several thousands of characteristic physicochemical values (Fig. 1).

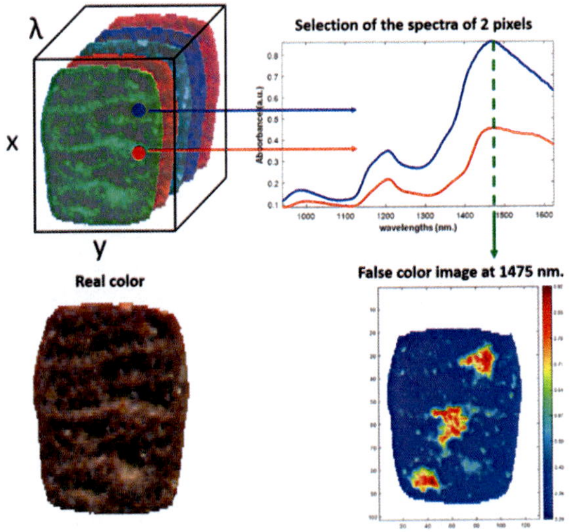

Fig. 1. Top left, hyperspectral image of a biscuit. Top right, the spectral information obtained at two different pixels. Bottom left, real color picture of the biscuit. Bottom right, the analytical information obtained at a certain wavelength shows defects on the biscuit surface.

The richness in spatial and spectral information, together with the advances in hardware and Machine Learning, promoted their usage in almost all scientific fields. The most used configuration in food processing industrial scenarios is the hyperspectral cameras working in the visible-Near Infrared spectral range. Two examples of this breakthrough are, for instance, the possibility of classifying nuts for protecting the denomination of origin (Fig. 2a) or the surface quantitation of bread being stale with time for quality assurance of the bread before releasing it to the market (Fig. 2b).

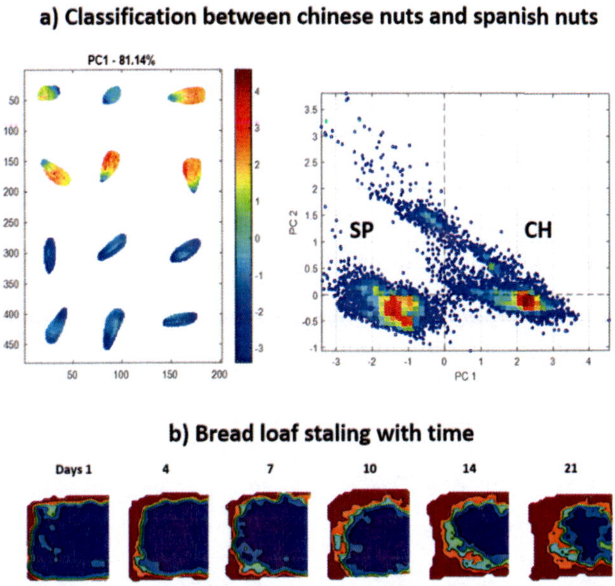

Fig. 2. (a) Classification of Chinese pine nuts (CH) and Spanish pine nuts (SP) using a NIR hyperspectral device [7]. (b) Monitoring of the chemical reactions occurring inside a bread loaf to assess the staling status of the bread. In red, the surface of the bread being stale. In blue, the surface of the bread is fresh and moist [8].

The major bottleneck of the full implementation of hyperspectral devices in industrial scenarios is the high maintenance costs, high measurement speeds, and the lack of robustness of the machine learning algorithms make producers reluctant of implementing this technology. In this context, this article presents the work in progress in developing a low-cost multispectral camera prototype based on electronic and mechanical systems that are simple to assemble and program. The system can be replicated by educational centers or researchers who do not have a high budget to acquire a high-performance hyperspectral camera for the price it entails. The prototype developed and presented here is currently in the validation and comparison phase with commercial hyperspectral cameras, so the images shown in the results section are examples of the images obtained with the first version of this prototype and subsequently analyzed using the software HYPER-Tools v.3 [9] (freely available at https://www.hypertools.org/, last accessed November 2021).

2 Background and State of the Art

Encouraged by the strong regulatory framework of the Pharmaceutical Industry and the launch of the Quality by Design program [10], food production industries started to implement smarter production systems. They must encompass non-destructive and fast measuring technologies together with dedicated data analysis (Process Analytical Technologies) to give a tailored and timely response of the quality of the product in all the stages of the production line [11]. In this regard, visible and near infrared (NIR) hyperspectral systems have been proposed showing excellent results in specific cases like finding foodborne pathogens in foodstuff [12], sorting plastics and other non-desired elements [13], or even assessing the content of ingredients in pharmaceutical preparations [14].

Despite this apparent success, the major problem is that most of the applications have been performed in controlled environments without a clear transition to industrial scenarios where the speed in the measurement and the changing environmental conditions (temperature or humidity) makes the hardware (camera) and the software (models) unstable, imprecise and inaccurate to a certain level.

Concerning the hyperspectral cameras normally used, there are four major configurations available in the market (Fig. 3)[15]:

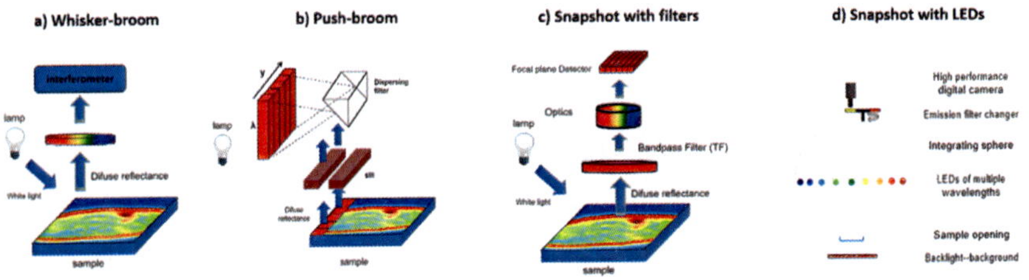

Fig. 3. Hyperspectral cameras configurations.

Whisker-Broom Configuration: This measuring modality implies that one spectrum is collected for each measurement in what is called a pixel. Then, the sample moves in an X–Y direction in order to obtain a hyperspectral image. This configuration is not an imaging method but a scanning one. Despite its high spectral quality and the possibility of measuring up to 10 microns pixels, the measurements are extremely slow. Therefore, it is not a configuration that is applied at the industrial level.

Push-Broom Configuration: This is the preferred industrial configuration since a complete line of spectra is acquired in each measurement. Therefore, the sample only moves in one direction. With this configuration, it is rather easy to imagine the samples passing under the cameras with a conveyor belt and being scanned in real-time. The major drawback of this configuration is that integrating the mechanics, the measurements and the data analysis normally conveys a high price.

Snapshot Configuration: The snapshot configuration is, indeed, the only configuration that is considered to be a real imaging technique, since one image is taken at a specific wavelength. The major benefit of the snapshot cameras is that the price is normally lower than the two previous configurations, and the hardware is normally easier to maintain. The drawback is that a limited amount of spectral information is obtained. That is the reason for being called multispectral cameras instead of hyperspectral. In this regard, two technologies are mostly used:

(a) Filter-based technologies: In this configuration, a rack of filters is normally inserted between the sample and the detector. The rack contains specific band-pass filters that sequentially obtain an image at the selected wavelengths of the filters. The major benefits of this configuration are using standard illumination sources and the easiness of changing the filters to specific wavelengths. Nevertheless, the rack needs to spin in each snapshot creates an extra movement in the camera that can negatively influence the acquisition of the images.
(b) LED-based technologies: The LEDs act illumination sources at specific wavelengths is the preferred snapshot technology. The LEDs are easy to maintain, have an affordable price, and offer a constant and reproducible amount of light. The drawback of the LEDs is that no external illumination source can arrive at the sample, and therefore, the sample needs to be in a closed environment for the measurement.

All configurations have benefits and drawbacks, making difficult to find the ideal configuration for solving general problems. While whisker-broom and push-broom configurations give the best spatial and spectral resolution, they result in expensive and difficult to maintain cameras. On the other side, the limited amount of spectral information of snapshot cameras hampers the plausible analytical possibilities of these configurations.

HYPERCam v1.0 aims to offer a low-cost LED-based camera that will serve as a platform to find specific analytical solutions in an easily replicable Do It Yourself (DIY) solution that can be easily adapted to specific situations. Moreover, HYPERCam v1.0 comes with an arm and a rotatory platform that take snapshot multispectral images at different sample positions, so a 3D multispectral image is obtained. This configuration is not available in the market, and it is extremely interesting in issues like, for instance. finding fungal infection in fruits and vegetables.

3 HYPERCam V1.0 Description

3.1 Software Design

The operational procedure of the HYPERCam v1.0 is very simple. The objective is to provide the user with the possibility of taking images of the sample under analysis using up to 17 different LEDs (Table 1), being able to place the camera in 3 relative positions: 90° (vertically with respect to the sample), 45° and 0° (in the horizontal of the sample). In addition, the user will have the possibility of rotating the plane on which the sample is located 360°. The option of rotating the sample and taking images from different angles offers the user the possibility of making a hyperspectral 3D image during the processing stage.

In a first step, the user will be able to modify certain camera parameters, such as brightness, contrast or ISO, among others. During this process, the user is offered a real-time video image of the captured image to see the effect of each of the modified configuration parameters. Once the camera is configured to your preferences, you can

Table 1. Available LEDs classified by wavelenth.

LED 370E	LED 630E	LED 910E
LED 450L	LED 660L	LED 930L
LED 465E	LED 750L	LED 940E
LED 525E	LED 780L	LED 970L
LED 570L	LED 840L	LED 1050E
	LED 890L	LED WHITE

calibrate it by taking 17 images of a reference sample for white calibration and another for black. If you want to use any other reference color palette in this calibration step, you can do that as well.

After the configuration and calibration steps, the user can access the measurement menu (Fig. 4) and configure the position of the camera, rotate the sample or not, and select the LEDs to be used when taking images. For example, the user can select the 570 nm and 1050 nm LEDs, and the 90° and 45° positions. The image acquisition sequence will be: place the camera in the 90° position, turn on the 570 nm LEDs, take the image, turn off the LEDs and turn on the 1050 nm LEDs to take images again. The LEDs turn off, the camera is placed at 45°, and the process is repeated. On this screen, the user is also shown the camera configuration made during the calibration phase.

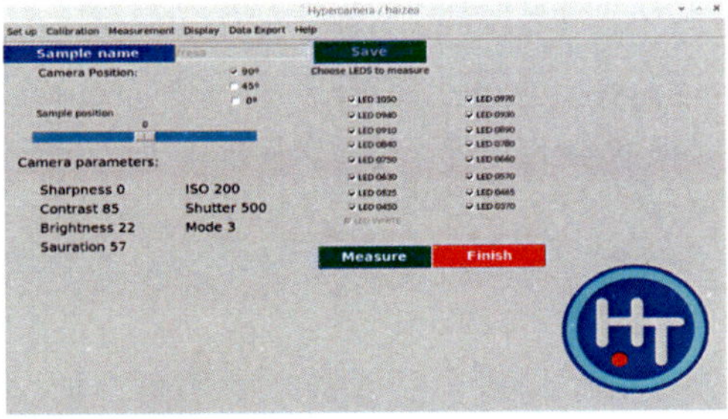

Fig. 4. Measurement user interface

A user may need to take images of different samples using different positions and different camera settings and positions during a session. That is why the system created includes a simple and intuitive directory tree (Fig. 5), so that it is easy for the user to identify which sample, camera position and type of LED corresponds to each captured image. The camera configuration is also stored as metadata in a txt file and the results of the images obtained during the calibration process (black and white). These images are necessary for the subsequent signal processing to take them as a reference.

The measurement process can be repeated by the user as many times as necessary. Once finished, the user will access the visualization of the captured images through the Display menu (Fig. 6). In this menu, you will be offered a dialog box to navigate through

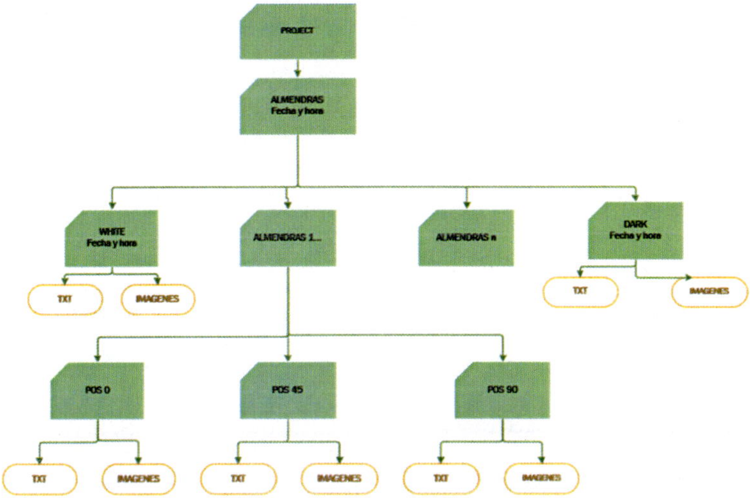

Fig. 5. Folder tree structure.

the directory tree and select the directory that contains the images to display. All the images are shown when selecting the directory, each one identified with the LED used during its acquisition. Furthermore, the user can select one of them, and it will be shown maximized.

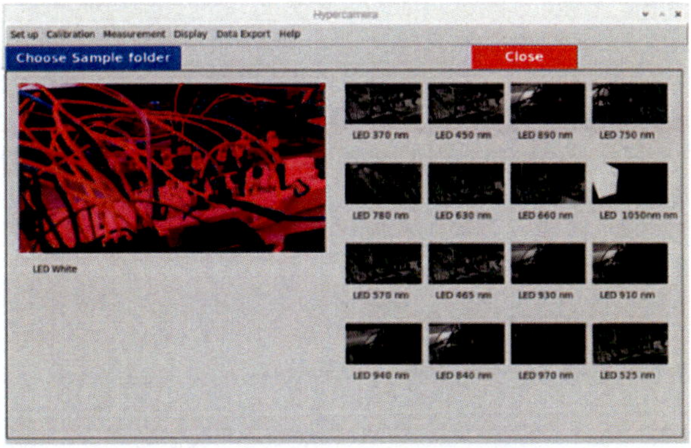

Fig. 6. Display of acquired images.

Finally, the user can export the images he selects through a USB connection. All the software is developed in Python using the Tkinter package for the development of the UIs.

3.2 Hardware Design

The hardware architecture of this version of the prototype is made up of several main elements. The main element of the system is a Raspberry pi 4. This embedded system is responsible for:

- Through its GPIO terminals, control the on and off of the LEDs according to the selection made by the user.
- Control the NIR camera in charge of acquiring images.
- Execute the communications protocol with the Arduino Uno system in charge of controlling the servomotor and stepper motor that allows moving the arm where the camera is positioned and the base on which the sample is located, respectively.

To control the LEDs, an electronic board has been designed to tune the LEDs according to the necessary light intensity characteristics. It is switched on and off by a series of transistors connected to the GPIO outputs of the Raspberry Pi 4. The LEDs, the NIR camera, and a diffuser in charge of providing create a better distribution of light inside the camera. The camera has its own communications interface, although, through the picamera.py library, its configuration and control through Python is very simple. An Arduino Uno system in combination with a CNC Shield module is used to control the engines. The communication between the Arduino and the Raspberry is done through a comprehensive communication protocol. This protocol has been designed for controlling the rotation of each of the engines. This design is based on a finite state machine.

The 3D parts necessary to create the rotating support where the LEDs and the camera are attached, and the rotating base attached to the stepper motor where the sample is placed, have been expressly designed for this prototype version. The final assembly and assembly that appears in Fig. 7 is provisional, and a final solution is currently being worked on to obtain an industrialized product with appropriate usability and operability analysis. The interaction with the user interface is done through the keyboard, mouse, and using the available touch screen.

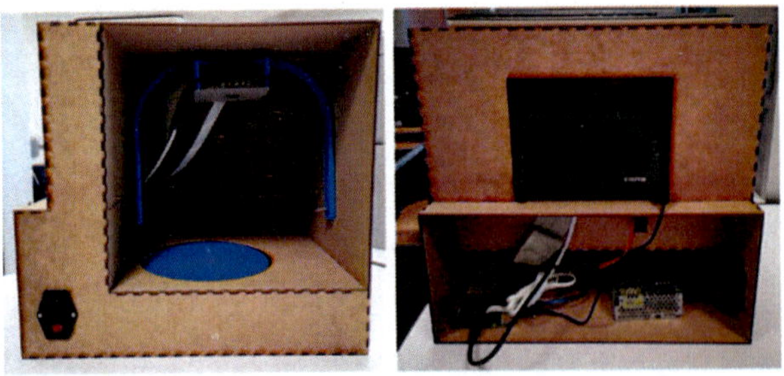

Fig. 7. HYPERCam v1.0.

4 Results

As one example of the usefulness of the HYPERCam v1.0. a simple oxization monitoring exercise was developed. This exercise consisted of taking multispectral images of a slice of an apple in a time span of 35 min during 5 min. In this regard, the camera was calibrated with the following parameters (metadata provided by the camera): camera_position: 90; sample_position: 0; iso: 200; sensor_mode: 1; sharpness: 0; contrast: 0; brightness: 50; saturation: 0; shutter_speed: 1000.

With these parameters, a multispectral image of the white reference (spectralon plate of 99.9% of reflectance) and of the dark reference (camera covered by a lid) were taken to correct the intensity of the different LEDs.

After taking the calibration images, the slice of apple was placed in the camera and multispectral images were taken every 5 min for 35 min. A multispectral image of dimensions (300 × 400 × 16) was obtained every 5 min. Each individual multispectral image was corrected following Eq. 1:

$$\text{For each wavelength } R_c = \frac{R - D}{W - D} \tag{1}$$

where, for each LED, the corrected reflectance of the sample (R_c) is the ratio between the reflectance of the sample (R) minus the reflectance of the dark reference (D), and the reflectance of the white reference (W) minus the reflectance of the dark reference (D). The RGB images obtained for 5 of the measured times can be seen in the Fig. 8 (top).

After the correction, the multispectral images of the corrected samples were concatenated, forming a fourth-order tensor of dimensions (9 × 300 × 400 × 16) corresponding to 9 measuring times, 300 × 400 pixels and 16 wavelengths. This fourth-order tensor was analyzed by using Principal Component Analysis (PCA). The results obtained can be seen in Fig. 8.

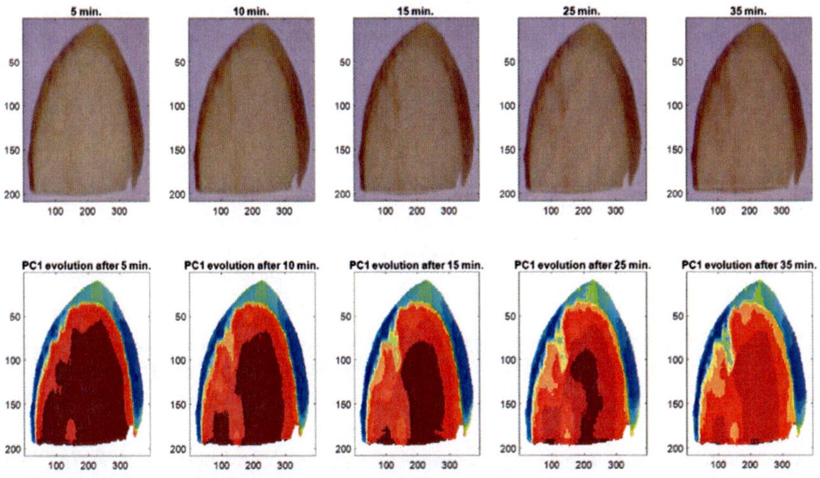

Fig. 8. Evolution of the oxidation of an apple slice. Top, the RGB images taken every 5 min. Bottom, First principal component score surface obtained.

The top line of the Fig. 8 displays the RGB images obtained of the apple slice during 35 min. Certain changes in color are appreciated in the surface due to the oxidation procedures that are involved. Nevertheless, these changes are merely subtle if the are compared to those observed after the analysis of the multispectral cube. The first principal component score surface is shown in the bottom line of Fig. 8. A strong decreasing of the dark red color is observed with time. This color is correlated with the lost of water in the surface with time, from the external part towards the center of the apple slice. At the same time, the increasing of the oxidized products is observed in the yellow and orange color, evolving from the outside of the surface to the internal part of the apple, indicating a time evolution that could be parametrized and that could give an indication of the storage time that an apple can hold before losing certain properties.

5 Conclusions and Work in Progress

The case presented above is just a small example of the utility of HYPERCam v.1.0. More works are being developed, specially in the 3D multispectral scanning of elements with commercial interest.

Due to the DIY character of HYPERCam v1.0. and the full integration with an analytical software like HYPER-Tools, HYPERCam will become a piece of integrated analytical hardware + software that any analytical laboratory could have at an affordable price. It is also expected that this prototype can be accessed remotely. The final goal is to build a remote laboratory that can be used by different institutions, thus sharing the images obtained through the camera and that each research group can apply their image processing algorithms.

References

1. Caballero D, Ríos-Reina R, Amigo JM (2021) Chemometrics and food traceability. Compr Foodomics 387–406
2. Patel KK et al (2012) Machine vision system: a tool for quality inspection of food and agricultural products. J Food Sci Technol 49:123–141
3. Land EH, McCann JJ (1971) Lightness and Retinex theory. J Opt Soc Am 61(1):1–11
4. Goetz AFH, Vane G, Solomon JE, Rock BN (1985) Imaging spectrometry for earth remote sensing. Science 228(4704):1147–1153
5. Khan MJ, Khan HS, Yousaf A, Khurshid K, Abbas A (2018) Modern trends in hyperspectral image analysis: a review. IEEE Access 6:14118–14129
6. Landgrebe D (1999) Information extraction principles and methods for multispectral and hyperspectral image data. Inf Process Remote Sens 82:3–37
7. Ríos-Reina, R., Callejón, R.M., Amigo, J.M.: Feasibility of a rapid and non-destructive methodology for the study and discrimination of pine nuts using near-infrared hyperspectral analysis and chemometrics. Food Control 130(108365) (2021)
8. Amigo, J.M., del Olmo, A., Engelsen, M.M., Lundkvist, H., Engelsen, S.B.: Staling of white wheat bread crumb and effect of maltogenic α-amylases. Part 3: spatial evolution of bread staling with time by near infrared hyperspectral imaging. Food Chem 353(129478) (2021)
9. Mobaraki N, Amigo JM (2018) HYPER-Tools. A graphical user-friendly interface for hyperspectral image analysis. Chemometr Intell Lab Syst 172:174–187

10. Davies R (2015) European Parliamentary Research Service. PE 568:337
11. USFDA, US Health and Human Services: PAT – a framework for innovative pharmaceutical development, manufacturing and quality assurance (2004)
12. Michael M et al (2019) Hyperspectral imaging of common foodborne pathogens for rapid identification and differentiation. Food Sci Nutr 7:2716–2725
13. Caballero D, et al (2019) Application of hyperspectral imaging and chemometrics for classifying plastics with brominated flame retardants. J Spect Imag 8(a1)
14. Ravn C et al (2008) Near-infrared chemical imaging (NIR-CI) on pharmaceutical solid dosage forms—comparing common calibration approaches. J Pharmaceut Biomed Anal. 48:554–561
15. Amigo JM (2020) Hyperspectral and multispectral imaging: setting the scene. Data Handling Sci Technol 32:3–16

Monitoring System for Glider Flight Parameters

Sebastian Pop[1], Marius Cristian Luculescu[1], Luciana Cristea[1], Florentina Custura[2], Attila Laszlo Boer[1], and Constantin Sorin Zamfira[1](✉)

[1] Transilvania University of Brașov, Eroilor 29, 500036 Brașov, Romania
{pop.sebastian,lucmar,lcristea,boera,zamfira}@unitbv.ro
[2] Romanian Aeroclub, Brașov, Romania
florentina.custura@aeroclubulromaniei.ro

Abstract. Training a pilot involves considerable time and resources, especially for glider pilots. The piloting of such an aircraft (without an engine) has some peculiarities of preparation because the lack of the engine does not allow errors, especially during the take-off and landing stages. Classically these gliders contain the minimum navigation equipment mechanically implemented. The paper presents a mechatronic system capable of acquiring and monitoring the flight parameters of the aircraft. Data from the aerial unit are displayed on a screen in the glider and transmitted to a ground unit that can monitor several aircraft simultaneously. Weather sensors information (temperature and humidity) are very useful for the pilot jhkkfor identifying and exploiting the presence of thermodynamic currents. The system represents an essential tool for the trainers to analyze the pilot mistakes and learn how to correct them in the future, improving the training process and increasing the safety of flights. In addition, in the case of a crash, the system data can be used as a BlackBox.

Keywords: Flight parameters · Navigation sensors · Monitoring system · Glider · Mission viewer

1 Introduction

1.1 Context

Gliders are non-motorized aircraft, which launched from a certain height will fly with a continuous descent slope. These aircraft were and are good for civil aviation, as well as military aviation in some states for the training of military pilots, before making the switch to motorized aircraft [1].

Training a pilot involves considerable time and resources, especially for glider pilots. The piloting of such an aircraft without an engine has some peculiarities of preparation because the lack of the engine does not allow the existence of errors, especially during the take-off and landing stages. An important aspect in correcting these skills is flight analysis. This is done naturally through the experience of instructors.

M. E. Auer et al. (Eds.): REV 2022, LNNS 524, pp. 252–260, 2023.
https://doi.org/10.1007/978-3-031-17091-1_26

1.2 Problem Formulation

The paper aims to present how a mechatronic system (Glider Flight Parameters through Radio Network - GFPRN) was created, which supports the process of pilot training. The GFPRN can monitor and collect data on the aircraft's flight parameters and transmit them in real-time on the ground. By monitoring these parameters, the aim is to collect data from the flight exercises performed on the track tour and advanced exercises that involve the exploitation of thermodynamic rising currents [2]. This data is displayed in real-time both onboard the glider and centrally on the ground. The proposed system does not replace or interfere with mandatory certified equipment (Transponder or ADSB) to transmit flight data to airspace management authorities.

A significant advantage of the GFPRN system is that for its installation, no structural changes have to be made to the glider, which would require equipment certification activities.

At the same time, an important aspect was taken into account, that of cost. This system is made with standard sensors used in robotics projects that use development modules such as Arduino or RasberryPi and Open Source software modules. This type of approach allows the optimization of the GFPRN for any type of glider. It also allows the equipment to be easily upgraded both in terms of software and hardware.

Unlike motorized planes, glider flight differs a lot, as its evolution is based on the exploitation of thermodynamic currents. For this reason, classically, gliders designed to train gliders are equipped with a minimum of navigation equipment (speedometer, altimeter, variometer, turn and slide indicator, magnetic compass).

Figure 1 shows the dashboard with the classic equipment of a two-seater glider type IS 28B2, produced by IAR Ghimbav - Romania and used both as a school or training glider.

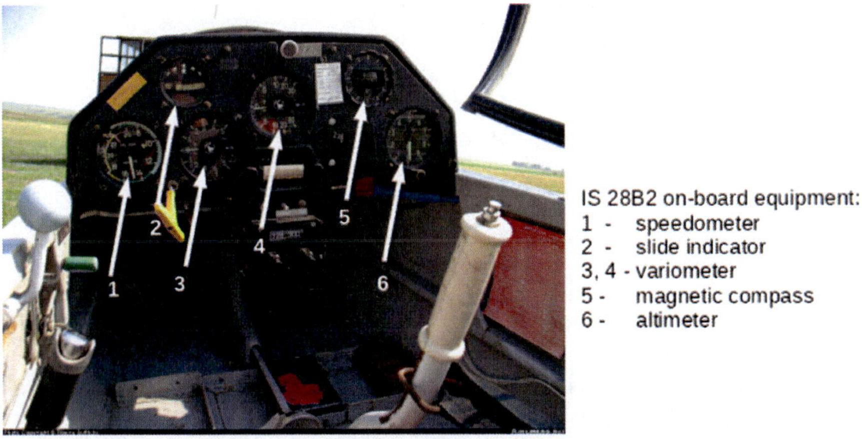

Fig. 1. Classic IS28B2 onboard equipment

The operation of this type of device is exclusively mechanical, having in its composition: aneroid capsules, open capsules, membranes, pressure sockets, Bourdon tubes, and harmonic capsules. This type of analog device does not have a very high sensitivity in detecting variations in speed and altitude, the data being displayed late. The data is displayed on analog dashboard clocks.

To benefit from modern technology, a solution used by most glider pilots was the use of PDA or Smartphone devices. However, this solution solves only a small part of the problem, namely that of navigation, with the help of digital maps and GPS positioning.

The solution proposed by GFPRN solves all these problems by integrating into a single platform both the inertial sensors and the navigation system.

Besides the main purpose for which it was made, the system can also be used to manage the technical resource of the glider on which it was mounted. GFPRN will provide the technical staff with the data and meteorological conditions in which the glider was operated. This information is currently recorded manually, and it is necessary for planning maintenance work.

At the same time, the system can be configured to transmit other data onboard the glider, such as biometric data of the pilot, or the amount of oxygen in the cylinders, in the case of flights made in undulating rising currents. In such flight conditions, the altitude of 3500 m is exceeded from where it is necessary to use oxygen, and the temperatures can reach up to −40 °C.

2 The Monitoring System for Glider Flight Parameters

2.1 The Structure of the System

The system's design started from the premise that it should be as small and easy to install onboard the glider. The installation can be done rigidly with screws or with adhesive. The choice of the sensors and the microcomputer was made so that the energy consumption is minimal, and the power supply could be made from a power bank or any existing voltage source onboard the glider.

The installation procedure involves an initial calibration of the sensors performed on the mounting bench according to the aerodynamic specifications of the glider. No further calibration of the equipment is required during operation.

The system built by our team is made up of two components: an air module installed onboard the aircraft and a ground module installed in the flight control center, as shown in Fig. 2.

A permanent telemetry radio connection between the two systems creates a network with several air modules.

2.2 The Aerial Unit

The aerial module is designed to monitor, display and transmit two types of data: Data collected by navigation sensors (GPS, Air speed, Gyro, Compass, Acc), as well as weather data (Temperature and Humidity), the latter being handy for the operation of thermodynamic currents.

The construction of the air module (see Fig. 3) was done with a Small Board Computer (configured with 32bit STM32F427 Cortex-M4F core with FPU, 168 MHz/252 MIPS, 256 KB RAM, 2 MB Flash) [3].

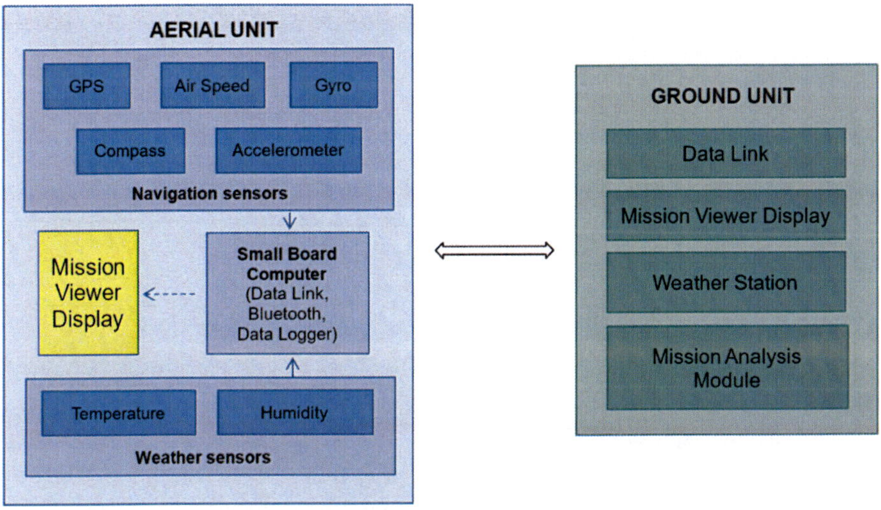

Fig. 2. The structure of the monitoring system

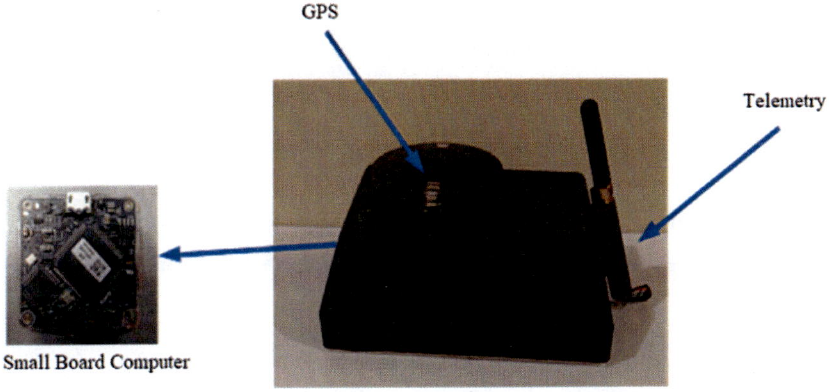

Fig. 3. Aerial unit

The position data is read with a Ublox M8P GPS module, capable of receiving data from several constellations of satellites (GPS L1C/A, GLONASS L1OF, BeiDou B1I), with a reading rate of 8 Hz. Glider inertial data is read using an IMU type ICM20948. Temperature and humidity data were collected with a DHT22 sensor. The whole assembly has been placed in a box, which allows its quick assembly and relocation.

The data display onboard the glider is done through the open-source application Mission Planner (see Fig. 4) installed on any Android OS device via Bluetooth connection with an HC05 module.

The interface used to display glider edge data is derived from the open-source Mission Planner application. It has been configured to have only the data needed by the glider pilot. The advantage of this type of interface is that it has an audio variometer

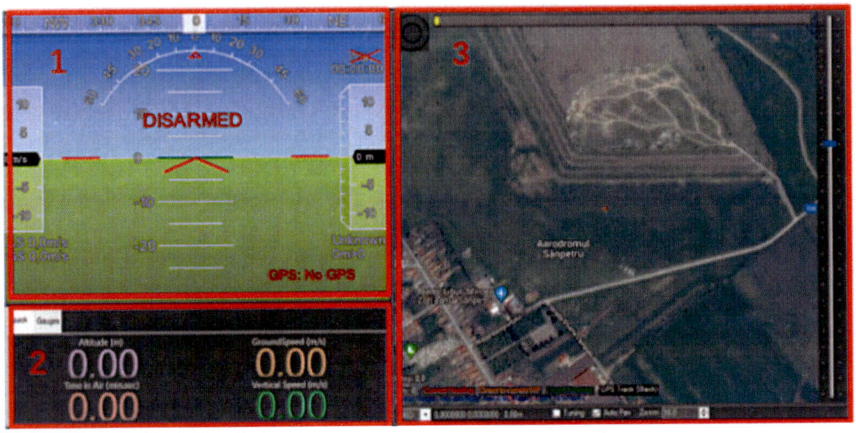

Fig. 4. MissionPlaner onboard software.

implemented. The interface consists of three screens delimited with red color and numbered from 1 to 3 (Fig. 4): the artificial gyro-horizon screen (1), the digital data window (2), and the navigation window (3).

Ground data transmission was performed using RFD868 data link modules. In addition, by implementing the RFD900x-Multipoint Sik firmware, it was possible to create a multipoint radio network [4]. Thus, several gliders in flight could be monitored simultaneously in the same software interface on the ground station.

2.3 The Ground Unit

The soil module is made up of four subsystems: the RFD868 multipoint data link module, the Mission Planer module, weather station, as well as mission post-analysis software. The data link module is made with RFD868 multipoint for simultaneous reception of data from several aircraft. This module is easily configurable, with a serial connection, data reception, and encrypted AES 256 bits.

For communication, it uses the MAVlink protocol and a two-antenna system with an automatic selector of the best reception feature. The software module (see Fig. 5) for displaying data collected onboard aircraft can be installed on any platform running a Windows operating system, receiving data on the serial interface from the RFD868 module [3].

The ground station interface is similar to onboard software, with the difference that it can display data from multiple aircraft simultaneously. When selecting a particular aircraft, it can be read in real-time: aircraft speed (1), flight altitude from the take-off point (2), tilt angle (roll) and pitch (pitch) (3), data provided by inertial sensors (4) as well as the position of the aircraft in the WGS84 coordinate system (5). It can also view the route of the aircraft (6) and the position on the map of the operational aircraft (7).

These parameters are stored both onboard the aircraft via SBC and on the ground via the Mission Planer interface. Through the Log analysis utility, this data can be transformed into formats such as *.kmz/kml, *.shp, or in tabular form [5].

Fig. 5. MissionPlaner ground unit software

3 Results and Interpretation

Thus the data collected on one of the monitored flights are represented in graphical form in the following figures. These data are correlated in the following images: flight path (see Fig. 6), variation of pitch angle (pitch) (see Fig. 7), roll angle (roll) (see Fig. 8), variation of altitude (see Fig. 9).

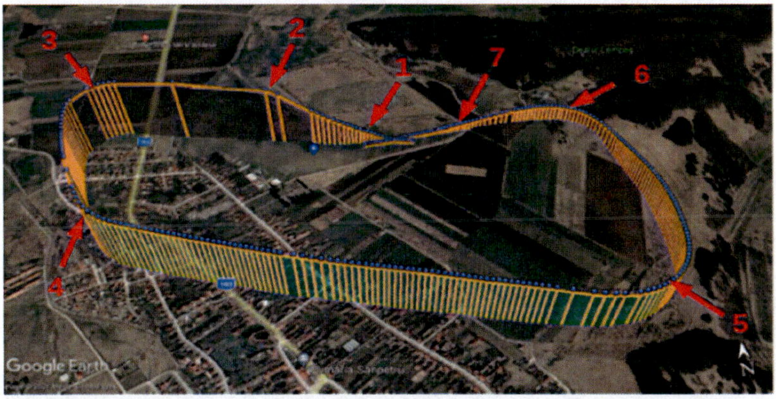

Fig. 6. Flight path

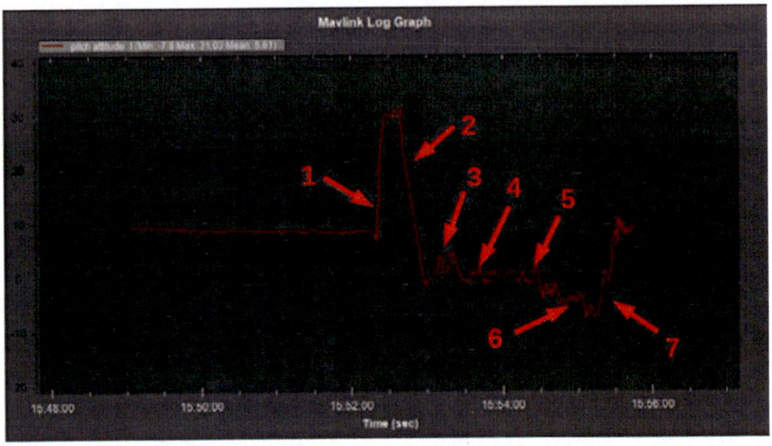

Fig. 7. Variation of pitch angle

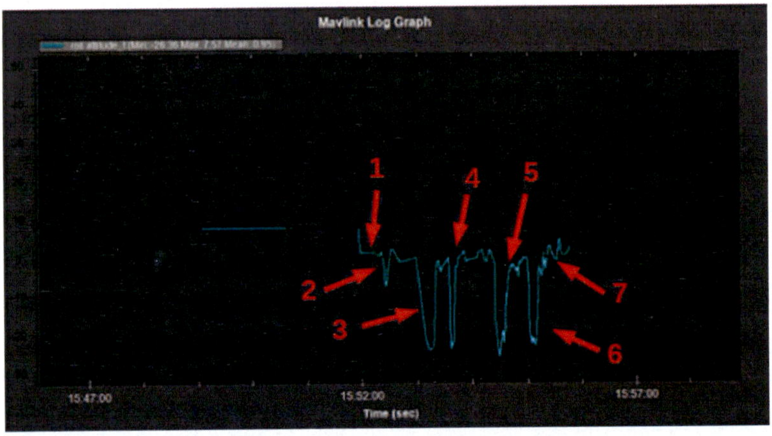

Fig. 8. Variation of roll angle

Thus, on the flight performed, from the data recorded with the help of GFPRN and graphically transposed in Google Earth can be identified the main points of the runway made by the glider, namely: take-off with ascent slope (1), trigger point (2), performing the first turn (3), make the second turn (4), trajectory to turn three (5), make turn four and enter the landing slope (6).

The flight analysis must be performed by analyzing several onboard parameters to see the correctness of the commands given by the pilot student [6].

The pitch angle variation (pitch) (see Fig. 7) would show us if the pilot student observed the optimal flight slope in the execution phases of the runway.

It can be seen that after take-off, the glider had an ascent slope (1) of 33°, followed by a correct trigger (2) to the tow cable up to an angle of -2°. Later it can be seen that in

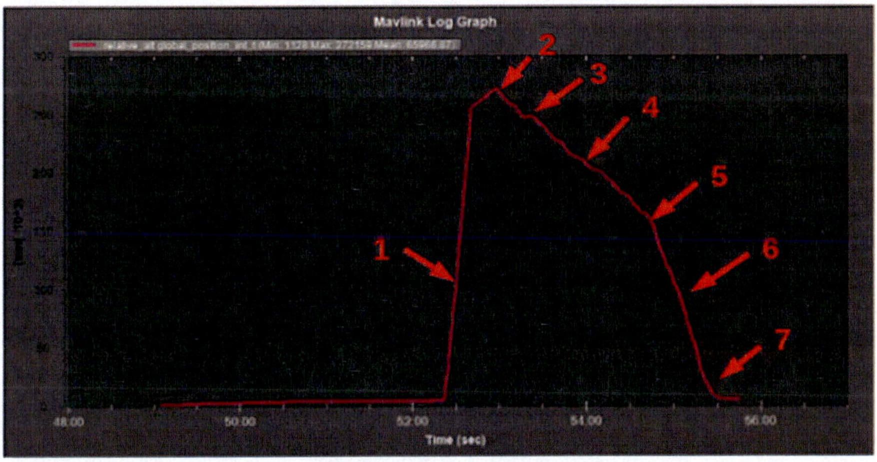

Fig. 9. Altitude variation

the other stages of the runway, the variation of the pitch angle was small and relatively constant, which indicates that we performed a correct flight.

Figure 8 shows the data recorded by the GFPRN regarding the movement of the aircraft around the longitudinal axis. These variations correspond to the angle of lateral inclination (roll), and as can be seen on the graph, the angle of 28° corresponds to the four turns related to the track lap. The roll angle was within normal flight parameters for the rest of the journey.

The analysis of altitude variation must be correlated with the other factors analyzed. It can be seen that the altitude of 280 m reached by the glider in tow (1) corresponds to the pitch angle of 33°, followed naturally by the trigger (2). The glider's free flight (3, 4, 5) corresponds to its standard gliding slope. The accentuation of the descending slope (6,7) corresponds to the landing maneuvers [7].

In this case, we analyzed a flight in which the ascent slope during towing was performed correctly. Turns 1, 2, 3, and 4 met the requirements specified by the airport's operating procedures. In addition, the landing slope, and the contact with the ground, were realized in safe conditions.

As the glider is not equipped with a powertrain, the weather influences its flight path. Thus, the correct interpretation of the data obtained from the GFPRN cannot be achieved without meteorological data. In this sense, a weather station was installed at the base of the track.

4 Conclusions

Currently, there are EASA (European Union Aviation Safety Agency) certified digital onboard devices on the market mounted on the newest gliders. There is also a low-cost software solution installed on phones or tablets, but they are more focused on providing GPS positions and are limited to the sensors installed on these devices. The approach presented in this paper is innovative. It is flexible in terms of the installed sensors, and it can transmit the full range of information collected onboard the glider to the Ground Unit.

An important outcome is the low-cost feature of the system due to standard sensors used in robotics projects. In addition, the system implementation does not require changes in the glider structure. Anticipated outcomes aim to monitor several gliders on the same Ground Unit, offering an essential tool for the trainers to analyze the pilot mistakes and learn how to correct them in the future. In the case of a crash, the system data can be used as a BlackBox.

At the same time, the system is an excellent tool for the technical management of the glider. It records all the data regarding the operating time and the meteorological conditions in which it evolved.

In the tests performed with this system, the following parameters were analyzed:

- The proper execution of the maneuvers within the take-off procedure;
- The proper execution of the two launch procedures;
- The observance of the procedure in the execution of the track tour;
- The track lap analysis especially turns three and four;
- The analysis of the landing slope;
- The contact with the ground.

By using the above-presented system, the training process is improved, and at the same time, the safety of flights is increased.

We intend to miniaturize the system in future development directions by making our interface boards and using other communication technologies, including 5G.

References

1. European Union Aviation Safety Agency EASA: Sailplane Rule Book. https://www.easa.europa.eu/document-library/easy-access-rules/sailplane-rule-book-easy-access-rules. Accessed 10 Nov 2021
2. Rene Heise, Herold W-D, Hertenstein R (2009) Weather Forecasting for Soaring Flight, Edition: WMO-No. 1038: World Meteorological Organization, Hindman. ISBN 978-92-63-11038-1
3. Pisano W, Lawrence D, Gray P (2007) Autonomous UAV control using a 3-sensor autopilot. In: Conference: AIAA Infotech@Aerospace 2007 conference and exhibit https://doi.org/10.2514/6.2007-2756
4. Lebreton J, Murad N, Lorion R (2016) Radio frequency mapping using an autonomous robot: application to the 2.4 GHz band. In IOP conference series: materials science and engineering. IOP Publishing, 2016, IEEE Radio and Antenna Days of the Indian Ocean (IEEE RADIO 2015), Conference 2016. LNCS, vol 9999, pp 1–13. Springer, Heidelberg
5. Kun P (2015) Applying open GIS solutions in glider competitions https://doi.org/10.13140/RG.2.1.3338.6969.
6. Popovici D (2009) Proceduri operaționale, Aeroclubul României, București
7. Landell-Mills N (2021) The physics of gliding and how dynamic soarin could enable perpetual flight. https://doi.org/10.13140/RG.2.2.18506.52162/1.

Augmented Reality as a Support to Solve Pallet Loading Problem

Vule Reljić(✉), Slobodan Dudić, Želimir Trišić, Branko Jekić, Vladimir Jurošević, and Ivana Milenković

Faculty of Technical Sciences, University of Novi Sad, Trg Dositeja Obradovića 6, 21000 Novi Sad, Serbia
vuketa90@uns.ac.rs

Abstract. This paper describes one of the methods for solving the problem of packaging transportation boxes on an auxiliary transportation equipment such as a flat pallet. An augmented reality application has been developed for this purpose. In the application, real boxes have their virtual twins that can be handled before they are physically arranged on a flat pallet. The application uses a smartphone camera to detect the flat pallet and allows the worker to see the stacked virtual boxes in real time, so that the worker knows which box should be stacked next. The order of stacking boxes is determined by an algorithm that runs in the background of the application. The current disposition of the boxes on the pallet is kept until the worker clicks and adds a new virtual box. In this way, by applying augmented reality for the purpose of stacking boxes on a pallet, possible stacking errors are reduced. While augmented reality continues to evolve, this paper is a contribution to its practical application in the field of material handling.

Keywords: Augmented reality · Pallet · Stacking

1 Introduction

Difficulties and problems that occur in production processes related to the handling of the workpieces, final packaging of finished products and the formation of collective packaging, lead to the intensification of scientific research and engineering approaches in finding concrete solutions in this area. One of the problems that arises is the process of proper formation of collective packaging by stacking the transportation boxes on auxiliary transportation equipment such as a flat pallet, or the pallet loading problem [1–5]. In this paper, a proposal for solving this problem by applying augmented reality is given. Augmented reality [6–10] enables the insertion of virtual objects into the real three-dimensional space of the observer. People experience these objects as if they were real. However, in this case, the virtual objects (virtual boxes) that appear on the smartphone screen should be distinguished from the real boxes that are physically stacked on the pallet. Virtual boxes are displayed on the smartphone, while real boxes are arranged on a flat pallet. The real boxes are arranged on a pallet according to the layout given on the smartphone. Each real box corresponds to one virtual box of identical dimensions, i.e.

M. E. Auer et al. (Eds.): REV 2022, LNNS 524, pp. 261–270, 2023.
https://doi.org/10.1007/978-3-031-17091-1_27

each real box has its own virtual twin. The real boxes are physically stacked on a standard flat EUR-pallet with the dimensions of 1200 mm × 800 mm. The flat pallet enables easy grouping, handling and distribution of products, which reduces the possibility of their damage. By scanning the QR code, the process of stacking one virtual box at a time is shown to the worker on the smartphone. In this way, the application of augmented reality in solving this practical task, increases efficiency and reduces errors [11].

1.1 Augmented Reality on Smartphones

The application of augmented reality was achieved with the help of a smartphone. Smartphones are widely used in everyday human life, so handling the application is intuitive and easy. The touch screen and a camera, which are indispensable parts of modern smartphones, are necessary for the successful functioning of the application. The instalation of the application is simple. After the application is downloaded to the phone, the user should run the downloaded file with the extension *.apk and click to confirm that he/she wants to install the application. Once installed, the application is ready to run.

1.2 The Brief Overview of the Paper

This paper is organized as follows: Sect. 2 gives a description of the developed augmented reality application. Section 3 is dedicated to a detailed description of the implementation of the developed application on a smartphone. Section 4 describes the testing process throught one example. Finally, Sect. 5 concludes the paper and introduces future work.

2 Development of an Augmented Reality Application

As part of the process of improving the exchange of data and goods within the plant or with the external environment, augmented reality is suitable for modern logistics operations in the warehouse. The use of augmented reality aims to increase the overall efficiency of production systems, which includes storage processes. About 20% of the total factory logistics processes go to warehousing. The process of taking over (finding, collecting) products from the warehouse based on customer requirements - Order picking, makes from 55 to 65% of all costs in the warehouse [12]. Based on these assessments, the importance of process sharing within the warehouse - within production and service systems - is confirmed. Although not explicitly stated, this procurement process ends with the grouping of products into transportation packaging (transportation boxes) and stacking of packaging (boxes) on one load-bearing unit - Unit Cargo. Packaging that is stacked on a pallet is grouped on the basis of a loading order formed by sales department. In order to make the packaging stacking process more efficient, an augmented reality application was introduced as a guide.

2.1 Algorithm Used

The virtual boxes that appear on the smartphone direct the worker where to place the real box on the pallet and how to orient it. An algorithm is executed in the background of the application, so that after each new click of the user, the application displays the position of the box which has to be stacked next. It should be noted that the whole algorithm is executed immediately, while the boxes on the screen are displayed sequentially; with each new click a new box is displayed. Namely, it would be confusing for the worker to see all boxes stacked on the screen. He would not see inner boxes and their placement. Therefore, the worker adds new boxes on the screen with a new click in accordance to coordinate the display of the boxes on the screen and the actual stacking of the physical boxes on the pallet. The algorithm referes to rectangular boxes, which are very often used for pallet stacking and transport. It is important to note that all boxes to be stacked have one dimension equal. This is a necessary condition for the algorithm function. The equal dimension can be any of three dimensions - length, height or width. In this way, the three-dimensional problem of stacking boxes on a pallet is reduced to two-dimensional. In this way layer are formed. Each row is a layer. The process of stacking rectangular boxes comes down to placing a rectangle on a surface which size is equal to the size of the active surface of the pallet. Each of the rectangles within the algorithm is the base of one box lying on the surface of the pallet or the previous row of boxes. The algorithm was created with the idea of minimizing the free space between the boxes, i.e. the "slits" that appear between the rectangles. When the arrangement of a new row of boxes begins, in order to minimize the free space, the box with the largest surface of the base is chosen. When one active pallet surface is filled (it cannot fit a new rectangle, even the smallest one), then one row of boxes is stacked. Thus, all rows are arranged one after the other, until all the boxes provided by the loading order are assembled.

2.2 Augmented Reality Application

As a result of the implementation of the algorithm, the worker sees a three-dimensional content on the smartphone in the form of a box that is currently in the loading position. Developed application recognizes the appropriate marker (pallet) and enables a visual display of the box in a certain color. The Unity development platform was used to create the application for Android systems. The Vuforia plugin was used to create markers for pallet recognition. The marker is a CAD model of the EUR-pallet (Fig. 1). The smartphone camera should be oriented so that that the spatial orientation of the pallet agrees with the spatial orientation of the marker created on the basis of the CAD model. In other words, the view of the pallet obtained from the smartphone camera should correspond to the marker created in the application. A certain deviation from the complete match of the display from the camera and the marker in the application is allowed. When the marker is spotted using the correct camera orientation, the area inside the marker will still be recognized, although there is little change in the camera angle relative to the palette. The recognition angle in relation to the vertical axis is from $-180°$ to $+180°$, and the recognition angle in relation to the horizontal axis from $0°$ to $60°$. These data should be taken into account when recognizing and storing the recognized marker. It should be noted that the closest corner of the pallet to the camera is not damaged (truncated).

Until the pallet marker is successfully recognized, it will not be possible to display the stacking of virtual boxes on the screen. If the marker is lost during the operation of the application, the current status is saved. Redetecting the marker continues where it left off.

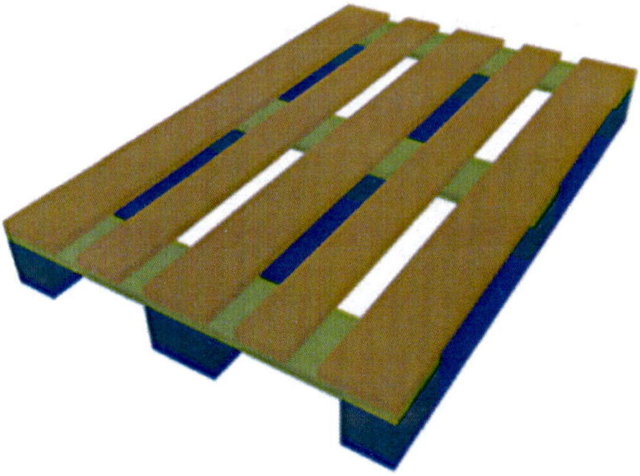

Fig. 1. Pallet recognition marker

A prerequisite for the successful operation of the application is that after starting the application, the smartphone camera scans the QR code that contains the information necessary for the algorithm function (number, dimensions and color of the boxes). Therefore, the procedure for successful operation is as follows: the worker needs to run the application on the smartphone, scan the QR code from the loading order and follow the instructions within the application. The QR code does not have to be scanned exclusively from a paper loading order, but can be done from screens such as monitors, tablets, televisions and similar devices.

3 Implementation of the Application on a Smartphone

As already mentioned, when the transportation packaging in the form of transportation boxes is prepared for stacking on a pallet, it is necessary to start the application and follow a certain procedure in order to successfully stack all transportation boxes. For ease of understanding, this procedure is shown in Fig. 2, as a graphical representation of the flow of augmented reality application. Arrows indicate the progress of the application.

The application steps appear in the following order: launch the application, scan the QR code, generate a sequence algorithm with the position of stacking the boxes on the pallet, and add each of the virtual boxes. The back arrow in the step of adding a virtual box, indicates that this step is repeated until all the virtual boxes on the displayed on the smartphone are stacked.

3.1 Visual Display of the Application on the Smartphone

The display of stacking and virtual boxes on the smartphone is realized through the application interface on the smartphone (Fig. 3).

The order of the steps is shown by the letters a to d. After successfully scanning the QR code (Fig. 3a), the user has the option to start generating the algorithm (Fig. 3b) according to the data received from the QR code. Confirming the algorithm generation

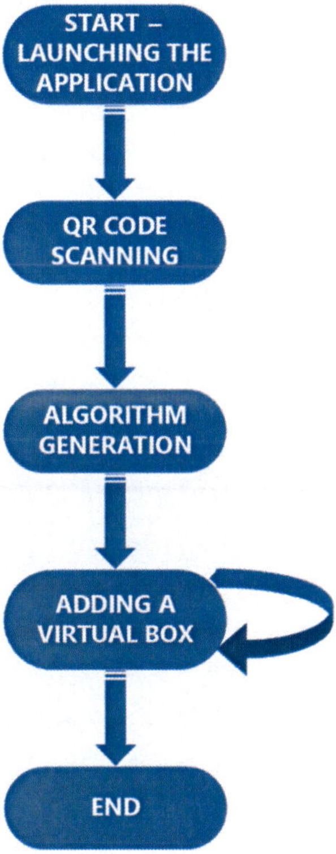

Fig. 2. Application flow diagram

option allows the user to add each individual box. It is advisable to click on the button to add the first box (Fig. 3c), in order to determine whether the application has recognized the pallet marker via the smartphone camera. Until the pallet marker is detected, virtual boxes will not be enabled.

As soon as the first box appears (Fig. 3d), it is clear that the pallet marker has been detected. This is not a problem for the worker, since he/she knows which box is the first and is sure that he/she has detected the pallet marker. As the first box is added, the procedure is repeated for the remaining boxes.

It should be emphasized that if the camera is directed to the other side in relation to the pallet, the display of the pallet and complex virtual boxes is lost. This happens because the pallet marker is lost. The current state of the stacked virtual boxes is saved, and the previously saved state is displayed again by detecting the pallet marker again.

It is important to distinguish the virtual boxes that appear on the smartphone from the real boxes that physically stack on the pallet. Namely, the display of virtual boxes is independent of physically stacked real boxes. The virtual boxes that appear on the smartphone are an element of augmented reality and different boxes have different colors. The colors of virtual boxes are sharp and clear, and generally differ from the colors of real boxes. To further point out the difference between the virtual boxes displayed on the

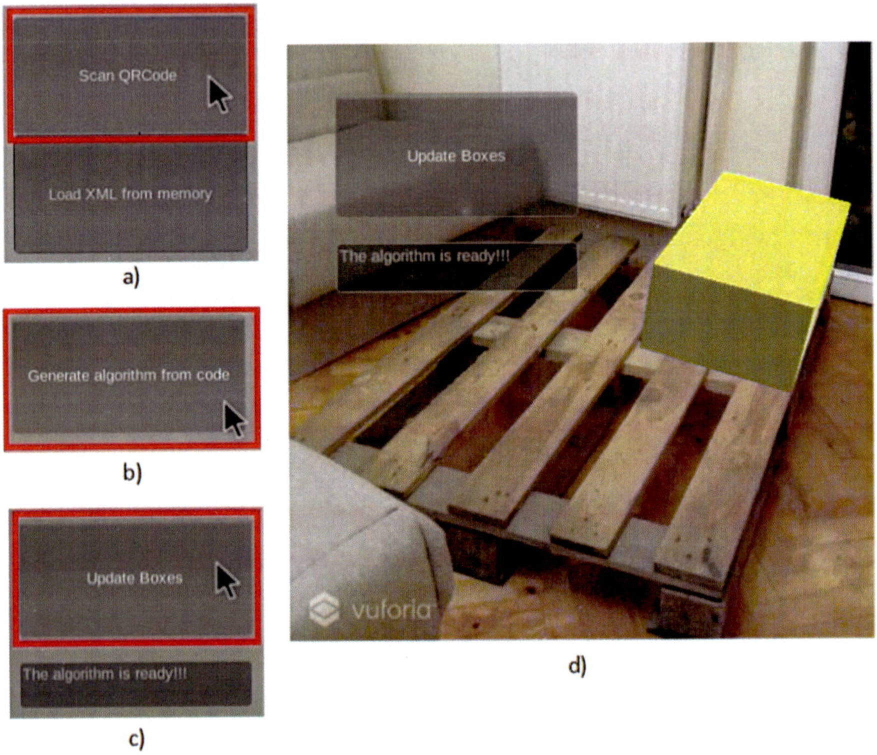

Fig. 3. Sequence of operations in the application

smartphone and the real ones, the worker can see the presentation of the stacked virtual boxes without physically stacking even single box on a pallet.

Based on this, the worker has two options: to display all virtual boxes on the smartphone, then to stack the real boxes, or to display an individual virtual box (or some of them) on the smartphone, and then to stack it (or them) on a pallet. The worker will know that he/she has finished stacking the virtual boxes, when the button titled "Update boxes" disappears. In this way, it is immediately clear to the worker that he/she cannot add new virtual boxes, since the button that enables that has disappeared. The process of stacking transportation boxes within a single loading order will be completed when the worker will physically stack the boxes on a pallet according to the display of the stacked boxes on the smartphone.

4 Testing and Results

The functionality of the application was tested on two markers. One marker is a paperback book, on which the application was initially tested. The second marker is a standard flat EUR-pallet with the dimensions of 1200 mm × 800 mm used to transport packaging in the plant. The application begins by scanning the QR code that contains the information necessary for the algorithm function. As previously stated, the required data are:

dimensions (width, length and height), number and color of the boxes. The dimensions of the boxes are given in mm. These are the dimensions that correspond to the real boxes that physically stack on the pallet. As also mention above, a necessary condition for the function of the algorithm is that all boxes have one dimension equal. Any of the three dimensions can be equal to each of the boxes. In addition, it is necessary to state that the boxes of the same dimensions are displayed in the same color on the smartphone.

The application functionality and algorithm were tested for several different sets of input data. The data required to generate the algorithm are contained in one data set as can be seen in Table 1.

Table 1. Example of a set of input data

Dimensions [mm]			Number	Color
Width	Length	Height		
250	175	100	10	RED
350	250	120	6	BLUE
420	250	140	8	GREEN
275	200	250	14	YELLOW
200	335	250	7	MAGENTA
250	100	80	18	GREY
450	375	250	4	WHITE

Each of the packaging boxes corresponds to one row in the table, which, in addition to the dimensions and color, also specifies the number of identical boxes. From the Table 1 it can be seen that the equal dimension for several boxes is 250 mm. The color names given in the table are used identically within the application. Based on the data set from the table, a QR code is generated (Fig. 4).

Fig. 4. QR code based on the set of input data

As already pointed out, all the data from the loading order necessary for the operation of the algorithm are in the QR code. Scanning the QR code via a smartphone camera allows the algorithm to run in the background of the application.

4.1 Displaying Stacked Boxes on a Book

For the set of input data from Table 1 and the generated QR code, the functionalities of the application and the algorithm on the paperback book were tested first. The virtual boxes are stacked on the front cover, as can be seen in Fig. 5.

Fig. 5. Stacking boxes on a book

It should be noted that the real dimensions of the book are smaller than the real dimension of the flat EUR-pallet. However, in order to stack virtual boxes to the cover of the book, the sizes of the boxes are proportionally reduced.

4.2 Displaying Stacked Boxes on a Flat Pallet

The main functionality of the application is reflected in the stacking of the transportation boxes on a flat pallet (Fig. 6). Unlike the book, the flat pallet is really intended for transporting packaging boxes. In this case, the virtual boxes retain their actual size and are displayed as such on the smartphone.

Empty space on the surface of a flat pallet occurs when the algorithm determines that even the smallest box can no longer be stacked. In this regard, the algorithm strictly specifies that the size of the boxes cannot exceed the range of the active surface of the pallet to which they belong. In other words, all the complex boxes are inside a rectangular prism, and the base of that rectangular prism is the upper surface of a flat palette. The height of that prism depends on the height at which the upper surface of the last stacked box is located. In the case of a smaller number of boxes in the last row, the worker can manually place them closer to the middle of the surface of the previous row. This ensures greater transportation stability.

Finally, the success of the algorithm for the input data set (Table 1) is shown through performance evaluation, which is given in Table 2.

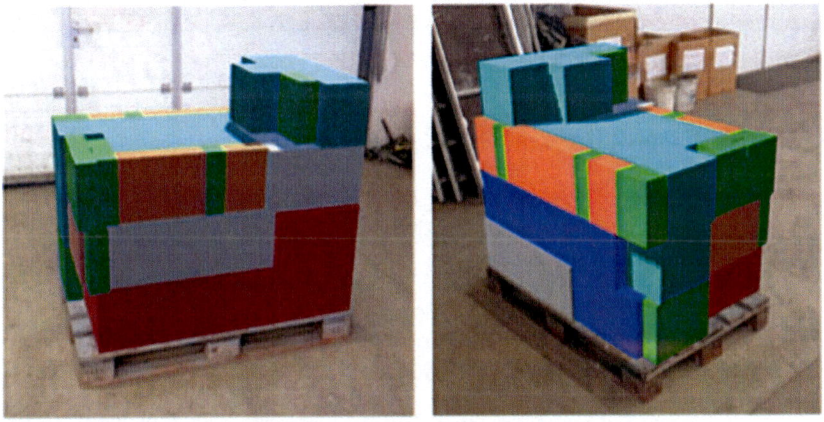

Fig. 6. Stacking boxes on a pallet

Table 2. Performance evaluation of the algorithm

Running time [ms]	Number of packed boxes	Number of given boxes	Unused space [%]
13	67	67	6

The data in Table 2 show that the algorithm was executed in 13 ms, that it set all the boxes and that the unused pallet space was 6%.

5 Conclusions

With practical implementation and stacking results, the functionality of the augmented reality application and the success of the algorithm for stacking transportation packaging (transportation boxes) on an auxiliary transportation equipment (flat EUR-pallet) in an industrial environment were confirmed. In addition, the functionality is tested on a pallet that has been exploited for a long time and is slightly damaged at certain ends.

Operating with the application is simple and intuitive, and no additional training of workers is required. The use of the augmented reality application when stacking transportation packaging reduces the time required to stack boxes on a pallet, and transporting a flat pallet with boxes becomes faster. In this way the time that elapses from the moment of receiving the request from the customer to the delivery of the finished product to the customer is reduced. This speeds up the transport and delivery of a larger number of grouped product packages, which also affects the lower cost of finished products due to reduced logistics costs of the plant.

In this paper the AR application is implemented on a smartphone. The manageability is a little difficult, due to simultaneously handling of smartphone and boxes with two hands. Our aim was to practicaly implement two main components – the algorithm and AR application on a widespread smart device. Further work will be directed to other portable devices with the Android platform such as smart glasses or screens for displaying content - smart TV's.

References

1. Terno J, Scheithauer G, Sommerweiß U, Riehme J (2000) An efficient approach for the multi-pallet loading problem. Eur J Oper Res 123(2):372–381. https://doi.org/10.1016/S0377-2217(99)00263-5
2. Martins GHA, Dell RF (2008) Solving the pallet loading problem. Eur J Oper Res 184(2):429–440. https://doi.org/10.1016/j.ejor.2006.11.012
3. Morabito R, Morales S (1998) A simple and effective recursive procedure for the manufacturer's pallet loading problem. J Oper Res Soc 49:819–828. https://doi.org/10.1057/palgrave.jors.2600588
4. Hodgson JT (1981) A combined approach to the pallet loading problem. A I I E Trans 14(3):175–182. https://doi.org/10.1080/05695558208975057
5. Ram B (1992) The pallet loading problem: a survey. Int J Prod Econ 28(2):217–225. https://doi.org/10.1016/0925-5273(92)90034-5
6. Paelke V (2014) Augmented reality in the smart factory: supporting workers in an Industry 4.0. environment, In: Proceedings of the 2014 IEEE Emerging Technology and Factory Automation (ETFA), Barcelona, Spain, pp 1–4. https://doi.org/10.1109/ETFA.2014.7005252
7. Reljić V, Milenković I, Dudić S, Šulc J, Bajči B (2021) Augmented reality applications in Industry 4.0 environment. Appl Sci 11:5592. https://doi.org/10.3390/app11125592
8. Fraga-Lamas P, FernáNdez-CaraméS TM, Blanco-Novoa Ó, Vilar-Montesinos MA (2018) A review on industrial augmented reality systems for the Industry 4.0 shipyard. IEEE Access 6:13358–13375. https://doi.org/10.1109/ACCESS.2018.2808326
9. Gattullo M, Scurati GW, Fiorentino M, Uva AE, Ferrise F, Bordegoni M (2019) Towards augmented reality manuals for Industry 4.0: a methodology. Robot Comput-Integr Manufact 56:276–286. https://doi.org/10.1016/j.rcim.2018.10.001
10. Masood T, Egger, J (2019) Augmented reality in support of Industry 4.0—implementation challenges and success factors. Robot Comput-Integr Manufact 58:181–195. https://doi.org/10.1016/j.rcim.2019.02.003
11. DHL Trend Research (2015) DHL Global Technology Conference Augmented Reality in Logistics, Dubai, 16 April 2015. https://na.eventscloud.com/file_uploads/b05d26158820d377ca7a022173486cb0_T.6_InnovationinPractise-AugmentedRealityinLogistics.pdf. Accessed 11 Oct 2021
12. De Koster R, Le-Duc T, Roodbergen KJ (2006) Design and control of warehouse order picking: a literature review. Eur J Oper Res 182(2):481–501. https://doi.org/10.1016/j.ejor.2006.07.009

Impact of Industry 4.0 as Perceived by the South African Railway Industry

K. Letlala, B. B. S. Makhanya(✉), J. H. C. Pretorius, and H. Nel

University of Johannesburg, Johannesburg, South Africa
bheki.makhanya@transnet.net, jhcpretorius@uj.ac.za

Abstract. The South African railway industry is in the process of aligning itself with Industry 4.0 trends. To stay competitive with today's customer demands, the industry has begun to digitize some of its services. Despite the benefits of the Industry 4.0 initiatives, employees in the railway sector are concerned that automation will replace humans in the workplace. Because of this negative perception, some people have expressed reluctance to fully embrace Industry 4.0 within the sector. The goal of this study was to determine the level of awareness of Industry 4.0 and its perceived impact, and some of the challenges that the South African railway sector has faced in adopting Industry 4.0. The study targeted a total population of 110 engineering practitioners within the South African railway infrastructure maintenance environment. The findings indicated that there was a high level of awareness of industry 4.0 in the South African railway sector. Respondents believed that Industry 4.0 would necessitate the acquisition of skills such as critical thinking, computer programming, agile management, soft skills, and project management. Some of the challenges that the South African railway sector faced in fully embracing Industry 4.0 were a lack of training and a lack of resources. Future research could build on this study by comparing the readiness, challenges, and lessons learned by different sectors in South Africa as they strive to embrace Industry 4.0.

Keywords: Industry 4.0 · South African railway

1 Introduction

The current market conditions have forced the South African railway industry to reconsider their business strategy if they are to maintain their commitment to providing high-quality services at lower costs. As a result, the South African railway industry is making an effort to align with Industry 4.0 trends. To keep up with today's customer expectations, the industry has begun to digitize some of its services. Industry 4.0 is seen as having great potential for the railway industry in terms of increasing asset reliability, availability, and maintainability, as well as safety, through digitalization and smart infrastructure [1–3]. Industry 4.0 introduces new technologies such as "big data and advanced analytics, the internet of things, digital modelling, and artificial intelligence" [4, 5].

M. E. Auer et al. (Eds.): REV 2022, LNNS 524, pp. 271–279, 2023.
https://doi.org/10.1007/978-3-031-17091-1_28

From research, it has emerged that most companies are finding it difficult to embrace and implement Industry 4.0 [5]. This is due to a significant gap between current technological advancement and the company's expectations. Furthermore, some companies argue that Industry 4.0 is too complex, too big, too unrealistic, and too much technology to absorb [6]. However, scholars argue that the key to unlocking Industry 4.0 is to empower human resources [7, 8]. This entails equipping the human resource to enable the interaction of humans and computers systems such as Cyber Human Systems (CHS).

According to Sung [5], self-monitoring and fault diagnosis are capabilities of components operating in the context of Industry 4.0. Furthermore, the author [Ibid.] claims that the components and systems provide greater insight into management by increasing self-awareness and self-productivity. Kunze [4] indicates that workers would have to acquire a new set of skills to operate and maintain modern technologies. The introduction of industry 4.0 results in customer-oriented manufacturing and machines that are autonomous entities that collect data and are self-customized to meet the needs of the customer [4]. According to the research, technology leads to economic growth, which increases workloads and, as a result, job opportunities [7]. Despite the benefits of the Industry 4.0 initiatives, employees in the railway sector are concerned that automation will replace humans in the workplace. Because of this negative perception, some people have expressed reluctance to fully embrace Industry 4.0 within the sector. The human role in Industry 4.0 is not clearly defined and there is little to no knowledge about the required skills for employees to operate effectively in Industry 4.0 in the South African railway sector. According to the reviewed literature, very little has been done to understand the level of awareness and perceived impact of industry 4.0 in the South African railway sector. The study's goal was to investigate the level of awareness of Industry 4.0, its perceived impact, and some of the challenges that the South African railway sector has faced in adopting Industry 4.0.

2 Research Method

This is exploratory research that used both qualitative and quantitative data [9]. In terms of research strategy, this study used a survey research method because it can reach a larger community at a lower cost and allows participants to respond at their leisure [10]. This research took place within the South African railway industry, specifically in the infrastructure maintenance environment. The study targeted a total population of 110 engineering practitioners within the South African railway infrastructure maintenance environment. In this study, the infrastructure maintenance organization's global email address was used to identify contact information for the target group.

The questionnaire, which included both closed-ended and open-ended questions, was chosen to elicit the views of engineering practitioners in the railway sector on the concepts of Industry 4.0. The first stage of this research was to review the literature to identify the concepts of interest and how other researchers have approached industry 4.0 investigations. The second stage entailed the creation of a questionnaire based on information gleaned from the literature. The third stage involved piloting the survey with a smaller sample group of 10 people from the target population. The pilot results were used to fine-tune the questionnaire and remove items determined to be irrelevant to the study; these findings were not included in the final report. The fourth stage was to distribute the main survey to all sample participants.

We had to apply for ethical clearance from the University of Johannesburg ethics committee before distributing survey questionnaires to the target population. The committee required proof that the researchers would not coerce people into participating in the study, that the study would not endanger the participants in any way, that it would provide some sort of benefit, and that there would be no injustice in the process of conducting the research. Following the submission of evidence in accordance with the guidelines of the ethics committee, the ethical clearance certificate to collect data was granted.

3 Results

This is a survey study, and the online survey questionnaire was distributed to a total population of 110 infrastructure maintenance engineering practitioners (Table 1). The rail network infrastructure maintenance organization's global e-mail address was used as a sample frame, and the study received 61 responses (55% response rate) over the course of 52 networking days. The 55% response rate represented the random sampling method because all members of the population were invited to participate in the study. To address precision issues, the study used a 95% confidence level, and the margin of error was calculated using tools built into the SurveyMonkey online platform. With a 95% confidence level, the true population scores ranged between −8 and +8% of the observed values.

Table 1. Study specifications

Characteristics	Description
Population	110 Engineering practitioners from the South African railway infrastructure maintenance
Sampling technique	Random sampling
Method of data collection	Online self-administered questionnaire
Unit of observation	Engineering practitioners
Sample size (N)	61
Confidence level	95% ($Z = 1.96$, $p = 0.5$, $N = 61$)
Margin of error	±8%
Data collection duration	27 November 2020 to 7 February 2021

Most respondents in this study (57%) were between the ages of 25 and 34, with the other bigger group (23%) between the ages of 35 and 44 (Table 2). The majority (41%) had a Baccalaureate degree as their highest qualification, while the other large group (38%) had postgraduate degrees as their highest qualification. 38% of those surveyed described themselves as engineers, while 34% described their position as principal engineers. When asked to rate their awareness of Industry 4.0, 95% indicated that they were aware, while only 5% indicated that they were not aware. We concluded that the individuals who took part in this study were knowledgeable about the subject under investigation and had a good level of education to provide insight into the subject in their working environment.

Table 2. Demographic information (N = 61)

Variables	Frequency	Percent
Age group		
18–24	3	5%
25–34	35	57%
35–44	14	23%
45–54	5	8%
55 and above	4	7%
Totals	61	100%
Highest Qualifications		
Post matric certificate	13	21%
Baccalaureate degree	25	41%
Post graduate degree	23	38%
Totals	61	100%
Job Title		
Technician	16	26%
Engineer	23	38%
Principal Engineer	21	34%
Executive Manager	1	2%
Totals	61	100%
Have you ever heard of industry 4.0?		
No	3	5%
Yes	58	95%
Total	61	100%

3.1 Level of Awareness of Industry 4.0 Concepts

In this section, participants were asked to rate their level of familiarity with the concept of Industry 4.0 (Table 3). The scale had a high internal consistency (Cronbach's Alpha = 0.916), indicating that the items were highly correlated. Cloud computing was identified as the most well-known concept (weighted mean = 4.08), followed by simulations (weighted mean = 4.03), system integration (weighted mean = 4.02), and big data analytics (weighted mean = 4.01). Some of the concepts that the participants were unfamiliar with included additive manufacturing (weighted mean = 3.12), cyber-physical systems (weighted mean = 3.20), and the Internet of things (weighted mean = 3.43).

Table 3. Level of awareness of Industry 4.0 concepts

Items	Not aware	Somewhat not aware	Neither aware nor aware	Somewhat aware	Aware	Weighted mean
Cloud computing	2%	8%	16%	28%	46%	4.08
Simulation	6%	7%	13%	26%	48%	4.03
System integration	3%	10%	13%	30%	44%	4.02
Big data analytics	7%	2%	9%	47%	35%	4.01
Autonomous robots	8%	10%	12%	31%	39%	3.83
Internet of Things	5%	15%	10%	36%	34%	3.79
Augmented reality	13%	9%	25%	28%	25%	3.43
Cyber physical systems	8%	21%	26%	33%	12%	3.20
Additive manufacturing	20%	7%	24%	39%	10%	3.12

Respondents were also asked to describe Industry 4.0 in their own words in this section. Thematic coding was used to analyze the emerging themes from the respondents, which were then grouped as shown in Fig. 1. The respondents described Industry 4.0 in terms of ten themes, with digitalization and big data analysis appearing more frequently (18%), followed by innovation (16%), and systems integration and automation (14%). Other respondents looked to Industry 4.0 as a source of efficiency improvement, which appeared in 10% of the data. There were 4% of groups that defined Industry 4.0 as a source of change and uncertainty.

3.2 Industry 4.0 Tools Used in South African Railways

In this section, respondents were asked to describe industry 4.0 tools and technologies that they used daily at work. Respondents identified five tools and technologies that they believe are Industry 4.0 tools and have become a part of their daily work life (Fig. 2). According to the responses, the smartphone (22%) appears to be the most important industry 4.0 tool in the South African railway industry, followed by machine learning (21%), 3D printing (20%), laptop (20%), and blockchain (16%).

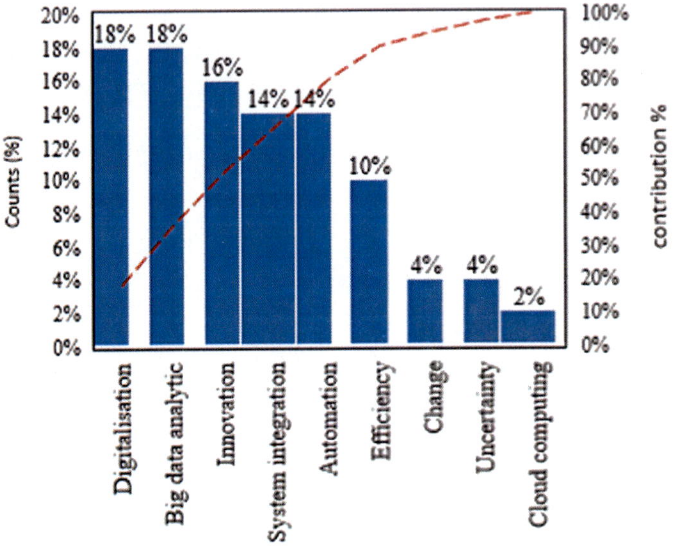

Fig. 1. Themes describing Industry 4.0

3.3 Skills Perceived to Be Required for Industry 4.0

In this section, respondents were asked to describe the skills they would need to operate effectively in the Industry 4.0 era. The respondents identified five skills (Fig. 3) that they believe will be necessary for them to function effectively in the era of Industry 4.0, with critical thinking (24%) topping the list. Other skills which were identified from the results included computer programming (23%), agile management (22%), soft skills (17%), and project management (14%).

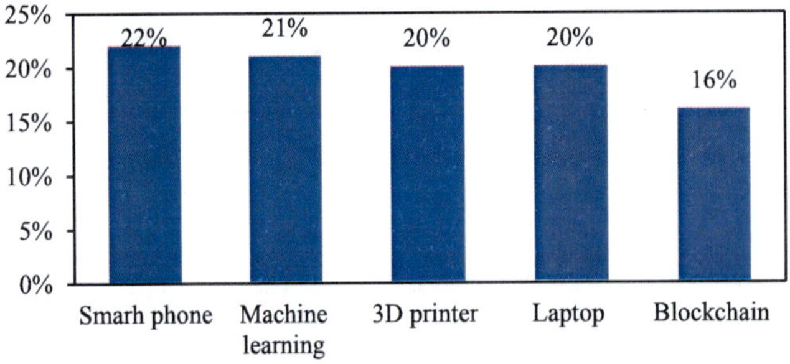

Fig. 2. Industry 4.0 tools used in the South African railways

3.4 Readiness to Operate in Industry 4.0 Era

Respondents were asked to rate their readiness, companies, and industries to operate in the Industry 4.0 era in this section. Respondents believed they were prepared to

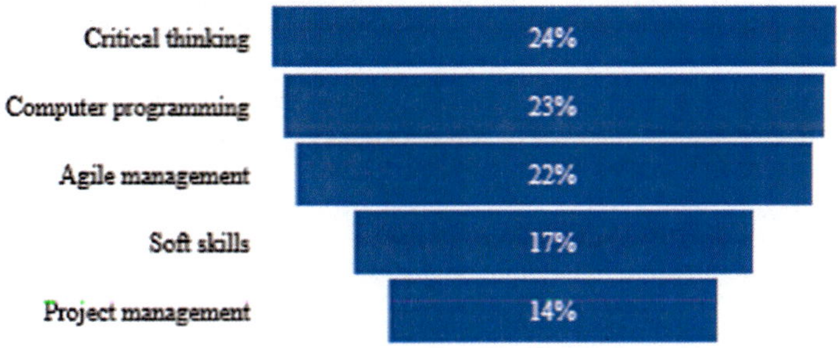

Fig. 3. Skills perceived to be required for Industry 4.0 in the South African Railways sector

operate in the Industry 4.0 era (weighted mean = 3.85), but there was uncertainty about the readiness of their companies (weighted mean = 3.21) and the industry as a whole (weighted mean = 3.44) (Table 4).

Table 4. Perceived readiness for Industry 4.0

Variables	Strongly disagree	Disagree	Neither disagree nor agree	Agree	Strongly agree	Weighted mean
I am ready for industry 4.0	4%	5%	23%	38%	30%	3.85
The railway industry is ready for industry 4.0	5%	10%	31%	44%	10%	3.44
The company I work for is ready for industry 4.0	4%	19%	37%	32%	8%	3.21

3.5 Obstacles to Adopting Industry 4.0

Respondents were asked to rate the difficulties they face in fully embracing Industry 4.0 in this section. The lack of training (weighted mean = 4.27) was identified as the most significant barrier to the adoption of industry 4.0 in the South African railway sectors, followed by a scarcity of skilled resources (weighted mean = 4.19) (Table 5). The lack of commitment from top management (weighted mean = 3.44) was identified as an area that requires further investigation.

Table 5. The railway industry's obstacles to adopting Industry 4.0

Variables	Strongly disagree	Disagree	Neither disagree nor agree	Agree	Strongly Agree	weighted mean
Lack of training	2%	3%	24%	8%	63%	4.27
Shortage of skilled resources	4%	5%	20%	10%	61%	4.19
Lack of top management commitment	5%	20%	33%	10%	32%	3.44

4 Discussion and Conclusion

The purpose of the study was to investigate the level of awareness of Industry 4.0, its perceived impact, and some of the problems that the South African railways sector has had in embracing Industry 4.0. The findings indicated that there was a high level of awareness of industry 4.0 in the South African railway sector, with only a small group of people (5% of respondents) indicating that they were unaware of Industry 4.0. Cloud computing, simulation, system integration, and big data analytics were among the well-known industry 4.0 concepts, with augmented reality, cyber-physical systems, and additive manufacturing among the unfamiliar concepts. Industry 4.0 was defined by the respondent as digitalization, big data analysis, innovation, system integration, and automation, which corresponded to the definitions found in [6, 11, 12]. Other respondents see Industry 4.0 as a source of efficiency improvement, changes, and uncertainties, confirming the theory that Industry 4.0 is interpreted differently among different role players [5, 12].

Respondents believed that Industry 4.0 would require them to acquire skills such as critical thinking, computer programming, agile management, soft skills, and project management to operate effectively in the Industry 4.0 era. The findings are comparable to those found in [12], where the authors identify programming skills, technological skills, and interpersonal skills as some of the skills required to operate in industry 4.0. Given the study findings, we recommended that companies invest in the competencies identified as skills required to operate effectively in the Industry 4.0 era.

When asked to rate their readiness, the readiness of their companies, and the readiness of the industry, respondents indicated that they were ready to operate in Industry 4.0, but there was uncertainty about the readiness of the company and the industry. Industry 4.0 has rendered the traditional approach to business management obsolete, to remain relevant, businesses should adopt new ways of doing business and embrace Industry 4.0 trends. Lack of training and a lack of resources were some of the challenges that the South African railway sector faced in fully embracing Industry 4.0. As a result, it was recommended that the sector invests in training programs related to the skills required to effectively operate in industry 4.0 and benchmark with industry 4.0 leading companies for information sharing. Future research could expand on this study by comparing the readiness, challenges, and lessons learnt by various sectors in South Africa in the quest to embrace Industry 4.0.

References

1. Britain Railway (2012) The future Railways. In Technical Stratergy Leadership Group, Britain
2. Miguel AN, Carlos S, Jose C (2017) Reliability analysis in railway repairable systems. Int J Qual Reliab Manag 34(8):1373–1398
3. Lohmeier C (2014) Big data? Qualitative approaches to digital research. In The researcher and the never-ending field: reconsidering big data and digital Ethnography, pp 75–89
4. Kunze A (2017) World Economic Forum, 15 June 2017. https://www.weforum.org/agenda/2017/06/digitization-can-mean-a-bright-future-for-rail-travel-heres-how/. Accessed Dec 2017
5. Sung T (2017) Technological focus and social change. Republic of Korea, Kyanggi University, pp 1–3
6. Rojko A (2017) Industry 4.0 concept: background and overview. Int J Interact Mob Technol (iJIM) 11(5):77–88
7. Sung TK (2018) Industry 4.0: a Korea perspective. Technol. Forecast Soc Change 132:40–45
8. Andrea Benešováa JT (2017) Requirements for education and qualification of people in industry. In: 27th international conference on flexible automation and intelligent manufacturing, FAIM 2017, no 11, pp 2195–2201
9. Karanja E, Zaveri J (2013) A comprehensive review of survey-based research in MIS. J Syst Inf Technol 15(2):159–188
10. Uma S, Roger B (2016) Research methods for business. John Wiley and Sons, Chichester
11. Barbara M, Gabriele B, Stefano U, Domenico S (2017) How will chnage the engineers' skilld in the Industry 4.0 framework? A questionnaire survey. In: 27th international conference on flexible automation and intelligent manufacturing (FAIM 2017), Modena, Italy
12. Maisiri W, Darwish H, Dyk LV (2017) An investigation of Industry 4.0 skills requirements. South Afr J Ind Eng

Remote Learning: Implementing IIoT and Industry 4.0 Technologies Using PLCs

Zhen Gao, Dan Centea(✉), and Ishwar Singh

McMaster University, 1280 Main Street West, Hamilton, ON L8S 0A3, Canada
{gaozhen,centeadn,isingh}@mcmaster.ca

Abstract. Rapid technology advances related to Internet of Things (IoT), Industrial Internet of Things (IIoT) and Industry 4.0 technologies have led to a need of software and hardware knowledge that students must learn and apply in an academic environment so that, after graduation, they can accelerate the adoption of these technologies in industrial and commercial workplaces. The purpose of the paper is to present the remote learning approach in teaching ways to implement these technologies using Programmable Logic Controllers (PLCs) and to describe the open-source Information Technologies (IT) and Operational Technologies (OT) options and choices that can be implemented in a remote learning mode. A remote delivering strategy of the PLC courses has been successfully developed for both lectures and labs without compromising the quality. The PLC courses were re-designed with high-quality and large quantity of practice modules and were synchronously delivered with the aims of cultivating students' technical competency to solve the real problems in industry and paving the foundation for their future professional career. The paper describes some of these courses, focuses on the courses related to teaching Industry 4.0 and IIoT technologies, and provides a detailed description of how remote learning has been implemented in automation courses and projects. The experience that students gained via PLC courses is applicable for senior courses, capstone projects, co-op employment, and full-time jobs in manufacturing and automation industry.

Keywords: Remote learning · PLC · IIoT · Industry 4.0

1 Introduction

Industry 4.0 is where IoT technologies and associated enablers are applied to improve manufacturing processes based on data collection/analysis and a smart decision making and execution system [1, 2]. IoT consists of a network of computing and sensory devices that gather and share large scale data, yet not necessarily big data. Through the integration of the methodologies of IoT and Industry 4.0 in an industrial environment, IIoT can be deployed by connecting agile nodes and configuring intelligent machines with advanced analytics to monitor, collect, exchange, examine, and deliver valuable and meaningful results [3, 4]. The data can be sent to a cloud-based server which is then shared with various end users in a more productive way [5]. By combining machine-to-machine communication, cloud/edging computation, and secured networks, IIoT not

M. E. Auer et al. (Eds.): REV 2022, LNNS 524, pp. 280–287, 2023.
https://doi.org/10.1007/978-3-031-17091-1_29

only allows for increased productivity, but also increases efficiency and performance of the workforce. The implementation of IoT in industry enables businesses to produce predictive analytics, whether it be to detect possible defects in machinery or by providing real-time production data for asset tracking.

The enablers of Industry 4.0 and IIoT are shown in Fig. 1 [6–8].

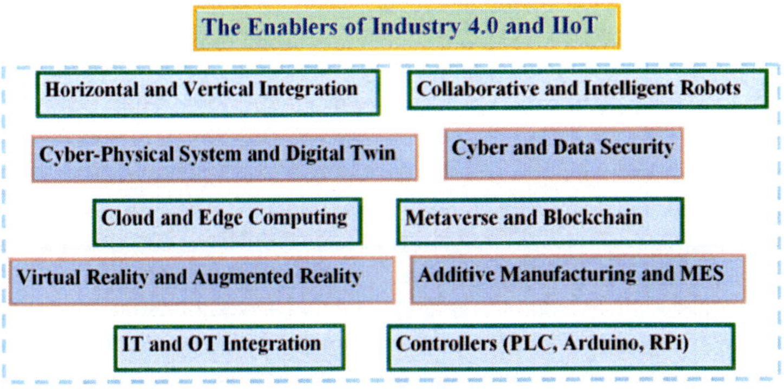

Fig. 1. List of enablers for Industry 4.0 and IIoT

By integrating with other enablers, PLCs are shaping the traditional landscape of industrial automation and continue to contribute to the ecosystem of Industry 4.0 and IIoT [9, 10]. Therefore, for education in automation engineering and technology, the importance of PLC training should not be reduced due to the emergence of other popular topics, such as Cyber-Physical Systems (CPS), virtualization and digital twin, horizontal and vertical system integration, big data and analytics, collaborative and autonomous robots, cloud/fog/edge computing, underneath the umbrella of Industry 4.0.

Although the global pandemics has been bringing challenges of teaching and learning in engineering, the remote delivering strategy of the PLC courses at McMaster University were successfully developed for both lectures and labs without compromising the quality. To advocate the pedagogy of practice-based education as part of the strategic planning of the W Booth School of Engineering Practice and Technology (SEPT) at McMaster University, the PLC courses were re-designed with high-quality and large quantity of practice modules which were synchronously delivered with lectures, aiming to cultivate students' technical competency to solve the real problems in industry to pave the foundation for their future professional career [11].

2 Industry 4.0 and IIoT

Industry 4.0 and IIoT are the promise of the future as these technologies can lead to lower costs of production along with flexibility to switch over to new products in the production line. IIoT has evolved from the inception of IoT to where one can connect devices from everywhere, obtain their status, and control them in an effective visual environment. In industry, the status of operations and visualization have been a very much a local operation. IoT concept extended to the plant floor using PLCs have become the hallmark of IIoT. While things are changing with IIoT, the PLC remains at the heart of industrial control. Thanks to advances in technology, PLCs have steadily shrunk in size and cost and can operate in harsher environments. IIoT brings new considerations as to whether centralized or decentralized control is the best option. The choice of the best control option, along with other operational issues and challenges that propelled the evolution of IoT technologies to industrial operations, have led to the evolution of IIoT. Moving forward with the vertical and horizontal integration of IT and OT technologies in the supply chain, IIoT has become the basis implementation strategies for Industry 4.0. These technologies allow to connect many machines, in multiple factories, in any part of the world.

These rapid advances have led to the development of a plethora of software tools and hardware devices that automation students must learn so that they can accelerate the adoption of these technologies in the industrial and commercial workplaces.

Businesses all over the world are struggling to find workers with digital skills as the pandemic continues to accelerate technology adoption and automation. A recent KPMG survey in Canada found the inability to find and retain talent was identified as the No. 1 threat to the growth prospects of businesses across the country. The poll found nearly 80% of businesses stating that the pandemic changed the way they work, and they now need more workers with IT skills.

The skill sets are required for the implementation of digital strategies in the industrial workplace include: programming microcontrollers and sensors for IoT applications; PLC programming and visualization; networking concepts and technologies; MQ Telemetry Transport (MQTT) publish/subscribe protocol; database programming and internet technologies, Open Platform Communications (OPC) Data Access (DA) and Unified Architecture (UA), cloud computing, condition monitoring and predictive maintenance, data analytics and Artificial Intelligence (AI) application development.

Besides being a remote lab, a pocket lab kit can be used for PLC courses. By following the standard of IEEE/IEC 61131, a microcontroller such as Arduino UNO with a size of credit card or the single board computer (SBC) such as Raspberry Pi can be adopted as a replacement of a regular-sized PLC from vendors such as Allen Bradley, Siemens, Omron, and ABB. The pocket lab is with low-cost kit that contains a microcontroller/SBC, input/output components such as push buttons, switches, proximity sensors, infrared sensors, LEDs, mini servo, and step motors, protoboard, jump wires, resistors, beeper, and battery pack. The pocket lab kits could be either purchased by students or by the institution. Such version of lab is suitable for both individual work and group work, ideally group of two. In other words, pocket lab is a version of conducting PLC lab anywhere and anytime by using lightweight and low-cost lab kits that complies the IEEE/IEC 61131 standard of ladder diagram programming.

Figure 2 shows the open-source IT and OT technology options and choices that are available and can be implemented in a remote learning mode. In addition, they deserve future exploration on how remote learning has been implemented in various automation courses and projects PLCs and basic networking; IoT and IIoT; and Internet technologies and databases.

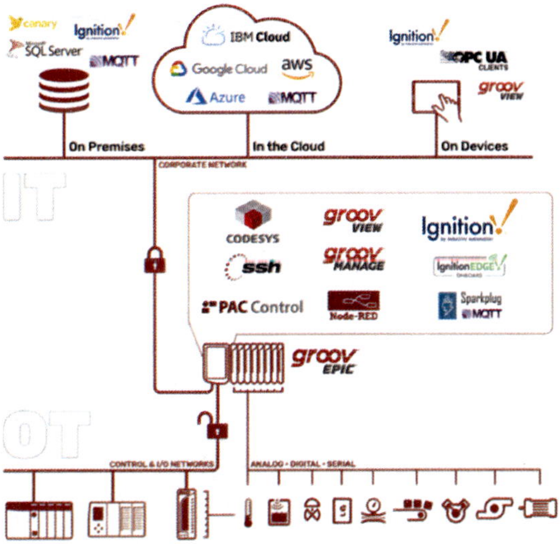

Fig. 2. *groov*EPIC software architecture [12]

3 Remote Learning of PLCs

There are two undergraduate PLC courses at Automation Program of SEPT, namely PROCTECH 2PL3 (PLC-1) and PROCTECH 3PL3 (PLC-2). Allen Bradley's RSLogix 5000, ControlLogix and 1756 I/O modules have been deployed as the main programming environment and hardware respectively for onsite labs. To shift all labs online, LogixPro and Connected Components Workbench (CCW) have been utilized as the main platforms for PLC-1 and PLC-2, respectively.

Examples of practice modules in PLC-1 include: (1) Simple relay logic circuits; (2) Logic basics of motor control; (3) Silo control; (4) Conveyor belt control; (5) Traffic light control; (6) Building a digital clock with display of hours, minutes, seconds, AM and PM; and (7) Batch process control. All programming activities were based on Ladder Diagram that follows the open international standard IEC 61131–3. The lectures and labs were synchronized in terms of content (fundamental and practice), and it kept going with this paradigm throughout the entire course. Besides, OpenPLC and LDmicro with Arduino Mega were tested and considered as alternative low-cost lab kits if students were expected to conduct non-virtual labs to gain hands-on experience with physical lab components during pandemics.

Examples of online labs of PLC-2include: (1) Sequential control of batch process application based on CCW Micro800-Simulator; (2) Introduction of Function Block Diagrams (FBD); (3) Elevator re-design based on FBD - up, down, 3rd floor, and timed/sensing doors; (4) Structured Text (ST) programming to convert an electromechanical process into an equivalent ST programming; (5) Analog input and output using CCW; (6) PWM and PID Control - using PWM to control the temperature of an incandescent light bulb manually and applying PID instruction to setup an automatic temperature control system; and (7) Human Machine Interface programming based on CCW and VTScada. Similar with PLC-1, the content of weekly lecture and lab/project of PLC-2 were well coupled. By doing so, the learning of the fundamental part and practical part can consolidate to achieve deeper and reinforced learning objectives.

Then for senior courses and projects, topics of industry networking, I/O interfacing, data visualization, HMI, OPC, and system integration will be covered via onsite, virtual, or hybrid delivery approaches, based on the updated policy from the University and government.

4 LogixPro: A Real-World Simulator of PLC Ladder Diagram

LogixPro as a professional simulator based on Allen Bradley's Rslogix 500 PLC. It provides the beginner of PLCs with a user-friendly visible environment that is enriched with various industrial and real-world applications, including but not limited to garage door control, process compressor, and elevator level control. The software was offered at an affordable price so that all students were encouraged to obtain licenses which they could use permanently.

As shown in Fig. 3, a Google Street view on a local one-way street could be used as the real-life example to engage the students to take the PLC courses virtually. Students were asked to solve the traffic light control problem by applying 6 timers that they learnt from previous lecture, with the following constraints:

- assume no left/right turn option. Assume people do not stop when facing yellow light. And assume the lights work 24/7 with fixed duration.
- to avoid collision, there are two overlaps when the red lights in both directions are illuminated with a few seconds.
- to make the practice more interactive, students would fill up the timetable that showed the desired sequence of lights' illumination in a loop.

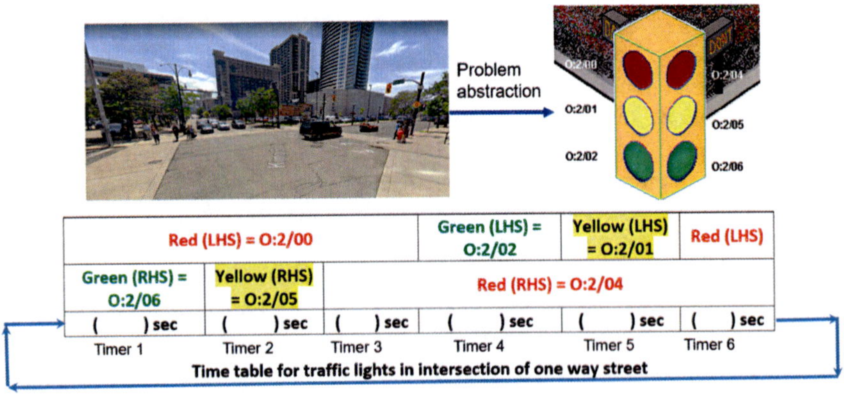

Fig. 3. Traffic light control on simulated one-way street

Besides, Factory IO provides 3D simulation of real scenarios of process automation industry where various PLC models are embedded in the virtual environment that is appropriate as a training kit for seniors or graduate students [13].

5 Lightweight Lab Kit for Remote PLC Training

5.1 Arduino-LDMicro

Conducting labs and projects with simulations apparently is not the only option for delivering PLC courses virtually. Arduino as a low-cost microcontroller has been applied for hundreds of thousands of projects in areas of automation and mechatronic systems. LDMicro as an open-sourced program editor, it allows the students to implement ladder logic programming without using the original PLC software and hardware. As shown in Fig. 4, Arduino UNO based lab kits can be shipped to students' residential addresses. By doing so, students not only enhance their PLC programming skill, but also gain their understanding about 1) compatibility of different controllers, 2) the working principles of ladder diagram which is the main PLC programming language among five languages that comply with IEC 61131–3 standard.

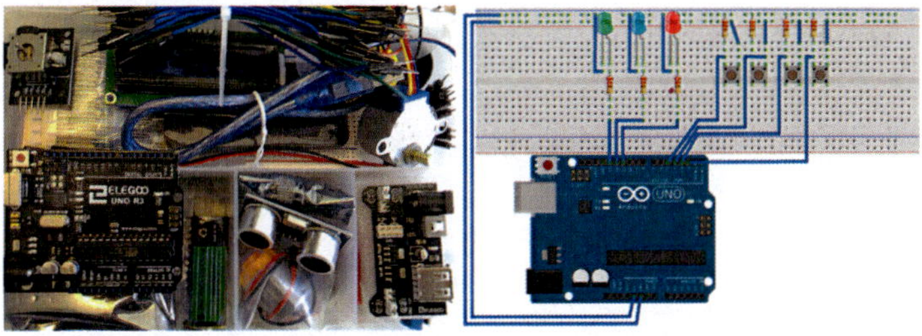

Fig. 4. Low-cost Arduino lab kits for PLC programming and practice

Hardware simulation about bottle filling was developed as one of the labs, with the instructions of XIO, XIC, Timer, OTE, etc. Sensors (such as Photoeye and limit switch) and actuators (such as step motors) which are available in the lab kit can be engaged for such lab. As shown in Fig. 5, the style of LDMicro based ladder diagram is similar to those from the major vendors such as Allen Bradley and Siemens.

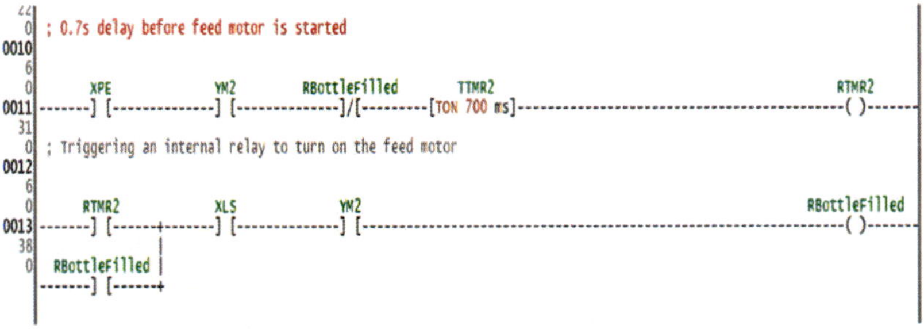

Fig. 5. Sample code for LDMicro based ladder diagram

The IO pins on microcontroller can be associated within the LDMicro editor instead of manually within the Arduino code. This can be easily achieved by double-clicking on the instruction name and select the appropriate pin.

5.2 Raspberry Pi for IoT-PLC

In addition to Arduino, Raspberry Pi (RPi) as a single-board computer has the high potential to replace regular PLC with expandable I/O, various communication protocols, high-performance microprocessor, and data storage capability. Besides, RPi serves a powerful local controller in the context of Industry 4.0 and IIoT. It is foreseeable that RPi will be used for more applications in digital manufacturing and automation since it has great compatibility with other enablers.

In a virtual learning environment, RPi based lightweight lab kit is an alternative consideration to build IoT-PLC blocks of industrial control systems and industrial cyber-physical systems with great integration of IT and OT [14]. Raspberry Pi can also function as a node for edge and fog computing to implement publish-subscribe protocol such as Message Queuing Telemetry Transport [15, 16]. Cyber and data security for RPi system needs to be investigated in an open architecture environment [17, 18].

6 Conclusions

This paper explores remote learning approaches of PLCs in the context of Industry 4.0 and IIoT. It concludes that remote learning does not mean total virtualization of the teaching environment. Low-cost lightweight lab kit is available for students to deal with hands-on labs and projects. Although automation engineering curriculum was constantly changed to enhance the learning skills needed to implement these technologies, the challenges created by the pandemic became the main reason behind the change in practice towards learning these skills in a remote learning environment. Whenever switching back to in-class learning, the materials of remote learning can be considered as additional and accessible supporting resources. The experience that students gained via PLC courses is applicable for senior courses, capstone projects, co-op employment, and full-time jobs in manufacturing and automaton industry. Furthermore, the lesson learned in implementing open-source software packages in remote learning environments can be used by faculty members and professionals interested in these topics.

Acknowledgements. This project is supported by the Future Skills Centre, Canada.
The authors thank Adam Sokacz for testing Arduino PLC labs.

References

1. Silvestri L, Forcina A, Introna V, Santolamazza A, Cesarotti V (2020) Maintenance transformation through Industry 4.0 technologies: a systematic literature review. Comput Ind 123:103335. https://doi.org/10.1016/j.compind.2020.103335
2. Amjad MS, Rafique MZ, Hussain S, Khan MA (2020) A new vision of LARG manufacturing — a trail towards Industry 4.0. CIRP J Manuf Sci Technol 31:377–393. https://doi.org/10.1016/j.cirpj.2020.06.012
3. Bu L, Zhang Y, Liu H, Yuan X, Guo J, Han S (2021) An IIoT-driven and AI-enabled framework for smart manufacturing system based on three-terminal collaborative platform. Adv Eng Inform 50:101370. https://doi.org/10.1016/j.aei.2021.101370

4. Javaid MK, Haleem A, Singh RP, Rab S, Suman R (2021) Upgrading the manufacturing sector via applications of Industrial Internet of Things (IIoT). Sens Int 2:100129. https://doi.org/10.1016/j.sintl.2021.100129
5. Liu P, Liu K, Fu T, Zhang Y, Hu J (2021) A privacy-preserving resource trading scheme for Cloud Manufacturing with edge-PLCs in IIoT. J Syst Archit 117:102104. https://doi.org/10.1016/j.sysarc.2021.102104
6. Ashima R, Haleem A, Bahl S, Javaid M, Mahla SK, Singh S (2021) Automation and manufacturing of smart materials in additive manufacturing technologies using Internet of Things towards the adoption of industry 40. Mater Today Proc 45(6):5081–5088. https://doi.org/10.1016/j.matpr.2021.01.583
7. Tobe F (2022) Why co-bots will be a huge innovation and growth driver for robotics industry. IEEE Spectr. https://spectrum.ieee.org/collaborative-robots-innovation-growth-driver. Accessed 11 Jan 2022
8. Liu B, Zhang Y, Zhang G, Zheng P (2019) Edge-cloud orchestration driven industrial smart product-service systems solution design based on CPS and IIoT. Adv Eng Inform 42:100984. https://doi.org/10.1016/j.aei.2019.100984
9. Bellmunt OG, Miracle DM, Arellano SG, Sumper A, Andreu AS (2006) A distance PLC programming course employing a remote laboratory based on a flexible manufacturing cell. IEEE Trans Educ 49(2):278–284. https://doi.org/10.1109/TE.2006.873982
10. Niang M, Riera B, Philippot A, Zaytoon J, Gellot F, Coupat R (2020) A methodology for automatic generation, formal verification and implementation of safe PLC programs for power supply equipment of the electric lines of railway control systems. Comput Ind 123:103328. https://doi.org/10.1016/j.compind.2020.103328
11. Gao Z, Wanyama T, Singh I (2020) Project and practice centered learning: a systematic methodology and strategy to cultivate future full stack artificial intelligence engineers. Int J Eng Educ 36(6):1760–1772. https://www.ijee.ie/1atestissues/Vol36-6/05_ijee3986.pdf
12. *groov* EPIC: the world's first Edge Programmable Industrial Controller. https://www.opto22.com/products/groov-epic-system/groov-epic-software. Accessed 11 Jan 2022
13. https://factoryio.com. Accessed 11 Jan 2022
14. Mellado J, Núñez F (2022) Design of an IoT-PLC: a containerized programmable logical controller for the Industry 4.0. J Ind Inf Integr 25:100250. https://doi.org/10.1016/j.jii.2021.100250
15. Amanlou S, Hasan MK, Abu Bakar KA (2021) Lightweight and secure authentication scheme for IoT network based on publish–subscribe fog computing model. Comput Netw 199:108465. https://doi.org/10.1016/j.comnet.2021.108465
16. Nasir M, Muhammad K, Lloret J, Sangaiah AK, Sajjad M (2019) Fog computing enabled cost-effective distributed summarization of surveillance videos for smart cities. J Parallel Distrib Comput 126:161–170. https://doi.org/10.1016/j.jpdc.2018.11.004
17. Mudaliar MD, Sivakumar N (2020) IoT based real time energy monitoring system using Raspberry Pi. Internet Things 12:100292. https://doi.org/10.1016/j.iot.2020.100292
18. Urrea C, Kern J (2021) Design and implementation of a wireless control system applied to a 3-DoF redundant robot using Raspberry Pi interface and User Datagram Protocol. Comput Electr Eng 95:100250. https://doi.org/10.1016/j.compeleceng.2021.107424

Distributed IoT System for Indoor Air Quality Monitoring

Marius Cristian Luculescu, Luciana Cristea, Constantin Sorin Zamfira(✉), Attila Laszlo Boer, and Sebastian Pop

Transilvania University of Braşov, Eroilor 29, 500036 Braşov, Romania
{lucmar,lcristea,zamfira,boera,pop.sebastian}@unitbv.ro

Abstract. Air quality matters for all of us. Even we discuss indoor or outdoor, our health status and comfort depend on the air that we breathe. The negative effects can be immediate or long-term. To maintain the air quality, we need to monitor the values of the parameters that influence it. The paper presents a distributed IoT system for indoor air quality monitoring. The acquisition process from humidity, barometric pressure, ambient temperature, and gas sensors is accomplished by a small board computer integrated into each module that communicates with a server to upload data automatically at certain time intervals. The software keeps a history of the monitored parameters that can be displayed in numerical or graphic format for each room, allows the setting of predetermined intervals for these parameters, and generates alarms when the limits are exceeded. The Indoor IoT Air Quality Monitoring Module is low-cost and has low-power consumption. It was designed as an open platform, ready to be changed to monitor outdoor air quality. For doing this, other sensors can be connected, such as a GPS that would identify the geographical location to generate maps for monitored parameters, respectively other communication modules, such as GSM, to connect from places where the WiFi signal is not present. Including a MicroSD module, an independent data-logger can easily be obtained. The system is implemented in university rooms where the educational process takes place, offering an image of indoor air quality.

Keywords: IoT · Air quality monitoring · Sensors

1 Introduction

1.1 Context

Air pollution is one of the most significant environmental risks to health. The World Health Organization (WHO) Air Quality Guidelines: Global Update 2021 assess health effects of air pollution and thresholds for health–harmful pollution levels [1]. The studies show that in 2019, approximately 99% of the world population was living in places where the WHO air quality guidelines levels were not met. In addition to outdoor air pollution, indoor smoke can be considered a severe health risk for some 2.6 billion people who cook and heat their homes with biomass, kerosene fuels, and coal [1].

M. E. Auer et al. (Eds.): REV 2022, LNNS 524, pp. 288–299, 2023.
https://doi.org/10.1007/978-3-031-17091-1_30

Air quality matters for all of us. Even we discuss indoor or outdoor, our health status and comfort depend on the air that we breathe.

The air quality can be measured outdoor or indoors. If we are discussing outdoor, there are official measurement stations, but they record air quality only in their immediate proximity. While these stations measure only outdoor parameters, the indoor concentration is largely unknown.

Indoor Air Quality (IAQ) considers the air quality within buildings related to people's health, comfort, and work performance. The negative effects can be immediate or long-term [2]. For example, indoor ventilation problems can lead to increasing pollutant levels if the air does not flow properly. Also, high temperature and humidity levels can increase the concentration of some pollutants. There are different ways that the outdoor air can enter and leave a building. Some are uncontrolled, like infiltration through cracks, floors, ceilings, openings, and around windows and doors due to wind and air temperature differences between inside and outside. The others can be controlled, like natural ventilation through doors and windows that can be opened or closed, or mechanical ventilation using specific devices. To maintain the air quality, we need to monitor the values of the parameters that influence it.

1.2 Problem Formulation

The Transilvania University of Brasov is the most prominent university in the central region of Romania. The preoccupation for ensuring comfortable conditions for the educational process development is a priority for those who carry out their activity in the university buildings.

One of the projects started at the Department of Product Design, Mechatronics and Environment of this university aims to create a distributed system of IoT modules that monitor indoor air quality. There are already many commercial systems on the market for this purpose. Still, the students and teachers of the Mechatronics bachelor program wanted to create their platform by highlighting the knowledge gained in disciplines such as Microcontrollers, Sensors, and Programming Languages.

1.3 The State of the Art

IAQ refers to the environmental characteristics inside buildings that include the concentrations of pollutants in indoor air, air temperature, and humidity [3].

Air pollutants can be generated from indoor sources or enter the building as microscopic particles within outdoor air. Gases and particles can be produced, for example, by tobacco smoking, cooking, heaters, stoves, molds, and bacteria that grow indoors on damp surfaces. Other pollutants are the Volatile Organic Compounds - VOC emitted by some building products, furnishings, and consumer products. People can also be sources of indoor air pollutants. Considering the pandemic period of COVID-19, when persons are encouraged to stay at home for a higher amount of time, IAQ becomes more important. Poor indoor ventilation might lead to different health problems.

An overview of IAQ in schools, based on papers published in peer-reviewed journals, was made by Lawrence Berkeley National Laboratory [3]. The review focuses on allergens, volatile organic compounds (VOCs), particles, dampness and mold, and ventilation with outdoor air in schools.

There are also different studies on monitoring the indoor air quality of the laboratories of higher education establishments [4] or in schools worldwide [5, 6], identifying the potential air pollutants [7].

There are different solutions for taking information about the IAQ, commercial or specially designed for research. Some of them include independent systems that measure and display temperature, humidity, air pressure, dust, gases, etc. [8–10]. Others point to intelligent distributed systems based on Internet of Things (IoT) modules that monitor these parameters, compute specific values, trigger alarms, and control devices that improve the IAQ [11–15]. Data can be sent through a local network or to remote destinations over the Internet.

2 The System for Indoor Air Quality Monitoring

2.1 System Requirements

The System for Indoor Air Quality Monitoring (SIAQM) was designed to meet the following requirements:

a. from the **hardware** point of view: it has to contain temperature, humidity, pressure, and gas sensors; the sensors should be connected to a microcontroller development system (Small Board Computer – SBC) that can communicate via a wireless interface with a server or mobile device, making from that module an IoT device; it has to contain elements for local visual and audio alarm; it has to be compact and as small as possible;
b. from the **software** point of view, the system must allow: the acquisition of data from sensors; automatic transfer of information to the server, at certain set time intervals; the calculation of the Index for the Air Quality (IAQ); keeping a history of the monitored parameters that can be displayed and processed by the user; setting of predetermined intervals for these parameters and generating alarms when the limits are exceeded; display the monitored buildings and rooms with the possibility of visualizing the data in numerical or graphic format, respectively;
c. **other conditions**: to be a low-cost system; to have low-power consumption; to be an open platform, ready to monitor outdoor air quality with the addition of other sensors, such as a GPS sensor that would identify the geographical location to generate maps for monitored parameters, respectively the addition of other communication modules, such as a GSM module to connect from locations where the WiFi signal is not present.

2.2 The Structure of the System for Indoor Air Quality Monitoring

An analysis was performed starting from the above hardware requirements. The structure of the resulted system is presented in Fig. 1. It consists of distributed Indoor IoT Air Quality Monitoring Modules (IIAQMM) that communicate with a server through the Internet.

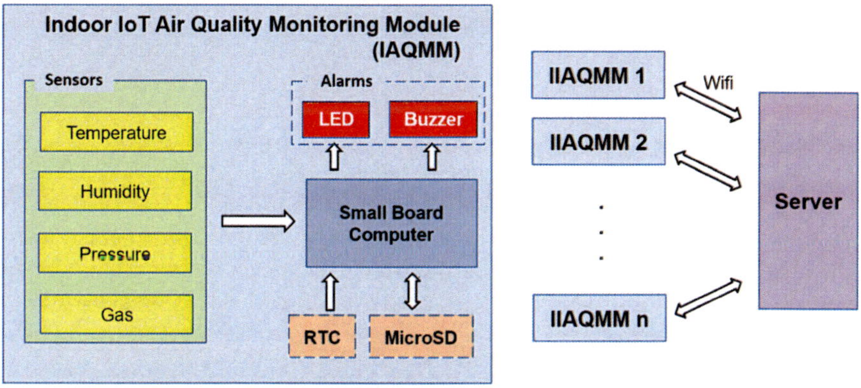

Fig. 1. The structure of the system for indoor air quality monitoring

The Small Board Computer

Each module has a Small Board Computer (SBC) responsible for the data acquisition, processing, storing, and sending to the server.

As SBC, different microcontroller development systems were considered: Arduino, ESP32, ESP8266, Raspberry Pi. Requirements such as low cost, low-power consumption, small dimensions, and wireless interfaces led to choosing the ESP32 as an SBC, namely the ESP32 DEVKIT V1 (Fig. 2).

Fig. 2. The SBC used for IIAQMM

ESP32 has a 32-bit dual-core architecture based on Tensilica Xtensa Dual-Core 32-bit LX6 microprocessor, running at 160 or 240 MHz. It has 512 KB RAM and built-in wireless communication interfaces: WiFi (2.4 GHz up to 150 Mbits/s) and Bluetooth (BLE - Bluetooth Low Energy and legacy Bluetooth). The board that we used contains 30 pins and has a wide variety of peripherals, like an interface with DMA that includes capacitive touch, ADCs (Analog-to-Digital Converters), DACs (Digital-to-Analog Converters), I^2C (Inter-Integrated Circuit), SPI (Serial Peripheral Interface), UART (Universal Asynchronous Receiver/Transmitter), I^2S (Integrated Interchip Sound), RMII (Reduced Media-Independent Interface), PWM (Pulse-Width Modulation).

ESP32 can be programmed in different programming environments like Arduino IDE, Micropython, Espressif IDF (IoT Development Framework), JavaScript, LUA, etc.

The Alarm Indicator

In case of exceeding the limits of the intervals for monitored parameters, a visual alarm using an RGB LED is implemented. Different colors and blinking combinations are coding the alarms for a specific parameter. In addition, there is also an audio alarm implemented with a buzzer. For this, different combinations of beeps are coding the exceeded parameter. Each alarm can be enabled or disabled by software, as it will be presented later in the paper.

The Sensors

An analysis was also performed regarding various sensor types, single or in combination. As mentioned in the system requirements, temperature, humidity, pressure, and gas parameters have to be monitored.

The following types of sensors were analyzed: temperature sensors, LM35DZ and Dallas 18B20; temperature and humidity sensors DHT-11 and DHT-22; barometric pressure/temperature/altitude sensor Bosch BMP180; barometric pressure sensors Bosch BMP280 and BMP390; humidity, barometric pressure, and ambient temperature BME280; humidity, barometric pressure, ambient temperature and gas (VOC) BME680; MQ Series gas sensor, etc.

Due to compact design and low-power consumption conditions, a solution with ESP32 and BME680 was preferred (Fig. 3), defining an Indoor IoT Air Quality Monitoring Module (IIAQMM).

Fig. 3. The BME680 sensor module used in the testing phase

BME680 is the first small-size low-power consumption gas sensor manufactured by Bosch that integrates high-linearity and high-accuracy gas, pressure, humidity, and temperature sensors [16]. It measures relative humidity, barometric pressure, ambient temperature, and gas (VOC - Volatile Organic Compounds). It can communicate by I^2C and SPI interfaces.

VOCs are detected by adsorption on the sensitive layer of the sensor. The BME680 measures the sum of nearly all VOCs/contaminants in the surrounding air generated by paint, furniture, garbage, cooking, food consumption, exhaled breath and/or sweating. The sensor resistance is inversely proportional to the concentration of gases. The raw signal is also influenced by parameters other than VOC concentration (e.g., humidity level).

Table 1 presents the essential characteristics of the BME680.

Table 1. BME680 characteristics [16].

No.	Parameter	Operating range/monitoring interval			Unit	Condition
		Min	Typ	Max		
1	Temperature	−40	25	85	°C	Operational
		0		65		Full accuracy
2	Humidity	−40	25	85	°C	Operational
		0		100	% r.H	
3	Pressure	300		1100	hPa	Full accuracy
4	Index for Air Quality (IAQ)	0		50	–	Excellent
		51		100		Good
		101		150		Lightly polluted
		151		200		Moderately polluted
		201		250		Heavily polluted
		251		350		Severely polluted
		351		500		Extremely polluted
5	Response time		92		s	Ultra-low power mode
			1.5			Low power mode
			0.75			Continuous mode

Using libraries developed by Bosch Software Environmental Cluster, the raw values of the resistance are transformed to an index for air quality (IAQ) by smart algorithms that automatically calibrate and adapt themselves to the typical environments where the sensor is used. The IAQ scale ranges from 0 (clean air) to 500 (extremely polluted air).

Other Features of the IIAQMM

The IIAQMM is a versatile device. If a MicroSD module is included, it can also be used as an independent data-logger in the absence of an Internet connection. In this case, a Real-Time Clock (RTC) module is needed to store the data together with the acquisition date and time.

The IIAQMM is also designed to be an open platform, ready to monitor outdoor air quality with the addition of other sensors. For example, a GPS sensor can be connected for identifying the geographical location. Thus the system will allow generating maps for monitored parameters. Furthermore, for areas where the WiFi signal is not present, the addition of other communication modules, such as a GSM module, is possible.

2.3 The Structure of the Software

The software of the System for Indoor Air Quality Monitoring consists of two parts:

- The routines for the IIAQMM that are stored in the memory of the Small Board Computer;
- The server application.

2.3.1 The Routines for the IIAQMM

These routines were written using the Arduino IDE programming environment.

They allow the following tasks:

- Connect the module to the WiFi network
- Acquire data from the sensors (temperature, pressure, humidity, staticIaq, co2Equivalent, breathVocEquivalent)
- Extracts date and time from the Internet using a special function
- Send data to the server for storing in the database and displaying as real-time values of the monitored parameters
- Send data to the Serial Monitor in Arduino IDE.
- Trigger the visual and audio alarms if they are enabled, and the parameter limit is exceeded
- Implement a Webserver for the IIAQMM to which you can connect to see the real-time values or make different settings

The Webserver interface displays the location where the module is installed; IAQ Status (from Excellent to Extremely polluted – see Table 1); Date and time for the last read value from the sensors; Module ID (the MAC of the ESP32); IP address of the ESP32; current values of the temperature, humidity, pressure, IAQ, CO2 Equivalent, and Breath VOC (Fig. 4).

Also, the following settings can be made through another option of the Webserver interface:

- The name for the location where the module is installed
- The time interval for the data to be sent to the server (default setting is 2 min)
- The time interval for the data to be refreshed in the webserver window (default setting is 5 s)
- Option for Enable or Disable the monitoring (check-boxes for each one of the six parameters)
- Set the limit values of the interval for temperature, humidity, and pressure (Min. Value and Max. Value), respectively IAQ, CO2 Equivalent, Breath VOC (only Max. Value)
- Enable or disable the alarm for each parameter (check-boxes)

The visual and audio alarm codes are presented in Table 2.

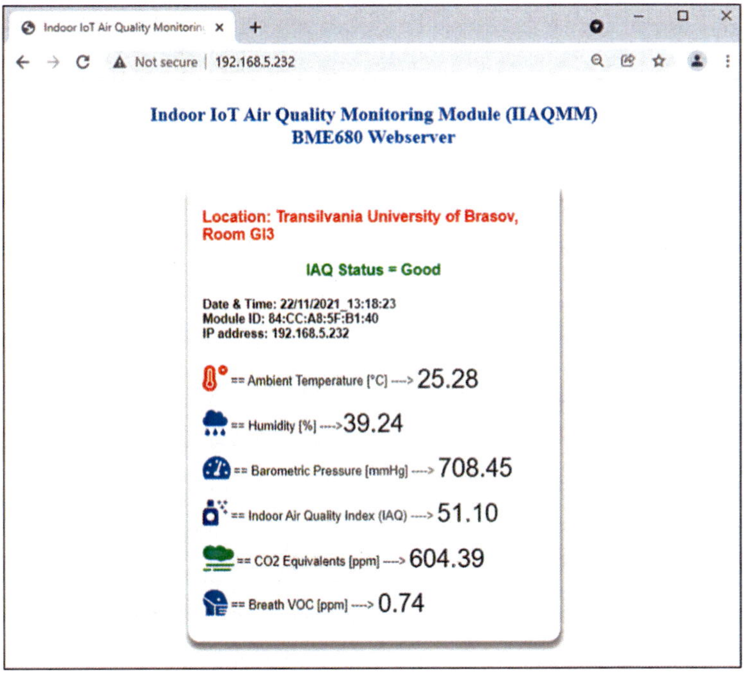

Fig. 4. The Webserver interface for displaying real-time values

Table 2. Alarm codes

No	Parameter name	LED color	Visual alert code	Audio alert code
1	Temperature	Yellow	1 blink	1 beep
2	Humidity	Blue	1 blink	2 beeps
3	Pressure	Orange	1 blink	3 beeps
4	IAQ	Red	1 blink	4 beeps
5	CO2 Equivalent	Red	2 blinks	5 beeps
6	Breath VOC	Red	3 blinks	6 beeps

2.3.2 The Routines for the Server

These routines were written in PHP, while the database is in MySQL.

They allow the following tasks:

- Receiving data from the IIAQMM modules
- Append data into the database
- Display last 100 records
- Keep a history of the monitored parameters for each IIAQMM module
- Display data in table format and generate different graphs for each IIAQMM module

The structure of the database includes a table with the following fields: Module ID (the MAC of the ESP32); Location (a string containing the location name); Time (date and time for the received record when data were acquired from the sensors); Temperature (ambient indoor temperature); Humidity (relative humidity); Pressure (barometric pressure); IAQ (Static IAQ – Index for Air Quality); Carbon (CO2 equivalents); VOC (Breathe Volatile Organic Compounds); TempMinVal (Minimum value of temperature for alarm trigger); TempMaxVal (Maximum value of temperature for alarm trigger); HumMinVal; HumMaxVal; PresMinVal; PresMaxVal; IAQMaxVal; CarbonMax, VOCMax (Fig. 5).

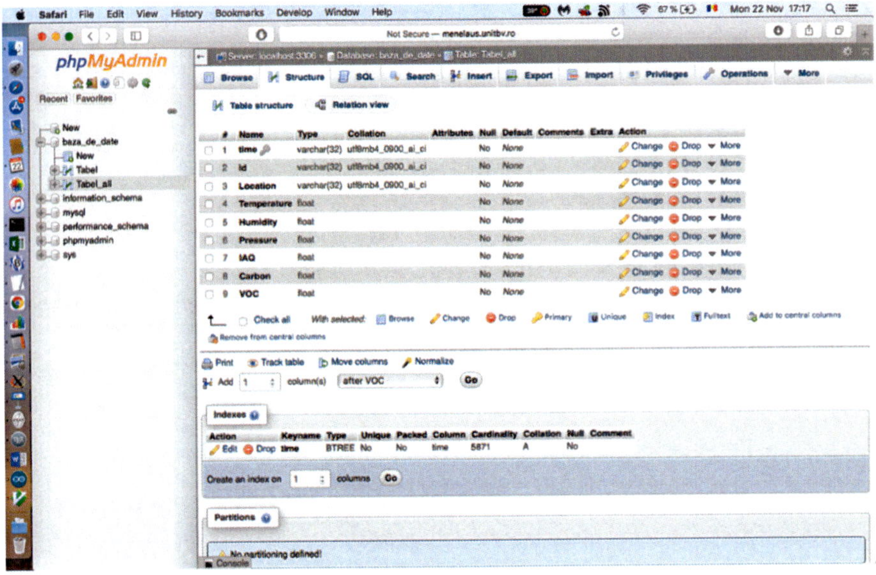

Fig. 5. The design of the server database for the system for indoor air quality monitoring

3 Tests and Results

IIAQMMs are connected by WiFi to the server that runs the monitoring application, exploiting the benefit of the Campus WiFi network that connects all Transilvania University of Brasov buildings and rooms.

Each IIAQMM is programmed with the SSID and password for that network, and on the back of its case, on a label is written the MAC address.

If we want to connect to the IIAQMM Webserver to see the current values of the monitored parameters or to make settings, we need the IP address of the Webserver. It can be obtained after scanning the network using the "arp –a" command to see a list of the connected hosts. Then, the associated IP address can be found by looking for the MAC address. (Fig. 6).

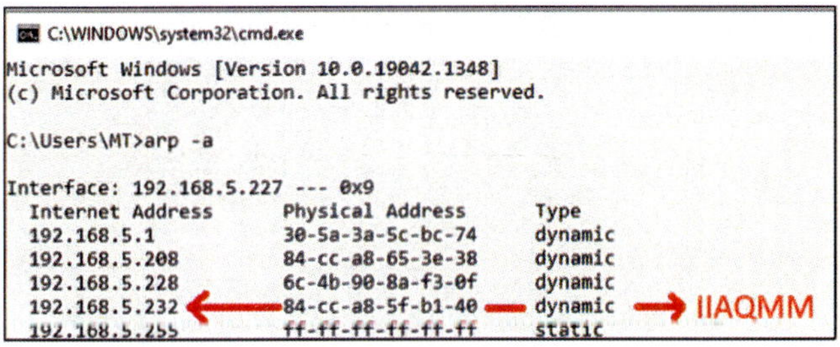

Fig. 6. How to find the IP address of an IIQAMM

For programming the IIQAMM, the USB cable connection is used. In our tests, for an IIQAMM installed in room GI3 at the Transilvania University of Brasov, acquired data are displayed into the Webserver (Fig. 4), transmitted to the Serial Monitor of the Arduino IDE, as shown in Fig. 7, and sent to the server and displayed by it (Fig. 8).

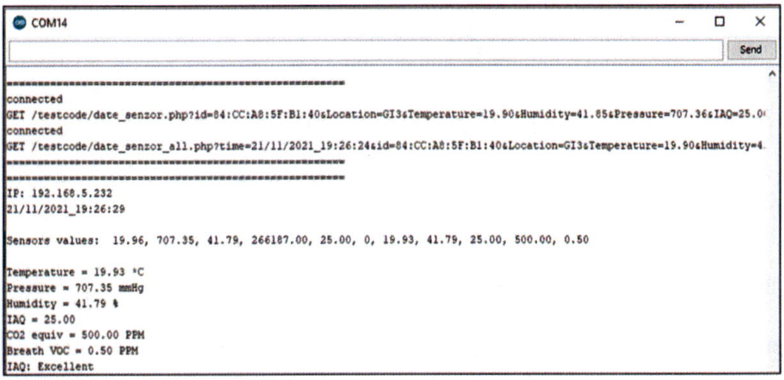

Fig. 7. Data sent to the serial monitor

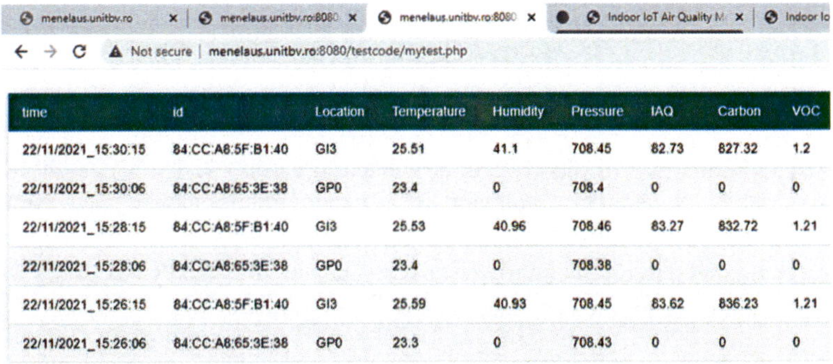

time	id	Location	Temperature	Humidity	Pressure	IAQ	Carbon	VOC
22/11/2021_15:30:15	84:CC:A8:5F:B1:40	GI3	25.51	41.1	708.45	82.73	827.32	1.2
22/11/2021_15:30:06	84:CC:A8:65:3E:38	GP0	23.4	0	708.4	0	0	0
22/11/2021_15:28:15	84:CC:A8:5F:B1:40	GI3	25.53	40.96	708.46	83.27	832.72	1.21
22/11/2021_15:28:06	84:CC:A8:65:3E:38	GP0	23.4	0	708.38	0	0	0
22/11/2021_15:26:15	84:CC:A8:5F:B1:40	GI3	25.59	40.93	708.45	83.62	836.23	1.21
22/11/2021_15:26:06	84:CC:A8:65:3E:38	GP0	23.3	0	708.43	0	0	0

Fig. 8. The server displays the last 100 records from the database

Figure 9 presents the IAQ graph generated on the server.

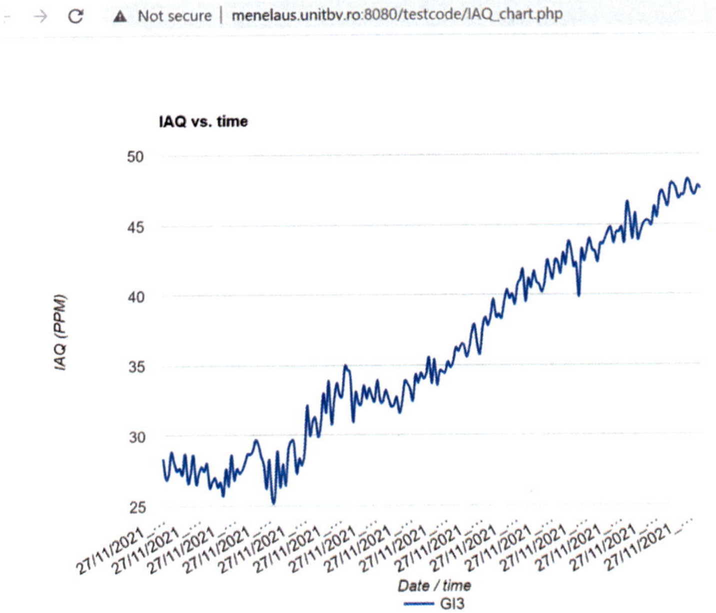

Fig. 9. The IAQ graph generated on the server

Similar graphs can be generated for all the other parameters.

4 Conclusions

A distributed indoor IoT air quality monitoring system was developed and implemented in some rooms of the Faculty of Product Design and Environment, Transilvania University of Brasov, Romania. Each device monitors relative humidity, barometric pressure, ambient temperature, and gas, computing the Index of Air Quality. They are connected to the Campus WiFi network that connects all the buildings and rooms in the university. Information is sent to a server at certain time intervals. A software application collects, processes, and displays data, raising alarms if the predefined interval limits are exceeded and offering an image of the air quality inside the rooms where the educational process takes place.

The actual outcomes of the system are related to the low-cost and low-power consumption modules, scalability – you can add as many modules as you need, computing the Index for the Air Quality, alarm triggering, etc.

The anticipated outcomes aim to versatility – the modules can be adapted to monitor outdoor air quality or be used as independent data-loggers.

References

1. World Health Organization - Ambient (outdoor) air pollution. https://www.who.int/news-room/fact-sheets/detail/ambient-(outdoor)-air-quality-and-health. Accessed 11 Nov 2021

2. U.S. Environmental Protection Agency - Indoor Air Quality (IAQ). epa.gov/indoor-air-quality-iaq/introduction-indoor-air-quality. Accessed 11 Nov 2021
3. Indoor Air Quality (IAQ) Scientific Findings Resource Bank (IAQ-SFRB) - Lawrence Berkeley National Laboratory. https://iaqscience.lbl.gov. Accessed 11 Nov 2021
4. de Carvalho JM, Ferreira A, Figueiredo JP, Loureiro A (2020) Indoor air quality in classroom labs. In: Arezes PM et al (eds) Occupational and Environmental Safety and Health II, vol 277. SSDC. Springer, Cham, pp 285–292. https://doi.org/10.1007/978-3-030-41486-3_31
5. de Gennaro G et al (2014) Indoor air quality in schools. Environ Chem Lett 12(4):467–482
6. Rajagopalan P, Andamon MM, Woo J (2021) Year long monitoring of indoor air quality and ventilation in school classrooms in Victoria, Australia. Archit Sci Rev 65:1–13 (2021)
7. Becerra JA, Lizana J, Gil M, Barrios-Padura A, Blondeau P, Chacartegui R (2020) Identification of potential indoor air pollutants in schools. J Clean Prod 242:118420 (2020)
8. Demanega I, Mujan I, Singer BC, Andelkovic AS, Babich F, Licina D (2021) Performance assessment of low-cost environmental monitors and single sensors under variable indoor air quality and thermal conditions. Build Environ 187:107415
9. Saini J, Dutta M, Marques G (2021) Sensors for indoor air quality monitoring and assessment through Internet of Things: a systematic review. Environ Monit Assess 193(2):1–32. https://doi.org/10.1007/s10661-020-08781-6
10. Adochiei FC (2020) Electronic system for real-time indoor air quality monitoring. In: 2020 international conference on e-health and bioengineering (EHB)
11. Kodali RK, Pathuri S, Rajnarayanan SC (2020) Smart indoor air pollution monitoring station. In: IEEE 2020 international conference on computer communication and informatics, pp 535–539
12. Esfahani S, Rollins P, Specht JP, Cole M, Gardner JW (2020) Smart city battery operated IoT based indoor air quality monitoring system. In: IEEE sensors
13. Chojer H, Branco PTBS, Martins FG, Alvim-Ferraz MCM, Sousa SIV (2020) Development of low-cost indoor air quality monitoring devices: recent advancements. Sci Total Environ. 727:138385
14. AbdulWahhab RS (2019) Air quality system using IoT for indoor environmental monitoring. In: Proceedings of the 2019 5th international conference on computer and technology applications, pp 184–188
15. Zhi SD, Wei YB, Cao Z, Hou C (2017) Intelligent controlling of indoor air quality based on remote monitoring platform by considering building environment. In: IEEE international conference on systems and informatics, pp 627–631
16. Gas sensor BME 680. https://www.bosch-sensortec.com/products/environmental-sensors/gas-sensors/bme680/. Accessed 11 Nov 2021

Inventory Control for Liquid Products by Using Internet of Things

Noha Mostafa[1,2](✉), Sara Sherif[1], and Alaa Adel[1]

[1] Mechanical Engineering Department, The British University in Egypt, Cairo 11837, Egypt
noha.mostafa@bue.edu.eg

[2] Industrial Engineering Department, Zagazig University, Zagazig 44519, Egypt
namostafa@eng.zu.edu.eg

Abstract. Many sectors are confronted with numerous problems and flaws when it comes to supply chain management and especially inventory management. The objective of inventory management is to identify the right amount of inventory to meet demand while avoiding overstocking. A methodology for inventory management system with integrating Internet of Things (IoT) technology is proposed in this paper to monitor the conditions of a resembling inventory prototype. The construction of the proposed system was explained starting with the mechanical design of the tank and proceeding with the electrical and coding part and their assembly all together. A mobile application was developed for continuous state monitoring of the used prototype that simulates an inventory system.

Keywords: Supply chain · Internet of Things · Inventory management · Mobile application

1 Introduction

The first and second industrial revolutions that took place in Europe and the united states have initiated the development in manufacturing processes and were followed by subsequent revolutions which aimed to eliminate the complications and human inefficiencies by relying more on the automated manufacturing [1].

The term "Internet of Things" was first introduced in 1999 by Kelvin Ashton from P&G who developed a system that represented the communication between computers in the manufacturing process in order to reduce the human error in any process, this has also led to higher levels of control and organization over the whole life cycle of products which will later have a positive impact on the customer's satisfaction as well as a smoother operation throughout different stages of production [2].

IOT can be explained as a large network of systems that are connected and communicate through standard protocols [3]. It simply helps create a bridge between both digital and virtual worlds by being more dependable on machines and advanced objects interaction to make sure that if anything happens; an on-time response would occur with no human errors or delays which can turn a traditional industry field into an intelligent one through proposing techniques such as self-maintainability, self-optimization, self-cognition, and self-customization [4].

M. E. Auer et al. (Eds.): REV 2022, LNNS 524, pp. 300–311, 2023.
https://doi.org/10.1007/978-3-031-17091-1_31

2 Literature Review

IoT applications in the supply chain (SC) provide easy means of monitoring and connecting goods and packages via Global Positioning System (GPS) as well as other technologies. The warehouse conditions for goods may also be tracked, which increases quality control across the supply chain. The integration of IoT in SC is a recent interesting topic and many researchers have been trying to explore how implementing this would impact the supply chain. In this section we will be reviewing some of these papers.

In [5], the impact of IoT on the social and economic aspects when implemented in a warehouse was studied. IoT can be beneficial when implemented in an area such as warehousing as it can add smart features like monitoring and tracking which will eventually lead to a cut down in the overall cost. To gain a better understanding of the IoT systems, the paper has discussed the four main layers included in the architecture of such systems which are: the perception layer, the transmission layer, the computation layer, and the application layer. In [6], it was mentioned that to gain maximum outcome, this approach should be implemented on a real warehouse in future research and the benefits should be compared with the actual current status.

Another study conducted in [7] where the effect of implementation of the IoT on the revenue of a two-level perishable product supply chain was investigated. Radio-Frequency Identification (RFID) technology was used to conduct the investigation. Two revenue models were built for the supply chain, before and after the implementation of IoT. After data collection about the aquatic product which was studied in this paper, the two models were solved by a computer simulation which was based on Colonial Competitive Algorithm. It was concluded that the implementation of IoT for large-scaled enterprises is beneficial revenue-wise as it helped reducing the overall cost by reducing the logistics cost, recall cost, and the cost of product losses.

In [8], a statistical analysis was proposed by using the software NVivo to be conscious of the effects of IoT on the supply chain of the Australian retail industry. The study revealed that IoT made a significant difference internally in the retail industry, as well as with vendors and consumers, it strengthens logistics processes. For greater retail supply chain incorporation, added features such as auto-capture, visibility, intelligence, and knowledge sharing were introduced. As a consequence, supply chain efficiency in terms of expense, consistency, and delivery increases.

A similar research was developed in [9], where the various aspects of SCM, ERP, IoT, and Industry 4.0 were discussed. A conceptual model has been created to connect the important requirements needed by each of the areas of supply chain. The paper has outlined few drawbacks though, such as the need to invest in technology in order to reap the benefits of Industry 4.0 so it will cost more as a capital. Furthermore, transitioning to Industry 4.0 necessarily requires appropriate IT security technologies for security, a workforce with the requisite skill sets, and information exchange in an interconnected setting with clients.

3 Methodology

After reviewing multiple papers in the literature review that discussed the implementation of IoT systems and how they are connected to supply chains, it was clear that one of the most important areas in supply chain, for which the application of IoT is important, was warehouse and inventory management [10]. Inventory Management is used at many phases of the supply chain. In this chapter we will be discussing the potential applicable methods in order to implement a demonstrative IoT system that can perform adequately in the realms of inventory management. A common problem in inventory management occurs for production processes that involve liquid materials; the problem is that it is often hard to measure the stock for these products. For example, if the product is solid or packaged in boxes, it is usually measured and scanned by methods such as barcode scanners, but for liquid products this is an uprising issue. In this paper, an idea of measuring the fluid in a container was proposed; this liquid can be present in inventories of various industries such as detergents, carbonated beverages, and other products in healthcare facilities [11]. The base of this idea was such a solid proposition for us to implement IoT techniques and incorporate them into supply chain. Figure 1 gives the flow chart of the design methodology of the proposed prototype in this paper.

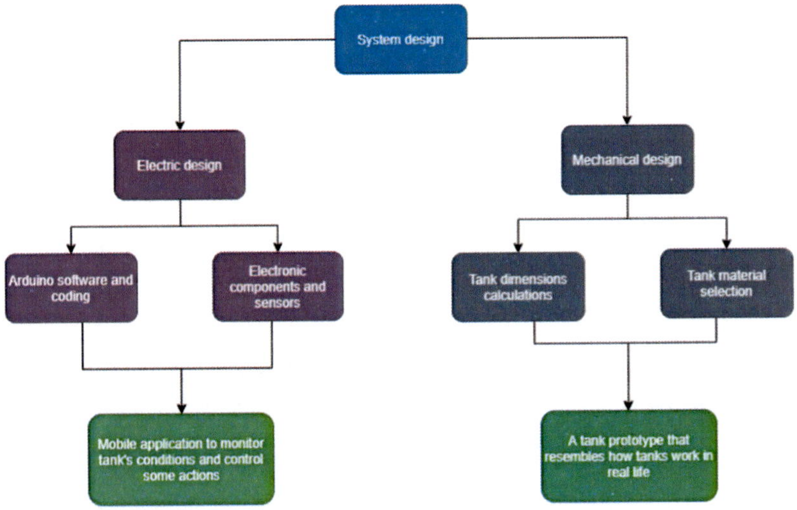

Fig. 1. Flow chart of the design methodology of the prototype

3.1 Mechanical Design

The desired prototype consists of two main stages; the first one is the mechanical part of the prototype and the second one is the electrical part of the prototype. Starting with the mechanical part, it is divided into the following segments:

3.1.1 Tank Material Selection

A tank was designed to resemble the tanks used in industrial factories for storing liquids used in the manufacturing process, the two options that were available were wooden tank with good sealing and a tank made out of steel, and as both of them would do the same function without any drawbacks if any of these two options were chosen, it was decided to go with the cheaper option which is the wooden tank with paying more attentions to the sealing applied at the corners.

3.1.2 Tank Dimensions Calculation

In this step the only constraint that had to be taken into consideration was the maximum distance that the ultrasonic sensor can work with which is 400 cm, so the depth had to be less than that for the ultrasonic sensor to work properly and do its function of measuring the amount of liquid in the tank by the calibration between the measured distance (cm) and the quantity of liquid in the tank (Liters) as any dimensions would have served the same purpose of constructing a prototype to hold all of the electric equipment and simulate as close as possible the function of industrial tanks. However, there were some arbitrary considerations when designing the wooden tank, the first was not to make it very compact to avoid being too crowded with all of the electric equipment, and at the same time not to make it too large to avoid requiring more time with filling and dispensing the liquid. Table 1 was constructed to illustrate the material and dimensions of the designed tank.

Table 1. Summary of material and dimensions of designed tank

Name	Specification
Container type	Stationary container
Tank material	Wooden with well-sealed edges
Capacity	33 L
Base	450 * 400 mm
Height	400

3.2 Control System and Electrical Components

This part of the developing the prototype is divided into the following sections.

3.2.1 Hardware and Electrical Components

There has been multiple electrical components and sensors used in the process of developing the prototype, all of which are summarized in Table 2.

Table 2. Summary about the used electrical components and sensors

Part name	Main function	No. of units used	Price (EGP)	Image of the component
NodeMCU or ESP28266	Considered as the brain of the system and connects it to Wi-Fi	1	140	
Ultrasonic sensor	Measure distances from solid surfaces	1	50	
Humidity sensor	Measure humidity in the air	1	75	
Waterproof temperature sensor	Measure water temperature	1	140	
Solenoid valve	Dispense liquid based on electrical signals	1	140	
Relay module	Closes and opens an electric circuit based on given orders with certain given current	1	5	
7805 Voltage regulator	Reduce supplied voltage to 5V to provide consistent low voltage for certain components	1	5	
Printed circuit board (PCB)	Connect all of the electronic components on a board rather than with wires	1	75	
Opto-coupler pc123	Acts as a switch and insulates some electric components from unwanted high voltages	1	5	
12V Power supply	Supply the designed circuit with the needed voltage	1	95	

3.2.2 Coding and Simulation

Arduino is an accessible open-source platform for skilled programmers to build on. Anyone who is interested in learning more about the syntax can use C or C libraries [12]. The libraries of both languages are such an easy coding languages as they smoothly interpreted by humans. They are also simple for those who want to learn more about the technical aspects by switching from Arduino to the Automatic Voltage Regulator (AVR) and C language. Using the Arduino code and software among the hardware components will lead to the creation of the autonomous part of our project which is the mobile application that will succeed in demonstrating the IoT technique in autotomizing the process of inventory management in supply chain. In the next section the experimental work will be previewed with the detailed steps of how the prototype will come together. The flow of the used code is illustrated in Fig. 2.

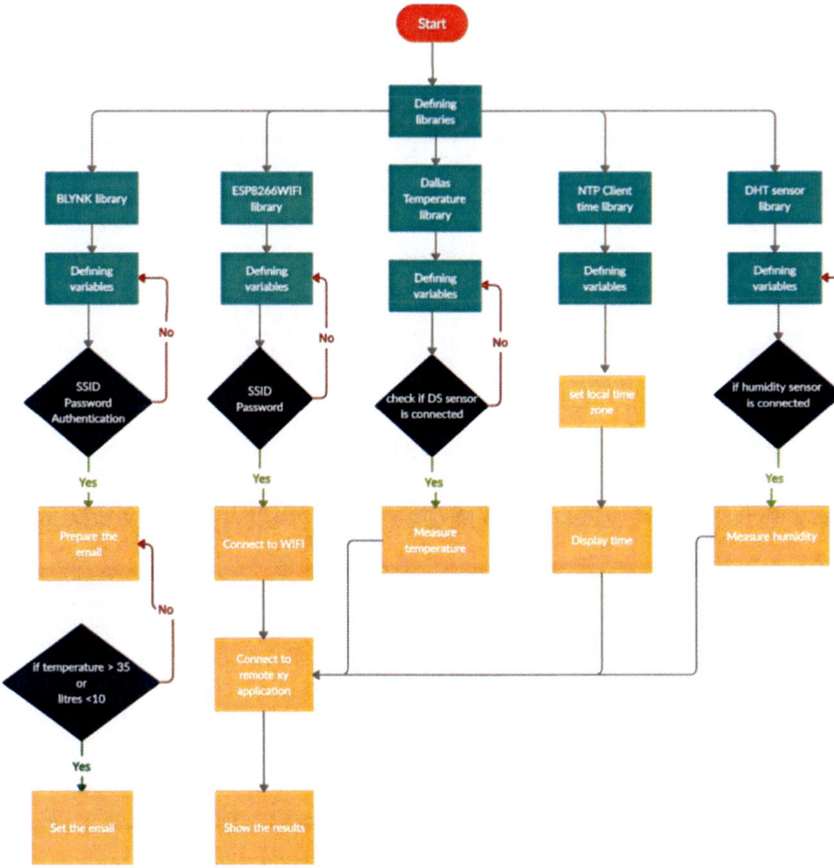

Fig. 2. Flow chart of the developed code

First thing in constructing the code the libraries have to be identified and inserted into Arduino interface, after that each variable in each library must be introduced for easier access of each one later on throughout the programming process. In our case we have used 5 main libraries as mentioned prior. The first library which is BLYNK makes sure that the password, SSID and authentication to cloud is achieved, if they are both not achieved the loop returns to the set variables phase which resembles a failure in logging in and therefore a failure in setting up the email which will update the user with the condition of the tank, otherwise, if the password is correct it proceeds to prepare an email for updates. The system then checks the conditions of the tank before sending an email, if the temperature is above 35 °C or the liters are above 10 then an email is sent, if otherwise the loop goes back to the phase of setting up the email.

The second library which is ESP8266WIFI also checks first for the password and authentication to cloud, if this isn't achieved it returns back to the library set variables, if it is true it proceeds to connect to the WIFI and afterwards it automatically connects to the application that will then be used by the user to display the results. The third library which is responsible for temperature measurements is called DS sensor library, this library firstly checks whether the sensor is connected or no, if not it returns in the loop, if it is connected it tends to measure the temperature and then also connects to the application that will then be used by the user to display the results. The fourth library in the code is the NTP time which manages the time in the application. This library starts by setting an initial local time zone and then displays this time on the application. The fifth and last library is the DHT sensor library and its role is to measure the humidity in the tank, it initially starts by checking if the humidity sensor is connected or not, If not the loop returns back, but if so it proceeds to measure the humidity and begins to display those readings on the application.

4 Prototype Development

4.1 Tank Construction

First the wooden tank was constructed by local workshop with the previously stated dimensions, all of the wooden pieces were cut according to the dimensions and had assembled together with nail gins and all of the corners were sealed properly with a sealant.

4.2 Drawing the Electrical Circuit

The circuit was first designed to allow connecting various electrical components together virtually and testing their function and the objective of the whole circuit with the Arduino syntax. The circuit was built and was checked for its function before building it in real life for the prototype.

4.3 Writing the Arduino Code Syntax and Testing the Electric Components

All of the electric components were connected on a bread board and each component was individually tested to confirm that it is working properly and there are no errors in the connections of the circuit or the components.

4.4 Construction of PCB Board

A PCB board was constructed for the previously explained reason which is to have a secure connection for all of the electric components and preventing any chances of error coming from these loss connections. After constructing the PCB board, all of the electrical components were wired to it by soldiering as shown in Fig. 3.

4.5 Assembly of the Mechanical and Electrical Parts

After all of the above steps, the tank and the circuit with all of the components have been put together with a silicon gun as shown in Fig. 4.

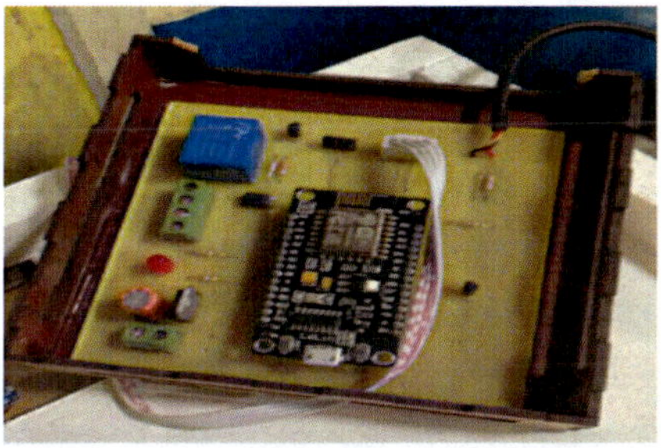

Fig. 3. Illustration of wiring all of the electric components to the PCB board

Fig. 4. Illustration of the assembly between mechanical and electrical parts

5 Testing and Results

The main objective of this project was to have an IoT system to monitor the condition and the quantity of the liquid in the tank, the intention for the designed system that it will work remotely using Wi-Fi and have a mobile application for continuous condition monitoring. The main two outcomes of the system were the mobile application and have an email sent for the owner whenever a certain condition comes true.

5.1 Mobile Application

A mobile application was established that is able to continuously monitor the conditions of the tank via three main interfaces:

a. The first main interface included three windows for continuous displaying of the condition of temperature, humidity, and quantity with a bar that represents the liquid level inside the tank and a real time clock and a green light indicator for the valve state as shown in Fig. 5.
b. The second interface manual control of the threshold of the three monitored parameter, so that the used could easily set the threshold for each parameter according to the monitored process. Figure 6 illustrates the second interface of the mobile application
c. The third interface had the option of controlling the solenoid valve manually as well as for the option of sending emails whenever a certain monitored parameter exceeds the set threshold, so that the user could decide for himself whether to have the option of sending the mail or just stay with the mobile application notification a shown in Fig. 7.

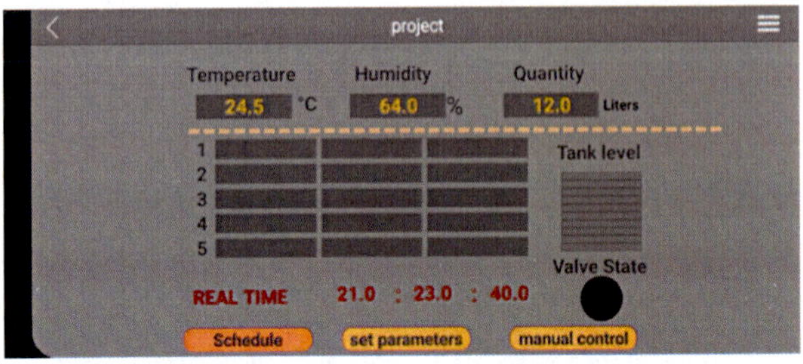

Fig. 5. Developed mobile application interface part (1)

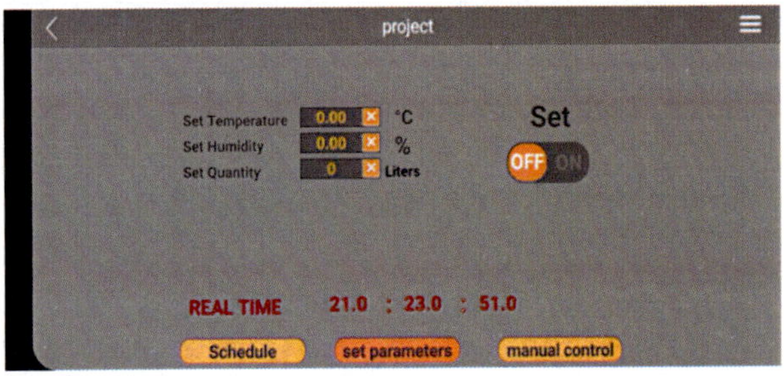

Fig. 6. Developed mobile application interface part (2)

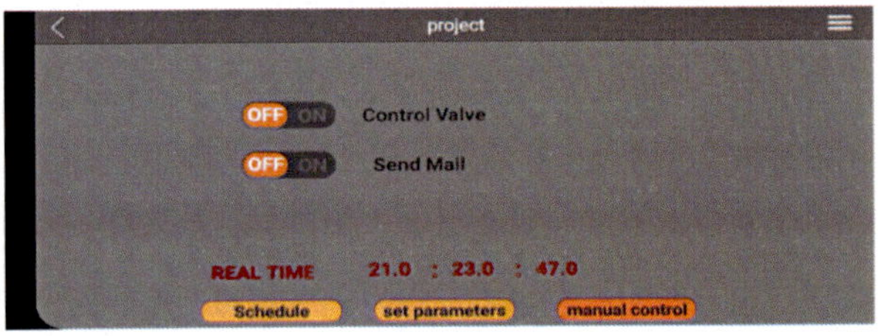

Fig. 7. Developed mobile application interface part (3)

5.2 Email Functionality

As the one of the most important outcomes of the designed IOT system is to receive an email whenever a certain threshold is exceeded, the team has successfully achieved this goal by receiving two separate emails for each of the parameters temperature and quantity as shown in Fig. 8.

As for even further showing the advantages of the developed system, a test was conducted to show that the measurements from the system are relatively accurate, so a thermometer was used to measure the temperature and compare it with the measurement obtained by the system and the reading of the thermometer shoed 38.8 C and the measurement of the system was at that time 36.5 C, so the error was to be calculated, the error would be 0.8% which is very low and shows the accuracy of the system. The test is shown in Fig. 9.

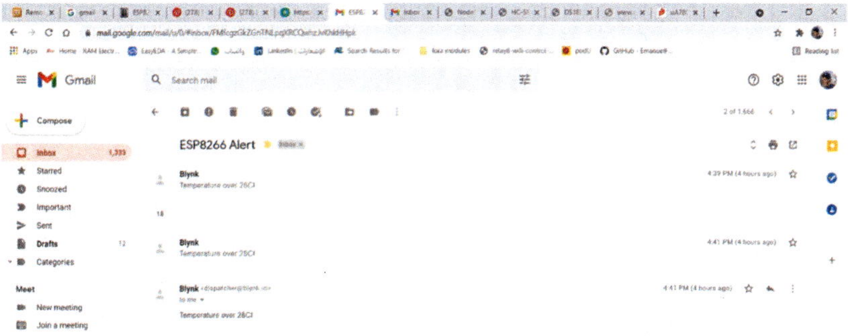

Fig. 8. Email functionality for temperature threshold

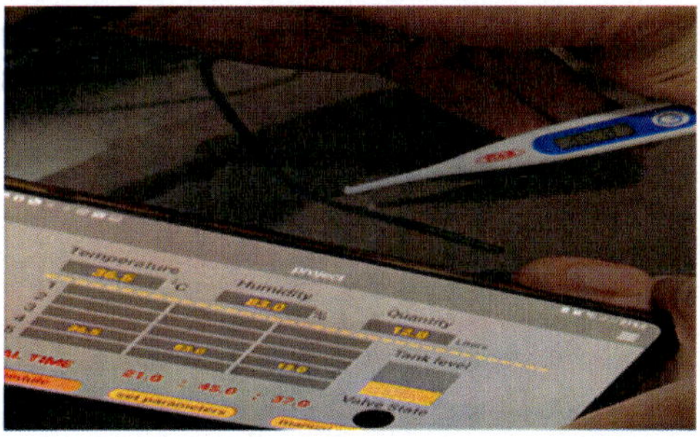

Fig. 9. System accuracy test

6 Discussion

After reviewing each step of the designing process, it is safe to say that the main objective of an IoT-based system design is achieved by the proposed and developed prototype. The following goals were obtained by the designed IoT inventory system:

- Continuous monitoring of the conditions of the reservoir tank (temperature, humidity and quantity) by developing a mobile application with graphical demonstration in the app about the liquid level in the tank
- Notification of a threshold exceeded for one of the parameters via application notification or by email
- Ultimately, ease of flow of information and less human interaction were achieved by the designed IoT system, which are the number one priority for any IoT-based system.

Also to father more prove the efficiency of the proposed design cost wise, cost analysis and power consumption were conducted to illustrate the fixed and running cost of the proposed design. The total energy consumption per month was found to be 4.4316 Kwh/month, while the total price of consumption per month was found to be 7.09 EGP. As a conclusion of both analysis (cost and power consumption), it was concluded that the proposed design offered many helpful and desired features with low cost compared to the offered achievements that the system has got. However, there were some difficulties that appeared in some of the designing steps such as:

- The difficulty in sending an email whenever a threshold is exceeded because it was difficult to find a working mediator that will connect between the ESP chip and the mail to be sent for, however it was overcome by an application that had available cloud to send the email,
- The difficulty to choose the perfect water proof temperature sensor for our system, so it was used another separate temperature sensor, and the humidity and temperature sensor was only left for measuring the humidity in the tank,
- The solenoid valve need high current and voltage (12 v), and it could not be controlled directly by the microcontroller, so a relay was used to act as a switch to help controlling the valve.

7 Conclusions and Further Research

This paper has proposed a methodology for designing an IoT system for inventory control, a step by step procedure of the design was explained and each of the used components were explained, and successful outputs have been obtained of continuous monitoring of the conditions as well as having a manual control over their thresholds and also having an alarming email sent automatically whenever a threshold is exceeded. This system is designed to be inexpensive and accessible. After the smart system is installed, it becomes easier not only to monitor the conditions of fluid products in any inventory but also to react instantly in case of any disruption or shortage. The system is environment-friendly, as it uses very less amount of power. The created system is efficient since it uses a web database server with continuously updated data and real-time operations mostly from computer system. Users may quickly monitor the reemergence of any liquid products with accurate results and little effort thanks to the installation of a customer user experience. The system can also work as a replacement for a huge part in the jobs of some daily workers which might lead to a reduction in the total expenses paid to labor. As for further research, more work can be done to find a way for the measurements to be stored on a cloud, also using different methods of coding programs could attain new features that could be integrated in the inventory management system which will enrich its functionality.

References

1. Qin J, Liu Y, Grosvenor R (2016) A categorical framework of manufacturing for industry 4.0 and beyond. Procedia CIRP 52:173–178
2. Ashima R, Haleem A, Javaid M, Rab S (2021) Understanding the role and capabilities of Internet of Things-enabled additive manufacturing through its applications. Adv Ind Eng Polym Res 5:137–142. https://doi.org/10.1016/j.aiepr.2021.12.001
3. Hozdić E (2015) Smart factory for industry 4.0: a review. J Mod Manuf Syst Technol 7(1):28–35
4. Lampropoulos G, Siakas K, Anastasiadis T (2019) Internet of Things in the context of industry 4.0: an overview. Int J Entrep Knowl 7(1):4–19

5. Mostafa N, Hamdy W, Alawady H (2019) Impacts of Internet of Things on supply chains: a framework for warehousing. Soc Sci 8(3):84
6. Mostafa N, Hamdy W, Elawady H (2020) An intelligent warehouse management system using the Internet of Things. Egypt J Eng Sci Technol 32(1):59–65. https://doi.org/10.21608/eijest.2020.42338.1009
7. Yan R (2017) Optimization approach for increasing revenue of perishable product supply chain with the Internet of Things. Ind Manag Data Syst 117(4):729–741
8. de Vass T, Shee H, Miah SJ (2020) IoT in supply chain management: a narrative on retail sector sustainability. Int J Logist Res Appl 24(6):605–624
9. Manavalan E, Jayakrishna K (2018) A review of Internet of Things (IoT) embedded sustainable supply chain for industry 4.0 requirements. Comput Ind Eng 127:925–953
10. Hamdy W, Mostafa N, Elawady H (2018) Towards a smart warehouse management system. In: Proceedings of the 3rd North American IEOM conference, 27–29 September 2018, Washington, USA, pp 2555–2563
11. Kaistha N (2021) Liquid level control in a recycle loop. J Process Control 104:11–27. https://doi.org/10.1016/j.jprocont.2021.05.014
12. Arduino Technical specs. www.arduinocc/en/Main/arduinoboarduno. Getting Started with XBee RF Modules Version 1.0 by Martin Hebel and George Bricker with Daniel Harris

3D Printing - An Important Industry 4.0 Tool for Online and Onsite Learning

Andre Espach[1] and Kapil Gupta[2(✉)]

[1] Mechanical Engineering Science, University of Johannesburg, Johannesburg, South Africa
[2] Mechanical and Industrial Engineering Technology, University of Johannesburg, Johannesburg, South Africa
kgupta@uj.ac.za

Abstract. Additive manufacturing (AM) which is commonly named as 3D printing has been found as an important tool for teaching and learning in engineering education. For both, onsite and online learning, it is exploited to facilitate students in project works, trainings, and prototype development. Fused deposition modeling (FDM) is one of the important material additive manufacturing technologies used most. It is a rapidly growing manufacturing process that offers many opportunities for research in advancing its capabilities for the enhanced performance to obtain better product quality, higher productivity, and sustainability. This paper presents a review of 3D printing with special emphasis on use of FDM for online and onsite learning. It also includes a discussion on optimization strategies in FDM for ease of part fabrication for various applications. Tool path and topology optimization strategies have been found the most prominent for part structures made by FDM. The aim of this paper is to facilitate the teachers, scholars and researchers by providing knowledge on 3D printing and encourage them to harness the potential of digital technologies for the development of students.

Keywords: Additive manufacturing · 3D printing · Industry 4.0 · Learning · Optimization

1 Introduction

The fourth industrial revolution (or Industry 4.0/4IR) has drastically improved business processes with the intervention of technologies such as internet of things, additive manufacturing (3D printing), augmented and virtual reality, cloud computing, big data, cyber security, and autonomous systems [1]. Engineering education is one of them who got facilitated by Industry 4.0 tools and techniques. 3D printing is defined as a process of product manufacturing by adding raw material (wax, polymers, metal powders) layers and building a 3D object [1, 2]. The addition of material is usually carried out layer upon layer as opposed to subtractive, or traditional, manufacturing. Resource (material and energy) efficiency reduced process planning requirements and ideal time, low human interventions errors, flexibility, flexible design and rapid production, light weight parts, and low waste etc. are some of the advantages offered by 3D printing-based techniques over conventional manufacturing. Interest in additive manufacturing started when

M. E. Auer et al. (Eds.): REV 2022, LNNS 524, pp. 312–322, 2023.
https://doi.org/10.1007/978-3-031-17091-1_32

engineers first explored the rapid prototyping techniques to produce a physical object from 3D models in 1980s [3]. Selective laser sintering (SLS) was patented in 1986 followed by the patents of Fused deposition modelling (FDM) in 1989. As the technologies improved, interest kept growing and further practical implementation of additive manufacturing with research, development, and innovation has been continued by the date. The widespread applications of additive manufacturing are in prototype building, rapid tooling, and fabrication of parts for scientific, industrial, and educational purposes. A typical set of steps in any additive manufacturing-based technique is as follows-

1. Geometric modelling: The enclosed volume is modelled geometrically using CAD software. Solid modelling is preferred to 2D modelling for a better representation of all features of geometry.
2. Tessellation of the geometric model: A series of triangular facets are used to process CAD model to approximate the surfaces. The most common tessellation format used in additive manufacturing is STL. It's a simple conversion of CAD model of parts to be fabricated into 3D meshes of triangular elements.
3. Model slicing into layers: The parallel horizontal layers are formed in the x-y plane and the layering in the z-axis direction after slicing of model which is in STL file format. The corresponding part program file loaded with instructions is generated for specific AM machine.
4. Product fabrication: Actual fabrication of part is accomplished using part program which may be followed by postprocessing operation to obtain the required shape, finish, and size.

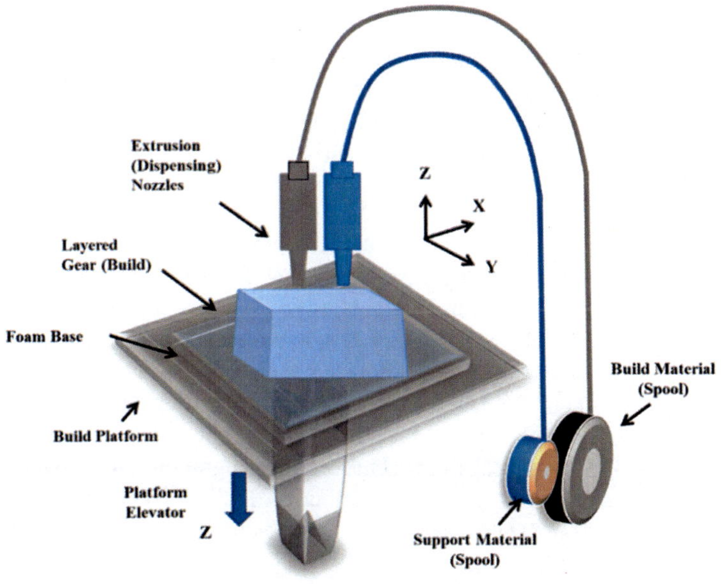

Fig. 1. Working principle of FDM.

Starting or raw material forms in additive manufacturing include liquid polymers, wax, metals, or ceramic which are cured into bonded solid layers of physical parts [3]. The layer formation process can be achieved using a variety of combinations of raw material and forming technologies. Based on that additive manufacturing may be classified into Stereolithography, Selective Laser Sintering, Fused deposition modeling, Laminated Object Manufacturing etc. [1, 2].

Fused deposition modelling (FDM) is a more precise term to describe an additive manufacturing technique where a polymeric material is deposited layer-by-layer to produce a physical 3D model. Figure 1 illustrates the working principle of FDM where raw polymer material (filament) from the spool is drawn to the extruder head where it is heated and extruded to the table to build the part layer by layer. The important process parameters of FDM are- filament type and diameter, printing bed size, nozzle temperature, layer height, cooling fan speed.

2 3D Printing in Engineering Education

The use of 3D printing technology is not only limited to the mechanical or manufacturing technology education, but also extended to chemical engineering type branches [4]. This technology enables students to test their ideas with ease, speed, and accuracy.

One such case is reported in [5], where various parts and prototypes were made to explaining the concepts of course topics. The library of University of Johannesburg has started an initiative *makerspace* that helps undergraduate and postgraduate students to develop and test their prototypes for projects and research [6]. This center is equipped with 3D printing and scanning facilities, robots, and computers. It is one of such initiatives of the university that embraces 4IR. It provides training with an aim to prepare future entrepreneurs and encourages students to share their ideas, concepts, and designs. It also contributes towards batch production of products which can be used in the university for various purposes such as face shields for protection against COVID-19, small equipment and instrument to be used in labs and offices. Similarly, 3D printing based FabLab was established in Italy with an objective of promoting digital culture in teaching and learning [7]. It is also one of the important tools being used in problem- and project-based learning, where students use it to develop models as per the problem or project given and test them accordingly [8]. In an important work reported in [9], 3D printing was named as 'learning by doing' technique for marine electrical engineering education. A positive response from the students was reported. Other than FDM, different 3D printing technologies such as selective laser sintering etc. have also been used for the enhancement of mechanical engineering education [10]. An interesting work reveals the integration of computer-aided design and 3D printing to make students' contributions towards United Nations sustainable development goals (SDGs) [11]. In this work, the students of electrical and electronics engineering programme were tasked to design the projects solving problems related to various SGDs. They converted their project designs into reality using 3D printing technology which ensured a better understanding and learning.

3 Optimization Strategies in FDM

As discussed in the previous section, there is a growing trend of the use of FDM from biomedical and robotic engineering to engineering education [12, 13]. FDM has been used mainly for demonstrations, presentation models, visual aids, and education, these applications account for 25% of customer use in the additive manufacturing industry [13]. A high degree of automation and the ability to create complex geometry has provided unique potential for FDM in low volume production [14]. Some parameters affect the structural performance of FDM manufactured parts more than others.

Experiments conducted by Kuznetsov et al. [15] show that layer height affects structural performance considerably. Moreover, as the layer height increases, structural strength decreases. Material mechanical characteristics may vary between filament material brands. The experimental study also explains how extruder nozzle diameter had a significant influence on interlayer cohesion. As the material is fused layer-by-layer, there are many variables that influence how well the previous layer and the new layer will fuse. Improperly fused layers may contribute to areas of high stress concentration as the effective cross-sectional area may be lower at areas with less 'working' material. It was showed that an increase in interlayer cohesion increases the part's structural performance. Experiments conducted by Ćwikła et al. [16] compared the effects of various parameters on the structural performance of FDM manufactured parts. These parameters include infill density, shell thickness, extrusion multiplier, and infill pattern. With the aim of studying FDM's potential in more structurally demanding scenarios, extensive studies have been done on improving structural performance by altering the printing orientation, raster angle, layer thickness and air gap size [12].

Structural optimization is an umbrella term for mainly three areas of structural design optimization. These are sizing, shape, and topology optimization. With the omnipresent increase of computing power and capacity, structural optimization is becoming a major tool for engineers in the automotive, aerospace, and other relevant industries. FDM along with all its other advantages is a well-suited technology to take even further advantage of structural optimization. Structural optimization deals with the optimal design of load-carrying mechanical structures.

Figure 2 depicts three main strategies of structural optimization [17]. A simply supported beam is used to illustrate the different disciplines. Sizing optimization targets dimensional constraints like the cross-sectional area and other internal dimensions of the beam. The aim is to improve structural stiffness while minimizing the weight of the structure. The application of this discipline is common in the aircraft industry [18]. Shape optimization optimizes the overall shape of the beam, as the name implies. The method targets the predetermined specified boundaries of the beam. In a finite element model, the ideal shape is achieved by optimizing the node placements of the mesh. Topology optimization is used most often in structural optimization. The aim is to obtain a solution that optimized the total material distribution. The shape, size and topology of a structure are defined using a set of distributed functions defined on a fixed design domain. The design formulation of topology optimization is obtained by parametrizing the stiffness tensor of the continuum [19]. A typical approach to solving topology optimization problems is to discretize the problem using finite elements.

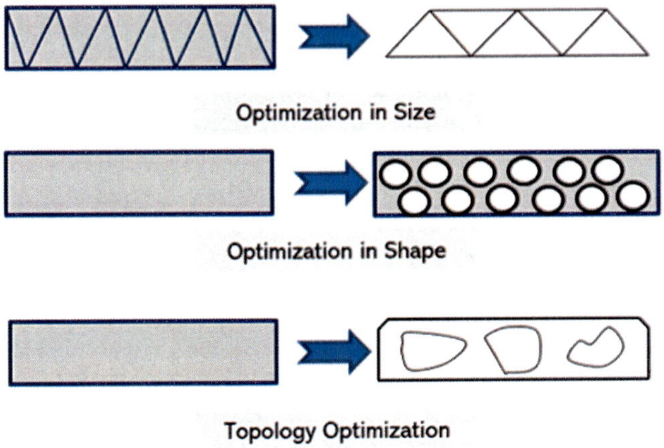

Fig. 2. Structural optimization strategies for FDM part manufacturing.

3.1 Effect of Parameters

Product fabrication for functional or rapid prototyping, using FDM 3D printers is considered a multiparametric process. Many of these factors may affect the mechanical behavior of the 3D printed parts. Any level of structural application of FDM parts will reveal the anisotropic characteristics of this manufacturing technique. FDM process parameters such as raster angle, raster width, raster-to-raster air gap, nozzle diameter and layer height affect the performance of the fabricated part [20].

An experimental study conducted on the dynamic mechanical properties of FDM parts have shown that one of the more effective parameters to alter to increase mechanical rigidity and strength is that of the number of shell parameters. The work done also shows that infill can be used as a reinforcement if oriented correctly for a specific loading direction on the FDM part [20]. Previous research attempts have shown that the raster angle has a definite effect on the tensile strength of FDM parts. Showing that raster angles of 45°/45°, with reference to the angle of the applied tensile load, will yield higher tensile strength values, followed by 30°/60° and lastly 90°/90° [21]. Increasing the layer height may improve the tensile strength of a part. This is presumably due to less airgaps being present in the specimens resulting in a higher effective cross-sectional area.

Raster width shows a similar effect on tensile strength than that of layer height. Decreasing the raster width will result in more material extrusion paths and an increase in the number of airgaps present in a part. This will reduce the effective cross-sectional area of the part and one can expect a decrease in tensile strength. The effective bonding will also decrease as the layer height of raster width increases. This may be an effect of non-ideal bonding of the filament material because of melting point limitations with conventional FDM 3D printing [21].

Figure 3 shows five different tensile strength test specimens after failure. The dashed red line indicates the crack propagation at failure. The angles refer to the raster angle relative to the applied tensile load. The subscripts indicate the number of layers. For raster angles 30° to 90°, the crack that led to failure is at the same angle as the raster angle and in between the layer. This is expected as the FDM part is at its weakest where material is bonded together during the extrusion process. The most commercial 3D printing slicers available today don't work on the principle of extrusion of every layer in the same direction. The reason for this is that, in tensile strength tests, results have shown that by altering the extrusion angle for every layer the specimen will fail at a higher tensile load. Thus, increasing the effective yield and ultimate tensile strength [22].

In case of functional 3D printed parts, one may be interested in how FDM manufactured parts will behave under cyclic loading. It makes sense to plan for a design that will experience fatigue loading and the engineer may thus be interested in knowing which parameters will increase the part's lifespan. Keeping in mind that fatigue failure is caused by cyclic loads that cause microscopic imperfections, it can be reasoned that process parameters like raster width-and-height will affect the fatigue failure lifespan of a mechanical part. A part with small raster width-and-height will have more airgaps compared to a part with larger raster width-and-height values. A study has shown that the infill type and pattern also affect the lifespan of a part that will be subjected to cyclical loading. Results have shown that the best type of infill to use for these parts is the Honeycomb infill pattern [23]. Past work has also revealed that 3D printed parts will perform better, under static and dynamic loading, in cases where the material filament fibers are in the direction of the force [24]. It is not always practical to design a part that can be 3D printed in such a way that the fibers of the material filament will all be in the direction of the expected load.

3.2 Tool Path Optimization

Parts fabricated by FDM are best described as laminates due to the varying mechanical behavior in the x, y, and z coordinates. There are many ways in which a g-code file can be created with regards to filament feed rate, number of perimeters, infill density and shape, and layer height. The path that is used to deposit material to the 3D printer's printing surface is called the tool path. Most 3D printer slicer software creates the infill pattern randomly or by use of some algorithm to produce an infill that follows the user's input parameters. It is therefore a worthwhile idea to try and manipulate the infill pattern in a way that will contribute to the part's mechanical behavior. Tool path optimization may also contribute to the time taken to produce an FDM part as the optimized path may require less changes in direction and filament extrusion rate which adds to production time [25].

In an important investigation, a genetic algorithm that uses a combination of Asymmetric Traveling Salesman Problem and Assignment Problem was introduced to solve it [25]. The tests and experiments conducted show that the technique may only be useful in cases where the problem size is low, therefor complex geometries and with/without complex loading scenarios may not be practical with their path optimization technique.

Comparative studies conducted by Brooks et al. [26] show that optimized tool path for FDM manufactured parts may offer large savings in production time. With conventional 3D printing being a timeous process, this technique will offer large production plants some potential in increasing their turnover due to being able to produce more parts in the less time.

An interesting approach to optimizing the tool path is with taking into consideration the principal stress directions that an FDM part will experience due to known loads and loading directions. Experimental studies conducted show that it is an effective technique [27]. They were able to test the fracture toughness of an FDM fabricated part that has optimized material deposition layouts based on the known principal stress direction. Their experiments show that the optimized part show up to 20% increase in strength when compared to conventionally 3D printed parts that are produced as sliced.

3.3 Topology Optimization

Some early work done on improving the material weight to structural performance ratio of FDM manufactured parts include that of the Fill Compositing technique. This technique involves reducing material weight by the addition of hollow voids in the part's internal infill structure and filling these voids with resin [28, 29]. The resin will exhibit isotropic material property behavior and may add to the part's mechanical performance [29]. The resin acts as an internal reinforcement. It was found that a completely hollow 3D printed part that is filled with epoxy resin may increase the overall yield strength of the part with 24% and stiffness with 25%. A strength and stiffness to weight ratio improvement of 13.6% and 16.1% respectively, was obtained with the resin filled part. As known, shape optimization is a technique that is most likely more practical in FDM fabricated parts. This is because an FDM machine can easily produce a part with hollow sections. More complex geometries are also easier to produce with an FDM machine as the manufacturing process is done layer-by-layer at a minimum of 0.1mm layer heights. It is essential to take anisotropic properties into account to exploit the full potential of topology optimization. Depending on the application, a stiffness-based topology optimization algorithm may be applied to FDM manufactured parts and still yield acceptable results.

Mirzendehdel et al. [30] used non-homogenous failure criteria, specifically Tsai-Wu, to formulate a topological sensitivity. Tsai-Wu was shown to be an effective optimization criterion in their comparative experimental studies. The experiment involved a topology optimized C- bracket that was shape optimized using von Mises and Tsai-Wu as optimality criterion, respectively. The models were then 3D printed and tested in both tension and compression. An interesting approach to decreasing the amount of material used but keeping/increasing the mechanical performance is studied by Zhang et al. [31]. The study aimed to measure the effectiveness of a custom infill style in 3D printed FDM manufactured parts. The technique involved creating a spine, referred to as the medial axis, for the 3D model and then have material branches that attaches the 3D model's outer shell to the inner spine-like structure. The spine-like structure as well as the branches are optimized based on applied loads. The models showed large material savings due to their application of a hexagon shell pattern that reduced the amount of material used to produce an outer shell. The study claimed that the 3D printed parts showed higher mechanical strength compared with conventional 3D printing methods.

A possible issue for shape optimization and 3D printing is that of increased manufacturing time due to rough shape optimized areas in a part. Depending on the mechanical hardware - specifically stepper motor type, driver and operating protocol, the acceleration times for the extrusion nozzles movement in the x-and-y directions may lead to an overall increase in manufacturing time. Shape optimized shapes may lead to very rough boundaries due to mesh sizing. The addition manufacturing time may make the process of using topology optimization more costly in these cases compared to high infill FDM parts.

There has been work done on the smoothing of the boundaries of the shape optimized meshes. This has a massive practical implication for an FDM 3D printing as the less rough the boundary layer of a shape optimized part; the less time is required to manufacture [31]. Liu et al. [32] have devised a way for decreasing the amount of control points in a topology optimized shape boundary. In the study, the skeleton of the topology results is extracted, and a filtering method applied. An adaptive B-spline fitting is proposed and results in a much smoother boundary for the optimized shape. Figure 3 shows the optimization of the medial axis of the test specimen, this structure represents a typical result that is used for the manufacture of an optimized part. The optimization of the tool path is shown in Fig. 4, that results in the reduced manufacturing time.

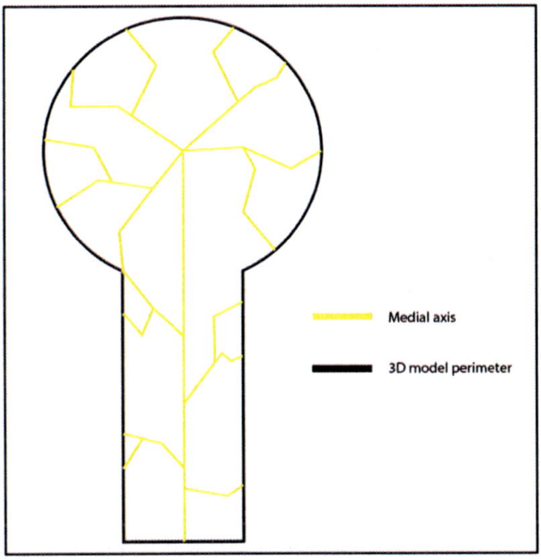

Fig. 3. Medial axis optimization.

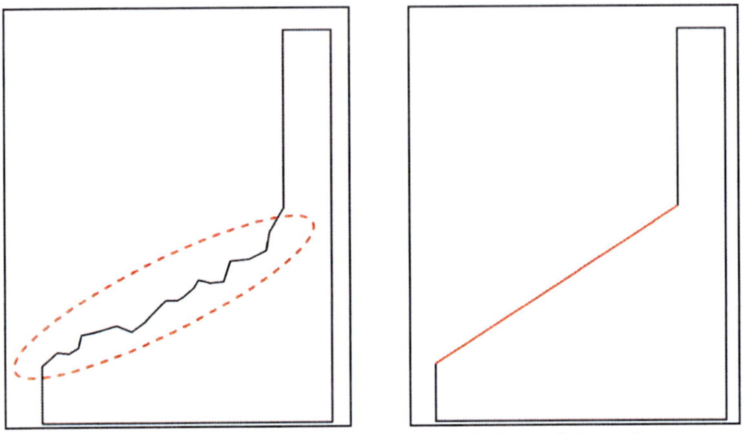

Fig. 4. Smoothed boundary of FDM G-code.

4 Conclusion

This paper has shed light on utilization of 3D printing as an important tool for the present engineering education. It also highlighted some optimization strategies for better utilization of FDM process. It can be concluded that 3D printing specially FDM has significantly complemented engineering education from training of teachers to prototype development for projects and research. This digital technology also reduces the onsite presence requirements and thereby ensures continuous learning remotely.

With regards to the optimization, most of the past research focusing on improving mechanical performance remain largely inconclusive and the adoptability of most of these techniques is low due to additional complexities during the manufacturing and post-processing of FDM parts. Printing parameters such as layer height and infill density affect the mechanical performance greatly. Common parameters that directly influence the manufacturing time include printing speed, infill density and layer height. Merely increasing the printing speed and the layer height and reducing the infill density is not the best option for maintaining acceptable printing quality and mechanical performance. The important future research directions are: investigating the effects of topology optimization on the consumption of energy and manufacturing time, evaluating the effects of topology optimization on mechanical performance of FDM parts, assessing the sustainability of FDM process subjected to topology optimization. Furthermore, sincere attempts are also required to resolve copyright issues, online security challenges, and post processing requirements of 3D printed parts.

Acknowledgment. The work reported is based on the funding provided by the Royal Academy of Engineering - UK, Grant number TSP1332

References

1. Manjaiah M, Raghavendra K, Balashanmugam N, Davim JP (2021) Additive manufacturing: a tool for industrial revolution 4.0, 1st edn. Elsevier
2. Pou J, Riveiro A, Davim JP (2021) Handbook of additive manufacturing, 1st edn. Elsevier Inc. ISBN 9780128184110
3. Gibson I, Rosen D, Stucker B (2014) Additive manufacturing technologies: 3D printing, rapid prototyping, and direct digital manufacturing, 1st edn. Springer, Heidelberg. ISBN 978-1-4939-2113-3. https://doi.org/10.1007/978-1-4939-2113-3
4. Stavarek P (2018) Paving the future: introduction of 3D printing in chemical engineering. In: PRINTCR3DIT: process intensification through adaptable catalytic reactors produced by 3D printing. Project ID: 680414
5. Srivastava S, Pande S (2019) Pedagogical evaluation of mechanical engineering education using additive manufacturing. In: Chandrasekhar U, Yang L-J, Gowthaman S (eds) Innovative design, analysis and development practices in aerospace and automotive engineering (I-DAD 2018). LNME. Springer, Singapore, pp 167–172. https://doi.org/10.1007/978-981-13-2697-4_18
6. https://www.uj.ac.za/news/uj-library-makerspace-develops-3d-printed-face-shields-in-fight-against-covid-19/
7. Baldassarre F, Ricciardi F (2017) The additive manufacturing in the industry 4.0 era: the case of an Italian FabLab. J Emerg Trends Mark Manage Bucharest Univ Econ Stud 1(1):105–115
8. Ali N, Khine MS (2020) Integrating 3D printing into teaching and learning. Koninklijke Brill NV, Leiden, The Netherlands
9. Kuzmanic I, Vujovic I, Kulenovic Z, Petkovic M (2021) Introduction of 3D printing into marine electrical engineering education - a case study. Mar Eng Educ Pedagogika-Pedagogy 93(6):263–272
10. Billings C, Siddique Z, Liu Y (2020) Enhancement of mechanical engineering education with additive manufacturing projects. In: Proceedings of the ASME 2020 international mechanical engineering congress and exposition, volume 9: engineering education, virtual, online, 16–19 November 2020. V009T09A039. ASME. https://doi.org/10.1115/IMECE2020-24568
11. Zaman MHM, Mokhtar MHH, Ibrahim MF, Huddin AB, Beng GK (2020) Integration of 3D printing in computer-aided design and engineering course. Int J Eval Res Educ 9(4):934–938
12. Rankouhi B, Javadpour S, Delfanian F, Letcher T (2016) Failure analysis and mechanical characterization of 3d printed abs with respect to layer thickness and orientation. J Fail Anal Prev 16(3):467–481. ISSN 1547-7029
13. Bakerjian R, Mitchell P (1992) Design for manufacturability, in tool and manufacturing engineers handbook, 1st edn. Society of Manufacturing Engineers. ISBN 9780872634022
14. Pham DT, Gault RS (1998) A comparison of rapid prototyping technologies. Int J Mach Tools Manuf 38:1257–1287. ISSN 0890-6955
15. Kuznetsov V, Solonin A, Urzhumtsev O, Schilling R, Tavitov A (2018) Strength of PLA components fabricated with fused deposition technology using a desktop 3d printer as a function of geometrical parameters of the process. Polymers 10(3):313. ISSN 2073-4360
16. Ćwikła G, Grabowik C, Kalinowski K, Paprocka I, Ociepka P (2017) The influence of printing parameters on selected mechanical properties of FDM/FFF 3D-printed parts. IOP Conf Ser Mater Sci Eng 227:012033. ISBN 1757-8981
17. Pham D, Dimov SS (2001) Rapid manufacturing, 1st edn. Springer, New-York. ISBN 978-1-4471-0703-3. https://doi.org/10.1007/978-3-662-05086-6
18. Cavagna L, Ricci S, Riccobene L (2012) Structural sizing, aeroelastic analysis, and optimization in aircraft conceptual design. J Aircr 48(6):1840–1855

19. Bendsoe MP, Sigmund O (2004) Topology optimization - theory, methods and applications. Springer, New York. ISBN 978-3-662-05086-6
20. Domingo-Espin M, Borros S, Granada A, Reyes G (2014) Influence of building parameters on the dynamic mechanical properties of polycarbonate fused deposition modeling. Printing Addit Manuf 1(2):70–77. ISBN 2329-7662
21. Rajpurohit SR, Dave HK (2018) Impact of process parameters on tensile strength of FDM printed crisscross PLA. Int J Mater Metall Eng 12(2):52–57
22. Kiendl J, Gao C (2020) Controlling toughness and strength of FDM 3D-printed PLA components through the raster layup. Compos Part B Eng 180:107562. ISBN 1359-8368
23. Gomez-Gras G, Jerez-Mesa R, Travieso-Rodriguez JA, Lluma-Fuentes J (2018) Fatigue performance of fused filament fabrication PLA specimens. Mater Des 140(15):278–285. ISSN 0264-1275
24. Ziemian CW, Ziemian RD, Haile KV (2016) Characterization of stiffness degradation caused by fatigue damage of AM parts. Mater Des 109:209–218. ISSN 0264-1275
25. Wah PK, Murty KG, Joneja A, Chiu LC (2002) Tool path optimization in layered manufacturing. IIE Trans 34(4):335–347
26. Brooks HL, Rennie AEW, Abram TN, McGovern J (2011) Variable fused deposition modelling - analysis of benefits, concept design and tool path generation. In: Innovative developments in virtual and physical prototyping. CRC Press, pp 511–517. ISBN 978-0-415-68418-7
27. Gardan J, Makke A, Recho N (2016) A method to improve the fracture toughness using 3D printing by extrusion deposition. Procedia Struct Integrity 2:144–151. ISSN 2452-3216
28. Espach A, Gupta K (2021) An investigation on achieving sustainability in fused deposition modeling via topology optimization. Int J Recent Contrib Eng Sci IT 9(3):4–19
29. Belter JT, Dollar AM (2015) Strengthening of 3D printed fused deposition manufactured parts using the fill compositing technique. PLoS ONE 10(4):e0122915
30. Mirzendehdel AM, Rankouhi B, Suresh K (2018) Strength-based topology optimization for anisotropic parts. Addit Manuf 19:104–113. ISSN 2214-8604
31. Zhang X, Xia Y, Wang J, Yang Z, Tu C, Wang W (2015) Medial axis tree - an interesting internal supporting structure for 3D printing. In: Computer aided geometric design, pp 35–36, 149–162. ISBN 0167-8396
32. Liu S, Li Q, Liu J, Chen W, Zhang Y (2018) A realization method for transforming a topology optimization design into additive manufacturing structures. Engineering 4(2):277–285. ISBN 2095-8099

Smart Education for Sustainable Digital Era

Integration of Business and Education in the Training of Engineers for Industry 4.0

Irina Makarova(✉), Polina Buyvol, Eduard Mukhametdinov, and Vadim Mavrin

Kazan Federal University, Syuyumbike prosp., 10a, 423812 Naberezhnye Chelny, Russia
kamIVM@mail.ru

Abstract. The fourth industrial revolution (Industry 4.0 - IR 4.0) led to a paradigm shift and changes in production and related processes, increased requirements for the competence of engineers, and therefore the business is experiencing difficulties in recruiting. Dynamic changes in the field of engineering and technology have radically changed the requirements for the engineering education system, since its inertia does not allow satisfying the needs of business for personnel with creative thinking, eco-consciousness and skills to work in the context of digital transformation. At the same time, there is a decline in interest in the profession of an engineer in the world. All this requires a paradigm shift and an increase in the sustainability of the engineering education system in a dynamically changing environment, the goal of which will be to prepare an engineer "for the future", that will meet the real needs of the business. The internationalization and digitalization of the economy and new work functions make digital competencies critical. At the same time, new and promising technologies should be adapted and included in the educational process in order to facilitate the adaptation of university graduates in production. The purpose of the article is to consider topical issues of engineering education related to interdisciplinarity, new standards and technologies for engineering education, digital educational environment and online technologies, the interaction of engineering education with high-tech business and industry.

Keywords: Digital competencies · Engineering education · Industry 4.0

1 Introduction

The rapid speed of technology and technique development requires a radical paradigm change from a rather inert educational system, since, according to analysts, the fourth industrial revolution (IR 4.0) will force humanity to reconsider attitudes towards work and the operate process: cardinal changes will affect more than 35% of work skills. Since Industry 4.0 implies cyber-mechanical systems that are tuned to perform new tasks: analyzing customer requirements, changing the technological processes, as well as serving themselves, a completely new engineering approach will be required: enterprise digital models [1, 2]; engineers and programmers who will work with them; smart devices [3, 4] that can communicate with each other and the digital model. The most important is

M. E. Auer et al. (Eds.): REV 2022, LNNS 524, pp. 325–336, 2023.
https://doi.org/10.1007/978-3-031-17091-1_33

the management system modeling and the intelligent labor's share growth. At the same time, the need for maintenance personnel and equipment downtime will decrease, and its utilization rate will increase, which will lead to the optimization of technological [5] and logistic processes and an increase in production efficiency. In addition, IR4.0. Involves the digital twins (DT) creation, and the creation of a system of virtual simulators will allow personnel to practice real actions in virtual workplaces. The changes also affect related processes - the most important of which is Logistics 4.0 [6], since transport provides not only the people mobility (including workers in any industry), but also the delivery of raw materials, materials and products to the manufacturer and the buyer [7, 8].

The university should become a platform where the technologies of the future are not only used, but already "live". The university, like production, should be "digital" - these are digital textbooks and teaching materials, electronic document management, digital online courses, digital logistics of the educational process. A transition to the "University 4.0" model is necessary, which will train an "engineer of the future" who can not only formulate a problem in a professional language, solve it with the help of new intelligent technologies, but also be able to create fundamentally new technologies and products.

The educational system must be continuously improved in order to provide opportunities not only for the professional competencies formation in future engineers, but also for the development of their environmental awareness and social responsibility to future generations. A transition to a creators' society is necessary. This can be realized only as a result of systemic interaction and integration in the training process of all its participants efforts and business representatives.

The article describes the concept of creating such a system in relation to the logistics 4.0, shows the first steps implemented in this direction on the example of organizing the interaction of the education system with the automaker.

2 Background: Trends in Logistic 4.0 and Its Realization in Engineering Education System

2.1 Industry 4.0 and Logistics 4.0 and Their Impact on the Development of Engineering Education

IR4.0 technologies bring increased productivity, reduced costs and delays to enterprises affecting all processes. Due to the complexity of analyzing all processes in aggregate, article [9] focuses on logistics and supply chain, in particular exploring the potential for introducing logistics 4.0 (LOG4.0) and supply chain 4.0 (SCh4.0) in the automotive industry in Bulgaria. A survey conducted showed that production transparency and blockchain technology are of the greatest importance for logistics and the supply chain. The main problems are cybersecurity and incorrect data.

The LOG4.0 paradigm means the optimization of inbound and outbound logistics through achieving the highest degree of automation supported by intelligent systems embedded in software and databases, and the necessary information is provided and disseminated through IoT systems [10].

The document [11] aims to address the changes that will occur for logistics operators in the IR4.0 era, describing two possible scenarios for LOG4.0 (or smart logistics) in a real industrial context and its impact on human labor. Deeper exploration of new competencies required by employees can also help companies train and engage employees in logistics transformation.

Decision makers in the IR4.0 era must create the appropriate information systems, physical tools and technologies of various types that can meet the future economy needs in changing the paradigm LOG 4.0 and SCh4.0. Document [12] aims to provide a balanced viewpoint of the view's diversity of physical Internet professionals to identify the main challenges (technological, social, business paradigms) of LOG4.0 to support IR4.0. It also requires a completely new organization of logistics activities, which cannot be realized without the broad participation of industry, since large investments are required in loading units, handling and transport vehicles. Both IR 4.0 and LOG4.0 are complex systems that entail large technological changes and changes in the business environment and their further development will be gradual and collaborative.

2.2 Business Needs for Personnel Competencies and New Trends in Competency Management

Technical developments for IR4.0 in digital transformation frame of are changing the requirements for employee competencies. The document [13] goal is to define the competencies that play a central role in IR4.0 and are divided into professional, methodological, personal and social. The analysis shows that the competencies "willingness to learn", "interdisciplinarity", "comprehensive understanding of the process" and "communication skills" are becoming more and more important. In specialized areas of logistics and manufacturing, there is a tendency to improve skills, acquire IT competencies and understand processes, due to the expansion of data sets, their analysis and operational interpretation.

Modern companies need to have a certain skill and visibility of their human capital in order to provide a better competency management strategy to increase the flexibility of supply chain management. The article [14] goal is to present an approach that will allow companies to assess and control their human capital (competencies) to improve the productivity of the supply chain, in accordance with the needs of companies through appropriate indicators. This will allow companies to properly classify their teams and serve as the basis for training plans, hiring or changing teams, which is useful for both production managers and the human resources department. In the future, it will be possible, according to the authors, to expand the system through qualitative indexes, including both individual and collective competencies.

The article [15] proposes mathematical models and approaches to optimization for solving the "People-Job" (PJ) problem, which has five goals. In doing so, each employee is assigned to perform a job based on the level of competence appropriate for that job. It then looks at employee satisfaction in terms of a list of preferences and rating points. Decision-makers (DMs) find the right employee to effectively define the job assignment and get the job done, based on a trade-off between overall competency and employee satisfaction.

Nowadays, competence is gaining special importance, and the interests of the organization are shifting from work to personality. The article [16] proposes a new approach to the ontological's competence model based on the Deming wheel, which facilitates the transfer and understanding of information to improve the management of resource's competencies within the company. Ontology is a suitable tool for defining and managing competencies and knowledge in companies due to its suitability for knowledge modeling. The authors proposed a model for improving information processing and exchange that describes competency management as an alternative to job and skill management.

The development of human resources for regional strategic industries is becoming an increasingly important area for the development of industries with growth potential. Article [17] identifies the competencies and knowledge levels required for personnel in the port and logistics industry. The study consisted of interviews with experts to design a competency survey. Six clusters of competencies and 16 sub-competencies were developed. This study was useful in developing an educational program and suggests ways to use regional human resource development (RHRD).

As new technologies change logistics processes, the role of people and the tasks they perform, companies will need to develop competency models. The article [18] develops a future-oriented competency model for employees, considering the changes caused by the digitalization of automotive logistics, which combines future competencies for operational activities and planning. The created competency model consists of professional, methodological, social and personal competencies, taking into account the implementation processes and future adjustments, which will contribute to strategic personal planning. The integration of the logistics competency model into the strategic workforce planning structure, taking into account the future qualitative and quantitative competencies of the automotive logistics employees, will facilitate strategic workforce planning.

Modern logisticians need to integrate business management and administration skills with process design, IT systems, and other engineering fields. However, according to research on university curricula and competence standards in the field of logistics, the engineering aspect is not fully represented. Document [19] proposes a competency profile for a "logistics system engineer" that introduces a system engineering perspective into logistics by synthesizing systems engineering competencies with modern logistics engineering, creating a "Logistics systems development" profile what reinforces the engineering aspect. Another term suitable for this area is intelligent logistics, borrowed from intelligent transport systems, extending to the cross-functional nature of logistics. We live in an era when more and more automation and autonomous systems are becoming in logistics, therefore current and planned developments require people with the proper education, and this is the basis of modern logistics engineering.

2.3 Trends in the Development of Modern Engineering Education

Article [20] explores the interests of engineering students in relation to working in global companies. According to the authors, interest in working abroad should motivate students to develop global competencies. In addition, students' interest in international careers, according to social cognitive career theory, may be influenced by personal and environmental factors. The results show a significant interest of students in traveling or working abroad, although they differ both by discipline and by educational institution. In future studies, according to the authors, it is necessary to study the influence of global work interests on motivation to study global competencies, as well as the influence of goals at the work place and expectations regarding results on the sources of students self-efficacy.

The article [21] focuses on the role of Experimental Learning Spaces (ELS) for task-based learning (CBL) in industrial engineering education, which arises from the need to define "where to teach and learn" in an active learning setting, because unlike experimental learning that ELS can support, CBL needs further adaptation. The document also provides a case study of the Social Laboratory for Sustainable Logistics at Tecnologico de Monterrey in Mexico. Learning spaces include areas of social interaction where students, faculty and academic partners undertake learning activities. To assess the effectiveness and contribution of ELS to improving the quality of education, a statistical analysis of students' opinions, their academic achievements, including the development of their disciplinary and personal competencies, is required.

The study [22] is focused at analyzing the impact of the engineering professors competence on their lecturer performance according to students and graduates of the Brazilian Federal State University, which was carried out using exploratory factor analysis (EFA) and other methods. As a result of the survey, three core competencies (namely, content-pedagogical knowledge, innovative attitude and emotional skills) were identified that are important for engineering professors, since in order to motivate students to innovate, it is necessary to develop innovative competencies not only among students, but also among professors.

Article [23] describes a real-world case study of Project Based Learning (PBL) using a methodology that takes into account the UN Sustainable Development Goals (SDGs) to develop them in students. The goal is to develop in future engineers an understanding of the needs of companies and the impact importance of their decisions on people and the environment. Sustainability and socially responsible engineering is achieved through a methodology that integrates the SDGs into real projects, which contributes to the awareness of their mission by future engineers, teaching them how to advance towards SDGs 4 and 5. However, because establishing and maintaining relationships with partners, as well as the project coordinating increases the workload teachers, an incentive system is needed for those who will implement such projects.

The study [24] proposes a method for predicting student performance by comparing different algorithms. To assess the learning quality (to determine academic performance), indicators such as academic performance, satisfaction, acquisition of skills and competencies, persistence, achievement of learning goals and career success are used. Data mining applications include various techniques: naive Bayesian classifiers, support vector machine (SVM), logistic regression, decision trees, neural networks, and deep learning (DL). The authors found that the Gradient Boosted Trees algorithm is most appropriate for measuring student performance, but researchers must compare the algorithms before using them to predict student performance.

3 Results and Discussion

3.1 New Methods and Tools for Training Engineers

The COVID-19 pandemic has shown the need to improve Learning Management Systems (LMS). For example, LMS, as an online learning management system, allows users to communicate with a friend from anywhere in the world using the Internet. It can be used to create digital classrooms and connect teachers with students, and in any office or organization to train staff. Nowadays, many organizations are starting to use LMS because of its many benefits.

Smart learning is based on smart devices and technologies that create an effective digital learning environment. Essentially, e-learning is a set of intelligent tools to support learning and as a means to stimulate learners to develop thinking and problem solving. The use of online technologies such as the Internet fosters collaboration and encourages teamwork. Students can work through online discussion groups in a multithreaded format asynchronously or synchronously via chats. The advent of Web 2.0 applications has changed the way people communicate and access knowledge (weblogs (blogs) and wikis). Microblogging services (e.g. Twitter) are also used for near-synchronous discussion, resource sharing, and the development of personalized personal learning networks (PLN). Mobile technology and wearable smart devices are becoming essential tools for expanding learning opportunities virtually anywhere.

Technologies such as virtual and augmented reality, 3D printing and chatbots help improve the quality of the educational process. So, the ability to create 3D prototypes for better visualization of models and concepts and 3D printing, educators can stimulate the interest of automotive students who can print small batches or modified samples of parts for testing. In addition, while working on 3D printing projects, students develop creativity, problem-solving skills, and critical thinking.

Virtual reality can create extreme environments and scenarios, allowing students to test systems and learn without any negative consequences. This not only reduces risk, responsibility and cost, but also allows users to practice solving real-world problems. AR allows you to overlay graphics on objects in the real world, facilitating visualization and immersion in the subject.

AR headsets make learning interactive, ensuring a safe learning environment and the ability to recreate many real-life situations, motivating students to learn difficult topics. In manufacturing and logistics, AR gives step-by-step instructions to workers.

Using language learning platforms, it is possible to participate in a semi-conversational environment in an instant messaging format using a chatbot. Chatbots can also send reminders, provide feedback, track student progress, and complement training and encourage employees to apply their knowledge and skills to real-world work situations.

3.2 The System of Training Logistics Engineers for the Automotive Industry

The automotive industry, like any high-tech industry, needs highly skilled engineers who can generate and implement innovative ideas. These are the problems of efficiency and environmental friendliness of vehicles, the creation of an energy efficient and intelligent transport, reduction of the negative impact on the environment at all stages of the vehicle life cycle. In addition, the engineer must be able to create intelligent transport systems, digital vehicle's models, virtual models of its production, as well as digital twins of service systems.

For more than 10 years we have been interacting with a full-cycle global automotive company in training engineers for all stages of the vehicle's life cycle, as well as with others companies engaged in logistics, service, management and transport safety [25]. That company has not only an R&D center, but also a logistics center, as well as a branded service system, including in a number of countries, We took into account that from the point of view of education digitalization, two main directions appear definable: instrumental and technological, associated with the use of new opportunities of digital and information technologies to improve the education system's efficiency [26, 27]; and substantive, associated with the formation of a new content of the educational process itself.

Taking into account the business demands, the first stage was devoted to streamlining the goals and the means of achieving them. We have systematized the tasks that engineers will have to solve at their workplaces. To identify the most significant competencies of graduates for successful work and career growth, a survey of specialists from partner companies was conducted. The survey results were processed and grouped into the appropriate categories (see Fig. 1).

Competency groups	Competencies	Functions of enterprises			Competency groups	Competencies	Functions of enterprises		
		1*	2**	3***			1*	2**	3***
Engineering	Fundamental knowledge	90	60	72	Interpersonal and communicative	Teamwork ability	80	90	90
	Engineering knowledge	90	56	78		Knowledge of business communication methods	90	75	85
	Digital competencies for solving professional problems	70	60	65		Ability for multilingual communication	90	70	70
	Understanding the problems on the product life cycle stages	90	48	63		Ability to work successfully in the company	95	95	95
Personal	Creative and critical thinking	80	45	60	Professional	Engineering thinking	98	56	68
	Initiatively and Leadership Role	90	85	80		Ability to solve professional problems	95	95	92
	Ability to continually improve	90	90	90		Systems thinking	95	48	71
	Ability for self-purpose and career planning	95	94	90		Ability to search and analyze information	95	85	80
	Ethics and responsibility	98	95	95		Engineering Trends Awareness	98	60	70

The main company activity: 1 * - Design and manufacture of vehicles and vehicles' intelligent systems; 2 ** - Management in transport & logistics systems; 3 *** - Organization of transport processes and transport systems safety

Fig. 1. Results of a survey of engineers from partner companies

After processing the questionnaires, were discussed the requirements for engineers with the HR department and heads of other departments. The specificity is that the main emphasis is placed on information, environmental, communication components, and business satisfaction with the quality of graduates is a criterion for the quality of education.

Now one of the main requirements of almost all enterprises for a young specialist is competence in the field of digital and intelligent technologies [28]. The formation of digital literacy is provided by a set of disciplines, educational situations and practices that simulate real professional tasks, for the solution of which software tools are used related to the specifics of the industry and a specific workplace [29].

To harmonize professional and educational standards, we have identified general training courses that all engineers must study (the first, general technical block), as well as training courses that are unique for each workplace (the second, special professional block). Then, matrices of key competencies were developed that correspond to the professional standards of the industry. Training courses for engineers for different companies and types of professional activities were included in the curricula, which ensured orientation towards a specific type of professional activity.

Considering that logistics processes are typical for all stages of the product life cycle, we examined the competencies typical for the logistics system and the supply chain in the context of the tasks to be solved, taking into account the concepts of IR4.0 and logistics 4.0. At the same time, the competencies in demand at a specific workplace were compared with a competency's matrix corresponding to the curriculum, including for adjusting the programs of disciplines, during the study of which this competence can be formed.

The specificity of the personnel training system lies in the fact that students are gradually "integrated" into the professional environment: in the junior courses they undergo practical training at the enterprise, and then combine their labor activity in the positions of engineers with the educational process. At the same time, they get access to information resources of the enterprise, educational content of the corporate university and software, that necessary for the project's implementation. So, the study of the first block disciplines is carried out at the university, and the second block - at the enterprise (at the basic departments). This makes it possible to resolve the issue of "separation" of management functions in two learning management system (LMS) [30].

The second stage of creating a training system is associated with the development of educational content, the choice of a method and means of educational process' effective implementation. At the third stage, the effectiveness of the proposed system was tested. For this, experimental students' groups, that learned in this system were formed. At the same time, the main emphasis was on digitalization and the use of the computer in different contexts (Fig. 2).

The discipline name from the curriculum	Competency groups 1	2	3	4
Intelligent Vehicle Systems Design Training Courses				
UG NX system for Computer-aided design (CAD) and Computer-aided manufacturing (CAM)	■		■	
Microprocessors in the automotive industry	■	■	■	■
Design and construction of intelligent vehicle systems	■		■	
Reliability and safety of electronic systems structure	■		■	
Vehicle testing. Equipment and test methods	■	■	■	
Methods and systems of engineering judgement and engineering analysis	■		■	
On-board computers and communication facilities V2V, V2I	■	■	■	■
Connected and unmanned vehicles	■	■	■	■

The discipline name from the curriculum	Competency groups 1	2	3	4
Transport Systems Safety Learning Courses				
Hardware systems for active safety of transport systems	■		■	■
Safety Basic Principles of Transport Systems	■	■	■	
Communication and information support of the transport process		■	■	■
Software Systems for Transport Systems' Safety	■		■	
Satellite positioning systems for transport systems	■			
Fundamentals of the reliability and diagnostics theory	■		■	
Environmental friendliness of vehicles	■		■	
Transport Systems Safety Standards	■		■	

The discipline name from the curriculum	Competency groups 1	2	3	4
Learning courses on the transport process organization				
Computer-aided design systems for the transport processes organization	■		■	
Theory of transport processes and systems	■		■	■
Maintenance and repair of vehicles	■	■	■	■
Modelling of transport processes	■			
Methods for optimizing traffic flows	■			
High-quality technical service in transport	■	■	■	
Geoinformation technology in transport	■	■	■	■
Automation and telemechanic in transport	■			

Fig. 2. Training courses for different educational programs

3.3 Examples of the System's Implementation

Markets globalization has made important the issue of judicious organization of logistics processes. The engineers involved in cargo delivery must know the means and methods of cargo shipment planning and managing of motor vehicle fleets, methods of building logistical chains accounting for interaction of different kinds of transport. They must also be trained in how to select the optimal routes with account for different factors and risks.

These aspects are studied by doing such courses as statistical data analysis, methods of planning and forecasting, theory of constraint, systems analysis methods, theory of experiment planning, management in logistical systems, and the theory of transport flows. The course and graduation projects include the issues of constructing of logistical systems, selecting of optimal logistical chains, rational warehousing logistics, building of optimal routes, and motor vehicle fleet management.

The ITS area most in need of use engineers in different activity lines is that of transport systems control because it is here that the greatest scope of tasks is solved. The engineers, creating programs for intelligent onboard systems, study the synthetic vision systems and pattern recognition techniques. In order to comprehend the structural principles of such structures, they do study projects and test the adequacy of proposed solutions in practice (Fig. 3a).

The issues of transportation planning and simulation are dealt with in different companies. There are companies creating the infrastructure, building and reconstructing the street-road networks, determining the location of sites for public transport stops, petrol filling and service centers. Others are involved in planning of traffic along the street-road networks and of traffic control with the purpose of enhancing its safety and effectiveness.

Engineers in these companies must know how to obtain adequate information and operate with great data volumes, analyze them and make strategic and operative decisions. Training for these positions envisages instruction in the systems theory, intelligent control systems, statistical data analysis, the theory of experiment planning, and methods of optimization. In this case, both the course and graduation projects of the students are to include such points as constructing of models of a street-road network site, revealing the causes of hindered traffic in a model, and searching for variants of optimizing the traffic parameters at the site in question (Fig. 3d).

Constructing of public transport routes requires exploring the consumer demand for transportation both within and outside the city. Engineers employed in companies dealing with passenger transportation must be aware of demand analysis techniques, principles of rational constructing of routes, methods of bus fleet management, and ecology-minded? driving. This is taught at such courses as statistical data analysis, theory of experiment planning, methods of optimization, Technosphere safety, systems modeling, management of passenger transportation, and of transport flows (Fig. 3b).

Alongside with other issues, the course and graduation papers of these students include constructing of correspondence matrixes, identifying of gravity centers and places of passenger flows formation, principles of building the macro-models of street-road networks, methods of determining the transport and passenger flows density, methods of constructing rational routes and managing of motor vehicle fleets (Fig. 3c).

Fig. 3. Examples of student projects: a) Project of a road sign recognition system; b) Macromodel of a road traffic system; c) System for managing logistic processes; d) Optimization of traffic at an intersection using a micromodel.

As can be seen in the presented examples of students' projects, the students solve real problems which help them to understand the problems of the industry and professional tasks already during their study time. This approach facilitates professional adaptation.

By participating in real projects, the students got a better idea of requirements imposed on modern vehicles and developed them by employing new design technologies. Their graduation papers were later extended in their master's degree studies.

4 Conclusion

In the era of transition to a digital society, the engineering education system needs to create conditions for training an engineer "for the future": socially responsible, with creative thinking, and eco-consciousness. In our opinion, developing of a high-quality study content in the simulation environment later to be used at real workplaces will improve both the quality of engineer training and the learning process management in realization of the "life-long learning" concept.

Here are the advantages of the proposed system over the traditional methods: (1) For business, it gives the advantage of shorter time spent on a graduate adaptation in real-work environment, a higher level of graduates' training, and new-coming employees with already developed business-relevant competences. (2) For the student, it promises higher interest in academic disciplines, a possibility to become involved in real problem solving, and personal competitive advantages at the labor market. (3) For the education system it means higher occupational prestige for engineers, a possibility of closer contact with business and higher quality of education.

Moreover, interaction with business will give an opportunity for the teachers to learn new technologies and apply them in teaching of students.

References

1. Buyvol P, Makarova I, Sizikin V, Badriev A (2020) The service center simulation model as a part of a reverse logistics system. In: Proceedings of the international conference on developments in esystems engineering, DeSE, pp105–110
2. Makarova I, Shubenkova K, Buyvol P, Shepelev V, Gritsenko A (2021) The role of reverse logistics in the transition to a circular economy: case study of automotive spare parts logistics. FME Trans 49(1):173–185
3. Gabsalikhova L, et al (2020) Connected vehicles fleet expanding problems. In: Proceedings of the 6th international conference on vehicle technology and intelligent transport systems, VEHITS 2020, pp 642–650
4. Makarova I, Shepelev V, Mukhametdinov E, Pashkevich A (2020) Changing the maintenance and repair system while expanding the connected vehicles fleet. In: Proceedings of the 6th international conference on vehicle technology and intelligent transport systems, VEHITS 2020, pp 622–633
5. Ostroukh A, Karelina M, Filippova N, Marusin A, Evtyukov S (2021) Intelligent system for digital substation control. Transp Res Procedia 57:385–391
6. Galkin A, Popova Y, Chuprina E, Shapovalenko D (2019) Interaction of logistics 4.0 and consumer oriented marketing using ICT. In: Proceedings of the 33rd international business information management association conference, IBIMA 2019: education excellence and innovation management through vision 2020, pp 6751–6760
7. Vakulenko K et al (2019) Designing optimal public bus route networks in a suburban area. Transp Res Procedia 39:554–564
8. Popova I, Abdullina E, Danilov I, Ruchkina I, Shemyakin A (2021) Application of the RFID technology in logistics. Transp Res Procedia 57:452–462
9. Markov K, Vitliemov P (2020) Logistics 4.0 and supply chain 4.0 in the automotive industry. IOP Conf Ser Mater Sci Eng 878:012047
10. Barreto L, Amaral A, Pereira T (2017) Industry 4.0 implications in logistics: an overview. Procedia Manuf 13:1245–1252
11. Cimini C, Lagorio A, Pirola F, Pinto R (2019) Exploring human factors in Logistics 4.0: empirical evidence from a case study. IFAC-PapersOnLine 52(13):2183–2188
12. Maslarić M, Nikoličić S, Mirčetić D (2016) Logistics response to the industry 4.0: the physical internet. Open Eng 6:511–517
13. Sapper S, Kohl M, Fottner J (2021) Future competency requirements in logistics due to industry 4.0: a systematic literature review. In: 2021 10th international conference on industrial technology and management (ICITM), pp 94–105
14. Ikram B, Chahinaze FB, Naoufal S, Hamid A (2020) Competencies assessment approach based on multiple criteria decision making. In: 2020 IEEE 13th international colloquium of logistics and supply chain management (LOGISTIQUA), pp 1–9
15. Asawarungsaengkul K, Laoraksakiat W (2019) Optimization approach to multi-objective person-job fit job assignment problem. In: 2019 research, invention, and innovation congress (RI2C), pp 1–6
16. Benbrahim CF, Jaber E, Hajira B, Cheikhrouhou N, Sefiani N, Reklaoui K (2018) Towards an ontological approach to company compenteces management. In: 2018 international colloquium on logistics and supply chain management, pp 57–62

17. Ahn Y, McLean GN (2008) Competencies for port and logistics personnel: an application of regional human resource development. Asia Pacific Educ Rev 9:542–551
18. Kohl M, Heimeldinger C, Brieke M, Fottner J (2020) Competency model for logistics employees in smart factories. In: Karwowski W, Trzcielinski S, Mrugalska B (eds) AHFE 2019, vol 971. AISC. Springer, Cham, pp 133–145. https://doi.org/10.1007/978-3-030-20494-5_13
19. Niine T, Koppel O (2015) Logistics systems engineer – interdisciplinary competence model for modern education. Int J Eng Pedagogy (iJEP) 5(2):54
20. Bielefeldt AR (2018) Engineering student interest in future work abroad. In: 2018 IEEE frontiers in education conference (FIE), pp 1–5
21. Salinas-Navarro DE, Garay-Rondero CL (2020) Requirements of challenge based learning for experiential learning spaces, an industrial engineering application case. In: 2020 IEEE international conference on teaching, assessment, and learning for engineering, pp 1–8
22. Gomes GD et al (2018) Competencies and performance of engineering professors: evidence from a Brazilian public university. Ingeniería e Investigación 38(3):33–41
23. Terrón-López M-J et al (2020) Preparing sustainable engineers: a project-based learning experience in logistics with refugee camps. Sustainability 12(12):4817
24. Arifin M, Widowati, Farikhin, Wibowo A, Warsito B (2021) Comparative analysis on educational data mining algorithm to predict academic performance. In: 2021 international seminar on application for technology of information and communication (iSemantic), pp 173–178
25. Makarova I, Shubenkova K, Buyvol P, Mavrin V, Mukhametdinov E (2018) Interaction between education and business in digital era. In: 2018 IEEE industrial cyber-physical systems (ICPS), pp 503–508
26. Makarova I, Shubenkova K, Tikhonov D, Pashkevich A (2017) An integrated platform for blended learning in engineering education. In: Proceedings of the 9th international conference on computer supported education, vol 2, pp 171–176
27. Makarova I, Pashkevich A, Shubenkova K (2018) Blended learning technologies in the automotive industry specialists' training. In: 2018 32nd international conference on advanced information networking and applications workshops, pp 319–324
28. Makarova I, Shubenkova K, Antov D, Pashkevich A (2019) Digitalization of engineering education: from e-learning to smart education. In: Auer ME, Langmann R (eds) REV 2018, vol 47. LNNS. Springer, Cham, pp 32–41. https://doi.org/10.1007/978-3-319-95678-7_4
29. Makarova I, Shubenkova K, Bagateeva A, Pashkevich A (2018) Digitalization of education as a new destination of e-learning. In: 2018 international symposium ELMAR, pp 31–34
30. Ribón JCR, Villalba LJG, Kim T-H (2013) Virtual learning communities: unsolved troubles. Multimedia Tools Appl 74(19):8505–8519

Impact Analysis of Changing Education Forms on the Future Engineers' Motivation in Connection with COVID-19

Irina Makarova, Larisa Fatikhova, Polina Buyvol(✉), and Gleb Parsin

Kazan Federal University, Syuyumbike prosp., 10a, 423812 Naberezhnye Chelny, Russia
skyeyes@mail.ru

Abstract. Currently, the situation with COVID-19 has changed our daily lives, creating serious problems in all spheres of society. This has also affected the system of engineering education, new forms of e-learning have appeared. In the current conditions, the integration of digital technologies into university education has become an integral component that meets the challenges of global digitalization. In this context, an important issue is the study of students' attitudes to new educational formats. This article presents some accumulated experience in the use of distance learning technologies by students of engineering specialties of the university. The main directions of influence on the increase of educational motivation in the process of digital learning are determined. The advantages and difficulties arising in the implementation of online forms of education, as well as socio-economic factors affecting the increase of educational motivation of engineering students are identified. The article considers virtual environments that have the greatest efficiency in terms of process management, quality of learning, student engagement and increasing their motivation.

Keywords: COVID-19 · Remote learning · Student motivation

1 Introduction

Improving the quality of training of graduates of higher educational institutions of a new type who are able to work creatively in the digital economy, think independently and make responsible decisions, and have a dominant motivation for success, requires the mandatory use of modern forms and methods of teaching in the organization of the educational process, allowing to continuously stimulate the educational motivation of students.

In these conditions, the issues of students' attitude to new learning formats were also actualized. The traditional model of education is disintegrating, and it is also impossible to focus only on distance learning technologies, it is necessary to search for new technologies, tools and opportunities for further development and quality assurance in the engineering education system.

This research was conducted in order to analyze the accumulated experience of using distance learning technologies by students of engineering specialties of the university,

M. E. Auer et al. (Eds.): REV 2022, LNNS 524, pp. 337–348, 2023.
https://doi.org/10.1007/978-3-031-17091-1_34

and to determine the directions of influence on the growth of educational motivation in the process of digital learning. We have formulated the following research questions: 1) to assess what difficulties arise when implementing online forms of education, and what its advantages help to increase the motivation of engineering students; 2) how socio-demographic factors affect the success of mastering and the perception of distance learning; 3) which virtual environments have the greatest effectiveness in terms of process management, quality of learning, student engagement, increasing their motivation; 4) find out which forms of classes are best suited for online learning.

2 State of the Problem: Distance Learning Technologies and Educational Motivation

2.1 Distance Technologies in Modern Engineering Education

Digitalization of vocational education and the creation of electronic educational environments were a lifesaving response to the challenges of the pandemic, providing opportunities for continuing education remotely for most students. Such a forced international experiment on the introduction of distance learning has become unprecedented in its scale, in terms of urgency and completeness of isolation of students, teachers and university territories [1]. Distance learning and related technologies have become an integral part of modern higher education [2]. The authors of the article [3] believe that good technologies are necessary for the successful implementation and support of distance learning. In their research, they propose a new technological solution for video conferences, using WebRTC technology as a replacement for applications that still use flash memory technology. According to the results of the tests carried out, the system is user-friendly and can be recommended for use.

The article [4] describes the transition from full-time to blended learning in the engineering master's program by adding a parallel online group, which had a positive impact on the learning outcomes and ensured the continuity of the master's program. The live broadcast of lectures allowed online students to interact with students in the classroom as in a normal classroom lesson. Interaction between the teacher and students, as well as between students, is provided through online chat, interactive training programs, face-to-face lectures and discussion forums. In addition, collaboration was expanded using modern open BIM-based software, which improved the quality of communication and teamwork of students, as well as their interaction with the teacher. The analysis of the final grades showed that the new methodology significantly improved the average grades of students, although no significant differences were found between the two study groups. The proposed approach to learning "at any time and in any place" with auxiliary materials and mentoring of students contributed to the improvement of the educational process.

The author of the study [5] suggests using the Moodle system in distance learning, in which theoretical lecture material is duplicated. The proposed new approach combines two types of education: classical (in classrooms and laboratories of the university) and self-education (distance learning outside the university). Theoretical materials and local tests on each topic uploaded to Moodle contribute to the successful mastering of knowledge and skills by students in any place and time convenient for them.

A hybrid teaching method combining a massively open online course (MOOC) and personal work of students is proposed by the author of the article [6]. An important component of this approach is that with the help of a mixed MOOC laboratory, teachers have information about the progress of students in the learning process, which allows them to provide timely support to the student during training or to improve the learning material for effective assimilation.

Currently, new forms of distance technologies in education are being formed and proposed. Among the most popular are: micro-learning, chatbots, adaptive learning, virtual and augmented reality, video content, peer-to-peer, content curation.

2.2 Motivation of Students in Distance Learning

Motivation of students is important for all forms of education, but especially for distance learning. The problem of motivation has become particularly relevant with the mass transfer of education at all levels to a distance format in the context of a pandemic. Undoubtedly, the digital educational environment, affecting the various senses of students, should stimulate their interest and develop motivation to study, but this is not always possible. The authors of numerous scientific and practical studies study various aspects of educational motivation, ways and mechanisms of its improvement.

The article [7] describes a new educational model for improving cognitive learning and motivation of undergraduate students in industrial engineering education. To assess the attitude of students to the computer program proposed by the authors, as well as its usefulness in understanding the basic concepts and theory of processing, the authors used both qualitative and quantitative assessments. For these purposes, a comprehensive questionnaire with open questions was developed. According to the results of the study, the authors come to the conclusion that the educational model and the computer program associated with it have proved their effectiveness and efficiency in stimulating the interest of students, as well as in expanding and deepening students' understanding of the main concepts of the discipline being studied.

The article [8] discusses motivational factors affecting students of higher educational institutions at the Faculty of Technology at the University of Portsmouth. The most important pedagogical issue addressed by the authors concerns the expediency of structuring teaching, for a deeper understanding and development of students. Another important aspect is the lecturer's potential, his ability to inspire and motivate students and influence their perception of information. An interesting feature of this work is the possibility of disagreements between the teacher's expectation and the student's experience. This study suggests that in order to promote active learning while observing academic standards, it is necessary to stimulate students' internal motivation, while preserving opportunities for independent learning and self-development.

The authors of the article [9] identified the conditions of the educational space of an engineering university aimed at motivating students to study, presented the characteristics of pedagogical support for the education of students of an engineering university and analyzed the human resource. An integrative indicator of pedagogical support of students' education is their involvement in the scientific and educational process of an engineering university.

An attempt to expand the learning opportunities for undergraduate students of mechanical engineering is made in the article [10]. The authors describe various interactive activities for the main course, stimulating students' participation in the learning process, contributing to a better understanding of the material and the development of critical thinking. The authors also believe that the Faculty of Engineering, which is open to any kind of active learning, should also be prepared for potential obstacles, such as student resistance and dissatisfaction with the transition from traditional lectures to active learning. The study allowed the authors to implement feedback and explore opportunities to improve the proposed activities and increase educational motivation.

The article [11] presents the basics of motivation theories: their impact on students' learning and their importance for teaching engineering disciplines. It is shown how the provisions of the theory can be used for teaching and the formation of positive learning motivation.

2.3 New Distance Technologies for the Formation of Students Motivation

Improving the quality of training of graduates of higher educational institutions of a new type who are able to work creatively in the digital economy, think independently and make responsible decisions, and have a dominant motivation for success, requires the mandatory use of modern teaching methods and forms of control in the organization of the educational process, allowing to continuously stimulate the educational motivation of students.

Currently, there are many tools and techniques in the field of engineering education. Technologies such as virtual and augmented reality, digital twins, smart education, allow improving the educational environment, increasing students' motivation and preparing them for life in the digital age [12]. The latter becomes especially important as both the cars themselves [13–15] and the processes that ensure their operability [16–19] become more intellectualized.

Augmented and virtual reality (VR) technologies are gaining popularity very quickly. By superimposing graphics and images, you can diversify the learning process. 3D modeling technologies are in high demand in construction companies and in oil production. If the model is built and supplemented with the necessary data, then it allows you to optimize the processes of making managerial decisions. Elements of virtual reality are used to teach the necessary skills, for example, in medicine or safety [20].

The article [21] describes the gamification technique in the study of geosciences as an innovative teaching method for engineering students. When using this technique, students are divided into small groups and perform group tasks during the semester. Tasks were evaluated in points, and a table was compiled showing the current academic performance of each group. At the end of the semester, questionnaire surveys of students were conducted, the answers of which showed that gamification can improve the educational motivation of students and increase interest in the subject. Gamification is increasingly used in the educational process and in e-learning.

Game methods in the educational process contribute to a deeper understanding of the essence of real processes, the growth of interest and motivation of students, a detailed analysis of technological processes and risks, and also contribute to a faster adaptation of young professionals in the workplace.

The article [22] describes the use of the VISIR (Virtual Instrument Systems In Reality) remote laboratory in the educational process, based on a network of virtual cooperation between universities. VISIR is a modern laboratory using real equipment, measurements on which are made using real instruments, with the possibility of remote control. VISIR systems have been installed in thirty different higher education institutions from eight countries (Argentina, Austria, Brazil, Georgia, India, Portugal, Spain and Sweden) and have been successfully integrated at different levels of education. The PILAR project (a form of laboratory integration based on the VISIR architecture), by providing access to a remote laboratory to other educational institutions, expands the possibilities of practical training at a higher level of digital integration.

Of particular interest in modern engineering education is modeling based on the integration of augmented and virtual reality (AR/VR) with traditional engineering simulators. The study [23] proposes a simple and flexible methodology for integrating AR software with technical content, such as animation, 3D models and engineering modeling. The digital tool developed in this work is available in the online version.

The analysis of applied forms of learning shows that blended learning has advantages over traditional and electronic learning. For its successful implementation, a unified information and educational platform is needed that allows developing educational content, managing the educational process and providing the opportunity for virtual communication [23].

3 Studying the Impact of Changing Forms of Education on the Motivation of Students

3.1 Difficulties and Advantages of Distance Learning

After the introduction of lockdown, announced in the spring of 2020 against the background of a pandemic, our university, as well as other educational institutions in the world, had to make a forced transition from traditional forms of classroom instruction to remote forms. Currently, training is conducted in a hybrid format, when some students "participate" in the educational process remotely in connection with the problem of crossing borders or self-isolation due to illness. Therefore, since the beginning of the pandemic, experience has accumulated, problems have been identified, as well as solutions that can help in the future. To analyze students' perception of various forms of education, a survey was conducted, as well as the results of the examination session after the outbreak of the pandemic were analyzed. The survey of students was conducted using online questionnaires in GoogleForms. 344 full-time students (85.1% of bachelors and 14.9% of masters) took part in the pilot study of students' attitude to online digital learning technologies. The age structure of the respondents: 20 years and younger - 25.6%, 20–25 years - 70.3%, 26–30 years - 1.7% and over 30 years - 2.3%.

The survey results showed that distance learning has led to a number of difficulties for students. Only 51% of the students surveyed noted that they had successfully adapted to the distance learning format. In addition to the technical difficulties that arise in the process of distance learning, there are also problems associated with the specifics of the distance format itself and the lack of skills necessary for successful learning.

The main difficulties in the distance learning format, called students: dependence on technical means (47.7%); the need to spend a lot of time at the computer (45%); lack of personal interaction (40.5%); it is difficult to take into account the individual characteristics of students (32.6%), etc. (Fig. 1).

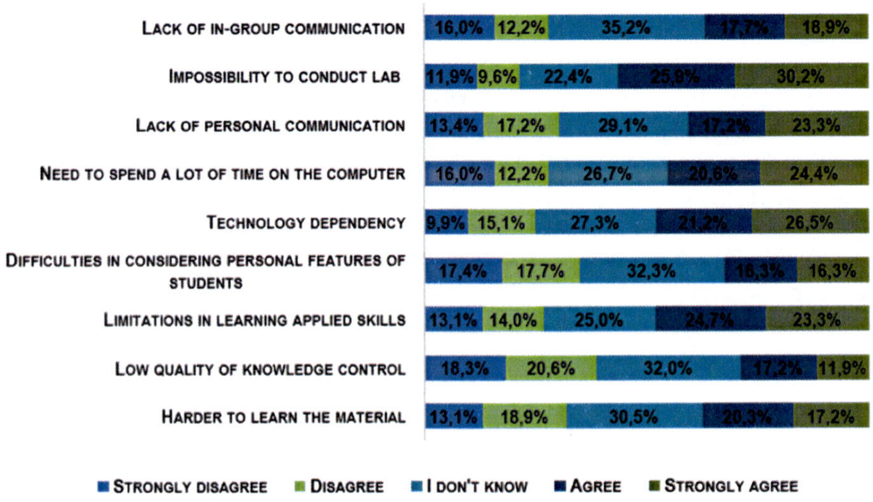

Fig. 1. Challenges in distant learning

The results of the questionnaire revealed difficulties with self-control and organization of independent work, as well as a lack of free time to complete tasks and untimely feedback from the teacher (Fig. 2).

Fig. 2. Challenges in Students individual work

As advantages of the distance learning format adopted at the university, students highlighted: the possibility of combining study with work; improving the skills of independent work; availability of materials and the ability to use them at a convenient time and in a familiar environment. Disadvantages are noted: the need to spend a lot of time at the computer and an increase in the time spent on mastering disciplines; lack of personal communication, both with teachers and with fellow students; limitations in obtaining practical skills and the inability to conduct laboratory classes; dependence on technical means and a decrease in the quality of training.

3.2 The Effectiveness of Using Communication Platforms and Programs for Educational Purposes

The main methods for interacting with teachers in the distance learning format, students note the use of Microsoft Teams (56.7%), "Virtual Audience" (26.2%), Zoom, Skype (13.4%) and third-party platforms (3.7%).

As shown above in Fig. 1 the main limitations of remote platforms are the inability to conduct laboratory classes (56.1%); limitations in obtaining practical skills (49%). The analysis of the session also showed that the expediency and effectiveness of online classes is higher for a lecture course, and personal communication between a student and a teacher is necessary to obtain practical skills. The average score in the theoretical (lecture) part of the natural sciences turned out to be higher than for performing practical tasks (Fig. 3).

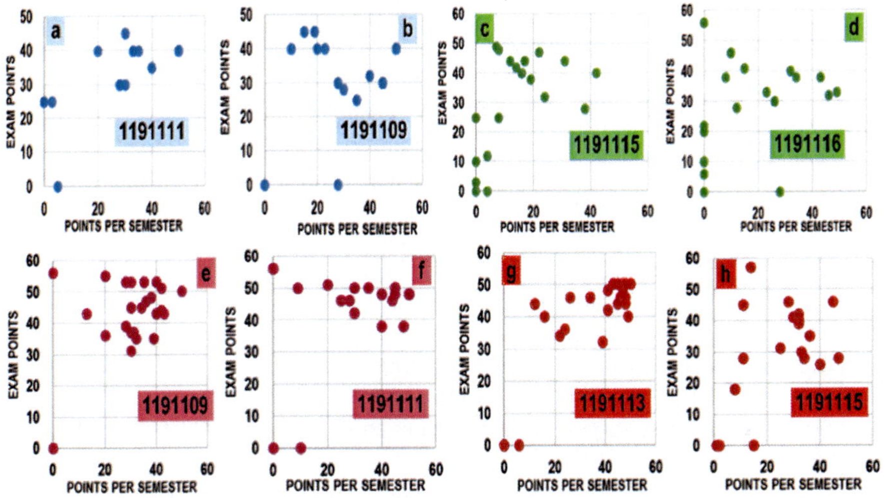

Fig. 3. Exam results in maths (a, b - teacher 1, c, d - teacher 2) and in structural materials technology (e, f - teacher 3, g, h - teacher 4)

The questionnaire survey also showed that students had problems with the search and use of teaching materials, which indicates the need to address questions about the effectiveness of the use of distance technologies in teaching disciplines that include the acquisition of practical skills. Students of the first cluster note problems with the use of teaching materials, which forces them to solve the issue of using online technologies for courses related to obtaining practical skills. The most popular methods of mastering educational materials were the methodological guidelines of the University (62.8%), methodological guidelines on the Internet (56.4%) and video materials provided by the university (40.4%) (Fig. 4).

At the same time, teachers - 33% of students helped to find the listed educational resources and methods of their development, colleagues-students - 32% shared their experience, mastered at an intuitive level - 32.2% and 2.8% of students answered "failed to master".

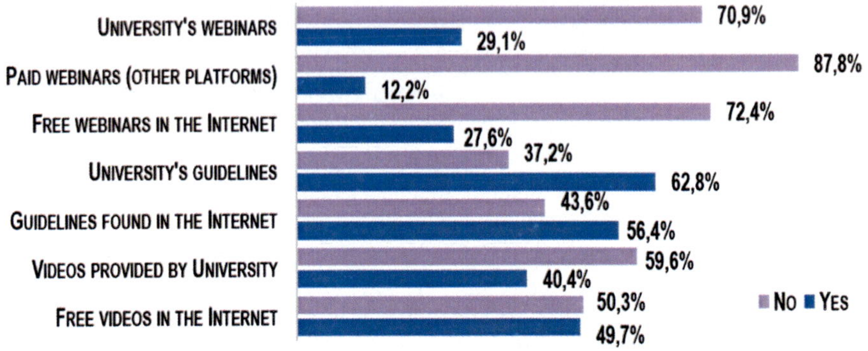

Fig. 4. Distant learning methods

3.3 The Influence of Socio-demographic Factors on the Success of Information Perception in Distance Learning

In order to clarify the readiness of students in technical and technological terms for the transition to a distance learning format, a number of socio-demographic questions were asked. Among the most significant technical problems faced by respondents were technical interruptions in the process of reproducing the material (25%) and the need to have stable access to the Internet (24.9%), then low Internet speed (19.8%), lack of feedback on the platform, lack of technical means and other reasons were highlighted. Internet connection when connecting to classes, the remaining 35% noted its poor quality. At the same time, only 74.4% of students were satisfied with the power of personal hardware communication. 49.4% of students named a laptop (computer) as the most frequently used devices for distance learning, 49.1% - a smartphone (tablet). The data show that not all students are provided with stationary computers and laptops for distance learning.

The data of the online survey do not allow us to get a full assessment of the technical component of the inequality in access to learning in a distance format. However, differences in the technical conditions of students' education indicate the existence of a problem of "digital inequality" of students, which may be related to both the material inequality of students' families and the lack of access to broadband Internet in their places of residence in distance learning.

3.4 Assessment of Satisfaction and Motivation with the Educational Process When Changing the Form of Education

The student's educational motivation is determined by various factors: the educational system, the organization of the educational process, the subjective characteristics of the student, the specifics of academic disciplines. It is possible to increase the motivation of a student's educational activity with the help of specially organized independent work and an objective assessment of his activities. At the same time, a necessary condition for students' interest in learning activities is the opportunity to show independence and initiative in the learning process. The more active the teaching methods, the easier it is to interest students in them. But maintaining motivational activity is the most difficult component when using remote technologies. The success of the educational process directly depends on the level of involvement in the educational process. Therefore, further in the study, we wanted to find out the degree of influence of distance technologies on the educational motivation of students.

In the course of the study, answering the question "What tasks do distance learning technologies solve in your opinion?", respondents noted: the opportunity to engage in education at a convenient time and in a convenient place - 22.6%; forms skills to work with large amounts of information and independence in finding the necessary information - 22.1%; increases motivation - 16.9%; increases my level of applied skills - 14.8%; the opportunity to expand the circle of communication - 11.1%. 8% of respondents expressed a negative attitude to remote technologies and 4.5% found it difficult to answer.

Students' vision of using distance technologies in the educational process of the university (Fig. 5).

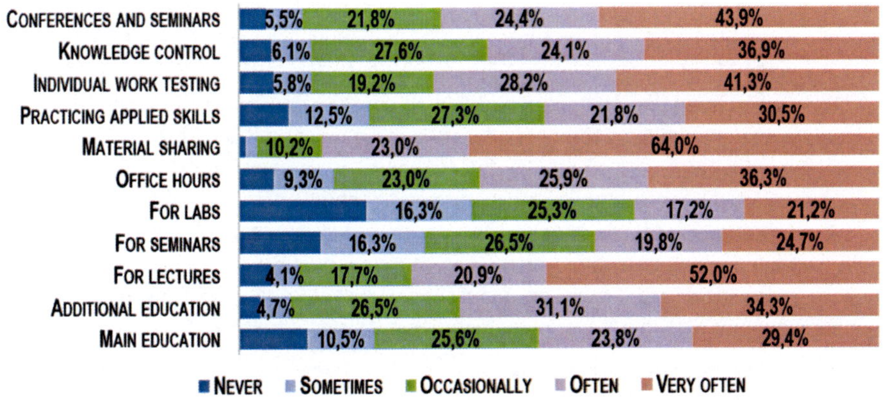

Fig. 5. Students attitude towards frequency of distant learning tools in the educational system

Students generally have a positive attitude to distance learning technologies in the educational process, but at the same time they consider lectures, placement of educational materials, issuance and verification of independent assignments, conferences and seminars, as additional training, etc. to be the most acceptable types of classes in this format.

Further in the study, we proposed a set of statements related to the motivation and satisfaction of students with the educational process in the distance format. A 5-level Likert scale was used to answer the questions (Fig. 6).

The majority of students agree with the statement that modern education should combine distance learning technologies and traditional educational practices (68.9%), and the skills acquired in the distance learning format will be useful to them in their future professional activities (57.8%). In general, 65.7% of the students who took part in the study confirmed their positive attitude to distance learning: almost 58% noted positive changes in their ability to study independently, 52.9% stated an increase in

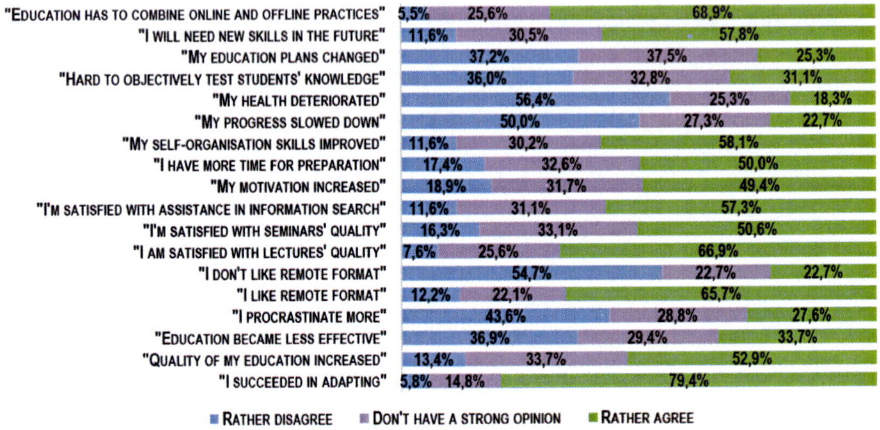

Fig. 6. Students experience with distant learning.

the quality of education, almost 50% noted the appearance of free time and increased learning motivation.

Our accumulated experience in studying the problems of motivation of university students when using various forms of education in a pandemic has allowed us to identify two factors affecting motivation. We defined the first factor as an external factor: the organization of the process and the structure and content of the discipline. The second factor, internal: personal parameters (orientation, interests, etc.). Both categories can have both positive and negative effects on motivation. Also, this study showed that intrinsic motivation is the main component in distance and blended learning and plays a more important role in motivating students than extrinsic motivation.

4 Conclusions

The task of developing, improving and optimizing the forms and methods of teaching in engineering education has always been one of the urgent problems. Modern technologies make it possible to solve the problems of personalizing learning, increasing its productivity and intensity, accessibility and openness, optimizing knowledge transfer and increasing student motivation.

The study showed that the majority of students have a positive attitude to the use of distance learning technologies in higher education, however, they prefer this format only if it comes to lectures and the placement of manuals and textbooks. The overall level of digital literacy directly affects the results of academic performance in the subject, since the student does not have to solve the problem of mastering information and communication technologies in parallel.

The analysis of the review results showed that the existing digital learning tools are insufficient to stimulate learning motivation, especially for students with a low initial level of training. In addition, in engineering education, online training cannot completely replace laboratory and practical classes held in the traditional form. The educational content existing at the time of the lockdown could not provide the proper level of quality of education and understanding of the essence of real processes, although with a certain level of elaboration, practical classes can be replaced by VR laboratories, therefore the development of specialized virtual and remote laboratories is necessary.

References

1. Kiryakova AV, Kargapoltseva NA, Belonovskaya ID (2020) Scientific and pedagogical projections of the transformation of university education. Vysshee Obrazovanie v Rossii 29(8–9):155–167
2. Makarova I, Shubenkova K, Antov D, Pashkevich A (2019) Digitalization of engineering education: from e-learning to smart education. In: Auer ME, Langmann R (eds) REV 2018, vol 47. LNNS. Springer, Cham, pp 32–41. https://doi.org/10.1007/978-3-319-95678-7_4
3. Alimudin A, Muhammad A (2018) Online video conference system using WebRTC technology for distance learning support. In: 2018 international electronics symposium on knowledge creation and intelligent computing (IES-KCIC), Bali, Indonesia, pp 384–387
4. Martínez P et al (2020) Transitioning from face-to-face to blended and full online learning engineering master's program. IEEE Trans Educ 63(1):2–9
5. Filipova-Petrakieva S (2019) Some problems in distance learning on theoretical electrical engineering in Technical University of Sofia. In: 2019 16th conference on electrical machines, drives and power systems (ELMA), Varna, Bulgaria, pp 1–5
6. Plaza P (2019) Portable blended MOOC laboratory. In: 2019 IEEE learning with MOOCS (LWMOOCS), Milwaukee, WI, USA, pp 15–20
7. Fang N, Stewardson GA, Lubke MM (2007)Work in progress - an innovative instructional model for improving manufacturing engineering education. In: 2007 37th annual frontiers in education conference - global engineering: knowledge without borders, opportunities without passports, Milwaukee, WI, pp S4D-19–S4D-20
8. Savage N et al (2011) Motivation of engineering students in higher education. Eng Educ 6(2):39–46
9. Makarova I et al (2018) Improving the quality of engineering education by developing the system of increasing students' motivation. Adv Intell Syst Comput 716:150–161
10. Takemoto I (2017) Autonomous learning applying for expectancy motivation theory on English education. In: 2017 IEEE 9th international conference on engineering education (ICEED), Kanazawa, pp 86–89
11. Finelli CJ et al (2013) Factors that influence faculty motivation of effective teaching practices in engineering. Paper presented at 2013 ASEE annual conference & exposition, Atlanta, Georgia. https://peer.asee.org/19604
12. Makarova I et al (2020) Usage of microscopic simulation to estimate the environmental impact of road transport. Trans Res Procedia 44:86–93
13. Gabsalikhova L et al (2020) Connected vehicles fleet expanding problems. In: VEHITS 2020: proceedings of the 6th international conference on vehicle technology and intelligent transport systems, pp 642–650
14. Makarova I et al (2020) Changing the maintenance and repair system while expanding the connected vehicles fleet. In: VEHITS 2020: proceedings of the 6th international conference on vehicle technology and intelligent transport systems, pp 622–633
15. Braniš M, Balint G, Takacs J, Šulík M, Galkin A (2020) Shared electric scooters like a tool of a micro-mobility in cities. In: SGEM 2020: international multidisciplinary scientific geoconference surveying geology and mining ecology management, pp 631–638
16. Ostroukh A (2021) Intelligent system for digital substation control. Transp Res Procedia 57:385–391

17. Galkin A (2019) Interaction of logistics 4.0 and consumer oriented marketing using ICT. In: Proceedings of the 33rd international business information management association conference, IBIMA 2019: education excellence and innovation management through vision 2020, pp 6751–6760
18. Makarova I, Shubenkova K, Buyvol P, Shepelev V, Gritsenko A (2021) The role of reverse logistics in the transition to a circular economy: case study of automotive spare parts logistics. FME Trans 49(1):173–185
19. Popova I (2021) Application of the RFID technology in logistics. Transp Res Procedia 57:452–462
20. Uvarov AU (2018) Virtual reality technologies in education. Sci Sch 4:108–117
21. Hasan M (2017) Improving students' motivation to learn through gamification. In: 2017 7th world engineering education forum (WEEF), Kuala Lumpur, pp 642–647. https://doi.org/10.1109/WEEF.2017.8467059
22. Castro M et al (2019) Workshop: teaching practices with VISIR remote lab: technical, educational and research fundamentals from the PILAR Project. In: 2019 IEEE world conference on engineering education (EDUNINE), Lima, Perupp, pp 1–3
23. Solmaz S et al (2021) A practical development of engineering simulation-assisted educational AR environments. Educ Chem Eng 35:81–93

Virtual Reality Laboratory as a Factor in Increasing Engineering Students' Motivation

Irina Makarova, Gleb Parsin, Aleksey Boyko, and Polina Buyvol(✉)

Kazan Federal University, Syuyumbike prosp., 10a, 423812 Naberezhnye Chelny, Russia
skyeyes@mail.ru

Abstract. Globally, young people have a decline in interest in obtaining engineering specialties. Therefore, it is necessary to create a learning environment that will motivate students to learn engineering professions. It is necessary to use all possible resources, as well as stimulate and encourage the creativity of both teachers and students. Therefore, the purpose of this study is to create a method for increasing motivation based on the use of new tools and approaches in training future engineers. Since virtual reality technologies allow a deeper understanding of real processes' essence, demonstrate the main steps of the process and possible critical situations use of these technologies are very relevant. We put forward the assumption that the process of developing educational content itself can be used as an incentive motivating force in the educational process. We have developed virtual reality laboratory, with the help of which students are able to study the work of stands used in vehicle repair. They can master structure and operation principles of the balancing machine, study and practice of the technological process of balancing vehicle tires and disks. In order to maximize usage of students' and teachers' creative, it was proposed to attract computer students to develop the content of a virtual reality laboratory for the study of vehicle repair processes. A survey conducted among students showed that this approach allows to get a higher motivation level compared to the use of traditional teaching.

Keywords: Engineering education · Remove laboratory · Motivation increasing

1 Introduction

The digitalization of the economy and society determines the replacement of direct human control of production systems and urbanized territories with autonomously controlled smart factories and intelligent transport systems. However, their deployment and operation place new demands on staff competencies. At the same time, on a global scale among young people, there is a decline in interest in obtaining engineering specialties. Therefore, significant efforts are required to find ways to increase motivation for obtaining engineering specialties. It is necessary to create a learning environment that will motivate students to study for mastering engineering professions.

Business needs a graduate who already possesses the required engineering competencies, knows the peculiarities of modern equipment functioning and operation. A

M. E. Auer et al. (Eds.): REV 2022, LNNS 524, pp. 349–359, 2023.
https://doi.org/10.1007/978-3-031-17091-1_35

limiting factor in the formation of a competent specialist is the lack of funding with the constant updating and expansion of the equipment and tools model range. In such conditions, it is impossible to ensure the relevance of the material and technical base for laboratory and practical training. This widens the gap between the level of competencies available to graduates and the requirements of the business. The lack of modern technical equipment reduces the motivation of students, since one of the important issues when taking training courses is the realism of the considered situations, the degree of their correspondence to the real production and technological process. To overcome the existing limitations, it is necessary to use all possible resources, as well as stimulate and encourage the creative potential of not only teachers, but also the students themselves. Therefore, the purpose of this study is to create a methodology for increasing motivation based on the use of new tools and approaches in the training of future engineers.

2 Methods and Technologies of Engineering Education: An Overview of the State

2.1 Requirements for the Competence of a Future Vehicle Maintenance Engineer

Speaking to the participants of the II online conference "Modern training of engineers" [1] Minister of Science and Higher Education of the Russian Federation Valery Falkov emphasized that over the past decades, engineering has made great strides, so in the 21st century it is impossible to train engineers in the same way as in the middle of the 20th century.

At the same event, Petr Shchedrovitsky formulated a number of theses, which, in fact, reflect the requirements for describing the competencies of future engineers:

- "The training and education of the modern engineer must take into account the context of the new industrial revolution and the forward-looking practical challenges that face different areas.
- Training and education of a modern engineer should, in the course of the educational process, ensure an imitation of an expanded system of labor division (training of engineering teams).
- The result of the educational process should be considered the formation of "cross-cutting" competencies that allow an individual to be included in situations, processes and systems of jointly-distributed thinking activity.
- The need to understand literacy as a level of mastering (possession) of the basic semiotic systems (2–3 natural languages, mathematics, logic, economics, "drawings", infographics, "schemes").
- The content of modern engineer training should be the development of the basic technologies of engineering thinking: design, engineering, research and "programming" (activity), and each next stage includes elements of technologies of previous industrial revolutions.
- The content of modern engineer education should be the development of not only the traditional picture of the world, but a broader activity-natural ontology (a weak version of this shift is the growth of the "humanitarian" component in education programs for engineers)."

It is impossible to disagree with each of the points listed. At the same time, special attention should still be paid to increasing digital competence. Digital technologies are firmly entering the field [2, 3], along with this, the importance of computational thinking skill among engineers is growing. However, the results of the conducted research show that students demonstrate low abilities in abstraction and algorithmic thinking [4]. Therefore, when determining the trajectory of training for automotive specialists, it is necessary to pay more attention to the development of skills in working with computer and engineering graphics programs, to teach the basics of algorithmization and programming.

Classic courses should be complemented by digital subjects such as mathematical modeling, programming tools and software, digital twins, algorithms and data structures, control systems and advanced sensors, data analytics and preventive maintenance, cybersecurity, digital business models [5].

In the context of the accelerating digitalization of modern society, rapidly changing technologies, when the number of typical template tasks decreases, the engineer has to solve non-standard tasks that require a creative approach and creative critical thinking. The high level of creativity in the engineering profession is confirmed in their study by Frey and Osborne [6]. After assessing the susceptibility of professions to computerization, they found that the likelihood of replacing people in engineering positions in mechanical engineering with computerized intelligent systems is low. Software robots and artificial intelligence are not able to compensate for the creative component.

Without diminishing the importance of analytical thinking and creativity in defining key competencies, it should be borne in mind that nowadays, more than ever before, it becomes true that the initial education received at school, university or during professional training is not enough for a lifetime. The need for training is increasing, and the speed with which the acquired knowledge must be replaced and the corresponding competences must be re-studied increases from decade to decade [7]. At the same time, the task of the education system is not only the acquisition by the student of relevant knowledge, skills and abilities, mastering specific competencies, but also the ability to independently build their own learning path in the future, manage and maintain the necessary level of self-motivation in order to look for sources of information and acquire additional knowledge.

In recent decades, there has been a transformation of the vehicle fleet. The share of vehicles using environmentally friendly fuels is growing: the popularity of electric and natural gas vehicles is increasing. Following connected cars, there is a likelihood of autonomous vehicles appearing on public roads with the creation of an appropriate ecosystem. Also, with the reduction of the ice cover in the northern hemisphere, there is an active development of the Arctic territories and an increase in the importance of the northern sea route. As a result, the share of the vehicle fleet operating in extremely low temperature conditions is growing. Changing the fleet requires changing the service network [8]. At the same time, the competence of the operation engineers should also be transformed: they should have ideas about the features of operation in a cold climate [9], the operation of a digital platform for controlling connected and autonomous vehicles [10], the technology of servicing computer vision modules [11].

2.2 Problems and Challenges of the Educational Process in the Context of the COVID-19 Lockdown

The COVID-19 pandemic has made significant changes in the usual way of life, forcing either to suspend the educational process, such as in some educational organizations in India [12], or to move from traditional forms of face-to-face education to a remote format. Many educational organizations were not prepared for the latter case, did not have accredited distance learning programs [13] and faced various kinds of difficulties. The results obtained by the authors [14] for California State Universities can be extrapolated to many educational institutions. Issues they identified include: logistic/technical issues, learning/teaching issues, privacy and security issues, and lack of sufficient hands-on training. For example, more than half of students reported lack of activity in the classroom, difficulty maintaining concentration, and Zoom fatigue after attending multiple online sessions.

In general, the problem of a decrease in the level of motivation for young people when switching to a distance mode has become aggravated. As it was shown in work [15], when implementing this form of training, a student has to show great self-discipline, self-motivation and self-regulation in order to achieve learning outcomes similar to full-time ones. The absence of a teacher and one-group students in the immediate physical environment reduces external motivation and control. Considering that all over the world before the pandemic there was a decrease in the interest of young people in obtaining engineering education in many countries [16], the development of a system for increasing motivation is becoming urgent.

As the authors showed in their study [17], the use of the new format had a direct impact on academic performance. Not all online education was equal: students who were given recorded online lessons from outside more qualified teachers had higher exam scores than those whose lessons were recorded by teachers from their own school. The educational benefits of distance learning were similar for rural and urban students, but students who used a computer for online learning had better exam scores than those who used a smartphone. At the same time, the greatest benefit from distance learning is received by children with a low level of academic performance.

Another problem is related to the organization of practical and laboratory classes. The main task that remains to be solved is to teach future engineers to work with real equipment, which is not available at home. Implementation of digital twin concept for industrial equipment can partially solve the arisen problem associated with the inaccessibility of equipment in the process of remote learning, as well as improve the quality of professional training after the end of the pandemic [18]. At the same time, starting with the use of digital twins of separate equipment, it is advisable to then combine them into a single technological process.

In this sense, simulation models can also help to study the processes occurring in the systems under study. For example, work [19] describes the methodology and presents the structures of simulation models of the road network sections to study the impact on the environment of emissions from auto-mobile transport. Also, simulation models can help in the study of logistic [20, 21] and service [22] processes in the system of firm vehicle maintenance.

Despite all the obvious extremely negative aspects of the COVID-19 pandemic, the crisis led to the discovery of new dimensions in online teaching of engineering courses and allowed the use of various distance learning tools for teaching and assessment [23].

2.3 Virtual Training Laboratories

Without educational laboratories, it is impossible to implement the formation of practical skills in any direction of engineering education. They allow students to engage in various stages of experiential learning, including conceptualization, experimentation, followed by reflection, analysis, and interpretation of the findings.

In the context of a lockdown due to COVID-19, educational institutions were faced with the task of transforming educational material from a traditional form of presentation to a remote one, while maintaining the quality of its development. Since augmented and virtual reality technologies allow a deeper understanding of the essence of real processes, demonstrate the main steps of the process and possible critical situations [24], new approaches and achievements in the field of online education based on the use of these technologies are not only relevant, but the only possible to ensure the required quality of the educational process.

After the removal of travel restrictions, the prospects for the development of virtual laboratories did not decrease, since they are often a good alternative to physical laboratories in the traditional form of face-to-face training, especially in the case of the potential danger of the experiments being carried out. It succeeds, on the one hand, to increase student engagement, and, on the other, to provide a safer learning environment. The introduction of a virtual simulator system allows students to practice real actions on virtual workstations.

A lot of evidence has appeared in the press about the successful implementation of virtual reality technologies in the educational process during a pandemic and lockdown. For example, [25] describes an example of how the intensive practical, group project engineering course "Design and Innovation of Medical Devices" was effectively adapted. All previously scheduled face-to-face lectures were delivered remotely via Zoom. Instead of personal tours, medical device companies conducted virtual tours of their facilities, showcasing their design, manufacturing, and validation processes. 82% of the students surveyed were satisfied with the quality of the remote lectures. Also, feedback from students showed that it is more difficult to establish personal communication in the format of distance learning. Communication and networking difficulties were particularly evident in virtual tours of medical device companies, where 44% of students said they lacked opportunities to meet and communicate with engineers. However, the main problem from the point of view of a full-fledged specialist formation is still not in the absence of communication.

Laboratory and practical exercises had to be rethought and replaced. Practice and internships at the sites of employers have almost disappeared. And although some part of the educational sector perfectly coped with this variety of problems with the help of such things as digital artifacts, virtual laboratories, remote control through a proxy [26], for the most part, the problem remained unresolved.

Despite the great relevance of this direction, great attention from the international scientific and educational communities and a large number of publications on research results that have appeared, many questions remain unresolved. Mainly, in many subject areas, a full-fledged replacement of practical exercises through the use of virtual laboratories has not been implemented. So, after conducting a search, we could not find works in which examples of the introduction of virtual laboratories into the educational process that implement an experiment with a full-fledged technological process of vehicle repair are presented. Therefore, the creation of such laboratories is the agenda of many educational institutions around the world. The analysis of the research shows that the authors focus only on the elaboration of the technical side of the issue and the perception of the received digital instrument by the students. We assumed that the active involvement of students in the very process of developing educational content is promising and ensures the growth of students' motivation.

3 Results and Discussion: Development of a Virtual Training Laboratory as Part of a Methodology for Increasing Motivation

3.1 The Essence of the Proposed Method

To increase the level of students' mastery of professional and digital competencies at the Naberezhnye Chelny Institute of Kazan Federal University, a methodology was formulated according to which, in order to develop the content of a virtual reality laboratory for the study of vehicle repair processes, it is necessary to involve computer students in addition to teachers.

To train practical skills in working with equipment during training engineers for operation of road transport in the traditional form "face-to-face", it is necessary to have machines: tire fitting, balancing, wheel alignment. It is expected that as a result of the collaboration, a virtual reality laboratory will be developed, with the help of which students will be able to study the work of stands used in the vehicle repair. Thus, it becomes possible to master, consolidate and test the professional competencies of bachelor's training in the areas of "Operation of transport and technological machines and complexes" and "Service" in distance learning mode in the absence of real equipment, for example, in a situation of isolation or blended learning.

Currently, a standard model of equipment for balancing work has been developed, and laboratory work with its use has been tested in the educational process. In the course of the implementation of this work, students study the purpose of the balancing machine, the device and the principle of its operation, work out the steps of the technological process of balancing vehicle tires and disks. When performing these types of work, students install a virtual wheel on a balancing stand, set the size of the disc and the distance from the disc to the machine, check the degree of imbalance of the wheel and, adding balancing weights to the wheel rim, achieve the required values within the tolerance (Fig. 1). The Unreal Engine game engine was used as a software platform for the implementation of a virtual laboratory (Fig. 2).

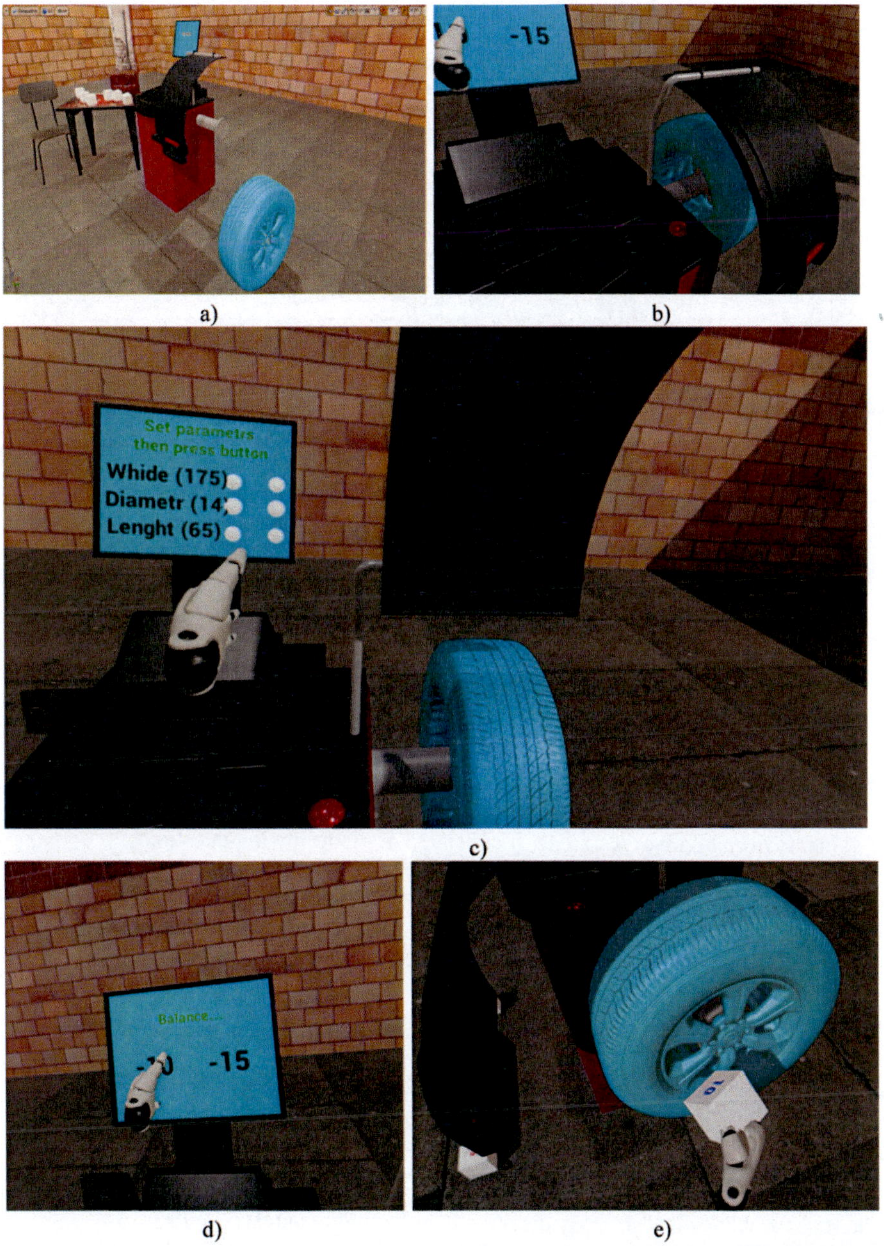

Fig. 1. View of virtual laboratory: a) before the wheel is installed on the balancing bench; b) after installation; c) wheel settings; d) wheel needs balancing: ten grams on the left and fifteen on the right; e) installing weight on the right rim.

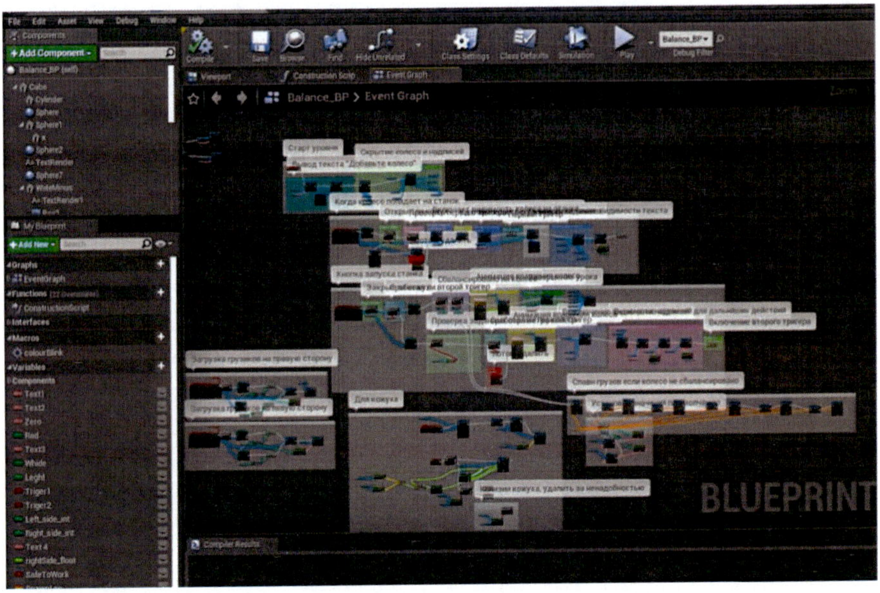

Fig. 2. Setting the logic of the balancing machine's virtual model in the Unreal Engine development environment.

3.2 Evaluation of the Proposed Methodology Effectiveness

In order to compare the proposed teaching method based on a virtual laboratory with the traditional teaching method using real physical equipment, we conducted a survey of students from all involved areas. It consisted of the following questions:

1. I am more motivated by the prospect of acquiring new knowledge, skills and abilities in the learning process than by the grades I can get.
2. It doesn't matter what the result of the work is, I get satisfaction, because feel like I've gained a new experience.
3. I enjoy attending classes in the subject and try not to miss them without a valid reason.
4. I am not distracted by extraneous objects in the course of the work.
5. I believe that the subject needs to be expanded with new modern practices.

Students were asked to rate the degree of agreement with each statement on a 5-level Likert scale, where 1 - "strongly disagree", 5 - "strongly agree."

A survey among students showed that this approach allows to get a higher level of motivation (Fig. 3) and, as a result, academic performance (Fig. 4) compared to the use of traditional teaching materials and real repair equipment for students in the automotive direction "Operation of transport and technological machines and complexes" and "Service" and educational topics for students of the computer direction "Information systems and technologies".

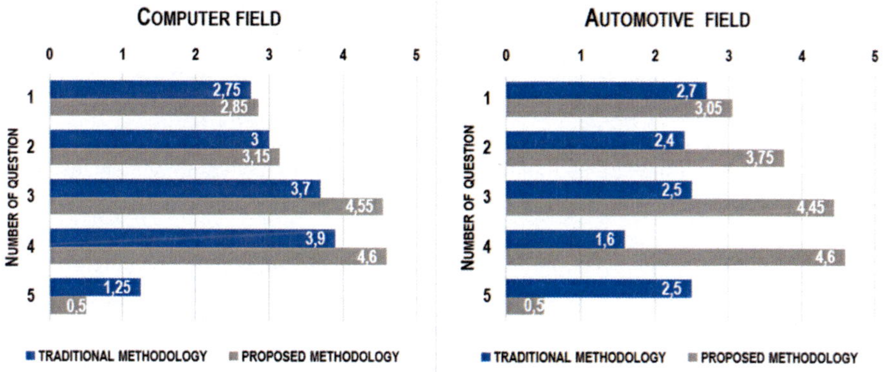

Fig. 3. Survey results

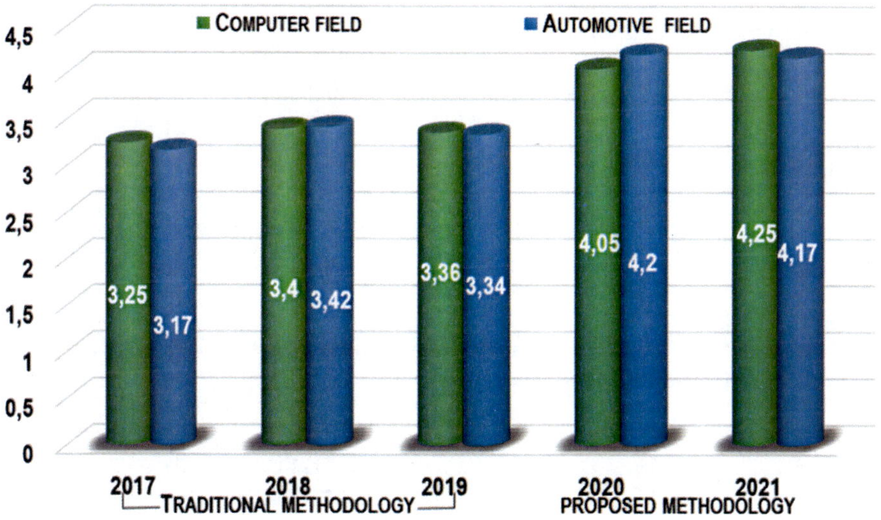

Fig. 4. Exam results - average score obtained by students

The increase in motivation and academic results among IT students can be explained by the fact that the creation of virtual models of operating equipment in a modeling environment used in enterprises has increased the interest of students. The increase in the performance of future vehicle maintenance engineers is due to the study of current service processes using modern virtual reality technologies. The proposed methodology stimulates students' enthusiasm for learning and develops the ability to independently work with various software, allowing maximum use of the creative potential of students and teachers.

4 Conclusion

Currently, society is moving towards the digitalization of all life spheres. One of the reasons for this was the COVID-19 pandemic, which radically affected the world and all aspects of education, leading to social distancing and technological change in the form of knowledge transfer and practical skills training. When developing new teaching methods, consideration must be given to how best to meet the academic intellectual and emotional needs of engineering students.

Using virtual reality technologies, it is possible and necessary to transform the very process of transferring educational information, during which students will already acquire digital competencies. The proposed method allows, during the development of a virtual reality laboratory, to create a learning environment that maintains the required level of motivation. As part of the collaboration of teachers and students of several directions, experience and knowledge are exchanged in an environment similar to a real professional one. Using a simulation software environment similar to that used in real production conditions, working on a joint actual project, achieving goals in a team contributes to the growth of students' motivation, which allows them to successfully master not only professional skills, but also communication skills.

References

1. The second online conference "Modern training of engineers" gathered more than 1500 participants. https://nticenter.spbstu.ru/news/7924. Accessed 2 Nov 2021
2. Galkin A, Popova Y, Chuprina E, Shapovalenko D (2019) Interaction of logistics 4.0 and consumer oriented marketing using ICT. In: Proceedings of the 33rd international business information management association conference, IBIMA 2019: education excellence and innovation management through vision 2020, pp 6751–6760
3. Vakulenko K et al (2019) Designing optimal public bus route networks in a suburban area. Transp Res Procedia 39:554–564
4. Bilbao J, Bravo E, García O, Rebollar C, Varela C (2021) Study to find out the perception that first year students in engineering have about the Computational Thinking skills, and to identify possible factors related to the ability of Abstraction. Heliyon 7(2):e06135
5. Feise HJ, Schaer E (2021) Mastering digitized chemical engineering. Educ Chem Eng 34:78–86
6. Frey CB, Osborne MA (2017) Ow susceptible are jobs to computerization Technol. Forecast Soc Chang 114(C):254–280
7. Levine A (2018) The "Just in time" learner and the coming revolution in higher education. Chang Mag High Learn 50(3–4):27–29
8. Makarova I, Shepelev V, Mukhametdinov E, Pashkevich A (2020) Changing the maintenance and repair system while expanding the connected vehicles fleet. In: VEHITS 2020 - proceedings of the 6th international conference on vehicle technology and intelligent transport systems, pp 622–633
9. Safiullin R, Pyrkin O, Marusin A, Safiullin R, Boryaev A, Ruchkina I (2021) Method for the level optimization of vehicle parameters when using fuels of different quality in cold climates. Transp Res Procedia 57:581–590
10. Ostroukh A, Karelina M, Filippova N, Marusin A, Evtyukov S (2021) Intelligent system for digital substation control. Transp Res Procedia 57:385–391

11. Gabsalikhova L et al (2020) Connected vehicles fleet expanding problems. In: VEHITS 2020 - proceedings of the 6th international conference on vehicle technology and intelligent transport systems, pp 642–650
12. Dhanalakshmi R, Anuja M, Shrijith D, Vijayaraghavan N (2021) A study on Covid-19 – impacting Indian education. Mater Today Proc. ID: covidwho-1142122
13. Milićević V et al (2021) E-learning perspectives in higher education institutions. Technol Forecast Soc Chang 166:120618
14. Asgari S, Trajkovic J, Rahmani M, Zhang W, Lo RC, Sciortino A (2021) An observational study of engineering online education during the COVID-19 pandemic. PLoS ONE 16(4):e0250041
15. Park M, Park JJ, Jackson K, Vanhoy G (2021) Online engineering education under COVID-19 pandemic environment. Int J Multidiscip Perspect High Educ 5(1):160–166
16. Makarova I, Shubenkova K, Tikhonov D, Buyvol P (2018) Improving the quality of engineering education by developing the system of increasing students' motivation. In: Auer ME, Guralnick D, Simonics I (eds) ICL 2017, vol 716. AISC. Springer, Cham, pp 150–161. https://doi.org/10.1007/978-3-319-73204-6_19
17. Clark AE, Nong H, Zhu H, Zhu R (2021) Compensating for academic loss: online learning and student performance during the COVID-19 pandemic. China Econ Rev 68:101629
18. Rassudov L, Korunets A (2020) COVID-19 pandemic challenges for engineering education. In: 2020 XI international conference on electrical power drive systems (ICEPDS), pp 1–3
19. Makarova I, Buyvol P, Magdin K, Pashkevich A, Boyko A, Shubenkova K (2020) Usage of microscopic simulation to estimate the environmental impact of road transport. Trans Res Procedia 44:86–93
20. Buyvol P, Makarova I, Sizikin V, Badriev, A (2020) The service center simulation model as a part of a reverse logistics system. In: Proceedings of international conference on developments in eSystems engineering, DeSE, pp 105–110
21. Makarova I, Shubenkova K, Buyvol P, Shepelev V, Gritsenko A (2021) The role of reverse logistics in the transition to a circular economy: case study of automotive spare parts logistics. FME Trans 49(1):173–185
22. Makarova I, Shubenkova K, Pashkevich A, Buyvol P, Mavrin V, Abeshev K (2020) Improvement of automotive service management by means of computer simulation. Transp Res Procedia 44:160–167
23. Khan ZH, Abid MI (2021) Distance learning in engineering education: challenges and opportunities during COVID-19 pandemic crisis in Pakistan. Int J Electr Eng Educ 0(0):1–20
24. Opriş I, Ionescu CS, Costinaş S, Nistoran DEG (2020) Challenges and opportunities to overcome. The impact Of COVID-19 pandemic on power engineering education. TEM J 9(4):1687–1691
25. Liu Y, Vijay A, Tommasini SM, Wiznia D (2021) Hands-on engineering courses in the COVID-19 pandemic: adapting medical device design for remote learning. Phys Eng Sci Med 44(1):195–200. https://doi.org/10.1007/s13246-020-00967-z
26. FEATURE: 5 ways the Covid-19 pandemic has changed engineering forever. https://www.imeche.org/news/news-article/feature-5-ways-the-covid-19-pandemic-has-changed-engineering-forever. Accessed 4 Oct 2021

Language in Artificial Neural Network Training

Jamila Mustafina[1](✉), Leysan Musina[1], Khayrullin Askar[1], Tara McKeown[2], and Dmitry Jdanov[1]

[1] Kazan Federal University, Kazan, Russia
muss_jane@mail.ru
[2] Liverpool John Moores University, Liverpool, UK

Abstract. Now, in the era of Industry 4.0, the connection between engineering and language is especially topical. We can talk not only about machine translation of a text from one language to another, but about a live conversation between a person and a robot. With the advent of modern technologies, the problem of human speech recognition by a robot appears. Electronic computers generally understand formalized, unambiguous words and expressions, logical commands, however, for an everyday user it is much more accustomed to have a more natural dialogue, with various kinds of synonyms, interjections, pauses, emotions.

Engineering linguistics is a branch of applied linguistics that studies the automation of processes such as machine translation, voice search, and the creation of tools for human-machine communication. Voice assistant is artificial intelligence, it not only follows pre-written algorithms, but also is able to learn and make its own decisions. A neural network is one of the ways to implement artificial intelligence, which resembles the human brain in its functions, simulating our nervous system. As a neural network is created on the basis of the human brain, its training is based on similar principles as human training.

Keywords: Neural networks · Natural language processing · Artificial intelligence · Engineering linguistics

1 Introduction

The research contains the analysis of the latest information concerning the AI and its development. Big amount of data on neural network function has been collected, so that to understand what has been already integrated and what is in the process of study and is an issue of interest. It has been found out, that the problem to teach AI the language, to give right responses and moreover emotional feedback is on the agenda of well-known innovative companies, such as Google and Yandex. To teach robots to process language, it is necessary to involve linguists, psychologists and psycholinguists, that is, those specialists who determine how the process of understanding a language works [1, 2]. For example, one of the important stages of information recognition is sound processing. Here, voice assistants work on the principle of dividing sounds into phonemes - fragments. This phenomenon was studied by the Leningrad phonological school, which investigated the connection between phoneme and speech, linguistic perception.

M. E. Auer et al. (Eds.): REV 2022, LNNS 524, pp. 360–365, 2023.
https://doi.org/10.1007/978-3-031-17091-1_36

The creation of artificial intelligence without language is considered impossible, since it loses all its significance, as language is one of the most important features of the human brain. Today, teaching computers the basics of the language is one of the topical tasks. It becomes obvious that for this it is necessary to reproduce the model of human learning, perception and understanding of the language. It is important to involve deep linguistic research, the basis for understanding and perceiving language. Language must be studied in a cognitive aspect, in other words, how the language process occurs in our head [3].

2 Methods

Natural language processing, (NLP), is one of the most well-known areas of data science. Over the past decade, it has gained a lot of popularity in both industrial and academic circles.

NLP is a field of science that has become especially popular over the past 10 years. However, this is not a completely new area; the human desire for the computer to understand and reproduce human speech has arisen a long time ago. People have tried to teach machines natural language and have gained some experience. A natural language is absolutely any human language, such as Russian, English, Turkish, etc. [4]. The human desire for computers to understand our language has existed since their inception. Those old computers that could hardly run several programs at the same time still managed to get acquainted with the complexity of natural languages. How difficult it is to teach a computer to recognize and understand a natural language depends on its structure. Moreover, the way we speak, our accents, the expression of thoughts are different, even if we speak our native language. If a person understands us intuitively, then a machine needs a certain algorithm, which is prescribed by programmers. Programmers, in turn, work closely with linguists, and psychologists [5].

Engineering linguistics is a science that directly deals with the processing of natural language and the automation of processes such as voice search, machine translation, and the work of a voice assistant. In terms of grammatical structure, there are many ways to combine words into one sentence. One approach to teaching a voice assistant is similar to teaching a child - dividing a sentence into a set of words that can be processed individually. When processing a word, the voice assistant determines its part of speech in order to understand what role it plays in the sentence. The problem is that everything works well if the sentences are simple and straightforward enough. If the sentence contains ambiguous words or neologisms, it will be difficult for the computer to understand what the speech is about. Natural language processing covers many areas where neural networks and data from many conversations between people are commonly used. The goal of the developers is to train the computer to understand human speech, both spoken and written. Here it is important to understand not only the logic of the language, but also to learn how to determine the emotional color of the sentence.

There are 5 main methods that help a computer analyze text:

1. N-grams are a type of probabilistic language model used to predict the next element in a sequence of words.

2. Tokenization is the process of breaking a sentence into separate words or tokens: verbs, nouns, pronouns, etc.;
3. Stemming is the process of finding the stem for a given word, for example swimming => swim.
4. POS tagging - a text processing process, the task of which is to determine the part of speech of a word and assign a corresponding tag to it;
5. Lemmatization is the process of bringing a given word to its dictionary form [6, 7].

After the text has been analyzed and the necessary response has been selected using neural networks, the voice assistant transmits it to the person. However, it is also important with what voice, timbre, with what emotional coloring the answer will be received. We may not have previously thought to teach AI empathy, but in Industry 4.0 we see a growing number of automated systems (from voice assistants which communicate with us both at home and in the car, to self-driving devices). Almost every company has a voice assistant on the telephone line that can perform almost any action, instead of a call center operator. It is important for a person not only to get the right answer, but also to receive some emotional feedback, since we perceive the language assistant as a real interlocutor [8].

3 Results

The daunting task of creating emotional AI is getting easier with the advent of machine learning. On the basis of several hundred thousand statements compared with the "emotional alphabet", the neural network learns to guess emotional markers. Then the AI will be able to categorize expressions according to similar emotional connotations, and in the future, learn to select statements based on the received data. Here it is necessary to involve a large team of linguists and psycholinguists to analyze the linguistic coloring of the language, identify interjections, expressive vocabulary, epithets, and other tools that give speech one or another emotional color. Thus, in connection with the need to train artificial intelligence, a profession such as a digital linguist is getting more and more demanded. A digital linguist is a specialist who develops systems for semantic translation (contextual and semantic), analyzes textual information, and also studies ways of effective human-computer interaction. A digital linguist is a link between a machine and a person, he sets an algorithm for neural networks to learn from the point of view of language. The neural network works according to the basic principle of machine learning - deep learning. The more information it receives, the more patterns it distinguishes between linguistic units. For a more correct operation, the neural network must be taught the laws of text construction, but not only from the point of view of grammar, but also from the point of view of expressiveness. Computer linguists have been able to teach artificial intelligence a lot, but a lot of effort is needed to make the voice assistant a real conversationalist.

If no request fits any of the information search scenarios, then it is redirected to a separate module of informal conversations. Very often, users use voice assistants not to search for specific information, but to just talk about abstract topics. Here it is important to teach the voice assistant not only to support the conversation, but to make it interesting using various literary techniques, jokes, etc. This task is accomplished using pre-prepared responses. The editorial team analyzes hundreds of the most popular questions and prescribes various answers to them. The answers should meet the expectations of the person, moreover, they should attract him/her, so that the next time he/she wants to talk with the voice assistant again [9, 10]. Also, voice assistants use neural networks trained on a huge database of books, texts, articles.

The next problem faced by computational linguists is that we miss a lot in colloquial speech. If a person understands us by context, then for a voice assistant this will become a certain problem. Therefore, perhaps a new task for computational linguists should be teaching a computer to recognize a natural language by context [11, 12].

Sentiment analysis is another popular direction in digital linguistics, in other words, it is the analysis of the emotional load of a text. It is necessary to teach the machine to analyze the emotional load of a written or oral text in order to correctly give feedback. This area is also interesting for commercial organizations that intend to study the attitude of customers to their activities, as well as the internal relationship of employees. A positive or negative attitude can be revealed in one way or another of evaluative statements, as well as contextually, in a hidden way. If the evaluative statement can be analyzed and entered a neural network database, then understanding the context is more difficult. Teaching artificial intelligence to do this is another goal of digital linguistics [13, 14].

4 Conclusions

Since in the era of Industry 4.0, people not only want to assign tasks to machines using alphabetic or digital text, but also want to have a real oral dialogue with a machine, such science as engineering linguistics has become especially important and advanced. We see that when training AI, it is necessary not only to conduct research in the field of programming and mathematics, but also it is important to study linguistic, psycholinguistic phenomena and methods, and apply them in teaching neural networks [15, 16].

We believe that in order to train a voice assistant and to create a specific feedback algorithm of a voice assistant for a person's speech, it is necessary to turn to some methods of linguistic analysis. In other words, to teach the neural network to analyze speech, it is necessary to accumulate an extensive database. This will help make the speech of the voice assistant more literate, livelier and bring it closer to human dialogue.

1. The method of component analysis is the analysis of words by means of highlighting the main components of meaning. For example, in the verb "to come", components such as "movement", "independence" can be distinguished. When studying the meaning of a word, linguists often turn to the analysis of dictionary definitions, that is, interpretations of the meaning of a word in various dictionaries. These interpretations explain the component composition of the word, as well as the possibility of its use in speech.
2. Contextual analysis is used to describe the functionality of a word and its meaning. This is an analysis of a text or a fragment in which a given word is used and determining to what extent the context of the entire text depends on it.
3. The method of linguistic modeling is used to compile various models or samples of linguistic units - sentences, phrases, words.

4. Statistical method – is a method of calculating the frequency of certain words use in a sentence, the typical use of language units for a particular language style.
5. Method of transformational analysis - analysis of the transformation of linguistic elements, for example, "They watch a movie" - "What they do is a movie watching".
6. Stratification method - this method studies linguistic differences between representatives of different social groups, examines phraseological phrases, dialect expressions and other features of communication of various social groups [17, 18].

These methods will help improve the speech of the voice assistant, make it more adaptable to different situations.

Acknowledgements. The reported study was funded by RFBR, project number 19-013-00679.

References

1. Bolshakov I, Gelbukh A (2004) Computational linguistics: models, resources, applications. Carretera Picacho, Mexico
2. Chernyavskaya V, Beliaeva L, Nefedov S (2018) Language worker in the framework of Information 4.0. In: Conference: 18th PCSF 2018 - professional culture of the specialist of the future, pp 608–609
3. Lacroix, F.: Writing for the 21st century. In: Towards a European Competence Framework tekom-Jahrestagung und tcworld conference, Stuttgart, pp 102–106
4. Gollner J.: Information 4.0 for Industry 4.0. In: Towards a European Competence Framework tekom-Jahrestagung und tcworld conference, Stuttgart, pp 93–94 (2016)
5. Zhou S, Mondragon RJ (2004) The rich-club phenomenon in the internet topology. IEEE Commun Lett 8:180–182
6. Paul T (2009) Text-to-speech synthesis. Cambridge University Press, Cambridge
7. Roemmele M, Gordon AS (2018) Automated assistance for creative writing with an RNN language model. In: Proceedings of the 23rd international conference on intelligent user interfaces companion, pp 1–2
8. Aditya J, Gandhar K, Vraj S (2018) Natural language processing. Int J Comput Sci Eng 1(6):161–167
9. Wang X, Zhao Y, Pourpanah F (2020) Recent advances in deep learning. Int J Mach Learn Cybern 11(4):747–750. https://doi.org/10.1007/s13042-020-01096-5
10. Salakhutdinov R, Hinton G (2012) An efficient learning procedure for deep Boltzmann machines. Neural Comput 24(8):1967–2006
11. Young T, Hazarika D, Poria S, Cambria E (2017) Recent trends in deep learning based natural language processing. IEEE Comput Intell Mag 13:55–75
12. Smagin M, Filatova E (2017) Computer linguistics, theory and practice. Success Chem Chem Process 31(14):364–365
13. Katranov S, Kuznetsov I (2016) Principles of training translators in the field of professional communication in the system of additional education in higher education, Philological sciences. Quest Theory Pract 5(59):187
14. Leontyeva N (2006) Automatic understanding of texts: systems, models, resources. ITS Academia, Moscow
15. Volkova I (2006) Introduction to the computer linguistics. Publishing Department of MGU, Moscow

16. Aleksandrovich N (2016) Stylistic analysis of a literary text: theory and practice. Flinta, Moscow
17. Goldman A (2015) The strategy and tactics of text analysis. Flinta, Moscow
18. Bortnikov V (2020) Linguistic analysis of the text. Ural Federal University, Ekaterenburg

Automatic Curriculum Scheduling in the Context of Learning Analytics

Jamila Mustafina[1](✉), Lenar Galiullin[1], Aigul Khanova[1], Evgeniya Maklakova[1], and Ahmed J. Aljaaf[2]

[1] Kazan Federal University, Kazan, Russia
muss_jane@mail.ru
[2] Liverpool John Moores University, Liverpool, UK

Abstract. Today, there are many educational institutions that have their own Internet resources. There are quite a few advantages of having a website for an educational institution: convenient access to information for employees and students, as well as accessibility for almost all applicants, so that in the future they can become website visitors and potential students. An automated information system for class planning should have a convenient interface for the user and administrator. The system must implement all the functionality required to manage the schedule. An automated scheduling information system must be built using some of the most current development tools used by developers of various qualifications around the world. It was decided to divide the development of the system into two stages, each of which is a separate project. The first step involves developing a lesson planner to assist with scheduling. The second stage includes the development of a convenient display of the lesson schedule for each day for students and teachers.

Keywords: Schedule · Learning analytics · Automation

1 Introduction

The experience of the massive transition to distance learning revealed many problems in the organization of the educational process and proved the need for good systems of educational analytics for each educational organization [1].

The development of a class schedule is an important process for an educational organization, but very time consuming for a compiler [2]. It is also important that the process of displaying the timetable is convenient for everyone. In this regard, optimization is required: a convenient system for drawing up and displaying the schedule in the context of the educational analytics system [3]. The object of research in this work is the system of educational analytics. The subject of research is automated scheduling [4].

2 Methods

Today there are about 10 known scheduling systems. For comparison, consider 3 of them:

M. E. Auer et al. (Eds.): REV 2022, LNNS 524, pp. 366–371, 2023.
https://doi.org/10.1007/978-3-031-17091-1_37

- 1C system: Automated scheduling;
- "AUTO Schedule" system;
- "NIKA" system.

1C: Automated scheduling is a standalone software product designed to solve the problems of automated scheduling and operational room management in educational institutions.

With its help, you can create a schedule in automatic, manual and mixed modes, taking into account many restrictions and conditions. In this case, it is possible to build both a permissible schedule and an optimized one, in which the number of windows or the number of rooms used is reduced.

The product implements the functionality of scheduling in an automated mode, in addition to the manual mode.

The scheduling process is divided into the following stages:

- input of primary information: courses, groups, disciplines, teachers, premises;
- entering the curriculum for the semester, which indicates who, for whom, what lesson and in what volume should be conducted;
- introduction of restrictions and preferences for teachers, students, premises;
- drawing up a training schedule.

Pros:

- the decision is being worked out for a specific organization;
- integrates with other solutions based on 1C.

Cons:

- high price;
- there is no easy-to-integrate curriculum development program.
- AVTOR software product allows:
- build a schedule without "windows" for classes (study groups);
- to optimize teachers' windows in the schedule;
- take into account the required range of days/hours for classes, for teachers and for classrooms;
- take into account the nature of the work and the wishes of both full-time employees and part-time employees;
- optimally place classes in classrooms (audiences), taking into account the characteristics of classes, subjects, priorities of teachers and the capacity of cabinets;
- enter the call schedule;
- to set the time of transition (moving) between educational buildings;
- to optimize the number of transitions from office to office and from body to body;
- it is easy to connect any classes (study groups) into streams for any classes;
- to divide classes (study groups) when conducting classes in a foreign language, physical culture, labor, computer science (and any other subjects) into any number of subgroups (up to ten);

- to introduce combined lessons for subgroups (such as “foreign/informatics”) in any subject;
- to introduce (in addition to the main subjects) special courses and electives;
- to optimize the uniformity and complexity of the schedule;
- easy and quick to enter and correct the initial data;
- have any number of schedule options;
- automatically convert schedules when changing the database;
- it is easy to save in archives, copy and send by e-mail complete databases and schedule options (the size of the archive of the full base of the schedule of a high school - 10–30 Kbytes, a large university – 50–70 Kbytes);
- quickly make any necessary adjustments to the schedule;
- find replacements for temporarily absent teachers;
- automatically control the schedule, excluding any “overlaps” and contradictions;
- display schedules in the form of convenient and visual documents: text, Word, HTML, as well as dBase files and Excel books;
- Expose ready-made schedules in the local network and on Internet pages for general access.

Pros of use: produced by the compilers of the “Shakhty” program, that is, it can be integrated with the curricula drawn up in this system (GosInsp). Cons of using: high cost.

The program takes into account:

- the presence of a second shift, five or six-day forms of training;
- features of the cabinet fund;
- methodological days and unwanted working hours for teachers.

The program provides:

- division of groups into students for five or six days;
- “floating” start of classes;
- splitting groups into subgroups (up to four), including by different subjects;
- doubling a given number of hours;
- scheduling taking into account the characteristics of the existing auditorium fund;
- combining groups into streams (including with further division into groups);
- teaching two or more subjects by one teacher;
- display and print of the timetable of classes in five forms, including individual timetables of teachers and timetables of individual classes;
- export of the schedule for display on the website and on information kiosks;
- export of the schedule to Excel;
- a convenient tool for hourly control of the schedule of individual teachers, as well as a “friendly semi-automatic” for manual editing of the calculated schedule.

Based on the data obtained, in the course of the work, the goal of the work was formulated [5]. It has also been found that each scheduling system is unique in its own way [6].

The purpose of this work is to improve the efficiency of scheduling management by using web technologies [7].

To achieve this goal, it is required to solve the following tasks:

- make it convenient to display the schedule of classes;
- help the scheduling planner to optimize his processes;
- update the site interface.

3 Results and Discussion

JavaScript (JS) language was chosen to develop an automated information system for lesson planning. This language is the most popular programming language on the web.

It was decided to divide the development of the system into two stages, each of which is a separate project.

The first stage includes the development of a lesson planner, which helps the head of the educational unit in drawing up a timetable.

The second stage includes the development of a convenient display of the schedule of classes for each day for students and staff.

The following solutions were chosen as development tools.

During the development of the project, the MySQL database management system is used. MySQL belongs to the type of relational database management systems and is one of the most popular today.

To effectively interact with the DBMS, Navicat Premium 15 was used, which is a powerful database administration tool. Navicat has a clear interface that allows you to create and edit the structure of the database. This tool is suitable for both professionals and beginners in the field of database management systems.

A database called "student-DB" was created containing three tables: accounts; teachers; study.

The "accounts" table contains information about the logins and passwords of the lesson planning system administrators.

The table "teachers" contains information about the names of teachers, about which subject or subjects each teacher teaches, as well as the classrooms assigned to teachers.

The table "study" contains information about the educational process, where the days of the classes are indicated, as well as data for each day of the classes: number of groups; Course of Study; names of teachers; names of items; room numbers.

This project is based on a standard layout using HTML and CSS.

The HTML markup language is the basis of the site; it is used to create a frame of the page that is displayed in the browser.

CSS is a style language that extends the capabilities of HTML. CSS helps you set the background, font style, change colors, and more.

JavaScript is added to HTML and CSS to make web pages dynamic.Node.js – этокроссплатформенная среда с открытым исходным кодом, обеспечивающаявыпол году она стала невероятно популярной и сегодня занимает лидирующиепозиции в обла

Node.js is built on the basis of the V8 JavaScript engine developed in C++ by Google and in addition to node.js, which is used as the engine at the heart of the Chrome browser, Node.js can work with external libraries, this platform is most often used for web development. Servers, but the scope of its application is far from limited by them.

The advantage of using Node.js is that, due to its modularity, this platform makes scaling tasks much easier. Despite the fact that the code is executed in one thread, due to the use of the event loop and the mechanism of deferred actions, it is possible to organize work with asynchronous logic, which allows a very convenient way to implement any external interactions, such as requests to third-party services, timers, user input and other asynchronous interactions.

In this project, using Node.js, a project development environment is provided, in which development, debugging, assembly and preparation for publication of the application source code are carried out.

Vue.js is the main framework for developing an application. Vue.js is an open source framework for building custom interfaces.

In this project, it was decided to use Vue.js, since it allows you to customize the structure of the application in accordance with your own requirements, and Vue.js is also easier to learn than, for example, AngularJS. To work with this framework, you can only use knowledge of JavaScript and HTML. The framework is well documented and responsive, which makes Vue.js state management pretty simple and intuitive.

LESS is a CSS preprocessor. LESS is an open source product as well as a nested metalanguage: valid CSS will be a valid LESS program with similar semantics.

The LESS language provides the following CSS extensions: variables, nested blocks, mixins, operators, and functions.

LESS can run client-side (Internet Explorer 6+, WebKit, Firefox) or server-side running Node.js or Rhino.

LESS is used in the project to modify the design of interface elements in the project as a whole.

VS Code was used in the project as an IDE for working with environment code. VS Code is a source code editor developed by Microsoft. Positioned as cross-platform for web and cloud applications. Includes a debugger, refactoring tools, syntax highlighting.

VS Code also has a navigation menu and the ability to quickly switch between project files, which allows you to see the entire project and quickly make changes to the code.

The tab bar at the top is convenient in that it allows you to close files that are not needed for work at the moment.

4 Conclusion

Based on the data obtained during the work, the goal of the work was formulated [8]. It has also been found that each scheduling system is unique in its own way [9]. The aim of the work is to improve the efficiency of schedule management in the context of educational analytics systems by using web technologies [10]. To achieve this goal, you need to solve the following tasks: make it convenient to display the schedule of classes; help the scheduler optimize its processes.

In the course of the work, the scheduling process was analyzed, the main activities of the process, input and output data were identified. The analysis of three existing scheduling systems was carried out: 1C systems: Automated scheduling; system "AUTOSchedule"; systems "NIKA". Based on the analysis results, custom requirements were generated, a use case diagram was developed, and custom requirements specifications were described. Functional and non-functional development requirements were developed based on user requirements. Some of the most relevant web development tools have been selected.

In the course of the work, an analysis of the schedule management systems market was carried out. An analysis of existing solutions was also carried out.

The automated lesson planning information system was created using some of the most current development tools used by web developers of various qualifications around the world.

Acknowledgements. The reported study was funded by RFBR, project number 19-013-00679.

References

1. Romero C, Ventura S (2010) Educational data mining: a review of the state-of-the-art. IEEE Trans Syst Man Cybern Part C Appl Rev 40(6):601–618
2. Baker RS, Inventado PS (2014) Educational data mining and learning analytics. In: Larusson JA, White B (eds) Learning analytics: from research to practice. Springer, New York, pp 61–75. https://doi.org/10.1007/978-1-4614-3305-7_4
3. Dutta Borah M, Jindal R, Gupta D, Deka GC (2011) Application of knowledge based decision technique to Predict student's enrolment decision. In: Proceedings of international conference on recent trends in information systems, pp 180–184
4. Okoli C, Schabram K (2010) A guide to conducting a systematic literature review of information systems research. Sprouts Work Pap Info Syst 10(26)
5. Sarker F, Tiropanis F, Davis HC (2014) Linked data, data mining and external open data for better prediction of at-risk students. In: 2014 international conference on control, decision and information technologies, pp 652–657
6. Avella J, Kebritchi M, Nunn S, Kanai T (2016) Learning analytics methods, benefits, and challenges in higher education: a systematic literature review. Online Learn J 20(2):1–17
7. Haykin S (2008) Neural Networks and Learning Machines, 3rd edn. Pearson, Upper Saddle River
8. Srivastava N, Salakhutdinov R (2012) Multimodal learning with deep Boltzmanns machines. In: Proceedings of advances in neural information processing systems 25 (2012)
9. Stahl F, Jordanov I (2012) An overview of the use of neural networks for data mining tasks. WIREs Data Min Knowl Disc 2:193–208
10. Blikstein P (2013) Multimodal learning analytics. In: 3rd international conference on learning analytics and knowledge, pp 102–106

Some Techniques for Smart Engineering Education

Kapil Gupta(✉)

Mechanical and Industrial Engineering Technology, University of Johannesburg, Johannesburg, South Africa
kgupta@uj.ac.za

Abstract. In this era of smart education, several novel and smart techniques are available to facilitate both instructors and learners. The role of these techniques is prominent for easy and effective content delivery, better understanding of course contents, and deeper and enhanced learning. This paper sheds light on few of such smart engineering education techniques. It mainly focuses on use of technology, innovative pedagogy, knowledge decolonization, and novel assessment practices. The paper aims to facilitate instructors who delivers engineering education courses by providing some brief knowledge on smart engineering practices and techniques.

Keywords: Engineering education · Learning · Teaching

1 Introduction

It is said that teaching with passion and keeping respect and emotion for students with an objective to engage and empower them, helps to involve them enthusiastically to gain knowledge, learn, develop and refine understanding to succeed. Based upon the discussion with peers and feedback from the students, one can keep on improving teaching style that is a continuously evolving practice. The success of teachers can be met when all students have become knowledgeable, critical thinker, problem solver, and industry ready. For learning, development, and success of the learners, it is very important that teaching and learning (T&L) should be grounded on a certain educational theory and equipped with smart techniques. Educational theory-based T&L promotes development, growth and progress of the learners, instructors/teachers, and the institution as well. There are various educational learning theories that work for engineering education as well, behaviorism, cognitivism, and constructivism theories are mostly followed or adopted [1, 2]. Each theory has a rationale behind it and explanation that how is knowledge acquired, how does learning take place, and how to teach to enhance learning and development. Cognitivism (or cognitive learning) focuses on how to receive, organize, store and retrieve information, and the consistent use of information to attain learning. The acquisition of knowledge is a mental activity in cognitivism, and structuring, organizing and sequencing of information are important. The role of the teacher is to prepare environment for learning. Mnemonics, concept or mind mapping, information chunking, analogies, demonstrations, animations, videos and illustrative examples are

M. E. Auer et al. (Eds.): REV 2022, LNNS 524, pp. 372–382, 2023.
https://doi.org/10.1007/978-3-031-17091-1_38

the learning strategies under cognitivism. In other words, as far as cognitivism is concerned, explanation of concepts with creative examples and through pictures, posters, and videos, facilitates information processing and interpretation (overall cognition) and thereby ensures learning.

The discipline and nature of the subjects one deals with, is also one of the grounds behind the selection and adoption of any learning theory. For example, subjects belong to manufacturing engineering require the knowledge of engineering materials, manufacturing operations, processes mechanisms, systems, and machines to be given to the students. For these types of subjects, lack of understanding of the contents is a major problem, as the traditional lecture notes and simple power point-based lectures merely solve the purpose. Therefore, following the cognitivism theory, using animations and illustrations for a better explanation of the course contents, and understanding of the students is a better way to teach them. Digital interventions promote a creative learning environment that connects the students with the topic and enables them to visualize the process/operation/mechanism/system being taught. Cognitive learning-based teaching philosophy and practice are of great assistance to remain operational during pandemic lockdown phase.

Hands on skills and laboratory practical work promote constructivism. Constructivism and experiential learning are major tools to develop industry ready engineers. Interventions of innovative pedagogy skills are required to promote deep and meaningful learning of the students. There are many ways to evaluate the effectiveness of innovative teaching techniques used for smart engineering education, such as, by conducting students' interviews, conducting feedback from them, analyzing their performance in exams etc. Regular evaluation of teaching and supervision by peers and students should be intended to refining teaching and supervision practice, pedagogical skills, and facilitation approaches. Knowledge decolonization, information & communication technology (ICT) interventions (or 4IR based digital techniques), and social inclusion are the curriculum imperatives should be evident in pedagogy [3]. This paper introduces some important teaching practices for smart engineering education and presents few examples for the benefits of prospective teachers. These practices and techniques are important to facilitate students for a better understanding and learning. While conducting the pedagogical research using these techniques, the possible objectives could be- to investigate the effectiveness of smart engineering education techniques, to evaluate the students' response and impact on performance, to assess technique sustainability and interrelationship with other techniques.

2 Smart Engineering Education

2.1 Teaching Practice

As discussed, discipline and nature of the subjects determine the learning theory to be adopted. However, there are some common teaching practices and relevant techniques which can be adopted by instructors involved in engineering education. These techniques are used to course material development, planning of tutorials and laboratory work, and conducting assessments. The use of videos, animations, power point slides embedded with pictures and video links, and virtual and augmented reality based digital learning material and space etc. are important tools and techniques. The lab work and practical should be designed in such a way that every student to get maximum opportunity for laboratory exploration, best practical understanding of the contents studied in the lecture classes, and to engage in experimental learning by realistic experience that enable life-long learning. The overarching goal of any course material design should be to facilitate the students and enable them to suggest, justify, analyze, and interpret the solutions of the broadly defined engineering problems and provide valid conclusions.

2.2 Facilitation Methods

For an effective contribution to the professional qualification programme purpose, teaching and assessment both are required to be aligned with the learning outcomes of the module. The purpose of facilitation methods is to develop students who can demonstrate detailed knowledge with an ability to use this to solve broadly defined engineering problems for societal, environmental, and economical benefits. Following key strategies, can be kept in practice for engagement and facilitation [4]:

Teacher Must Act as a Facilitator

To try best to understand the limitations and psychology of the students, and provide many opportunities to the students to explore, come out of fear, and learn. Rather having a traditional tendency of being an authority figure and having power to control the class or atmosphere of the class system, it is needed to play a role of facilitator and try best to provide enough opportunities to the students to engage, explore, and understand.

Defining Objectives and Establishing Ground Rules in Advance or in the Beginning

To communicate the objectives and rules of engagement, as well as about various aspects of the module/course to the students well in advance. It helps to make them feel included and engaged. Individual engagement and consultation are also important to be included into practice.

Adopting Service-Learning Strategy

Under this, it is necessary to expose students to the service-learning sites, and hands-on skills and labs, to break their comfort zone, and have real-life learning experience. It helps them to analyse the gap between theory and practice and makes them industry ready.

From a social inclusion perspective, participation, accountability, non-discrimination, and empowerment while engaging with and facilitating the students are important factors. Providing feedback is also an important facilitation method, and it helps the students to bridge the gap between their actual performance and the desired learning outcomes. Therefore, timely and constructive feedback is essential for students to extend and refine understanding and make revisions. A feed-forward technique which focuses more on future development than on rating them or judging their performances should be followed based on the following principles:

- Future focus: for students to improve and perform better in next assessments.
- Inform rather than attack: instead of finding faults, to provide proper information and suggestion for future work.
- Focus on person: to focus on underperformed students and counsel for respective concerns.

2.3 Assessment Practices

Some common assessment strategies are worth discussing:

- Discussing review questions and practice test opportunities with prompt feedback.
- Allowing consultation beyond the prescribed schedule.
- Informing students about type and timing of assessments well in advance.

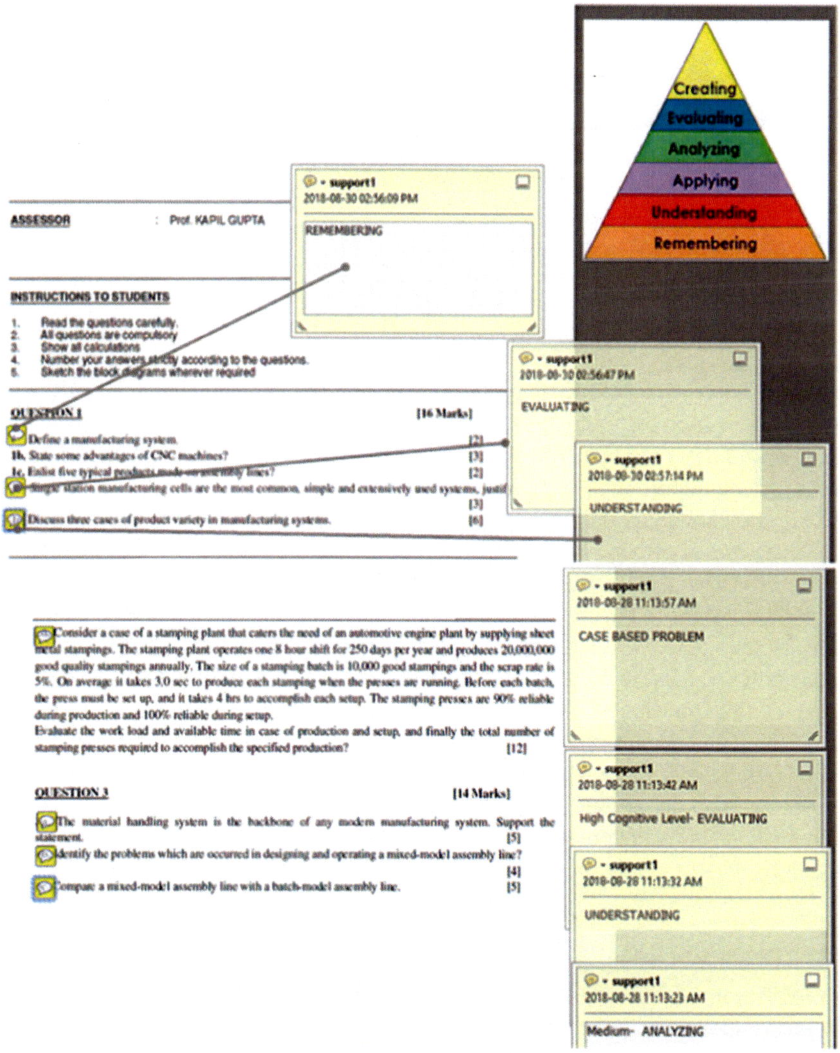

Fig. 1. Example of an assessment based on Bloom's taxonomy.

Going beyond the aforementioned common strategies, there are some smart assessment techniques namely bloom's taxonomy-based assessment questions and mediated learning theories [5, 6]. The mediated learning theory characteristic 'feeling of competence' implies to conduct practice tests and quizzes. Following the mediated learning theory characteristic 'challenge of novelty', while drafting questions for the tests, exams, and assignments, it implies to refer bloom's taxonomy and develop questions accordingly [5]. Another mediated learning theory that can be applied is 'goal seeking and achieving', which implies to make all the learning objectives of the modules or subjects, units, and topics clear to the learners, and plan lectures and lessons accordingly. The students are informed about objectives, and they study accordingly to achieve them.

Bloom's taxonomy was developed to facilitate educators to conduct assessment practices grounded on learning objectives that have a varying level of complexity [5]. As per this, the learning and its depth is assessed on a variety of cognitive levels whose hierarchy is as shown in the Fig. 1. The main focus is on intellectual skills such as critical thinking, problem solving, and creating a knowledge base. By designing such assessments, students' abilities to remember, understand, apply, analyze, evaluate, and create, are assessed.

2.4 Use of Technology

Technology enabled teaching is the requirement for better communication and understanding of the contents of the subjects and to remain operational remotely. Information and communication technology tools such as digital lectures, animations and videos, power point slides, drop box, WhatsApp and various other virtual and augmented reality based tools are important. These tools facilitate students to have a better understanding and gain the knowledge in comparatively less time. Blogs and discussion forums are very effective and provide ease to the students to ask questions and have discussions. The use of ICT in teaching practice is a mean to facilitate learning and its assessment through integration with face-to-face teaching. The role of technology is more prominent and its effective use is essential in the circumstances like pandemic lockdown phase to provide remote access to the students for continuation of their studies. During this phase, recommending e-books, access to virtual labs and learning space etc. are effective ways for continuation of teaching and learning. An example of a manufacturing engineering virtual learning space developed at author's institute is illustrated in Fig. 2. This virtual space provides a comprehensive information on contents of various modules related to manufacturing engineering. This virtual space is equipped with theoretical details, animations, videos, and quizzes, and facilitates learners to stay learnt remotely [7].

2.5 Decolonization of Knowledge

Decolonization of knowledge implies transformation to affirm local knowledge and cultural traditions in universities [8]. For knowledge decolonization, efforts significantly more than changing curriculum and reading lists are required to be made, so that students can also become locally relevant along with staying internationally competitive. Relevant workshops and seminars are required to be attended for a preliminary formal or informal training of knowledge decolonization. It can also be initiated by incorporating examples and case studies in local contexts and exposing students to the scholarly thinking of local scientists and engineers working in the local industries. It helps them to understand and prepare to meet the local requirements and deal with challenges. Minor modifications in the curriculum can also be planned.

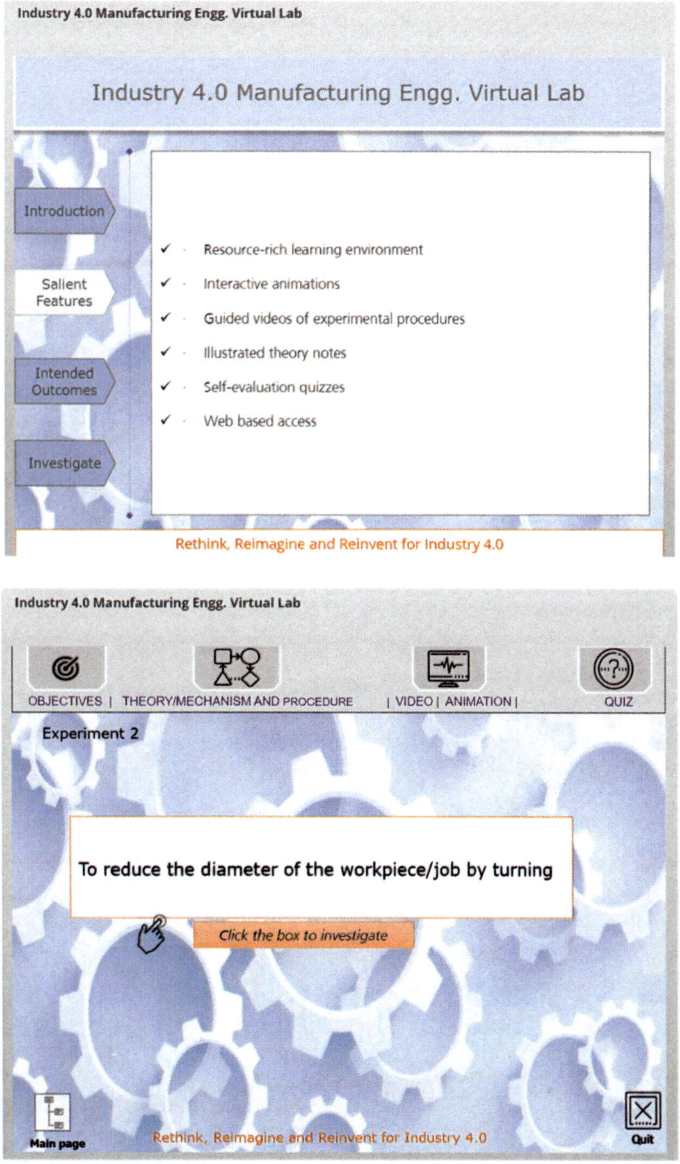

Fig. 2. An example of virtual learning space developed for manufacturing related modules.

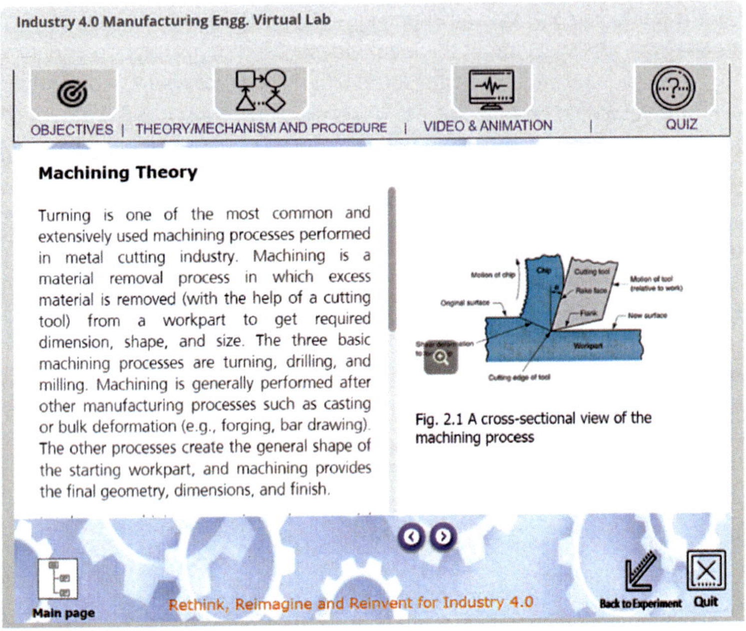

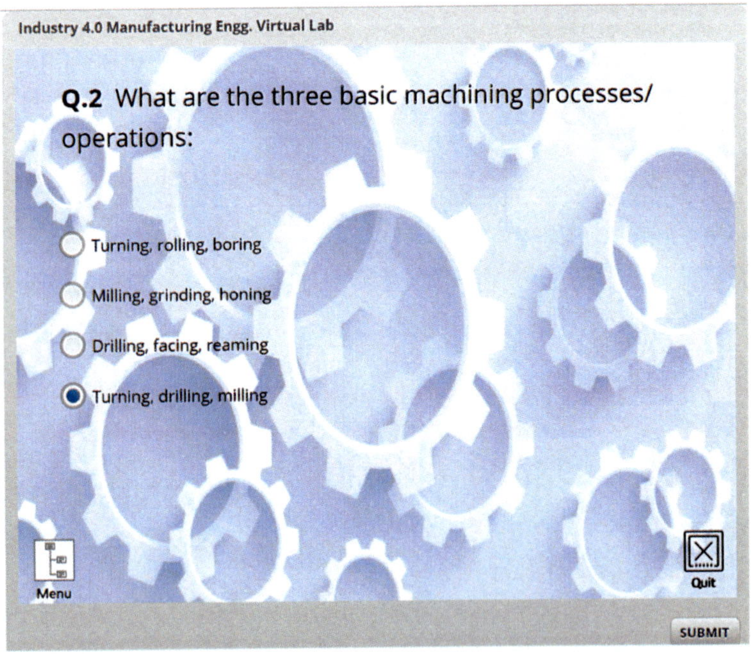

Fig. 2. continued

Fig. 2. continued

2.6 Approach to Supervision

For effective supervision of postgraduate and project students, it needs to provide different supervision environments to students as they move from one stage to another. To identify their readiness level and needs at the beginning, and matching supervisory behaviour accordingly are important tasks. There is a directive supervision model of developmental approach which can be followed in the beginning [9]. Considering the importance of demonstration at the beginning stage, sharing skills and knowledge with students and reinforcing them to build competencies and gain expertise to achieve the established goals as regards to their postgraduate studies are essential. Exposing students to new information and making use of counselling skills to facilitate their progression to the next stage fosters the development of necessary critical skills. By the time, collaboration and interaction with them is important to make research plans and their execution, analysis and discussion of the results, and set future directions towards finalizing their research work.

2.7 Sustainable Development Goals (SDGs) and Engineering Education

Teaching and learning is being embedded with activities contributing to achieving United Nation's sustainable development goals [10]. SDG 17 out of seventeen global goals encourages collaborations or partnerships to work towards other goals. Based upon the engineering field or branch, teaching, learning, and research strategies can be formalized to make contributions towards SDGs. For example, advanced and sustainable manufacturing can fit in SDG-9 industry, innovation, and infrastructure, and can assist to achieve other goals as well. Industrial engineering branches such as ergonomics and safety engineering can promote SDG 8 decent work and economic growth and help to achieve SDG 3 good health and well-being. Trainings, projects, seminars, and workshops can be possible mediums to disseminate the concept of SDG based engineering education and to encourage teachers and learners to make contributions.

3 Review of Past Work

Table 1 summarizes some of the past work on teaching and learning techniques used for smart engineering education.

Table 1. Use of innovative teaching and learning techniques for smart engineering education

Researcher	Technique used	Outcome
Richard et al. [11]	Virtual lab for fluid mechanic education	Satisfactory feedback by the learners
Dobrzanski and Honysz [12]	Virtual lab for material science	Satisfactory feedback by the learners
Gupta et al. [13]	Project-based learning	Successful completion of two workshop projects with learners' training on digital design and manufacturing
Fernandes [14]	Project-based learning	Students demanded to employ this technique for other courses and modules
Lai et al. [15]	Game-based learning	Winning in games attracted students towards course study
Mavromihales et al. [16]	Game-based learning	Students who were taught by this were found having better understanding of computer aided design assembly, compared to those taught in conventional way
Setyowidodo et al. [17]	Problem-based learning	Creativity enhancement of mechanical engineering students
Lutsenko [18]	Problem-based learning	Students learnt teamwork, problem solving skills, and communication skills via problem-based learning

4 Conclusion

Some important techniques pertaining to smart engineering education have been discussed in this paper. For the success, it is essential to have an effective engagement and involvement by both instructor and learner. Bloom's taxonomy-based assessment practice, learner facilitation methods, decolonization of knowledge, and use of technology are found very important in literature. Regular feedback from the learners and peers for continuous improvement is essential. Moreover, continuous professional development as a teacher including scholarship of teaching and learning also play major role. The attempts can be extended to incorporate engineering education 4.0 based teaching and learning practice along with the techniques discussed in this paper. The future work can be explored in the following domains:

- Applying experiential and project learning based teaching techniques and investigating their impacts.
- Hybridizing two or more smart techniques and assessing the results.
- Flipped classroom teaching, problem-based learning, game-based education, triggered virtual labs etc. are some of the fields that need more sincere future attempts.
- Incorporating UN's SGDs into teaching, learning, research.

Acknowledgement. The work reported is based on the funding provided by the Royal Academy of Engineering- UK, Grant number TSP1332.

References

1. Hergenhahn BR, Olson MH (2005) An introduction to theories of learning, 7th edn. Pearson/Prentice Hall, Upper Saddle River
2. Schunn CD, Silk EM (2011) Learning theories for engineering and technology education. In: Barak M, Hacker M (eds) Fostering human development through engineering and technology education. Sense Publishers, Rotterdam, pp 3–18
3. Gupta K (2020) Some insights on engineering education 4.0. In: 5th North American international conference on industrial engineering and operations management proceedings, Detroit, pp 19–25
4. Fry H, Ketteridge S, Marshall S (2003) A handbook for teaching and learning in higher education enhancing academic practice, 3rd edn. Routledge, London
5. Gogus A (2012) Bloom's taxonomy of learning objectives. In: Seel NM (ed) Encyclopedia of the sciences of learning. Springer, Boston, pp 469–473. https://doi.org/10.1007/978-1-4419-1428-6_141
6. Seng TO (2003) Mediated learning and pedagogy: applications of Feuerstein's theory I twenty-first century education. REACT 1:53–63
7. Gupta K (2020) Evaluation and analysis of students feedback on Industry 4.0 virtual lab developed for manufacturing engineering education. In: 5th North American international conference on industrial engineering and operations management proceedings, Detroit, pp 9–18

8. Hall BL, Tandon R (2017) Decolonization of knowledge, epistemicide, participatory research and higher education. Res All 1(1):6–19
9. Glickman CD (1980) The developmental approach to supervision. Educ Leadersh 38(2):178–180
10. https://www.un.org/sustainabledevelopment/sustainable-development-goals/
11. Richards CD, Thompson BR, Richards RF (2015) A remote access laboratory for fluids education in mechanical engineering. In: 122nd ASEE annual congress and exposition, Seattle, USA, 14–17 June 2015, Paper ID-11430
12. Dobrzanski LA, Honysz R (2007) Materials science virtual laboratory as an example of the computer aid in materials engineering. J Achiev Mater Manuf Eng 24(2):219–222
13. Gupta K, Mashinini M, Mukhawana D (2019) A project based learning tool for Industry 4.0 manufacturing engineering education. In: Proceedings of the international conference on industrial engineering and operations management, Toronto, pp 978–984
14. Fernandes MAC (2015) Project-based learning laboratory for teaching embedded systems. Math Probl Eng 2015. http://dx.doi.org/10.1155/2015/236749. Article ID 236749
15. Lai CH, Lee TP, Jong BS, Hsia YT (2012) A research on applying game-based learning to enhance the participation of student. In: Park J, Jeong YS, Park S, Chen HC (eds) Embedded and multimedia computing technology and service. LNEE, vol 181. Springer, Dordrecht, pp 311–318. https://doi.org/10.1007/978-94-007-5076-0_36
16. Mavromihales M, Holmes V, Racasan R (2019) Game-based learning in mechanical engineering education: case study of game-based learning application in computer aided design assembly. Int J Mech Eng Educ 47(2):156–179
17. Setyowidodo I, Sutanto, Handayani AD, Mahmudi H (2019) Problem based learning in mechanical engineering to train student's creativity. J Phys Conf Ser 1280:052072
18. Lutsenko G (2018) Case study of a problem-based learning course of project management for senior engineering students. Eur J Eng Educ 43(6):895–910

How Oversea University Students Use Mobile Devices to Support Their Foreign Language Learning and Their Learning Motivation

Binbin Zhu[1](✉) and Ying Sun[2]

[1] Technische Universität Berlin, Straße des 17. Juni 135, 10623 Berlin, Germany
zbbcau@hotmail.com

[2] Friedrich-Schiller-Universität Jena, Leutragraben 1, 07743 Jena, Germany

Abstract. International students have more difficulty than host nationals in dealing with a wide range of academic and social situations. The common challenges reported by students studying abroad encompassed the themes of language. Learning a new language is difficult for many students, who study abroad and are dependent on language to find their way around. If asked to identify the most significant influence on language learning, motivation would probably be high on most teachers' and learners' lists. Language teachers constantly use the term 'motivation' when they describe successful or unsuccessful learners. However, there is no doubt that motivation is the most complex and challenging issue faced by language learners, especially in such a digital age. A significant number of recent studies have been done on investigating the effectiveness of computer-assisted language learning, and the literature on psychology in language learning is very considerable. Relatively little work has been conducted about the psychological factor within a mobile-assisted language learning environment. Furthermore, few studies have explicitly focused on international students' perceptions of their use of mobile devices in motivating their foreign language learning. This paper is intended as an investigation of Chinese oversea university students' experience in Germany of using the MALL applications in motivating foreign language learning. The research investigated whether and with what tools or activities they perceive can affect their motivation of learning foreign languages in informal settings. The reasons behind the mobile devices motivating language use and specific types of motivation are also examined.

Keywords: Foreign language learning motivation · Mobile assisted learning · Social-psychological linguistic

1 Motivation: One of the Most Significant Influences on Language Learning

If asked to identify the most significant influences on learning, motivation would probably be high on most teachers' and learners' lists [1]. Many research pieces found that the motivation element has the most significant impact on students' achievement in learning a second language. Positive attitude, motivation, and efforts in mastering the target

M. E. Auer et al. (Eds.): REV 2022, LNNS 524, pp. 383–394, 2023.
https://doi.org/10.1007/978-3-031-17091-1_39

language are crucial elements that can determine the success of language students [2]. More broadly, research has consistently shown that engagement arises from students' motivation [3]. Some researchers [4] regarded passion as the most significant factor in the formation of motivation. In many cases, learners with sufficient motivation can achieve a working knowledge of a foreign and second language regardless of their language aptitude or other cognitive characteristics [5].' Motivation remains a significant factor to consider in any form of learning.

1.1 Motivation Defined

"Motivation refers to the choices people make as to what experiences or goals they will approach or avoid and the degree of effort they will exert in this respect" [6]. Motivation in psychology is a wide range of phenomena, encouraging people to do something. It is a concept that has been defined in various ways and dimensions by researchers in psychology and other scientific disciplines [7]. Therefore, the motivation concept includes a strong definition. Longman Dictionary of Language Teaching and Applied Linguistics explains that 'motivation stands for the driving force in any situation that leads to action.' The term "motivation" is described as the inner motor of human beings that leads to behavior [8], a result of conscious and unconscious factors that stimulate and energize people in continually doing something to attain a preset goal [9]. In a research [10], he once referred to a finding from Keller's [6] study that states motivation is a choice made by an individual on experience, objective, and effort in achieving or avoiding one's goal. However, motivation is more than merely arousing interest—a desire to do something. It also needs to sustain that interest and invest time and energy into putting in the necessary effort to achieve the specific goals [1]. Motivation is defined as the boost to provide action and retain intention in achieving the desired goals [11]. Williams and Burden [1] define general motivation as more concerned with initiating motivation in three stages: 1) Reason for doing something, 2) Deciding to do something, 3) Sustaining the effort.

When the learning target is mastery of a foreign language, it is not simple to define motivation. Motivation to learn the second language cannot be built easily as it covers a multitude of aspects such as cognitive, affective, and behavior as well as external wants. Motivation cannot be gauged using any scale [7]. General motivation in foreign and second language learning includes three elements that address the relationship between motivation and goals [10]: 1) Intensity of motivation, 2) Desire to learn the language, and 3) Attitudes towards learning the language.

1.2 Motivation on Language Learning

Regarding language learning motivation, three elements were addressed about the relationship between motivation and goals, including intensity of motivation, desire to learn, and attitudes towards learning. However, learning a foreign language involves far more than learning grammar or skills, but an alteration in self-image, social-psychological oriented, the adoption of new social and cultural behaviors and ways of beings and therefore has a significant impact on the social nature of the learner. Thus, motivation to learn a foreign language covers multifaceted aspects such as cognitive, affective and behavior as well as external wants, which echoes the statement that motivation involves conscious and unconscious factors of human mental activities and is affected by cognitive, affective and behavioural processes [12].

The researcher explained different types, roles and conditions of motivation, and finally stated key factors of motivation in language learning through reviewing some of the most considerable previous studies. It includes integrative and instrumental motivation [13]. Integrative motivation concerns a positive attitude towards the foreign language community or culture and the desire to communicate with, which involves attitudinal factors and goal-orientated behaviors, while instrumental motivation concerns external or practical factors for language learning such as good grades, passing exams, or furthering higher education. Besides, from the perspective of cognitive psychology and why people choose to act in specific ways, motivation is divided into intrinsic and extrinsic [1]. Intrinsic motivation is associated with the individual's identity, and refers to satisfaction, whether due to the development of knowledge, self-achievement or gaining exciting experience. Self-efficacy is central to enhancing intrinsic motivation. In comparison, extrinsic one comes from outside the individual and activity, which aims for the sake of rewards, diplomas, occupations, and prizes. However, a student's motivation is most frequently a combination of intrinsic and extrinsic motivation and external rewards can either increase or decrease intrinsic motivation. The learner's intrinsic motivation could be increased by providing them with a learning experience that meets their needs for competence, relatedness, and enjoyment. Moreover, based on the source of motivation, it can be considered in internal and external motivation [5]. Internal motivation means the job itself motivates the learner, and is the urge to pursue something because of their desire. It contributes to obtaining the pleasures of learning, which relates to learners' interests, excitement and self-esteem. In contrast, external motivation means the environment motivates that individual and refers to being motivated from rewards that come from the surrounding environment of individuals and the human interaction with its environment. It is based on incentives, penalties, and other types of stimulations that are either direct or inhibit the behavior. Therefore, the multifaceted motivation construct was affected by target language factors such as CL facet, what the learner brings to the learning activity such as cognitive aspects, and learning situation perspectives such as social constructivist characteristics.

2 Foreign Language Learning in a Mobile Assisted Language Learning Environment

Recent technological advancements in cloud computing and nanotechnology have enabled computing devices to shrink in size and increase their functionality, which is driving changes in the social and pedagogical milieu. The review indicated that mobile devices are becoming more and more pervasive in today's world for both personal use and educational purposes. In order to provide a general overview of the various stages of technologies enhanced language learning, especially the flourishment of mobile language learning, [14] built four stages of CALL based on [15] three stages of CALL, proposing a fourth stage for the second decade of the new millennium: ecological CALL with a reflection of ecological perspective of second and foreign language learning [16], and with a language acquisition goes beyond classroom boundaries, as technologies nowadays can be accessed anytime and anyway. Besides, the traditional definitions of literacy have been extended to digital literacies or multiliteracies [17]. Language teaching practices should involve the active construction of digital artifacts, literacies and multiliteracies, which can be refined gradually supported by the online community [18]. The scope of language learning moves from the cognitive and sociocognitive view to the symbolic and intercultural competence, which is "the ability not only to approximate or appropriate for someone else's language, but to shape the very context in which the language is learned and used" [19]. The principal use of computers develops from the drill and practice and communicative exercises to the use in authentic discourse and for global communication.

With the rapid development of mobile technologies, like smartphones, personal digital assistants (PDA), laptops, and tablets, learning via mobile devices would play a significant role in the educational realm, given that 98% of the university students possessed cell phones [20]. The educational use of mobile devices was referred to as mobile learning, focusing on facilitating and extending learning and teaching [21]. Mobile learning has been acknowledged as a useful approach in language training, particularly for English as Foreign Language (EFL) students [21]. In detail, the ecological system of college English teaching is integrated with mobile learning. The results testified the feasibility and practicality of applying mobile learning to college English teaching and raised the guidelines for constructing the ecological system of college English education. Similarly, it is investigated the attitudes and self-efficacy of using mobile learning devices for college students in a language class by employing task-based instruction. Results showed that most students agreed that their motivation for English learning was enhanced and had positive attitudes towards mobile learning. In another study, Mobile-Assisted Seamless Chinese Learning Questionnaire is used to measure students' motivation and the learning strategy they use in a seamless learning environment [22]. The Confirmatory Factor analyses (CFA) were conducted to validate the psychometric properties of the instrument and the findings showed that students' motivation predicted the learning strategy they used in a seamless learning environment. Despite the emergence of studies on seamless language learning, there is a lack of instruments specifically designed to measure students' motivation and learning strategy in such technology-enhanced environments [23]. Mobile technology facilitates holistic and perpetual learning experiences that bridge different locations, times, technologies or social settings.

Based on what was found from the literature, the following issues are important to address as the plans for future studies in foreign and second language learning.

- The existing literature has mainly focused on technology-related factors to examine language learning adoption. However, few studies have investigated this topic from the users' motivation perspective and even fewer from the oversea university learner perspective;
- More research needs to be conducted in the less explored language skills areas such as speaking, listening, and culture rather than imbalanced studies on writing, reading and vocabulary acquisition;
- Computer technologies, apps need to be based on relevant language pedagogical or design principles for more effectiveness.

3 Problems Met by Oversea University Students

With the development of economics and globalization, more and more students have the opportunity to study abroad. Learning a new language is difficult, especially for students who study abroad. They usually need to get a language certificate such as TOEFL and IELTS when applying to overseas university. They are also highly dependent on language to find their way around. Therefore, they spent much time learning a language to satisfy the language requirements of studying and living overseas.

The international students see their residence abroad period as a unique chance to develop oral fluency, and they look for interactions with native speakers. However, they have more difficulty than host nationals in dealing with a wide range of academic and social situations. The common challenges reported by them encompassed the themes of: language, which included the technical aspects of language, speaking and listening, insufficient interaction in tutorial group discussion, and perceptions of ethnocentrism [24]. They reported particular difficulties in dealing with tutorials of not knowing how to behave and feeling less confident about the language proficiency to express their opinions. Learning in this sense is not just a case of cognition or something that only happens in a classroom, but a process by which humans survive in an unpredictable and constantly changing environment, which are frequently brimming with emotionally destabilizing encounters. This broad range of sentiments can lead students to question who they are and how they must act to gain access to and legitimately participate with peers in local host communities.

Furthermore, international students hope to improve listening and speaking proficiency dramatically, but quickly lose their motivation because of their shaky grammar and lack of procedural knowledge [25]. They often focus on linguistic shortcomings instead of successes and are often unable to see themselves as anything but deficient communicators [26]. Therefore, fears of being unable to produce native-like language can keep them from using the language and becoming relevant local participants. In this modern age of technology, a sizeable possible cause for the apparent low level of international students' motivated behavior might be related to the new means of communication and entertainment, which allowed them to stay connected with their social networks. It seems to have reduced the effort they put into making the most of their international study in language learning. Due to globalization and new means of electronic communication, 'immersion is increasingly a matter of choice, and perhaps in some cases a locus of struggle. Language learning in study-abroad settings requires a more profound and durable commitment than has been needed in the past [26]. These results emphasize the importance of investigating the role of motivation and intended language learning effort in study-abroad contexts.

4 Research Questions and Research Methods

4.1 Research Questions

The central aim is to identify how the MALL environment motivates Chinese oversea university students in Germany to learn a foreign language. In detail, the research intends to investigate what they perceive MALL can affect their motivation to learn language listening, speaking, reading, speaking, and vocabulary. Investigating students' motivation and its source helps us recognize the insights, to help the language learning stakeholders better understand how affective experiences shape how students construct and engage with learning context, when- and how- it is appropriate for educators and teachers to intervene. The research is designed to investigate three research questions:

- Do Chinese oversea university students perceive MALL to affect their motivation of learning foreign languages (including listening, speaking, reading, writing, and vocabulary learning) in informal settings?
- What tools or activities do Chinese oversea university students perceive MALL to affect their motivation to practice specific language skills?
- What types of motivation do Chinese oversea university students perceive that can motivate them to learn foreign languages in the MALL?

4.2 Research Methods

This research aims to examine Chinese oversea university students' experience of how the MALL can affect their motivation to learn a foreign language in informal settings. The study conducted a survey using quantitative methods, which is designed to investigate the research questions of the dominantly quantitative character of whether and why they think MALL influences their language learning motivation.

Sample. The population is the Chinese oversea university students who are studying in Germany. Most of them need to attend an authorized language test (e.g., IELTS, TestDaf) and submit the results when applying to the University. It means they need to spend much time and effort to learn a foreign language such as German or English, which requires great motivation to learn. As a member of several Chinese students' online groups in Berlin, I request formal permission to answer the questionnaire by sending a message with the questionnaire link to these students. Besides, I asked them to introduce their classmates or friends. Students in other German universities are also accepted as long as they have experience in using mobile technology to learn a foreign language in an informal context. The selection of the students is based on a volunteer effort. The informants would be notified of the right to refuse to participate or withdraw consent to participate at any time without reprisal.

Quantitative Method. Since Dörnyei claimed that "motivation is abstract and not directly observable," observation is not an appropriate technique to measure, judge, or document motivation. Thus, the survey adopted the questionnaire to obtain experiences and perceptions of students. Quantitative methods are research techniques used to gather quantitative data dealing with numbers and anything measurable, which involves the use of formal and structured questionnaires. Using a questionnaire has certain advantages. Firstly, 'the knowledge needed is controlled by the questions, and it affords a good deal of precision and clarity [27]. Secondly, the questionnaire is convenient for collecting survey information, especially for numerical data.

5 Research Findings

5.1 The Participants' Background

There are 394 valid questionnaires to be analyzed, including 179 male and 215 female students. Nearly half of them (42.39%) are between the ages of 26 to 29, and 25.13% of respondents are between 22–25 and 18.78% between 30–33. Besides, 61 respondents (15.48%) read language-related subjects such as German language and literature, English language and literature, etc., and 333 participants (84.52%) read non-language related subjects. Regarding their language proficiency, more than half of them have B2 or above level in German (62.19%) and English (74.37%).

5.2 The Preference of Mobile Devices Use

Regarding students' preference of devices used to support foreign language learning, compared with traditional desktop computer-assisted language learning, overwhelmingly 82.23% of participants using notebook computer assisted foreign language learning, including listening, speaking, reading, writing, and vocabulary. And 89.09% of respondents stated that they use smartphones to support foreign language learning, which is the biggest group. Surprisingly, the first choice of foreign language learning for oversea students is smartphones instead of laptops.

5.3 Which Language Skill Does the MALL Affects Students' Motivation

MALL Motivated Different Language Skill Learning. Firstly, for research question 1, most students (73.86%) expressed that they were motivated to learn foreign language vocabulary when doing activities. Besides, 69.29% of students were encouraged to learn listening when doing activities, followed by reading, of approximately 64.47%. Compared with a high proportion of participants obtaining the motivation in language vocabulary, listening, and reading, fewer respondents gain motivation in speaking (38.58%) and writing (29.19%) with the help of MALL.

Different Motivation Intensity of Specific Language Skill Learning Among Different Student Groups. Regarding different study subjects, the Independent T-test shows a significant difference between the proportion of language- and non-language-related subjects about listening activities affecting the learning motivation ($t = 4.350$, $p = 0.0000 < 0.01$). The average of the language-related subject students (4.39) is significantly higher than those of non-language-related majors (3.88). On the contrary, there is no significant difference between the different subject groups about their motivation of speaking, reading, writing and vocabulary learning. Through the sample variance analysis, there was no significant difference between the proportion of gender, age, about whether MALL activities focused on listening, speaking, reading, writing and vocabulary can affect the learning motivation.

5.4 What Tools or Activities Does the MALL Affect Students' Motivation

The research needs to get some detailed insights about what specific learning activities they did in the MALL motivated them to practice their specific language skills. These questions are related to research question 2 'how oversea students perceive MALL can help them practice specific language skills?' and give more insights into how oversea students feel motivated to learn foreign languages in the MALL.

Firstly, over 87% of participants stated that watching foreign language authentic movies/TV programs motivates them to practice their language listening. More than 65% of students proposed that listening to foreign music, and audio clips helps them gain motivation to practice listening. Approximately half of them stated that communicating synchronously (47.21%) (e.g. video chatting) motivates them to practice listening. Some respondents (44.67%) were motivated to practice listening by using social media (e.g. Facebook). Secondly, regarding language speaking, most respondents (64.97%) thought that watching authentic foreign movies or TV series programs motivates them to practice speaking, which is the same with listening motivation. Nearly 60% of participants gain motivation to practice language speaking by communicating synchronously (e.g. video chatting). Besides, more than one-third of respondents were motivated to practice their speaking by utilizing social media.

Thirdly, regarding language reading, nearly four-fifths of participants stated that reading authentic materials (e.g. online newspaper) motivated them to practice reading. More than 65% of students proposed that having access to multimedia reading materials is helpful for them in gaining motivation to practice reading. Approximately half of them stated that the use of electronic dictionaries or translation apps motivates them to practice reading. Almost one-third of respondents were motivated to practice their reading by communicating with people synchronously or asynchronously. Fourthly, regarding language writing, more than half of respondents thought reading authentic materials motivates them to practice writing. Nearly half of them proposed that getting immediate feedback after writing; such as getting corrected by software or getting a response from an online teacher, is helpful for them gaining motivation to practice writing. And more than one-third of respondents were motivated to practice their writing through communicating with people asynchronously and having access to multimedia reading materials. Fifthly, regarding vocabulary learning, nearly four-fifths of participants stated that watching authentic movies, or TV programs motivates them to practice vocabulary. More than 67% of students proposed that reading authentic materials (e.g. online newspaper) is helpful for gaining motivation to practice reading. More than half of them stated that listening to foreign music, podcasts, and authentic audio clips motivates them to practice vocabulary practice.

5.5 To What Extent Do Students Perceive Mobile Devices to Affect Their Motivation for Learning Foreign Language

Generally, most respondents agree that the MALL environment motivated their foreign language learning more or less. To be more specific, they can be motivated to learn based upon the various elements that MALL brings. Most of them (80.46%) felt motivated to learn because they can learn at any time and place. Having a chance to communicate asynchronously with native speakers or other learners (e.g., email) made a large number of participants (77.41%) feel motivated. Communicating synchronously with native speakers or other learners (e.g., chatting by WhatsApp, WeChat) motivated the oversea students (73.35%) to learn. It means communication, both synchronously and asynchronously, brings participant motivation to learn. The majority of respondents (75.38%) felt motivated to learn because they can choose what they need to learn, which is followed by many students (71.58%) think that learning a foreign language is useful in a real context (e.g., when travelling in other countries). Furthermore, more than 70% of students felt motivated to learn because they can get to know more cultures of target language-speaking countries. Also, nearly 70% of participants felt motivated to learn because they could determine their learning objectives and pace. More than two-thirds of respondents felt motivated to learn because they think many valuable websites are written in target languages, even with a mobile version. Some students (62.19%) felt motivated to learn because they can imitate the authentic accent and speak closer and closer to the native speaker. Nearly three-fifths of participants get motivated to learn because the authentic or similar to real-world learning materials could be found via mobile devices, which has the same respondents because of the abundant interesting learning resources to choose.

6 Conclusion and Discussion

By focusing on Chinese oversea university students' perceptions and preferences in the MALL, the research investigated whether and with what tools or activities these students perceive can affect their motivation to learn foreign languages including listening, speaking, reading, writing, and vocabulary in informal settings. The reasons behind the mobile devices motivating language use and the emotions the learners experienced are also examined during the learning activities. The primary purpose of this research was to survey Chinese oversea university students' foreign language learning experience in the MALL, including its motivation for specific language skills, motivational activities. It also tried to help foreign language learning stakeholders and decision-makers determine what mobile devices use were most fruitful to engage and motivate students and learners and help shape future university students, especially oversea university students, mobile technologies initiatives. The research findings reveal a rich and diversified learner-initiated learning experience, the learner's views and concrete evidence in the MALL context. The attitudes of foreign language learning the oversea university students toward utilizing mobile devices were mostly reported as positive, and mobile technologies are becoming an indispensable social, cultural artefact and tool to support language learners' learning.

On the one hand, most university students perceived that they were motivated to learn language vocabulary, listening and reading in the informal MALL settings. In contrast, fewer students felt motivated to learn speaking and writing. The results indicated that the learners overall feel motivated to learn a foreign language in the MALL, which is coherent with researchers' findings. Furthermore, students are more motivated within the input and receptive type learning activities (reading, listening and vocabulary) than output and productive type language learning (speaking and writing). However, there is an interdependence between the language input and the output. The four skills are interconnected to serve the learner's interlanguage development. The input is insufficient for foreign language acquisition, and the Output Hypothesis is also necessary [28]. Language learners are encouraged to produce comprehensible language to notice the 'gaps' in their language knowledge that could urge them to improve accordingly [29]. Since fewer students get motivated for their speaking and writing activities, compared with the number of students who get motivated in listening, reading and vocabulary, the MALL is still not a robust environment for Chinese oversea university students to improve speaking and writing.

On the other hand, the study discussed what activities or tools motivate the learners to improve specific skills in such a digitalization age. Firstly, most students perceived that watching authentic movies or TV motivates them to practice listening, speaking and vocabulary learning. The luxurious linguistic setting offered by authentic video with an engaging and unfolding storyline and embedded target language patterns could create real-life communicative contexts to learn the language and a rich learning context similar to when a child receives authentic input from adult speakers. The intercultural course that integrated foreign authentic video materials could be applied as an instructional tool, and students had real improvement in developing intercultural motivation, attitudes, knowledge, and awareness [30]. Many students get motivated to practice listening and speaking by listening to their chosen and preferred music and podcasts and communicating synchronously that include unique expressions and various voices. Secondly, reading authentic materials motivated learners to practice language reading, writing, and vocabulary. The platforms with social attributes help form the learning community for learners with common or complementary goals and benefit students by motivating them to become involved in authentic projects and read and write for a real audience instead of merely composing for the teacher. The learning process could not just result in promoting reading and writing but a powerful sense of user engagement with multimedia texts. Thus, the various applications and activities motivate learners to develop specific language skills development, which echoes the statement of Motiwalla [31], who suggested two categories of mechanisms in mobile learning, including the push and pull mechanisms. The two mechanisms include several critical functions that could be applied in mobile learning to offer personalized content and support collaborative learning and interaction with peers. However, Motiwalla argued that most learners felt it difficult to read and type because of the screen and keypad size and navigate through pages. Also, some emerging technologies could raise metalinguistic awareness to evaluate word choice and sentence construction by viewing authentic examples in a rich linguistic context. However, traditional drill and practice programs did not provide sufficient authentic multimedia materials and allow enough authentic communication of much value. Instead, a considerable number of students get motivated to practice writing by communicating asynchronously and getting timely feedback through their engagement in more accurate, sophisticated, formal, and longer utterances communicative practice to increase learners' grammatical, lexical, and syntactic competence when compared with synchronous interactions. The research reveals a significant factor that students get motivated by using mobile technologies: the support of personalized learning. The learners' characteristics (including motivation and learning styles and beliefs, strategic knowledge) are taken into account. Thus, the learner's conscious personalized choice regarding what and how to learn could be satisfied. The learning applications developer should focus on improving these primary motivators for supporting specific language skill learning in the MALL. The portability of mobile devices is believed to bring new methods that can shape learning styles to become more individualized and allow learners to learn on the move [32]. However, the most crucial principle is still to understand the positive and negative features of a particular application or activity while deploying appropriate practices to achieve specific learning goals.

Note. Both authors contributed equally to the preparation and writing of this article, and should be considered as co-first authors.

References

1. Williams M, Burden R (1997) Psychology for language teachers: a social constructivist approach. Cambridge University Press, Cambridge pp 111–125
2. Mat SSC, Yunus MM (2014) Attitudes and motivation towards learning English among FELDA school students. Aust J Basic Appl Sci 8(5):1–8
3. Tseng SC, Tsai CC (2010) Taiwan college students' self-efficacy and motivation of learning in online peer assessment environments. Internet High Educ 13(3):164–169
4. Gilakjani AP, Leong L-M, Sabouri NB (2012) A study on the role of motivation in foreign language learning and teaching. Int J Mod Educ Comput Sci 7:9–16
5. Dörnyei Z (2001) Motivational strategies in the language classroom. Cambridge University Press, Cambridge, pp 5–25
6. Keller JM (1983) Motivational design of instruction. In: Reigeluth CM (ed) Instructional design theories and models. Erlbaum, Hillsdale, NJ, pp 386–433
7. Kamaruddin R, Sha'ri SN, Hamdan R (2017) Intrinsic and extrinsic motivation as extralinguistics factors in second language learning among foreign students in five universities of Malaysia. J US China Public Adm 14(1):26–37
8. Burghart I, Stehr K, Ulferts E, Vahl C (2011) Sprechmotivation im fach Deutsch an niederländischen Schulen. Universiteit Utrecht
9. Kea H (2008) How are NASA engineers motivated? An analysis of factors that influence NASA Goddard engineers' level of motivation. Ph.D. dissertation in Antioch University. https://etd.ohiolink.edu/
10. Gardner RC (2005) Integrative motivation and second language acquisition. Canadian association of applied linguistics/Canadian linguistics association joint plenary talk. The University of Western Ontario, London, Canada
11. Ames C, Ames R (1989) Research in motivation in education. Academic Press, San Diego
12. Söyler S, Gökalp Y, Şahin SK, Yiğit P (2019) Effect of internal & external motivation and burnout levels of nurses on work performances: cross-sectional analysis. Int J Bus Ecosyst Strategy 1(1):63–69
13. Krashen SD (1988) Second language acquisition and second language learning. Prentice-Hall International
14. Chun DM (2016) The role of technology in SLA research. Lang Learn Technol 20(2):98–115
15. Warschauer M (2004) Technological change and the future of CALL. In: Fotos S, Brown C (eds) Language learning and technology, vol 19, no 2, pp 156–176
16. Kramsch C, Whiteside A (2008) Language ecology in multilingual setting: towards a theory of symbolic competence. Appl Linguist 29(4):645–671

17. Kern R (2015) Language, literacy, and technology. Cambridge University Press, Cambridge
18. Chun DM (2019) Current and future directions in TELL. Educ Technol Soc 22(2):14–25
19. Kramsch C (2011) The symbolic dimensions of the intercultural. Lang Teach 33(3):354–367
20. Yang S (2012) Exploring college students' attitudes and self-efficacy of mobile learning. Turk Online J Educ Technol 11(4):148–154
21. Qi C (2019) A double-edged sword? Exploring the impact of students' academic usage of mobile devices on technostress and academic performance. Behav Inf Technol 38(12):1337–1354
22. Ma Q (2017) A multi-case study of university students' language-learning experience mediated by mobile technologies: a socio-cultural perspective. Comput Assist Lang Learn 30(3–4):183–203
23. Chai CS, Wong L-H, King RB (2016) Surveying and modeling students' motivation and learning strategies for mobile-assisted seamless Chinese language learning. Educ Technol Soc 19(3):170–180
24. Shakya A, Horsfall JM (2008) ESL undergraduate nursing students in Australia: some experience. Nurs Health Sci 2(3):163–171
25. Dekeyser R (2010) Monitoring processes in Spanish as a second language during a study abroad program. Foreign Langu Ann 43(1):80–92
26. Kinginger C (2008) Language learning in study abroad: case studies of Americans in France. In: Modern language journal monograph, vol 1. Blackwell, Oxford
27. McDonough J, McDonough S (1997) Research methods for English. Language teachers. Arnold, New York
28. Swain M (2005) The output hypothesis: theory and research. In: Hinkel E (ed) Handbook of research in second language teaching and learning. Routledge, London, pp 471–483
29. Rhalmi M (2019) Input and output in second language acquisition. My English pages. https://www.myenglishpages.com/blog/input-and-output-in-second-language-acquisition/. Accessed 07 Apr 2020
30. Chao TC (2013) A diary study of university EFL learners' intercultural learning through foreign films. Lang Cult Curric 26(3):247–265
31. Motiwalla LF (2007) Mobile learning: a framework and evaluation. Comput Educ 49(3):581–596
32. Ros i Solé C, Calic J, Neijmann D (2010) A social and self-reflective approach to MALL. ReCALL 22(01):39–52

Billing and Booking System for Distributed IoT Laboratories

Davide Reverberi[1] (✉), Valentin Kammerlohr[2,3], Giovanni Esposito[1], Ankita Shankar Ghorpade[2], Giovanni Romagnoli[1], and Dieter Uckelmann[2]

[1] University of Parma, Parma, Italy
{davide.reverberi,giovanni.esposito,giovanni.romagnoli}@unipr.it
[2] Hochschule fur Technik Stuttgart, Stuttgart, Germany
{valentin.kammerlohr,dieter.uckelmann}@hft-stuttgart.de
[3] Auburn University, Auburn, USA

Abstract. In the last decades, the sector of non-traditional labs (NTLs) has been really active in creating new initiatives. Also, Everything-as-a-Service (XaaS) is one of the key concepts, adapted from software business models, which is moving the product-based revenue models to service-based revenue models, characterized by the sale of fee-based services and hence periodically recurring revenues. Having these two concepts, the implementation of a billing and a booking system within a NTLs network is becoming a priority. This research has been carried out within a study of the cross-institutional project DigiLab4U, which aims at the creation of a network of NTLs. The idea behind is to combine network of laboratories and XaaS in order to make the laboratories available to stakeholders who can use the available technologies for their own purposes. To do so, a booking system is required, and a platform is required to bill the laboratory usage according with the DigiLab4U policy. The combination of these two points, lead to the exigency to select the most suitable solutions for both booking and billing systems accordingly with the requirements of the DigiLab4U network. Starting from a literature research, solutions on the market and requirements for selections have been identified. The analysis has been limited to ten solutions for the billing and four software for the booking. The requirement analysis led to two software fitting the DigiLab4U needs. These platforms have been implemented and tested with specific use cases of the DigiLab4U according to technical and organizational perspectives.

Keywords: Booking & billing · Non-traditional laboratories · IoT · XaaS · Everything-as-a-Service

1 Introduction

Digitalization is changing the business world because of even more increasing connected devices via IoT technologies [1, 2], which are leading to business model innovation for reinventing the way companies create and capture value for their customers [3–5]. Everything as a Service (XaaS) is one of the key concepts, adapted from software business models, which is moving the product-based revenue models who are based on one-off

M. E. Auer et al. (Eds.): REV 2022, LNNS 524, pp. 395–405, 2023.
https://doi.org/10.1007/978-3-031-17091-1_40

payments to service-based revenue models. The service-based revenue models are characterized by the sale of fee-based services under which a periodically recurring revenues are realized [2]. Here, Wortmann has demonstrated that this revenue models also ensure profitable cumulative cash flow on the long period. This disruptive business approach can involve players otherwise traditionally out of business, because of different (i) mission, (ii) positioning on the market, and a (iii) traditional inability to reach potential customers. In particular, Esposito *et al.* [6] have proved that business possibilities are disclosing for Universities and Research Institutes, through digitalization and networking of their laboratories (labs), through the provisioning of lab access and usage by remote. As a consequence, the present paper wants to discover whether is actual feasible to develop a stable system for using labs by remote and to generate revenue by the provisioning of this service. In doing it, two research question (RQ) have been set as follows: (**RQ1**) "Which requirements needs to be full field by modern billing and booking system?", and (**RQ2**) "How can a billing and booking system be integrated into a shared resource environment?". The cross-institutional project DigiLab4U acts as a case study to test the developed solutions. DigiLab4U pursues the objective to provide users within the IoT-environments with practical, digitalized and networked labs independently of their location. Thereby labs are functioning as IoT device and users as customers.

The reminder of the paper is as follows. In Sect. 2 a literature review regarding billing and booking systems is conducted. A brief description of the DigiLab4U case study follows to analyze possible pricing mechanisms of IoT applications, and then a requirements analysis is conducted as a result of the literature review. Section 4 introduces the prototypical implementation of the system, and then Sect. 5 summarizes the findings. Finally, Sect. 6 address conclusions.

2 Literature Research for Billing and Booking Systems

This literature review has been conducted by querying different databases until December 2019. First of all, the Hochschule für Technik Stuttgart (HFT) and the University of Stuttgart libraries, providing a good access to many physical books, EBooks and Journals. Moreover, IEEE database, and Google-Scholar, among the largest databases that hold millions of articles from different fields and allows to search more specifically of scholarly literature across different sources, including books, journals, articles, and conferences. The review has followed a three-stage process, according to Tranfield's [7]. We have previously analyzed the title of 187 documents, retrieved using combinations of keywords as (i) 'payment' or 'billing' or 'pricing' systems, and (ii) 'remote' or 'virtual' labs and Industry 4.0 or Internet of Things and relative forms (e.g., IoT). We then selected the 58 documents that met the following inclusion criteria: (i) source available, (ii) year of publication from 2015 (for meeting the Industry 4.0 perspectives), and especially (iii) relevance of the topic. Next, we have selected 12 documents skimming through the abstract, according to the previous inclusion criteria and the language of the document in English. Finally, we have reviewed the papers, and the important information related to booking and payment system is provided in the next sub-section.

2.1 Analysis of the 12 Full Papers and Outlooks for Requirements

The highlights of the analyzed documents are presented in *italics* in the next.

Costas-Perez, Lago, Farina, and Rodriguez-Andina [8] described solution designed to be used by both *non-experienced students,* who act as plain users provided with predesigned library modules, and *advanced students* who can demonstrate the design skills they have learnt by developing their own LabVIEW applications. If the experiment is free, student is given exclusive use of it during *predefined period of time*, which depends on the *complexity* of the experiment. Once the assigned time has finished, student must *stop working* on that experiment to allow others to use those resources. Web based system shows the user the list of already confirmed reservations and user can choose one of the available *time slots.* Apart from students, professors have same access rights as advanced students that they can define *access restriction, modify* and *cancel queues* or even cancel current works and provide online documentation. They can access the *stored data* related to student's experiments and corresponding reports for evaluation purpose.

Uran, Hercog, and Jezernik [9] represents Moodle as booking system used to reserve remote experiments in remote web lab. In the original booking system, experiment administrator must provide *the booking name, booking description, the maximum number of students that can work on experiment at the same time, experiment url address and the time interval when the booking link will be available during booking creation process.* By selecting the booking link a 7-day calendar pop ups. The calendar page displays the local time (at user's time zone) which is different from server time. This *time conversion* takes place automatically, so that remote users can see free or booked time slots according to their local time zones.

Spilakova and Schauer [10] described RemLabNet, a remote lab management system, and its new booking module for experiment. Module is created which consist of calendar, tables, and elements for making the reservation. Remote user is allowed to view daily, weekly, monthly calendar. Reservations are displayed by coloring the respective time slots with short information inside this element. Booking data is represented in the form of a table. This table includes *experiment ID, experiment name, description, username, duration, start date, end date respectively.* User can *search* the entry dynamically. User can enter start date, end date, *choose the experiment* from drop down and add a *description* to book the particular experiment. User checks the availability of the experiment and book it by filling the fields in booking form. If user leaves any field empty or tries to book reserved experiment, the user will be *notified,* and reservation will not be performed. Once confirmed the booking data is sent to database and booking system which displays in calendar and table. Management tool for deleting or modifying the booking is also found in the booking module. User can book any available experiment using this module in RemLabNet.

Hercog, Gergic, Uran, and Jezernik [11] presented a framework for rapid remote experiment implementation in the field of automatic control. In Digital signal processing based remote control lab, each experiment contains experiment link to book it. After selecting this link and login process, a new web page with booking table displays. User can then book experiment for 1 h and when booked time slot becomes current time, *user gains a valid link to the remote experiment page.*

Restivo, Mendes, Lopes, Silva, and Chouzal [2] mentioned the 6 remote experiments offered by the remote lab are accessed through web pages. The presented booking system has been developed as an extension of the Moodle platform. It provides 1 h time slots. If user does not arrive within given slot, Moodle extension will *free the previously reserved time slot*. Moodle brings additional advantage of storing tutorials on subjects, as short experiment description and how to use it through booking application. For evaluation, user is asked to post answers to a few questions correlated with experiment. These answers are used as report later accessed by teachers. At the end, this survey has been sent to all students in order to evaluate the impact of use of remote labs.

Gardel Vicente, Munoz, Galilea, and del Toro [12] presented a remote lab for industrial automation for different programmable logic controller (PLC) manufactures. The lab architecture makes up the remote automation laboratory (RAL) and it depends on Moodle server master PC (MPC) which manages the booking system of PLC benches in the lab. It implements scripts and plugins into Moodle for booking PLCs and configuring the local devices appropriately. The server manages the access to the Virtual machine cluster taking into account booking system on Moodle- meeting room booking system (MRBS).

Salzmann, Govaerts, Halimi, and Gille. [13] described the smart device specification for remote lab which enable interoperability with client applications and external devices such as booking service. The lab repository, which is a portal where teachers can search online lab and resources to use in their courses. Booking systems provides *common user interface* for teachers to reserve remote lab. In addition to enable user interaction with remote lab equipment, the smart device uses existing mechanism for booking a lab. When booking is needed, user retrieves an *authentication token* from booking system which she can authenticate to smart device. Smart device has authentication component that validates tokens with booking system. In Go Lab booking system, token-based authorization is used which can be modeled with swaggers API key type as the booking token is a sort of temporary API key for the duration of the booking. When remote user books a lab, Go Lab booking system provides an authentication token. At the reserved time the user can connect to smart device with this token. Smart device then contacts booking system to validate to check if user is allowed to access smart device. Thus, integrating the booking service in the Smart Device requires small effort compared to providing its own authentication and booking mechanisms.

Lowe, and Orou [14] presented scheduling schemes for managing access to limited Remote Access Laboratories (RAL) resources, which supports booking or queuing. RAL software system has come to be referred as Sahara system supports two scheduling methods to handle remote user requests: queuing and booking. *Priority queuing* supports on demand requests, providing soonest available access. Another is *calendar booking* which allows users to make reservation that provides guaranteed access to experiments at specified time. Experiment page appears with available time slots that can be booked and showing maximum allowed reservation time for selected experiment. Once time slot is selected then reservation is created, a *confirmation message* displays and *email* is sent to user with reservation details. The above case study highlights an unanticipated consequence of the interplay between booking and queuing schemes. It demonstrates that considerable care must be taken in the design of scheduling algorithms if we are to optimize performance measures of RAL systems, including aspects such as overall level of utilization and queue waiting times. Movable *booking times* involves modifying bookings so that they represent an approximate time rather than an absolute time for example like a booking to see a doctor where you arrive at the booked time and may need to wait a short period. Consider there are 1 h booking at 3 pm and 4 pm and one user waiting in the queue. User with 3 pm booking does not log in and so the session times out at 3:10 pm. This leaves the 50 min till next booking and the queued user will not be allocated. If we can move the 4 pm booking back to 4.10 pm then the queued user can be allocated. Another option is when *shorter time window becomes available* the users in the queue are asked whether they would like to accept this shorter period. Sometimes they may know that they are able to complete the experiment quickly and would be willing to accept the shorter time slot in preference to waiting for a slot that guarantees their maximum time allocation.

Luís Gomes and Javier García-Zubía [15] provided overview on several aspects of remote labs development and usage and their impact on teaching and learning process. Access server prevents unauthorized users from accessing the experiment by *login and password*. This also contains booking system that manages access from different students. The evaluation of remote lab has been divided in 3 phases. Pre-lab phase consist of activities needs to be performed by student before each experiment such as tutorials to train students to use instrumentation, background information of experiment. Central phase consists of execution of experiment. Post-phase stabilizes and confirm student's learning with assessment activities. Distributed lab allows equipment sharing consist of lab server where register of all lab resources is stored in database. Authorized teachers can log on to lab server and book lab session via web interface. They also set number of seats in each session with start and end date of the course, list the emails of enrolled students by specifying identification of university LMS. To perform the experiment, student log on to the courses and select the lab session. LMS then sends student email address and *list of equipment* to be used to lab server. Address of this server, manipulator, and other devices are entered into lab client software module. This module is downloaded on student's machine and student then can start experimenting.

El-Ghareeb, H. [16] presented e-Learning ecosystem evolution. Third platform components and how they fit into e-Learning system is presented in detail with example of applied research. Integrate cloud based virtual lab with LMS to provide an interactive virtual training environment for ICT training. Utilize the cloud computing 'Pay as you go' business model so users can only *pay for the time they use virtual* lab, thus providing cost effective access to large number of computing resources. Partner schools can integrate with Coursera using *shibboleth* or SAML2 to secure their private on campus courses.

Stich, Schumann, Beverungen, Gudergan, and Jussen [17] described that subscriptions in combination with *pay-per-use* can also be offered to customers. Depending on their complexity, the modular services can be billed on a monthly basis or be completely independent of usage. Simple tools can be provided by a one-time payment. The prices should be transparent for the customer, especially when using a store. Furthermore, it makes sense *to offer free services to customers to get started with the platform*. These make it easier for customers to access the platform, so that a larger circle of customers can be addressed and show through the ease of use what added value can be created. After a certain period of time, these free services can be cancelled or are converted into paid long-term services.

Khali, Hasna, Benammar, Chaabane, and Amar [18] presented remote lab for electrical engineering experiments. Booking system is implemented to avoid access conflict and help the student to organize their time and activities. The management room booking system (*MRBS*) is used to provide access to remote user. Small modification in MRBS php file redirects the user to the experiment after checking the user details against identity and time slots schedule stored in Moodle database. The experiment web page is viewed and accessed for the duration fixed by Moodle administrator after validation.

Six main requirements have come out of the review: (i) open authentication, (ii) billing system, (iii) platform integration, (iv) notification functionalities, (v) open source, and (vi) booking system. These requirements have been adopted as evaluation criteria for the different solutions compared in the next chapter.

3 Case Study DigiLab4U: Billing and Booking Within Shared Laboratory Resources

The cross-institutional research project DigiLab4U intends to offer digitized lab environment that enables a real IoT learning marketplace. The goal is to develop an integrated, hybrid learning and research environment consisting of a large variety of lab technologies for a digital offering that can be used by any kind of user, especially students. The DigiLab4U learning marketplace consists of both digital labs from several suppliers, and users who access to the labs. The cooperation of universities, research institutions and industry allow to pool the suppliers so that users have access to a larger variety of digital courses based on different IoT labs. Essential steps in the project are the integration of the labs in an automated booking and billing system that allocates digital lab slots, logs and records usage, and can generate corresponding invoices. The IoT service to be implemented refers to digitalization and connection features arranged in Romagnoli *et al.* [1]. More in detail, the general architecture of the DigiLab4U [19, 20], is showed in Fig. 1 for highlighting the access to the resources (local/remote), the nature of labs (real/virtual), and the possible cross-institutional nature of organizations (i.e., one or multiple).

The users are considered the focal point of the network, and the booking and billing system are the tools which link the user with the labs.

3.1 System Requirement of Billing and Booking Applied to Market Solutions

Market solutions for billing and booking software have been carried out, and 14 possible software and platforms for billing and booking have been retrieved. The evaluation of the software solutions is based on their core features reflecting the six requirements of the literature review, declined by the DigiLab4U consortium as follows:

1. *Open Authentication:* User validation (reliable), open API, third party software integrability (e.g., Shibboleth).

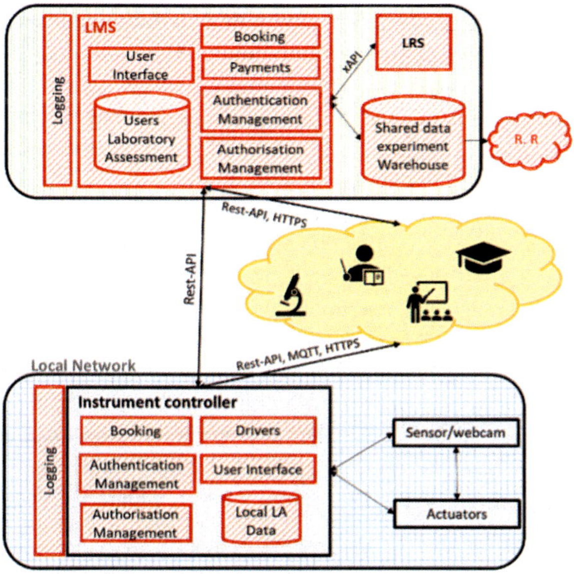

Fig. 1. DigiLab4U general architecture [20]

2. *Integrated billing system:* billing functionalities, pay-per-use pricing model, multi-currency support, accounting and invoicing.
3. *Easy platform integration:* Learning Management System Integration (i.e., Moodle), common user interface, integration support for booking and billing systems.
4. *Notification functionalities:* notification of bookings, lab information, used lab time, payment information via email.
5. *Open Source:* Cost free use of system.
6. *Integrated booking system:* predefined period of time for user to access experiments and recurring interval, calendar export, modifiable bookings, time conversion for different countries.

The degree of fulfilment of each software with requirements is measured on a scale from 0 (not available) to 5 (very good), and a focus group of 4 experts ranked solutions. The result of that analysis is provided in Table 1. Scores have been given as collective score by the focus group. All examined solutions have their advantages and disadvantages. First of all, it must be said that none of the systems cover both the pricing function and the booking function, and hence the integration requirement is missing. This is probably due to the fact that these are often considered separate solutions. All in all, 10 billing solutions have been analyzed: (1) NEON, (2) ANC CO, (3) Wave, (4) Manager, (5) Akaunting, (6) Invoice Plane, (7) jBilling Community, (8) jBilling Enterprise (9) Kill Bill, (10) BoxBilling; and 4 booking solutions: (I) Tutorial Booking, (II) EJS App, (III) Bookkit, (IV) Agendo. As shown in Table 1, 3 billing solutions scored more than 10 out of 30 points. A main reason is the missing possibility of (i) open authentication like Shibboleth and (ii) multi role support, and the missing notification functionalities for the user and the provider of the request. The Akaunting solution is an open-source solution which furthermore has no platform integration. The NEON software has a well possibility of platform integrations, and a highly flexible pricing mechanism support like usage dependent pricing and multicurrency, it is not an open-source solution. The billing systems with highest scores are BoxBilling and jBilling with community/enterprise version. Though Boxbilling provides most of the needed functionality, but it is not an active solution in the market. Hence, it is excluded from the selection process. jBilling community version is an open-source system with no integration support. However, jBilling enterprise version provides integration support at some defined cost. Similarly, out of 4 booking systems found in the research, the Tutorial booking and EJS App are Moodle plugins. Whereas Bookkit and Agendo are independent booking systems. The EJSApp booking system is removed out of the further research due to lack of requirements matched. The system does not provide any support for integration of other systems such as billing system and not allows accessing secured functionality in open-source version. Bookkit is one best solution which matches all the requirement but with no integration support. However, Agendo system is one of highest scored system but there is large cost structure applied to access the functionality. Hence, the Tutorial booking plugin has been selected.

Table 1. Requirements analysis for Billing (1–10) and Booking (I–IV) solutions

Requirements	Billing										Booking			
	(1)	(2)	(3)	(4)	(5)	(6)	(7)	(8)	(9)	(10)	(I)	(II)	(III)	(IV)
Open autentication	2	0	0	0	0	2	2	2	0	0	4	4	1	5
Billing system	5	5	5	5	5	5	5	5	5	5	0	0	3	3
Platform integration	4	3	0	0	0	0	0	5	0	2	4	4	0	5
Notification functionalities	0	0	0	0	0	0	0	0	0	0	5	5	5	5
Open source	0	0	2	5	5	5	5	0	5	5	5	3	5	0
Booking sytem	0	0	0	0	0	0	0	0	0	0	5	5	5	5
Total	**10**	**8**	**7**	**10**	**10**	**10**	**12**	**12**	**10**	**12**	**23**	**21**	**19**	**23**

Both selected systems differ fundamentally, since the payment solution is the central focus for jBilling and the booking solution is for Tutorial Booking. The two applications have both undergone to an implementation and evaluation procedure in order to define their effectiveness. The outcomes are discussed in the next section.

4 The Billing and Booking Implementation in the DigiLab4U

Beyond technical aspects of software installation and architecture, which does not concern this paper, the system implementation in both cases has been carried out referring to 3 user stories for defining possible use cases of the systems and hence set the relative workflows:

The first user story relates to professors and lab managers, who manage the course, the lab utilization, and arrangement of content (documentation, equipment) for performing the experiments. The user story has been designed as a professor in information logistics at the Stuttgart University of Applied Sciences. This user offers lectures on Robotics, RFID in Logistics and Information Logistic Processes. For the RFID in Logistics lecture, the course has a special RFID measurement chamber and a laboratory set up specifically for this chamber. In this laboratory, exercises are offered to the students to accompany the lectures and provide them with practical knowledge. Through the DigiLab4U project of the Stuttgart University of Applied Sciences, this laboratory is now also available online and can be operated remotely. The use would like the students from Information Logistics to complete the laboratory in a fixed timespan. For this purpose, the user creates corresponding timeslots in the Moodle Scheduler for the students, which then can be booked by them. If, for whatever reason, he does not agree with one of the student's appointments, he can revoke it. The user also cares about service provisioning to people from outside the university. The user is finally in charge for invoicing customers for service utilization.

The second user story relates to researchers who access and use the digital online labs or deal with supporting the professors when managing the courses and lab accesses. The user story has been designed as a researcher working on RFID technology. The user registered on DigiLab4U platform (on Moodle) and needs to book an exercise in the RFID laboratory in a date that suits relative business commitments. A fee for lab utilization is due or not on the basis of the institution the user belongs and the institution owning the lab needed.

The third user story relates to students and people external to the organization providing the digital online labs, who want to access and to use the provided content. The user story has been designed as a visiting student, so to cope with both role of student and guest. The user is a visiting student of information logistics. Therefore, he is taking part in the exercise offered by the professor in the RFID laboratory this semester. The user books the lab on the basis of its availability and must pay a charge according to institutions relationships and agreement.

Matching all these 3 user stories, possible scenarios to face for both booking and billing practices have been simulated, from routine scenarios to system breakdown.

5 Results and Discussion

The implementation of the systems referring to the user stories and the possible scenarios when they use the platform, has followed a Deming cycle "Plan-Do-Check-Act" approach. The technical installation has been verified up to the convergence of the system with stability. From organizational perspective, namely the effectiveness of the system as set, has been discussed in a DigiLab4U workshop attended by a panel of 8 experts, and the implementation of the system on the basis of the user stories has been successfully validated. Only three user stories have been set since limits in the number of user typologies for some platforms. Furthermore, the professor user story as also lab manager is a forcing of the design, since lab managers usually are non-teacher staff only caring about management and setting of equipment and other material. Similarly, professors do not directly deal with payment procedures. However, this 'super user' can be played by professors, or otherwise can be accessed by staff of the institutions dealing with their business.

As a result, the booking part enables the possibility for all the users to schedule an experiment, or experience, in a laboratory of the network by considering all the constraints linked to its usage. The billing system enables a charging system which could be based both on the hours of laboratory usage, especially for the partners within the network, or on a monetary fee based on the policy applied.

The system as set, has been then tested in a 'simulated' environment, in which 6 users, from different organization in the network from different geographic are, played each one a role in the booking and billing systems coping with all the user stories. The 'simulation' achieved positive results, and the players appreciated the easiness of use of the platforms, and the implementation of the possible scenarios within the systems. The only gap highlighted has been the missing integration of the two software, but this limit was discussed and accepted during the software selection phase.

6 Conclusions

In this paper a requirement analysis has been carried out for defining booking and billing software and platforms suitable for digital online lab consortium and their transition towards Internet of Things systems of Industry 4.0. A practical use case of the DigiLab4U has been proposed, and a new approach in which the viability of the solutions has been proved by means of suitable user stories. The prototype of the implementation has been verified and validated and is now approaching the Go Live phase.

The study has some limitations. Firstly, software and platforms considered has been not evaluated on a basis of rigorous benchmarking or gap analysis. Moreover, a-priori ranking and the further verification of the degree of consistency of judgements would have been more robust for evaluating the software, than the collective score by experts' focus group. However, due to the simplicity of the use case, a leaner approach has been adopted. Secondly, no stress test has been carried out for testing user stories, hence the verification step does not stick to huge number of scenarios that certainly guarantee for the convergence of the system with its effectiveness. Finally, the systems work in prototypical environment, and needs confirmation from the official released version.

Acknowledgements. This research has been carried out within the 42-month DigiLab4U project, funded by the German Federal Ministry of Education and Research (BMBF), research grant FKZ 16DHB2112.

References

1. Romagnoli G, Esposito G, Rizzi A, Zammori F, Bertolini M, Uckelmann D (2020) Lab networks in engineering education: a proposed structure for organizing information. Int J Online Biomed Eng 16(5): 41–70. https://doi.org/10.3991/IJOE.V16I05.11891

2. Restivo MT, Mendes J, Lopes AM, Silva CM, Chouzal F (2009) A remote laboratory in engineering measurement. IEEE Trans Ind Electron 56(12):4836–4843. https://doi.org/10.1109/TIE.2008.2011479
3. Henry J (2002) Using the modern chemical engineering laboratory at a distance. In: ASEE annual conference proceedings, pp 841–849
4. Marincean S, Scribner SL (2020) Remote organic chemistry laboratories at university of Michigan-dearborn. J Chem Educ 97(9):3074–3078. https://doi.org/10.1021/acs.jchemed.0c00812
5. Gitelman DR (2002) ILAB: a program for postexperimental eye movement analysis. Behav Res Methods Instrum Comput 34(4):605–612. https://doi.org/10.3758/BF03195488
6. Esposito G, Kammerlohr V, Reverberi D, Rizzi A, Romagnoli G, Bisaschi F (2021) Business model validation for a marketplace of lab network initiatives. In: Proceedings of the 26th summer school "Francesco Turco"
7. Zutin DG, Auer M (February 2014) A simple LabVIEW based framework to facilitate the deployment of iLab batch lab servers. In: Proceedings of 2014 11th international conference on remote engineering and virtual instrumentation, REV 2014, pp 328–331. https://doi.org/10.1109/REV.2014.6784182
8. Costas-Perez L, Lago D, Fariña J, Rodriguez-Andina JJ (2008) Optimization of an industrial sensor and data acquisition laboratory through time sharing and remote access. IEEE Trans Ind Electron 55(6):2397–2404. https://doi.org/10.1109/TIE.2008.921687
9. Uran S, Hercog D, Jezernik K (2007) Remote control laboratory with moodle booking system
10. Spilakova P, Schauer, F (2015) Remote laboratory management system remlabnet and its booking system. In: 2015 4th international conference one-technologies and networks for development ICeND 2015, pp 80–84. https://doi.org/10.1109/ICeND.2015.7328544
11. Hercog D, Gergic B, Uran S, Jezernik K (2007) A DSP-based remote control laboratory. IEEE Trans Ind Electron 54(6):3057–3068
12. Vicente AG, Muñoz IB, Galilea JLL, del Toro PAR (2010) Remote automation laboratory using a cluster of virtual machines. IEEE Trans Ind Electron 57(10):3276–3283
13. Salzmann C, Govaerts S, Halimi W, Gillet D (2015) The smart device specification for remote labs. In: Proceedings of 2015 12th international conference on remote engineering and virtual instrumentation (REV), pp 199–208
14. Lowe D, Orou N (2012) Interdependence of booking and queuing in remote laboratory scheduling. In: 2012 9th international conference on remote engineering and virtual instrumentation, REV 2012, pp 1–6. https://doi.org/10.1109/REV.2012.6293100
15. Kazmierkowski M, Liserre M (2008) Advances on remote laboratories and e-learning experiences (Gomes, L. and Garcia-Zubia, J., Eds.) [Book News], vol 2, no 2
16. El-ghareeb HA: Third platform and e-learning ecosystem
17. Stich V, Schumann JH, Beverungen D, Gudergan G, Jussen P (2019) Digitale Dienstleistungsinnovationen. Smart services agil und kundenorientiert entwickeln. Springer, Berlin. https://doi.org/10.1007/978-3-662-59517-6
18. Khali A, Hasna M, Benammar M, Chaabane M, Ben Amar C (2009) Development of a remote lab for electrical engineering program. In: Proceedings of the - ICELIE 2009, 3rd IEEE international conference on e-learning in industrial electronics, pp 40–44. https://doi.org/10.1109/ICELIE.2009.5413214
19. Kammerlohr V, Pfeiffer A, Uckelmann D (2020) Digital laboratories for educating the IoT-generation - heatmap for digital lab competences. In: Online engineering & internet of things: proceedings of the 16th international conference on remote engineering and virtual instrumentation REV 2020, pp 11–27
20. Galli M, Mezzogori D, Reverberi D, Uckelmann D, Ustenko M, Volpi A (2020) Digilab4u: general architecture for a network of labs

Hotel Process Simulator as a Tool for Increasing the Level of Managerial Skills – An Educational Application

Petra Poulova[1](✉), Martin Kocourek[1], Filip Maly[1], Tomas Kozel[1], Pavel Kriz[1], Blanka Klimova[1], Jan Han[2], and Zdenek Ulrych[3]

[1] Faculty of Informatics and Management, University of Hradec Kralove, Rokitanskeho 62, 500 03 Hradec Kralove, Czech Republic
{petra.poulova,martin.kocourek,filip.maly,tomas.kozel, pavel.kriz,blanka.klimova}@uhk.cz

[2] Institute of Hospitality Management in Prague, Svídnická, 181 00 Prague, Czech Republic
han@vsh.cz

[3] University of West Bohemia, Univerzitni 8, 306 14 Pilsen, Czech Republic
ulrychz@kpv.zcu.cz

Abstract. Simulators are a standard part of training for a variety of jobs. There is a rising emphasis on teachers in passing their knowledge and experience on students in order to get students more engaged into the process of education. Hotel Process Simulator prepared in frame of project Innovation of Tourism Management Systems Using the Process Management Tools and their possibilities will be described in this work. The aim of this article is to describe the possibilities of simulation in the training of hotel staff.

Keywords: Application · Blended learning · e-learning · Education · Game · Process · Simulation

1 Introduction

Computer games and simulations are used at all levels of education. Most of the educational games try to introduce students to new knowledge and help them acquire and improve their skills and abilities. [1].

Perkins [2] defined the simulations as "imitating any real thing, state or process".

A simulation game is usually referred to as a technology that enables a user to interact with a simulated environment. By using simulation technologies, it is possible to create the illusion of the real world. It is a visual, auditory, tactile or other experience creating a subjective impression of reality using computer imaging equipment.

Simulators are a standard part of training for a variety of jobs. There is a rising emphasis on teachers in passing their knowledge and experience on students in order to get students more engaged into the process of education. Current students expect more interaction and communication in the online world [3].

M. E. Auer et al. (Eds.): REV 2022, LNNS 524, pp. 406–412, 2023.
https://doi.org/10.1007/978-3-031-17091-1_41

2 Use of the Simulation Games in Hotel Management Education

As in many other areas, the simulators can also be used in the field of hotel management education. There are currently several applications on the market that can be used in hotel management training.

Computer hotel simulations were generally designed for college students in the hotel industry. The simulation should provide managerial experience in accounting, problem solving and decision making.

Significant development of computer systems that help manage production and logistics in restaurants and hotels could be seen during the last decade. It is essential that potential hotel managers were informed about the principles and rules involved in the development and use of computer applications, both in accommodation and food sectors of the hotel industry.

An important area is also Revenue Management.

2.1 Revenue Management

Economic pressure on the bottom line, ever-increasing costs, especially personnel costs and other influences are forcing hoteliers to continuously increase revenues. In today's globalized world, sales strategy is the alpha and omega of success for almost every company. In the hotel industry, this is certainly true as well.

Revenue management is a management tool that is used by hoteliers to achieve the desired results, particularly in terms of occupancy and the achieved price per available room. This tool is based on the analysis of the amount of data available to the hotel. It is used not only in the hotel industry, but also to a large extent in other service sectors that are characterized by fixed capacity. The main idea is to sell identical services at different prices according to the predetermined criteria [4].

Revenue management is a comprehensive and analytical view of management that not only uses disciplines, such as statistics, mathematics and marketing, but also uses technology. Its implementation and successful use are made possible mainly by a significant change in the understanding of revenue management by business managers themselves. This is the prioritization of profitability over occupancy [5].

The existing research on the importance of Revenue management in the hotel industry has mainly focused on the potential of the Revenue management tool and its effectiveness from several perspectives.

1. the influence of the size of the accommodation facility and its accommodation capacity [6] on the potential of the Revenue management is tool and its effectiveness
2. the impact of the level of the accommodation facility as declared by the number of stars [7]. Alternatively,
3. whether the level of the Revenue management tool is affected by whether or not the hotel is integrated into a hotel chain [8].

Demand forecasting is another important area of Revenue management application capability issues. Demand forecasting is at the very beginning of the Revenue management process and is a key [9] in terms of the whole follow-up process concerning the very basis of Revenue management which is: to offer the right price, to the right customer, at the right time.

2.2 Modelling and Simulation of Processes in the Hotel and Catering Industry

The prerequisite for effective process management is a sufficient level of understanding and setting of individual processes of the organization and the system as a whole. The knowledge of individual processes is then the basis for their possible innovation and optimization, i.e., a positive development of the organization and increasing the level of its competitiveness. There are many ways of understanding the processes and behavior of sub-enterprise systems today. We will focus mainly on process modelling and simulation.

Not only does process modelling allow for an increased level of process understanding, it is also a powerful tool for corporate communication and education. The simulation is an even more effective method of understanding the properties of individual processes and the behavior of sub-systems. It is based on the existence of established process models and its outputs are a powerful source of data for further decision-making on changes in the setup of sub-business processes and systems.

During the simulation it is possible to observe under variable conditions the following:

- time requirements of individual processes and sub-activities (including delays),
- numbers of units served and not served (guests, rooms, etc.),
- the level of use of individual resources (especially staff) and facilities,
- queue lengths, etc. [9]

However, the simulation is not only a method used for innovation and optimization of business processes and systems, it has also been well received in training and human resource development. Using simulation games representing the behavior of selected business systems, markets, etc., it is possible to learn a range of managerial knowledge and skills in a very effective way.

In the hotel and catering industry, this is still in its infancy. While it is already possible to take parts of courses using simulation games at some hotel schools and other educational institutions around the world, optimizing business processes in practice using simulation experiments is definitely not yet that common. Modelling and simulation prove to be one of the effective ways of improving the quality and efficiency of the provided accommodation and catering services, and thus increasing guest satisfaction.

3 PROTUR Hotel Simulator

The application PROTUR is developed by the authors´ team for training students of the hotel school and for training hotel staff.

The following figure schematically demonstrates the frame algorithm simulator (Fig. 1).

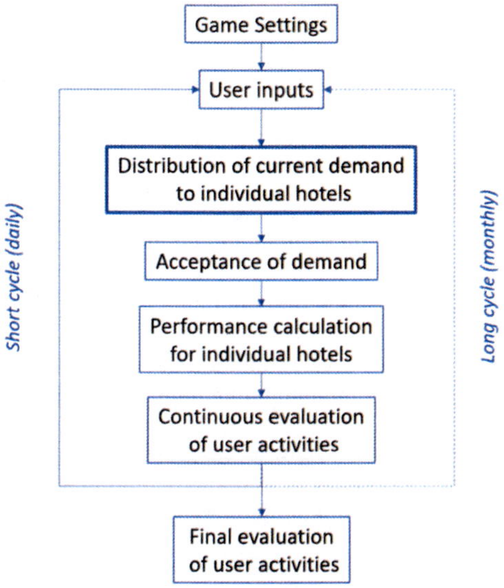

Fig. 1. Frame algorithm.

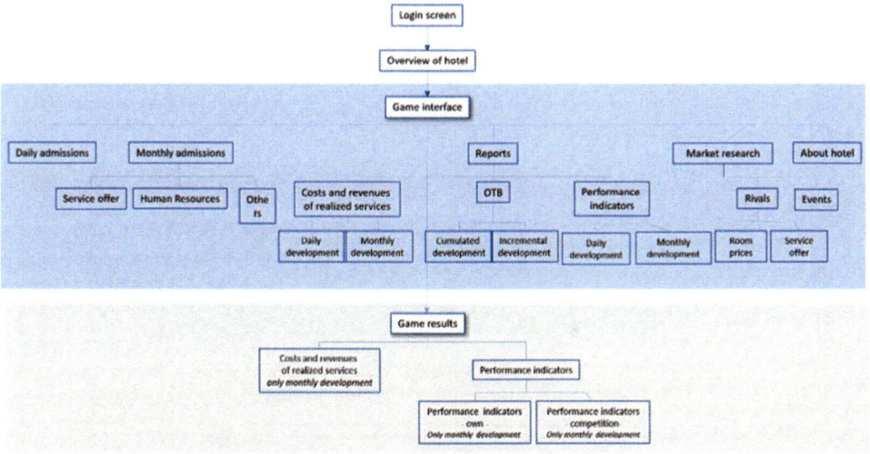

Fig. 2. Player user interface structure

The simulator is prepared in two basic modes. In the case of the on-line mode, players compete with each other on the market for current demand and players also control the operating side of the hotel (see Fig. 2).

In the off-line mode, players do not compete with each other in the markets, everyone has the full amount of demand regardless of the games of the others, and as for the hotel's operational side, it is firmly assigned by the game manager. In this case, players decide only revenue management processes.

The current version of the hotel process simulator allows the following activities:

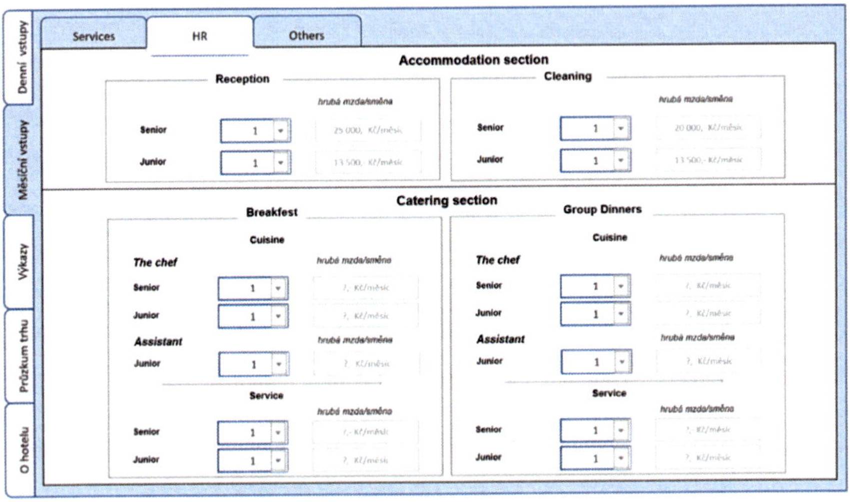

Fig. 3. Monthly admission

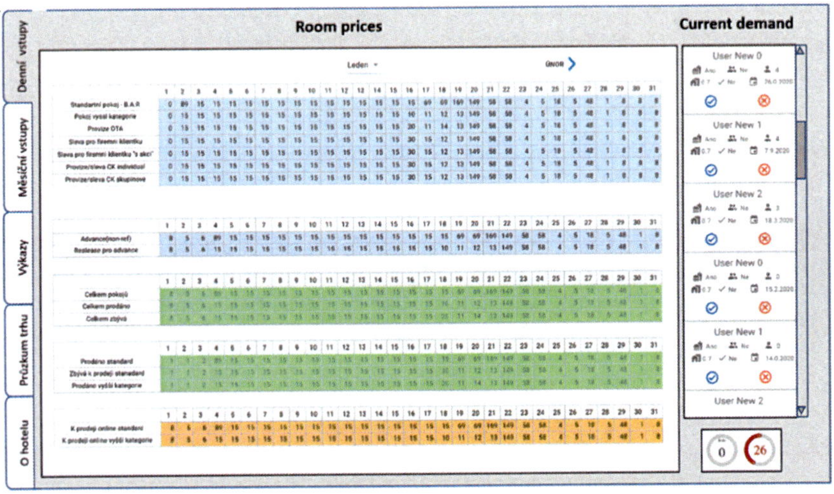

Fig. 4. Game

- setting the initial game parameters of the simulation (see Fig. 3)
- starting the simulation (game)
- timing the simulation, i.e., counting the time for each game round (see Fig. 4)
- receiving client inquiries generated by the system on the basis of pre-prepared curves
- displaying daily and monthly statistics and economic results. (see Fig. 5)

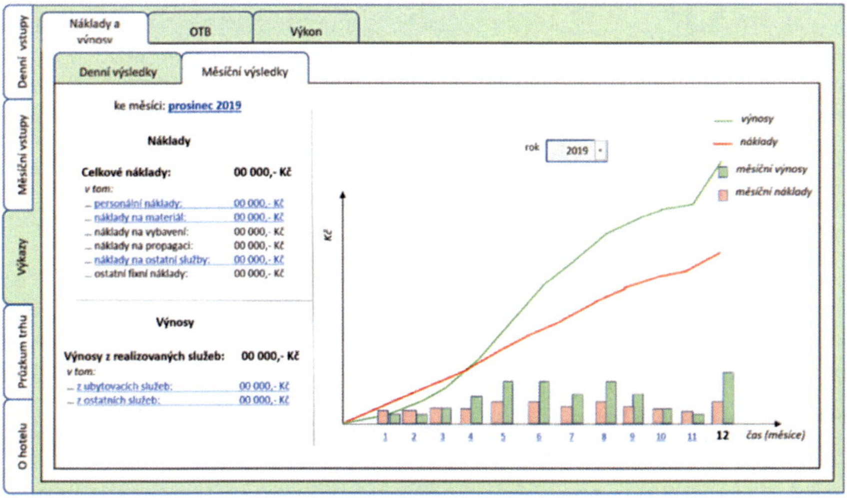

Fig. 5. Costs and revenues

4 Conclusion

The hotel simulators offer excellent tools that can be used in educational activities. The aim is to provide students with a gentle and playful way technique enabling them understand the basic principles and improve their decision-making skills.

Currently, there is a large number of all kinds of games on the market, which relate to the hotel industry. The development of the game market constantly moves forward. Therefore, it is important to be able to accept and work with new information. Automated world forces us to invent new innovations that ultimately make out life easier.

Acknowledgments. This study is supported by the TACR project TL01000191 Innovation of Tourism Management Systems Using the Process Management Tools and SPEV project 2021, run at the Faculty of Informatics and Management, University of Hradec Kralove, Czech Republic.

References

1. Bouki V, Mentzelopoulos M, Protopsaltis A (2011) Simulation game for training new teachers in class management. In SIGDOC 2011 proceedings of the 29th ACM international conference on design of communication. ACM, New York
2. Perkins (2007) Simulation in resuscitation training, resuscitation 73:202–211
3. Oblinger D (2003) Boomers, gen-xers a millennials: understanding the new students. EDUCAUSE Rev 38(4):37–47
4. Lee SH, Bai B (2014) Hotel discount strategies on consumer responses: the role of involvement 69(4):284–296
5. Talón-Ballestero P, González-Serrano L (2013) Yield revenue management in the hotel sector: an empirical analysis of its application and results in Madrid, Spain. In: Matias Á, Nijkamp P, Sarmento M (eds.) Quantitative Methods in Tourism Economics. Physica, Heidelberg, pp 213–231. https://doi.org/10.1007/978-3-7908-2879-5_12

6. Ivanov SH, Ayas C (2017) Investigation of the revenue management practices of accommodation establishments in Turkey: an exploratory study. Tour Manag Perspect 22:137–149
7. Ivanov S, Del Chiappa G, Heyes A (2021) The research-practice gap in hotel revenue management: Insights from Italy. Int J Hosp. Manag. 95:102924
8. Ivanova M, Ivanov SH, Magnini VP (eds) (2016) Pricing and revenue management in hotel chains. In: The Routledge handbook of hotel chain management, pp 262–273. Taylor and Francis
9. Hán J, Petříček M, Chalupa Š (2020) Business process management approach in the hospitality Industry 24:145–158

Data Science and Online Engineering

Work-in-Progress: Design and Development of an Advanced Technological Platform for the Diagnosis and Simulation of Electric Power Consumption

Mario Figueroa de la Cruz(✉), Vicente Chibilisco, and Mauricio Monteros

Facultad Regional Tucumán, Universidad Tecnológica Nacional, Tucumán, Argentina
mariofigueroa@frt.utn.edu.ar,
vicentechibilisco@doc.frt.utn.edu.ar,
mauriciomontero@alu.frt.utn.edu.ar

Abstract. At present, the rational use of energy and energy efficiency within the framework of an adequate energy management system, represent a step towards sustainable energy planning, understood as that which makes good use of natural resources. Cares for the environment and incorporates low-carbon energy sources [1].

Knowing that per capita consumption is disaggregated by sector [2], added to the impact produced by COVID-19 in these times of pandemic [3], it seeks to bring, through an intelligent web platform, the possibility that anyone can carry out a energy audit, knowing their consumption and helping to identify behaviors [4, 5], which can improve use in favor of the consumer.

Keywords: Energy efficiency · Energy consumption · Smart systems

1 Introduction

Framed within the research project PID UTI5407 - Study and Development of a Lighting System with Natural Light (Phase II) [6], the study of energy efficiency through different means is one of the mainstays of the current research project.

Within this scheme and understanding the importance of a responsible use of efficient energy consumption, and predictability in the face of such consumption and costs, the development of an advanced technology software platform that facilitates the diagnosis and simulation of electricity consumption in different environments is proposed.

This development is carried out in conjunction with the Institute of Energy and Sustainable Development (IEDS – Arg.), which has the mission of promoting the use of energy in a sustainable way.

Currently the energy survey and audit are done manually with an enormous work of specialists in each installation, which presents a great challenge for the development of this platform, as well as the assembly of databases with a national base, which are very basic, and that are the ones that will serve as support for the analysis and subsequent studies.

M. E. Auer et al. (Eds.): REV 2022, LNNS 524, pp. 415–420, 2023.
https://doi.org/10.1007/978-3-031-17091-1_42

2 Objectives

Development of an intelligent software platform, which is operational through a cloud environment on the web, which will allow auditing of electricity consumption and gas consumption, suggesting recommendations to improve the use of a property's equipment.

Said platform will be used for academic purposes, for professional training, and to facilitate the calculation of the consumption of homes, public institutions and other applications that will emerge from the joint studies, presenting suggestions to improve the use of household appliances in buildings. These processes will be validated with IEDS experts.

3 Stages

The stages proposed in this development were decided in conjunction with the IEDS. Analyzing the needs to carry out the development of the platform, where the following stages were defined:

- Stage 1: Analysis, Development and Implementation of the Authentication and Authorization module.
- Stage 2: Analysis, Development, and Implementation of the Property Management module.
- Stage 3: Analysis, Development, and implementation of the Electricity and Gas Invoice Management module.
- Stage 4: Analysis, Development, and Implementation of the Building Equipment Management module.
- Stage 5: Analysis, Development, and Implementation of Results and Recommendations modules.
- Stage 6: Validation and Verification with experts in the IEDS area.

These stages are intended for the first phase, which is the development of a Minimum Viable Product of the platform.

In a later phase, support for mobile devices, image recognition, consumption prediction and recommendations will be carried out, and functionality will be extended to expert users and buildings of any type. Table 1 represents the different phases of the project.

Table 1. Project phases

Phase 1: Analysis, design and development of the web platform	*Phase 6: Validation and Verification*
Phase 2: Mobile development	
Phase 3: Image recognition and recommendations prediction.	
Phase 4: Adaptation to Expert Users	
Phase 5: Support for buildings of any category	

4 Work Methodology

An incremental model is used for the design process, where the requirements are defined, it is analyzed and structured in conjunction with the experts, and a basic prototype is developed for field tests (see Fig. 1).

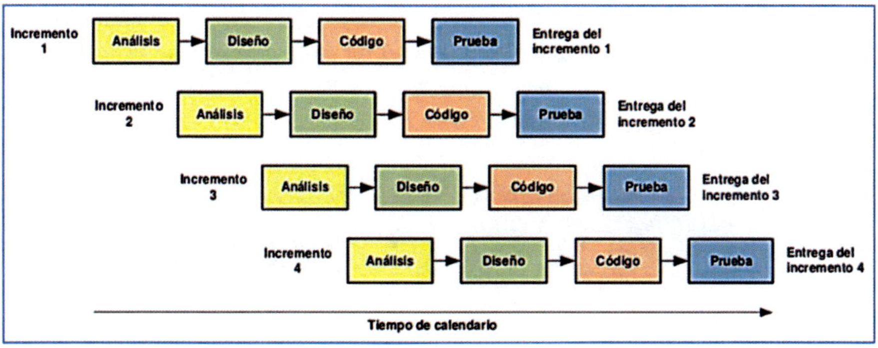

Fig. 1. Example of incremental model

5 Platform Architecture

The platform has a client-server architecture, which allows the client to be totally independent from the server; enabling it to be any application that wants to communicate, under the REST API interface with said server, which allows integrating the mobile application into the solution in a much easier and faster way.

The server is made up of two services:

- Service that receives requests from client applications.
- Service that performs the processing of equipment and invoices to generate the results. It is processed asynchronously.

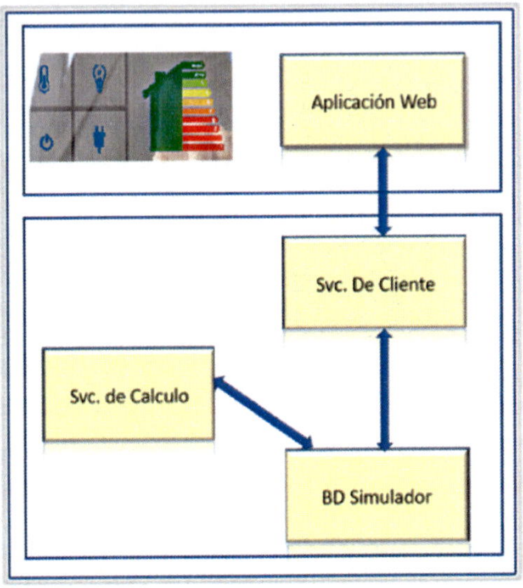

Fig. 2. Deployment diagram

In Fig. 2 you can see the deployment of the different applications.

In later phases, a mobile platform will be included, an image recognition module where convolutional neural networks will be used for this purpose, a prediction module also using neural networks, and recommendations using expert systems, all combined with each other (see Fig. 3).

Fig. 3. Interaction of all the modules of the complete platform

6 Technology Used in Development

The decision of the technology to use was made after an analysis based on the existing problems and the previous knowledge of the development team, leading to research and application of new technologies, allowing the professional growth of the members.

- The frontend is developed with the React library and relying on the ANTD library to speed up the development of the user interface.
- The backend is made up of two services, one that attends to client requests and the other that processes data to obtain results, and a database server.
- The service that attends customer requests is developed in Java 11, with Spring and Hibernet.
- The service that processes the data to obtain results is developed in Go.
- The database server is MySql 8.

All this technology is mounted on Docker (see Fig. 4).

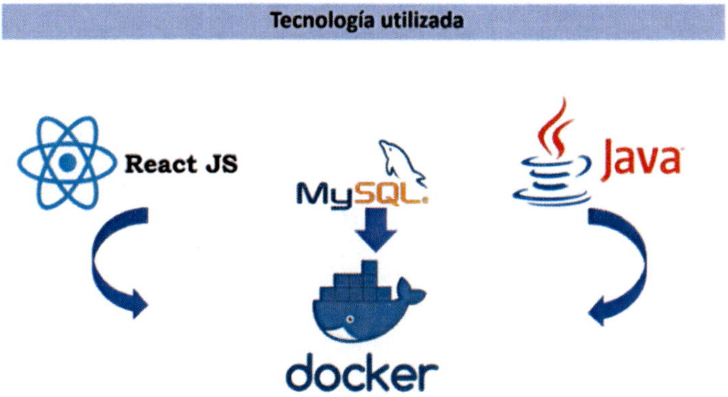

Fig. 4. Technologies used in Phase 1 of the project

7 Conclusions and Future Work

It is important to make a more rational and efficient use of energy. This behavior benefits users financially by reducing the costs of their bills, preserves valuable natural resources for future generations and reduces greenhouse gas emissions.

This platform seeks to raise awareness of the importance of taking care of consumption, and propose changes in behaviors that generate excessive consumption, bringing an easy and intuitive application to people. Also have a tool that could contribute to the reduction of energy demand through the analysis, prediction of behavior and implementation of energy efficiency measures.

Currently the project is in Stage 5 of Phase 1.

The process of analyzing the recommendations is being carried out to find the best possible solution with the data currently available.

The processing of equipment and invoices, and presentation in graphics that are relevant for decision making, have been completed.

In the coming months it is planned to present the platform, to be used by a select group, who will give feedback on it, and thus be able to make corrections for errors that may arise.

All this project will allow to have an instrument of great utility, of national development and free use, endorsed by national entities that promote energy efficiency, generating training of human capital for professionals and students.

References

1. Álvarez M, Baragatti A, Bergallo J, Bourgues C, Casabina G (2016) Hacia el uso racional y eficiente de la energía en la Administración Pública Nacional. IEDS
2. Pasquevich DM (2011) La creciente demanda mundial de energía frente a los riesgos ambientales. Asociación Argentina para el Progreso de las Ciencias en la sección "Energía y Ambiente"
3. Energy Efficiency Division of the International Energy Agency (2020) Covid-19 and energy efficiency. Int Energy Agency
4. Buranyi S (2019) The air conditioning trap: how cold air is heating the world. Guardian
5. Dongellini M, Marinosci C, Morini GL (2014) Energy audit of an industrial site: a case study. Elsevier Ltd.
6. Cesar PG, Stela EPO, de la Cruz Mario F, René BN, Hernán GD, Claudia S (2019) Ahorro de Energía Eléctrica Mediante la Iluminación con Luz Solar Transportada por Fibras Ópticas. Tercer Congreso de Investigación y Transferencia Tecnológica en Ingeniería Eléctrica CITTIE 2019 Octavo Seminario Nacional de Energía SENE 2019

User Stance Detection and Prediction Considering Most Frequent Interactions

Doaa S. Elzanfaly[1,2], Zeyad Radwan[2], and Nermin Abdelhakim Othman[1,2](✉)

[1] Department of Information Systems, Faculty of Computers and Artificial Intelligence, Helwan University, Cairo, Egypt
{doaa.elzanfaly,nermin.othman}@bue.edu.eg

[2] Faculty of Informatics and Computer Science, British University in Egypt, Cairo, Egypt

Abstract. Detecting and predicting users' stances on social media platforms is a challenging problem. There are many applications in detecting and predicting users' stances in many fields like politics, digital marketing and social science. Most research works in this area are using the users' textual contents like tweets or posts as indicators for their stances in a given topic. Recently, some works started to consider users' online activities like networks of accounts that users interact with in mentions, replies or retweets as indicators of their stances. There are challenges in using users' online activities to detect or predict their stances in given topics. One of these challenges is that most of users' interactions might not be relevant to the given topics. Moreover, the large number of users' interactions might have a significant impact in detecting their stances efficiently. The main objective of this paper is to overcome these challenges by considering the most relevant users' online interactions to improve the stances prediction and detection tasks. To meet this objective, we selected only the most relevant interactions to given topics using a filtration method based on the users' interaction frequency. Our results show, in most cases, that considering the most frequent and relevant users' interactions outperforms the research works that are including all the interactions of users.

Keywords: Stance detection · Stance prediction · Natural language processing · Social media mining

1 Introduction

User stances detection and prediction is a part of an academic field called Computational Social Science (CSS). This academic field is concerned with dealing with social sciences using big data and social online activities in order to model, visualize and predict social phenomena. User stance is an opinion, feeling, attitude or judgment that the user are expressing on some topic or situation. Recently, many researchers are interested in the area of automatic user stances detection/prediction. It is an important topic and has wide range applications in many areas like politics, social science, user behaviour analysis and social media analysis. In order to detect or predict user stance, many research works used users' text statements as the only indicator for the user stance. These works faced

M. E. Auer et al. (Eds.): REV 2022, LNNS 524, pp. 421–433, 2023.
https://doi.org/10.1007/978-3-031-17091-1_43

the problem of detecting the stance of 'silent users', who tend to express their stances using interactions rather than writing texts. To solve this problem, some researches started to use users' interaction with other users such as mentions, replies and likes as features. By using interactions as features in detecting stances, two new challenges have emerged. Firstly, not all interactions of the user may be relevant to the given topics. Secondly, the features of interactions, like the list of mentioned accounts or the list of following accounts, are multi-valued features. They are features that have multiple different values per each instance. In our case, these values represent the accounts that each user interacts with. The number of these values is large and many of them are irrelevant to the classification target. The highly common deal with this problem is to transform the number of values in a multi-valued feature into the same number of binary features and apply a traditional feature selection model [33]. However, this solution has many disadvantages like increasing the dimensionality of the dataset, especially when dealing with a large number of multiple values in the same attribute [32]. Other works tried to deal with the multi-valued attributes by using custom classification algorithms. This work aims to enhance user stances detection and prediction by solving the problem of multi-valued features, in the form of multiple interactions, and considering the most frequent ones. Our hypothesis in this research is that using the most common accounts of users' interactions as features, would be more effective, in detecting their stance, rather than considering all users' interactions.

The rest of the paper is organized as follows: Sect. 2 presents the related work. Section 3 illustrate the proposed technique. Section 4 discuss the experimental results and discussions. Section 5 Section for conclusion and future work.

2 Related Work

There are many attempts to solve the user stances detection problem using different approaches with different perspectives. In [13], they divided these works into two main approaches: content-based approach and network-based approach. The content-based approach depends mostly on classifying the text contents of the users' posts or comments in order to detect their stances. This approach represents the majority of the works in the literature. While the network-based approach depends on users' online interactions and connections on social media. Network-based approach is mostly used in predicting stances of prior events [7]. The techniques under this approach are based on the principle of homophily [14]. This principle states that individuals tend to interact and connect more with people with similar ideologies and opinions. There are some hybrid approaches that are using both the text content and network interaction in order to achieve a higher performance [5, 6, 21–23].

A. Content-Based Approach – User Texting

Most of the research on detecting user online stances rely on the text where the user is expressing his stance directly like [1] and [3]. In [13], they divided the content-based approach into two general methods: linguistic features and users' vocabulary. The first method, linguistic features, is based on the user text's linguistic features, which can detect his stance. For example, using n-gram modelling of words is one of the most common techniques of this method [15, 16] and [17]. According to [2], using n-gram modelling of text has achieved best F-score in SemEval-2016 stance detection task. Another technique that is used in linguistic features is sentiment polarity of the text [12, 18, 19]. However, according to [18] and [15], using sentiment features was not sufficient in the task of predicting users' stances as they are made for sentiment prediction. Another technique is mapping text according to some predefined topics (topic modelling). For example, in [12], they used the concept of frames and sentiment analysis of the text as additional features rather than using the content of the text only in order to achieve a better f1 score. The second method in the content-based approaches is the user's vocabulary, this method is based more on user's choice of words [11, 20–23]. This method depends on the hypothesis that users with the same stance would probably use the same words in order to express their opinions [11]. For example, [20] has used a generative model besides the Naïve Bayes and SVM classifiers in order to detect users' stances based on their vocabulary. Another work, [21] has used users' interactions in a word-generative model in order to generate a set of words representations for each stance. Also [22] has used the same approach of using users' interactions to build words representations of each stance. In addition, [23] also used users' interactions and text post embedding by using one embedding vector for pro-stance and another one for the con-stance. Recently, most of the works in user's vocabulary area have used users' interactions to enhance the performance of their models.

B. Network-Based Approach - User Interactions

When it comes to social media platforms, there are many other features, beside the users' text, that can be used to indicate their stances. [13] has grouped network features into two groups, namely, users' behavioral data [6, 11, 24] and users' meta-data attributes [25].

Users' behavioral data are used in stance detecting based on the social phenomena of homophily [14]. Homophily states that social media users tend to interact with people that have the same believes and stances which makes what is called 'echo-chambers' in social media platforms [8]. In addition, after facing the problem of the 'Silent User' (who use more interactions such as shares, likes, mentions, replies and comments), many researchers started to use these interactions as features and attributes in order to perform a higher performance in detecting user stance. One of the most used interaction elements is the hashtag; it has been used in many works to measure the similarity between users in order to predict their stance. For instance, [26] and [6] both measured the similarity between the users who are using the same hashtags. The former used the soft cuisine similarity feature [26] while the second used the graph reinforcement technique [6]. The work in [5] divided network features into three groups, namely, interaction network, preference network and connection network. The interaction network represents interactions where users interact directly with other users like retweets, mentions and replies. This type has achieved the best performance in stance detection task compared with the two other types. The second type, preference network, is the network of the users who post or are mentioned in posts the users like. This type helps when the user has limited direct interactions with other users. Finally, the connection network, which contains the friends and followers lists of the users. Another study in [27] has built a model of multi-layer graph in order to represent users' profiles and extract communities.

Using network features improved the ability to predict users' stances in future events (stance prediction) as predicting users' stances is dependent on users' past behaviour, which can be extracted from network features [4]. For instance, [28] has used users' similarity components as features to predict users' stances after Paris attacks. The similarity between users was calculated using interaction elements like mentions, retweets, replies, URLs and hashtags. On the other hand, [29] has used heterophily in order to measure the dissimilarity between the users based on the user's tendency to reply to the opposed stance. It used a rebuttal value to represent if the user's reply is against the author's post, this value depends on the degree of the opposition between the author's post and the user's reply.

C. Multi-valued Features Problem

When using users' interactions as features in stances classification models, we face an issue of what is called 'Multi-Valued Features Problem' where a feature can have multiple values for each instance. There are many works tried varied approaches in order to handle this problem. The highly common deal with this problem is to transform the number of values in a multi-valued feature into the same number of binary features and apply a traditional feature selection model [33]. However, this solution has many disadvantages like increasing the dimensionality of the dataset, especially when dealing with many multiple values in the same attribute [32]. Some previous works tried to have a solution of this problem like making some changes on widely used categorical attributes measures -like information gain and gain ratio to measure these attributes importance in classification [9]. [10] has proposed a new decision tree algorithm formula (AMDT) specially to deal with this kind of datasets with multiple values attributes. A recent work of [32] proposed a feature selection solution to select attributes from comprising multiple values using utility-scale. Using utility-scale in selecting the optimal set of values is based on the importance of their contribution rather than their frequency of occurrence. For instance, in market data, using profit gained from an item is more relevant than using frequency of this item occurrence (selling a car is more profitable that selling 100 car badges).

Most of previous research works in users' stances detection are focused on analysing the contents of the user-generated text. Recently, many research works started to use users' interactions on social network platforms as an indicator of users' stances on some topic. Using users' interactions shows new challenges like the huge number of each user interactions lists. Also, not all interactions of the user may be relevant to the stance of the user on specific topic. Some research works used users' interactions to calculate the similarity between users, which help in detecting users' stances based on homophily. The problem here is that users may have high similarity, due to having similar opinions and interactions in many topics but have different stances in a specific topic. This may result in detecting false stances for users who have different stances than users with high similarity. Therefore, our motivation in this work is to enhance using homophily concept by using users' interactions with focusing more on interactions that are relevant to a specific topic.

Regarding the problem of multi-valued attributes, our approach is similar to the approach by [32]. The key difference between the two approached is that our approach is based on frequency of values rather than values themselves. As in our case of users' interactions, the frequency of the values is more relevant than the value itself.

3 Proposed Technique

In this section, we discuss our proposed solution to users' stances detection problem. This contains discussing data collection, data cleaning, and most importantly, the preprocessing steps including the handling of multivalued feature, and finally the classification step to identify the stances. Our solution is using user-level features while the dataset used (SemEval 2016) [2] is labelled on tweet-level. In order to compare our solution to other related works, we apply our detection method on tweet level. This means giving the same result to all tweets written by the same user. [5] has argued that it would not be a problem based on the assumption that user would not change his stance on a given topic within a short period of time. Moreover, even if the user changes his stance; he would change his network interactions.

A. Dataset

Our proposed solution is using the benchmark dataset of The SemEval 2016 users' stances detection task [2]. The full dataset of SemEval contains 4063 tweets for five political, religious and social topics. These topics are Atheism (A), Climate Change (CC), Feminist Movement (FM), Hillary Clinton (HC), and Legalization of Abortion (LA). 2814 (70%) tweets are for training and 1249 (30%) are for testing. Regarding network interactions, we used network interactions provided by [5] which was collected using Twitter REST API of each user ID in SemEval 2016 dataset. [5] has collected two timelines interactions for each user in the dataset. The first timeline is Home timeline (IN), which is used to build user's interactions network from his home timeline. It contains the list of accounts the user retweet for, reply to or mentioned in his timeline (IN@) and the list of web domains the user links in his timeline (INDM). The second timeline is the Likes timeline (PN), which contains the interactions extracted from tweets the users' like. It contains list of accounts retweeted for, replied to or mentioned in the tweets the user likes (PN@) and list of web domains linked in the tweets that user likes (PNDM). We used (IN) and (PN) timelines interactions data in [5]. Moreover, we have collected hashtags from users' previous tweets. Nearly 25% of the full dataset users accounts was deleted or suspended. Therefore, we have used a smaller dataset, 1969 training instances and 906 testing instances. Table 1 Shows the distribution of the tweets and users for the five topics covered in the used dataset compared to the full dataset of SemEval 2016 [5].

B. Interactions Filtration Method

As shown in Fig. 3, each user has multiple lists of replies, retweets, mentioned accounts and linked websites for both home (IN) and preference (PN) timelines in addition to hashtags from previous tweets. Regarding detecting user stances for a given topic, most of these interactions are not relevant to the topic under investigation. In addition, it would increase the dimensionality of the data if we represent all of the accounts and websites in raw dataset as features and label for each user if he has interacted with this account

Table 1. Distribution of tweets and users for the five topics compared to the full dataset [5].

	Full dataset		Existing Users	
Topic	Train	Test	Train	Test
Atheism (A)	513 (434)	220 (196)	380 (302)	170 (146)
Climate change is a real concern (CC)	395 (347)	169 (145)	317 (269)	144 (120)
Hillary Clinton (HC)	639 (556)	295 (250)	447 (364)	223 (178)
Feminist movement (FM)	664 (620)	285 (256)	354 (312)	170 (141)
Legalization of abortion (LA)	603 (496)	280 (228)	471 (365)	199 (147)
Total	2814 (2453)	1249 (1075)	1969 (1612)	906 (732)

or not (0 or 1). Some previous works used these interactions to measure the similarity between users and use this similarity to detect users' stances based on homophily [6, 26, 28]. However, this approach may lead to some false detections, as users may be similar in many opinions except in each topic. Our approach is based on filtering the most frequent interactions for both favor and against stances in each topic. Some previous works used filtration technique to handle multi-valued attributes problem [32] and in the scope of users' stances detection to achieve better results [15]. The work of [15] has used filtration to use only hashtags that would help in predict users' stances. They gave each hashtag a score by a simple equation and considered only hashtags with score more than 0.6. The difference between our method and the previous one is that they filter interactions that would not help in detecting users' stances and consider the rest. On the other hand, we filter only the most frequent interactions to consider them and neglect the others. Figure 3 shows main steps that explain our filtration method (Figs. 1 and 2).

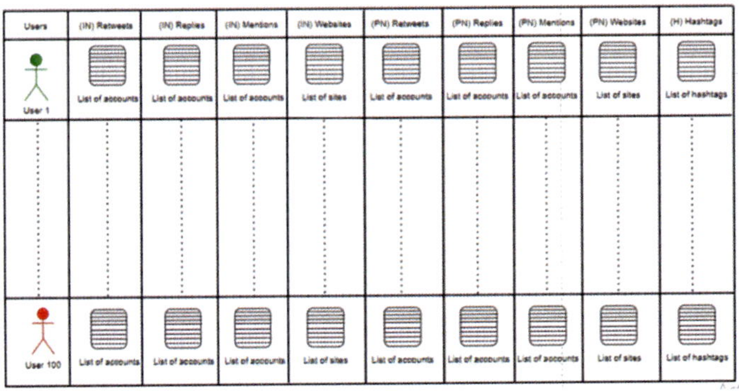

Fig. 1. Multi-valued interactions for each user.

The steps are as follows: for each interaction, i.e. replies, insert all unique interactions for all users in two lists: One list for favour users and the other for against users. Then, using a fixed threshold *x* (40 in our case) and get the most frequent x interactions among favour users. This will result in 2*x* interactions or less, if there are intersections, for each interaction element (IN_{Reply}, $IN_{Retweet}$, $IN_{Mention}$, $IN_{Websites}$, PN_{Reply}, $PN_{Retweet}$, $PN_{Mention}$, $PN_{Websites}$ and Hashtags). Table 1 Contains examples of most frequent interactions for each interaction element in the five topics datasets.

```
DECLARE List_favour_Interaction:ARRAY [0:N] OF STRING

DECLARE List_Against_Interaction:ARRAY [0:M] OF STRING

DECLARE Reply_Favour:ARRAY [0:N] OF STRING

DECLARE Reply_Against [0:M] OF STRING

FOR each user in List_favour_Interaction

   ADD DISTINCT interaction INTO Reply_Favour

ENDFOR

FOR each user in List_Against_Interaction

   ADD DISTINCT interaction INTO Reply_Against

ENDFOR

SELECT DISTINCT interaction as FI FROM List_favour_Interaction WHERE COUNT(*)>= Threshold (h)

SELECT DISTINCT interaction as AI from List_Against_Interaction WHERE COUNT(*)>= Threshold (h)

UNION FI and AI into output

Ouptut is presented as feature
```

Fig. 2. Main stepts of the proposed interactions fillteration technique

C. Interaction Filtration Technique

- *Data Pre-processing*

As noticed in Table 2, hashtags are significant indicators for detecting users' stances directly. In addition, replies and retweets interactions in both timelines contain accounts for people who are directly relevant to given topics. On the other hand, mentions and linked websites interactions in both timelines do not have any correlation to given topics. All of these interactions are about news channels or famous websites (i.e. Facebook, YouTube, NY times). Accordingly, we decided to neglect mentions and linked websites interactions in both timelines and use replies, retweets and hashtags.

Recent work of [5] has created second baseline by neglecting 'none' labelled classes and train the machine learning model on only 'favour' and 'against' classes. The reason behind this is the argument that there is no complete natural stance towards a given topic, as people tend to position themselves with or against a given topic [30]. Although this approach results in larger number of false positives to 'against' and 'favour' classes, it was proven in [5, 31] that this approach achieves better performance. As neglecting 'none' class in training the classifier results in larger number of false positives to 'against' and 'favour' classes, it results in larger numbers of true positives for both classes. This approach improves recall results with some reduction in precision results and overall improvement in average F-score. Accordingly, we decided to follow this approach and neglect 'none' labelled stances in the classifier training step.

Table 2. Stance prediction (using interactions) results.

Models	(A)	(HC)	(CC)	(FM)	(LA)	$f_{Against}$	f_{Favour}	f_{Avg}
Results of [5]	67.65	62.64	42.86	55.87	63.93	79.07	64.04	71.56
Filtration + SVM	58.30	59.76	**72.06**	55.85	52.97	77.0	57.0	67.05
Filtration + Random Forest	58.30	**68.47**	70.64	**60.12**	56.12	79.0	62.0	70.57

- *Classifier Models*

  The textual content of tweets in dataset were represented using n-grams (n = {1, 2, 3}) method with *sklearn.feature_extraction.text.CountVectorizer*() function. We have built two baseline models, one for detecting users' stances using features of both textual content and interactions. The second baseline model is for predicting users' stances using only features of interactions without any textual content. Each baseline model uses two classifier models, namely, support vector machine classifier with linear kernel (SVM) and random forest classifier (RF). In order to evaluate the performance of our methods, we have used the official SemEval-2016 macro average of $f_1$score for only 'favour' and 'against' labelled stances as used in previous works [2, 5].

4 Results and Discussions

A. Stance Prediction Results

Table 3 reports the results of the first baseline of our approach, stance prediction, compared to the best results of [5]. Stance prediction is predicting users' stances using only their interactions as features without using textual content revealing their stances explicitly. We have applied this technique for each topic in the dataset independently and for the full dataset contains the five topics. We also have applied two types of classifiers (Linear Kernel SVM and Random Forest) which show the best performance.

Table 3. Stance detection (using TXT + interactions) results.

Models	$f_{Against}$	f_{Favour}	f_{Avg}
Results of [5]	78.31	66.67	72.49
Filtration + SVM	76.0	58.0	66.94
Filtration + Random Forest	**79.0**	64.0	72.15

Compared to the results of work in [5], our approach using random forest classifier achieved better results in most of the topics of the datasets (three out of five) which are Hillary Clinton (HC), Climate Change (CC) and Feminist Movement (FM). The other two topics, Atheism (A) and Legalize Abortion (LA) showed lower performance than the work of [5]. Regarding the overall dataset contains the five topics together; our approach using random forest classifier has achieved results almost the same as to the results of [5]. We have achieved almost the same results in the f_1 score of against stances and 2% lower than the results of [5] in the f_1 score of favour stances. The macro average f_1 score of both against and favour stances is 1% lower than the results of [5].

B. Stance Detection Results

Our second baseline is combining textual content of users' tweets with users' interactions used in previous baseline. We also have recorded the result of two types of classifiers, which showed the best performance (SVM and Random Forest). Table 4 reports the results of our technique applied in both SVM and Random Forest classifiers, compared to the results of [5] in users' stances detection. This table reports only the results of stance detection in the full dataset contains the five topics together.

Compared to the results of [5], our approach using random forest classifier has achieved better results in f_1 score of against stances. On the other hand, it has achieved lower f_1 score of favour stances compared by [5] results. These results lead to almost similar results in macro average f_1 score of both against and favour stances (0.3% lower than [5] results). It has been noticed in the two baselines that random forest classifier outperforms SVM classifier. These results show that using only the most frequent users' interactions could achieve similar results -and better in some cases- of using all users' interaction in detecting and predicting their stances.

Our results of stances detection and prediction depends on some factors like dataset balance and used threshold of filtration.

Table 4. Examples of the most frequent interactions in five topics datasets.

	(A)	(CC)	(HC)	(FM)	(LA)
IN_{Reply}	@Gr8Darwinia @godFreeWorld @ atheistic_1 @RichardDawkins	@kylegriffin1 @acoyne @UN @LeoDiCaprio	@HillaryClinton @BarackObama @DineshDSouza @SpeakerRyan	@ChelseaClinton @HuffPostWomen @UN_Women @JessicaValenti	@NancyPelosi @Alyssa_Milano @NARAL
$IN_{Retweet}$	@Scripture_Truth @Women_Of_Chri st @AtheistRepublic @Atheist_Bot	@ClimateReality @RealJamesWoods @ACLU @PrisonPlanet	@realDonaldTrump @shannonrwatts @AdamParkhomenko @ChrisMurphyCT	@krassenstein @JessicaValenti @TakedownMRAs @larryelder	@samswey @seanhannity @AmyMek @CecileRichards
$IN_{Mention}$	@POTUS @PressSec @NewYorker	@BernieSanders @globeandmail @guardian	@realDonaldTrump @WhiteHouse @FoxNews	@thehill @guardian @CNN	@PressSec @TheAtlantic @BuzzFeedNews
$IN_{Websites}$	wordpress twitter independent	independent fb thegatewaypundit	theguardian twitter google	dailymail amazon google	instagram wordpress wikipedia
PN_{Reply}	@iamAtheistGirl @JoyAnnReid @LibAthe80 @MrOzAtheist	@cathmckenna @PierrePoilievre @MissionClimate @SpencerFernando	@jaketapper @politico @DonaldJTrumpJr @TheRickWilson	@EverydaySexism @Lauren_Southern @chrissyteigen @AVoiceForMen	@JackPosobiec @DrJenGunter @LilaGraceRose
$PN_{Retweet}$	@ahartforGod @acatholicprayer @CharityIdeas @Godly_Life	@OceanChampions @CLIMATEMAMA @healthandsafety @ActNowForCJ	@JimmySpencerNBA @kasujja @SexyLiberal @oceana	@iamfearlesssoul @logansrun98 @ladyhawkerfinds	@TheWomensWa tch @ProWomanCho ice @prochoicemass
$PN_{Mention}$	@CNN @WhiteHouse @YouTube	@CNN @nytimes @WhiteHouse	@realDonaldTrump @nytimes @CNN	@netflix @Twitter @FoxNews	@Jack @FoxNews @NPR
$PN_{Websites}$	blogspot time google	nytimes youtube medium	dailymail washingtonpost theguardian	theguardian dailymail youtube	theatlantic twitter blogspot
Hashtags	#Christians , #Pray #GOP , #Truth	#ParisAgreement , #Climate , #China	#Benghazi , #USA #DrainTheSwamp	#MeToo , #FF #WomenAgainstFem inism	#ProChoice #prolife #Catholic

5 Conclusions and Future Work

In this work, we present a new approach for detecting and predicting users' stances on social media. Many previous works used users' online activity, like mentions, replies and likes, as indicators for users' stances based on homophily concept [14]. This concept states that users tend to interact with people who have similar points of view. Our approach aims to enhance using homophily with focusing only in most relevant interactions to given topics. This approach filters the most relevant interactions based on frequency and represents them as features. We used two timelines interactions elements (interaction network IN and preference network PN) which was provided by [5] in addition to previous used hashtags. The first timeline, interaction network (IN), includes account that user interacted with directly (reply, retweet and mention) and website links that user shared. The second timeline, preference network (PN), includes account and websites links in the tweets user likes. We examined this approach on the benchmark dataset of The SemEval 2016 [2]. We have built two baselines for both stances prediction and detection, used two types of classifiers in each baseline (SVM and Random Forest) and compared the results with the recent work of [5]. Results shows that using only most relevant users' interactions in given topic perform similarly -or even better in some cases- than representing all users interactions as features.

We concluded that considering the most relevant interactions instead of all users' interactions leads to better results in most of the cases and lower results in some cases. However, these lower results can be enhanced by using balanced datasets covering more topics and trying more filtration thresholds.

Regarding future work, the filtration algorithm could be improved more and previous tweets written by the user could be used. In addition, this filtration technique could also applied on textual content in order to represent only most relevant words to given topics. Also using different datasets contain different topics might be considered since SemEval-2016 dataset is not balanced and labelled on tweet-level structure. More machine learning algorithms could be applied with more hyper-parameters tuning.

References

1. Augenstein I, Vlachos A, Bontcheva K (2016) USFD at SemEval-2016 task 6: any-target stance detection on Twitter with autoencoders. In: Proceedings of the 10th international workshop on semantic evaluation (SemEval-2016)
2. Mohammad S, Kiritchenko S, Sobhani P, Zhu X, Cherry C (2016) SemEval-2016 task 6: detecting stance in tweets. In: Proceedings of the 10th international workshop on semantic evaluation (SemEval-2016)
3. Mohtarami M, Baly R, Glass J, Nakov P, Màrquez L, Moschitti A (2018) Automatic stance detection using end-to-end memory networks. In: Proceedings of the 2018 conference of the north American chapter of the association for computational linguistics: human language technologies, vol 1, pp 767–776
4. Himelboim I, McCreery S, Smith M (2013) Birds of a feather tweet together: integrating network and content analyses to examine cross-ideology exposure on Twitter. J Comput Mediat Commun 18(2):40–60
5. Aldayel A, Magdy W (2019) Your stance is exposed! Analysing possible factors for stance detection on social media. In: Proceedings of the ACM on human-computer interaction, vol 3, pp 1–20
6. Darwish K, Magdy W, Zanouda T (2017) Improved stance prediction in a user similarity feature space. In: Proceedings of the 2017 IEEE/ACM international conference on advances in social networks analysis and mining 2017
7. Garimella K, De Francisci Morales G, Gionis A, Mathioudakis M (2017) Reducing controversy by connecting opposing views. In: Proceedings of the tenth ACM international conference on web search and data mining
8. Colleoni E, Rozza A, Arvidsson A (2014) Echo chamber or public sphere? Predicting political orientation and measuring political homophily in Twitter using big data. J Commun 64(2):317–332
9. Tasca M, Zadrozny B, Plastino A (2014) Relevance measures for multivalued attributes in multiclass datasets. In: Bazzan ALC, Pichara K (eds) IBERAMIA 2014, vol 8864. LNCS (LNAI). Springer, Cham, pp 344–355. https://doi.org/10.1007/978-3-319-12027-0_28
10. Yi W, Lu M, Liu Z (2011) Multi-valued attribute and multi-labeled data decision tree algorithm. Int J Mach Learn Cybern 2(2):67–74
11. Darwish K, Stefanov P, Aupetit M, Nakov P (2019) Unsupervised user stance detection on Twitter

12. Elfardy H, Diab M (2016) CU-GWU perspective at SemEval-2016 task 6: ideological stance detection in informal text. In: Proceedings of the 10th international workshop on semantic evaluation (SemEval-2016)
13. ALDayel A, Magdy W (2021) Stance detection on social media: state of the art and trends. Inf Process Manag 58(4):102597
14. Bessi A et al (2016) Homophily and polarization in the age of misinformation. Eur Phys J Spec Top 225(10):2047–2059. https://doi.org/10.1140/epjst/e2015-50319-0
15. Mohammad SM, Sobhani P, Kiritchenko S (2017) Stance and sentiment in tweets. ACM Trans Internet Technol (TOIT) 17(3):26
16. Anand P, Walker M, Abbott R, Tree JEF, Bowmani R, Minor M (2011) Cats rule and dogs drool! Classifying stance in online debate. In: Proceedings of the 2nd workshop on computational approaches to subjectivity and sentiment analysis. Association for Computational Linguistics, pp 1–9
17. Sobhani P (2017) Stance detection and analysis in social media, Ph.D. thesis, Université d'Ottawa/University of Ottawa
18. Elfardy H (2017) Perspective identification in informal text. Columbia University
19. Mohammad SM (2017) Challenges in sentiment analysis. In: Cambria E, Das D, Bandyopadhyay S, Feraco A (eds) A practical guide to sentiment analysis. Socio-affective computing, vol 5. Springer, Cham, pp 61–83. https://doi.org/10.1007/978-3-319-55394-8_4
20. Klebanov BB, Beigman E, Diermeier D (2010) Vocabulary choice as an indicator of perspective. In: Proceedings of the ACL 2010 conference short papers. Association for Computational Linguistics, pp 253–257
21. Rui D, Yizhou S, Lu W, Yupeng G, Yuan Z (2017) Weakly-guided user stance prediction via joint modeling of content and social interaction. In: CIKM17, Singapore
22. Benton A, Dredze M (2018) Using author embeddings to improve tweet stance classification. In: Proceedings of the 2018 EMNLP workshop W-NUT: the 4th workshop on noisy user-generated text, pp 184–194
23. Li C, Porco A, Goldwasser D (2018) Structured representation learning for online debate stance prediction. In: Proceedings of the 27th international conference on computational linguistics, pp 3728–3739
24. Thonet T, Cabanac G, Boughanem M, Pinel-Sauvagnat K (2017) Users are known by the company they keep: topic models for viewpoint discovery in social networks. In: Proceedings of the 2017 ACM on conference on information and knowledge management. ACM, pp 87–96
25. Pennacchiotti M, Popescu A-M (2011) Democrats, republicans and starbucks afficionados: user classification in Twitter. In: Proceedings of the 17th ACM SIGKDD international conference on Knowledge discovery and data mining. ACM, pp 430–438
26. Dey K, Shrivastava R, Kaushik S, Mathur V (2018) Assessing the effects of social familiarity and stance similarity in interaction dynamics. In: Cherifi C, Cherifi H, Karsai M, Musolesi M (eds) Complex networks & their applications VI. COMPLEX NETWORKS 2017. SCI, vol 689. Springer, Cham, pp 843–855. https://doi.org/10.1007/978-3-319-72150-7_68
27. Fraisier O, Cabanac G, Pitarch Y, Besançon R, Boughanem M (2018) Stance classification through proximity-based community detection
28. Darwish K, Magdy W, Rahimi A, Baldwin T, Abokhodair N (2017) Predicting online islamophopic behavior after# ParisAttacks. J Web Sci 3(1)
29. Trabelsi A, Zaiane OR (2018) Unsupervised model for topic viewpoint discovery in online debates leveraging author interactions. In: Twelfth international AAAI conference on web and social media
30. Kockelman P (2012) Stance: sociolinguistic perspectives - edited by Alexandra Jaffe. J Linguist Anthropol 22(2):E105–E108

31. Siddiqua UA, Chy AN, Aono M (2018) Stance detection on microblog focusing on syntactic tree representation. In: Tan Y, Shi Y, Tang Q (eds) DMBD 2018. LNCS, vol 10943. Springer, Cham, pp 478–490. https://doi.org/10.1007/978-3-319-93803-5_45
32. Lnc Prakash K, Anuradha K (2018) Optimal feature selection for multivalued attributes using transaction weights as utility scale. In: Bhateja V, Tavares JMRS, Rani BP, Prasad VK, Raju KS (eds) Proceedings of the second international conference on computational intelligence and informatics, vol 712. AISC. Springer, Singapore, pp 533–545. https://doi.org/10.1007/978-981-10-8228-3_49
33. Hall MA, Holmes G (2003) Benchmarking attribute selection techniques for discrete class data mining. IEEE Trans Knowl Data Eng 15(3):1437–1447

Pests' Attacks Prediction Using Sensor Fusion in Green Houses

Laila Hatem Mahmoud Hammam(✉) and Ihab Adly

Electrical Engineering Department, The British University in Egypt, Cairo, Egypt
{laila.hammam,ihab.adly}@bue.edu.eg

Abstract. Tomato is considered one of the most important crops not only because of its nutritional value but also due to its critical role in many fields of industry. Considering all of the crops' losses' factors like plants' diseases or stresses of weather conditions, pests and insects are considered one of the major threats that attacks the tomato crops, causing a lot of losses. This work considers the use of sensor fusion techniques combined with convolutional neural networks (CNN) on field programmable logic gate arrays (FPGAs). The use of FPGAs facilitates the reconfiguration of suitable CNN, adapting to the environmental conditions based on the sensors fusion results. A case study has been devised, consisting of two blocks, first the sensor fusion block by using sensors' integrated data from the soil pH, the temperature and the humidity level. The second block is the convolutional neural network block comprising both AlexNet and VGG16 models. On comparing the two models, the VGG16 proved to be more efficient in disease detection. Consequently, for the detection of the tomato mosaic virus disease, the yellow leaf curl tomato virus and the healthy tomato leaves, seven VGG16 neural network models were trained on different combinations of these three classes. By taking advantage of the sensor fusion, a higher accuracy for tomato diseases' detection has been achieved meanwhile reducing the workload of the neural networks' computations.

Keywords: Convolutional neural networks · Sensor fusion · Tomato plants' diseases · AlexNet · VGG16

1 Introduction

Among all kinds of crops that are known nowadays. One of the most popular, highly consumed and planted crops is the tomato crop. Tomato is considered one of the most important crop not only because of its nutritional value but also due to its critical role in many fields of industry. Considering all of the crop losses' factors like plants' diseases or stresses of weather conditions, pests and insects are considered one of the major threats that attacks the tomato crops, causing a lot of losses, damages and also deeply affects the crops quality [19]. Annually in Brazil the loss in the agricultural production due to insects and crops' pests reaches an average of 7.7 percent. This percentage means that the resources of biofuels, food or fiber has faced a huge reduction of twenty-five million tons. This leads to reduction in the country's economy by almost seventeen million sterling

M. E. Auer et al. (Eds.): REV 2022, LNNS 524, pp. 434–445, 2023.
https://doi.org/10.1007/978-3-031-17091-1_44

pounds [1]. One of most dangerous threats to the tomato crops is the Hubner pest that causes reduction in the quantity of the yield that may reach up to 55% annually, leading to degradation in meeting the increasing demand for the crops' quality and quantity. In addition to crops' losses this pest causes degradation in the height, weight and number of leaves, decreasing the yield of the tomato crop [2]. Another critical threats to tomato crops are the tomato plants' diseases. In the tropical and sub-tropical countries where the temperature and humidity levels are often high that can reach to 40 °C or even more. The tomato crops are planted and grown usually in summer or spring, and in such cases will be exposed to unusually high temperature and humidity. In such weather stresses the plants' infection with the tomato yellow leaf curl virus became a possibility. This virus is a very popular tomato crops' disease that infects the tomato plants' worldwide. It is very well known by the yellowing of the small and lower plants' leaves and in critical cases, the severity of this disease can cause entire loss in the tomato yield [3]. One more tomato disease that will be discussed is the tomato mosaic virus disease. It is one of the most highly infectious, widely prevalent viruses that can easily be spread by plants' contact. It has a great ability of survival and can last for long periods on contaminated equipment and in crop debris. This disease can generally cause uneven ripening, curling and mosaic leaves, or even varying fruits' internal browning [4].

In this work sensory data about the weather and the soil are gathered from different types of sensors. The main challenge that is faced by any sensor-based management system is that the sensors' data is very sensitive to changes in the soil type, the properties of the crops regarding its growing stages and processes or even usual changes in the climate. This makes the sensors' readings very uncertain and consequently taking decisions related to crops management or diseases' control will be inaccurate and complex. An efficient solution to solve the uncertainty problem is to use combination of different sensors' reading from various sources. This integration of the resultant data can provide promising results regarding accuracy in results' estimations. This technique is called sensor fusion at which data from different sources like soil pH or temperature degree are collected from different types of sensors and the fusion of these results can lead to more accurate decisions and confident results [5].

Due to the significant rise in convolutional neural network in many fields and the remarkable success in the tasks related to classification, recognition and detection. One of the work's targets is to take the advantage of convolutional neural networks' efficiency in order to accurately detect tomato plants' diseases. Concerning only the AlexNet and VGG16 neural network models, the main target of this work is to detect the mosaic virus, the yellow leaf curl virus tomato disease in addition to the detection of healthy tomato leaves. An essential condition to consider is to detect the plants' diseases with minimum neural network power and workload. So, during detection process, it is desired to decrease the intensive and expensive computations of the CNNs. The convolutional neural networks' detection process will be supported by primary estimations based on the sensors' collected data with the knowledge of tomato diseases' pathogens and symptoms. This work mainly focuses on two techniques; first the sensor fusion, of sensory data obtained from different environmental sensors, second the use of convolutional neural networks in order to detect the plants' diseases based on the inserted images' extracted features. Taking advantage of the detection accuracy using these two techniques can reach a higher accuracy for tomato diseases' detection. In addition, this can provide a more efficient technique to reduce the workload of the neural networks' intensive computations. This paper demonstrates the proposed architecture and design in Sect. 3, and the results from the proposed design architecture are discussed and analyzed in Sect. 4.

2 Literature Review

2.1 Fusion of Sensor Data for the Detection of Plant Diseases

Currently, in the case of fungal pathogens that infect cucumber plants, visual assessment is the most commonly used method to determine the pathogen occurrence. On plant diseases' detection, various technologies and sensor methods have been implemented to better study the effect of pathogens on plant's physiological parameters. There are different drawbacks in the current disease handling methods. For example, the disease assessment by human raters may be affected by subjective bias, also it is time consuming, in addition molecular diagnostic methods are expensive as well as time-consuming. The diagnostic and detection ability of the proposed methods can possibly be improved, through combining the information of the different measuring sensors in a sensor data fusion approach with a simple discriminant analysis. It is obvious that a combination of sensor data or of different vegetation indices is a highly successful approach to achieve an early detection of plants' diseases [6].

2.2 Tomato Plants' Diseases and Viruses

In many countries tomatoes are considered one of the most important crops. Unfortunately, this crop's production can be affected by many disorders and plants' disease infections [7]. Around the world, the reported losses in the tomato crops have been raised up from 20% to 90% due to the presence of more than twenty different viruses. The tomato mosaic virus is one of the most highly infectious viruses that can easily be spread by plants' contact. The symptoms of this disease were recorded in the seeds and leaves of the tomato plants by an average of 25.49% and 29.79% respectively that can cause serious reduction in the tomato yield by an average of 25%. As a final conclusion, it was found that there is positive correlation between the severity of the tomato mosaic virus and the temperature degree if the temperature lies between 20 °C and 31 °C. On the other hand, if the relative humidity level (RH) lies between 55% and 70%, a reduction in the disease development and severity will be noticed [4]. Another disease to be considered is the tomato yellow leaf curl virus (TYLCV). This disease is a series limiting factor for the production of tomato crop and loses due to its infection can frequently reach up to 100% in various regions. Since the year 1960, the tomato crops have been severely affected by this viral disease in various regions like Africa, Far East, Central America and Europe [3].

2.3 Artificial Neural Networks for Disease Severity Detection

This paper represents an artificial neural network model for predicting the severity of Septoria Tritici. The proposed neural network uses specific field data as an input parameter with the values of disease severity and corresponding growth stage recorded with fusing weather data in trials over ten years across Europe from 2008 till 2018. After data processing, a feed forward neural network with two hidden layers under two implementations is trained in order to define the relationship between weather data and the disease severity. The first approach is implementing of a two-layer feed forward neural network, and the second approach is the implementation of an advanced model that uses techniques like cosine annealing, cyclic learning rate, and stochastic gradient descent. The first approach achieved 0.17 mean absolute error, and 0.06 mean square error, regarding the second approach with improvements, it achieved 0.1 mean absolute error, and 0.02 mean square error [8].

3 Proposed Architecture and Design

The concepts of sensor fusion and convolutional neural networks are the two main bases on which this work relies on. First the sensor fusion block, it is believed that this concept mainly depends on gathering sensors' information from different sources in order to reach more reliable and accurate results. Gathering of sensory information from different sources is way better than depending on only a single source. Specifically, in this work the concept of sensor fusion is used as follows, three types of sensors were used temperature sensors, humidity sensors and soil pH sensors. For each of the previously mentioned tomato diseases, the diseases' pathogens were specified and according to the sensors' readings about the environmental conditions at a specific moment of time, it is possible to determine which disease is more liable to occur. For example, if there is a prior knowledge that the mosaic virus only occurs at a temperature degree greater than thirty degree Celsius, if the temperature sensors' reading is greater than thirty degree Celsius, this means that as a primarily prediction the disease can possibility be the mosaic virus disease. Otherwise, if the temperature sensors' reading was less than thirty degree Celsius, this means that with a high level of confidence the disease is not the mosaic virus disease. Moving to the second block that is the convolution neural network block. On this block it is desired to predict the tomato diseases from the three mentioned classes, the mosaic virus disease, the yellow curl virus and the healthy leaves class.

Any implemented neural network mainly depends on two processes, the training process and the inference process that is the prediction process. In the training process the neural network works on samples of data that are annotated in order to create a primary model. In the training process the two steps of forward propagation and back-propagation are implemented. For the forward propagation the created model starts to gain knowledge about the inserted dataset, and tries to make relationships between the images' features and the associated annotation for each image. In the back-propagation the model tries to interactively enhance the model's parameters, for example the model's convolution weights in order to enhance the power of implemented model's prediction. After completing the training process, the inference process will start the classification process for each image in the testing set. For a reliable neural network testing, it is

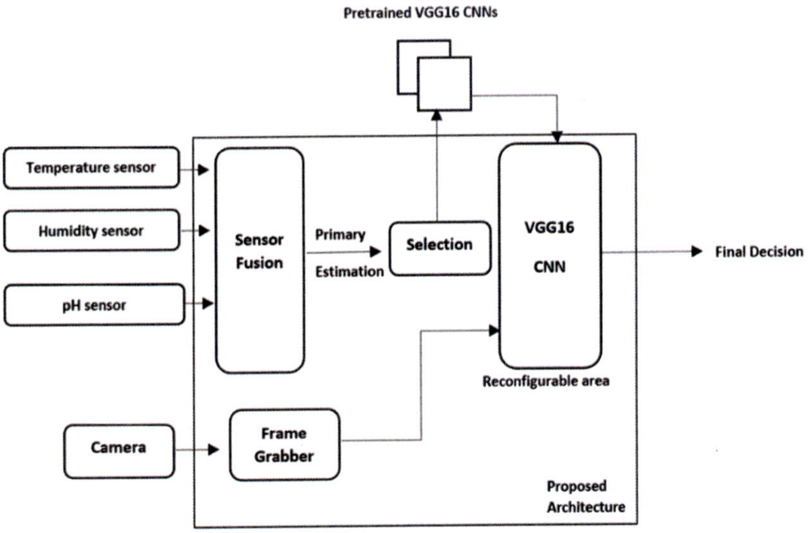

Fig. 1. Schematic diagram to represent the proposed design

important to be aware that the testing set should be a set of new data images for the same training data samples but never the same identical images. As an overview note, the training process for a neural network model is only implemented once on large GPU or FPGA clusters. On the other hand, the inference or classification process is implemented each time a new data sample is needed to be classified [9].

In this work the implementation of the neural network block mainly focuses on only the tomato crop diseases, so three models were chosen for the detection process. First the AlexNet convolutional neural network, second a modified AlexNet model, and third the VGG16 convolutional neural network. As a first step the three networks were trained on 38 classes of plants' diseases including the healthy classes. On comparing the performance of the three networks, the VGG16 model has proved to be the best with training accuracy and loss of 0.8801 and 0.3731 respectively, and with validation accuracy and loss of 0.8933 and 0.3331 respectively after two epochs. Consequently, the VGG16 was chosen in this work due to its high accuracy in disease detection and the layers' implementation in the VGG16 neural network architecture is better than the AlexNet architecture. As previously mentioned, this work focuses on three classes that yields seven combinations. For each combination of classes, a VGG16 neural network model was implemented and trained. So, each neural network will detect and classify a different set of classes. As shown in Fig. 1, one way to take the advantage of the sensors' information according to the temperature, humidity and soil pH sensors, it is possible to make a primarily prediction about the types of possible diseases that can occur in different environmental situations and consequently on a specific situation with a prior knowledge of the possible diseases that may occur, it is possible to choose one of the pre-trained neural network versions that are trained on different combinations of tomato plant diseases in order to speed up the prediction process, decrease the intensive workload of the neural network, decrease the resources specifically the power utilization, and finally to increase the accuracy of the neural network diseases' detection.

3.1 The Neural Network Block

3.1.1 The First AlexNet Neural Network Model

One of the most common and popular convolutional neural networks is the AlexNet convolutional neural network. This model was firstly originated by Alex Krizhevsky. After competing with the AlexNet convolutional neutral network in the ImageNet Large Scale Visual Recognition Challenge in the year 2012, this network became very popular by its high efficiency and detection accuracy. The layers' depth of the neural network was the main reason for the model's high performance and speed during both training and inference processes. The following section will include more detailed information about the AlexNet layers' implementation. The models' whole architecture consists of eight layers arranged as follows; there are five convolutional layers, three max-pooling layers, and two normalization layers. Finally in the output layer, there are two fully connected layers and one softmax layer [10]. It is important to mention that in the model's architecture the max pooling layers are only following the first, second and fifth convolutional layers. For each of the implemented convolutional layers there are convolutional kernels or filters with a nonlinear ReLU activation function. The pooling layers are implemented after the convolutional layer in order to mathematically extract the greatest values among a set of values that represent the output of the convolution layer. The existence of the fully connected layer is the reason for having a fixed sized input images. In most of the AlexNet models, the input image size is $224 \times 224 \times 3$ [11]. In this section a neural network example for an AlexNet convolutional neural network is being trained for the detection and classification of plants' diseases especially the tomato plant's diseases [12]. The model is trained first on 38 classes of different plants' diseases including the healthy crops' leaves classes. The following table represents the networks' training accuracy and loss with the validation accuracy and loss during the ten epochs of the model's fitting step (Table 1 and Fig. 2).

A group of images were chosen as a testing set in order to evaluate the trained model. Unfortunately, most of the results from the chosen testing set were wrongly estimated. As it is clearly illustrated from the previous plotted graph for the AlexNet CNN model's training accuracy and loss along with the model's validation accuracy and loss. It is obvious that the validation accuracy is high but the training accuracy is very low. On the other hand, on the loss graph, the training loss is very high in contrary with the validation loss. Such problem can be referred to as under-fitting where the neural network model struggles in the training step and cannot entirely fits the images in the training set. This problem can be solved by increasing the depth of the model, meaning to increase the number of the implemented layer in the model or another solution is to increase the number of neurons in the model or to increase the number of images in the chosen training set.

Table 1. The overall fitting step of the AlexNet CNN model including ten epochs

Model's epochs	Training accuracy %	Training loss %	Validation accuracy %	Validation loss %
Epoch 1	0.0588	7.2336	0.9707	0.0863
Epoch 2	0.0549	6.2280	0.9689	0.0919
Epoch 3	0.0582	5.6187	0.9652	0.1099
Epoch 4	0.0574	5.2555	0.9614	0.1255
Epoch 5	0.0764	4.8570	0.9601	0.1374
Epoch 6	0.0781	4.7476	0.9570	0.1522
Epoch 7	0.0748	4.6910	0.9567	0.1626
Epoch 8	0.0802	4.5734	0.9566	0.1720
Epoch 9	0.0745	4.5681	0.9546	0.1838
Epoch 10	0.0588	4.5785	0.9523	0.1957

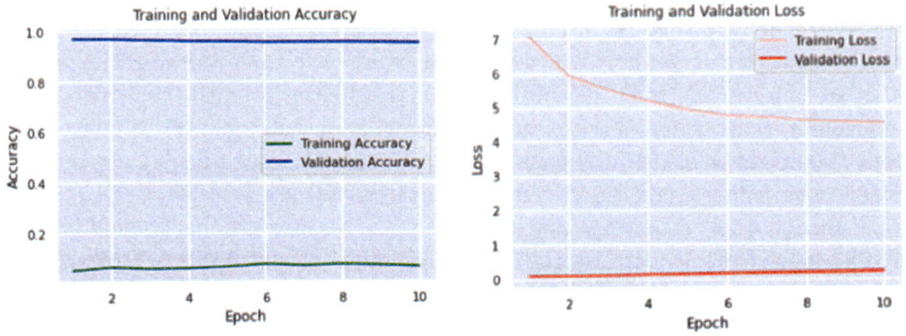

Fig. 2. The training and validation accuracy and loss of the AlexNet CNN model

3.1.2 The Second AlexNet Neural Network Model

In this section another AlexNet model is being trained and evaluated for the same training and testing set previously mentioned for the purpose of detecting the crops' diseases [13]. This model is a modified model to the previously mentioned AlexNet model. The enhancements are implemented in the modified weights' file that was loaded to the network and the increased number of epochs to enhance the classification accuracy of the implemented convolutional neural network. The following table represents the model's training accuracy and loss with the validation accuracy and loss for all of the twelve epochs of the neural network's fitting step (Table 2 and Fig. 3).

For the chosen testing set, this modified model showed better performance. As a comparison with the first AlexNet model, this model was actually capable of detecting and classifying more plant's diseases correctly. It is clear from the plotted graphs that this model has better training and validation accuracy. On the other hand, the model's training and validation loss is relatively small compared to the first AlexNet model.

Table 2. The overall fitting step of the modified AlexNet CNN model including twelve epochs

Model's epochs	Training accuracy	Training loss	Validation accuracy	Validation loss
Epoch 1	0.9660	0.1008	0.9754	0.0729
Epoch 2	0.9691	0.0968	0.9758	0.0708
Epoch 3	0.9691	0.0929	0.9758	0.0706
Epoch 4	0.9696	0.0923	0.9759	0.0698
Epoch 5	0.9710	0.0914	0.9764	0.0696
Epoch 6	0.9714	0.0880	0.9761	0.0697
Epoch 7	0.9716	0.0883	0.9766	0.0692
Epoch 8	0.9698	0.0929	0.9766	0.0691
Epoch 9	0.9704	0.0905	0.9767	0.0690
Epoch 10	0.9693	0.0911	0.9767	0.0689
Epoch 11	0.9708	0.0884	0.9770	0.0686
Epoch 12	0.9716	0.0880	0.9767	0.0686

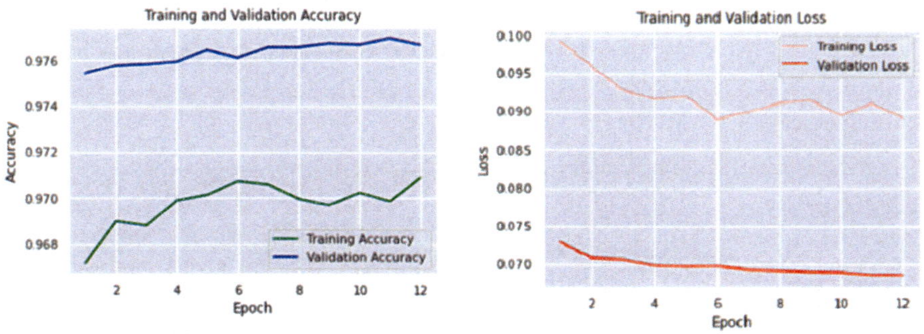

Fig. 3. The training and validation accuracy and loss of the modified AlexNet CNN model

3.1.3 The VGG16 Neural Network Model

Another very popular neural network is the VGG16 convolutional neural network. It is a deep convolutional neural network with sixteen implemented layers. It was mainly used at Oxford University by the group of visual geometry in order to achieve impressive results in the competition of ImageNet Large Scale Visual Recognition Challenge (ILSVRC) in the year 2014 [14]. As the depth of the neural network increases, the accuracy in the features' extraction and detection will also increase. The size of the input images that are passed to the VGG16 neural network is $224 \times 224 \times 3$ [20]. The main reason for the VGG16 model's depth is that it consists of thirteen convolutional layers divided into five convolutional layer groups, each layer is implemented using small sized filters. The filters' size is constant in all of the five layers that is 3×3 filters. In the VGG16 convolutional neural network the padding is preserved to one pixel for thirty-three filters and the stride in all of the convolution layers is fixed to one pixel. Each convolution layer group is followed by a max-pooling layer in order to extract the greatest values from the convolutional layer results. In the VGG16 model the max pooling operation is performed with stride of 2 on 22 different windows. Finally, after all of the implemented five convolutional layers groups, there are three implemented fully connected layers followed by a single soft-max layer. The output result from the soft-max layer is the classification probability for the inserted input image. In this work, the VGG16 neural network was chosen for the purpose of detecting and classification of plants diseases' images [15]. Focusing on the tomato plants' diseases. This network has achieved great improvement over the AlexNet neural network due to the replacement of the filters with relatively large sizes with smaller sized filters. It replaced both the eleven sized filters in the first convolutional layer and the five sized filters in the second convolution layer with 3×3 filters implemented one after the other. Using of multiple stacked small sized filters is better than using large sized filters due to the increase in the number of the non-linear layers and so increasing the depth of the network. More non-linear layers increase the capability of the network to extract the more complex feature from the input image leading to enhancement in the detection and classification accuracy with relatively low computational cost, because the earlier layers of the implemented network will detect the simple image's features and the deeper layers will detect the more complex features [11] (Table 3 and Fig. 4).

Table 3. The overall fitting step of the VGG16 CNN model including two epochs

Model's epochs	Training accuracy	Training loss	Validation accuracy	Validation loss
Epoch 1	0.7904	0.7059	0.9074	0.2831
Epoch 2	0.88801	0.3731	0.8933	0.3331

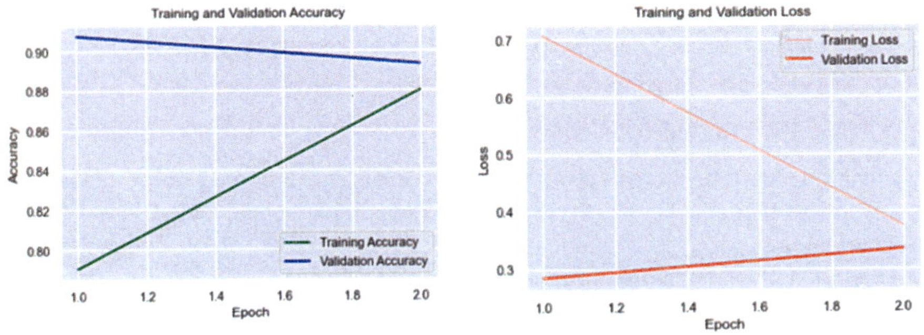

Fig. 4. The training and validation accuracy and loss of the VGG16 CNN model

After two epochs, it is clear that the chosen VGG16 model has high training and validation accuracy. On the other hand, the loss in the training process and validation process is quite small. Due to the depth of the network that improved its detection accuracy, this model was chosen in this work for tomato plants diseases' detection regarding the three chosen classes; tomato yellow leaf curl virus (TYLCV), the tomato mosaic virus (ToMV) and the tomato healthy leaves' class.

3.2 The Sensor Fusion Block

In this block the main focus is only on three tomato crop classes concerning the healthy class, the yellow leaf curl virus disease and the tomato mosaic virus disease. The temperature degree, the humidity level and the soil pH are the three main metrics that were chosen in this work and consequently the occurrence of the tomato diseases will be detected regarding these three metrics [4, 16, 17, 18]. By using temperature, humidity and soil pH sensors' readings, also with a prior knowledge about the diseases' pathogens regarding the tomato mosaic virus disease and the yellow leaf curl virus tomato disease, it is possible to determine in a specific moment whether a specific disease is liable to occur or not (Table 4).

Table 4. The conditions at which each class might occur

Class	Temperature (°C)	Relative humidity level (%)	Soil (pH)
The healthy tomato leaves	10 °C to 30 °C	60% to 70%	5.6 to 5.8
The tomato mosaic virus	20 °C to 31 °C	>70%	7.0
The yellow leaf curl virus	More than or equal to 40 °C	>75%	>7.0

4 Results and Analysis

The results of the prediction and training time are a clear prove that the neural networks trained with a large number of classes takes the longest time in the training and inference processes. As shown in the detection accuracy table, as the number of classes increases, the prediction accuracy of the neural network model decreases and vice versa (Tables 5 and 6).

So, one way to decrease the workload of the expensive computations of the convolutional neural networks, is to choose a neural network version that best suits the current situation. For example, if the primary estimations stated that the plant's disease is the tomato yellow leaf curl virus or mosaic virus, choosing the neural network that detect only these two diseases will be the best choice in order to achieve high accuracy in the disease classification and also to decrease the intensive neural network computations as possible.

Table 5. The VGG16 CNN prediction and training time

Network (VGG16) combinations	Average prediction time	Network training time
Tomato yellow curl virus	0.6386 s	43.33 min
Tomato mosaic virus	0.715 s	34.4 min
Healthy tomato leaves	0.743 s	72.2 min
Tomato yellow curl virus Tomato mosaic virus Healthy tomato leaves	1.70714 s	157.33 min
Tomato yellow curl virus Tomato mosaic virus	0.999964 s	78.7 min
Tomato mosaic virus Healthy tomato leaves	1.1174 s	80.2833 min
Tomato yellow curl virus Healthy tomato leaves	1.1092 s	81.9833 min

Table 6. The VGG16 CNN accuracy in disease detection

VGG16 networks	Mosaic virus	Yellow curl virus	Healthy leaf
Tomato yellow curl virus	0/20 = 0%	20/20 = 100%	0/20 = 0%
Tomato mosaic virus	20/20 = 100%	0/20 = 0%	0/20 = 0%
Healthy tomato leaves	0/20 = 0%	0/20 = 0%	20/20 = 100%
Tomato yellow curl and mosaic viruses	19/20 = 95%	9/20 = 45%	0/20 = 0%
Tomato mosaic virus Healthy tomato leaves	16/20 = 80%	0/20 = 0%	20/20 = 100%
Tomato yellow curl virus and healthy tomato leaves	0/20 = 0%	13/20 = 65%	19/20 = 95%
Tomato yellow curl, mosaic viruses, and healthy tomato leaves	12/20 = 60%	10/20 = 50%	19/20 = 95%

5 Conclusion and Recommendation for Future Work

On using both techniques of sensor fusion and convolutional neural network, it is possible to increase the detection accuracy of the tomato plants' diseases. Also, on having seven VGG16 neural networks trained on different combinations of classes, it is possible to use the sensor fusion results in order to choose the most suitable neural network to best detect the plants' diseases with the minimum computation workload, because the training and inference time increases as the number of classes to which the neural network is trained increases. The results of the prediction and training time are a clear prove that the neural networks trained with a large number of classes takes the longest time in the training and inference process that was 157.33 min and 1.70714 s respectively. For the detection accuracy, as the number of classes increases, the prediction accuracy of the neural network model decreases and vice versa. For the detection accuracy of the mosaic virus, the accuracy increased from 60% to 80% when the number of classes decreased. So, one way to decrease the workload of the expensive computations of the convolutional neural networks, is to choose a neural network version that best suits the current situation.

References

1. Oliveira C (2014) Crop losses and the economic impact of insect pests on Brazilian agriculture. Crop Prot 56:50–54
2. Singh N (2017) Assessment of crop losses due to tomato fruit borer, Helicoverpa armigera in tomato. J Entomol Zool Stud 5:595–597
3. m. eelm (2000) Tomato yellow leaf curl virus, an emerging virus complex causing epidemics worldwide. Virus Res 71:123–134
4. Imran M (2021) Influence of environmental conditions on tomato mosaic virus disease development under natural condition. Pak J Phytopathol 25: 117–122
5. Adamchuk VI (2011) Sensor fusion for precision agriculture. In: Sensor fusion-foundation and applications, Rijeka, InTech, pp 27–40
6. Berdugo CA (2014) Fusion of sensor data for the detection and differentiation of plant diseases in cucumber. Plant Pathol 63: 1344–1356
7. Gleason M (2006) Tomato diseases and disorders. Department of plant pathology, IOWA state university, Iowa
8. Shankar P (2020) Data fusion and artificial neural networks for modelling crop disease severity. In: 2020 IEEE 23rd international conference on information fusion (FUSION), Rustenburg
9. Abdelouahab K (2018) Accelerating CNN inference on FPGAs: a survey. HAL, Clermont-Ferrand
10. Krizhevsky A (2017) ImageNet classification with deep convolutional neural networks. Commun ACM 60:84–90
11. Yu W (2016) Visualizing and comparing AlexNet and VGG using deconvolutional layers. In: Proceedings of the 33rd international conference on machine, New York
12. Prajwal10031999 (12 May 2021) Plant-diseases-classification-using-AlexNet. https://github.com/Prajwal10031999/Plant-Diseases-Classification-using-AlexNet
13. Bhattarai S (29 November 2018) Plant diseases classification using AlexNet. https://www.kaggle.com/vipoooool/plant-diseases-classification-using-alexnet
14. Liu S (2015) Very deep convolutional neural network based image classification using small training sample size In 2015 3rd IAPR Asian conference on pattern recognition (ACPR), Kuala Lumpur
15. Kumar A (20 May 2020) Plant disease detection using VGG16. https://www.kaggle.com/amitkrjha/plant-disease-detection-using-vgg16
16. Anfoka G (2016) Tomato yellow leaf curl virus infection mitigates the heat stress response of plants grown at high temperatures. Sci. Rep 6:1–12
17. Adams SR (2001) Effect of temperature on the growth and development of tomato fruits. Ann Bot 88:869–877
18. Snyder RG (2019) Greenhouse tomato handbook, mississippi: mississippi state university extension service
19. Savary S (2012) Crop losses due to diseases and their implications for global food production losses and food security. Food Secur 4:519–537
20. Tindall L (2017) Plankton classification using VGG16 network

Personality Traits of Twitter Bullies Using Deep Learning

Alaa Hesham(✉) and Abeer Hamdy

Faculty of Informatics and Computer Science, British University in Egypt, Cairo, Egypt
Alaa.hesham@bue.edu.eg

Abstract. Research in the phycology domain pointed out a relationship between the personality traits and the ability of a person to bully or be bullied. This paper aims at analyzing the personality traits, in terms of the Big Five model (BFM), of cyberbullies on Twitter communities. To achieve our objective, two deep learning-based models were trained to detect the BFM traits from a text; the first of them is a multi-label CNN based classifier while the second includes five independent binary CNN classifiers (one for each trait). The trained models were applied to a tweeter dataset of cyberbullying people, in order to identify their BFM personality traits. Experimental results showed that the characteristics of cyberbullies are high neuroticism, high extroversion, and low agreeableness. Moreover, modelling the BFM personality prediction problem using five binary classifiers achieves higher accuracy than utilizing one multi-label classifier.

Keywords: Big five model · Personality traits · Cyberbullying · Tweeter

1 Introduction

Cyberbullying is using online platforms as a method of spreading hatred or producing offence towards other users. It had influenced users in a negative manner and unfortunately it is escalating [1]. Its consequence on victims often varies between sadness, depression and other psychological behaviors (depending on the situation) [2]. In some cases, cyberbullying has made its victims encouraged towards suicidal thoughts [2]. In extreme cases, these suicidal thoughts were turned into action. A live case study spotted a student (15 years old) committed suicide caused by cyberbullying [3]. Therefore, an action must be taken to handle this manner. As a gesture of handling cyberbullying, artificial intelligence (AI) techniques could be applied to detect it on a consistent daily matter. Moreover, AI could be utilized in identifying the various personality characteristics of bullies and victims; which may lead to insights into the causes of the bullying behavior and consequently improving the techniques of predicting this act and fixing it.

An efficient way to characterize the personality of the bullies is through the big five model (BFM); which is a psychological model that classifies people personalities into five classes which are: Openness, Extroversion, Agreeableness, Neuroticism, consciousness [4]. BFM is used in various fields; for instance: medical fields, job hiring agencies. BFM is selected in this paper for personality analysis as it analyses the

M. E. Auer et al. (Eds.): REV 2022, LNNS 524, pp. 446–455, 2023.
https://doi.org/10.1007/978-3-031-17091-1_45

personalities from different aspects including emotion expression, social engagement, responsibilities, behaviors.

1.1 Objectives and Contributions

The main aim of the paper is to find a correlation between the characteristics of the cyberbullies and BFM personality traits. This objective is accomplished through the following:

1- Developing a deep learning (DL) [5–8] based model for predicting the categories of BFM personality traits from text.
2- Applying the trained BFM DL Model on cyberbullying tweets to identify the characteristics of a cyberbully.
3- Find the correlation between cyberbullying characteristics and personality traits.

The paper is organized as follows: Sect. 2 introduces the BFM and deep learning models. Section 3 surveys the previous work. Section 4 discusses the proposed methodology. Section 5 discusses the experiments and results. Finally, Sect. 6 concludes the paper.

2 Background

2.1 Big Five Model

Big five model (BFM) is one of the popular psychological models to evaluate the personality traits. BFM is also called the ocean model; it represents 5 major characteristics that forms a complete personality assessment. This psychological model was initiated in 1936 by relating words into certain categories and later in 1940 the model was formed with 16 traits that was later reduced and narrowed down into the big five model, consisting of 5 general traits [4]. The personality assessment usually goes as a test or a group of questions each targets a trait and based on the answer a number is generated between 0 and 100 to record how high or low each characteristic is found while generating the assessment. The BFM personality characteristics are summarized in Table 1.

2.2 Deep Learning Models

Deep learning (DL) [5] is advance era of machine learning that is applied to various fields including text classification [6–8]. Convolutional neural network (CNN) and Recurrent neural networks (RNN) are two popular types of DL models. CNN was largely utilized with images, while RNN was utilized with text due to its memory mechanism. However, the performance of CNN models was showed in the literature to be competitive to the RNN models on sequential data.

Table 1. BFM personality characteristics.

Trait	description
Openness (OPN)	It demonstrates how the person is opened to try new experiences, which indicates creativity, braveness etc.
Conscientiousness (CON)	It defines the behavior of being organized and always ready with plans. People with high scores are considered hardworking, discipline and pay attention to details
Extroversion (EXT)	It targets the emotional expression where it examines the social act around the community. People with high score are always the one to initiate conversations. They feel relaxed talking to others
Agreeableness (AGR)	It focuses on the morals as well as the human-being part. High scores reveal that this person is trustworthy and often patient and empathetic also, their main feature is helping others
Neuroticism (NEU)	It puts the spotlight on the mental health and confidence of the personality. People with high scores are often found anxious, stressed, and exposed to negative environment

3 Literature Review

Some previous psychological studies examined the relationship between the BFM traits and the cyberbullying problem [9, 10]. Alonso and Romero [9] used the BFM to study the personality profiles of each of the aggressors and victims of cyberbullying and bullying. The authors discussed the opinions, of some previous studies, on the BFM model comprehending cyberbullying. The studies showed that the common scores between bullies are (low agreeableness, low conscientiousness high neuroticism, high extroversions). On the other hand, victims' characteristics were (high neuroticism, high agreeableness, low openness, and low conscientiousness). Meanwhile other research showed difference in victims' results for instance some showed engagement of low extroversion was related while others showed that the high extroversion was the one related, other disagreed and stated that high neuroticism is the key feature [11]. To avoid this confliction, Alonso and Romero [9] prepared a questionnaire about the BFM to investigate its relation to cyberbullying and bullying. The results showed a correlation between bullying and cyberbullying. In addition to, cyber-victims score is higher in neuroticism and openness, and agreeableness.

Zhou et al. [10] discusses the relationship between BFM and cyberbullying in terms of a definition called moral disengagement. The authors firstly investigated the relationship between cyberbullying, gender, and age. They found out that gender is not strongly relevant. According to a conducted questionnaire they found out that bullies score is low in agreeableness, conscientiousness, and high in neuroticism. While, victims score is low in extroversion, conscientiousness and high in neuroticism. It appeared that the consistency of doing that act on purpose to its effect, is related to weak attachment or obligations to self-morals or ethics, this is the definition of moral disengagement. Furthermore, moral disengagement was discovered to be affected by extroversion, neuroticism, and openness. Cyberbullying detection with psychological models. Balakrishnan et al. [12] used Random forests classifier to predict cyberbullying in twitter platform. They found out that feeding the classifier with the BFM and dark triad features improved the prediction accuracy. Recently, studies were conducted that use word embedding and deep learning models for predicting the BFM traits from text [13, 14]. Majumder et al. [13] utilized Word2vec embedding and CNN networks; while Kazameini et al. [14] utilized BERT network for word embeddings and bagged SVM for classifying the text to one of the five traits.

4 Proposed Methodology

4.1 Overview

The proposed approach for analyzing the personalities of cyberbullies in relation to the BFM includes two main stages. The first stage is training a deep learning-based model for predicting BFM personality traits from a text. Then, the trained model is applied to cyberbullying Twitter dataset, for estimating the BFM personalities of cyberbullies. The following subsections discuss these two stages in details.

4.2 BFM Personality Traits Prediction Model

The problem of BFM personality traits prediction from a text is a multi-label classification problem; as every text could be classified to one or more BFM traits. The proposed model (as depicted by Fig. 1), starts with pre-processing the text using natural language processing to extract the key words and remove the noise. Then the sentiment of each record in the data is analyzed, as the same concept can convey different meanings to people of different personality types; neutral sentences are removed. Every key word is represented using a vector generated by a word embedding model. Every text is represented as a fixed sized vector of words. Padding is used to keep the text vectors of fixed size. Finally, the embedding vectors are fed to a CNN based classification model. In this work we experimented with two CNN models for classification. The first model is a CNN multi-label classifier, which has five outputs; while the second model includes five independent binary CNN classifiers, each of them targets a single personality trait.

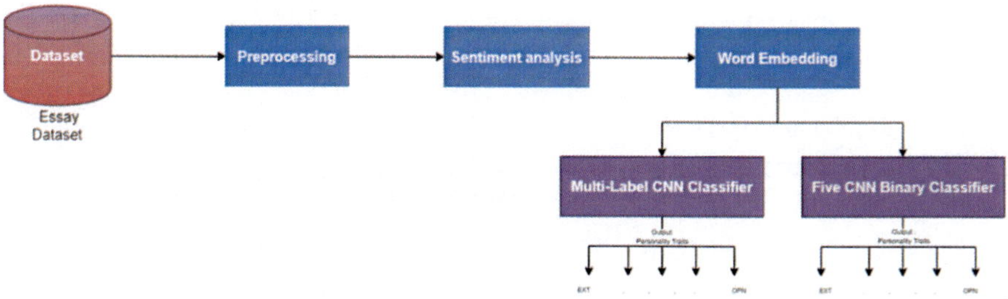

Fig. 1. BFM personality prediction models

Preprocessing
Natural language processing was used at this stage. This procedure is similar to the preprocessing procedure discussed in [15–20]. It starts with "tokenization" to split each text into a set of words, this step is performed using NLTK library. This step is followed with removing stop words and punctuations, then stemming and normalization to reduce the noise and the size of the input features to the classifier.

Sentiment Analysis
Sentiment analysis is conducted to predict the polarity of a text; Vader (Valence Aware Dictionary for Sentiment Reasoning) [21] which is a lexicon and rule-based sentiment analysis tool was used in this work. Vader determines the polarity and intensity of each text, through assigning a compound score, ranges between -1 and $+1$, that expresses how positive, negative or neutral is the text. Lexicon dataset is a compiled list of lexical words that are labeled with positive or negative polarity based on the semantic meaning. According to the lexicon dataset, the polarity of a sentence is evaluated through the evaluation of the polarity of its selected words. However, Vader assigns for each text a mathematical compound scores which are calculated based on a knowledge-based system.

Embedding
Embedding is a technique for learning the representation of entities words or sentences as real-valued, fixed-length vectors in a high-dimensional space. The vectors are learned such that words/sentences that have similar meanings are close to each other in the vector space [22]. Embedding provide a more expressive representation for text than classical methods like VSM, where semantic similarity between words are ignored [20]. In this work we used the pre-trained word embedding model Glove [23]. Glove includes a huge number of words and is provided in several vector dimensions [50, 100, 300], we used vectors of size 100. Every record in the dataset is represented as a vector of keywords. Padding was used at this stage to make the vectors of fixed length (150 keywords) before they are fed to the CNN based classifiers.

Multi-label /Binary CNN Classifiers
The objective of this stage is classify the text to one or more of the BFM traits. We experimented with two CNN models, the first of them is a multi-label CNN classifier, which is depicted by Fig. 2. It consists of 6 cascaded layers which are: embedding, convolutional, max pooling layer, flatten and two dense (fully connected) layers. The last dense layer has five output nodes 5 representing the BFM traits individually. The second classification model consists of five CNN binary classifiers, each one of them has the same architecture as the multi-label CNN model, except that the second dense layer has only one output node. We experimented with different values of hyper parameters, Table 2 lists the hyper parameters that achieved the results listed in the paper.

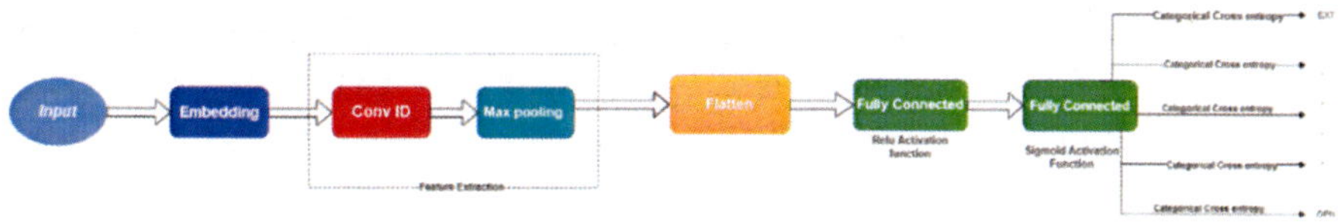

Fig. 2. BFM multi-label CNN classifier

Table 2. Hyper parameters values for BFM models

Hyper parameter	Nu. of filters	Kernel size	Batch size	1st-Dense layer	Loss	Optimizer
Multi-label CNN	512	3	8	60	Categorical	Adam
Binary CNN	200	3	8	60	Binary	Adam

4.3 Cyberbullies and BFM Traits Correlation

The aim of this stage is to characterize the twitter cyberbullies with respect of the BFM traits. Figure 3 depicts the steps of this stage, which starts with analyzing the sentiment of every twitter text and removing the neutral ones to eliminate the noise. Then, preprocessing and embedding every twitter text as previously discussed. Finally, the text vectors of the cyberbully tweets are fed to the previously trained BFM multi-label/binary CNN classifiers to estimate the BFM traits for every cyberbully tweet.

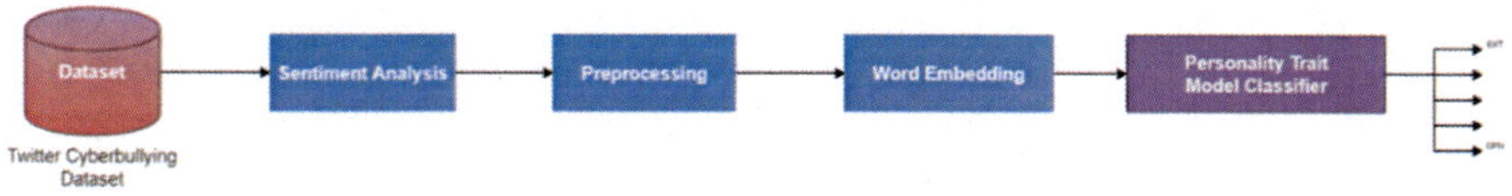

Fig. 3. Model for estimating correlation between cyberbullies and BFM traits.

5 Experiments and Results

5.1 Data Sets

Two datasets were used in the experiments. The first of them is the Essays dataset [24] to train the BFM model. While, the second dataset is the cyberbullying tweets dataset to characterize the cyberbullies in relation to the BFM.

Essays Dataset

It was constructed by a group of researchers on several phases. The authors of this dataset interviewed group of students where they expressed themselves or how their day was. Data was written in English and some of the texts were written in informal English.

Later this dataset was modified to express the BFM traits. Thus, 5 labels were added to each record in this dataset that represent the five BFM traits. The Essays include 2468 record, their statistical distribution over the traits is listed in Table 3.

Table 3. Statistics of the Essays dataset with respect to the BFM traits

Trait	EXT	NEU	AGR	CON	OPN
Yes	1269	1226	1300	1250	1267
No	1184	1227	1153	1203	1186

Cyberbullying Tweets Dataset
It includes 11091 records of tweets labeled as "offensive" and "non-offensive", indicating cyberbullying and regular tweets. The number of offensive records in this dataset is equal to 2953 records. However, the sentiments of these records were analyzed and the records that were marked, by the sentiment analysis model, as neutral were removed; which resulted in 2210 records of offensive tweets, to be characterized by the personality trait models.

5.2 Research Questions

RQ1: what are the performance of the BFM personality trait prediction models?
RQ2: How is the cyberbullying behavior related to the BFM personality traits?

5.3 Results

Answer to RQ1
Table 4 lists the accuracy of the multi-label and binary CNN classifiers in predicting the BFM traits. A multi-label classifier using LSTM was also trained and its accuracy is listed in Table 4. As could be observed the performance of the Multi-label CNN classifier is better that the multi-label LSTM by 8%. But the accuracy of the binary CNN classifiers is better than the multi-label CNN. This could be attributed to the fact that the multi-label classifier has to converge to a complex mapping function that predicts 5 classes simultaneously. While, the five independent binary CNN classifiers converge to simpler mapping functions. The precision, recall and F1-measure values of the multi-class CNN BFM model are listed in Table 5.

It was observed that the highest accuracy among the 5 traits in the binary CNN models is the NEU. This can be justified as neuroticism is fed on negative input. Thus, in classification it was easy to recognize sentences with negative outcome as per shown in the sentiment analysis. The second highest score was extroversion. Furthermore, comparing Agreeableness, conscientiousness, openness as the variation between their values was insignificant. Agreeableness resulted the highest among conscientiousness and openness.

Table 4. Comparison among the accuracy of the developed BFM personality trait models

Algorithm	Multi-label CNN	Binary EXT	Binary NEU	Binary AGR	Binary CON	Binary OPN	Multi-label LSTM
Accuracy	61.2%	82.7%	85.5%	76.0%	74.3%	74.9%	53.2%

Table 5. Precision, recall and F-measure of the multi-label CNN BFM traits classifier.

	EXT	NEU	AGR	CON	OPN
Precision	0.48	0.52	0.48	0.50	0.50
Recall	0.82	0.84	0.82	0.84	0.84
F1 measure	0.65	0.68	0.65	0.67	0.67

The reason for that is, Agreeableness classification is based on emotionally charged situations that are most likely to be positive, thus, recognizing the emotionally positive records was straightforward. On the other hand, conscientiousness was the lowest accuracy observed. The justification for that could be that, the act of conscientiousness is more related to a moral behavior which can be challenging to detect in text as it would be dependent on naturally emotional texts.

Answer to RQ2

The trained BFM prediction model is used to identify a pattern for personality traits of a tweet cyberbully. When feeding a tweet to the trained BFM prediction model, five decimal number will be generated representing the scores of each tweet in terms of the 5 traits. On preforming statistical analysis, it was concluded that the cyberbully is high on neuroticism, which indicates instable emotional state or consistent stress and anger tamper. The second feature of cyberbully is low agreeableness which represents morals or ethical principles, and this seems logical. Additionally, cyberbully is high on openness as cyberbullies could be adventurous and open to try new methods. Furthermore, cyberbully is high on extroversion as they are brave to contact others and express their feelings wildly.

6 Conclusion

The significance of Psychology has introduced models that helps understanding the personalities of each human. It also has provided a guide on how to deal with each characteristic. Some models can detect good and bad characteristics like the big five model (BFM) and some would focus on detecting the dark patterns like dark traits models. BFM composes any personalities as a combination of traits (extrovert, openness, agreeableness, neuroticism, and conscientiousness). Meanwhile, the escalating growth of the cyberbullying had affected many victims. An action towards limiting this behavior needs to be taken. Thus, this paper presents the BFM perspective of recognizing cyberbullies.

In this paper, deep learning based models were developed to analyze the personalities of cyberbullies according to the BFM. Two models were developed, the first is a multi-label CNN, while the second includes five binary CNN classifiers, one for each single trait individually. It was found that the binary models have better accuracy than the multi-label model. The highest accuracy achieved was for estimating the Neuroticism trait as it represents stress and anger. The reason is that its classification was based on detecting negative emotions which was easy to detect. The lowest accuracy was observed to be for predicting conscientiousness. It was reasonable as this trait is based on moral behavior. It was found that the common characteristics of a cyberbully are high neuroticism, high extroversion, and low agreeableness.

As a future work, for a deeper understanding to the personality of the cyberbullies and a better cyberbullying detection; a pattern of the characteristics of a cyberbully can be determined by studying their behaviors with respect to the dark traits personality model. It is expected to elaborate more as dark traits model targets negative behaviors, thus detecting an act like cyberbullying can be more accurate.

References

1. Chun J, Lee J, Kim J, Lee S (2020) An international systematic review of cyberbullying measurements. Comput Hum Behav 113:106485
2. Hinduja S, Patchin JW (2020) Cyberbullying identification, prevention, and response. Cyberbullying research centre (cyberbullying.org)
3. Ryalls E (2012) Demonizing "mean girls" in the news: was phoebe prince "bullied to death?" Commun Cult Crit 5(3):463–481
4. Lim AGY The big five personality traits. Personality theories
5. Goodfellow I, Bengio Y, Courville A, Bengio Y (2016) Deep learning, vol 1. MIT Press, Cambridge
6. Hamdy A, Ezzat G (2020) Deep mining of open source software bug repositories. Int J Comput Appl 1–9
7. Hamdy A, Tazy M (2020) Deep hybrid features for code smells detection. J Theor Appl Inf Technol 98(14)
8. Hamdy A, Ezzat G (2021) Deep embedding of open source software bug repositories for severity prediction. In: Research and evidence in software engineering
9. Alonso C, Romero E (December 2017) Aggressors and victims in bullying and cyberbullying: a study of personality profiles using the five-factor model. Span J Psychol 20:E76
10. Zhou Y, Zheng W, Gao X (2018) The relationship between the big five and cyberbullying among college students: the mediating effect of moral disengagement. Curr Psychol 38(5):1162–1173. https://doi.org/10.1007/s12144-018-0005-6
11. Çelik S, Atak H, Erguzen A (2012) The effect of personality on cyberbullying among university students in Turkey. Egitim Arastirmalari Eurasian J Educ Res 49:129–150
12. Balakrishnan V, Khan S, Fernandez, T, Arabnia HR (2019) Cyberbullying detection on Twitter using big five and dark triad features. Personal Individ Differ 141:252–257. ISSN 0191-8869. https://doi.org/10.1016/j.paid.2019.01.024
13. Majumder N, Poria S, Gelbukh A, Cambria E (2017) Deep learning-based document modeling for personality detection from text. IEEE Intell Syst 32(2):74–79. https://doi.org/10.1109/MIS.2017.23
14. Kazameini A, Fatehi S, Mehta Y, Eetemadi, S, Cambria E (2020): Personality trait detection using bagged SVM over BERT word embedding ensembles. ArXiv

15. Hamdy A, El-laithy A (2019) SMOTE and feature selection for more effective bug severity prediction. Int J Software Eng Knowl Eng 29:897–919
16. Hamdy A, El-Laithy A (2020) Semantic categorization of software bug repositories for severity assignment automation. In: Jarzabek S, Poniszewska-Marańda A, Madeyski L (eds.) Integrating research and practice in software engineering. SCI, vol 851. Springer, Cham, pp 15–30. https://doi.org/10.1007/978-3-030-26574-8_2
17. Hamdy A, Elsayed M (2018) Topic modelling for automatic selection of software design patterns. In: ICGDA 2018
18. Hamdy A, Elsayed M (2018) Towards more accurate automatic recommendation of software design patterns. J Theor Appl Inf Technol 5069–5079
19. Hamdy A, Elsayed M (2018) Automatic recommendation of software design patterns: text retrieval approach. J Softw **13**, 260–268
20. Hamdy A, El-Laithy A (2020) Multi-Feature approach for Bug severity assignment. Int J Open Sour Softw Proces 11(2):1–15
21. Hutto CJ, Gilbert EE (June 2014) VADER: a parsimonious rule-based model for sentiment analysis of social media text. In: Eighth international conference on weblogs and social media (ICWSM-14), Ann Arbor, MI
22. Mikolov T, Sutskever I, Chen K, Corrado GS, Dean J (2013). Distributed representations of words and phrases and their compositionality. ArXiv, abs/1310.4546
23. Pennington, J, Socher R (2014) Manning CD GloVe: global vectors for word representation. Standford university
24. Pennebaker JW, King LA.: Linguistic styles: language use as an individual difference. J Personal Soc Psych 77(6):1999, pp. 1296–1312

A Novel Approach for Financial Markets Forecasting Using Deep Learning with Long Short Term Networks

Somashekar Manjunath(✉) and Prathap Halasuru Manjunath

M S Ramaiah University of Applied Sciences, Bengaluru, India
soma1902@gmail.com

Abstract. Financial markets are highly unpredictable owing to their volatile nature. Due to this complexity, Financial time series forecasting is one of the most challenging fields of study. In the past 20 years many approaches were used to solve time series forecasting. However in the recent years, the implementation of Deep Learning algorithms for financial market predictions has proven to yield better accuracy and returns in the field of financial analysis when used with the rightful knowledge.

In this paper, we propose a novel method for forecasting stock price movements using LSTM. Long Short Term Networks (LSTM) are highly efficient for time series predictions due to their ability to store historical price information, which is important for our approach as previous prices of a stock is essential for predicting its future price. We choose 3 stocks from NIFTY 50 index, deploy our LSTM networks and train on its historical prices and evaluate the performance of the trained model.

Keywords: Financial analysis · Deep Learning · LSTM

1 Introduction

The world's stock markets encompass enormous wealth. In 2019, the value of global equites surpassed $85 trillion (Pound, 2019). As long as markets have existed, investors have searched for ways to acquire knowledge about the companies listed in the market to improve their investment returns [1].

In India during the pandemic from April 2020 to Jan 2021, around 10.7 million new demat accounts were opened. According to a survey around 95% of Indian traders lose money. Many institutional traders use the algorithmic approach known as Quant Trading which also benefits from the features of Artificial Intelligence, such systems are built and run on state of the art machinery. A retail investor cannot cope against such state of the art Quant algorithms. Technical Analysis is based on the premise "History repeats itself", we are building our model based on this premise.

A time-series is a sequence of data values taken at equally spaced successive points in time. Financial time-series data is frequently used to predict future stock prices. However, predicting future stock prices using financial time-series data is very challenging.

M. E. Auer et al. (Eds.): REV 2022, LNNS 524, pp. 456–462, 2023.
https://doi.org/10.1007/978-3-031-17091-1_46

This is because the price of a stock is influenced not only by the fundamentals of the company but also by the mood or sentiments of short-term and long-term investors Positive sentiments can drive prices up and vice versa. Some researchers have adopted techniques from natural language processing to study the impact of sentiments on stock prices. Some researchers proposed using deep recurrent neural network (RNN) along with data preprocessing techniques to reduce and denoise the input data [2]. RNN has the ability to remember its previous states. The algorithm of choice here is the LSTM (Long-Short term memory) network. It's a type of recurrent network that has proved very successful on a number of problems given its capability to distinguish between recent and early examples by giving different weights for each while forgetting memory it considers irrelevant to predict the next output. In that way, it is more capable to handle long sequences of input when compared to other recurrent neural networks that are only able to memorize short sequences [3].

The objective of this paper is to apply LSTM networks, on various stocks instruments and predict their price movements helping the retail investors/traders to test and analyze their investment strategies, Assess their performance in terms of accuracy and other metrics.

2 Background and Related Work

Technical Analysis is the study of the market action, primarily through charts for the purpose of forecasting price trends. The important premise for technical analysis is "History Repeats Itself". The current price of an asset always reflects all previous information available for it instantly.

There is also the Random-walk hypothesis which claims that a stock price changes independently of its history, in other words, tomorrow's price will only depend on tomorrow information regardless of today's price. Those two hypothesis determine that there are no means to accurately predict a stock price [3].

A variety of methods for predicting and modeling stock behavior have been object of study of many different disciplines, such as economics, statistics, mathematics and computer science. It's worth mentioning that in 2012, it was estimated that approximately 85% of trades within the United States' stock markets were performed by algorithms [4].

Long Short Term Memory (LSTM) networks, which are used in this paper are a deep and recurrent model of neural networks. Recurrent networks differ from the traditional feed-forward networks, implying that they not only have neural connections on a single direction, neurons can pass data to a previous or the same layer. Data doesn't flow in a single direction and the practical effects for that is the existence of short term memory, in addition to long term memory that neural networks already have in consequence of training. LSTM were introduced and it aimed for a better performance by tackling the vanishing gradient issue that recurrent networks would suffer when dealing with long data sequences. It does so by keeping the error flow constant through special units called "gates" which allows for weights adjustments as well as truncation of the gradient when its information is not necessary [3] (Fig. 1).

LSTM networks are composed of an input layer, one or more hidden layers, and an output layer. The number of neurons in the input layer is equal to the number of

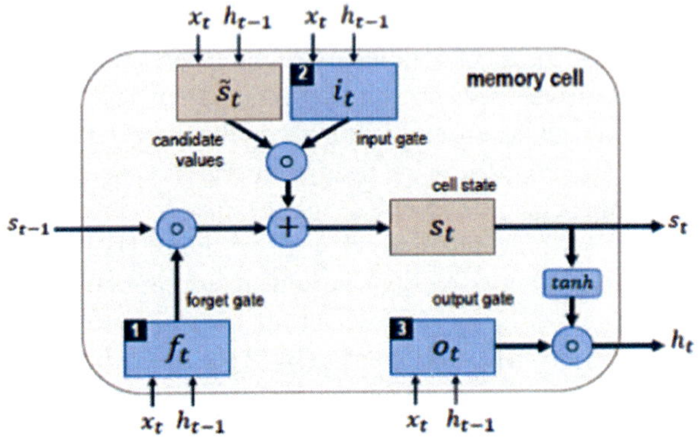

Fig. 1. Structure of a LSTM memory cell [4].

explanatory variables (feature space). The number of neurons in the output layer reflects the output space, The main characteristic of LSTM networks is contained in the hidden layer(s) consisting of memory cells. Each of the memory cells has three gates maintaining and adjusting its cell state s_t: a forget gate (f_t), an input gate (i_t), and an output gate (o_t). The structure of a memory cell is illustrated in the above figure. At every timestep t, each of the three gates is presented with the input x_t (one element of the input sequence) as well as the output h_t-1 of the memory cells at the previous timestep(t – 1) [4]. Hereby, the gates act as filters, each performing the following functions:

1. Forget Gate: Defines which information to remove from the memory (cell state).
2. Input Gate: Defines which information to add to the memory (cell state).
3. Output Gate: Defines which information from the memory (cell state) to use as output.

In this paper we propose a LSTM based model for forecasting the stock prices. Hence proving that LSTM based model will yield better results over traditional feedforward neural networks.

3 Methodology and Development

A classification algorithm is designed using LSTM network for stock price forecasting. In this section we discuss various stages in designing our system.

3.1 Data Preprocessing

Historic data of 3 different stocks (Tata Motors, SBIN, Reliance) from the NIFTY 50 index are collected such as Open price, Close Price, High, Low & Volume in a day's timeframe .csv file format. For this paper we have collected data from 1996–2021 of these stocks.

Exploratory Data Analysis is performed on the collected stock data. Data normalization is performed using Min-Max Scaling to stabilize the data along time series, it is expressed as

$$X_{std} = \frac{X - X_{min}}{X_{max} - X_{min}}$$

$$X_{scaled} = X_{std} \times (max - \min) + min$$

We split the data in train and test datasets by 70% & 30% respectively and preprocess the data such that the previous data is taken into account using timesteps.

3.2 Model Development

- We use Google's TensorFlow to build our model. In this stage, the data is fed to the neural network and trained for prediction assigning random biases and weights.
- Our system is designed as a sequential model of, the input LSTM layer which takes the close prices and feeds it to 2 sequential LSTM layers.
- Dropout Layer is present along with the LSTM layers to avoid overfitting of data.
- And an output dense layer with Sigmoid activation.
- The model consists of 139,019 total trainable parameters.
- Model is entirely developed online using Kaggle and various open sourced python libraries (Fig. 2).

Model: "sequential"

Layer (type)	Output Shape	Param #
lstm (LSTM)	(None, 100, 128)	66560
lstm_1 (LSTM)	(None, 100, 64)	49408
lstm_2 (LSTM)	(None, 50)	23000
dense (Dense)	(None, 1)	51

Total params: 139,019
Trainable params: 139,019
Non-trainable params: 0

Fig. 2. Developed model summary.

4 Results

4.1 Model Evaluation

For evaluating our model we use the root mean squared error (rmse),

$$\mathrm{RMSE} = \sqrt{\frac{\sum_{i=1}^{N}(x_i - \hat{x}_i)}{N}}$$

The overall model performance for selected stock data (Table 1) (Figs. 3, 4 and 5):

Table 1. Evaluation of model loss for different parameters.

Stock	Number of Epochs	Training RMSE	Testing RMSE
Tata Motors (Close)	50	4.1032×10^{-4}	0.0020
Tata Motors (Close)	100	2.1577×10^{-5}	0.0011
Reliance (Close)	50	5.4608×10^{-3}	0.047
Reliance (Close)	100	3.3861×10^{-3}	0.036
SBIN (Close)	50	2.8371×10^{-4}	0.0029
SBIN (Close)	100	3.1508×10^{-5}	0.009

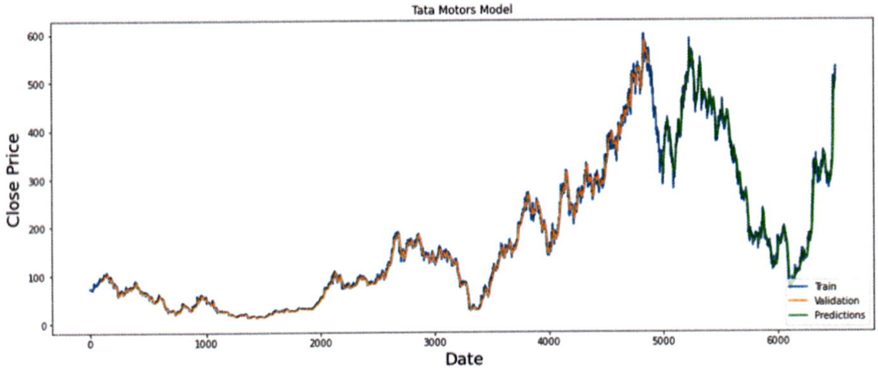

Fig. 3. Model validation, testing and prediction curve on the original data curve for Tata Motors.

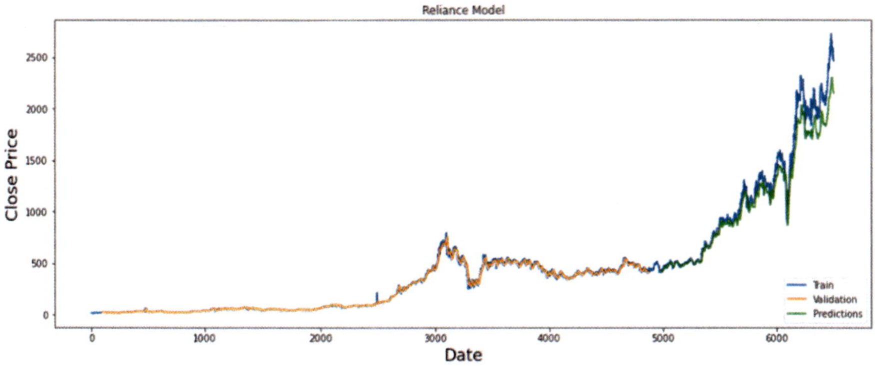

Fig. 4. Model validation, testing and prediction curve on the original data curve for Reliance.

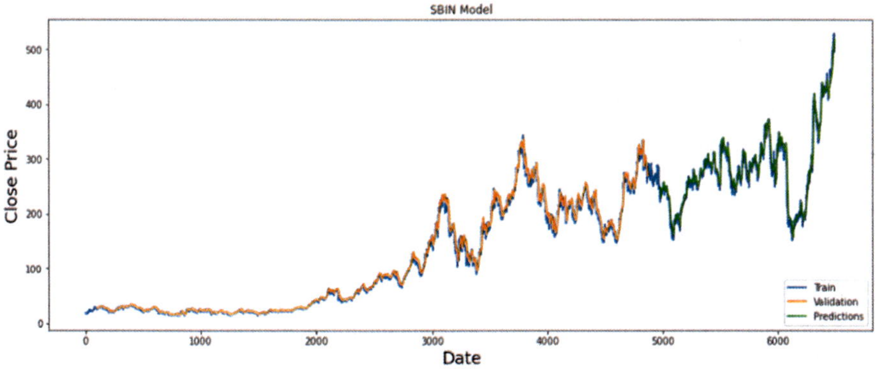

Fig. 5. Model validation, testing and prediction curve on the original data curve for SBIN.

5 Conclusions

A retail investors/traders can use our model to test and analyze their investment strategies, gauge the accuracy of the systems and predict the prices of the instrument. A classification model is designed based on LSTM to perform stock price movement predictions. Long Short Term Networks (LSTM) are highly efficient for sequence predictions. We have trained, validated and tested our model on 3 different stock data and compare the obtained predicted values to the actual historical data of the stocks and plot them simultaneously. We have also looked at our model's efficiency for with different learning rates and the number of epochs and proved that LSTM networks work best with time series data models.

The integration of Deep Learning using open sourced tools into financial analysis is going to be a good supplement to the retail investor in order to test out their strategy on the model and predict the future price which helps in analyzing the market conditions and scenarios with respect to the price movements.

With the inclusion of several other parameters we can design a real world working model to predict prices of various instruments.

6 Future Work Scope

We intend to explore ways to improve our model's prediction accuracy by using various approaches such as improved data pre-processing or using several other features or efficiently studying the design of neural network architecture.

The designed model can only be a supplement to the traders/investors, and in order to build a more sophisticated model we will have to factor in the many more features such as Candle movements, Candle Patterns, Volume, Open Prices and integrate various models like sentiment analysis, in order to achieve a Sustainable System that is useable in the real world scenario.

References

1. Strader TJ, Rozycki JJ, Root TH, Huang YHJ (2020) Machine learning stock market prediction studies: review and research directions. J Int Technol Inf Manag 28(4):63–83 Article 3 (2020)
2. Althelaya KA, Mohammed SA, El-Alfy E-SM (2021) Combining deep learning and multiresolution analysis for stock market forecasting. IEEE Access 9:13099–13111. https://doi.org/10.1109/ACCESS.2021.3051872
3. Nelson DMQ, Pereira ACM, de Oliveira RA (2017) Stock market's price movement prediction with LSTM neural networks. In: 2017 international joint conference on neural networks (IJCNN), pp 1419–1426. https://doi.org/10.1109/IJCNN.2017.7966019
4. Fischer T, Krauss C (2017) Deep learning with long short-term memory networks for financial market predictions. FAU Discussion Papers in Economics 11/2017, Friedrich-Alexander University Erlangen-Nuremberg, Institute for Economics

What are App Store Reviews on Mobile Apps in Retail About? Insights from the German Market

Atilla Wohllebe(✉) and Thomas Stoyke

MATE Hungarian University of Agriculture and Life Sciences, Kaposvár Campus, Kaposvár, Hungary
atilla.wohllebe@gmail.com

Abstract. App reviews drive of app downloads and are highly relevant to app developers because they contain feedback from app users about a mobile app. Research shows that many app reviews are not only related to functional aspects but often rather generic. Findings regarding mobile apps in retail do not exist yet. This exploratory case study examines the content of 300 reviews of 12 apps from stationary retailers in Germany. The results show that a majority of the reviews refer to the app functionalities. Negative reviews are mainly related to functional aspects while positive reviews are often not specified. A surprisingly large proportion of reviews do not refer to the app at all, but to the retailer itself or its stores. App reviews are therefore not only relevant for app developers but may also provide valuable insights for other departments in retail organizations.

Keywords: App reviews · Mobile apps · Stationary retail · App store · Case study

1 Introduction

Mobile apps are steadily gaining in importance in stationary retail as digitization increases [1]. Such digital offerings enable retailers to increase customer satisfaction and generate additional sales [2, 3]. The provision of an online store via an app plays only a downstream role. In particular, customers expect functions to digitally extend their physical in-store shopping experience, for example through additional information and services provided via mobile apps [4–6].

Of central importance for the successful marketing and operation of a mobile app in general, and in retail in particular, are mobile app reviews, which allow users to evaluate apps on the app marketplaces (especially Google Play for Android and App Store for iOS) [7, 8]. In this respect, they should not only be seen as a consequence of the work of app developers, but also an important component of app marketing in terms of App Store Optimization (ASO) [9, 10]. In addition, the use of appropriate keywords and the number of downloads are considered important factors for an app's placement in the charts, categories, and search [11].

M. E. Auer et al. (Eds.): REV 2022, LNNS 524, pp. 463–472, 2023.
https://doi.org/10.1007/978-3-031-17091-1_47

In App Reviews, users mainly report bugs and make suggestions for improvement [12]. App reviews help developers to gain information about user experience, consider it in future development, and thus meet user expectations [13].

In this respect, the relevance of app reviews is on the one hand of a direct technical nature and on the other hand also due to the influence of reviews on app marketing.

While research involving the Technology Acceptance Model primarily emphasizes the importance of technological components (feature set, stability, bugs), other studies indicate that the willingness to install and use mobile apps in retail is also highly associated with the retailer itself, i.e., its brand and service, as well as with customer satisfaction with the retailer [6, 14–19].

Conversely, this finding raises the question of whether consumers really limit themselves to providing feedback on an app when writing app reviews or whether they also include other aspects in them. This applies in particular to aspects that relate, for example, to the app publisher, i.e., in the case of retail, to the retailer, for example, its brand, services or the range of products offered.

In fact, while on the one hand research on app reviews, especially of a quantitative nature, is already very widespread, on the other hand the qualitative examination of the content of reviews represents a research gap [20–22].

This research gap and the previously discussed question of the extent to which app reviews actually focus exclusively on mobile apps give rise to the objective of this article. Based on a primarily qualitative examination of 300 reviews, this article aims to provide an overview of the content of reviews of mobile apps in the retail sector, taking the German market as an example. In particular, it aims to find out to what extent users relate their reviews to the retailer and its locations apart from the app.

First, an overview of the existing scientific literature is given. Then, the material and methods of the article are described. The results show app reviews in German retail first by means of selected descriptive statistics. This is followed by a qualitative examination of app reviews, assigning them to different categories by hand. The summary and discussion show potential for further research.

2 Literature Review

The academic papers selected for the literature review are found by searching for "mobile app store review" on ResearchGate, ScienceDirect, and Google Scholar in Summer 2021, and reflect selected findings of current research in this context.

Overall, many papers primarily emphasize the technical relevance of app reviews in terms of app development [12, 13, 23]. However, in addition to the widespread quantitative description of app reviews, there is also a lot of research - independent of retail - that deals with app reviews in a broadly qualitative sense, or at least tries to approach a qualitative discussion [21, 22]. The research findings discussed below are arranged chronologically in ascending order by year.

In a highly quantitative study, Vasa et al. [22] examine, among other things, the relationship between the length of a review text and the rating given. Among other things, they are able to show that many app review texts tend to be short. They quantify an average review with M = 117 characters (SD = 156). In fact, the content of many app reviews might be difficult to evaluate qualitatively. According to a study by Hoon et al. [20], only 37% of all words in English app reviews also appear in the English dictionary. Words that express a certain sentiment are used most frequently. The range of words used to express a negative opinion is much wider than for a positive review.

To match this, Khalid et al. [24] examine 6,390 negative reviews of 20 free iOS apps. They identify a total of 12 different complaint types, with functional errors, feature requests, and app crashes being particularly frequently addressed. On average, app reviews that refer to privacy, ethical issues or hidden costs have the greatest negative impact. Only 11% of the reviews explicitly refer to a recent update, so generic reviews related to the app are the most common. A study by Mojica Ruiz et al. [25] adds to these findings and finds, based on data from 10,000 Google Play apps, that while updates improve the average app rating, the accumulation of ratings over the overall lifetime of the app does not adequately reflect current user satisfaction.

Grano et al. [26] also examine the thematic relevance of app reviews. From 280,000 reviews of 395 apps, the authors extract 606,000 sentences and investigate to which topics these sentences refer. In line with expectations, the topics of features and functions as well as the app as a whole are the most present. The relevance of app reviews for identifying technical errors is also confirmed by Deshpande & Robke [27], The company behind the app plays a role only sporadically (in 1,672 sentences) in the study by Grano et al. [26]. Intention is distinguished between problem discovery (57,952) and information giving (42,192). It is noteworthy, however, that a large proportion of app reviews are characterized as "other topic" (189,784) and with "other intention" (465,876).

Nayabi et al. [28], comparing app reviews and app-related tweets, find that tweets are more critical but also more objective than app reviews. Also, tweets about apps often include feature requests and bug reports and thus are thematically comparable to app reviews. Jha and Mahmoud [29] study only non-functional requirements (NFR) in their research. Based on 6,000 app reviews, they can show that 40% of reviews contain at least one NFR.

The high relevance of mainly functional comments is also confirmed by a study of drone apps. 162,250 reviews of 1,825 apps show that the most important reasons for complaint are mainly functional errors, compatibility problems, and problems with connection and synchronization [30].

In summary, the review of scientific literature shows that it can be assumed that app reviews contain functional, technical, and user feedback directly related to the app [24, 26, 30]. Nonetheless, it must also be noted that qualitative engagement with app reviews at the content level is challenging and many user reports are uncharacterized or difficult to characterize [20, 26].

With regard to the research gap, it must be noted that previous research rarely if ever focuses on individual app categories. In particular, no findings exist on retail apps.

3 Material and Methods

In line with the previously identified research gap, this paper provides an insight into app reviews in the German retail sector. In particular, it will examine the extent to which app reviews do not focus on the app itself, but on the retailer and its locations. For this purpose, the first 12 apps from retailers with a strong stationary focus, e.g., no pure online stores or voucher portals, will be selected in the Google Play Store Shopping category. The data collection will be conducted in May 2021. The most recent 25 reviews of each of the 12 apps will be evaluated. They are manually examined to determine the extent to which they belong to one of the three superordinate topics or to a further specified sub-topic:

- App-related: Reviews related to the retailer's app in the broadest sense, e. g. referring to features, app-specific deals or coupons provided via the app
  - App function: Reviews related to the functionalities / features of the app, e.g. crashes, search function etc.
- Retailer-related: Reviews related to the retailer in general, e.g. referring to quality of service of weekly deals
  - Store: Reviews related to the stationary retail stores, e.g. friendly employees or clean stores
- General/undefined: Reviews not able to be classified based on previous categories, e.g. generic comments such as "very good" or "okay".

The following list shows the assignment of reviews to the individual categories and subcategories based on a few examples (translated from German):

- App-related: "Very good app, coupons change and there is always something for everyone. You can save good money." (5 stars)
  - App function: "Dark Mode is missing. Thanks" (3 stars)
- Retailer-related: "Bought a Sony TV with over 1,000 €, which was broken after 3 weeks. The store from Nuremberg did not want to do anything with, got a phone number from the manufacturer and [...]" (1 star)
  - Store: "Very nice employees, despite the current situation." (5 stars)
- General / undefined: "Very good" (5 stars)

Figure 1 shows the retailer apps examined and the average rating of the 25 most recent reviews at the time of the study.

The investigation of the app reviews is conducted in an exploratory manner. The observations are in many parts based on the existing literature and are placed in the appropriate context during the discussion.

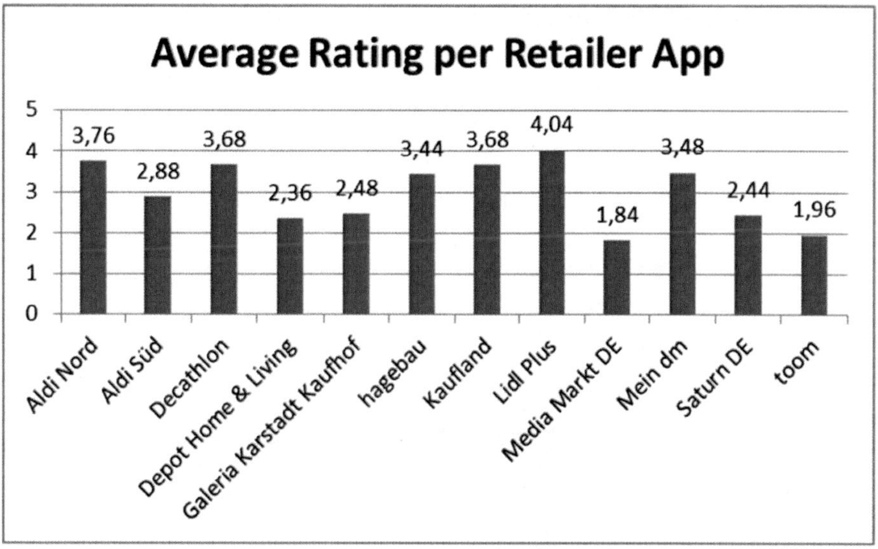

Fig. 1. Average rating per retailer app.

4 Results

The results of the study provide various insights into reviews of apps in the retail sector in Germany.

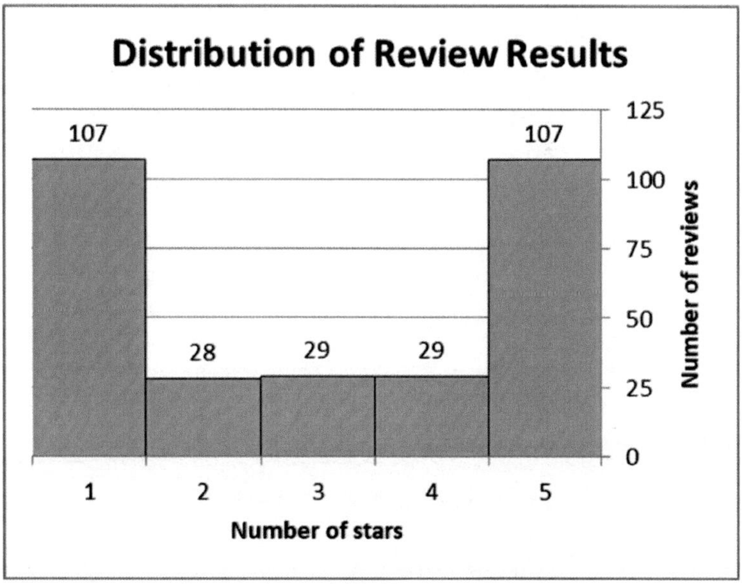

Fig. 2. Number of app reviews per rating.

First, an overview of the evaluated data is obtained. It can be seen that the ratings behind the app reviews average around 3 out of 5 stars (M = 3.003, SD = 1.747). The distribution of ratings on the 5-point Likert scale provided by the Google Play Store shows a clear tendency toward the extreme values of 1 and 5 (cf. Fig. 2). Around 71% of all submitted app reviews rate the respective app as either 1 or 5 [31, 32].

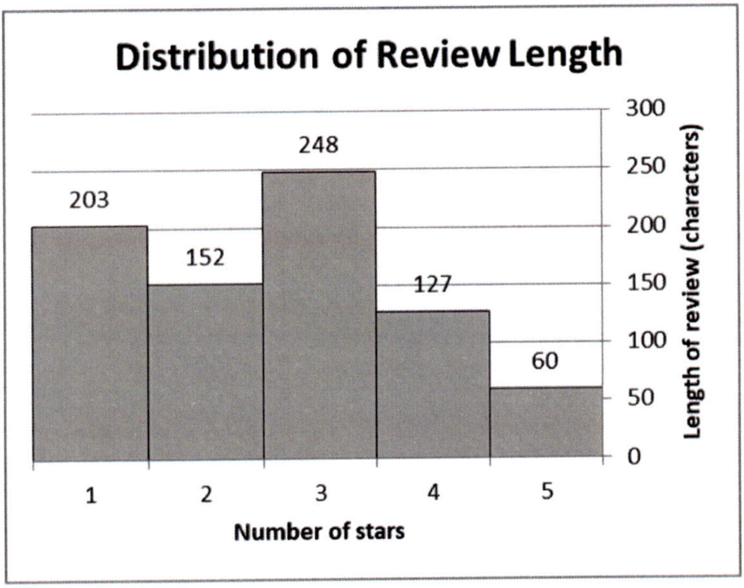

Fig. 3. Average number of characters per rating.

The average length of the text of the app reviews is around 144 characters (M = 144.35, SD = 137.23). The distribution of length by rating shows that reviews with three stars are the longest (M = 248), while very good ratings with 5 stars are very short (M = 60). Also, the distribution suggests a negative correlation of rating and text length (cf. Fig. 3).

Since there is no normal distribution (Shapiro-Wilk: stars W = .986, $p = .005$; lentextDE W = .864, $p < .001$), this correlation is tested using Spearman [33] correlation [34]. The result shows a strong negative correlation (rs = -.498, $p < .001$).

If one assumes a causal relationship between rating and text length and assumes that a poor rating is accompanied by a corresponding amount of criticism, this can be tested with a linear regression ($F(1, 298) = 70.0224$, $p < .001$, $R^2 = .190$). As the rating increases, the length of the text decreases significantly (β = -34.26, 95% CI [-42.31, -26.20]).

To determine the extent to which consumers' feedback in the app reviews also relates to the retailer as a whole and its locations in particular, the 300 reviews are analyzed manually in terms of the topics they cover. In addition, the reviews are divided into bad (1–2 stars) and good (4–5 stars) (cf. Table 1). If a review covers several topics, it may appear more than once in the overview.

Table 1. Number of app reviews per topic and subtopic.

	Total	1–2 stars	4–5 stars
Number of reviews	300	135	136
Topic: App broad	183	105	52
Subtopic: App functionality	165	97	42
Topic: Retailer broad	55	25	29
Subtopic: Store	19	4	15
Topic: Non-specific	73	13	58

Among all reviews, 61% refer to the app in the broadest sense (app broad, N = 183). Around 90% of these refer to functionalities of the app. It is also striking that 57% of all app-related reviews are poor (1–2 stars, N = 105), while only 28% are positive (4–5 stars, N = 52).

Away from the expected relevance of the app in app reviews, around 18% of the texts examined do not or not only refer to the app, but also or exclusively to the retailer (N = 55). Here, however, only 45% are negative (1–2 stars) (N = 25), while 53% - and thus the majority - are positive (4–5 stars) (N = 29).

It should be mentioned in more detail that 34.5% of the reviews that refer to the retailer in the broadest sense (N = 55) make specific reference to the retailer's individual stores (N = 19). Feedback on the stores was positive in 15 out of 19 cases - 78.9%.

No specific topic can be assigned to 24.33% (N = 73) of the reviews. Of these, however, 79.5% (N = 58) were positive with 4 or 5 stars and included texts such as "Everything okay" or "Super".

5 Summary and Discussion

In summary, the results show that the majority of app reviews (61%) actually refer to the app being reviewed. As a rule, these reviews are apparently rather critical and contain concrete feedback on individual functions and errors. These findings confirm the widely mentioned role of app reviews as a feedback channel for app developers [12, 13, 23, 26, 27, 29].

Also, the results show that 18% of the reviews directly refer to the retailer itself. These reviews even seem to tend to be slightly positive. The proportion of retailer-related reviews represents a clear difference to the research of Grano et al. [26], where only about 0.275% of all investigated sentences in app reviews also refer to the app publisher. In this respect, the results presented here show that app reviews can also be an important feedback channel outside of app development. Rather, they are considered part of customer feedback as a whole and are therefore also potentially relevant for marketing, sales and customer management.

The distribution of reviews presented at the beginning clearly tends towards the extremes, although the scale has an odd number of expressions with 1 to 5 [35]. This can probably be attributed to the anonymously perceived evaluation situation [31].

Regarding the length of the reviews, the review texts analyzed here have on average significantly more characters (M = 144.35) than Vasa et al. [22], where the length is M = 117 characters. However, it must be considered here that the difference could also be due to the language of the reviews analyzed in this paper (German). The negative correlation between rating and text length, on the other hand, can be confirmed.

6 Conclusion

This paper analyzes 300 reviews of mobile apps from German retailers. In addition to basic descriptive data, it includes a qualitative, manual text analysis to determine the topics of the app reviews. In particular, it examines the extent to which app reviews in retail actually refer to the app as a technological product and the extent to which retailers in the broader sense are also mentioned and evaluated in them.

Due to the manual topic definition of the reviews and the first look at app reviews in the retail sector, this paper represents an important addition to the existing scientific literature around app reviews. In addition to confirming app reviews as a feedback channel for app developers, the paper additionally shows the role of app reviews also as a feedback channel for marketing, sales and customer management functions in retail companies.

Further research steps should aim in particular at validating the results on a larger data basis, for example internationally and across industries.

References

1. Deckert R, Wohllebe A (2021) Digitalisierung und Einzelhandel: Taktiken und Technologien. Praxisbeispiele und Herausforderungen. Springer Gabler, Wiesbaden
2. van Heerde HJ, Dinner IM, Neslin SA (2019) Engaging the unengaged customer: the value of a retailer mobile app. Int J Res Mark 36:420–438. https://doi.org/10.1016/j.ijresmar.2019.03.003
3. Verma DS, Verma D (2013) Managing customer relationships through mobile CRM in organized retail outlets. Int J Eng Trends Technol - IJETT. 4:1697–1701
4. Dacko SG (2017) Enabling smart retail settings via mobile augmented reality shopping apps. Technol Forecast Soc Change 124:243–256. https://doi.org/10.1016/j.techfore.2016.09.032
5. Kang J-YM (2017) Augmented reality mobile apps in fashion retail: expectancy-value judgments. In: international textile and apparel association annual conference proceedings, vol 74
6. Wohllebe A, Dirrler P, Podruzsik S (2020) Mobile apps in retail: determinants of consumer acceptance – a systematic review. Int J Interact Mob Technol 14:153–164. https://doi.org/10.3991/ijim.v14i20.18273

7. Fu, B., Lin, J., Li, L., Faloutsos, C., Hong, J., Sadeh, N.: Why people hate your app: making sense of user feedback in a mobile app store. In: Proceedings of the 19th ACM SIGKDD international conference on Knowledge discovery and data mining, Chicago, Illinois, USA, pp 1276–1284. ACM (2013). https://doi.org/10.1145/2487575.2488202
8. Lee G, Raghu TS (2014) Determinants of mobile apps' success: evidence from the app store market. J Manag Inf Syst 31:133–170. https://doi.org/10.2753/MIS0742-1222310206
9. Sadiq S, Umer M, Ullah S, Mirjalili S, Rupapara V, Nappi M (2021) Discrepancy detection between actual user reviews and numeric ratings of Google App store using deep learning. Expert Syst Appl 181:115111. https://doi.org/10.1016/j.eswa.2021.115111
10. Wohllebe A, Hillmers M (2021) towards a scientific definition of app marketing – a practice-oriented approach using scientific and grey literature. IJARBM 2:13–25. https://doi.org/10.51137/ijarbm.2021.2.1.2
11. Karagkiozidou M, Ziakis C, Vlachopoulou M, Kyrkoudis T (2019) App store optimization factors for effective mobile app ranking. In: Kavoura A, Kefallonitis E, Giovanis A (eds) Strategic Innovative Marketing and Tourism. SPBE. Springer, Cham, pp 479–486. https://doi.org/10.1007/978-3-030-12453-3_54
12. Malgaonkar S (2020) Prioritisation of requests, bugs and enhancements pertaining to apps for remedial actions - towards solving the problem of which app concerns to address initially for app developers. https://www.researchgate.net/publication/351010227_Prioritisation_of_requests_bugs_and_enhancements_pertaining_to_apps_for_remedial_actions_Towards_solving_the_problem_of_which_app_concerns_to_address_initially_for_app_developers
13. Genc-Nayebi N, Abran A (2017) A systematic literature review: opinion mining studies from mobile app store user reviews. J Syst Softw 125:207–219. https://doi.org/10.1016/j.jss.2016.11.027
14. Briz-Ponce L, García-Peñalvo FJ (2015) An empirical assessment of a technology acceptance model for apps in medical education. J Med Syst 39(11):1–5. https://doi.org/10.1007/s10916-015-0352-x
15. Kim E, Lin J-S, Sung Y (2013) To app or not to app: engaging consumers via branded mobile apps. J Interact Advert 13:53–65. https://doi.org/10.1080/15252019.2013.782780
16. Papadakis S, Kalogiannakis M, Zaranis N (2018) Educational apps from the Android Google play for Greek preschoolers: a systematic review. Comput Educ 116:139–160. https://doi.org/10.1016/j.compedu.2017.09.007
17. Peng K-F, Chen Y, Wen K-W (2014) Brand relationship, consumption values and branded app adoption. Ind Manag Data Syst 114:1131–1143. https://doi.org/10.1108/IMDS-05-2014-0132
18. Saare MA, Hussain A, Yue WS (2019) Conceptualizing mobile health application use intention and adoption among Iraqian older adults: from the perspective of expanded technology acceptance model. Int J Interact Mobile Technol (iJIM). 13:28–41. https://doi.org/10.3991/ijim.v13i10.11285
19. Wohllebe A, Ross F, Podruzsik S (2020) Influence of the net promoter score of retailers on the willingness of consumers to install their mobile app. Int J Interact Mob Technol 14. https://doi.org/10.3991/ijim.v14i19.17027
20. Hoon L, Vasa R, Schneider J-G, Mouzakis K (2012) A preliminary analysis of vocabulary in mobile app user reviews. In: Proceedings of the 24th Australian Computer-Human Interaction Conference on - OzCHI '12. pp. 245–248. ACM Press, Melbourne, Australia. https://doi.org/10.1145/2414536.2414578
21. Hoon L, Vasa R, Martino GY, Schneider J-G, Mouzakis K (2013) Awesome!: conveying satisfaction on the app store. In: Proceedings of the 25th Australian computer-human interaction conference on augmentation, application, innovation, collaboration - OzCHI 2013 , Adelaide, Australia. pp 229–232. ACM Press. https://doi.org/10.1145/2541016.2541067

22. Vasa R, Hoon L, Mouzakis K, Noguchi A (2012) A preliminary analysis of mobile app user reviews. In: Proceedings of the 24th Australian computer-human interaction conference on - OzCHI 2012, Melbourne, Australia, pp 241–244. ACM Press. https://doi.org/10.1145/2414536.2414577
23. Oh J, Kim D, Lee U, Lee J-G, Song J (2013) Facilitating developer-user interactions with mobile app review digests. In: CHI 2013 extended abstracts on human factors in computing systems on - CHI EA 2013, Paris, France, p 1809. ACM Press. https://doi.org/10.1145/2468356.2468681
24. Khalid H, Shihab E, Nagappan M, Hassan AE (2015) What do mobile app users complain about? IEEE Softw 32:70–77. https://doi.org/10.1109/MS.2014.50
25. Mojica Ruiz IJ, Nagappan M, Adams B, Berger T, Dienst S, Hassan AE (2016) Examining the rating system used in mobile-app stores. IEEE Softw 33:86–92. https://doi.org/10.1109/MS.2015.56
26. Grano G, Di Sorbo A, Mercaldo F, Visaggio CA, Canfora G, Panichella S (2017) Android Apps and user feedback: a dataset for software evolution and quality improvement. In: Proceedings of ACM conference, Washington DC, USA
27. Deshpande G, Rokne J (2018) User feedback from tweets vs app store reviews: an exploratory study of frequency, timing and content. In: 2018 5th international workshop on artificial intelligence for requirements engineering (AIRE), Banff, AB, pp 15–21. IEEE. https://doi.org/10.1109/AIRE.2018.00008
28. Nayebi M, Cho H, Ruhe G (2018) App store mining is not enough for app improvement. Empir Softw Eng 23(5):2764–2794. https://doi.org/10.1007/s10664-018-9601-1
29. Jha N, Mahmoud A (2019) Mining non-functional requirements from app store reviews. Empir Softw Eng 24(6):3659–3695. https://doi.org/10.1007/s10664-019-09716-7
30. Kalaichelavan K, Malik H, Husnu N, Sreenath S (2020) What do people complain about Drone Apps? A large-scale empirical study of Google Play Store reviews. Procedia Computer Science. 170:547–554. https://doi.org/10.1016/j.procs.2020.03.124
31. Bogner K, Landrock U (2015) Antworttendenzen in standardisierten UmfragenAntworttendenzen in standardisierten Umfragen. https://doi.org/10.15465/GESIS-SG_016
32. Likert R (1932) A technique for the measurement of attitudes. Arch Psychol 22:1–55
33. Spearman C (1904) "General Intelligence", objectively determined and measured. Am J Psychol 15:201. https://doi.org/10.2307/1412107
34. Smigierski J Korrelation In STATA berechnen. http://www.beratung-statistik.de/statistik-beratung-infos/stata-tutorial/stata-nachhilfe-korrelation/. Accessed 29 June 2021
35. Schnell R, Hill PB, Esser E (1999) Methoden der empirischen Sozialforschung. R. Oldenbourg, München

Temporal Data Processing with ASAMPL Programming Language in Mulsemedia Applications

Danyil Peschanskyi[1], Pavlo Budonnyi[1], Yevgeniya Sulema[1](✉), Frederic Andres[2], and Andreas Pester[3]

[1] Igor Sikorsky Kyiv Polytechnic Institute, Kyiv, Ukraine
sulema@pzks.fpm.kpi.ua
[2] National Institute of Informatics, Tokyo, Japan
andres@nii.ac.jp
[3] British University in Egypt, Cairo, Egypt
andreas.pester@bue.edu.eg

Abstract. The paper presents a new programming paradigm that enables simplifying temporal data processing which can be used as a core of mulsemedia and affective computing applications. The paper also shows the semantics and pragmatics of basic timewise operations for mulsemedia data processing in the ASAMPL programming language. The proposed approach can be used for the development of various applications which requires the processing of mulsemedia information. The ASAMPL program code development tools and the practical result of their use are presented in this paper as well.

Keywords: Mulsemedia · Affective computing · Temporal data processing

1 Introduction

Mulsemedia and affective computing [1–3] applications can require complex representing and simultaneous processing of multiple types of information perceived by human senses, including information about taste [4–9]. For example, in the food industry, mulsemedia applications can require a complex representation of visual data (video, images), olfactory data, gustatory data, and textual data (receipts, calorie content, diet information). This creates a need for new approaches to software development tools and methods for increasing computational efficiency especially taking into consideration a wide range of specialized hardware required for both recording and reproduction of mulsemedia data. The hardware of multimedia systems includes devices for recording and reproduction of three-dimensional visual information [10], devices for recording and reproduction of spatial acoustic information [11], devices for tactile interaction [12–15], devices for determining and reproducing odour [16], devices for simulating taste sensations [4–9], devices for creating other sensory effects.

M. E. Auer et al. (Eds.): REV 2022, LNNS 524, pp. 473–485, 2023.
https://doi.org/10.1007/978-3-031-17091-1_48

The motivation for the research, part of which is presented in this paper, is to develop a general approach for the development of simple procedures for representing and computing multiple mulsemedia data sets received from specialized hardware and presented in arbitrary formats, which requires synchronization and matching. This approach enables presenting a virtual or real object in the most comprehensive way and with strong consideration of time scale with timestamps as humans perceive real-world objects not only in three dimensions but also in terms of time.

The rest of the paper is organized in the following way. The second section analyses existing researchers in the field of mulsemedia. The third section presents a new programming paradigm that allows the development of a software code that realizes a complex data representation and temporal data processing for handling information about a real or virtual object with mulsemedia features. The fourth section demonstrates the result of the practical implementation of the proposed approach. The conclusion summarizes the research results presented in this paper and shows the directions of future work.

2 Related Works

Many research works in the field of mulsemedia are related to solving such tasks as the design of tools and frameworks for mulsemedia applications development [28–36], mulsemedia system design [37], synchronization of multimodal data [17–19], organization of human-machine interaction and creation of new user interfaces based on the use of mulsemedia data [8, 9, 20, 21], development of mulsemedia applications [22, 23], increasing the level of QoS and QoE for mulsemedia systems [24, 25], and other aspects of mulsemedia. Since the research presented in this paper concerns mulsemedia data processing and mulsemedia applications development, let us analyse the researches and software tools that have similar purposes.

In [28, 29], the authors present software tools that are designed to implement tactile interaction between the user and virtual objects. The HFX Studio [28] editor is based on three perceptual models: a model of pressure, a model of vibration, and a model of temperature, which enable a spatial tactile effect on the user's body. The proposed models can also be used to describe the capabilities of tactile devices.

The H-Studio editor [29] is designed to add tactile and motor effects to audio-visual material. There are three methods for creating motion effects: manually creating at key points, recording the trajectory using a tactile feedback device, and importing object motion data.

The distributed mulsemedia platform PlaySEM is presented in [30]. The authors propose a test scenario for a mulsemedia application and analyse the time efficiency of its work. The authors note that when remotely using actuators that reproduce sensory effects over a wired network, the delay in reproducing mulsemedia effects averages 27–67 ms, which is satisfactory for most applications. The PlaySEM platform is compatible with the MPEG-V standard and the ISO/IEC standards.

In [31], a software platform for creating mulsemedia applications is presented. It includes three components: the SEVino sensory effects editor [32], the SESim sensory effects simulator [33], and the SEMP sensory effects player [34]. The software system is based on the MPEG-V standard. The components of the platform are developed in Java (SEVino, SESim) and C++ (SEMP).

In [35], the authors present the advanced framework Play-SEM SER 2 [36] for working with heterogeneous applications and devices, in particular, they describe its architecture design, applied design patterns, services provided to promote integration for heterogeneous multimedia applications, interaction with timeline multimedia applications, and other aspects of working with the framework. The authors also present two case studies: (1) video clip enriched with mulsemedia effects and (2) smell-intensive system. The framework Play-SEM SER 2 is the most advanced among existing tools for mulsemedia applications development.

To support a concept DIY (Do It Yourself), the tutorial on mulsemedia system development is presented in [37]. The authors provide both a survey on the mulsemedia hardware (haptic, gustatory, and olfactory displays) and a detailed explanation on options of building a mulsemedia environment.

In [19], the authors propose temporal models for related mulsemedia documents (models based on points, events or intervals) and basic methods of data synchronization, offer solutions to numerous application problems, and analyse the possibilities of using specialized programming languages SMIL [26] and NCL [27] for synchronizing multimedia elements. SMIL and NCL programming languages are based on XML. SMIL programming language is designed to develop web-based multimedia applications and allows a programmer to operate multimedia elements such as text, images, video, and audio. The NCL programming language is intended for declarative specification of software applications based on the concept of hypermedia.

In [38], the authors discuss technology-assisted sensory marketing, present the cognitive and emotive aspects of virtual reality, and propose applications of virtual reality technologies to encourage sustainable consumption.

The challenges and opportunities for using mulsemedia technology in educational applications are discussed in [39]. In particular, the authors provide six "golden rules" based on their experience in technology-enhanced learning projects.

Analysis of software that supports mulsemedia technology shows the following:

- Lack of specialized programming languages designed for software development of mulsemedia data processing systems.
- The focus of the process of developing mulsemedia applications is on the use of editors of sensory effects (authoring tools), which allow adding mulsemedia elements to the audio-visual material by annotation.

In the case of creating mulsemedia software, its development is mainly performed according to the paradigm of object-oriented programming. However, new approaches for mulsemedia data processing can provide opportunities for simplifying the development of mulsemedia applications.

3 Multi-image Programming Paradigm

One of the factors that affect the efficiency of data processing in a computer system is the optimality of the software code, which depends on the programming style and computational model, which is determined by the programming paradigm.

The programming paradigm is a set of principles, methods and concepts that determine the way programs are designed. The programming paradigm defines an approach to programming based on mathematical theory or an agreed set of principles. Each paradigm supports a set of concepts, which makes it the best for solving problems of a certain class. Thus, object-oriented programming is aimed at solving problems characterized by a large number of related data abstractions organized in a hierarchy. Logic programming is designed to solve problems according to the rules of inference. Real-time programming is focused on solving problems that require a real-time response with clearly defined data processing time requirements.

In this research, we apply a multi-image programming paradigm. This paradigm is focused on solving problems of temporal processing the multimodal data which has the following features:

1. A multimodal data set to be processed is accompanied by timestamps for each data value or these timestamps can be obtained as a result of computing.
2. Data of different modalities that define a certain real or virtual object are considered and processed in conjunction with data of other modalities that define the same object.
3. Data are obtained from arbitrary sources and in arbitrary formats.

The multi-image programming paradigm, as a methodology for creating software, is that the developer needs to consider the relationship of data in terms of both time and modality, i.e. to operate such an entity as a multi-image of an object [40, 41].

The data structure for describing the multi-image should be defined in the main part of the program code, which begins the solution of the data processing problem. The logic of solving this problem should be based on the operations and relations defined for the multi-image as a mathematical object in the algebraic system of aggregates [40].

The program code should be constructed in such a way that the developer can abstract as much as possible from what is the data source, in which format this data is received for processing and what is a receiver of data during its reproduction. The tasks of data conversion between formats, preparation for playback with the use of specialized hardware are solved with the help of specialized libraries, as should be supplied by software developers of this specialized hardware.

The basic principles of the multi-image programming paradigm are the following:

1. The main entity for the presentation and processing of data in the program is a multi-image.
2. Synchronization and aggregation of data of different modalities is the basis of the computational model for processing multi-image data.
3. The processing process is based on considering the properties of data temporality.

4. Processing of multi-images is carried out using operations and relations defined in the algebraic system of aggregates.

In addition, the abstraction of the concepts of source and receiver of data is performed through the use of specialized libraries that allow a programmer to convert data from one format to another without the need to develop a program code for this conversion procedure.

The programming paradigm is realized through a computational model. The computational model is a formal system that determines how calculations are performed.

The computational model for the multi-image programming paradigm is based on the concept of a multi-image as an abstract data structure. A multi-image is an aggregate of an algebraic system of aggregates, the first tuple of which is a non-empty tuple of timestamps [40, 41].

In the computational model, the multi-image is a key abstraction, which is defined at two levels: (1) at the level of processing of individual temporal data (partial multi-images) and (2) at the level of formation of a single synchronized and aggregated set of temporal multimodal data (the object multi-image).

At the first level, the multi-image is defined as an aggregate consisting of two tuples: a tuple of time stamps and a tuple of data of a certain modality. At this level, the primary data synchronization takes place which consists in the synchronization of timestamps and the modality data for which they are defined. Thus, the structure of the first level multi-image is typical, but the second tuple of each of these multi-images belongs to a specific type of data.

At the second level, the multi-image defines the method of aggregation of all multimodal data in one data structure. The second level multi-image corresponds to some data representation model, for example, one of those proposed in [42].

Thus, the scheme of the basic computational model includes a two-level hierarchical procedure of synchronization and aggregation. At the first level, synchronization and aggregation take place for a pair of tuples "time–modality", and at the second level, the data of all modalities are synchronized and aggregated into a single ordered set of data, ready for further processing. This processing is the third level of the schema of the basic computational model.

Further data processing of the obtained multi-images is to be performed using a programming language such as the ASAMPL programming language, which got its name from "Algebraic System of Aggregates" and "Mulsemedia data Processing Language" [41]. ASAMPL is a cross-platform programming language that allows the development of any applications which operate with mulsemedia data; the language offers to developers a set of programming constructions, abstractions, standard functions, and data types that help to realize any mulsemedia data processing. To work with sensory effects devices, the language supports using special functions – handlers and renders – included in the standard library. If a developer uses their own sensory effects device, such special functions must be developed especially for this device. This can be considered as a limitation but with every next iteration of the ASAMPL project [43], the number of handlers and renders for existing mulsemedia devices is expected to grow.

The purpose of the ASAMPL programming language is to simplify the processing of multimodal, in particular, multimedia data, the main feature of which is that they are temporal. In addition, the processing of such data should consider the data modalities, their compatibility and the relationship (including semantics) between them. Therefore, the main principle of organizing the program in ASAMPL is that the execution of the program is based on two entities: time and modality of data, which corresponds to the principles of the multi-image programming paradigm.

A key concept in ASAMPL is the concept of a multi-image. To represent a multi-image of a real or virtual object in this programming language, special data structures are used which are tuples and aggregates. The processing of aggregates is performed in accordance with the rules specified in the algebraic system of aggregates [40].

The main features of the ASAMPL language, which corresponds to the principles of the multi-image programming paradigm, are as follows:

1. The relationship of the data processing process to the timeline.
2. Synchronization of data of different modalities.
3. Presentation of multimodal data using aggregates and tuples.
4. Focus on the simultaneous use of different sources of multimodal data and different formats.
5. Use of the algebraic system of aggregates for processing of multi-images and their components.

These features are realized through the implementation of specific commands for mulsemedia data processing such as TIMELINE, SEQUENCE, SUBSTITUTE FOR WHEN, DOWNLOAD FROM, UPLOAD TO, and RENDER WITH [41]. These commands are designed specifically to simplify basic procedures of data handling that simplify application development. Let us study some of these commands in detail.

As it was defined above, since any human-perceived information exists in terms of time, mulsemedia data are temporal. Therefore, such data needs to be processed with consideration of the strong relation of any mulsemedia data value to a specific timestamp. In universal programming languages, it is solved by using cycles (loop operators), however, in ASAMPL there is a specific operator which makes temporal data processing more natural and simpler for a programmer. This operator is TIMELINE and its modification TIMELINE AS and TIMELINE UNTIL.

The operational semantics of the TIMELINE operator is given in Fig. 1. This diagram shows the flexibility of the TIMELINE operator for the simultaneous fulfilment of different actions on different data sets.

The modification of this operator named TIMELINE AS allows the processing of data in specific moments of time (Fig. 2).

Another modification of the TIMELINE operator named TIMELINE UNTIL enables processing dependable on a certain external event or condition that makes it similar to some extent to loop operators like WHILE in other languages.

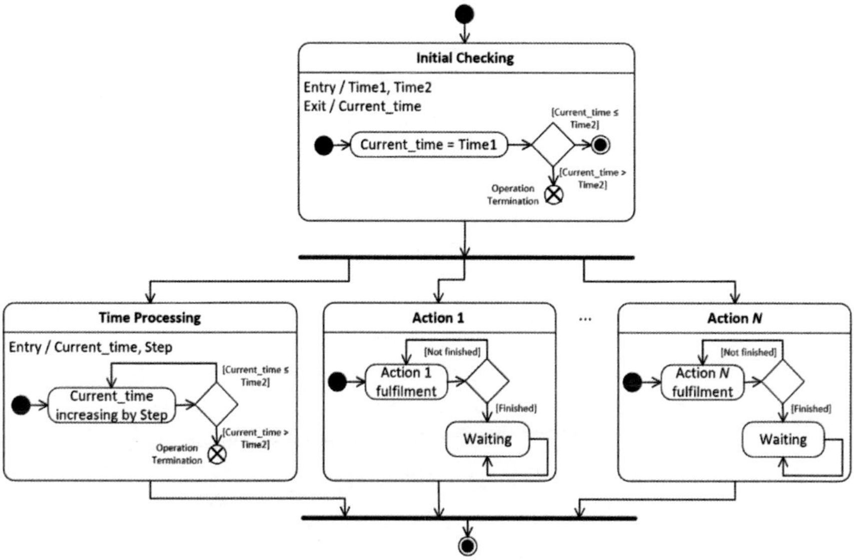

Fig. 1. Diagram of states for TIMELINE operator

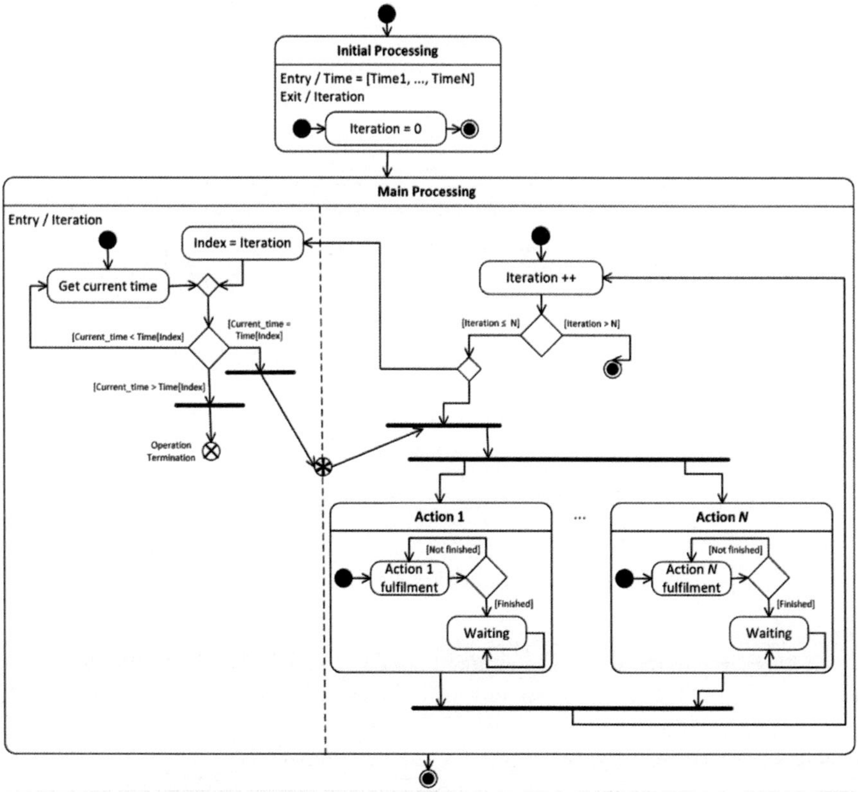

Fig. 2. Diagram of states for TIMELINE AS operator

4 Results and Discussion

To allow programmers to develop software applications and run ASAMPL code, the compiler, library, and IDE are developed as part of this research [43].

The ASAMPL language library allows a programmer to use built-in language types while developing a program code for processing mulsemedia data. The library implements functions to work with all types of data in the ASAMPL language and to convert them. The library also provides opportunities for interaction with the operating system, including input and output operations.

The existing standard library is not intended to provide tools for working with specific mulsemedia hardware and data formats; they must be implemented by developers of specific devices and formats. However, the library provides tools for working with audio and video as their formats are standardized.

Another feature supported by the developed library is the possibility to interact with a program code developed by using other programming languages, namely, to use functions and handlers. There are two interfaces for this: the machine code interface and the Python interface.

Interaction with machine code occurs through the mechanism of dynamic libraries. In the case of Windows, it is.dll files, in the case of Linux, it is.so. To be considered as ASAMPL library or handler, a dynamic library must export certain functions. The structure of such a file is shown in Fig. 3.

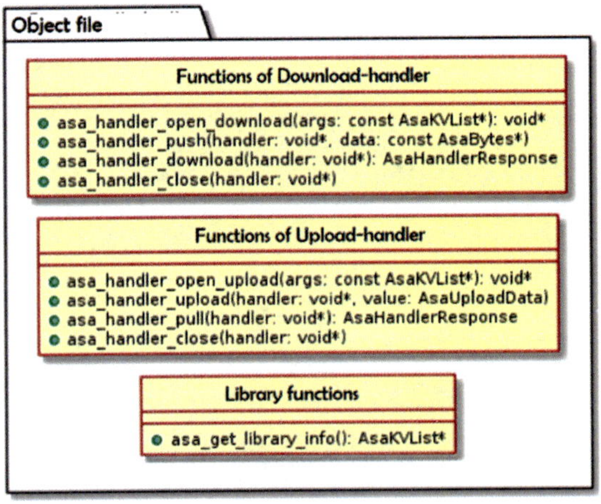

Fig. 3. The general structure of the object file that the ASAMPL language can interact with

To interact with the Python programming language, the ASAMPL interpreter uses boost-Python and boost-NumPy libraries. A library or handler written in Python looks like a regular Python library. All global values created in the imported file is available in ASAMPL. Most often these are functions, but there may be other global variables. The general view of the Python file that the ASAMPL language can work with is shown in Fig. 4.

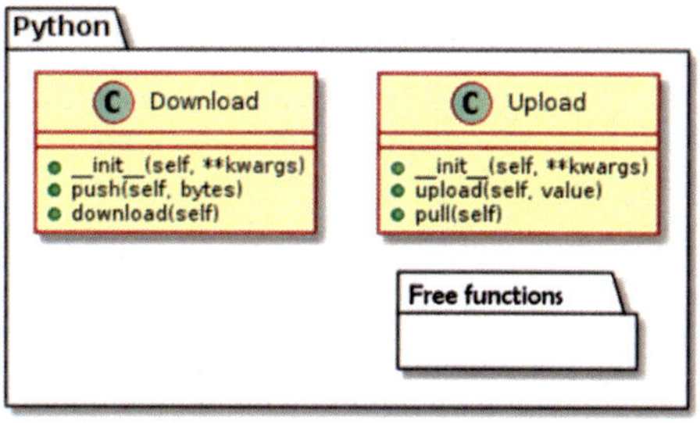

Fig. 4. General view of a Python file that ASAMPL can work with

To test the developed compiler, library, and IDE, a demo that demonstrates the processing of temporal mulsemedia data sets, such as visual data, gustatory data, and olfactory data, was developed (Fig. 5).

Fig. 5. Mulsemedia application with taste data processing developed in ASAMPL

The developed ASAMPL code uses the TIMELINE operator for simultaneous processing of mulsemedia data (Listing 1).

Listing 1. ASAMPL Code

```
import handler ffmpeg from "ffmpeg-handler"
import handler string_stream from "string-stream-handler"
fn main() { let tastes = [ "sweet", "sour", "salty", "bit-
ter", "umami" ];
    let images = t_map(tastes, |name| { s_concat("data/",
name, ".png"); });
    images = t_map(images, load_image);
    timeline { "downloads": [
             { "handler": ffmpeg,
               "source": "data/food.ogv"   },
             { "handler": string_stream,
               "source": "data/taste.txt"  }  ]  }  |img,
taste_value| {
        let offset = 0;
        let i = 0;
   while (i < t_len(tastes)) {
    if (s_contains(taste_value, tastes[i])) {
    i_overlay(img, images[i], { "x": 0, "y": offset });
            offset = offset + i_height(images[i]); }
            i = i + 1;   }
        img = scale_image(img, { "x": 300, "y": 500});
        show_image(img); };   }
```

For demonstration purposes, gustatory information is shown in the demo as added symbolic images of tastes (sweet, sour, salty, bitter, umami). In the case of using any hardware which can handle gustatory data, a corresponding action with this hardware is expected to be performed instead of a demonstration of these images.

5 Conclusions and Future Work

The proposed approach to the development of mulsemedia applications allows the development of any software for temporal data processing by using the ASAMPL programming language. It is a key difference between this approach of mulsemedia applications development and the use of authoring tools. The development of software code in the ASAMPL programming language allows a programmer to develop mulsemedia applications that perform any necessary processing of data and reproduce the result by using appropriate hardware. The current limitation for the wide use of ASAMPL for the mulsemedia applications development consists in a limited number of handlers and renders included in the standard library.

The future work will be focused on supporting a wider range of mulsemedia hardware, first of all, haptic devices. The language syntax also requires modification in order to simplify the syntactic constructions of the language and improve program code metrics. Another direction for the future advancement of the ASAMPL programming language is to align it with MPEG-V and other existing standards.

References

1. Ghinea G, Andres F, Gulliver S (2012) Multiple sensorial media advances and applications: new developments in mulsemedia. IGI Global, 344 p
2. Raheel A, Majid M, Anwar S (2021) Dear-Mulsemedia: Dataset for emotion analysis and recognition in response to multiple sensorial media. Inf Fus 65:37–49
3. Richardson S (2020) Affective computing in the modern workplace. Bus Inf Rev 37(2):78–85
4. Velasco C et al (2016) Multisensory approach for the design of food and drink enhancing sonic systems. In: Proceedings of the 1st workshop on multi-sensorial approaches to human-food interaction, Article 7, Tokyo, Japan, pp 1–7
5. Bruijnes M, Huisman G, Heylen D (2016) Tasty tech: human-food interaction and multimodal interfaces. In: Proceedings of the 1st workshop on multi-sensorial approaches to human-food interaction, Article 4, Tokyo, Japan, pp 1–6
6. Carvalho FR et al (2016) Sound-enhanced gustatory experiences and technology. In: Proceedings of the 1st workshop on multi-sensorial approaches to human-food interaction, Article 5, Tokyo, Japan, pp 1–8
7. Mroczko-Wasowicz A (2016) The unity and complexity of flavour perception: multisensory phenomenology and processing in human-food interactions. In: Proceedings of the 1st workshop on multi-sensorial approaches to human-food interaction, Article 6, Tokyo, Japan, pp 1–4
8. Narumi T (2016) Multi-sensorial virtual reality and augmented human food interaction. In: Proceedings of the 1st workshop on multi-sensorial approaches to human-food interaction, Article 1, Tokyo, Japan, pp 1–6
9. Sakurai K et al (2016) Saltiness and umami suppression by cathodal electrical stimulation. In: Proceedings of the 1st workshop on multi-sensorial approaches to human-food interaction, Article 2, Tokyo, Japan, pp 1–4
10. HoloVizio 3D. Technology principles. http://www.archive.holografika.com/Technology/Technology-Principles.html. Accessed 17 Jan 2022
11. 3D AudioSense. http://www.3daudiosense.com/uploads/2/4/7/2/24724752/zylia-audiosense-parp_polish_product_of_the_future.pdf. Accessed 05 Dec 2021
12. Fujitsu develops prototype haptic sensory tablet. https://www.fujitsu.com/global/about/resources/news/press-releases/2014/0224-01.html. Accessed 17 Jan 2022
13. VMG 30. https://www.virtualmotionlabs.com/vr-gloves/vmg-30/. Accessed 17 Jan 2022
14. Silva AJ et al (2009) Phantom omni haptic device: kinematic and manipulability. In: Proceedings of the IEEE conference "electronics, robotics and automotive mechanics conference CERMA, Cuernavaca, Mexico, pp 193–198
15. Butterfly Haptics. https://butterflyhaptics.com/products/system/. Accessed 17 Jan 2022
16. Ghinea G, Ademoye O (2021) The sweet smell of success: enhancing multimedia applications with olfaction. ACM Trans Multimed Comput Commun Appl 8(1):1–17
17. Abreu R et al (2019) Semi-automatic synchronization of sensory effects in mulsemedia authoring tools. In: Proceedings of the 25th Brazillian symposium on multimedia and the web webmedia 2019, Rio de Janeiro, Brazil, pp 201–208
18. Yuan Z, Bi T, Muntean G, Ghinea G (2015) Perceived synchronization of mulsemedia services. IEEE Trans Multimedia 17:957–966
19. Montagud M et al (2018) MediaSync handbook on multimedia synchronization. Springer, Cham. https://doi.org/10.1007/978-3-319-65840-7

20. Obrist M et al (2016) Touch, taste, & smell user interfaces: the future of multisensory HCI. In: Proceedings of the 2016 CHI conference extended abstracts on human factors in computing systems, San Jose California, USA, pp 3285–3292
21. Kim J, Lee C-G, Kim Y, Ryu J (2013) Construction of a haptic-enabled broadcasting system based on the MPEG-V standard. Image Commun 28(2):151–161
22. de Amorim MN, Saleme EB, Assis Neto F, Santos CAS, Ghinea G (2019) Crowdsourcing authoring of sensory effects on videos. Multimedia Tools Appl 78(14):19201–19227. https://doi.org/10.1007/s11042-019-7312-2
23. Ghinea G et al (2018) Towards a reference ontology on mulsemedia systems. In: Proceedings of the international conference on management of emergent digital ecosystems MEDES 2018, Tokyo, Japan, pp 23–30
24. Marconi C, Brovetto C, Mendez I, Perera M (2018) Multisensorial books: improving readers' quality of experience. In: Proceedings of the 13th Latin American conference on learning technologies LACLO, São Paulo, Brazil, pp 33–36
25. Mesfin G, Hussain N, Covaci A, Ghinea G (2019) Using eye tracking and heart-rate activity to examine crossmodal correspondences QoE in mulsemedia. ACM Trans Multimedia Comput Commun Appl 15(2):34:1–34:22
26. SMIL 3.0. https://www.w3.org/TR/SMIL3/. Accessed 17 Jan 2022
27. Nested Context Language. http://www.ncl.org.br/en/inicio. Accessed 17 Jan 2022
28. Danieau F et al (2018) HFX studio: haptic editor for full-body immersive experiences. In: Proceedings of the 24th ACM symposium on virtual reality software and technology, Article no. 37, Tokyo, Japan, pp 1–9
29. Danieau F et al (2013) H-studio: an authoring tool for adding haptic and motion effects to audiovisual content. In: Proceedings of the 26th ACM symposium on user interface software and technology UIST 2013, St Andrews, United Kingdom, 2 p
30. Saleme EB et al (2017) Time evaluation for the integration of a gestural interactive application with a distributed mulsemedia platform. In: Proceedings of the 8th ACM on multimedia systems conference MMSys 2017, Taipei, Taiwan, pp 308–314
31. Waltl M, Rainer B, Timmerer C, Hellwagner H (2012) A toolset for the authoring, simulation, and rendering of sensory experiences. In: Proceedings of the 20th ACM international conference on multimedia, New York, United States, pp 1469–1472
32. SEVino. https://sourceforge.net/projects/sevino/. Accessed 17 Jan 2022
33. SESim. https://sourceforge.net/projects/sesim/. Accessed 17 Jan 2022
34. SEMP. https://sourceforge.net/projects/semediaplayer/. Accessed 17 Jan 2022
35. Saleme EB, Santos CAS, Ghinea G (2019) A mulsemedia framework for delivering sensory effects to heterogeneous systems. Multimedia Syst 25:421–447
36. PlaySEM SER 2.0.0. https://github.com/estevaosaleme/PlaySEM_SERenderer/releases/tag/2.0.0. Accessed 17 Jan 2022
37. Saleme EB, Covaci A, Mesfin G, Santos CAS, Ghinea G (2019) Mulsemedia DIY: a survey of devices and a tutorial for building your own mulsemedia environment. ACM Comput Surv 52(3):58:1–58:29
38. Laukkanen T, Xi N, Hallikainen H, Ruusunen N, Hamari J (2021) Virtual technologies in supporting sustainable consumption: from a single-sensory stimulus to a multi-sensory experience. Int J Inf Manag 102455:1–5
39. Volpe G, Gori M (2019) Multisensory interactive technologies for primary education: from science to technology. Front Psychol 10:1076:1–1076:8
40. Sulema Ye, Kerre E (2020) Multimodal data representation and processing based on algebraic system of aggregates. In: Chakraverty S (ed) Chapter 5 in book mathematical methods in interdisciplinary sciences Wiley, USA, 464 p

41. Sulema Ye (2018) ASAMPL: programming language for mulsemedia data processing based on algebraic system of aggregates. In: Auer ME, Tsiatsos T (eds) IMCL 2017, vol 725. AISC. Springer, Cham, pp 431–442. https://doi.org/10.1007/978-3-319-75175-7_43
42. Sulema Ye, Dychka I, Sulema O (2019) Multimodal data representation models for virtual, remote, and mixed laboratories development. In: Auer ME, Langmann R (eds) REV 2018, vol 47. LNNS. Springer, Cham, pp 559–569. https://doi.org/10.1007/978-3-319-95678-7_62
43. ASAMPL Compiler and Library. https://github.com/orgs/Asampl-development-team/repositories. Accessed 17 Jan 2022

Adaptive Weighted Sum Bi-objective Bat for Regression Testing Optimization

Nagwa R. Fisal[1](✉), Abeer Hamdy[2], and Essam A. Rashed[3]

[1] Suez Canal University, Ismailia, Egypt
Nagwa.reda7@gmail.com
[2] British University in Egypt, El Shorouk City, Egypt
[3] Nagoya Institute of Technology, Nagoya, Japan

Abstract. Regression testing is a type of testing carried out during the software maintenance phase, to confirm the validity of a software system after any modifications. However, regression testing is expensive, and sometimes it cannot be carried out within the testing budget, due to the large size of a test suite. In order to reduce regression testing cost, the test suite should be reduced without losing its efficiency in terms of pre-defined criteria such as its capability of fault detection; this problem is known as test suite reduction problem (TSR). In this paper, the TSR problem was formulated as a bi-objective optimization problem using an adaptive-weighted (AW) sum method. Then an adapted binary Bat algorithm (AW-ABBA) was utilized to search for a Pareto-optimal set of solutions; allowing the decision-maker under various circumstances to choose the best solution from the proposed set. The efficacy of the AW-ABBA was assessed using three metrics, Cardinality ratio, IGD^+ and Diversity, over five test suites of different sizes. Experimental results showed that the AW-ABBA was able to efficiently approximate a reference Pareto-front.

Keywords: Multi-objective · Test suite reduction · Binary bat · Regression testing · Pareto front · NSGA-II

1 Introduction

Software testing is used to detect faults to ensure the release of high-quality software. However, testing is expensive, so software systems are usually released with several bugs, e.g., at the early releases of Mozilla and Eclipse, about 170 and 120 bugs respectively were reported daily [1–5]. During the maintenance phase, any modification will affect one or more modules of the software, which necessitates the re-execution of previous test cases in addition to new ones. Regression testing checks the validity of the entire software under test (SUT) after each change [6]. That makes regression testing costly to implement; it accounts for over half of the total cost of testing. Test suite reduction (TSR) technique used to reduce the cost of regression testing, the size of the test suite should be reduced without affecting its efficacy, measured in terms of pre-specified criteria [7], such as mutation testing [8]. TSR problem is a multi-objective combinational optimization problem, and it is an NP-complete. Different approaches for test suite reduction

M. E. Auer et al. (Eds.): REV 2022, LNNS 524, pp. 486–495, 2023.
https://doi.org/10.1007/978-3-031-17091-1_49

have been presented in the literature [9]. Heuristic algorithms proved effectiveness in addressing each of the single objective TSR (SO-TSR) [10, 11] and multi-objective TSR (MO-TSR) problems [9, 12–17]. Recently, Yang [18, 19] proposed the Bat algorithm (BA), which is a swarm intelligence search algorithm inspired by bat echolocation behavior. Yang [18] demonstrated the superiority of BA over each of the GAs [20] and Particle swarm optimization (PSO) [21, 22] over some benchmark functions. However, for some multi-modal functions, BBA occasionally fails to find the global optimal solution. The adapted Binary Bat algorithm ABBA [7, 23] uses an updated formula to the bat velocities different from the one used by the BBA [24], which enabled ABBA to converge to better solutions than the BBA. This paper aims at solving the MO-TSR problem using the ABBA. To achieve our objectives, a fitness function was formulated using the adaptive weighted-sum approach [25], which was proved to be effective in approximating Pareto-front trade-off solutions [26] for bi-objective optimization. The performance of the adaptive weighted-sum ABBA (AW-ABBA) was evaluated in comparison to a pre-determined weighted-sum ABBA (PW-ABBA) [7], pre-determined weighted-sum GA (PW-GA) and NSGA-II. The rest of this paper is organized as follows: Sect. 2 summarizes previous studies that addressed the TSR problem. Section 3 explains the AW-ABBA. The experiments and results are discussed in Sect. 4 finally, Sect. 5 concludes the paper and introduces future directions for this research.

2 Literature Review

Researchers proposed a variety of approaches for reducing the test suite size. Most of these approaches depend on greedy algorithms [27], with only a few depending on clustering algorithms [28] or hybrid algorithms [29]. Recently, heuristic algorithms have been used to solve single and multi-objective TSR [9, 11–17]. As mentioned by Khan et al. [9], the most used heuristics search-based for TSR problem are single-objective optimization, which represents 79% of them. Different studies have shown that some heuristic algorithms outperform greedy algorithms in solving the single and multi-objective TSR problems [11, 14]. The work of Yoo and Harman [14] is considered the first multi-objective optimization study to reduce the size of the test suite. The authors employed the NSGA-II algorithm and demonstrated its superiority over greedy techniques in an experiment. The NSGA-II method was also used by Geng et al. [12] and Gupta et al. [17]. Wang et al. [15] used different forms of weighted-sum method GA; their experiments showed that random weighted-sum GA outperformed NSGA-II and other common multi-objective algorithms. Wei et al. [16] compared six multi-objective optimization algorithms, including NSGA-II and several variants of the multi-objective decomposition-based evolutionary algorithm (MOEA/D). NSGA-II outperformed the MOEA/D in small programs, whereas the MOEA/D outperformed the NSGA-II in large programs. The approach proposed in this paper is a heuristic search-based approach, that utilizes the ABBA [23].

3 Adaptive Weighted Sum Bi-objective ABBA for TSR

3.1 TSR Solution Encoding

Each TSR solution is represented as bat's position vector X_i, i.e., each X_i represents a reduced test suite. X_i is encoded as a binary vector $X_i = (x_{i1}, x_{i2}, \ldots, x_{id})$, where, d is the total number of test cases in the original test suite TS, each bit x_{ij} corresponds to a test case tc_j, the value of bit is equal to "1" or "0". This means that tc_j is included/excluded in the reduced test suite, respectively.

3.2 TSR Bi-objective Fitness Function Formulation

Fitness function was formulated using two objectives, defined by Eq. (1). The first objective is to lower the test suite's cost. The second objective aims at maximizing the number of faults detected by a reduced test suite. The mutation score is used to express the reduced test suite's fault detection capability.

$$Objective1 = \frac{\sum_{|TS|} exec\_time_i}{\sum_{|RTS|} exec\_time_i} \quad Objective2 = \frac{|killedmu|}{|mu|} \tag{1}$$

where $|TS|$ is the original test suite size, $|RTS|$ is the reduced test suite size, $exec\_time_i$ is the execution time of tc_i, $|mu|$ is the total number of mutants and $|killedmu|$ is the number of mutants killed by the reduced test suite RTS.

To formulate the fitness function, we used the adaptive weighted-sum method for bi-objective optimization [25]; as defined by Eq. (2), the best solution is the one that maximizes the *fitness*.

$$fitness = (1 - \lambda) * nor(Objective1) + \lambda * nor(Objective2) \tag{2}$$

$$nor(x) = \frac{x}{x+1} \tag{3}$$

where, $nor(x)$ is a normalization function [30], $\lambda \in [0, 1]$, is a weighting factor that sets the relative importance of each objective. λ is not determined previously, it evolves according to the shape of the Pareto front of the problem.

The Adaptive weighted-sum method starts by applying the normal weighted-sum method, with a large weight step size, $\Delta\lambda$, to generate coarse representations of the solutions that form the Pareto-front. The algorithm proceeds according to the length of each segment (distance between neighboring solutions) such that closely overlapping solutions are deleted, while long-distance segments are refined by constructing additional inequality constraints in this region and performing sub-optimization with the weighted-sum method. When all the segments of the Pareto front reach a pre-specified resolution, the algorithm ends. The detailed procedure of the adaptive weighted-sum method is as follows.

Step 1: Use the normal weighted-sum approach with a small number of divisions, $n_{initial}$. The number of divisions determines the uniform step size $\Delta\lambda$ of the weighting factor λ as defined by Eq. (4).

$$\Delta\lambda = \frac{1}{n_{initial}} \tag{4}$$

Step 2: Calculate the segment lengths between all neighboring solutions. Delete solutions that are closely overlapping. When using the weighted-sum method, it is common to get several almost identical solutions. The Euclidian distances among these solutions are nearly zero, and the Pareto front can be represented by only one of them. So, if the distances among a set of solutions are smaller than a pre-specified distance value (ε), all these solutions are discarded except one.

Step 3: Determine how many more refinements each segment requires. The refinements are determined according to the relative lengths of the segments by Eq. (5).

$$n_i = round\left(c\frac{l_i}{l_{avg}}\right)\text{for the } i^{\text{th}} segment \tag{5}$$

where n_i is the number of further refinements for the ith segment, l_i denotes the ith segment length, l_{avg} is the average length of all the segments, and c denotes an algorithmic constant. 'Round' is a function that rounds to the nearest integer. If $n_i \leq 1$ the segment is not refined any further. If $n_i > 1$ for any segment i proceed to the next step for further refinement.

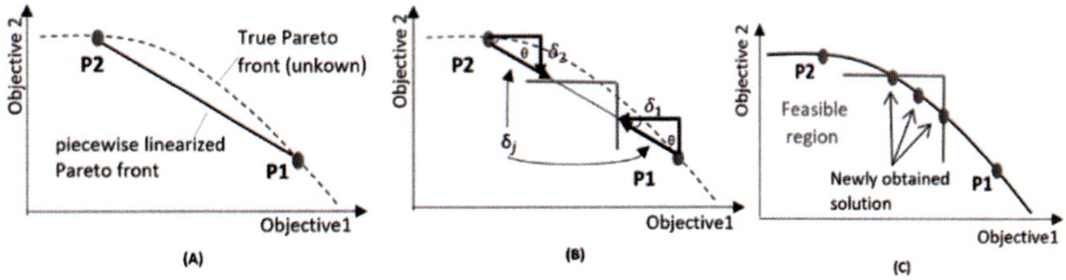

Fig. 1. Determining the offset distances, $\delta 1$ and $\delta 2$, based on δ_J

Step 4: Calculate the offset distances δ_1, δ_2 from the two endpoints of each segment. Firstly, link the endpoints, P1 and P2, to form a piecewise linearized secant line; as shown in Fig. 1 (a). Then, the user selects the offset distance along the piecewise linearized Pareto front, δ_J (Fig. 1 (b)); δ_J determines the density of the Pareto solution distribution. The angle θ in Fig. 1 (b) is calculated using Eq. (6) to find the offset distances parallel to the objective axes.

$$\theta = tan^{-1}\left(-\frac{p_1^y - p_2^y}{p_1^x - p_2^x}\right) \tag{6}$$

where p_i^x and p_i^y are the x(Objective1) and y(Objective2) positions of the end points, P1 and P2, respectively. Then, $\delta 1$ and $\delta 2$ are determined using δ_J and θ as follows:

$$\delta_1 = \delta_J cos\theta \quad and \quad \delta_2 = \delta_J sin\theta \tag{7}$$

Step 5: In each of the feasible regions, construct additional inequality constraints and conduct sub-optimization with the weighted-sum method. As shown in Fig. 1(b,c), the feasible region is offset from P1 and P2 in the direction of Objective1 and Objective2 by the distances of δ_1 and δ_2. Performing suboptimization in this region, the problem is formulated as:

$$\text{Max}(1-\lambda) * nor(Objective1) + \lambda * nor(Objective2) \tag{8}$$

$$s.t. Objective1 \leq p_1^x - \delta_1 Objective2 \leq p_2^y - \delta_2,$$

where δ_1 and δ_2 are the offset distances, p_1^x and p_2^y are the x and y positions of the endpoints. The uniform step size of the weighting factor λ for each feasible region is determined using Eq. (9):

$$\Delta\lambda_i = \frac{1}{n_i} \tag{9}$$

Step 6: Calculate the segment lengths between all neighboring solutions. Delete solutions that are almost overlapping. Terminate the process if the length of all segments is less than δ_J. Otherwise, go to Step 4 and iterate if there are segments in which their lengths are greater than the maximum length δ_J.

Four important parameters that must be set: the number of the Pareto front segments in the initial iteration ($n_{initial}$); the offset distance (δ_J); the Euclidean distance for specifying the nearly overlapping solutions (ε); the further refinement constant (c). Table 1 lists the parameter settings we experimented with.

Table 1. Parameter settings for adaptive weighted sum method

Parameter	δ_J	ε	c	$n_{initial}$
Value	0.1	$0.25 * \delta_J$	3	15

4 Experiments and Results

4.1 Data Sets

We experimented with five Java programs provided by Polo et al. [31]; Table 2, lists their characteristics which are: lines of code (LOC), test suite size $|tc|$, number of mutants $|mu|$, and execution time of the test suite (T_{exe}).

Table 2. Software under test (SUT).

	MBisectOk	MBubCorrect	MFindOk	MMidOK	MTringle
LOC	31	54	79	59	61
$\|mu\|$	44	70	179	138	239
$\|tc\|$	25	256	135	125	216
T_{exe}	25	256	135	125	216

4.2 Performance Metrics

Three metrics were used which are: Cardinality, Diversity and Inverted Generational Distance plus, to measure the convergence, diversity (spread and uniformity), and cardinality of the approximated solution set to the known Pareto-optimal front [32]. Assume, A is an approximated Pareto front and R is a reference Pareto front, which includes all non-dominated solutions that are already known. And m is the dimension of the objective space; the metrics are defined as follows:

(1) Cardinality Ratio (CR): The cardinality of a set A, $|A|$, can be defined as the number of NDS in this set. Cardinality Ratio of a set A, $CR(A)$, is the ratio between $|A|$ and the cardinality of R, $|R|$, as given by Eq. (10)

$$Cardinality\ Ratio\ CR(A) = \frac{|A|}{|R|} \tag{10}$$

(2) Inverted Generational Distance plus (IGD^+): IGD^+ [33] used to measure the convergence of A. It measures the distance to the reference Pareto front R using a distance measure that adopts Pareto dominance as given by Eq. (11, 12).

$$IGD^+(A, R) = \frac{1}{|R|} \sum\nolimits_{\vec{r} \in R} min_{\vec{a} \in A}\left(d^+\left(\vec{a}, \vec{r}\right)\right) \tag{11}$$

where,

$$d^+\left(\vec{a}, \vec{r}\right) = \sqrt{\sum\nolimits_{k=1}^{m} (\max\{r_k - a_k, 0\})^2} \tag{12}$$

(3) Diversity (Δ): It evaluates the spread and distribution of solutions on the known Pareto-optimal front. The spread refers to the range of values covered by the solutions in A, while the distribution is the relative distance among solutions in A. It's calculated by Eq. (13).

$$\Delta = \frac{d_f + d_l + \sum_{i=1}^{|A|-1} |d_i - \overline{d}|}{d_f + d_l + (|A| - 1)\overline{d}} \tag{13}$$

where d_i is the Euclidean distance between two consecutive solutions in A, d_f and d_l are the distances between the extreme solutions on the refernce Pareto front R and the boundary solutions of A. $\overline{d}$ is the average of all distances d_i, $i \in [1, |p| - 1]$. The value of Δ will be in the range of 0 to 1.

As R is a reference Pareto front (the true front is unknown), the Δ value of A should be compared with the Δ of R. So, in our work Δ of A computed as Eq. (15).

$$\Delta = \Delta_R - \Delta_A \tag{15}$$

Pareto fronts with a higher cardinality ratio, smaller IGD^+ and smaller spread are preferred.

4.3 Results

A set of experiments were conducted on the five Java SUT. In all the experiments, the maximum number of iterations was set to 100. Table 3 shows the predetermined values of λ that were used to generate ten non-dominated solutions (NDS) by the PW-ABBA and PW-GA.

Table 3. Values of λ used with the PW-ABBA and PW-GA

NDS	1	2	3	4	5	6	7	8	9	10
λ	0.6	0.676	0.744	0.804	0.856	0.9	0.936	0.964	0.984	0.9960

As could be observed from Table 4, NSGA-II has the largest CR over the five programs, which means that NSGAII was capable of discovering the largest number of NDS. However, CR is not enough to prove the superiority of a heuristic algorithm as the quality of the discovered NDS should be analyzed to determine their closeness to the true Pareto front (in terms of IGD+) and their diversity. As it could be observed, the performance of AW-ABBA is superior to the other algorithms, in terms of the IGD+, over three programs (MBisectOk, MMidOK and MTringle); where their IGD+ values are equal or close to zero, which means that the NDS discovered by the AW-ABBA are located on or very close to the reference front. Over Mfindok program, the solutions discovered by each of the PW_ABBA, AW-ABBA and NSGA-II are on the reference Pareto-front (IGD+ = 0). It could be observed from Table 6 that the diversity values of PW_ABBA and AW_ABBA are close to each other and superior to the NSGA-II over three programs. Finally, PW-GA has the worst performance results among all the algorithms; it approximates the pareto-front with the largest IGD+ and diversity values, i.e., the discovered NDS are located far from the reference front and their distribution and spread are not efficient. Figure 2 illustrates samples of approximate Pareto fronts discovered by the different algorithms. For MMid and Mtringle, whose reference fronts include the largest number of NDS, the figures show the efficiency of AW-ABBA over other algorithms in approximating the Pareto front (Table 5).

Table 4. CR of the approximated Pareto-fronts

	PW-ABBA	AW_ABBA	GA	NSGA-II
MBisectOk	2/2	1/2	1/2	**3/2**
MBubCorrect	1/3	2/3	1/3	**3/3**
MFindOk	1	1	1	1
MMidOK	2/5	5/5	5/5	**7/5**
MTringle	3/16	16/16	2/16	**17/16**

Table 5. IGD^{+} of the approximated Pareto-fronts

	PW-ABBA	AW_ABBA	GA	NSGA-II
MBisectOk	**0**	**0**	0.6	0.08
MBubCorrect	0.4	0.2	0.5	**0.08**
MFindOk	0	0	0	0
MMidOK	0.25	**0.06**	0.3	0.1
MTringle	0.1	**0.00**	0.1	0.1

Table 6. Diversity of the approximated Pareto fronts

	PW-ABBA	AW-ABBA	GA	NSGA-II
MBisectOk	**0**	**0**	1.00	−0.26
MBubCorrect	1.00	**0.32**	1.00	−0.26
MFindOk	1	1	1	1
MMidOK	0.53	**0.11**	0.24	0.38
MTringle	**0.01**	0.02	0.01	−0.003

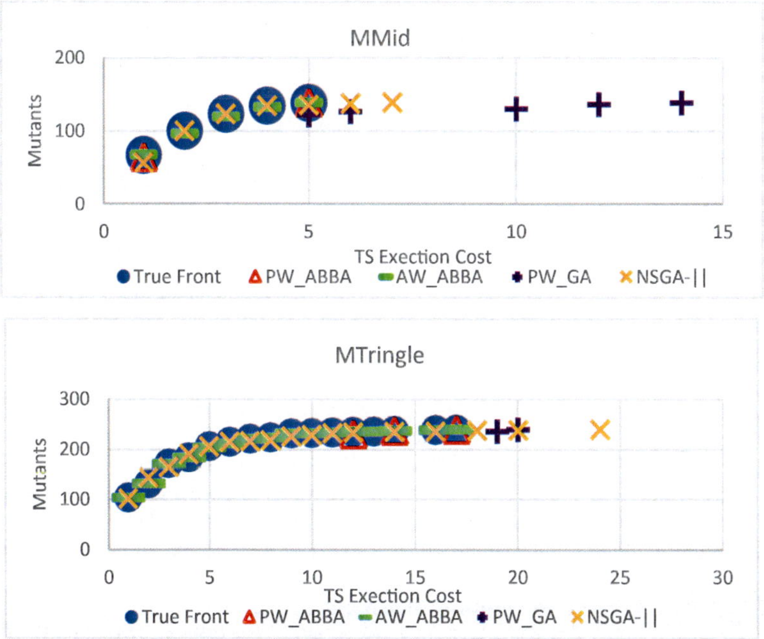

Fig. 2. Sample of the approximated NDS set of TSR by different algorithms.

5 Conclusion and Future Work

This paper proposed solving the test suite reduction problem, as a bi-objective optimization problem using the ABBA; which is one variant of the BBA and was reported to be more effective than the BBA and the PSO. An adaptive weighted-sum method was utilized to formulate the fitness function. The effectiveness of the adaptive weighted-sum ABBA (AW-ABBA) was evaluated over five Java programs and compared to a pre-determined weighted-sum ABBA(PW-ABBA), PW-GA and NSGA-II. The experimental results showed that the performance of the proposed AW-ABBA is superior to PW-ABBA and PW-GA. Although the NSGA-II could discover more non-dominated solutions than each of the AW-ABBA, PW-ABBA; the NDS discovered by AW-ABBA, over most of the programs, dominate some of the NDS discovered by the NSGA-II.

As a further extension for this work, more objectives such as branch coverage could be considered in the fitness function.

References

1. Hamdy A, Ellaithy A (2020) Multi-Feature approach for Bug severity assignment. Int J Open Source Softw Process 11(2):1–15
2. Hamdy A, El-Laithy A (2019) Smote and feature selection for more effective bug severity prediction. Int J Softw Eng Knowl Eng 29(06):897–919
3. Hamdy A, El-Laithy A (2020) Semantic categorization of software bug repositories for severity assignment automation. In: Jarzabek S, Poniszewska-Marańda A, Madeyski L (eds) Integrating research and practice in software engineering. SCI, vol 851. Springer, Cham, pp 15–30. https://doi.org/10.1007/978-3-030-26574-8_2
4. Hamdy A, Ezzat G (2020) Deep mining of open source software bug repositories. Int J Comput Appl 1–9
5. Hamdy A et al (2020) Deep hybrid features for code smells detection. J Theor Appl Inf Technol 98(14)
6. Rosero RH, Gómez OS, Rodríguez G (2016) 15 years of software regression testing techniques—a survey. Int J Softw Eng Knowl Eng 26(05):675–689
7. Reda N, Hamdy A, Rashed E-A (2022) Multi-objective adapted binary bat for test suite reduction. Intell Autom Soft Comput 31(2):781–797
8. Jia Y, Harman M (2010) An analysis and survey of the development of mutation testing. IEEE Trans Softw Eng 37(5):649–678
9. Khan SUR, Lee SP, Javaid N, Abdul W (2018) A systematic review on test suite reduction: approaches, experiment's quality evaluation, and guidelines. IEEE Access 6:11816–11841
10. Mohapatra SK, Mishra AK, Prasad S (2020) Intelligent local search for test case minimization. J Inst Eng (India) Ser B 101(5):585–595. https://doi.org/10.1007/s40031-020-00480-7
11. Sun J, Wang S (2010) A novel chaos discrete particle swarm optimization algorithm for test suite reduction. In: The 2nd international conference on information science and engineering, Hangzhou, China. IEEE, pp 1–4
12. Geng J, Li Z, Zhao R, Guo J (2016) Search based test suite minimization for fault detection and localization: a co-driven method. In: Sarro F, Deb K (eds) SSBSE 2016. LNCS, vol 9962. Springer, Cham, pp 34–48. https://doi.org/10.1007/978-3-319-47106-8_3
13. Mohanty S, Mohapatra SK, Meko SF (2020) Ant colony optimization (ACO-Min) algorithm for test suite minimization. In: Das H, Pattnaik PK, Rautaray SS, Li K-C (eds) Progress in computing, analytics and networking. AISC, vol 1119. Springer, Singapore, pp 55–63. https://doi.org/10.1007/978-981-15-2414-1_6

14. Yoo S, Harman M (2007) Pareto efficient multi-objective test case selection. In: Proceedings of the 2007 international symposium on Software testing and analysis, New York, NY, USA, pp 140–150
15. Wang S, Ali S, Gotlieb A (2015) Cost-effective test suite minimization in product lines using search techniques. J Syst Softw 103:370–391
16. Wei Z, Xiaoxue W, Xibing Y, Shichao C, Wenxin L, Jun L (2017) Test suite minimization with mutation testing-based many-objective evolutionary optimization. In: 2017 international conference on software analysis, testing and evolution (SATE), Harbin, China. IEEE, pp 30–36
17. Gupta N, Sharma A, Pachariya MK (2020) Multi-objective test suite optimization for detection and localization of software faults. J King Saud Univ Comput Inf Sci 34:2897–2909
18. Yang XS (2010) A new metaheuristic bat-inspired algorithm. In: González JR, Pelta DA, Cruz C, Terrazas G, Krasnogor N (eds) NICSO 2010. SCI, vol 284. Springer, Heidelberg, pp 65–74. https://doi.org/10.1007/978-3-642-12538-6_6
19. Yang X-S (2011) Bat algorithm for multi-objective optimisation. Int J Bio-Inspir Comput 3(5):267–274
20. Hamdy A (2014) Genetic fuzzy system for enhancing software estimation models. Int J Model Optim 4(3):227
21. Hamdy A, Mohamed A (2019) Greedy binary particle swarm optimization for multi-objective constrained next release problem. Int J Mach Learn Comput 9(5):561–568
22. Eberhart R, Kennedy J (1995) A new optimizer using particle swarm theory. In: MHS 1995. Proceedings of the sixth international symposium on micro machine and human science, Nagoya, Japan. IEEE, pp 39–43
23. Fisal NR, Hamdy A, Rashed EA (2021) Search-based regression testing optimization. Int J Open Source Softw Process 12(2):1–20
24. Mirjalili S, Mirjalili SM, Yang X-S (2013) Binary bat algorithm. Neural Comput Appl 25(3–4):663–681. https://doi.org/10.1007/s00521-013-1525-5
25. Kim IY, De Weck OL (2005) Adaptive weighted-sum method for bi-objective optimization: Pareto front generation. Struct Multidiscip Optim 29(2):149–158
26. Ngatchou P, Zarei A, El-Sharkawi A (2005) Pareto multi objective optimization. In: Proceedings of the 13th international conference on intelligent systems application to power systems, Arlington, VA, USA, pp 84–91
27. Assi RA, Masri W, Trad C (2018) Substate profiling for effective test suite reduction. In: 2018 IEEE 29th international symposium on software reliability engineering (ISSRE), Memphis, TN, USA. IEEE, pp 123–134
28. Coviello C, Romano S, Scanniello G, Marchetto A, Antoniol G, Corazza A (2018) Clustering support for inadequate test suite reduction. In: 2018 IEEE 25th international conference on software analysis, evolution and reengineering (SANER), Campobasso, Italy. IEEE, pp 95–105
29. Anwar Z, Afzal H, Bibi N, Abbas H, Mohsin A, Arif O (2018) A hybrid-adaptive neuro-fuzzy inference system for multi-objective regression test suites optimization. Neural Comput Appl 31(11):7287–7301. https://doi.org/10.1007/s00521-018-3560-8
30. Greer D, Ruhe G (2004) Software release planning: an evolutionary and iterative approach. Inf Softw Technol 46(4):243–253
31. Polo M, Piattini M, García-Rodríguez I (2009) Decreasing the cost of mutation testing with second-order mutants. Softw Test Verif Reliab 19(2):111–131
32. Falcón-Cardona JG, Coello CAC (2020) Indicator-based multi-objective evolutionary algorithms: a comprehensive survey. ACM Comput Surv 53(2):1–35
33. Ishibuchi H, Masuda H, Tanigaki Y, Nojima Y (2015) Modified distance calculation in generational distance and inverted generational distance. In: Gaspar-Cunha A, Henggeler Antunes C, Coello CC (eds) EMO 2015. LNCS, vol 9019. Springer, Cham, pp 110–125. https://doi.org/10.1007/978-3-319-15892-1_8

Studding the Benefits of BIM Plugins on the Efficiency of Architecture Practice

Rania Rushdy Moussa[1], Dina R. M. Moawad[2], and Samir Abou El-Seoud[3](✉)

[1] Architecture Engineering, The British University in Egypt (BUE), Cairo, Egypt
rania.rushdy@bue.edu.eg

[2] Environmental Sustainable Architecture Engineering, The British University in Egypt (BUE), Cairo, Egypt
dina.rushdy@bue.edu.eg

[3] Faculty of Informatics and Computer Science (ICS), The British University in Egypt (BUE), Cairo, Egypt
samir.elseoud@bue.edu.eg

Abstract. The exchange of information between the different organizations in the different stages of a building's life cycle has always been an important and difficult task at the same time. In each project, there are a vast number of participants with different views of the same physical structure who have to interact and exchange information throughout the whole building life cycle. In CAD data exchange, the product modeling and information exchange standards community have developed several times known as "Building Information Models" or "BIMs" in order to enable a more coherent exchange of data. Plug-in program have been created in order to create an invincible piece of software. Plug-in program help to improve the efficiency of the original program by making sub-programs strengthen the weaknesses in the original program and completing the original program if there was an uncompleted part in order to help the user complete their job by using one software and its plug-ins. In this paper, the researchers will conduct a comparison between the efficiency of one of the most popular program of BIM software, which is called the Archicad program, and the ArchiStair plug-in. The comparison was made in order to evaluate the efficiency of Archicad software and the efficiency of its plug-ins. This study revealed that drawing stairs using the Archicad program did not work as expected and that the plug-in is needed and basically cannot be dispensable.

Keywords: Building information model · Plug-in · Archicad software · Archistair software

1 Introduction

BIM (building information modeling) is now a very important possibility for developing efficiency and productivity in the design and construction profession sector as in the AEC (Architecture, Engineering and Construction) sector, professional practitioners could rely on the stability and time arrangement of a tridimensional information model providing data which count as interconnected, to the point and intelligent,

M. E. Auer et al. (Eds.): REV 2022, LNNS 524, pp. 496–506, 2023.
https://doi.org/10.1007/978-3-031-17091-1_50

in order to enhance designing, detailing and managing complex and more composite projects [1]. The acquiring and relying on BIM technology throughout its several stages could improve project culture which is a so far established in offices and construction firms as inserting the concept of incorporation and integrating into process interfaces, policies and technology [2], rather than designing in a way revolving around documentation, subsequent and fragmented, professional practitioners could achieve collaborative contributions throughout a complete model, combined with architecture, structural and MEP (mechanical, electrical and plumbing) system different disciplines, a full informatics model which authorize designers from different instances to join decisions making process, acknowledging the individual roles of involved others. These changes might need the support of BIM coordinators, professional practitioners who their specialty in modeling compatibility and coordination, in addition to performing the technology implementing into the process [3].

1.1 Traditional Design Phase Model

Figure 1 shows a swim-lane diagram separated horizontally, similar to a standard 2D CAD design phase information flow swim-lane diagram. Only the conceptual and schematic design phases are present vertically, as information coordination, sharing, and owner feedback occur during each of these phases and do not have to wait until the design is complete. The concept design phase begins with the creation of deliverables that are incorporated into the building information model and the development of the architectural concept in the BIM environment. The structural/civil and MEP design step differs from the usual 2D CAD design phase.

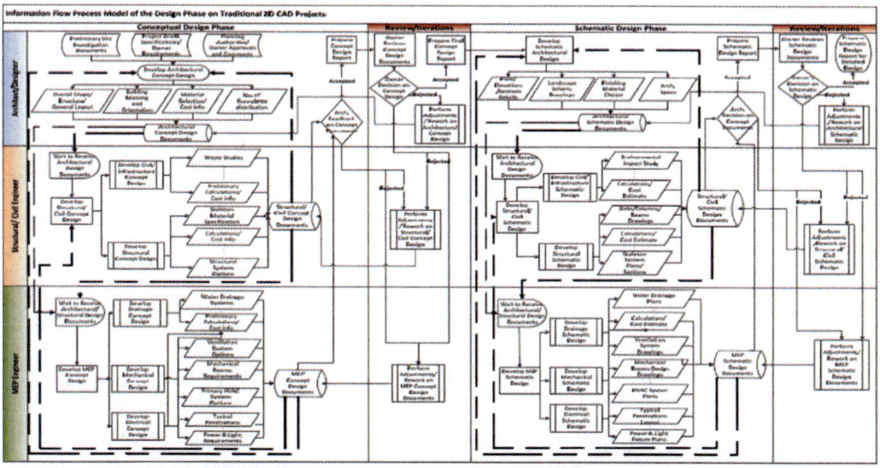

Fig. 1. Information flow process model of the design phase on traditional 2D CAD projects

The structural/civil and MEP engineers do not have to wait until the completion of the architectural design concept to proceed. Instead, early and easy data sharing is possible before data completion, thus the three cross-functional teams can develop their design concepts simultaneously as shown in Fig. 2.

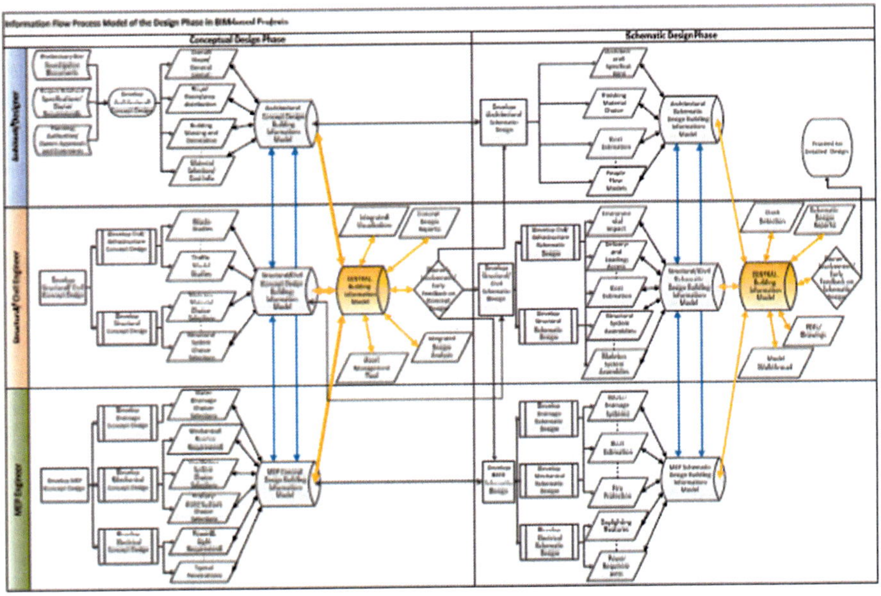

Fig. 2. Flow process model of the design phase in BIM programmers

A major aspect of BIM technology is the usage of external applications to the principal platform, these implementations are identified as plug-ins, whichever utilize the programming previously inhibited into the BIM software's in order to control several characteristics of the design environment, controlling specified objectives, and concentrating on adding versatility to already existing functionalities.

BIM professional practitioners could utilize the plug-ins to reach specific targets of any designing process, in addition to increasing their production, allowing the applications to perform unvaried repeating tasks, execute any calculations or execute sequential commands throughout programming routines, among other uses. The development process of applications plug-ins is now presented by the users' community of BIM technology, who are searching for particular solutions for several project situations [4].

Plug-ins gives the opportunity to custom tailor a software to suit particular project requirements in order to increasing range of potentials and exceed any platform limitations, even throughout tailoring a software to particular user's need has become newly practiced in architecture, and already extensively established in mechanical engineering and design [5]. Implemented plug-ins may also be beneficial with facilitating cost, managing and sustainable analysis for construction projects, so this provides a more simple replacement for other constructive system substitutes, accompanying the designing process and decision making an implemented plug-in can, as example, help increasing synchronization and automate processes by interrelating the model's existing information with other different software or servers which have data relevant to the task available, authorizing for active decision making and a lot more solutions possibilities [6]. Nevertheless, the vital role of sustainable design is established and spreading in the industry of construction, however, energy and performance analysis don't get the needed integration into BIM designing process, since that pace can typically get done after the architecture design is complete [7]. Plug-ins can also help increasing the possibility of evaluating several technological solutions throughout the early stages of the project, giving the chance for designers to achieve more positive impacts on building life cycle and energy efficient design [8]. Throughout plug-ins created in the Revit API (application programming interface), framework in order to help essentials to environmental certifications like LEED could be combined with the designing process [9]. Giving a chance for designers to implement contextual information within the building model so as to achieving specific certification credits, furthermore, the documentations submitted could be managed in obedience to designer's targets. Targeting particular fields of certifications like rainwater management [10], operational energy use, carbon emissions [11], and sustainable materials [12], Focusing on specific fields of certification such as for rainwater management [13, 14], operational energy use, carbon emissions [15–19] and sustainable materials [20–23], are applications too of plug-in implementation, whichever grants for more strategic design choices in certification processes.

This paper targets the knowledge gap in solution optimization and academic research about external applications such as one of the most popular programs of BIM software's which is called Archicad program and the ArchiStair plug-in which was made to serve the Archicad program, this study compares between Archicad program and its ArchiStair plug-in, in order to evaluated the difference between efficiency of the original software without external implementations and the efficiency of one of its plug-ins.

2 Archicad Plugins

Building Information Modeling is the new name for the virtual building technology Graphisoft developed over 25 years ago. Archicad is a BIM engine that pioneered the process of integrating information between the 3D model and construction documents. The Use of BIM to accurately derive quantities, synchronize coordination between trades, test constructability, and simulate building performance before construction even begins. Many architects have designed and constructed over a million buildings worldwide using Archicad proven technology. More than 20 plugin program has been seen lately for ArchiCAD software. This plugins are expensive and increasing the initial cost of the program in which it prevent many users for buying the software, in addition it increase the initial cost of the programmer in comparing to other BIM programs. May architects stated that ArchiCAD can operate efficiently without using these expensive plugins and these plugins only accelerate the design process. While others negotiated that the plugins are essential for operating ArchiCAD program and without the plugin the users will not be able to design their projects.

One of the most well-known plugins in ArchiCAD program is the StairMaker, A common problem faced by ArchiCAD users is the creation and management of stairs. The StairMaker Plug-in was developed to resolve this problem. This research intends to study the efficiency of designing stairs using the free ArchiStair.

Tool installed in ArchiCAD program and compare the cost and efficiency of using StairMaker Plug-in.

3 Case Study

The benefits of plug-ins in Archicad are well known as it was discussed before and how it provides new functions in addition to the standard characteristics. There was a negotiation about the effectiveness of the Archicad plug-in and if it was really needed in Archicad program and that what this paper is trying to prove it right now. There was an Interpreter explanation about for both the ArchStair which represent the original program and StairMaker which represent the plug-in of an original program.

3.1 Differences Between StairMaker and ArchiStair

In this section, the research will describe the interface and the differences between StairMaker and ArchiStair. StairMaker is the original drawing tool used by Archicad. While, ArchiStair is a professional plugin tool made to help the users in develop the drawing of stair cases.

3.1.1 StairMaker

More Flexible 2D Display of Stairs with Improved Interface

- Stair Settings offers many more parameters allowing you to adjust the Floor Plan symbol display of stairs to conform to local standards.
- Custom User Interface Settings (2D Symbol and Attributes) allow for flexible text display, railing/walking line symbols, handrail types, 2D symbol and 3D attributes.
- Text Display parameters similar to those of regular GDL stairs: Four New Standard Stair Types for StairMaker
- Single Winder at Lower End; Single Winder at Upper End; Single Winder U-Return; Double Winder U-Return.
- User defines number of risers in the winder; related parameters will be adjusted accordingly.
- User defines offset of risers in winders from the corner(s) of the stair.
- Flexible Options for Defining Location of Saved Stair, When the user saves a new stair created with StairMaker, you can browse any library location (Fig. 3)

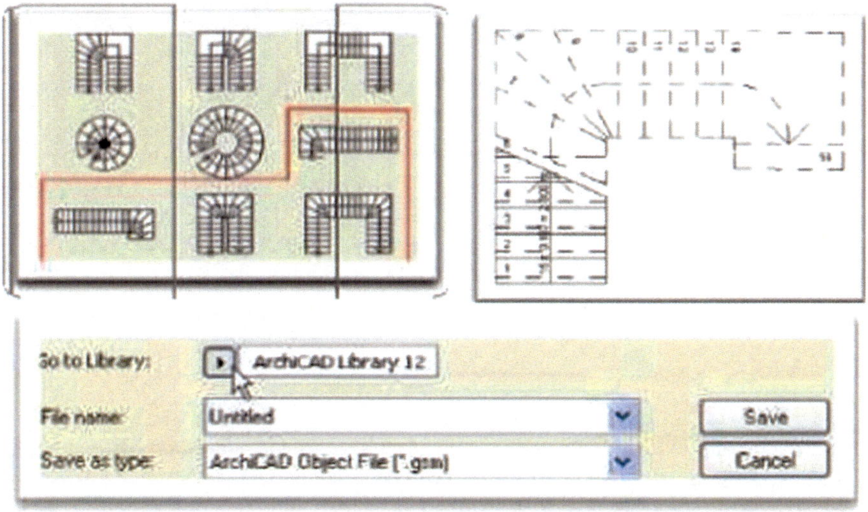

Fig. 3. Architerra interface of StairMaker

3.1.2 ArchiStair

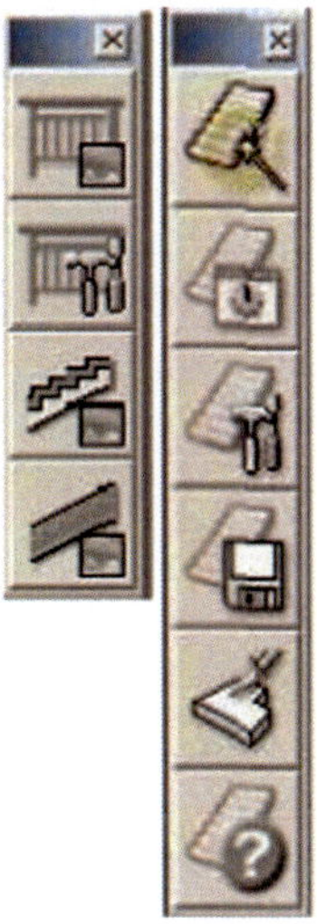

ArchiStair is an ArchiCAD extension that allows stairs to be created even when the design constraints of the new project or remodel prohibit them from having a regular shape. With its simple and rapid graphical interface, ArchiStair allows users to create stairs with treads and landings that have customized, creative shapes. ArchiStair is easy and fun to use.

ArchStair allow the user to either select one of the predefined stair types displayed in the main dialog window or use 2d ArchiCAD elements to draw the shape they desire.

Editable Hotspots allow treads and landings to be graphically modified both in the Floor Plan Window and in the 3D Window by simply selecting a node and dragging it to the desired position.

The treads of the stair can be curved and the risers can be either standard or angled. All other aspects of the stair are editable and customizable. Let's start with the handrails, The shape of the handrails and handrail ends can be chosen, as can the number and shape of the posts, panels and newels. Even custom objects can be used as elements. For the panels, both the material and thickness can be modified.

Users can add finishing touches such as skirting boards and stringers, which are also customizable both in terms of material and measurements. To facilitate 3D visualization, supporting walls can be placed under the stair, which can be edited for thickness and material. The user can even customize the 2D appearance of the stair and modify the scale with which it is viewed in the ArchiCAD worksheet. In addition, ArchiStair allows the users to cut the opening in the slab over the stair automatically as shown in Figs. 4.

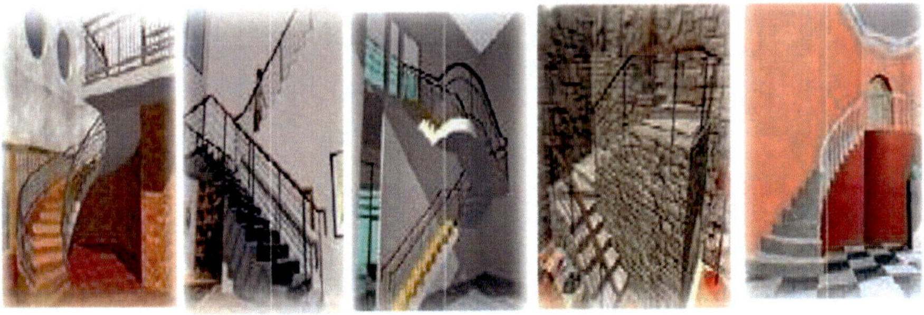

Fig. 4. The output of ArchiStair

As a special feature, the image of a person can be positioned in the Section Window or 3D Window, providing a means of controlling the height and comfort of the stair. Moreover, the settings of the stairs created can be saved and used again later to create a new stair.

3.2 Characteristics of Participants and Data Collection

The participants were chosen according to their level of experience, the participant were professional architects works in different local architecture firms that using ArchiCAD program to design their projects. 56 architects and landscape designers were willing to participate in the research. The sample started with more than 78 participants but the participants that completed the questionnaire were 56 out of 78. This questionnaire were conducted made April 2021. The distribution of the participants, social and demographic characteristics are presented in Table 1.

Table 1. Social and demographic characteristics of the research participants

Gender		Age		Education	
Type	%	Range	%	Education level	%
Male	42%	20–29	35%	PHD degree	15%
		30–40	32%	Master degree	20%
Female	58%	Above 40	33%	BSC of engineering	65%

A short questionnaire consists of 14 questions have been used in this research, the 14 questions were divided into 7 questions related to StairMaker while the other 7 questions targeting the efficiency of ArchiStair. The questions were measuring the speed of design (time saving), feasibility, the ability to modify the properties of chosen elements, visualization, software interface, rendering and efficiency. The participants were asked to indicate the rate of effectiveness of StairMaker and ArchiStair from 'totally disagree' (5 points) to 'totally agree' (1 point). Finally, questions were randomly ordered in the questionnaire to elicit more objective answers from the respondents.

3.3 Comparative Analysis

A questioner was made and distributed among architect and ArchiCad users to compare the original program with its plug-in by putting criteria in order to compare the effectiveness of Archicad program and its plug-in which was made for that program. 110 architects participated in this questionnaire some of them are working in international firms while other are fresh graduate.

A comparative analysis was conducted with the aim of comparing the original software with the implemented external plug-in, this comparison focuses on a specific criterion which is the effectiveness of the original software and its particular plug-in made for the same software.

Moreover, ArchStair is presented as the original programe, although the StairMaker is presented as the plug - in implemented in the original software. Nevertheless, the leading criteria used to conduct expressive results were: Speed, Accessibility, Design Verities, Efficiency, Visualization, Accuracy, and Communications.

The choices of answers were ranging from A to D. As "A" represents a percentage of 85%, "A−" represents 80%, "B+" represents 75%, "B" represents 70%, "B−" represents.

65%, "C+" represents 60%, "C" represents 55%, "C−" represents 50%, "D+" represents.

45%, and "D" represents 40%.

The answers were used to utilize the difference in qualifications between ArchStairs and StairMaker compared using the specified criteria mentioned above. Nevertheless, the ranging answers are presented in Table 2 as it is seen by my point of view.

Table 2. Comparison between the qualifications of ArchStairs and StairMaker.

Criteria		ArchiStairs	StairMaker
Speed		B	B
Accessibility		A	B+
Design Verities	Creative shapes	A	C
Efficiency	Editing	A	B+
	Removing	A	B
	Detailing	B+	C+
	Structural solutions	B+	C
Visualization	3D view	A	A
	Orbit	B	B
	Walk through	B	B
Accuracy	Different views	A	A
	Dimensions	A	B
Communications	Visually	A	B+
Rendering	Material	A	B

Answers were ordered to utilize and analyzed the results in form of comparative table as shown in Table 3 and conducted percentage charts as shown in Fig. 5.

Table 3. Analysis of the case study

Criteria	ArchiStairs	StairMaker
Speed	70	70
Accessibility	85	75
Design Verities	85	55
Efficiency	80	65
Visualization	75	75
Accuracy	85	75
Communications	85	75
Rendering	75	70

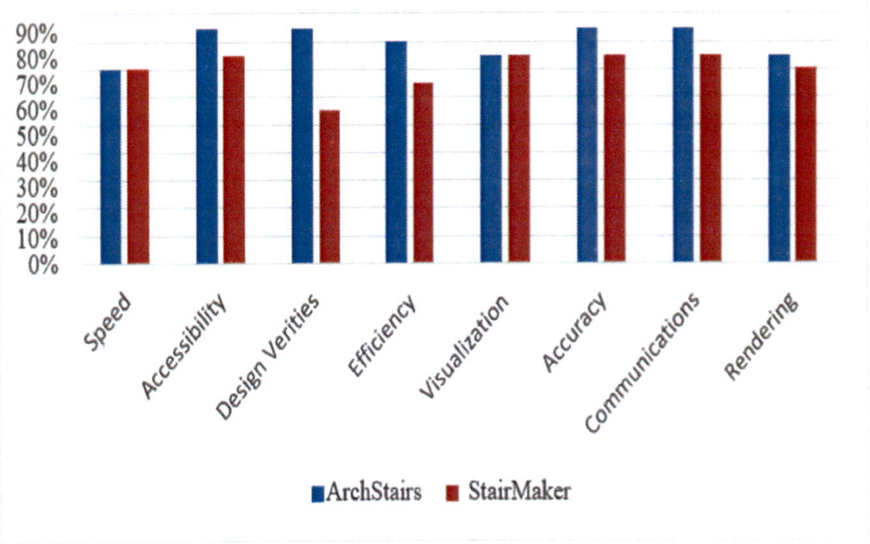

Fig. 5. Identifying the difference between ArchiStair and StairMaker

4 Conclusion

The comparison shows that there is a huge gap between the performance of the program with plug-in instead of use the program without plug-in, as an example the accessibility Archstairs has a very easy icons interface with helps any user to use it without having a good background for using it, while StairMaker have some complicated items which need users with a good background for using StairMaker in an efficiency way.

As in Structure Solutions the ArchiStair provides the users by a flexibility in add finishing touches such as skirting boards and stringers, which are also customizable both in terms of material and measurements, and supporting walls can be placed under the stair, which can be edited for thickness and material, with a complete structure system. While in StairMaker the user has only an opinion to choose the stair support which is not enough for civil engineering's that's why they chose a different program instead of Archicad.

As in Material the ArchStair gives the users a Varity in choosing the materials and change some of its properties as it offers a huge library which motivates the architectures to create a creative building while StairMaker has a very limited library with limited options and complicated.

As a conclusion the plug-in was very useful for Archicad and it really improve the qualification of Archicad program that encourages the users to choose Archicad program instead of others programs.

References

1. Eastman C, Liston K, Sacks R, Teicholz P (2008) BIM handbook: a guide to building information modeling for owners, managers, designers, engineers and contractors. Wiley, Hoboken
2. Moussa RR, Moawad DR (2020) Investigating the efficiency of building energy simulation software on architectural design process. In: ICSIE 2020: 2020 9th international conference on software and information engineering, Cairo, Egypt. Association for Computing Machinery, New York, pp 37–40. https://doi.org/10.1145/3436829.3436860
3. Succar B, Sher W, Williams A (2012) Measuring BIM performance: five metrics. Archit Eng Des Manag 8(2):120–142
4. Garbini MAL, Brandão DQ (2014) Proposta de Modelo para Implantação de Processo de Projeto Utilizando o Conceito BIM em Escritórios de Arquitetura. Gestão e Tecnologia de Projetos 9(1):7–24. (in Portuguese)
5. Autodesk (2016) Revit API developers guide. https://knowledge.autodesk.com/support/revit-products/learn-explore/caas/CloudHelp/cloudhelp/2014/ENU/Revit/files/GUID-F0A122E0-E556-4D0D-9D0F-7E72A9315A42-htm.html. Accessed 10 Nov 2016
6. Matcha H, Quasten G (2009) A parametric-typological tool: more diversity for mas produced single family homes through parametrized design and customized mass production. In: Design tool development. Presented at the 27th eCAADe Conference, Istanbul, Turkey, pp 409–416
7. Li J (2016) Integrating building information modelling (BIM), cost estimating and scheduling for buildings construction at the conceptual design stage. Master's degree thesis, Faculty of Engineering, Ottawa-Carleton Institute for Civil Engineering, University of Ottawa, Canada
8. Schlueter A, Thesseling F (2009) Building information model based energy/exergy performance assessment in early design stages. Autom Constr 18(2):153–163
9. Jalaei F, Jrade A (2014) An automated BIM model to conceptually design, analyze, simulate, and assess sustainable building projects. J Constr Eng 2014:1–21
10. Wu W (2010) Integrating building information modeling and green building certification: The BIM—leed application model development. Ph.D. dissertation, Design, Construction and Planning, University of Florida, Florida

11. Sanhudo LPN (2016) BIM for building sustainability assessment: development of a software tool for rainwater runoff mitigation. Master's degree dissertation, Curso de Engenharia Civil, Faculdade de Engenharia da Universidade do Porto, Porto, Portugal
12. Stadel A, Eboli J, Ryberg A, Mitchell J, Spatari S (2011) Intelligent sustainable design: integration of carbon accounting and building information modeling. J Prof Issues Eng Educ Pract 137(2):51–54
13. Zhang C, Chen S, Kang B (2015) LEED embedded BIM for construction sustainability analysis. Int J Sci Res Technol 1(1):95–101
14. Moussa RR, Hatem TM (2021) Estimating the potential of desalinate seawater using solar glass pyramid (SGP) in hot arid zones. Clean Eng Technol 4:100189. https://doi.org/10.1016/j.clet.2021.100189
15. Moussa RR, Mansour MMM, Yossif NM (2021) Statistic equation to estimate the amount of CO2 emission produced from high traffic density roads. WSEAS Tran Power Syst 16:78–86. https://doi.org/10.37394/232016.2021.16.8
16. Moussa RR, Dewidar KM (2020) Energy-scape web-based application is an effective tool to overcome the basic knowledge of architects toward renewable energy properties. Int J Emerg Technol Learn (iJET) 15(10):42–54. https://doi.org/10.3991/ijet.v15i10.13669
17. Alaa H, Khaled N, Moussa RR, Hatem TM (2021) Zero energy educational building: a case study of the energy and environmental engineering building in the British university in Egypt. In: 2021 IEEE 48th photovoltaic specialists conference (PVSC), 20–25 June 2021, pp 1913–1915. https://doi.org/10.1109/PVSC43889.2021.9518995
18. Moussa RR (2020) Installing piezoelectric tiles in children outdoor playing areas to create clean & healthy environment; case study of El-Shams sporting club, Cairo_Egypt. WSEAS Trans Environ Dev 16:471–479. https://doi.org/10.37394/232015.2020.16.48
19. Moussa RR, Ismaeel WSE, Solban MM (2022) Energy generation in public buildings using piezoelectric flooring tiles; case study of a metro station. Sustain City Soc 77:103555. https://doi.org/10.1016/j.scs.2021.103555
20. Abdullatif KG, Guirguis MN, Moussa RR (2020) Analyzing the structural properties of Fire Clay bricks after adding cigarette filters. WSEAS Trans Environ Dev 16:671–679. https://doi.org/10.37394/232015.2020.16.69
21. Micheal A, Moussa RR (2021) Investigating the economic and environmental effect of integrating sugarcane bagasse (SCB) fibers in cement bricks. Ain Shams Eng J 12(3):3297–3303. https://doi.org/10.1016/j.asej.2020.12.012
22. Guirguis MN, Moussa RR (2019) Investigation on utilizing garbage as a resource for a sustainable neighbourhood; case study of a neighbourhood in New Cairo. In: 2nd conference of the Egyptian affiliation of international building performance simulation association (BSCAIRO 2019) Cairo_ Egypt, 397-012018, 28–30 November 2019. IOP Conference Series: Materials Science and Engineering. https://doi.org/10.1088/1755-1315/397/1/012018
23. Dawoud N, Micheal A, Moussa RR (2020) A review on investigating the experimental process for partial replacement of cement with sugarcane bagasse in the construction industry. In: The international conference on civil and architecture engineering (ICCAE), Cairo_ Egypt. IOP Conference Series: Materials Science and Engineering

Collaborative Work in Virtual Environments

Connecting Spatially Separated Laboratory Environments by Combining Virtual and Augmented Reality Technology

Nils Höhner[1](✉), Anke Pfeiffer[2], Davide Reverberi[3], Mark Oliver Mints[1], and Julien Rodewald[1]

[1] University of Koblenz-Landau, Mainz, Germany
nhoehner@uni-koblenz.de
[2] Hochschule für Technik Stuttgart, Stuttgart, Germany
[3] University of Parma, Parma, Italy

Abstract. This paper proposes a technical solution to connect an existing Virtual Reality laboratory to a real-world laboratory by combining Augmented Reality glasses (Microsoft HoloLens 2) and a Virtual Reality Headset (HTC Vive Pro) in a single networking application. Furthermore, it investigates educational opportunities and challenges of combining both a virtual and a real lab for higher engineering education.

1 Introduction

The research project Digilab4U focuses on the development of Virtual, Remote and Mixed Reality (MR) laboratories. The aim is to bring real laboratories online and make them accessible on a platform, but also to create a network of connected and modular laboratories. We present a technical solution to connect an existing Virtual Reality (VR) laboratory to a real-world laboratory by combining Augmented Reality (AR) glasses (Microsoft HoloLens 2) and a VR Headset (HTC Vive Pro) in a single networking application. We investigate educational opportunities and didactical challenges of combining both virtual and real lab for higher engineering education.

The broader goals of our work are twofold. We want to enable collaboration by combining AR and VR technology in a meaningful way, as well as embedding these technologies in a didactical scenario. From the technological point of view this involves creating a software architecture capable of handling different hardware and rendering techniques that share the same context over network. Additionally, due to the completely different kind of user experience AR and VR headsets offer we had to develop device specific solutions to promote collaboration. This includes visualization techniques to share the view between both devices, creating ways to interact with each other by exchanging virtual objects, as well as utilizing video and audio-based communication channels. Our goal is to share our solution, discuss various technical approaches, their implications, as well as their influence on the didactical decision process.

M. E. Auer et al. (Eds.): REV 2022, LNNS 524, pp. 509–520, 2023.
https://doi.org/10.1007/978-3-031-17091-1_51

Related Work: Collaboration within the Reality Virtuality (RV) Continuum [11] includes different technologies as well as modalities such as synchronization and colocation. For our work we focus on synchronous collaboration and especially on the combination of AR & VR technology. A comprehensive overview of synchronous collaboration techniques spanning the entire RV continuum can be found in [16].

Already 1999 Kiyokawa et al. [9] proposed a combination of AR and VR technology based on special hardware that offered a shared virtual and a shared augmented reality environment allowing collaboration between colocated users. The users, however, had to choose a mode to collaborate and were able to switch it on demand. Similarly the magical book [1] lets users switch fluidly between the real and virtual worlds by initially augmenting its pages with augmented reality content. By flipping a switch, users are able to view the identical scene from VR. Oda et al. [12] demonstrate the effectiveness of combining AR and VR technology for collaboration by allowing an VR remote expert to annotate or move virtual replicas that are projected in real time for an AR user. Extending on this idea Le Chenechal et al. [10] proposed an interaction paradigm where both users sit on the same chair while the AR user is able to see the virtual hands of his guide. Wang et al. [17] combine VR and AR Glasses in a remote assembly task and share the VR users hands for the AR user to mimic the assembly process. Additionally a webcam view of the real world is streamed to the virtual environment letting the remote expert see his partner. The CoVAR system by Piumsomboon et al. [15] offers a unity based universal collaboration environment for colocated scenarios, supporting the HTC Vive as well as the Microsoft HoloLens. The AR user is able to share his reconstructed environment for the VR user to visit. Additionally multiple awareness cues as well as view modes are proposed. Similarly Pereira et al. [13] developed a framework for MR collaboration and discuss architecture design and challenges regarding different types of hardware.

System Overview and Contribution. Similarly to the CoVAR system [15] we focused on creating a universal collaboration application that can be integrated or used in different scenarios. As one can see in the schematic overview of our system architecture in Fig. 1 we split the application conceptually, as well as technically into an AR and VR part resulting in two non-colocated user spaces. The bridge between the two worlds is formed by a portal metaphor through which both users can look into the other world. While the AR user is able to look directly into the virtual environment, the VR user receives a first person video stream from his partner. Both users have their own space but are able to exchange virtual objects (like RFID-Chips) through the portal.

2 Connecting Spatially Separated Laboratories

The starting point for the development of the Mixed Reality learning environment are two independent, spatially separated laboratory environments. While the first laboratory at the Hochschule für Technik in Stuttgart (HFT) offers an RFID measurement cabinet where students learn the basics of RFID-technology by performing measurements under controlled conditions. The second laboratory

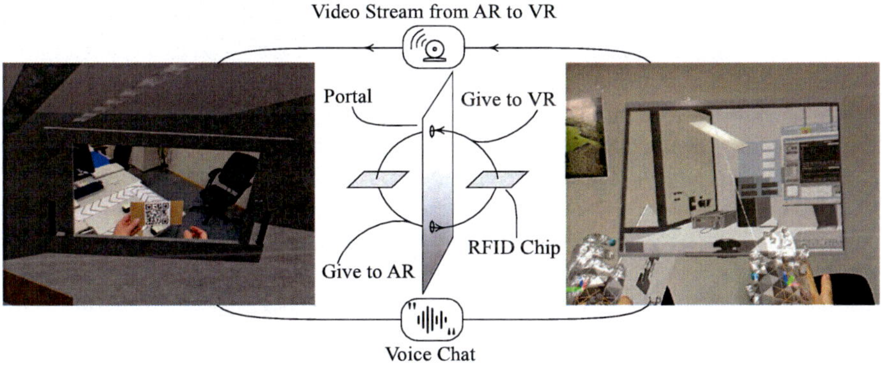

Fig. 1. Schematic Overview of our collaborative mixed reality application.

in Parma provides an industry standard conveyor belt setup where students are able to work on real life problems and experience the difference sensor noise can make in an uncontrolled environment. Given the thematical relation of both laboratories a didactical scenario was developed where students first utilize the RFID cabinet in Stuttgart to formulate hypothesises on how to optimize the experiment in the conveyor belt setup. Afterwards they have to verify their ideas by testing them in the real laboratory. In order to connect the two labs across spatial boundaries we decided to use a portal metaphor to connect an already existing VR twin of the RFID Cabinet [6] with the real laboratory. On a technical level this is done in a single Unity game engine scene where both students share the same virtual coordinate system and objects. The Augmented Reality user in the conveyor belt laboratory, however, only sees a cutout of the virtual environment through the portals surface, while the VR user is able to see the whole RFID cabinet and the first-person video stream of his or her partner. Both students share the same portal and can move it with their hands to change the field of view similar to moving a virtual camera that points into the other world. Furthermore, the student operating the conveyor belt is able to scan the real-world RFID-Chips via QR-Code to create virtual clones which can be transferred through the portal and measured in the Virtual Reality environment. This setup allows the students both to explore the real world together, as well as observing the virtual measuring results while discussing their ideas on the experiment.

2.1 Scenario Description and Didactical Potential

For a general understanding, the two laboratory scenarios are briefly explained below, followed by a more detailed description of connecting the laboratories from a didactical perspective.

The **RFID-Lab at the University of Parma** is a real laboratory reproducing an industrial like environment. The laboratory is used by the master students of management engineering in production and logistics to apply the knowledge acquired during the theoretical lessons in an industrial like environment. The

Reading Optimization scenario has the objective to teach the students to set the parameters in a way that optimize the readings of the RFID tags. Indeed, one of the main challenge for the use of RFID in industry, is to set the parameters to maximize the readings of the tags and minimize the errors. As well described by [2], the scenario is divided in two parts. The students run the experiment with different parameters, collecting the data, and then they perform an analysis. The analysis consists in the elaboration of the raw data and the calculation of Alpha and Beta Error (errors of first and second order). Having both the errors calculated, the students will be able to define the required combination of factors to minimize the errors. Moreover, to develop soft skills like collaboration, critical thinking, and to have a correct amount of work, the students are divided in groups of three or four.

The **RFID measuring chamber at the HFT Stuttgart** is a test environment for RFID UHF tags. RFID technology is a key technology in logistics, as it makes warehousing and the movement and traffic of goods transparent. To learn the correct use of the RFID measuring chamber students are given an industry-specific use case to learn the background knowledge, the practical use of the chamber and the interpretation of the results. The students work together in small teams and should gain multifaceted experiences, which support them to know, remember and explain the technical use and handling of the measuring instruments and to apply, analyse and evaluate selected RFID measurement values to optimize the use of RFID for a certain use case [14].

2.2 Connecting Laboratories for Teaching and Learning

Already in 2001, de Witt described that virtual environments can only play a role in educational processes when they are based on concrete experiences and gain meaning for future human activities [18]. Even though de Witt probably could not have foreseen the technical developments twenty years later, the statement is still relevant, especially in the context of laboratory-based learning. Lab-based learning is based on theory-practice transfer and on gaining professional experiences, which should be enabled in laboratory environments, whether they are real, remote or virtual [4]. In this sense, DigiLab4U has itself set the goal, to develop hybrid laboratory learning environments that enable the acquisition of laboratory based knowledge and skills and especially competences in the field of Internet of Things (IoT) and Industry 4.0 (I4.0). These fields of work are highly up-to-date in the Engineering Sciences and characterized by linking the real world and the virtual world. In order to allow students to experience the associated profound technical and economic challenges for production and business processes in a sustainable manner, these changes should be reflected in their learning processes in laboratories. The project DigiLab4U aims to connect labs not only to a shared digital interface (e.g. earning platform), but to interconnect them and make them available across institutions providing access to industry-related laboratory environments and equipment. Connecting the real laboratory with remote, augmented and virtual laboratory-applications, provides a hybrid (cross-media) learning environment that serves as learning content and learning medium at the same time. The theoretical foundations for the development of the scenario can only be outlined in this paper. For a further didactic consolidation of the presented scenario, the following articles are recommended [3]. The focus of the connected scenario is on theory-practice transfer, and here specifically on the acquisition of laboratory-based skills that are currently required for prospective specialists in the field of IoT and Industry 4.0 [5]. To provide students with the most extensive experimental time and experience possible, a use case approach was employed, which includes the task to find a suitable RFID tag for a company that transports cans and PET water bottles in cardboard boxes via a conveyor belt. The connected scenario proceeded as followed: At the beginning, the student from Stuttgart (student S) accessed the VR-Lab and familiarized herself with the surroundings while her peer student in Parma (student P) did the same with the AR glasses. The peer students observed the whole process via a shared VR-view on a computerscreen, that enabled them to support the VR- and AR-student with comments and hints. Immediately, when the student S entered the VR environment, she saw the real laboratory of the University of Parma through the portal and she was able to interact with the laboratory team of Parma. Via the portal, a virtual RFID tag was handed over from student P to student S. The latter inserted the digital tag in the VR measuring chamber and started the experiments. The measurement results were than explained by student S and discussed with the fellow students in Parma and Stuttgart. Subsequently, the selected RFID tag was tested in the real laboratory environment. For this purpose, a cardboard box was pasted with the RFID tags and a test run on the conveyor belt was initiated. The student P showed and explained the laboratory equipment and the experiment settings to the fellow students of Stuttgart, who observed the activities in the real laboratory.

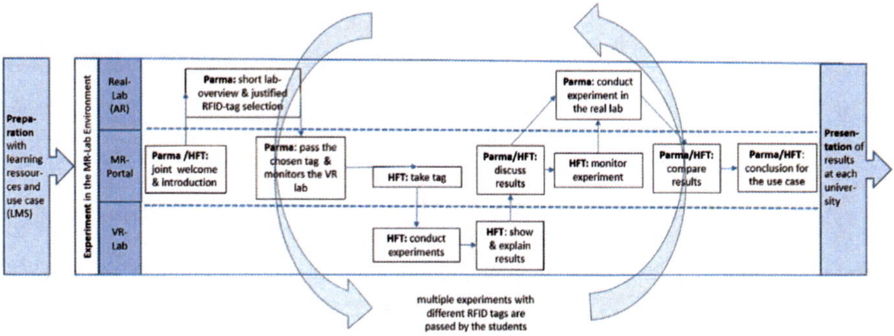

Fig. 2. Didactical process overview of mixing realities

In addition to the tag's reading data, the experiment with the conveyor belt also recorded the speed of the conveyor belt, the reading capability of the different antennas used, and possible sources of interference that could influence RFID reading results. This provides a wealth of data that can be jointly evaluated and compared to find an appropriate solution for the use case. The process shows the elements of the hybrid lab environment and clarifies the connection between the learning activities performed and the MR lab-environment. Here the students were given the opportunity to move freely and to switch between the different hybrid levels, to communicate and interact with their peers via the portal or just to test technical capabilities and boundaries of Mixed Reality.

3 Implementation

We designed and implemented a working prototype that is able to combine AR and VR hardware in a single application allowing two spatially separated students to work together on a common task. During development we tested and evaluated different visualization techniques which will be discussed in detail and summarized in a lessons learned area giving the reader the possibility to build upon our work.

3.1 Single Scene Approach

In principle, the Unity game engine works with so-called scenes to be able to divide games and applications into different sections. This serves both the clarity and the thematic and functional separation. When planning an application that can be separated into different sections as clearly as ours, one might first assume it makes sense to divide it into several scenes as well. For instance, into an AR and a VR version.

This has the advantage of being able to optimize each scene specifically for the underlying hardware and to completely decouple the interaction systems. At the same time, however, communication across different scenes is made more difficult by this very separation. Instead of separating at the scene level, we decided to logically divide all objects by ownership and control both interaction and visibility on this basis. Meaning that for both users the identical objects are present at all times and only the interaction and the visibility alters based on the users role. This is done by switching to AR or VR mode directly when the application is launched based on the underlying hardware. Suitable camera objects are loaded dynamically and render modes are set appropriately. Conveniently this allows us to use the Mirror networking framework to synchronize objects within the same scene over network.

Transferring Objects Through the Portal. The aim of the application is not only to provide a view into the virtual world, but to enable the interaction as well. For this reason, it is possible to pass all virtual objects through the portal (in our special case RFID-chips, which are scanned beforehand via QR code). From a technical point of view, a decisive advantage of the *Single Scene Approach* becomes apparent. Instead of transferring the object over the network, destroying it on one side in order to make it reappear on the other side, only the visibility and the ownership has to be altered. For example, in the *Render To Texture Approach*, the layer changes as soon as the center of the object crosses the surface of the portal.

3.2 Portal Rendering

The idea of the portal renderer is to present the AR user with a perspectively correct cutout of the virtual world to give the impression of looking through a window. In addition, the user should be able to move the portal to adjust the angle of view into the virtual world. In computer graphics, there are various approaches for displaying only parts of a virtual scene, which have various advantages and disadvantages, especially in combination with the HoloLens 2. In the following, our *Render To Texture* and *Stencil Buffer* variants will be presented. Table 1 summarizes the pros and cons of each approach.

Table 1. Pros and Cons of the presented portal rendering techniques.

	Render to texture	Stencil buffer
Pro	- Separation of visible objects - Easy to integrate	- Stereo vision works - Hologram stabilization - Single pass rendering - Performance
Contra	- No hologram stabilization - Stereo vision not correct - No single pass rendering	- Harder to integrate - 32-Bit depth buffer required

Render to Texture Approach. The Render To Texture approach utilizes the Unity layer system which allows the assignment of layers to game objects. These layers are used by the cameras culling masks in order to render only parts of the scene. In our case we created a portal layer which contains all objects that should only be visible through the portal, while the portal itself, as well as, all AR control elements reside on the scenes default layer. To convey the impression of a cutout view we defined two cameras inside the scene. One being the HoloLens main camera, while the other is a standard unity camera (portal camera) which mirrors the main cameras pose, as well as its resolution and projection settings. The portal camera is responsible to draw all objects from the portal layer onto a specified target texture, while the main camera draws everything else resulting in just the portal visible inside a seemingly empty scene. In order to visualize a cutout of the render texture the portal surface quad utilizes a special shader which maps the screen position to texture coordinates.

Even though this approach is easy to implement and integrate into existing applications it results in two renderpasses (one rendering to a texture, followed by the main render pass), which is not ideal for the HoloLenses optimization for single pass instanced rendering. Additionally the hologram stabilization solutions only work on objects rendered directly by the main camera resulting in a slightly skewed perspective, as well as a small asynchronicity during camera movements.

Stencil Buffer Approach. The stencil buffer is a special 8 bit integer buffer that stores a value for each pixel in the framebuffer. When enabling stencil testing the gpu will test against the given value in the stencil buffer to decide if the current pixel has to be drawn. The type of comparison (less, equal ...), as well as the reference value can be defined in Unity. For our portal we start by drawing the surface quad with enabled stencil testing, set a reference value of one and tell the renderer to replace the current values in the stencil buffer. This way we create a stencil mask which represent our portal window. In the next step all other virtual objects which should reside on the other side of the portal are drawn only if their pixels overlap with the stencil mask. In order to ensure that the portal surface is drawn first, the render queue order has to be altered manually.

In contrast to the texture approach, the stencil technique has the great advantage of requiring only one render pass. As a result, both the stereo display as well as the hologram stabilization work properly. The performance is significantly higher than with the multi-pass approach, but it is necessary to use a 24-bit depth buffer, since Unity only supports stencil operations in this mode. Unfortunately, this can also degrade performance, as the HoloLens is optimized for a 16-bit depth buffer. In our tests, however, the limitation was significantly greater with multiple render passes and the stencil variant could be used without any problems in the portrayed scenes. Nevertheless, the integration of a stencil based portal might be harder depending on the underlying application. The setup is associated with additional effort, since a special material with the correct stencil settings must be provided for each object. Furthermore, incompatibilities with already existing stencil rendering methods may occur.

4 Evaluation

For the evaluation of our system, we did a small real world stress test with two groups of students located in Italy and Germany to evaluate the technical and didactical design for the MR-environment, to observe the way the students communicate with each other and to identify potential technical and didactical limitations. Additionally, a controlled test with e-learning experts was done in a mock-up laboratory, followed by semi structured interviews.

Live Test (LT). Due to the paper restrictions, the didactical research procedure can only be presented in abbreviated form. The live test took place in the summer term 2021. Four students of the BA Information Logistics of the HFT Stuttgart and four students of the MA Management Engineering of the University of Parma were involved. The test was supervised and guided on site by a laboratory assistant and a professor. The presented preliminary analysis of the students activities in the cross-country test were related to the acquisition of competencies and here specifically to competences from the field of IoT and I4.0 with relevance for laboratory-based learning processes. To identify these competencies from the field of IoT and I4.0 a qualitative analysis of nine well-founded studies from the German-speaking area was conducted in advance. From this analysis, competencies were derived which, in a next step, were assigned to the competency dimensions of the German Qualification Framework, as this serves as a competence orientation for universities.

German Qualification Framework (DQR)		I4.0/-IoT Specification	Results of the qualitative Observation of the students actions in MR
Professional Competence	Knowledge	Knowledge of IT-related topics	Show basic knowledge in handling the equipment; Explain each other the experiments and results
		In-depth industry and work-specific expertise	Not observed
		Handling cyber-physical systems	Demonstrate their ability to act confidently in the MR environment
	Skills	Digital Literacy	Use various the MR-lab environment + technical systems for learning-/ communication-processes.
		Methodological knowledge	Analyze + compare measurements of different labs and optimize them in relation to a use case
		Dealing with Big Data	Transfer measurement results from the software into Exel data and prepare them for further processing
Personal Competence	Social Competence	Interdisciplinary thinking and acting skills	Not observed
		Intercultural competence	Can exchange and discuss professional information in English
		Networking capability	Not observed
		Teamwork skills	Conduct experiments and present results together
	Indepen-dence	Ability for self regulated learning and life long learning	Preparing for the experiment and analysing the results independently
		Leadership competence	Not observed
		Innovation capability	Not observed
		Act and think Creatively	Not observed

Fig. 3. IoT-/I4.0 competence analysis in the MR-Lab

As part of the DigiLab4U-project a previous work on building a competence grid was provided by Kammerlohr et al. 2020 [8]. To open up the lab exercise for further analysis in terms of skill acquisition, the students' activities and behaviour in the MR laboratory environment were videotaped and afterwards analyzed using an observation sheet to capture actions that could not be recognized at first glance, for example due to technical aspects which could bind the attention more strongly [7]. The observation of the MR-lab experiment was subsequently conducted using the video footage of the live test, with a total length of 02:36 h. The focus of the observation was on the students' actions and communication processes. The results of the observation are to be regarded as non conclusive, but they provide indications and tendencies which competencies can possibly be achieved and they will be used to further develop and expand the scenario in this respect. The analysis of the observation protocol shows that the scenario still has potential, particularly with regard to the acquisition of IoT/I4.0-relevant competencies. It is not yet sufficiently explored at what level competencies are acquired. Moreover, some competences were not observable in the current MR lab-experiment, but are to be expected, in higher semesters and or research projects (BA-/ MA Thesis).

Expert Tests (ET). For the expert test, we invited 4 experts and divided them into two teams who performed a mock-up experiment one after the other. For the mock-up experiment, the Conveyor-Lab from Chap. 2.1 was recreated using paper models. The task was to select suitable RFID transponders for the scenario and to decide to which side of the transport boxes they should be attached. The AR user showed the lab to his partner and they discussed the execution of the experiment together. Afterwards, existing RFID transponders were virtualized via QR code, transferred to VR and measured in the virtual rfid chamber. The in eye perspective, as well as the spoken word of all participants was recorded. Afterwards a semi structured interview with each participant was done. Interview results summarized from one perspective at a time.

4.1 Mixing Reality - Lessons Learned

The lessons learned are derived from the observations and the interviews of the **Live Test** (*LT*) and the **Expert Tests** (*ET*).

Navigation(*LT & ET:*) It could be observed, that after a short familiarization period, the users confidently navigated through the lab environment and used the portal for sharing the tags, for observation and discussion. Switching between the different hybrid levels, such as AR, MR-portal and VR was used very self-evidently by the students. This indicates a viable link between the laboratory environment and learning processes, that needs to be explored more in upcoming evaluations. **Perception** (*ET:*) Especially from the AR side the portal was described as a *"window to the other world"* with realistic depth perception. Most of the problems arose from the semi transparent holograms which are hard to see on cluttered background. For optimal use one should use the application in an empty room with bright walls. The VR side described the first person perspective from his partner as very helpful for following explanations but did not report the feeling of a window. Giving this statement we think it might be beneficial to also allow the VR user a view into the reconstructed environment the HoloLens provides. **Communication** (*LT:*) Working in the MR-environment requires increased communication discipline from all involved participants. Discussions that arise away from the MR environment can quickly lead to an interruption of the collaborative work. Disruptive situations could not always be prevented due to queries, technical difficulties and occasional language barriers. In addition, a time-out-moment could be tested next time, to give each University the chance to clarify their issues if required. This could provide space for planned interruptions without disrupting the collaboration too much. (*ET:*) In the more controlled environment the communication was more focused on the topic at hand. Interestingly the users tend to turn towards the portal while discussing which indicates that the metaphor works as kind of an anchor between the two worlds. **Collaboration** (*LT:*) Collaboration could be observed as fruitful, due to the fact that the different labs provided a good insight into the respective laboratory practice via the technical connection. Since the portal could be moved freely by both groups of students, irritating situations occasionally occurred, e.g., the portal was placed by the student wearing AR glasses directly in the field of view of the student in the VR measurement chamber, so that it obscured the view of the VR environment. Communicatively, these and similar situations could be quickly resolved. (*ET:*) The experts really enjoyed the virtualization of the real RFID-Chips and the exchange of them via gestures through the portal. They stated that this metaphor increases the immersion and motivates the users to collaborate as it offers a natural topic for a conversation. We think this demonstrates the importance of transferring real life communication cues to the virtual environment (instead of just teleporting the RFID-Chips per button click). **Interaction:** (*LT:*) While developing the MR-scenario, it was necessary to take into account, that in parallel with the MR lab environment, a videoconferencing system was used as an additional communication tool to record the live test, to screenshare the lab software or to observe students lab activities, by students that were at that point not in charge of VR or AR-activities. In this way, it could be ensured that the students of the respective universities could see the same processes in their observer role and they were constantly able to interact with their peer students if required. **Cooperation:** (*LT:*) Especially for the role of the observing students, journaling and observation tasks will be assigned in the upcoming runs to further involve them in the procedure of the MR environment and to foster engagement and cooperation. **Influencing technical issues:** *(LT:)* Some technical problems cannot be solved at the moment, for example, the overheating of the AR glasses. This must be taken into account in the didactic re-design of the scenario. In general, the connection of the laboratory environment was successfully implemented despite the long distance transmission, but initial problems with the WebRTC connection led to unplanned delays in the process. *(ET:)* Even though the experts really liked to scan the real RFID-Chips via QR code they had difficulties spawning them as the QR tracking on the HoloLens has a small delay and is influenced by the lighting conditions of the environment. Users have to keep in mind to provide enough light and print the QR codes as big as possible.

5 Conclusion and Future Work

The presented Mixed Reality technology could successfully be used to combine two thematically related laboratory scenarios from different courses allowing cross-university and in our case even cross-country collaboration between students. Moreover, it is plausible that the adoption of new technology together with the connection between foreigner students, increases the commitment and the involvement of the students which participate in the scenario. We emphasize that the presented software architecture, as well as the communication metaphors like the portal can be easily extended to other application areas.

Acknowledgement. Funded by the German Federal Ministry of Education and Research (BMBF), grants no. 16DHB2115 "DigiLab4U".

References

1. Billinghurst M, Hirokazu K, Poupyrev I (2001) The magicbook: a transitional ar interface. Comput Grap 25:745–753
2. Bottani E, Reverberi D, Romagnoli G, Ustenko M, Volpi A (2022) Batch experiment: A fruitful way of combining hands-on laboratory and e-learning. In: Auer ME, Bhimavaram KR, Yue X-G (eds) REV 2021, vol 298. LNNS. Springer, Cham, pp 244–255. https://doi.org/10.1007/978-3-030-82529-4_24
3. Burghardt M, Ferdinand P, Pfeiffer A, Reverberi D, Romagnoli G (2021) Integration of new technologies and alternative methods in laboratory-based scenarios. In: Auer ME, May D (eds) REV 2020, vol 1231. AISC. Springer, Cham, pp 488–507. https://doi.org/10.1007/978-3-030-52575-0_40
4. Feisel LD, Rosa AJ (2005) The role of the laboratory in undergraduate engineering education. J Eng Educ 94:121–130

5. Gottburgsen A, Wannemacher K, Wernz J, Willige J (2019) Ingenieurausbildung für die digitale transformation: Zukunft durch veränderung: Vdi-studie April 2019
6. Höhner N (2020) Integrating Virtual Reality in a Lab Based Learning Environment. In: Bourdot P, Interrante V, Kopper R, Olivier A-H, Saito H, Zachmann G (eds) EuroVR 2020, vol 12499. LNCS. Springer, Cham, pp 99–114. https://doi.org/10.1007/978-3-030-62655-6_6
7. Hubert K (2011) Die video-interaktions-analyse. In: Meyn C, Peter G, Dechmann U, Georg A, Katenkamp O (eds) Arbeitssituationsanalyse, pp 365–382. VS Verlag für Sozialwissenschaften, Wiesbaden
8. Kammerlohr V, Pfeiffer A, Uckelmann D (2021) Digital laboratories for educating the IoT-generation heatmap for digital lab competences. In: Auer ME, May D (eds) REV 2020, vol 1231. AISC. Springer, Cham, pp 3–20. https://doi.org/10.1007/978-3-030-52575-0_1
9. Kiyokawa K, Takemura H, Yokoya N (1999) A collaboration supporting technique by integrating a shared virtual reality and a shared augmented reality. In: IEEE International Conference on Systems, Man and Cybernetics (SMC 1999), pp 48–53
10. Le Chénéchal M, Duval T, Gouranton V, Royan J, Arnaldi B (Mar 2016) Vishnu: virtual immersive support for HelpiNg users - an interaction paradigm for collaborative remote guiding in mixed reality. In: 3DCVE 2016: International Workshop on Collaborative Virtual Environments, pp 1–5. IEEE, United States,
11. Milgram P, Takemura H, Utsumi A, Kishino F (Jan 1994) Augmented reality: A class of displays on the reality-virtuality continuum. In: Telemanipulator and Telepresence Technologies, vol 2351
12. Oda O, Elvezio C, Sukan M, Feiner S, Tversky B (2015) Virtual replicas for remote assistance in virtual and augmented reality. In: Proceedings of the 28th Annual ACM Symposium on User Interface Software & Technology - UIST 2015, pp 405–415
13. Pereira V, Matos T, Rodrigues R, Nóbrega R, Jacob, J (2019) Extended reality framework for remote collaborative interactions in virtual environments. In: 2019 International Conference on Graphics and Interaction (ICGI), pp 17–24
14. Pfeiffer A, Lukarov V, Romagnoli G, Uckelmann D, Schroeder U (2020) Experiential learning in labs and multimodal learning analytics. In: Ifenthaler D, Gibson D (eds) Adoption of Data Analytics in Higher Education Learning and Teaching. AALT. Springer, Cham, pp 349–373. https://doi.org/10.1007/978-3-030-47392-1_18
15. Piumsomboon T, Dey A, Ens B, Lee G, Billinghurst M (2017) Covar: mixed-platform remote collaborative augmented and virtual realities system with shared collaboration cues. In: Adjunct Proceedings of the 2017 IEEE International Symposium on Mixed and Augmented Reality, ISMAR-Adjunct 2017, pp 218–219
16. Schäfer A, Reis G, Stricker D (2021) A survey on synchronous augmented, virtual and mixed reality remote collaboration systems
17. Wan P et al (2021) 3dgam: using 3d gesture and cad models for training on mixed reality remote collaboration. Multimedia Tools Appli. 80:31059–31084
18. de Witt C (2001) Bildungsprozesse in interaktiven medialen umgebungen. In: Aufenanger S, Schulz-Zander R, Spanhel D (eds) Jahrbuch Medienpädagogik 1, pp 193–202. VS Verlag für Sozialwissenschaften, Wiesbaden

A Quantitative Research About MOOCs and EdTech Tools for Distance Learning

Serafeim A. Triantafyllou(✉)

Computer Science Teacher in School of Secondary Education of Greek Ministry of Education and Religious Affairs, Athens, Greece
ser.triant@gmail.com

Abstract. Educational technology is an integral part of modern life, research and teaching. MOOCs (Massive Open Online Courses) constitute a widely recognized mode of learning in distance education that can help to the dissemination of knowledge. Also, web-based digital tools such as Kahoot, Google Docs (Google Documents), TitanPad and Etherpad are edtech tools that let a group of learners work together to create and edit documents. These tools allow learners and instructors to ask questions, group them, create a text together and collaborate in real time. Nowadays, many engineering courses depend on group or team work. Each course must have documented learner outcomes that support the initial learning objectives. Attainment of these outcomes prepares learners to be sufficient to enter the professional practice of engineering. Instructors should encourage learners to work in groups or teams because in a team learners coordinate independent work toward a common goal. The purpose of our quantitative research is to study the degree of penetration of MOOCs and edtech tools in distance learning by better understanding MOOCs and other edtech tools such as Kahoot, Google Documents, TitanPad and Etherpad, as they can be used in professional practice of engineering.

Keywords: Educational technology · MOOCs · Edtech tools

1 Introduction

Each educational program must have documented learner outcomes that support the program learning objectives. Attainment of these outcomes prepares learners to be sufficient to enter the professional practice of engineering. Learners when participating in an educational program can learn to identify, formulate, and solve engineering problems by applying principles of engineering, science, and mathematics. They learn to work together and communicate effectively. Moreover, they learn to function effectively on a team whose members together create a collaborative and inclusive environment, set goals, plan learning activities and meet learning objectives. Also, it is very important that learners can construct their own knowledge by using learning strategies and apply new knowledge when needed in their everyday practice. Engineering team strategies should be taught, not assumed. Each strategy and every learning goal should be clear, observable and relevant to educational program [4, 10, 11]. For effective remote instruction the basic

M. E. Auer et al. (Eds.): REV 2022, LNNS 524, pp. 521–532, 2023.
https://doi.org/10.1007/978-3-031-17091-1_52

goals should be to enhance motivation, incorporate active engagement, design effective assessment, and adopt an inclusive mindset [8, 13, 14]. Adoption of inclusive mindset means that instructors should consider what they teach and how they teach during the learning process [1]. Self-Determination theory refers to motivation that drives an individual to act and is a theory of human motivation developed by psychologists Edward Deci and Richard Ryan. According to this theory, there are three innate psychological needs that drive motivation [3]. These needs are: (a) autonomy, (b) competence and (c) relatedness [2, 12]. Relatedness shows the importance of social presence. Isolation is a challenge that we face in online instruction. We should develop student social presence, build teacher social presence, and create connections to engineering practice and society. The need of autonomy is the need to experience behavior as voluntary and give learners autonomy choosing the content they want to study [9]. The need of competence means the need to experience behaviors as effectively enacted [9].

2 Methodology

The purpose of our quantitative research is to study the degree of penetration of MOOCs and edtech tools in distance learning. The methodology used to conduct our survey is the construction of a questionnaire. The completion of the questionnaire was optional and anonymous. What questions were selected to be included in the questionnaire was an extremely important issue. Two important attributes that the questions of the questionnaire should have been were reliability and validity. We also had to pay attention to some important points when formulating the questions, such as: (i) the questionnaire should have been simple, small in length and not include complex concepts, (ii) the questions should have been formulated in such a way as to motivate the respondent to answer them, (iii) to have a greetings message at the end of the questionnaire.

The type of questions we finally chose to include in the questionnaire was the structured questions and in particular: (i) **Dichotomous Questions**: Yes/No type questions that give the choice between the answer "Yes" or the answer "No", (ii) **Multiple Choices**: Questions accompanied by a number of alternative answers. It was important to report to the respondent that he/she could choose one option as an answer, (iii) **Rating Scales**: it is about questions with a rating scale in the answers. Specifically, we chose queries on a Likert scale of five levels from 1 (Not at all) to 5 (Too much).

Step 1: Collection of Responses

The final questionnaire sent in response contained eight questions and was divided into two parts. The first part concerned demographic data (gender, age and level of study). The second part contained Likert-type questions related to the use of MOOCs and edtech technologies in distance learning. The design of the questionnaire was done using Google's docs software. The collection of responses finally resulted within two months (October 2019–December 2019) after sending the online questionnaire to a sample of 50 learners. 29 of the learners, were male (58.00%) and 21 female (42.00%).

In order for the specific sample to be representative it was chosen at random, by random sampling.

Step 2: Coding Questions for Their Statistical Processing in SPSS (Statistical Package for Social Sciences)
For the statistical processing of the results the statistical package we selected was SPSS Statistics 20.0 (Statistical Package for Social Sciences). Below we quote in detail the encoding of the questions we have done to allow them to be entered into the SPSS database.

Part A: Demographic Data

Question A1: Your gender is:
Name of variable: GENDER
Measurement: nominal (dichotomical)
Label: Respondent gender
Values: 1 = "man"
2 = "woman"

Question A2: Your age ranges from:
(2) Name of variable: AGE
Measurement: ordinal
Label: Respondent age group
Values: 1 = «18–24», 2 = «25–31», 3 = "32–38", 4 = «39–45», 5 = "46–52", 6 = «53–59», 7 = "60 years and over"

Question A3: What is your level of study?
(3) Name of variable: EDUCATION
Measurement: ordinal
Label: Level of Studies
Values: 1' = "graduates of primary school", 2 = "junior high school graduates", 3 = "high school graduates", 4 = "tertiary education graduates"

Part B: MOOC and Edtech Tools

Question B1: Do you know the term MOOC?
(4) Name of variable: MOOC_KNOWLEDGE
Measurement: nominal (dichotomical)
Label: MOOC_KNOWLEDGE
Values: 1 = "Yes", 2 = "No"

Question B2: Do you use and to what extent the MOOCs?
(5) Name of variable: MOOC_USE
Measurement: ordinal
Label: MOOC_USE

Values: 1 = "Not at all", 2 = "Little", 3 = "Enough", 4 = "A lot", 5 = "Very Much"

Question B3: Do you know EdTech tools such as Kahoot, Google Documents, TitanPad and Etherpad?
(6) Name of variable: EDTECH_KNOWLEDGE
Measurement: nominal (dichotomical)
Label: EDTECH_KNOWLEDGE
Values: 1 = "Yes", 2 = "No"

Question B4: Which of the following are you considered the most popular EdTech tools?
(7) Name of variable: EDTECH_TOOLS
Measurement: ordinal
Label: Popular EDTECH_TOOLS
Values: 1 = "Kahoot", 2 = "Google Documents", 3 = "TitanPad", 4 = "Etherpad"

Question B5: Do you use and to what extent the EdTech tools such as Kahoot, Google Documents, TitanPad and Etherpad?
(8) Name of variable: EDTECH_TOOLS_USE
Measurement: ordinal
Label: EDTECH_TOOLS_USE
Values: 1 = "Not at all", 2 = "Little", 3 = "Enough", 4 = "A lot", 5 = "Very Much"

Step 3: Grouping of Questions and Presentation of Results
The questions are divided into two categories, each of which follows a different approach to the analysis of the results. These categories are:

Likert Type Questions
Questions, where there is a rating scale in the answers. The statistical analysis listed below follows the following structure: (i) Question and corresponding coding in SPSS, (ii) Tables depicting key statistics, (iii) Graphs, (iv) Observations and commentary on the results. Such type of questions are as follows: B2, B5.

Multiple Choice Questions
In these questions, we list the charts directly and make observations where necessary. Such type questions are as follows: A1, A2, A3, B1, B3, B4. Here is the quote of the results and annotation of the resulting statistical data. What follows is the analysis of the multiple choice questions with quoting the graphs and the necessary commentary.

Question A1: GENDER
Variables in SPSS environment:
GENDER - Respondent's gender: (i) Man, (ii) Woman

Remarks: With regard to the gender of the respondents, the male population (29 men) corresponds to 58.00%, while the female population (21 women) is 42.00% (see Fig. 1).

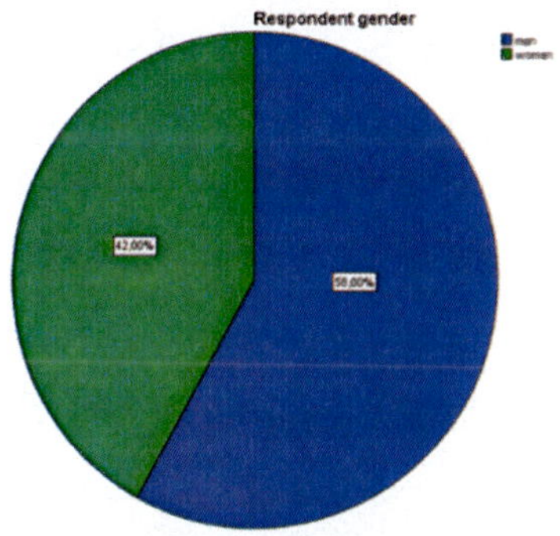

Fig. 1. Respondent gender

Question A2: AGE

Variables in SPSS environment:

(2) AGE

Age group of respondents: (i) 18–24 years old, (ii) 25–31 years old, (iii) 32–38 years old, (iv) 39–45 years old, (v) 46–52 years old, (vi) 53–59 years old, (vii) 60 years old and over

Remarks: The pie chart for the age groups in the sample shows that 42.00% of the respondents are in the age group 18–24, in the age group 25–31 it is 32.00% of respondents, in the age group 32–38 it is 6.00% of the respondents, in the age group 39–45, it is 8.00% of the respondents, in the age group 46–52 it is 8.00% of the respondents, in the age group 53–59 it is 2.00% of the respondents and in the age group of 60 years and over it is 2.00% of the respondents (see Fig. 2).

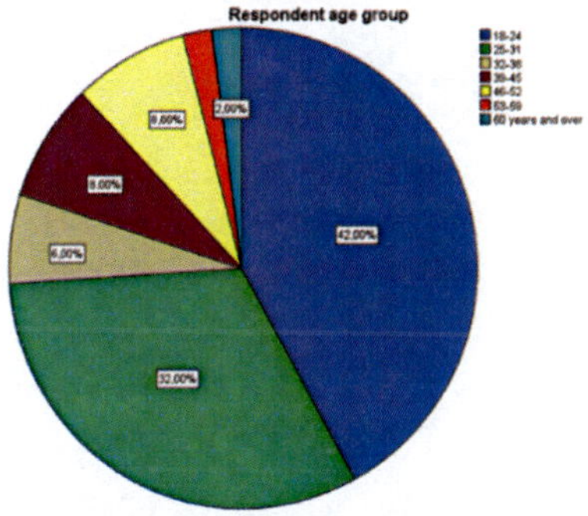

Fig. 2. Respondent age group

Question A3: Level of Studies

Variables in SPSS environment:

(3) EDUCATION.

Level of Studies: (i) Junior high school graduates, (ii) High school graduates, (iii) Graduates of tertiary education.

Remarks: From the sample of respondents 68.00% are graduates of tertiary education, 28.00% are high school graduates and 4% are junior high school graduates (see Fig. 3).

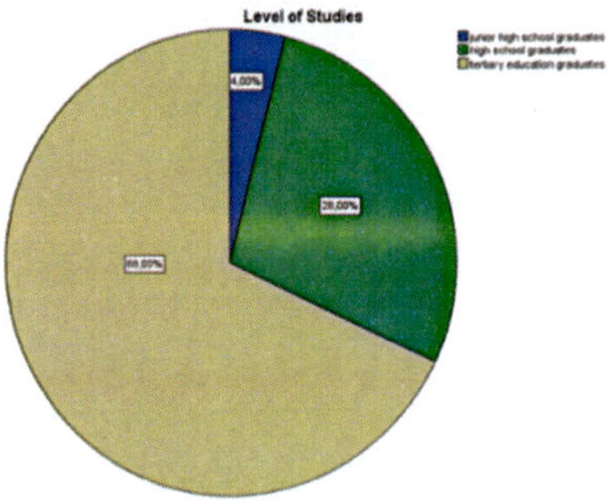

Fig. 3. Level of studies

Question B1: Do you know the term MOOC?
Variables in SPSS environment:
(4) MOOC_KNOWLEDGE

Remarks: According to the sample, 90.00% of the respondents are aware of the term MOOC, while 10.00% do not know this term (see Fig. 4).

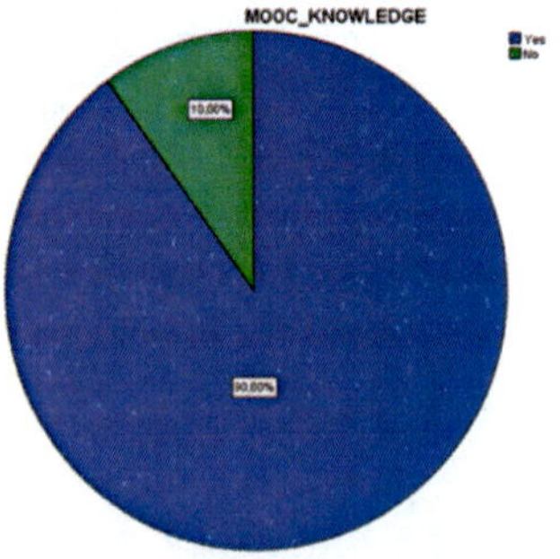

Fig. 4. MOOC_KNOWLEDGE

Question B3: Do you know EdTech tools such as Kahoot, Google Documents, TitanPad and Etherpad?
Variables in SPSS environment:
(6) Name of variable: EDTECH_KNOWLEDGE

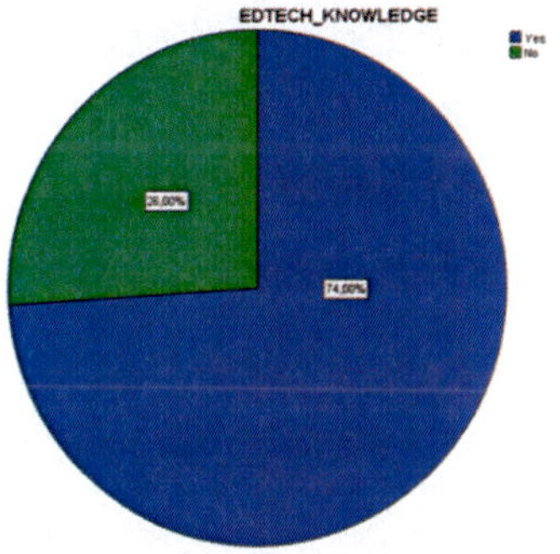

Fig. 5. EDTECH_KNOWLEDGE

Remarks: According to the sample, 74.00% of the respondents are aware of the EdTech tools such as Kahoot, Google Documents, TitanPad and Etherpad, while 26.00% do not know these edtech tools (see Fig. 5).

Question B4: Which of the following are you considered the most popular EdTech tools?

(7) Name of variable: EDTECH_TOOLS

Remarks: From the sample of respondents 42.00% chose Google Documents as the most popular EdTech tool, 34.00% chose Kahoot, 14% chose TitanPad and 10% chose Etherpad (see Fig. 6).

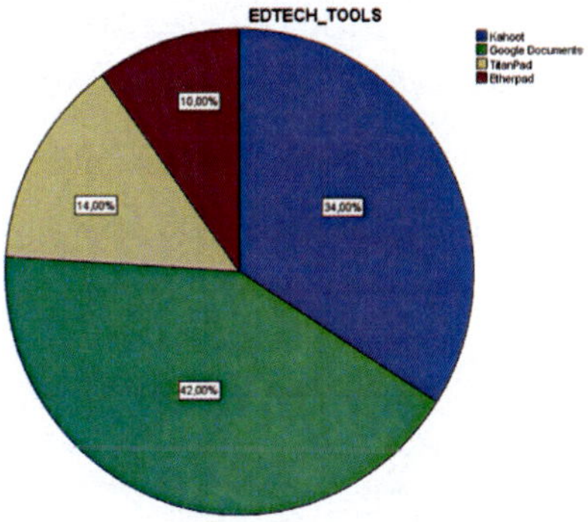

Fig. 6. EDTECH_TOOLS

Questionnaire Type Likert

Question B2: Do you use and to what extent the MOOCs?
(5) Name of variable: MOOC_USE

Table 1. Statistics MOOC_USE

Statistics		
MOOC_USE		
N	Valid	50
	Missing	0
Mean		3,640
Std. Error of Mean		,1393
Median		3,000
Mode		3,0
Std. Deviation		,9848
Variance		,970
Skewness		,126
Std. Error of Skewness		,337
Kurtosis		–,535
Std. Error of Kurtosis		,662
Range		4,0
Minimum		1,0
Maximum		5,0
Sum		182,0
Percentiles	25	3,000
	50	3,000
	75	5,000

Table 2. Statistics MOOC_USE

MOOC_USE					
		Frequency	Percent	Valid percent	Cumulative percent
Valid	Not at all	1	2,0	2,0	2,0
	Little	1	2,0	2,0	4,0
	Enough	27	54,0	54,0	58,0
	A lot	7	14,0	14,0	72,0
	Very Much	14	28,0	28,0	100,0
	Total	50	100,0	100,0	

Remarks (see Table 1, Table 2 and Fig. 7):
Mean: The responses of the sample regarding the extent of MOOCs use, show that they

use them “Enough” (mean: 3.640), **Median**: Learners seem to be using “Enough” the MOOCs (Median: 3.000), **Mode**: Most learners chose “Enough” (Mode: 3.0) in the use of MOOCs, **Standard Deviation**: The response of the sample is from the average of 0.9848 (Std. Deviation: .9848), **Skewness**: The sign of skewness is positive (.126). The distribution of responses is to the left and has its tail on the right. The distribution of responses show that learners use “Enough” the MOOCs, **Standard Error of Skewness**: The standard skewness error amounts to 33.7%, **Kurtosis**: Kurtosis has a negative sign (−.535) which makes the distribution a platycyte, with concentration of answers around the mean of “Enough”, **Standard Error of Kurtosis**: The typical Kurtosis error is 66.2%, **Range**: The answers cover the whole range from 1 to 5 and therefore the range remains 4.0, **Percentiles**: Percentiles tell us that 25% of respondents say they use “Enough” the MOOCs, 50% of respondents say they use “Enough” the MOOCs and 75% of respondents say they use “Very much” the MOOCs.

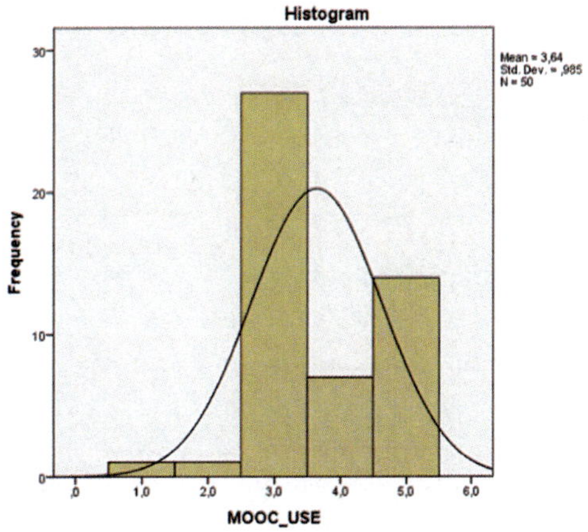

Fig. 7. MOOC_USE

Question B5: Do you use and to what extent the EdTech tools such as Kahoot, Google Documents, TitanPad and Etherpad ?

(8) Name of variable: EDTECH_TOOLS_USE

Remarks (See Table 3, Table 4 and Fig. 8): Mean: The responses of the sample regarding the extent of edtech tools utilization by learners, show that they use “Enough” the above edtech tools (mean: 2.940).

Median: Learners seem to be using “Enough” the edtech tools (Median: 3.000).

Mode: Most learners chose “Enough” (Mode: 3.0) in the use of edtech tools.

Standard Deviation: The response of the sample is from the average of 0. 5115 (Std. Deviation: .5115), **Skewness**: The sign of skewness is negative (−1.061). The distribution of responses is to the right and has its tail on the left. The distribution of responses show that learners use “Enough” the above edtech tools, **Standard Error of Skewness**: The standard skewness error amounts to 33.7%, **Kurtosis**: Kurtosis has a

Table 3. Statistics EDTECH_TOOLS_USE

Statistics		
EDTECH_TOOLS_USE		
N	Valid	50
	Missing	0
Mean		2,940
Std. Error of Mean		,0723
Median		3,000
Mode		3,0
Std. Deviation		,5115
Variance		,262
Skewness		−1,061
Std. Error of Skewness		,337
Kurtosis		4,598
Std. Error of Kurtosis		,662
Range		3,0
Minimum		1,0
Maximum		4,0
Sum		147,0
Percentiles	25	3,000
	50	3,000
	75	3,000

Table 4. Statistics EDTECH_TOOLS_USE

EDTECH_TOOLS_USE					
		Frequency	Percent	Valid Percent	Cumulative Percent
Valid	Not at all	1	2,0	2,0	2,0
	Little	5	10,0	10,0	12,0
	Enough	40	80,0	80,0	92,0
	A lot	4	8,0	8,0	100,0
	Total	50	100,0	100,0	

positive sign (4.598) which makes the distribution acute with concentration of answers around the mean of "Enough", **Standard Error of Kurtosis**: The typical Kurtosis error is 66.2%.

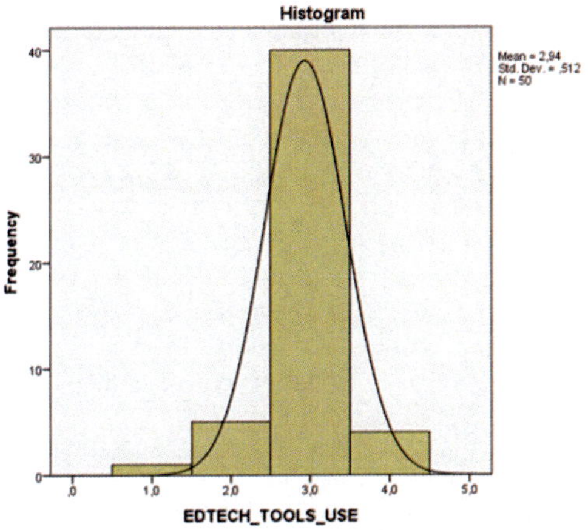

Fig. 8. EDTECH_TOOLS_USE

Range: The answers cover the whole range from 1 to 5 and therefore the range remains 3.0, **Percentiles**: Percentiles tell us that 25% of respondents say they use "Enough" edtech tools, 50% of respondents say they use "Enough" edtech tools and 75% of respondents say they use "Enough" edtech tools.

3 Discussion and Conclusion

The main aim of all education initiatives is to enhance the learning process and make learners participate actively during teaching [6]. We want to incorporate in our teaching sound pedagogical practices in order to develop responsible and autonomous learners [5, 7].

Through this piece of work, some conclusions are drawn. In the first place, distance education is a broad concept and involves many parameters. Over time, it turns out that it is not a stagnant process but changes, evolves and adapts to the needs of the learners. Now effective remote instruction is interwoven with motivation, incorporate active engagement, design of effective assessment, and the adoption of inclusive mindset. This is achieved with the help of the instructors and with the use of MOOCs and edtech tools. In general, educational institutions using MOOCs gain a significant competitive advantage. Distance Education with the proper use of MOOCs and edtech tools becomes more competitive and its long-term viability is achieved.

References

1. Ambrose SA, Bridges MW, DiPietro M, Lovett MC, Norman MK (2010). How learning works: seven research-based principles for smart teaching. In: Good practical strategies and the research that supports them. Appendix on verbs for learning objectives is very helpful. Jossey-Bass, San Francisco

2. Deci EL, Ryan RM (1985) Intrinsic motivation and self-determination in human behavior. Springer, Boston. https://doi.org/10.1007/978-1-4899-2271-7
3. Deci EL, Flaste R (1996) Why we do what we do: understanding self-motivation. Penguin Books, New York
4. Felder RM, Brent R (2016) Teaching and learning STEM: a practical guide. In: Many research-based strategies for designing and teaching courses and assessing students' learning. Draws on the authors' extensive backgrounds and decades of experience in STEM education and faculty development. Jossey-Bass, San Francisco
5. Holec H (1981) Autonomy and foreign language learning. In: First Published 1979, Strasbourg: Council of Europe. Pergamon, Oxford
6. Johnson DW, Johnson RT, Smith KA (2006) Active learning: cooperation in the college classroom, 3rd edn. Interaction Book Co., Edina
7. McCombs B (2012) Developing responsible and autonomous learners: a key to motivating students: teacher's modules. American Psychological Association, Washington, DC. https://www.apa.org/education/k12/learners
8. Nichols M (2003) A theory for eLearning. Educ Technol Soc 6(2):1–10. http://ifets.ieee.org/periodical/6-2/1.html
9. Niemiec CP, Ryan RM (2009) Autonomy, competence, and relatedness in the classroom: applying self-determination theory to educational practice. Theory Res Educ 7:133–144
10. Normark OR, Cetindamar D (2005) E-learning in a competitive firm setting. Innov Educ Teach Int 42(4):325–335
11. Piaget J, Roberts G, Piaget J (1973) To understand is to invent: the future of education. Grossman Publishers, New York
12. Ryan RM, Deci EL (2000) Self-determination theory and the facilitation of intrinsic motivation, social development, and well-being. Am Psychol 55(1):68
13. Selim H (2007) Critical success factors for e-learning acceptance: confirmatory factor models. Comput Educ 49(2):396–413. https://doi.org/10.1016/j.compedu.2005.09.004
14. Smith K (2016) Toward an understanding of training to teach online: a review of the literature. Pearson Efficacy & Research

Precovid, Covid and Post Covid Classes and Online Engineering

Thiyagarajan Anushalalitha(✉)

BMS College of Engineering, Bangalore, India
anusha.tce@bmsce.ac.in

Abstract. The covid pandemic has altered the way in which people live and the academic environment is no exception. Covid has changed the way in which classes are conducted and a lot of classes were held online and students were adapting faster to the changing needs and learning, projects and assessment can all happen online along with the regular pattern, which is a positive scenario. The paper is about the success story of the blended mode of education during the covid times and the transformation in the education sector due to the changes and subsequent developments in the field of education.

Keywords: Online classes · Remote learning · Online labs · Online and offline assessments · Course delivery · Management of digital resources for education

1 Purpose or Goal

The goal of this work is to study the outcomes of online and blended modes of learning during the covid times with respect to the courses handled by the faculty. The usage of online tools, classes, labs changed the way in which the system used to operate and there are lot of positive sides because of these changes, albeit the hiccups here and there. The courses handled are Project Management and Finance for the VIII semester, Satellite Principles and Applications for the VIII semester, Project Work II for the VIII semester, Sensors and Measurements for the III semester, Optical Fiber and Satellite Communication for the VII semester, Project for Community Service for the VII semester and Project Management and Finance for the next batch-VIII semester, Project Work II for the VIII semester, which are undergraduate Engineering courses.

For the first time classes were conducted using Online Tools. Students were made to join Google Classroom and link shared and classes conducted, recorded and shared. Quizzes were conducted online. Continuous internal examinations were conducted online or offline depending on the circumstances. Labs were handled online or also offline after the improved situations. Semester end examinations were conducted offline and assessments were done online or offline according to situations. Students performed alternate assessments and projects and presented online and assessments happened online or offline depending on the circumstances.

The outcome of the whole exercise is the increase of technology enabled classrooms, faculty and students connecting seamlessly over the network for classes or meeting or

M. E. Auer et al. (Eds.): REV 2022, LNNS 524, pp. 533–546, 2023.
https://doi.org/10.1007/978-3-031-17091-1_53

workshop or online courses and the alternate working from home environment which is relatively new for the academicians or working from college premises online connecting to students from various locations, all brought a massive change in the education during pandemic and attracted lot of interest from people away from the field also on the changing scenario. Newer tools came up to aid Teachers and Students and although initially it was a struggle to keep pace with the change, in due course, out of necessity and also interest, things started falling into place and online Engineering became the norm. The results are amazing and innovations in teaching methodologies came up and the fact that recorded materials are made available to students, Students could learn anytime of their convenience also. They could upload assignments on time and get notified of their marks. Labs were handled online and simulations and all submissions could be shared by the students and test conducted. Project presentations could be conducted online and demonstration shown and assessed online.

The objective of the paper is to highlight the different components of online learning in the courses that the faculty handled and the comparison of the quality of delivery, student learning, assessments and alternate assessments during such times and the take away of the positive improvements of the process and assessing the need for further improvements. The Pandemic has thrown lot of opportunities for digital learning and there is a huge scope for progress in future to make use of the trends and developments in this field.

2 Quality of Delivery

The faculty improved the quality of lectures many times compared to offline classes since the classes were recorded and the students could listen at their convenience later on also, which is a definite advantage. On the flip side, there was no proper control on the learning by the students due to remote learning. Recorded lectures can pave way for future learning also and dedicate more time for additional academic activities [17].

3 Student Learning

While the student would like to go classes online, wanting the teachers to showcase their performances, the literature shows the other way around also. Lecturers who have taught online are well prepared to use the e-learning system, which includes time management, class planning and online class design. However, lecturers who previously only taught face-to-face (do not have online teaching experience) are less prepared for class planning, time management and online communication [1]. The students of the digital era were more than comfortable logging in to online classes compared to physical classes. The Pandemic classes could drive them into a positive state of mind during the Pandemic period which is true beyond doubt [2]. The findings from literature support that online learning increases student motivation and satisfaction levels and decreases the number of online course dropouts. Additionally, literature has provided some suggestions that a balance can be created between online and offline activities [3]. Quality of instructor, course design, prompt feedback, and expectation of students positively impact students' satisfaction and further student's satisfaction positively impact students' performance and Quantitative study using four independent variables has led to a model proposed [4]. Although online learning is proving helpful in safeguarding students' and faculty's health amid COVID-19 pandemic, however, it is not as effective as conventional learning [5]. There are no significant effects on the measures of functional health in college students who are taking an online learning course [6]. For the teacher, it is recommended to use google classroom not only to face this pandemic but also to use it in regular class to integrate the technology into the teaching-learning process [7].

Migrating from a traditional face-to-face teaching approach to a fully functional virtual educational system will require time and expertise [8]. Timely and quality feedback should be provided for the successful online practical classes [9]. Although the social-distancing measures were necessary to slow the spread of COVID-19, they may have limited children's ability to engage in sufficient levels of physical activity (PA) to maintain health and prevent disease [10]. Technology, or technology variables, is a main focus in the survey to be used as an evaluation tool, for example, teachers' use of it, students' adaption to it, unstable Internet connections, and so on, which apparently is one of the factors to affect the effectiveness of online classes. The questionnaire focused on elements of online class including content, structure, motivation, feedback, interaction and learning strategies, but simply no mentioning of the content blocking [11], which highlights privacy concerns in online classes. Five constructs were identified (attitude, curriculum, motivation, technology and training). All the five constructs were found to have impact on the online classes. Motivation is one factor that has almost 50% (approx…) impact on the performance of online classes [12].

Financial demands, revenues collection, payment of the salaries of teaching, and academic staff are the biggest issues to go for online teaching [13]. There is a negative influence between the academicians and the students during the online learning platform in reference to the communication and its effectiveness [14]. The upside to ERT (Emergency remote Learning) during the pandemic is it will force us to consider what we have lost from in-person teaching that we could not compensate for [15]. Online teaching is a rather exciting concept of teaching for Indian teachers in general. Initially, there was a feeble attendance, but subsequently, its' race gained momentum, and the students were comfort- able in attending online classes [16].

4 Assessments

The continuous internal assessments (CIE) and Semester end examinations (SEE) which were generally held offline were conducted online/offline depending upon circumstances. Three sets of question paper were prepared for CIE and assessed. Lab exams were also held online. Project examination was also held online.

5 Alternate Assessments

Students submitted alternate assessment through online mode during Pandemic which were usually offline before.

5.1 Courses Handled During Covid

In the course project Management and Finance in VIII semester classes were initially held online and then later on after first test moved to online mode. Lectures were given online and students participated positively and submitted assignments, took quiz and submitted self-study presentations using project management software tools and took semester end exams offline and assessed online. The VIII sem project presentations were held completely online and internal and external assessments were conducted online with very good outcomes.

The Satellite principles and Applications institute elective course in the VIII semester was offered to students from different departments of the college and lectures delivered and students participated in the class and took quiz, submitted seminars and assessed online during CIE and offline exams were conducted for SEE.

Project Work II was conducted completely online except for the synopsis presentation.

In the next semester, VII semester department elective, Optical Fiber and Satellite Communication lectures were given online and students participated and took quiz, submitted seminars, attended tests and semester end examination was conducted offline and results announced.

For the III semester, Sensors and Measurements course was completely held online and a similar exercise was practised and could successfully conduct online classes and regular submission of lab experiments and miniproject and seminars and quiz were performed and assessments completed online CIE and Offline SEE and after pandemic labs were given access for students interested and now regular classes continued further offline. VII sem Project for Community service was also held online completely.

6 Approach

6.1 Project Management and Finance-VIII Sem

As the Pandemic started, the offline classes got disturbed and lockdown was announced. To avoid disruption of classes, online tools were researched and chosen for classes. Different faculty resorted to different tools and Google classroom was chosen as a common tool for all the classes for sharing online class links, sharing materials, conducting quizzes, tests, seminars, project presentations etc., GoToMeeting or Google meet platforms were mostly used for sharing online lectures. The students could join and listen and actively participate and follow the academic requirements online or offline as per the instructions.

Although few classes technical problems were faced, students found it interesting and responded positively. Assignment on risk management due to the pandemic, Assignment on the benefits/Challenges of online classes were submitted by the students during Covid times. Mock tests were conducted to train the students and faculty for online tests.

Apart from that, Students used project management tools for the self-study component and submitted detailed analysis of the management component of their projects and presented as a seminar for 10 Marks. This way they could practically apply various

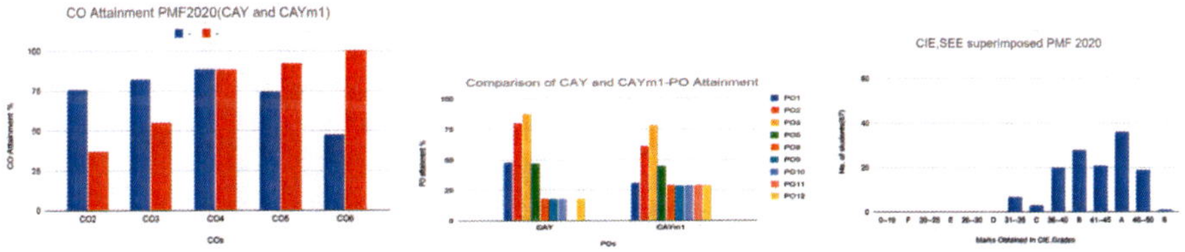

Fig. 1. Project management and finance (2020)

techniques of project management to their VIII sem projects using templates and learn the course in a simple and interesting manner (Fig. 1).

Positive Observation: PO1, 2, 3, 4 Attainment is relatively good, Active learning methods, Case study-based approach; Online Teaching and Assessment has helped learning PM concepts and mapping them to their project and Team building, Design skills, Feasibility studies, case study, PM software skills improved very much. Areas that need improvement: PO5 attainment needs to be increased, Higher POs Attainment has to be gradually increased.

6.2 Satellite Principles and Applications Students - VIII Semester

The Institute elective Satellite Principles and Applications students in VIII semester submitted Seminars on specific subsystems of the Satellite apart from the regular classes and assessments online (Fig. 2).

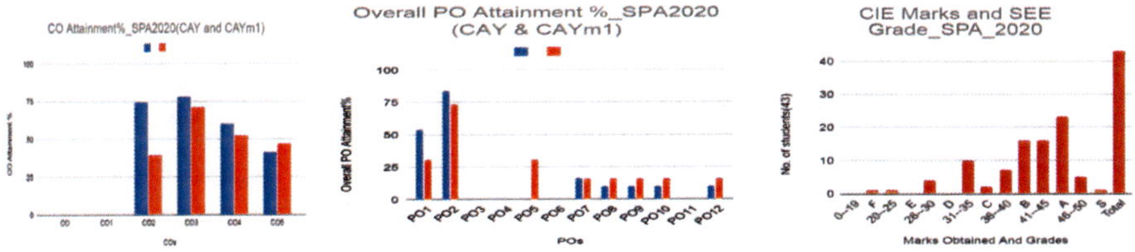

Fig. 2. Satellite principles and applications

Positive Observations: PO1, 2 Attainment is very good, Online Teaching and Assessment is done during Covid time. Areas that need improvement: PO5 can be increased, Higher POs Attainment has to be gradually increased [18].

6.3 Optical Fiber and Satellite Communication Students - VII Semester

The next semester, online classes started from the beginning, the Optical Fiber and Satellite Communication students of the VII semester submitted seminars on Indian Satellites for various applications apart from the regular classes and assessments online. VII semester project for community service was conducted and presentations were conducted online and assessed for CIE and SEE (Fig. 3).

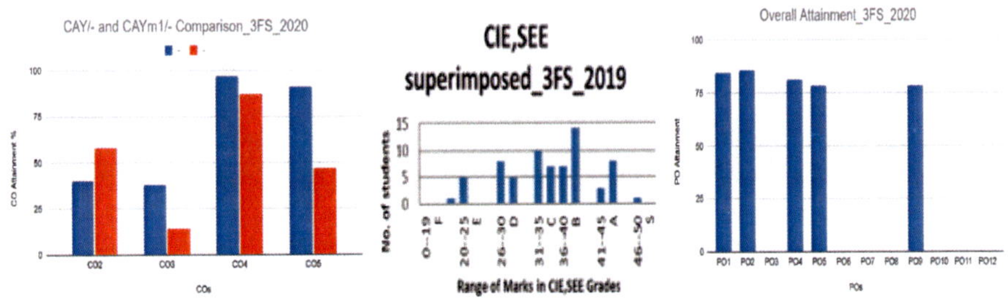

Fig. 3. Optical fiber and satellite communication

6.4 Project for Community Service-VII Sem

This is the VII sem project for Community service which was held online. This course was jointly handled by Prof. Archana. K as the Project Evaluation Committee (PEC) member (Fig. 4).

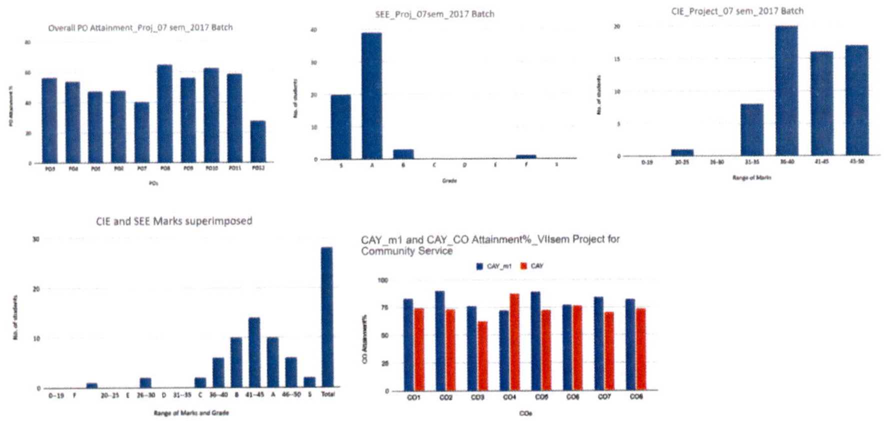

Fig. 4. Project for community service

The CAY CO attainment is slightly less compared to the CAY_m1.

6.5 Sensors and Measurements-III Sem

The Third semester students of Sensors and Measurements had to complete the semester online and then attended labs offline. They had submitted e-waste seminars, multisim online labs were conducted and lab submissions happened every week online and multisim mini-project was submitted in groups and quiz and CIE were conducted online and SEE conducted offline (Fig. 5).

The CO Attainment for the CAY is slightly less compared to CAYm1 in CO2, CO4, CO5 but CO3 attainment (Lab experiment conduction) is more compared to CAYm1 and CO6 (e-waste Presentation) there is no significant change at all. PO Attainment also is

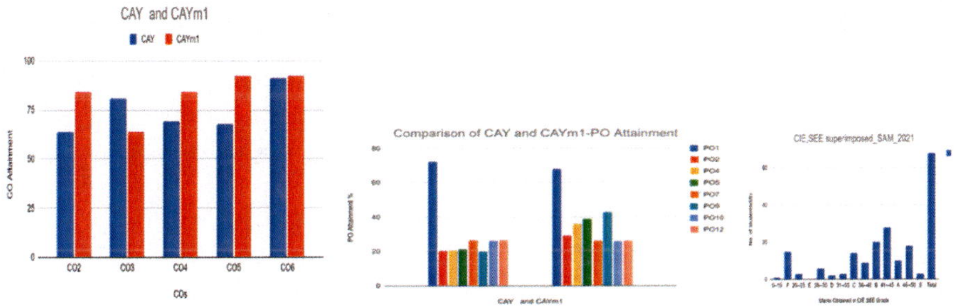

Fig. 5. Sensors and measurements

less compared to CAYm1 except CO5 (Multisim mini project). The interested students could do the lab experiments offline only after lockdown due to Pandemic, otherwise most of the labs were conducted online using Multisim [19].

6.6 Project Management and Finance - VIII Semester

The next batch of students were taught Project Management and Finance in the VIII semester and classes conducted online and the process repeated along with the regular and alternate assessment tools. Most of the project were based on Artificial Intelligence and web application (Fig. 6).

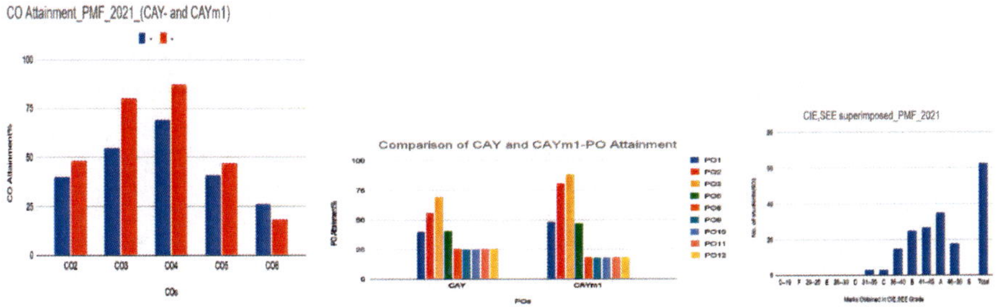

Fig. 6. Project management and finance (2021)

The CO Attainment is slightly less compared to CAYm1 except CO6 which is the self-study component. The PO Attainment is also slightly less compared to CAYm1 except for the self-study.

Internals	Self-study
Best 2 of 3 Tests	**10 M** (Assignment for each phase (3) + Final video (5) + Presentation (2)
40 M	Total CIE = **50 Marks**

6.7 Microcontroller Lab-IV Semester

Co-teaching Microcontroller lab for the IV semester, the students submitted lab experiments on Microcontroller 8051 and submitted miniproject presentations on Sensor applications using Tinker CAD, Arduino or 8051, which was assessed online and interfacing labs conducted offline after pandemic.

Overall, the teacher has become a digital Teacher, managing everything online as opposed to the conventional academic routine, which will be physical classes and assessments.

There is lot of scope for documenting everything online from classes, to materials to assessments and results which is a positive change for any further reference.

When the classes go online, there is lot of improvements in the content delivery, scope for innovations in TLP, be it asking questions, organising debates, conducting polls, usage of different tools such as whiteboard or any other scribbling pads, different changes in presentation styles, participation of students in the form of submission of assignments on a weekly basis and assessing their presentations and answer booklets although they were little bit time consuming in the beginning.

While there are loads of positive outcomes of online classes, there were quite a few hiccups too. There was total lack of physical classes for some time which was not the norm and it took time and resources to keep abreast of the digital technology for education both from the faculty side and the student side, lack of proper network or power or devices were always challenging and since there was no formal offline classes, the teachers were not very sure of how the information was received by the students. Still there is long way to go to ensure the online environment matches the conventional classes and also enhances and supplements the learning scenario in a bigger way in future, provide the demands of resources such as network, power and other device requirements are fulfilled. Since there is increased demand for network, the cost requirements also increased which was a challenge for some.

The details of specific success stories in various courses will be highlighted as the important component of the paper further.

6.8 Project Work II–VIII Semester

The VIII sem project was carried out by 18 batches and the entire Final Project presentations were held online and students could present seminars and show the demonstration and take viva voce online. Most of the projects were based on signal processing using Artificial Intelligence, Embedded system project using IOT, VLSI projects, some on communication etc. This course was jointly handled by Prof. Archana. K as the Project Evaluation Committee member (Fig. 7).

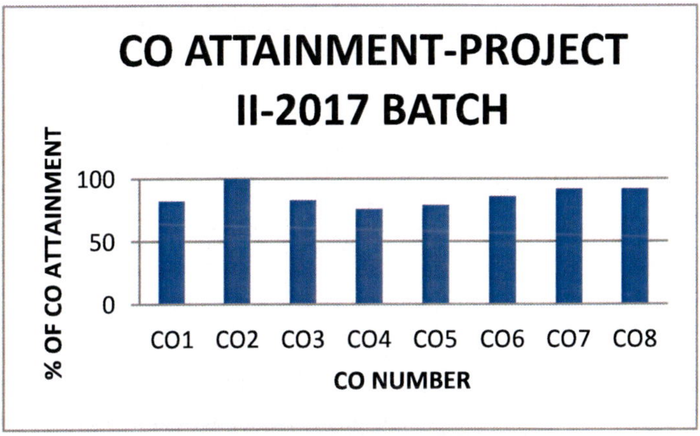

Fig. 7. Project work II

7 Detailed Analysis

7.1 Project Management and Finance - VIII Semester

Students submitted risk matrix for handling the Covid situation. By arriving at solutions, precautions for the real pandemic, they could learn about the risk analysis for crisis conditions for any project-35 students submitted assignment on Covid risk analysis which is a positive sign. Then Mock test were conducted online and students were submitting positive and negative aspects of Online tests and that of lockdown, before the actual CIE. These were ungraded. The graded Assessment was done as follows:

Internals	Self-study	Total
Best 2 of 3 Tests	Video presentation	CIE + AAT
40 M	10 M	50 M

Apart from that, 62 students submitted self-study for 10 marks on the project management aspects of their final year project using project management tools, in the form of video presentation. The comparison of PO Attainment in the course with the previous year which was offline is shown in Fig. 1.

7.2 Satellite Principles and Applications - VIII Semester

Apart from the quiz and regular CIE and SEE assessments, the students submitted video presentations on satellite subsystems of Indian satellites mostly on the following topics. Indian National Satellite System (Insat), Thermal Analysis Of The Electronics Of A Cubesat Mission, Astrosat, Chandrayaan-1, Military Satellites, The Communication Subsystem In Mission Starlink, Megha-Tropiques, Mos-1 (Marine Observation Satellite), New Horizons Mission, Case Study On Apollo 11, Kalpana-1/Metsat-1 (Meteorological Satellite-1), Space Segment, Satellite Communication, Satellite Power Subsystem, Cassini, Chandrayaan-2, Pslv-C48/Risat-2br1 (Radar Imaging Earth Observation

Satellite), Gsat-30 A Telecommunication Satellite. Students from departments such as CS, IS, EC, TC, EE, CIVIL, IEM, ME of the Institute took the course. The Assessment was done as follows:

Internals	Alternate assessment (AAT)	Quiz	Total
Best 2 of 3 Tests	Video presentation (10 M)	Online quiz	CIE + AAT + Quiz
40 M	Avg. to 5 M	5 M	Total 50 M

7.3 Optical Fiber and Satellite Communication-VII Semester

Apart from the regular quiz and CIE and SEE, the students submitted video presentation on Communication satellite GSAT 16, Chandrayaan 1-Indian Satellite, Oceansat and its Applications, Indian Remote sensing satellites, Communication satellite GSAT 16, Aryabhata Satellite, Indian Regional Navigational satellite system (IRNSS), Cartosat 2 Satellite, G Sat 30 Satellite, IRNSS (India Regional Navigation Satellite System), SCATSAT-1, this being a department elective. The Assessment was done as follows:

Internals	Best 2 of 3	40 M
QUIZ	Average of TWO	05 M
AAT	Seminar	05 M
Total		50 M

7.4 Project for Community Service-VII Semester

The project topics typically include Remote Weather Monitoring System, Sustainable Pesticide Usage In Agriculture, Voice Controlled Wireless Electronic Notice Board Using Android, Implementation Of Traffic Light Controller, Traffic Violation Detection and Number Plate Extraction of a Vehicle through Machine Learning Algorithms, Detection of Tumour Cells in Brain using Artificial Intelligence, Telebot, college campus Navigation using Augmented Reality, Driver Drowsiness Detection, Home Automation System, Electric Vehicle Efficiency Calculator, Secure Data Transfer Over Internet using Image Steganography, Enhanced Diffie Hellman Key Exchange using string comparison, Iris recognition system, Covid-19 Social Distancing system In Queue Based Services using Android App, AI Assistant Using Python, Smart Card based student database and Smart Waste Segregation System. This course was jointly handled by Prof. Archana. K as the Project Evaluation Committee member. The Assessment was done as follows:

Internals	Marks
Mid-sem evaluation	50
End-sem evaluation	50
Final	100

7.5 Sensors and Measurements-III Semester

The students submitted an ungraded Assignment on block diagram of a typical measuring instrument/system and identifying the components of the system such as input, signal conditioning and output. The Lab experiments were conducted using Multisim and continuous evaluation happened and exams conducted for regular experiments and datasheet verification and applications of sensors.

Students gave presentation in groups on e-waste. Soil Moisture Sensor, Touchless Tap, Burglar Alarm System, Hall Effect Sensor, Automatic Lighting Sensor, Motion Sensor, Sensor Circuit For Reverse Parking, Corrosion Sensor, Touch-Less Door Bell, Cell Phone/Mobile Detector, Rain Sensor, Ultrasonic Distance Sensor (Hc-Sr04), Fire Alarm Sensor, Solar Street Light Sensor, Electromagnetic field sensor, Infrared Sensor, Temperature sensor, Electronic Repellents, Emergency alarm, Infrared Transmitter are some of the topics for mini-project implemented using Multisim.

The Assessment was done as follows:

Tool	Remarks	Marks
Internals	Best 2 of 3	20
Lab	Two lab tests + Lab quiz	20
AAT	Multisim mini project + Presentation on E-waste	10
Total		50

7.6 Project Work II–VIII Semester

The Students did projects and all the reviews were conducted online and assessed. Most of the project were based on Artificial Intelligence and some on Web Design and Embedded System etc., Project Video and Poster presentation contest was conducted and three best projects were chosen on three categories selected by two distinguished Alumni from the department in the presence of PEC, Guides and students from the entire department. The Assessment was done as follows:

Internals	Marks
Mid-sem evaluation	50
End-sem evaluation	50
Final	100

Emotion Recognition From Facial Expression Using CNN, Sustainable Pesticide Usage In Agriculture, IOT Based Face Mask Detection And Body Temperature Security System, Classroom Attendance Using Face Recognition, Analysis Of Customer Churn Prediction In Telecom Industry Using ML, Music Genre Classification, Efficient Data Transfer In Software Defined Network(Using Similarity And Trust Score), An Optimized Hybrid Neural Network Model For Detecting Depression Among Twitter Users, Digital Library, Speech Enhancement Using Advanced Techniques, Smart Vacuum Cleaner,

A Smart Box For Courier Delivery, Automatic Pill Dispenser System, Bio-Medical Image Analysis On Covid-19, FPGA Based Implementation of Image Segmentation Technique, Aeye -Shades To Aid Blind, Safety Survey Robot etc., were some of the topics implemented for Project work. This course was jointly handled by Prof. Archana. K as the Project Evaluation Committee member.

7.7 Microcontroller Lab-IV Semester

This course was handled by Prof. Rajeshwari Hegde and I was the course co-faculty for Microcontroller lab in the Google Classroom. The students were taught labs online and Continuous Assessment of lab experiments were conducted and the students did miniproject on sensor applications using microcontroller using tools like Tinker CAD, Proteus, Arduino, Multisim, 8051 etc., and presented online seminars of their mini projects and demonstrated their applications. Interfacing experiments were demonstrated offline after pandemic.

Automatic street light controller using 8051-Tinker cad, proteus, Fire Alarm system using 8051-Tinker cad, proteus, Motion detection using PIR sensor-Proteus, Ultrasonic sensor using Arduino, Temperature based FAN Speed controller, Switching ON/OFF Appliances using IR Remote and Arduino, Automated cooling using Arduino, Digital clock using Arduino, Human detection using PIR sensor, Automatic water heater-Proteus, Traffic signal controller using 8051, Smart RADAR control using Arduino, Password based Door lock system using 8051, Password protected Lock using Arduino, Automatic Hand Sanitizer, Simple calculator using Arduino Uno, Social distancing and Alarm system, Fire Detection and alarm system, Automatic Water level control and monitor, Metal detector, Alcohol detecting system, Automatic door lock system using Tinker cad, Frequency variation circuit using 555 timer, Shadow Sensor Alarm etc., were some of the topics implemented for miniproject.

The Assessment was done as follows.

Project presentation	Project demo	Project report 1	Comparison of ICs report 2	Lab experiments submission	Total
4 M	6 M	3 M	2 M	10 M	25 M

8 Actual or Anticipated Outcomes

Technology for education is a boon or bane is debatable, since students are enabled for use of technology, online classes or blended classes would definitely prove to be vital tool for education of Tomorrow. This also opens up for further improvements in research and need for developments of tools in this online engineering system domain. There is a need to develop tools for greater engagement with the students, since online classes could not completely complement the actual classes. Faculties could indulge in research, students could learn online more and do their projects, internships also. Visiting industries and indulging in group activities physically in college were not possible, which greatly affected the regular college routines. Otherwise, Online classes could serve as an effective alternate for technological education anytime.

9 Conclusion

Technology for education became the new normal during pandemic times and changed the way in which technical education was offered to the students. Faculty became digital ambassadors of education, students and faculty utilised their time effectively with less travel, managing their faculty/student development programmes, workshops, internships, online courses, projects and other activities along with the regular academic requirements, which was quite a welcome step amidst the chaos of the pandemic. Post pandemic, things are started to change, with vaccinations, colleges are started to move offline again. The benefits of online and blended classes will continue to take education to new levels and create avenues for new tools for further changes in this area.

Acknowledgements. The author would like to acknowledge Teqip III, BMS college of Engineering, Bangalore and the faculty of the department, co-faculties who handled the courses with the author and staffs of the department who have contributed to the academic activities for supporting this work.

References

1. Junus K, Santoso HB, Putra POH, Gandhi A, Siswantining T (2021) Lecturer readiness for online classes during the pandemic: a survey research. Educ Sci 11:139. https://doi.org/10.3390/educsci11030139
2. Agarwal S, Kaushik JS (2020) Student's perception of online learning during COVID pandemic. Ind J Pediatr 87(7):554
3. Baber H (2020) Determinants of students' perceived learning outcome and satisfaction in online learning during the pandemic of COVID19. J Educ e-Learn Res 7(3):285–292 (2020)
4. Gopal R, Singh V, Aggarwal A (2021) Impact of online classes on the satisfaction and performance of students during the pandemic period of COVID 19. Educ Inf Technol 26(6):6923–6947
5. Adnan M, Anwar K (2020) Online learning amid the COVID-19 pandemic: students perspectives. J Pedag Sociol Psychol 1(2):45–51
6. Mangis, J.: Online learning and the effects on functional health: a pilot study. EWU Masters Thesis Collection vol. 386 (2016). http://dc.ewu.edu/theses/386
7. Okmawati M (2020) The use of google classroom during pandemic. J English Lang Teach 9(2):438
8. Sarwar H, et al (2020) Self-reported effectiveness of e-learning classes during COVID-19 pandemic: a nation-wide survey of Pakistani undergraduate dentistry students. Eur J Dent 14(suppl S1):S34–S43
9. Yu J, Jee Y (2021) Analysis of online classes in physical education during the COVID-19 pandemic. Educ Sci 11:3. https://doi.org/10.3390/educsci11010003
10. Dunton GF, Do B, Wang SD (2020) Early effects of the COVID-19 pandemic on physical activity and sedentary behavior in children living in the U.S. BMC Publ Health 20:1351. https://doi.org/10.1186/s12889-020-09429-3
11. He W, Xiao J (2020) The emergency online classes during COVID-19 pandemic: a Chinese University case study. Asian J Dist Educ 15(2):21–36
12. Zia A (2020) Exploring factors influencing online classes due to social distancing in COVID-19 pandemic: a business students' perspective. Int J Inf Learn Technol 37(4):197–211

13. Zulfiqar Hasan KM, Islam A (2020) Academic, financial and administrative issues of online teaching during corona pandemic: the scenario of Private Universities in Bangladesh. Int J Account Financ Rev 5(1):116–122
14. Maney KL, Saritha SR, Hegde S (2021) An exploratory study on the challenges faced by teachers in online classes during pandemic–COVID-19. Int J Res Eng Sci Manag 4(9)
15. Gurung RAR, Stone AM (2020) You can't always get what you want and it hurts: learning during the pandemic. Scholarsh Teach Learn Psychol. https://doi.org/10.1037/stl0000236
16. Mishra L, Gupta T, Shree A (2020) Online teaching-learning in higher education during lockdown period of COVID-19 pandemic. Int J Educ Res Open 1:100012. https://doi.org/10.1016/j.ijedro.2020.100012
17. Anushalalitha T (2020) Open education resources and new age teachers. In: Auer ME, Ram B. K (eds) REV2019 2019, vol 80. LNNS. Springer, Cham, pp 742–747. https://doi.org/10.1007/978-3-030-23162-0_67
18. https://bmsce.ac.in/home/Electronics-and-Telecommunication-Engineering-Programmes
19. https://bmsce.ac.in/home/Electronics-and-Telecommunication-Engineering-Syllabus

An Experience of Using a Handwriting Haptic Device to Fine-Tune the Children Motor Skills

Jezia Zakraoui[1], Sumaya Al Maadeed[1], Samir Abou El-Seoud[2](✉), Moutaz Saleh[1], and Jihad Al Ja'am[3]

[1] Qatar University, Doha 2713, Qatar
{s_alali,moutaz.saleh}@qu.edu.qa
[2] The British University in Egypt (BUE), Cairo, Egypt
samir.elseoud@bue.edu.eg
[3] Leb-French University, Tripoli, Lebanon
jihad.jaam@ulf.edu.lb

Abstract. Handwriting can help children to improve their learning of a language and fine-tune their motor skills. Every child needs to develop her/his handwriting skill to grasp new concepts appropriately and learn the language vocabulary. Therefore, in-hand manipulation of the traditional pen is highly important to develop pre-writing and transform the scribbled writings to legible ones at a later stage. In this paper, we evaluate the effectiveness of a customized haptic device in improving the children motor skills and their handwriting quality of Arabic letters. We use the Touch™ device from the 3D-Systems company with a controllable stylus that can be adapted to children needs. Fifteen pupils from the Deutsch International School in Doha, have participated in this experience after obtaining all necessary ethical approvals from concerned stakeholders. We conducted the experiments for a period of two weeks with the assistance of the school instructors and staff. Results show that there is an important increase of children motivation, and a good improvement of their motor skills and handwriting experience. The device can be used at home to learn independently during COVID-19 pandemic that continues to hit severely the whole world and enforces schools to adapt online teaching approach.

Keywords: Haptic · Handwriting · Learning skills · Arabic letters · Motor skills

1 Introduction

Learning to write vocabulary in any language is a challenging task especially for new learners and young children [1–3, 5]. The lack of proper knowledge of the language alphabetic shapes often causes issues in forming vocabulary correctly. For instance, Latin native children learning the Arabic language often mimic Latin letters when they write in Arabic. They start writing from left to right while in Arabic vocabulary should be written from right to left.

Regardless of the drastic advancement of digital technology, handwriting is still an essential factor for children to master the language. In fact, writing by hand using a

M. E. Auer et al. (Eds.): REV 2022, LNNS 524, pp. 547–558, 2023.
https://doi.org/10.1007/978-3-031-17091-1_54

pen allows the children to engrave the words in their brains so that they can remember their shapes and structure quickly. The technology of haptic devices can help children learn the language independently and efficiently. In fact, they can learn at home and repeat the task several times. Recently, research works have been carried out on how children with low handwriting skills react to different virtual reality environment (VR) writing tasks. Psychologists hypothesized that children with writing difficulties will be cognitively empowered while interacting in a VR during the learning process. VR offers positive re-enforcement and a better sense of engagement to keep children's attention awake. They are immersed in the VR environment and can grasp new concepts through virtual manipulation of objects. Other studies [6, 7] used haptic devices to help children learn the language independently. However, for young Latin native children, and to the best of our knowledge, there is no detailed studies published in the literature.

In this research, we target young children in the Deutsch International School in Doha, Qatar to explore the development and refinement of their handwriting skills in Arabic in an immersive VR environment. We use a haptic device with a stylus-like tool that allows them to physically touch the pen, feel the force and movement and write in a virtual space within a harsh environment to fine-tune their motor skills. Since, handwriting is difficult for children, writing in a 3D environment can help them learn better to write. In fact, they should maintain their hands in the VR space for a long time while tracing the letter. This may cause some fatigue, great efforts, and brain concentration. These factors can contribute to mastering the propre writing and strengthening the understanding of the language vocabulary. Children spend a lot of time to manipulate the pen, follow up the letter trajectory, set on the needed punctuation, and joining independent letters. This hard environment contributes to better learning and recalling of the proper manner to write letters. We investigate whether using a haptic device with a hard pen to practice handwriting can be an effective approach to resolve the difficulties in the writing process. We use the Touch™ haptic device from 3D-Systems company shown in Fig. 1 (left). It is a motorized device that applies force feedback on the child's hand, allowing her/him to feel natural and producing true-to-life touch sensations. Many parameters are set for a smooth touch such as force, velocity, and movement. We developed several control algorithms and adapted them to the children forces of writing. The device is connected to a laptop so that the child can see the letter trajectory and trace it in the VR space. We video recorded all the experiments for further analysis.

Fig. 1. The haptic device with a child writing the Arabic letter 'Alef'.

2 Experiments

In this section, we discuss the experiments settings including the children selection and training sessions.

2.1 Participants

Initially, we obtained all the IRB approvals from all the stakeholders (i.e., parents, school principal, research office, and ministry of education) to assess the haptic device with children in schools in Qatar. We selected fifteen children from the Deutsch International School in Doha where parents agreed to sign a detailed consent form. We explained to the parents the objective of this study, the experiments phases, the data collection procedures, and the confidentiality of the obtained results. The children were able to hold a pen and manipulate it appropriately. They were from different ethnicity and of an age between 5 and 11 years (n = 15, M = 7.2, SD = 1.6). We explained to them the details of the experiments, answered all their questions. We gave them a training session for around 35 min to write anything they like with the haptic stylus in the VR environment. Whenever we found that all children were able to write properly, we started the real experiments.

2.2 Dataset

The children produced many writings outcomes. However, we retained only 75 of them for the analysis since many attempts were exclude due to being either too noisy (i.e., illegible) or incomplete. Figure 2 shows a cleaned character from our dataset.

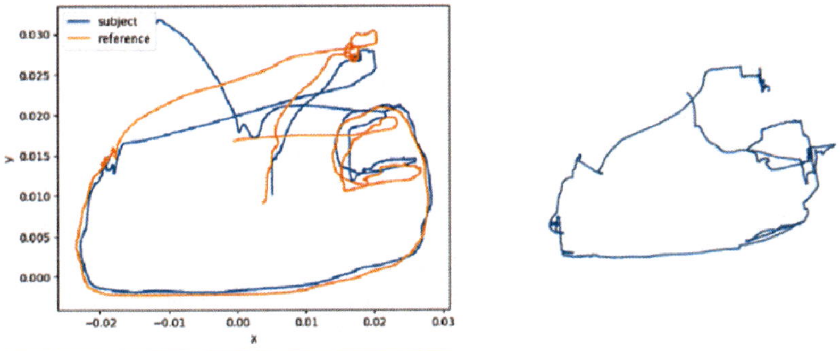

(a) (b)

Fig. 2. Samples of traced Arabic character *Faa*, via (a) guided mode, (b) free mode.

2.3 Task Procedure

We showed the children the image of an animal with an incomplete name on a computer screen. The child should complete the name and write the missing letters in guided and free modes. The haptic task procedure is illustrated in Fig. 3. After the pre-test, each subject completed three sessions: a training, a test session, and a post-test with 8 min for every task. Each participant spent around 40 min on average in the experiment.

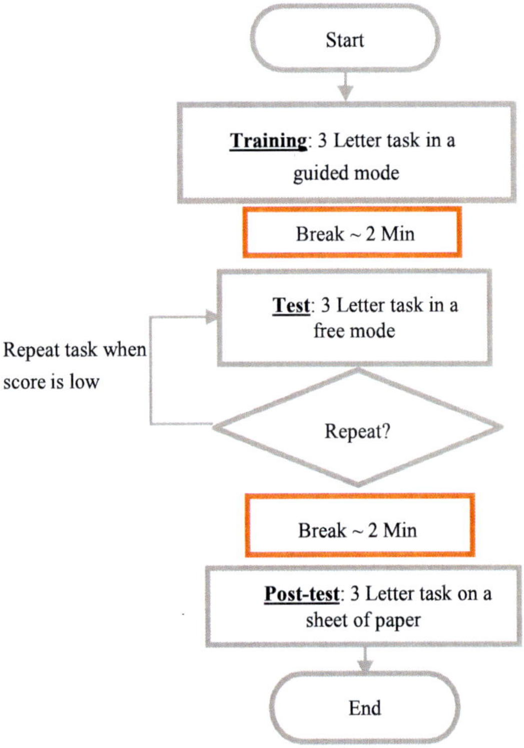

Fig. 3. The haptic task procedure, pre and post assessment.

2.4 Metrics

To assess the performance of participants, we used both subjective and objective metrics. Subjectively, we asked the children verbally, at the end of the experiment, whether they liked this approach of learning and to what extent. We noted their responses for analysis. The questionnaire was composed of the following questions:

Q1. "Do you like writing Arabic using this haptic device?"
Q2. "Have you enjoyed this activity?"
Q3. "Do you like to repeat it?"

Another subjective metric is the observations from the teachers of the children' behavior [3] while writing. Whenever, we noted that the child improved his/her handwriting, we concluded that the haptic device boosted her/his motivation and increased her/his engagement. Likewise, we noted, how many attempts the child did?

Did she stop the writing process and how many times? Did she try to finish her/his writing following the template as precise and as fast as possible to engage with the activity? All observations and remarks were recorded. Concerning the objective metric, we used the average tracing error [8] to analyze the subjects' performances.

3 Preliminary Results

We analyzed the video recorded handwriting of participants in the following section using qualitative and quantitative metrics. The collected data represent the 3D positional data, in addition to the 3D grip force data. Other data that include the user's performance (e.g., completion time, velocity, etc.) are also collected. All these objective data are collected in a time-stamp manner and converted into csv files for further analysis. Furthermore, the participants did a post-assessment similarly to the pre-assessment to measure their handwriting progress at the end of our activity.

3.1 Observations on Children Behavior with Qualitative Metrics

A teacher from the school observed the children's behavior (e.g., Table 1) while they were taking part of the experiments. When they completed the handwriting tasks, the teachers assessed their handwriting on a five-point Likert scale which ranges from 1, not at all, to 5, very much. In addition, teachers gave their feedback with regards to letter formation. Two young students showed significant improvement of their writing skill of the letter ف(Faa). Almost all children showed consistent motivation and engagement over the period of the experiment while also enjoying the handwriting task. Most of the participants (93%) liked the handwriting task using the haptic device.

Table 1. Teachers' observations

Nr	Teachers' observations	Pre-rating	Post-rating
1	Wrote more smoothly without thinking	4	5
2	Learned to distinguish letters, enjoyed, enhanced, has problems to adapt to the pen in virtual board, has language gap	2	3
3	Motivated and performing well, enjoyed, has LD, he was able to remember some letters, but still has difficulty	3	3
4	Motivated, engaged in letter activity, not aware of alphabet, she used down to up letter formation	1	2
5	Has language gap (because he missed a school year), aware of letter formation	2	2
6	Determined to improve, patient, no significant improvement	3	4
7	NA	1	2

(continued)

Table 1. (*continued*)

Nr	Teachers' observations	Pre-rating	Post-rating
8	NA	4	4
9	NA	3	1
10	NA	1	2
11	NA	4	5
12	NA	2	2
13	Significant improvement in letters *Alif*, *Faa*	2	1
14	Notable improvement in letter *Faa*, better letter formation	1	3
15	NA	2	4

At the end, the children were asked to give a rating for their experience with the haptic activity. Most of the participants (93%) liked the handwriting task using the haptic device. However, only 60% of them admitted repeating the experiment in the future. Figure 4 summarizes the results.

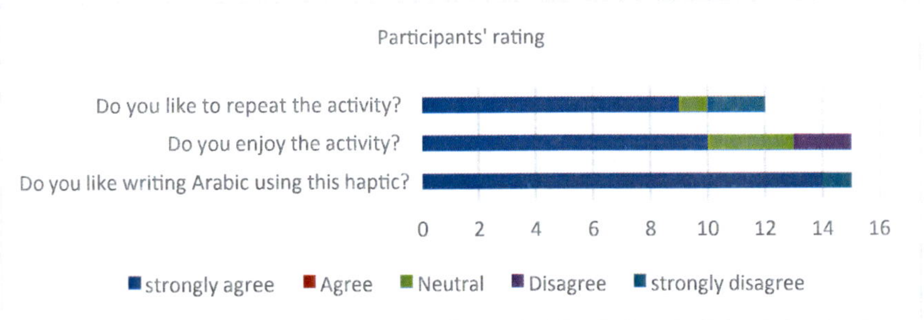

Fig. 4. Participant's survey responses.

3.2 Performance with Quantitative Metrics

Almost all children showed better handwriting performance in the guided mode than in the free tracing mode due to applied force feedback on the participants' hand. Consequently, we believe that our handwriting task assistance was effective for the tracing tasks using our haptic virtual system. Some children showed slight improvement over the time. Overall, most of the children showed higher tracing accuracy in the guided mode.

After training with the haptic guidance, the children wrote the 3 letters in the free mode and again in this test teachers observed the improvement in handwriting. With the haptic assistance, all children showed better tracing performance when made more attempts in the free tracing. Overall, all students showed fair improvement in the free mode session. Figure 5 shows the aggregated data for the task of a single letter tracing for the two different ages groups, notably, group 1 (age: 5–7) and group 2 (age: 8–11).

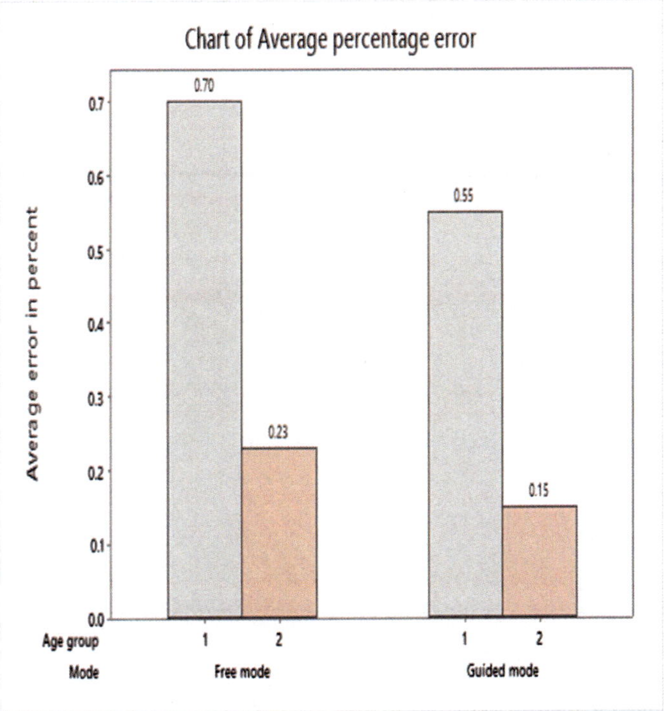

Fig. 5. Average percentage error rates for guided and free mode per age group.

The children' post handwritings were scored by the teachers on a five-point Likert scale where *Very much* (5), *A little*, *Neutral*, *Not much* and *Not at all* (1) are scales used to assess the handwriting outcome in letters trained on during the sessions. For the analysis, we check whether the participants had an overall learning outcome in writing letters for all sessions, we did a paired t-test (since our data is normal distributed) using Minitab software. The results are shown in Fig. 6 which indicates that post- assessment scores were significantly higher than pre-assessment scores (*p-value* = *0.036*). It's low enough to reject the null hypothesis (i.e., all means are equal) using the common significance level of 0.05. The alternative hypothesis indicates that post-assessment mean is significantly greater that pre-assessment mean. Subsequently, all students showed fair improvement in the post-assessment session.

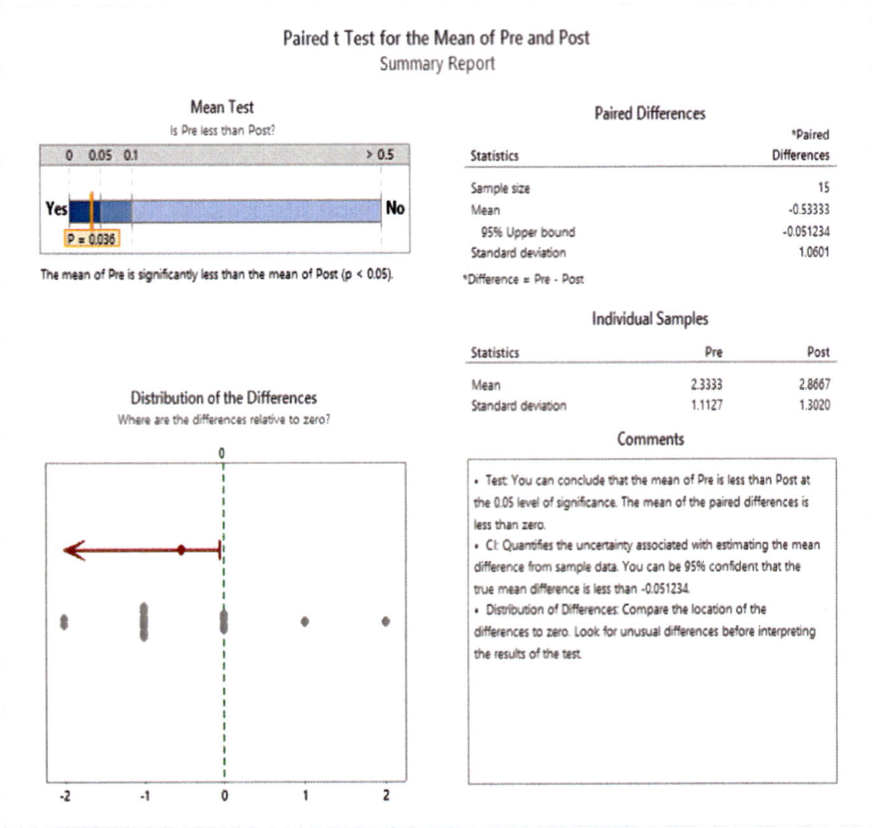

Fig. 6. Paired t test for all participants (using Minitab software)

We break down the teacher ratings by groups using a statistical significance method. First, we computed the differences between post- and pre-assessment as follows:

$$Difference = Post{-}assessment{-}Pre{-}assessment$$

Then, we apply a One-way ANOVA test on that computed difference. Figure 7 reports results which indicates that participants having slower than normal motor development (group 1) are able to improve in letter formation thereafter. This was demonstrated by an overall significant positive difference compared to group 2.

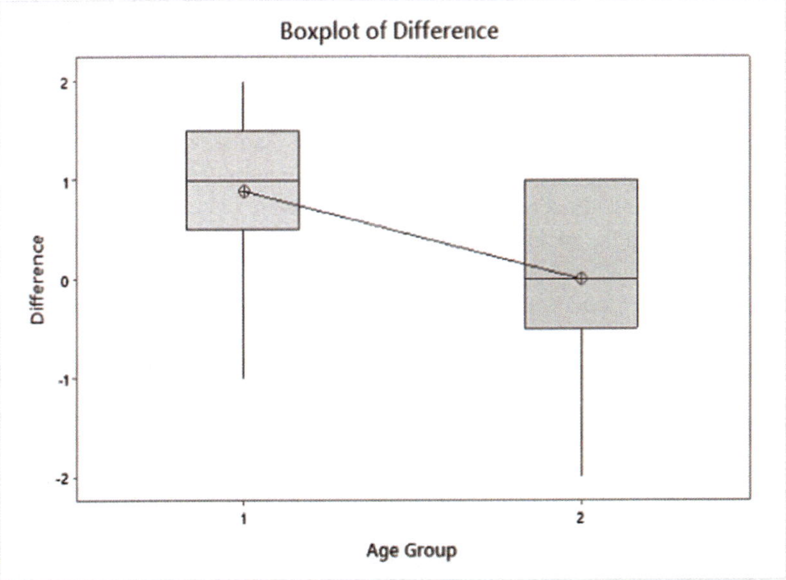

Fig. 7. Difference of post- and pre-assessment per age group (using Minitab software)

4 Discussion

We investigate in our study whether a VR environment that provides a haptic writing in Arabic could help assess and improve fine motor skills of young children. The collected objective data and the preliminary results confirmed of providing fine motor writing skill in an attractive and engaging way. The performance assessments and our initial results from this experience showed that the haptic task was adequate to study children' handwritings. The children did not face large difficulty in using the haptic device and the virtual system. In contrast, they found the training sessions interesting and easy to perform to some extent. The force feedback was helpful for the children to improve eye-hand coordination via haptic feedback. In particular, the results of this study showed a significant improvement for two young kids where both are at the age of five. These are participants #13 and #14 from Table 1. It is observed that, especially, those participants who have had some difficulty in tracing the correct shape and orientation were dedicated to trace the letters correctly. Furthermore, overall, the results indicate that using the haptic device had a significantly positive effect on children' writing performance. This is confirmed by measuring the difference between the scores of the pre- and post-test which showed an acceptable significance level, see also Fig. 6. The results agreed with the responses of the survey, where most children confirmed that the handwriting was enjoyable even if the grip and the applied finger pressure were uncomfortable and difficult for some participants. All children performed better in guided mode that was because a little bit of effort was required to hold the stylus and move smoothly along the visualized letter traces. Thus, the influence of the gravity against hand control was minimized, reducing thereby hand muscles fatigue.

While most of the participants performed well in the guided mode, their letter tracings in free mode have major issues and resulted in high error rates and poor performance. The relative error rate of 70% and 23% for group1 (age: 5–7) and group 2 (age: 8–11), respectively is reported for the free mode activity, while 55% and 15% for group 1 (age: 5–7) and group 2 (age: 8–11), respectively is related to guided mode activity, as shown in Fig. 4. The average relative error rate for young children was higher than the older ones in both modes. It was observed that the errors were either due to the participant writing a segment in the wrong direction, missing a segment, or missing a directional change completely and continuing in the same direction.

Likewise, many participants started the letter tracing from left to right or down to up which caused huge error rate. Some children missed the punctuations after finishing the letter tracing, while others lost the focus and had difficulties to control the grip of the pen. A 6-year child was holding the pen and she followed the full guided mode only and was unable to follow the instructions for continuing the test session. One participant played a lot with the pen, and one stopped earlier and had to quit the experiment.

We noted that, maintaining a precise eye-hand coordination was challenging for most of the participants. The stability of their hands' movement had impact on the grip strength, and grip precision where we observed that participants were moving their hands while writing which caused unnecessary noise. Certainly, at the begin and end of the trace, the participant who do not know the shapes were drawing without notice. However, we believed that the guided mode allowed using the force feedback is responsible of the results (e.g., improvement in the engagement, handwriting legibility and speed) observed here and presents a possible enhancement compared to traditional methods. This finding is in line with research outcomes [9, 10] where slight correction in movements, were found to influence the virtual task performance for each participant. Also, grip force precision decreased for both groups in free mode, which suggested that they were unable to maintain stable grip pressure all the time. Besides, the overall performance analysis demonstrated that older participants in both modes performed better than younger counterpart. These results indicated that the proposed metrics could be useful to quantify fine motor skills for younger participants. Yet, these metrics together with new ones might assist in assessing fine motor skills in an efficient and objective way. Whereas in other studies, the benefit for learning using advanced devices was demonstrated, in certain context, to be ineffective for younger audience, for an overview of these studies, see also [11]. The inconsistent findings might happen from methods e.g., unconventional methods, analysis e.g., qualitative metrics, and related issues e.g., small sample size. Besides, the inconsistent results also may occur from other participant variables (e.g., slower than normal motor development for early learning of letter knowledge). However, for young Latin children handwriting Arabic Alphabet, and to the best of our knowledge, there is no detailed studies published in the literature. So, our work can be considered as a starting position to shed light on the role of haptic devices and VR experience in early childhood for language learning.

Certainly, we learned also that ranking the handwritings between 1 and 5, as shown in Fig. 7 was indeed not showing the improvement in resulting performance of children, especially, who were initially unable to write at all. We found instances where the pre-test was not gradable because no letter was written at all. Some post-tests instances were rated low but comparably closer to the corresponding letter trace. For those cases, we clearly, remark that some improvements are shown, but both instances scored with low rank. To quantify the letter tracing performance efficiently, we suggest considering other objective metrics e.g. focus on shape quality. For that reason, we need to switch to other metrics in order to reliably assess the writing performance by focusing on visual perception and visuomotor coordination requirements, as experimented in [7, 10].

5 Conclusion

This study showed the feasibility of utilizing a haptic device to explore the challenges and the solutions while producing handwritings which resemble to a harsh learning environment. The impact of writing with haptic device to fine-tune the children motor skills and improve their focus was evaluated and discussed. It was observed that some children had difficulties to grip the pen in a 3D environment. If longer training time were given to them, this might help to master the pen properly. However, longer handwriting training during one session may cause out of focus, demotivation, and fatigue, especially by repetitive letter shapes that must be written with continuous segments.

Therefore, a tradeoff strategy should be applied. We noticed that the written letters were engraved in the children's memories. They spent a lot of time to trace them in the VR environment. This approach is based on the philosophy that what it was given easily can be forgotten quickly. Currently, we are testing this haptic device on a larger number of children from different background and ages. The ability of subjects to draw more complex patterns using haptic and VR sensations is also being explored.

Acknowledgement. This work was made possible by NPRP grant #10-0205-170346 from the Qatar National Research Fund (a member of Qatar Foundation). The statements made herein are solely the responsibility of the authors.

References

1. Wu Z, et al (2021) LightWrite: teach handwriting to the visually impaired with a smartphone. In: Proceedings of the 2021 CHI conference on human factors in computing systems, pp 1–15
2. Pilatásig M, Tobar E, Paredes L, Sánches Z, Naranjo G (2021) Virtual reality and haptic devices applied in the system of teaching learning with children of early education age. In: Botto-Tobar M, Montes León S, Camacho O, Chávez D, Torres-Carrión P, Zambrano Vizuete M (eds) ICAT 2020, vol 1388. CCIS. Springer, Cham, pp 33–43. https://doi.org/10.1007/978-3-030-71503-8_3
3. Kim Y, Collins M, Bulmer W, Sharma S, Mayrose J (2013) Haptics assisted training (HAT) system for children's handwriting. In: World haptics conference (WHC), pp 559–564. https://doi.org/10.1109/WHC.2013.6548469
4. Voutsakelis G, Kokkonis G, Kontogiannis S (2020) Haptic applications for people with limited motor skills. In: The 3rd world symposium on communication engineering (WSCE), pp 1–6. IEEE
5. Zhao H, Zheng Z, Swanson A, Weitlauf A, Warren Z, Sarkar N (2018) Design of a Haptic-Gripper virtual reality system (Hg) for analyzing fine motor behaviors in children with autism. ACM Trans Access Comput 11:4

6. Bowers L, Ryan H (2021) Creative haptics: an evaluation of a haptic tool for non-sighted and visually impaired design students, studying at a distance. Br J Vis Impair 39(3):214–230
7. Ozgur Arzu G, et al (2020) Iterative design and evaluation of a tangible robot-assisted handwriting activity for special education. Front Robot AI 7:29. https://doi.org/10.3389/frobt.2020.00029
8. Williams CK, Tremblay L, Carnahan H (2016) It pays to go off-track: practicing with error-augmenting haptic feedback facilitates learning of a curve-tracing task. Front Psychol 7:2010. https://doi.org/10.3389/fpsyg.2016.02010
9. Kim Y-S, et al (2013) Haptics assisted training (HAT) system for children's handwriting. In: Proceedings of the world haptics conference (WHC 2013), pp 559–564. IEEE, Daejeon
10. Zhao H, Zheng Z, Swanson A, Weitlauf A, Warren Z, Sarkar N (2018) Design of a Haptic-Gripper virtual reality system (Hg) for analyzing fine motor behaviors in children with autism. ACM Trans Access Comput 11(4):1–21
11. Herodotou C (2018) Young children and tablets: a systematic review of effects on learning and development. J Comput Assist Learn 34(1):1–9

The Effectiveness of Tourism in KSA Using Mobile App During COVID-19 Pandemic

Hosam F. El-Sofany[1,2] and Samir Abou El-Seoud[3(✉)]

[1] King Khalid University, Abha, Kingdom of Saudi Arabia
helsofany@kku.edu.sa
[2] Cairo Higher Institute for Engineering, Computer Science and Management, Cairo, Egypt
[3] The British University in Egypt (BUE), Faculty of Informatics and Computer Science (ICS), Cairo, Egypt
samir.elseoud@bue.edu.eg

Abstract. Tourism in Saudi Arabia varies to include the most attractive and active religious tourism due to the presence of the Two Holy Mosques, eco-tourism, and business tourism. Global tourism in Saudi Arabia has a more than distinctive character, for example, Saudi Arabia has many coastal destinations, and it has green spaces that cover its mountains, plains, and countless parks. In addition to many heritages, archaeological sites, and markets that extend Some of them date back to the pre-Islamic era. In recent years, the Government has developed the tourist areas in the Kingdom and optimized its services and facilities to attract the largest number of tourists from different countries. According to the government, the travel and tourism sector would be one of the most hit by the COVID-19 epidemic, with an expected 60–80 percent drop in international tourism in 2020. They acknowledge that COVID-19 may cause a paradigm change in the travel and tourism industry. Therefore, there are many efforts and support are targeting this sector to reach the indicators of tourism to the level of vision of 2030. With the increase of using mobiles globally, it has become increasingly necessary for tourists to use mobile apps for tourism. This research study proposes a mobile tourist app, it offers a variety of services to consumers including 1) providing the tourist with the nearest touristic locations to his present location, 2) providing the tourist with all information about nearest shopping centers, restaurants, coffee shops, hospitals, festivals, and malls, 3) offers hotel, restaurant, and movie theatre bookings to tourists. The paper shows a promising impact on using the mobile app for tourism. The research study uses Cronbach's alpha to assess the stability and validity of the questionnaire's content and step-wise multiple linear regression analysis and Spearman correlation to analyze the influence of the related factors on the outcome.

Keywords: Mobile tourist app · Tourist information system · Tourism sector · COVID-19 · Cronbach's alpha · Spearman correlation

1 Introduction

The travel and tourism sectors were among the first to be impacted by COVID-19. The crisis has had an impact on every aspect of the business, highlighting the fragility of

M. E. Auer et al. (Eds.): REV 2022, LNNS 524, pp. 559–570, 2023.
https://doi.org/10.1007/978-3-031-17091-1_55

the sector's enormous value chain. Since 2000, global travel and tourism revenues have nearly tripled, with the sector now accounting for 10.4% of global Gross Domestic Product GDP and employing one out of every ten people globally [1]. International travel has come to a complete halt as of mid-April 2020. According to the World Travel and Tourism Council (WTTC), international travel might decrease by more than 30% in 2020. According to the United Nations World Tourism Organization (UNWTO), all destinations have some form of travel restrictions in place, with 72% of governments prohibiting all international travel [2]. The tourism sector value chain impacts practically every part of a country's economy since it is a cross-cutting business. The repercussions will be felt most strongly through many channels.

The Saudi people have a strong belief in their country's cultural vision, and they have worked hard over the last year to make that vision a living reality. Despite significant challenges and social distancing measures, cultural and creative fields have proven their vital importance as a powerful unifying factor. Nowhere has this been more evident than in the sector's ability to adapt to the virtual space and utilize its various tools and platforms –as the industry found inspiration from within, it was met with abundant creative talent and inventive adaptation. The culture sector exemplifies the vibrancy of Saudi Vision 2030. The Kingdom is committed to developing and conserving national cultural heritage while also exploring new cultural boundaries. Over the past year, community involvement in cultural sub-sectors such as culinary arts, fashion, and the visual arts increased dramatically. Economic contributions from the architecture and design, film, and music sectors also witnessed a substantive uptick, in addition to positive strategic momentum in publishing, translation, and language, as well as cultural activities at large [3].

In response to the COVID-19 epidemic, governments have used a variety of tools to assist travel and tourist businesses. While some governments have offered assistance to enterprises across the economy and the tourist industry in general, others have targeted particular tourism sub-sectors such as hospitals, food services, hotels, and airlines. While debt financing appears to be the most popular form of government support, other instruments used include regulatory relief, tax, fee, and charge exemptions, cash grants and subsidies, and employment and training guidance [4].

Tourism has evolved into a highly competitive industry all around the world. The evolution of information technology and innovation is increasingly driving competitive advantage. Nowadays, the Internet is the principal source of information on tourism destinations for tourists. The Internet is rapidly evolving toward wireless usage, thanks to recent breakthroughs in hardware and software technology. New mobile apps operating on these devices give users easy and quick access to distant services that are available at any time and from any location and will soon make use of wireless networking's ubiquity to create new virtual worlds. Furthermore, artificially intelligent agents can operate on these devices and give travelers personalized help during their journeys. They build a Multi-Agent Information System with traditional information agents such as hotel broker agents, tour planning agents, and other diverse tourist resources to provide collaborative and intelligent help to visitors. [5].

In this study, we aim to use all of the above mentioned technologies to enhance tourist coordination and integration toward full support. The biggest challenge of our system is to provide efficient coordination and integration of disparate information and service resources at any time and from any location, as well as to provide personalized assistance and automation to tourists, each of whom has different preferences and support requirements that frequently change during the trip [6].

The research aim is to analyze, design, and implement a Mobile-based app, that is accessed via wireless devices (such as mobile telephones, and Personal Digital Assistants (PDAs, etc.). The following are some services provided by the proposed app:

- The app provides the tourist with the nearest touristic locations to his present location,
- The app provides the tourist with all information about nearest shopping centers, restaurants, coffee shops, hospitals, festivals, and malls,
- The app provides tourists with service of hotel, restaurant, and cinema reservations,
- The app delivers user-friendly mobile administration features

2 The Impact of Coronavirus Pandemic on the Tourism

The coronavirus (COVID-19) pandemic has triggered an unprecedented crisis in the tourism economy, giving a sudden and massive shock to the industry. The COVID-19 impact estimates a 60% drop in foreign tourism in 2020. If recovery is delayed until December, this estimate might reach 80%. International tourism in certain geographical areas. Domestic tourism, which accounts for over 75% of the countries' tourist income, is likely to rebound more swiftly.

The crisis's impact is being felt throughout the tourist ecosystem and reopening and restoring sites will need a collaborative effort. Tourism firms and employees are benefiting from broad-based economic stimulus packages, with many governments implementing tourism-specific policies. Governments and businesses are concentrating their efforts on:

- Lifting travel restrictions and collaborating with enterprises to get access to liquidity and financial support, implementing new health measures and protocols for safe travel, and to assist in diversifying.
- Restoring tourist trust and increasing demand through new safe and clean labels for the industry, tourist information applications, and domestic tourism promotion campaigns.
- Developing comprehensive tourist recovery strategies to rebuild destinations, encourage innovation and investment, and redefine the tourism industry.

These steps are critical, but more must be done in a coordinated manner to successfully restore the tourism industry and get companies up and running, as tourism services are highly interrelated. The travel and tourism sector, as well as governments, should continue to strengthen their coordination mechanisms to assist enterprises, particularly small firms, and workers. [7].

3 Using Mobile Technology for E-tourism

M-tourism refers to the delivery of tourism services using wireless devices such as portable computers, PDAs, mobile phones, and Tablet PCs. Our mobile tourist guide system includes capabilities for entering and organizing tourism items in a database and making them available to visitors and administrators via E-tourism management information systems (TMIS) portals that structure the tourism process and its operations. M-tourism object management solutions are now accessible with all types of user devices, including traditional desktop PCs as well as mobile devices like laptops, PDAs, and cell phones. [8–12].

- *The advantages:* The advantages of mobile tourism primarily benefit the mobile user: greater and better flexibility, the ability to improve tourist schedules, increased productivity during downtime, and just-in-time tourism.
- *Difficulties and limitations:* Nevertheless, they have relatively small screens, low memory capacity, and the huge diversity of mobile devices hinder a decent travel experience. In the M-tourism industry, we note that the tourist services must answer to specified conditions. We will have to work around technical constraints in order to provide effective tourism services. The use of video, audio, simple interfaces and segmented tourism services must all help. In addition, we will need to customize the material to the demands of the mobile consumer. Because the user's schedule is fragmented, we must ensure that the tourist objects are not overly lengthy. A solution is to divide the knowledge into smaller parts.

4 Social Implications of the Proposed App

The following facts make the suggested research project considerably significant:

- *For individuals*: The proposed app has a great effect on individuals who are the tourists that it support them with good information about the tourism places and services centers near the tourism places.
- *For organizations*: The proposed app organizes the ways of offering the information through this adopted system for tourism places, that it will be a trusted site.
- *For society*: The proposed system effect KSA society, that it makes the people use the information technology in getting the needed information regarding the tourism sector. Therefore, the app supports the tourism sector and encourages tourists to visit the tourist places that support the country's economy.
- Other benefits of using the proposed M- tourists app:-
  - The research is significant for researchers working on open, popular, and current research fields of the next-generation Web.
  - The research enables (new) employees and workers to utilize their wireless devices (PDAs, mobile phones, and Smartphones) at any time and from any location to simply and rapidly locate the destinations that they are looking for.
  - Tourists, particularly first-time visitors, will feel more at ease. This is due to the fact that just by pushing a few buttons, they will be able to learn everything they need to know about national landmarks, restaurants, and other interesting sites.
  - Tourists will be able to save a lot of time by learning the information they want without having to contact or spend time looking on the internet.

- First-time drivers and drivers from other countries will be able to navigate around quickly and easily.
- Hotels, hospitals, businesses, and other key sectors will not get unnecessary calls, such as inquiries about the location of the building or department phone numbers, and so on. This is because this sort of information will indeed be available on the proposed app.
- The application directs scholars and conference attendees on an easy tour in the country.
- The application directs people who have traveled from afar for festivals and Eid celebrations to places where they may enjoy themselves without wasting time.

5 The Proposed App Development

The proposed app is completely Mobile-based and requires no technical expertise from potential and possible users. The system access is easy and ideal for on-campus or off-campus users. The main three components of the proposed system architecture include the client, server, and database.

The requirement engineering process of the proposed app is a major step in determining the criteria of the system in detail. The aim is to reflect system goals into specified functions and operations of the proposed Mobile-based services at this stage in system development. A complementary series of diagrams such as use cases, sequence, class, and activity diagrams for the mobile tourist app was created. Each diagram offers a different view of the app and presents different demands and requirements. The requirements of the mobile tourist app are divided into a non-functional requirement and a functional requirement.

- Non-Functional requirements: The features and interfaces that are not directly related to the system's functional behavior, such as accessibility, supportability, reliability, security, and performance that have been checked and demonstrated by the authors.
- Functional requirements: Different users are allowed to benefit from the proposed app according to the specified privileges. There're two main users including administrators and tourists. The proposed app provides each user with the following services:

  - This system is a mobile app that is designed to support tourists who like to visit tourist places in KSA.
  - Firstly, we have to determine the services in this application that this application offers information about the tourism places in KSA such as photos, videos, and stories; Also, it offers information about the important services centers which are close to the tourism place such as restaurant, coffees, health centers, and hotels.
  - This application offers an evaluations service which was left by the tourists about the tourism places and the services centers to be reviewed by all tourists in the system, and any tourist can write a comment about the tourism place and service centers.
  - The *Admin* actor can manage the tourist user through activating or deactivating his account, also the admin can manage the tourism places, region, service center type, and the services centers nearby through adding, updating, and deleting.

– The *tourist* can register in this application to get the availability of logging into the app using a username and password. He can search for a tourism place and search for the nearest service center to this tourism place. He can view the comments about the tourism places and the service centers near the places, and also can add a comment about a tourism place and service center.

Figure 1 illustrates the data flow diagram (context diagram) that was initially created for the implementation of the proposed app.

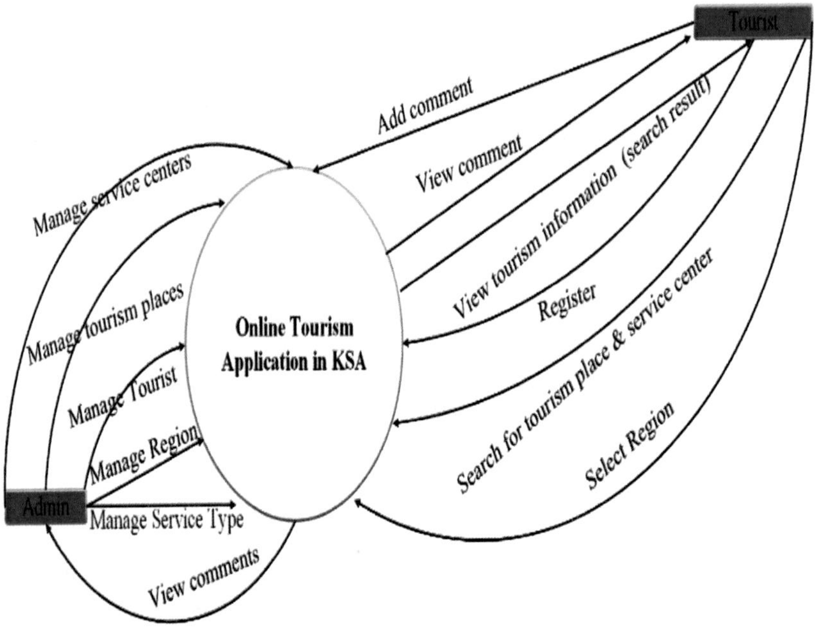

Fig. 1. Data flow diagram of the proposed app

The authors used object-oriented design methodologies to develop the proposed app. Besides, to create the required UML diagrams, the researchers implemented the following tasks:

- Identify the system' actors and their functions
- Identify the internal and external interactions of the proposed app
- Create system architecture
- Identify the principal classes of the system
- Identify the associations among classes
- Create the design models
- Develop users' interfaces

On the other hand, the authors design and implemented the app's database according to the defined requirements. We have used the MySQL Server database as a database management system (DBMS). The following entities are created in the implementation stage of the proposed app: Admin (adminId, username,password), Tourist (touristId, username, password, gender, age, telephone, country), TourismPlace (placeId, placeTitle, description, informationLink, latitude, longitude), ServiceCenter (serviceCenterId, centerName, description, video-photoLink, latitude, longitude), Region (RegionId, RegionName),CenterType (CenterTypeId, CenterType), and Comment (commentId, commentText).

The users can access the system by username/password authentication. If the user accesses the system as an Administrator, then he/she can manipulate whole system information such as add, update, and delete tourist (activate/ deactivate), tourism place, service center nearby the tourism place, region, and center type. If the user accesses the system as a Tourist, then he will be able to use the app to register, log in to the app, review comments, select regions, look for tourism places (view the search details), search for the available services centers, and add a comment. Figure 2 illustrates use case diagram of the proposed app.

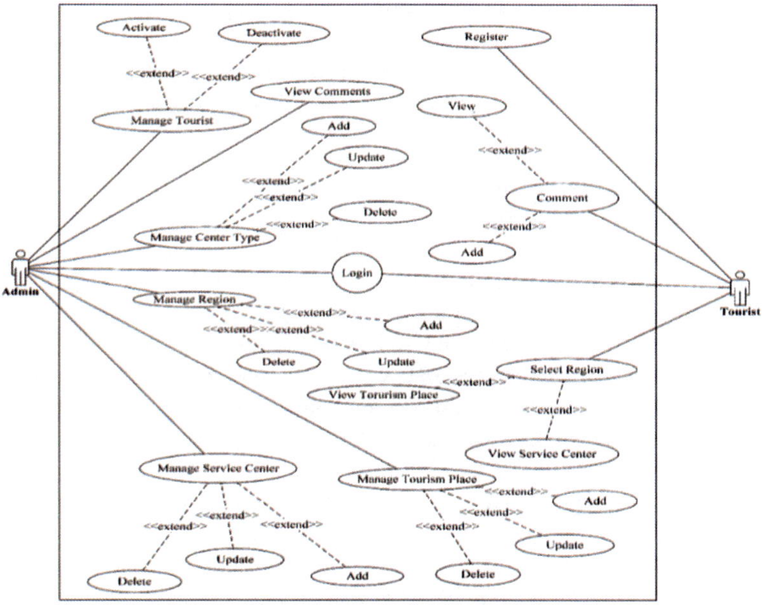

Fig. 2. Use case diagram

6 App Evaluation and Results Analysis

The authors focus on testing a key hypothesis to handle the dimensions of the research problem and objectives. The research key hypothesis states that " There is a positive impact of the different dimensions of dominating practices in using a mobile app on enriching the performance quality of tourism message". The research key hypothesis is divided into the following variables:

- *Independent vars*: include independent dimensions that affect the use of the mobile app for tourism.
- *Dependent vars:* variables that influence the trial outcomes and include user comments on the proposed app.

The authors use Spearman Correlation to identify and quantify the relationship and direction between research variables. Cronbach's alpha, on the other hand, will be used by the authors to assess the stability and validity of the study's questionnaire contents. The questionnaire is evaluated for its validity, completeness, and suitability for statistical analysis and data entry. SPSS software was used to do statistical computations on the experimental variables that were coded. Gender, age, region, and employment are all considered dependent variables. The authors extracted descriptive statistics (e.g., Arithmetic mean, standard deviation, coefficient of standard variation, etc.) for the selected dependent and independent variables that were associated with the study samples.

The findings in this research study are based on information and data from the authors' questionnaire, which was performed among a sample of 200 tourists (120 males and 80 females).

The Likert Scale evaluation for the survey responses to the given problem is described in the following results:

- 65% of tourists are "strongly agree",
- 20% of tourists are "agree", and
- 11% of tourists are "satisfied",
- 04% of tourists are "disagree",
- Cronbach's Alpha was calculated and determined to be (0.8546). That suggests that the research samples have such a significant degree of validity, i.e., the face validity of the suggested viewpoints, as represented in the study samples, has achieved a high level (0.8576). The authors used the descriptive statistics presented in this section to establish the characteristics of the research sample. In the study studies, the demographic variables descriptive statistics include gender, age, area, and job:

- Figure 3 illustrates the distribution of "gender"-based study samples. The female sample received 60%, while the male sample received 40%.
- Figure 4 shows the distribution of "age"-based study samples, which is divided into two groups: The "age" category of 40 to less than 70 years was the highest; 61%, followed by age groups from 25 to less than 40 years was 29%, followed by other age groups from 19 to less than 25 years was 10% of the research sample individuals.
- The distribution of the research samples by "region" is divided into four groups: the first is the "Asia" group, which had the highest; 55.25 percent, followed by the "Europe" group, which had 24.75 percent, followed by the "Africa" group, which had 11.35 percent, and finally the "Americans" group, which had 8.65 percent; as shown in Fig. 5.

On the other hand, the authors offered the independent variables descriptive statistics, which explained the data in the table. Table 1 displays the items with the highest or lowest agreement or disagreement based on the responses of the study sample.

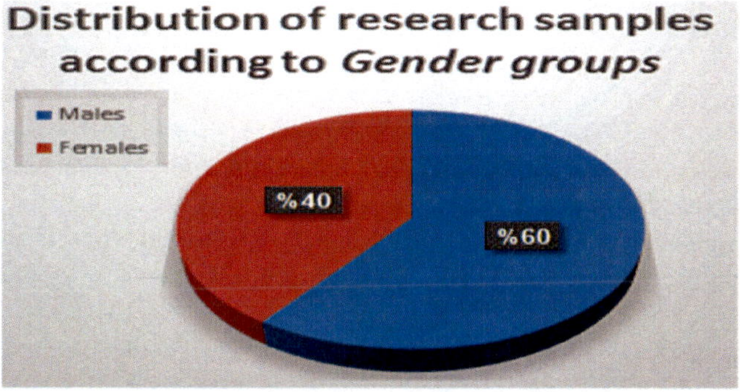

Fig. 3. Gender-based research samples distribution

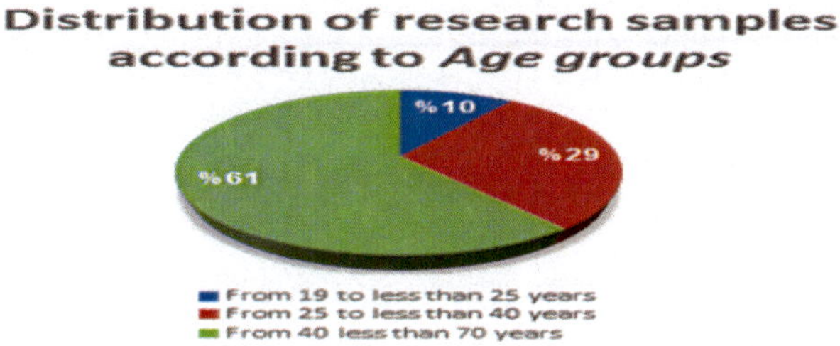

Fig. 4. Age-based research samples distribution

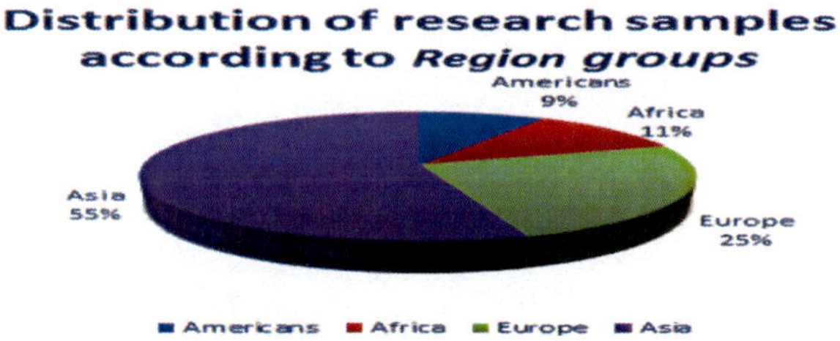

Fig. 5. Region-based research samples distribution

Table 1 represents the researchers' suggestions for the research samples in terms of "the usage of the mobile app for tourism." This questionnaire's data show that the arithmetic mean is (1.984) and the coefficient of standard variation is (53.75 percent). Among the selected items are:

1 Do you prefere to travel to KSA for make Umrah and Hajj?
2 Do you prefer to travel to tourist places in KSA other than Umrah and Hajj?
3 Do you have background information about the tourism places inside KSA?
4 Do you like to get information about the ancient ruins and stories about old cities in KSA?
5 Do you prefer an application supporting the tourists with all needed information during their travel to the tourism places?
6 Are you agree to develop a mobile app for supporting tourists in KSA?

7 Are you agree to provide the tourists with the locations and names of the tourism places in KSA?
8 Are you agree to provide the tourists with stories about the old places and cities in KSA?
9 Are you agree to provide the tourists with the famous service centers (restaurants, coffees, hotels, hospitals, etc.) in KSA?
10 Are you agree to provide the tourists with the nearest locations of the tourism places in KSA?

- The Spearman Correlation is used to detect and measure the relationship and direction between dependent and independent variables. Table 1 gives the following conclusions. There is a positive correlation between the two variables (independent and dependent) related to the research sample, indicating that there is a positive outcome with multiple dimensions for the use of the mobile app for tourism, at a level of significance less than (0.02), (0.05), demonstrating the validity of the research key hypothesis.

Table 1. The descriptive analysis of the effectiveness of tourism in KSA using the mobile app during COVID-19 pandemic

Items	Weighted Arithmetic Mean	Standard Deviation	Coefficient of Standard Variation	Order
1	1.84	1.022	55.72	3
2	1.83	0.898	48.76	2
3	1.99	1.182	53.82	6
4	1.72	0.971	51.90	1
5	2.13	1.201	57.99	8
6	2.16	1.270	56.97	9
7	2.03	1.141	59.27	7
8	2.38	1.242	55.75	10
9	1.94	1.177	57.71	5
10	1.90	0.604	44.66	4

7 Conclusion

This research study proposes a mobile tourism app that offers customers a variety of services including 1) providing the tourist with the nearest touristic locations to his present location, 2) providing the tourist with all information about nearest shopping centers, restaurants, coffee shops, hospitals, festivals, and malls, 3) provide tourists with service of hotel, restaurant, and cinema booking. The paper has shown a promising impact on using the mobile app for tourism. Cronbach's alpha was used in the research study to analyze the stability and validity of the questionnaire's content, as well as stepwise multiple linear regression analysis and Spearman correlation to examine the effect of relevant factors on the outcome. The results of the study are based on data from questionnaires from the authors' survey, which was conducted among 200 visitors. The results described by the Likert Scale evaluation and produce 65% of tourists were "strongly agree", 20% of tourists were "agree", 11% of tourists were "satisfied", and 4% of tourists were "disagree". Cronbach's Alpha was calculated and found to be (0.8546), indicating that the study samples have a high degree of validity, i.e., that the face validity of the proposed views that were reflected on the study samples validity has reached a high level (0.8576). The authors determined the characteristics of the study sample using descriptive statistics include gender, age, region, and job. The "gender" based research samples distribution is shown; the female sample got 60% and the male sample got 40%. The "age" based research samples distribution is shown and categorized into two groups: "age" group from 40 to less than 70 years was the highest; 61%, followed by age groups from 25 to less than 40 years was 29%, followed by other age groups from 19 to less than 25 years was 10% of the research sample individuals. The distribution of the research samples by "region" is divided into four groups: the first is the "Asia" group, which has 55.25%, followed by the "Europe" group, which has 24.75%, followed by the "Africa" group, which has 11.35%, and finally thc "Americans" group, which has 8.65%.

Acknowledgment. The authors extend their appreciation to the Deanship of Scientific Research at King Khalid University for funding this work through a research groups program under grant number (R.G.P.1–185/41).

References

1. https://www.unwto.org/news/covid-19-world-tourism-remains-at-a-standstill-as-100-of-countries-impose-restrictions-on-travel
2. https://www.wttc.org/about/media-centre/press-releases/press-releases/2020/latest-research-from-wttc-shows-an-increase-injobs-at-risk-in-travel-and-tourism/
3. Ministry of Culture, "Report on the State of Culture in the Kingdom of Saudi Arabia, 2020: Digitalization of Culture". Ministry of Culture. – Riyadh 1442H. Ministry of Culture, Riyadh (2020)
4. World Bank Group, "Building Tourism Competitiveness". Tourism response, recovery, and resilience to the COVID-19 crisis. Markets & Technology Global Tourism Team, July 2020
5. Dickinson JE, Karen G., Cherrett T, Speed C, Davies N, Norgate S (2014) Tourism and the smartphone app: Capabilities, emerging practice and scope in the travel domain. Curr Issues Tour 17(1):84–101
6. Hoehle H, Venkatesh V (2015) Mobile application usability: conceptualization and instrument development. Mis Q 39(2):435–472
7. Tan GW-H, Lee VH, Lin B, Ooi K-B (2017) Mobile applications in tourism: the future of the tourism industry?. Ind Manag Data Syst 117(3): 560–581
8. Liu X, Mehraliyev F, Liu C, Schuckert M (2019) The roles of social media in tourists' choices of travel components. Tour Stud 20:1–12

9. Gretzel U, Sigala M, Xiang Z, Ko, C (2015) Smart tourism: foundations and developments. Electron Mark 25:179–188
10. Wang D, Xiang Z, Fesenmaier DR (2014) Adapting to the mobile world : A model of smartphone use, vol 48, pp 2012–2014
11. Dickinson JE, Hibbert JF, Filimonau V (2016) Mobile technology and the tourist experience: (Dis)connection at the campsite. Tour Manag 57: 193–201
12. Lalicic L, Weismayer C (2016) Being passionate about the mobile while traveling. Curre, 3500:0–14

Students' Perception of English Language Learning in Virtual Discussion Boards

Wesam K. Morsi[1] and Samir Abou El-Seoud[2](✉)

[1] Faculty of Arts and Humanities, The British University in Egypt (BUE), P.O box 43, Suez Desert Road, El Sherouk City 11837, Cairo, Egypt
wesam.morsi@bue.edu.eg

[2] Faculty of Informatics and Computer Science, The British University in Egypt (BUE), P.O box 43, Suez Desert Road, El Sherouk City 11837, Cairo, Egypt
samir.elseoud@bue.edu.eg

Abstract. This study investigated students' perception of English Language Learning in virtual discussion boards (VDBs). A 5-likert scale questionnaire of 5 sections was piloted to 108 students in a private university. Results showed that VDBs were convenient tools for learning English and for promoting students' critical thinking skills. The general trend of results for all sections was towards the agreement with average mean for Sect. 1 "convenience of VDBs" (3.52) and the other 4 sections, namely interaction of instructors and students (mean = 3.14), English language learning (mean = 3.74), critical thinking (mean = 3.70) and topics of discussions (mean = 3.71). Pearson statistical measure showed positive significant correlation between Sect. 1 "convenience of VDBs" and all other sections at p-value 0.01 which confirms that VDBs are effective tools for learning. It is recommended that instructors use topics relevant to course syllabus and employ new strategies in VDBs to engage learners and make them value peer interactions.

Keywords: Students' perception · English language learning · Virtual discussion boards

1 Introduction

The outbreak of COVID 19 pandemic, according to the United Nations report in 2020, affected 94 percent of global students' population and forced stakeholders in formal educational institutions to reflect on new approaches to effectively deliver the in-class 'lesson plan' and ensure the sustainability of "teacher's monitoring" and "interactive learning", using virtual learning platforms. Since 2020, innovative educational digital tools have been suggested and experimented to replicate traditional methods of learning in a remote context. Instructors have developed their digital literacy skills and introduced learners to different online platforms. However, a number of educational institutions may not afford this educational software due to competition or tight finances in the time of the crisis. Therefore, taking advantage of already available tools in electronic Learning Management Systems (LMS), such as Virtual Discussion Boards (VDB) has

M. E. Auer et al. (Eds.): REV 2022, LNNS 524, pp. 571–582, 2023.
https://doi.org/10.1007/978-3-031-17091-1_56

been recommended. VDBs are online platforms that engage students and instructors in activities outside the traditional classroom, and it is time to consider how they can innovatively be used in all subject courses to reinforce delivering of teaching remotely and enhance learning [1].

1.1 Online Learning

Inclusion of online activities has become an integral part of students' learning experience in campus-based higher institutions. These activities aim at promoting meaningful learning through the use of interactive quizzes, piloting surveys or virtual discussions. Accordingly, hybrid or blended learning methods have been adopted to complement in-class traditional learning with online activities in virtual learning environments. Learning in the traditional classroom can be replaced by 45 to 80 percent of designated online activities [2]. With the digital transformation in the 21st century in all activities of life, various applications and online tools are developed every day to engage English language learners in activities outside the classroom. Research has shown the benefits of using technology in language learning; yet, there are still some concerns that should be addressed. These include students' attitudes towards face-to-face learning and online learning, their digital literacy skills and their familiarity with course curriculum. For instance, some students may feel anxious when participating in virtual learning environments, while others prefer interaction in online platforms [3]. Factors such as course design, quality of peer and instructor interaction, and convenience of technological tools affect learners' satisfaction of learning online. Recently, increasing number of researches confirm that when English language courses are well designed with adequate support to students, blended learning can significantly improve their language skills, academic performance and overall success of delivered courses. One of the most commonly used digital tools in online or learning is the asynchronous VDBs which allow for exchange of dialogues and materials that are relevant to English courses [4].

1.2 Virtual Discussion Boards

Virtual discussion boards (VDBs) are tools that are used frequently in the LMS of higher educational institutions to complement traditional instruction. They are e-platforms that enable teachers to post various topics for discussion, and students can reply, by sharing alternative viewpoints, developing new insights, building arguments and drawing conclusions that are supported by research evidence. Being the admin and moderator in VDBs, instructors can use this tool to enhance students' higher order thinking, improve their interpersonal and language skills, encourage collaborative learning and foster problem-solution or decision-making skills [5]. Asynchronous VDB has many useful features. First, they act as storage of students' messages, shared materials and comments in the virtual space. Second, they give students the chance to post a delayed response after thinking, reflecting and analyzing all input available at their own pace. During in-class time, some students may not be able to participate effectively to discussions due to shyness, hesitance or dominance of other students [6]. In English Language Learning (ELL), VDBs provide valuable opportunities to practice the language in meaningful activities outside the limited classroom time by engaging in written dialogue on almost a daily basis. Consequently, the continuous and sustained interaction in this virtual learning environment that is being monitored by instructors enable learners to better understand subject matter, construct knowledge and provide opportunities for collaborative learning that is integral part of their future life.

In spite of all these benefits, there is much disappointment reported by instructors with regards to the actual effectiveness of VDBs. Some students would only respond to receive credit on participation or complete a task, and that actually makes the discussion pointless. Some students post surface level responses with no quality contribution; they do not add any new information or express their viewpoints; instead, they repeat previous comments. Some may not abide to deadlines of posting or they just would rather ignore the task. Inappropriate language and incorrect grammatical structure may confuse the readers and disturb understanding of content. Consequently, more research has been conducted to examine convenience of VDBs as tools for learning collaboratively and whether student-student and instructor-student interaction can significantly develop critical thinking abilities and learners' academic performance positively. Other research has become concerned with developing best strategies for effective VDBs. Findings of the present research can contribute to the body of literature about the effectiveness of VDBs in ELL [7]. Using VDBs should aid learners develop positive study habits, participate and utilize course content and extra resources successfully to achieve better learning outcomes [8]. Therefore, gaining more insights into how students view the effectiveness of VDBs as a tool is essential to help promote their English language learning.

2 Review of Literature

2.1 Theoretical Background

The theoretical framework of this paper depends on three theories that are used in Computer Assistant Language Learning (CALL) research: the *Sociocultural Theory* and *Interactionist SLA*, and the *Technology Acceptance Model (TAM)*. The Sociocultural theory is originally developed by Vygotsky. It argues that contributing in activities via cultural artifacts and interacting with others can result in the development of higher mental functions of humans. Further, within this framework, communication between novice and experienced partners is also supported by Vygotsky's Zone of Proximal Development Theory [9]. The second widely used theory in CALL studies is the Interactionist SLA developed by Long in 1996; it signifies the role of integration and interaction for developing second language skills [10]. The last theory on which the theoretical framework is based is the Technology Acceptance Model developed by Davis in 1989 [11]. According to Davis, individuals' motivation to use any technological system are determined by its perceived usefulness, perceived ease of use and their attitude toward using this system. In this paper, individual students interact in collaborative dialogue and collaborative writing with their classmates to discuss various viewpoints on discussion board. Their tendency to engage in discussions are dependent on their attitudes toward e-learning management system and whether virtual boards are convenient and useful for them to perform assigned tasks successfully.

2.2 Precedent Studies

Asynchronous VDBs have been used for almost two decades. They are considered the oldest tools in CALL. Nowadays they have been used dramatically due to its convenience as part of LMS compared to other online applications; and their effectiveness have been reinforced after the outbreak of COVID-19. Researchers have been interested in investigating its benefits in different educational settings and whether learners' have perceived these benefits on the development of their academic performance, including foreign language learning. It has been agreed that VDBs have an effective role in fostering learning, collaboration and construction of knowledge among learners [1; 3; 4; 5; 12; 13].

Learner's cognitive processing functions by engaging in organizing their thoughts, writing them down to be presented to authentic audience and being exposed to wide range of views on different topics. Reference [13] conducted an experimental study to examine students' attitudes towards structured and unstructured discussions on 52 computer science students. Structured discussions are planned by instructors about specific topics related to the course with rules of participation, while unstructured discussions were mainly about responding to questions asked by participants or receiving feedback from students. Findings of closed and open-ended questionnaires, that were distributed to the control and experimental groups, showed that students' attitudes towards structured discussions were more positive because of clear rules of discussion, explanations, collaboration in assigned tasks and continuous involvement in the discussion to give feedback. Structured discussions had a significant effect on improving students' learning and on their perceptions of VDBs. In political science, [14] conducted a study to compare between traditional and online discussions with respect to students' academic performance. It was found that when VDBs were perceived as a tool for task completion and for finding answers, students were more likely to attain lower grades. However, students who perceived VDBs as tools to deepen their understanding of concepts tended to attain better grades.

In ELL, the researcher in [12] adopted Classroom Action Research methodology and collaborated with an English teacher to observe improvement of 8th grade learners' writing skills in recounting a text upon using VDBs. The study was implemented in 5 stages. After initial stages of brainstorming, prewriting and drafting in the classroom in pairs, students were assigned to revise, draft and publish their products on VDBs for feedback from their classmates and their teacher. In the final stage, students posted their individual recount texts to the discussion board. The researcher's observation checklist and field notes, students' questionnaires and scores of learners' recount texts showed the success of this learning experience based on 2 criteria: students' scores of final product and percentage of participation. Eighty two percent of students were actively engaged in the planned meetings of using the discussion board, and more than 81.4% scored B (65) in the final product. Also, 84.07% of students reported positive responses towards using discussion boards to improve their writing skill. Reference [15] confirms the above findings with regards to improving students' writing and critical thinking skills of 54 senior high school students by using posttests, perception surveys and interviews. It was found that the use of VDBs made students understand how to organize the structure of their essays effectively compared to students who did not use discussion forums.

to reach these results. Other findings showed that learners' reading, vocabulary and grammar performance were not improved.

Moreover, Korean EFL learners' experience of using online discussions were examined in [4] by allowing them to share their thoughts in a semi-structured interview and instructors' reflection journals. Students had assignments on discussion forums related to course lessons outside the traditional classroom. Every 14 or 21 days, students participated in a discussion forum topic by writing a primary post about 150 words and posting a reply of about 50 words to at least three classmates. The researcher also counted students' replies to the instructor in the following week. Results revealed students' understanding of effective learning as high participation in discussions with quality posts that reflect deep thinking. For example, a student commented, "Effective learning is.....when I get better at English and don't have to spend so much time thinking about what I want to write. I think my English is getting better, but it is still difficult at times to write exactly what I want to say", p. 14. Findings revealed convenience of the tool to practice the language outside the classroom for shy and part time working students. Similarly, in a small case study, 30 intermediate level Persian EFL students revealed positive attitudes towards learning English using VDBs [16]. The experimental group participated in discussions via Yahoo groupsite as an after-class activity, whereas students in the control group had paper and pencil reading and writing assignments. After 2 months, a writing test of three parts: short and long essay and sentence transformations was conducted. Learners in the experimental group were found to outperform others in the control group. Researchers believed that it could be the tool or the motivating effect it had on students that made a difference in students' performance. Results of the questionnaire showed that 29% believed that interaction in English was very effective followed by learning grammatical structures 23%, learning vocabulary 18% and viewing posts 15%, respectively. Students commented that they had time to review their posts for grammatical errors after reading their classmates' ones. They did the writing activities at their own pace and time and they were exposed to various viewpoints and broader range of vocabulary.

It is worth noting that discussion forums are not necessarily perceived by learners as useful or effective tools in improving ELL and communication skills. For some, it is only a storage of resources and a board for posting important announcements [8]. Instructors' observations even reveal that some learners ignore accessing LMS unless it is necessary. Consequently, more qualitative and quantitative data from students in different cultural backgrounds are needed to investigate how learners perceive utilization of VDBs in performing of tasks, discussing topics, and interaction with instructor/peers to maximize their effectiveness on learning. This could advise curricula developers and teachers on how to manage various activities rules of participation by using effective directed strategies that can aid in creating student-centred environment and ensure students' involvement to meet course objectives and learners' needs.

After the pandemic and the shift to online learning or hybrid learning, it has become a must in the university, where the researchers conducted this study, to use VDBs on a weekly basis to follow up on students' progress and engage them in activities that can support development of their English language and prepare them for final assessments (Lee, 2017). The purpose of this paper is thus to learn about Egyptian students' perceptions of using VDBs in an English language course.

The following are the research questions:

1. Do students perceive VDBs as useful online tools for learning?
2. From students' perspectives, what is the impact of contribution to VDBs on the development of their English language skills?

And so, the following null hypotheses are formulated:

i. Students do not perceive VDBs as useful online tools for learning. ii. Contribution to VDBs has no significance on promoting students' English Language Skills.

3 Research Methodology

3.1 Sample of Population and Procedures

The participants are 108 Engineering Prep. Year students in an Egyptian private university whose age ranges from eighteen to twenty years old. They are enrolled in an English for Academic Purposes module to improve their academic writing, argumentation and presentation skills. The semester is divided into 10 weeks. In the first 6 weeks, students were assigned topics to contribute to VDB each week. In the first two weeks, general topics were chosen for students to contribute to by sharing their personal experiences and supporting their views by information from reliable sources. In the next 4 weeks, students were assigned topics that are relevant to their course syllabus and assessments (e.g., one forum was about their viewpoints of single sex schools vs mixed schools, another one was about their viewpoints of aspects of similarities between learning online vs learning on campus, the third is about their viewpoints of euthanasia in preparation for their writing assignment and the fourth is about the advantages of online shopping). Students were informed that participation was essential to record their online attendance, but their contribution to the VDBs was not graded. In week 7 and 8 discussions were conducted in-class; finally, in week 9 students were required to fill out a questionnaire about their perception of VDBs. They were informed about the purpose of the questionnaire and that their perceptions would make a difference in how to design activities and assess learners' performance using VDBs in the next semester.

3.2 Data Collection and Analysis

For data collection, participants were required to answer a semi-structured questionnaire about their perception of using VDBs in the English module. It consists of 20 questions: 17 statements provided in a 5-likert scale (1 = strongly disagree to 5 = strongly agree) and 2 questions to collect data about students' gender and their ability to access VDBs on e-learning. Questions of the survey were developed by the researchers based on the dimensions mentioned in [4] and some questions were adapted from [17]. This study adopts the non-experimental quantitative method; descriptive analysis was administered by the SPSS 20 program to quantitatively analyse participants' responses in the questionnaire and to confirm that the results obtained are valid and reliable.

4 Results and Discussion

In this section, findings of the questionnaire distributed to Egyptian undergraduates are displayed and discussed. The questionnaire was given to students in a language that they can understand to avoid any misinterpretations. The questionnaire was made up of five sections: investigating VDB as a tool for learning, its influence on students' English language development, quality of interaction among students and instructor, its effect on critical thinking and problem-solution skills choice of topics and assessment of contribution. Statements in all these sections were based on a 5-likert rating scale. There were two "Yes or No" questions to collect data about gender and accessibility of VDB via e-learning, and one open-ended to ask students about their preferred mode of discussions -Face to Face or online. To measure whether variables for the total participants' responses were stable, "Alpha Cronbach", was used for checking the reliability. Results showed that the stability coefficient was (0.845); this indicated a high degree of validity for the study sample that was reflected also by its coefficient that is equal (0.919).

As illustrated in Table 1., 76.9% approved that VDB were easily accessible on e-learning, while 23.1% disapproved.

Table 1. Frequency Distribution of Yes/No Question of Accessibility of VDBs on e-learning

Do you find using virtual discussion boards on e-learning convenient for you with regards to accessing, posting replies and materials?	Frequency	Percentage %
Yes	83	76.9%
No	25	23.1%
Total	*108*	*100*

These results confirm previous research which highlights the importance of technology to enhance the quality of teaching and learning provided that there is proper instructional design of course curriculum that aids in delivering content effectively [18]. Further, students' web-based learning or e-learning makes them appreciate discussion forums as a tool to share hyperlinks, materials and post replies to their colleagues in a way similar to their use of other social media platforms. Discussion forums via email, videoconferencing, and live lectures are all possible through the web. Reference [19] illustrates that modern students can relate easily to discussion on e-learning because they replicate online social media habits. In Table 2., descriptive statistics for the research variables were conducted to show the sample's general trend of responses for each statement, the percentage coefficient of variation, f-variation, standard deviation, and mean. The average means for each section are shown in Table 2. below.

The trend of the sample for S1 (Convenience of the digital tool), indicated that it was towards the (Agreement), with mean of (3.52), Std. Deviation (0.62), and the coefficient of variation (17.64%), and differences in variance's ability (%82.36). The most agreeable statements were, (Contribution to forum discussions was a valuable experience for college students), (I have adequate digital literacy skills to contribute to discussion forums on e-learning), with coefficient of variation, (28.45%), (30.08%),

Table 2. Descriptive statistical analysis of average mean, standard deviation and coefficient of variation for the Questionnaire's Five Sections

Questionnaire's sections	Mean	Std	C.V
S1 convenience of VDB	3.52	0.62	17.64
S2 interaction of instructors & students	3.14	0.62	19.85
S3 critical thinking & communication skills	3.70	0.80	21.74
S4 english language learning	3.74	0.81	21.64
S5 topics of discussion & grades	3.71	0.72	19.37

respectively. The least agreeable statement was, (Keeping track of the discussion can sometimes be confusing online), with coefficient of variation, (36.27%).

For S2 Interaction of instructor and students, the trend of the sample was towards the (Neutral) and (Agreement), with mean of (3.14), Std. Deviation (0.62), and the coefficient of variation (19.85%), and differences in variance's ability (80.15%). The most agreeable statements were, (I like my instructor's presence and interaction in the discussion; it motivates me to participate), (It is exciting to interact and share opinions with many of my classmates online), with coefficient of variation, (24.37%), (32.50%), respectively. The least agreeable statement was, (I always feel embarrassed when my instructor or my classmates disagree with my opinion online), with coefficient of variation, (49.28%).

Reference [5] had similar findings; students believed that using forums was a valuable, distinguished experience in college. Similarly, in [16], students revealed that the forums were convenient tools for learning and practice English Language outside the classroom at their own pace and time because some were overwhelmed with busy schedules or working. Also, his results showed that all Persian learners had adequate techno-literacy skills to participate in online forums.

The trend of students' responses for S2, "Interaction of Instructor and Students" was neutral. It is interesting to notice that most students prefer the presence of their instructor which motivates their participation in the discussion and that they rarely feel embarrassed when others disagree with their opinion. This is confirmed in [3] in which "anonymity was of little importance" p. 40. This is in accordance with most of previous research findings [4; 13; 16]. According to [20], VDBs are powerful tools that highly motivate students and instructors. The ultimate goal of online forums is to share difference perspectives on various topics whether students are all from the same cultural background or from different cultural backgrounds - eastern or western. However, it is worth noting that in this study, contrary to [3] which discovered that instructors' feedback and evaluation are the most important dimensions in VDBs, the general trend was neutral for S2. This means that instructors are challenged to make learners participate effectively so that VDBs would not fall flat.

With respect to S3, (Critical thinking and communication skills), overall responses were towards the (Agreement), with mean of (3.70), Std. Deviation (0.80), and the coefficient of variation (21.74%), and differences in variance's ability (78.26%).The most agreeable statements were, (I learnt new strategies for solving problems during forum discussions), with coefficient of variation, (28.45%); whereas the least agreeable statement was, (I feel participation to discussion forums develops my critical thinking and communication skills), with coefficient of variation, (29.32%).

These results are emphasized in [14] who compared online discussion to face to face discussions in a political science course with respect to students' academic performance. It was found that students who perceived online forum discussions only as a tool to find answers and complete assigned tasks were likely to be the ones with lower overall grades compared to students with higher academic achievements who perceived VDB as a tool to deepen their understanding of concepts develop their critical thinking abilities, and accordingly received higher overall grades. [21] explained that critical thinking was required to be mastered by educational students in university level due to their needs in further educational settings. Students are required to take positions in essays, or debates, write a final thesis or formal reports, present materials in classroom discussion and master educational practitioners. Graduate students in Advanced Sociolinguistics course in the study were able to use advanced English to share their inferences, deductions, assumptions and evaluations of arguments clearly in logical, flexible ways that were relevant to topics of discussions. Thus, it is evident that asynchronous VDBs promote better critical thinking abilities through writing one's thoughts after enough time of reflection and further enhance the higher order thinking processes [6; 20].

S4 (English Language reading and writing skills) showed the sample's (Agreement), with mean of (3.74), Std. Deviation (0.81), and the coefficient of variation (21.64%), and differences in variance's ability (78.36%). The most agreeable statements were, (I usually check my English language structure, vocabulary and punctuation before posting my replies), with coefficient of variation, (27.84%). The least agreeable statement was, (I like that writing rather than speaking gives me time to think and express my thoughts exactly the way I want), with coefficient of variation, (36.22%). The last section, (Topics of Discussion and Grades), indicated (Agreement), with mean of (3.71), Std. Deviation (0.72), and the coefficient of variation (19.37%), and differences in variance's ability (80.63%). The most agreeable statements were, (Discussing course concepts and doing extra practice help me retain information and get better grades in exams), (I actively participate in virtual discussion forums when they are GRADED and the Assessment Criteria of quality and number of posts are clear), with coefficient of variation, (24.50%), (25.64%), respectively. The least agreeable statement was, (Participation to the forum gives me time to search for information to support my viewpoint rather than sharing it spontaneously in class), with coefficient of variation, (36.22%).

In this study, students were Prep. Year undergraduates of upper intermediate level to advanced English levels. They appreciated they had time to check their English language to use it effectively in the forum more than the fact of having time to think about their thoughts at their own pace and time before sharing them or even checking information and sources. As Prep. Year students with lower English language levels, they may not clearly understand how their higher order thinking processes were being developed gradually as they shared and received feedback on their arguments from their instructors and peers. Their language level may hinder many opportunities in which they would like to share their thoughts to the discussions.

Further, learners' most agreeable statements for S5 were for topics that were related to the course syllabus and that are graded as affirmed in [18]. In this study, findings about using online tools in which EduBingo was used to teach arithmetic problems on e- learning, "Only by targeting the assessment goal in the design, can we know what material is needed and what pre-designed activity is appropriate for the content sample. Striving to fulfill the assessment goal, a content sample should be designed and re-deigned, experimented and re-experimented in order to seek continual improvement" p. 345. It should be noticed that learners in this study may have perceived VDBs as tools for task completion rather than developing their critical thinking abilities. That could be another rationale for the above findings. Prep. Year students need more time in college to develop their higher thinking processes and completely understand the benefits of using VDBs. To conclude this section, Pearson statistical method was used to measure the correlation between S1 (Convenience of VDB as learning tool) X and the other 4 sections S2-S3-S4-S5 Y. Results showed positive relationship between X and Y in which the correlation coefficient for S2-S3-S4-S5 reached 0.581, 0.674, 0.616 and 0.583 respectively, at p-value level significantly less than (0.01), see Table 3. below. Thus, there is a significant effect of using VDBs on all other sections of the questionnaire.

Table 3. Pearson correlation between variable X (S1 Convenience of VDB) &Y (S2-S3-S4-S5)

Section Name & Number	r	Significance
S2 interaction of instructors and students	0.581	0.01*
S3 critical thinking & communication skills	0.674	0.01*
S4 english language learning	0.616	0.01*
S5 topics of discussion & grades	0.583	0.01*

** Significant level 0.01.

These results are emphasized in the discussion mentioned above. The highest correlation was for using the tool to developing learners' critical thinking and English language followed by topics of discussions and interaction among students and instructors. Therefore, Egyptian undergraduates perceived VDBs as convenient tools for promoting their English Language Learning. In [22], it was discovered that there was a correlation and multiple regression between students' number of posts contributed to the forum and prior academic performance, "Each new posting contributed three times as much to the final unit mark as its nominal assessment value" p. 847. This suggested that preparation for new postings aids students in completing some measurable assignments that are related to the unit. Passiveness and low-quality posting in the forum did not affect students' academic performance significantly.

5 Conclusion and Recommendations

To conclude, e-learning management systems have provided opportunities for blending innovative technologies and traditional teaching methods which are indispensable for online learning in this digitalized era. VDBs are examples of such technologies that can highly motivate students' learning and eventually lead to improvement in their academic performance. This study has proved that Egyptian undergraduates highly perceive VDBs as effective tools for developing their English language skills and critical thinking abilities. Topics that are related to course content and clear assessment criteria of participation to VDBs are highly perceived as important aspects that affect students' learning positively. Findings showed that asynchronous VDBs allowed learners opportunities to interact and reflect on each other's contributions at their own pace and time although the general trend was neutral. Since academic staff and peer presence are important aspects of successful online discussions, it is recommended to incorporate in the instructional design directed strategies on how to follow up students' presence. Therefore, threads of discussions should be appropriately incorporated in course curriculum design. Academic staff should employ different strategies and activities' ideas to engage learners and increase level of peer interaction (e.g., assign students role plays, choose student leaders for discussions, divide students into smaller groups), and set specific criteria for contribution evaluation. In short, VDBs are platforms for flexible communication that are essential to promote learning from instructors' feedback, peer review and one's own mistakes besides being a tool to save course content and accomplish collaborative tasks; that is why, they should be implemented effectively in course design to maximize their ultimate benefits on students' academic performance.

References

1. Kelly J, McNair V (2021) Re-imagining blended learning. an experience-led approach to accelerate student future skills development. In: ICL2021 – 24th International Conference on Interactive Collaborative Learning 22–24 September 2021, TU Dresden and HTW Dresden, Germany, pp 418–424
2. Tomlinson B, Whittaker C (2013) Blended Learning in English Language Teaching: Course Design and Implementation. London: British Council
3. Alzahrani M (2016) Dimensions affecting student participation in online discussion forums: a review of literature and a current investigation. Int J Instruct Technol Dis Learn 13(10):29–46
4. Lee A (2017) Korean EFL students' perceptions of the effectiveness of peer interaction in online forum discussion. SNU J Educ. Res. 26(2):1–23
5. Akmal A (2017) Impact of Web Based Learning on EFL: Using on-line discussion forum (ODF) to enhance students' writing skill. Univ J Educ Res 5(8):1345–1348
6. Krasnova T, Ananjev A (2015) Students' perception of learning in the online discussion environment. Mediterr J Soc Sci 6(6):202–207
7. Biriyai A, Thomas EH (2014) Online discussion forum: a tool for effective student teacher interaction. Int J Appli Sci Res Rev 1(3):111–116
8. Zekry D, McKee G, (2021) A socio-educational app for digitally transforming online learning. In: ICL2021 – 24th International Conference on Interactive Collaborative Learning 22–24 September 2021, TU Dresden and HTW Dresden, Germany, 2017–2026
9. Akayoğlu S (2019) Theoretical frameworks used in CALL studies: a systematic review. Teach English Technol 19(4):104–118
10. Chapelle, (2009) The relationship between second language acquisition theory and computer assisted language learning. Mod Lang J 93(741–753):2009
11. Chuttur Y (2009) Overview of the technology acceptance model: origins, developments and future directions. Work Papers Inf Syst **9**(37):9–37

12. Sholihah L (2017) Using discussion board as the medium to improve the skill in writing a recount text. J English Lang Teach 1(1):38–48
13. Tibi M (2018) Computer science students' attitudes towards the use of structured and unstructured discussion forums in online courses. Online Learn J 22(1):93–106
14. Bliuc A, Goodyer P, Ellis R (2010) Learning through face-to-face and online discussions: Associations between students' conceptions, approaches and academic performance in political science. Br J Edu Technol 41(3):512–524
15. Zhang T, Gao T, Ring G, Zhang W (2007) Using online discussion forums to assist a traditional English class. Int J E-Learn 6(4):623–643
16. Ghorbanpour A (2013) The effect of online forum participation on the EFL learners' writing performance and their attitudes towards learning English. BARNOLIPI J 2(5):207–237
17. Tiene D (2000) Online discussions. a survey of their advantages and disadvantages vis-a-vis Face-to-Face discussions. J Educ Multimedia Hypermed 9(4) 371–384
18. Matuga JM (2009) Self-regulation, goal orientation, and academic achievement of secondary students in online university courses. Educ Technol Soc 12(3):4–11
19. Tiwary A (July 13, 2020) Learn Why online discussion forums are beneficial. elearning industry. 6 Tips to Keep your online forum academy alive. https://elearningindustry.com/why-online-training-discussion-forums-beneficial
20. Mbati L (2013) Discussion Forums and Weblogs: A Means to Fostering Learning. Mediterr J Soc Sci 4(13):513–518
21. Fatimah F, Rosimela Y, Syarif H (2018) Students' critical thinking and writing ability reflected on online discussion forum. In: 7th international Proceedings on Languages and Arts, vol 301, pp 680–688. Atlantic Press
22. Palmer S et al (2007) Does the discussion help? The impact of a formal assessed discussion on final students' results. The British Journal of Educational Technology 39(5):847–858

Cross Reality and Cyber Physical Systems

Integrating Virtual and Augmented Reality Applications into a Connected Laboratory Environment

Nils Höhner(✉), Mark Oliver Mints, and Julien Rodewald

University of Koblenz-Landau, Mainz, Germany
nhoehner@uni-koblenz.de

Abstract. Modern Learning Management Systems offer a variety of possibilities for learners to interact with learning material, to schedule their courses as well or to track and plan their individual learning path. Given this advantages and the related success of such systems it is understandable that many external services aim for a seamless integration in the ecosystem of Learning Managemet Systems and vice versa. Augmented Reality and Virtual Reality applications, however, are hard to integrate as they are often times very diverse in both technical implementation and content.

We present a technical solution to tackle this problem by proposing a software architecture for a Mixed Reality Client application that is able to connect to the Moodle LMS API and works as a root service for different Augmented and Virtual reality scenarios.

1 Introduction

Today, there are many established ways to create, maintain and operate individual online labs. Increasingly, the networking of diverse labs and a coherent presence, for example in a shared web platform or a learning management system has become more important, which is mostly ensured by standardized interfaces.

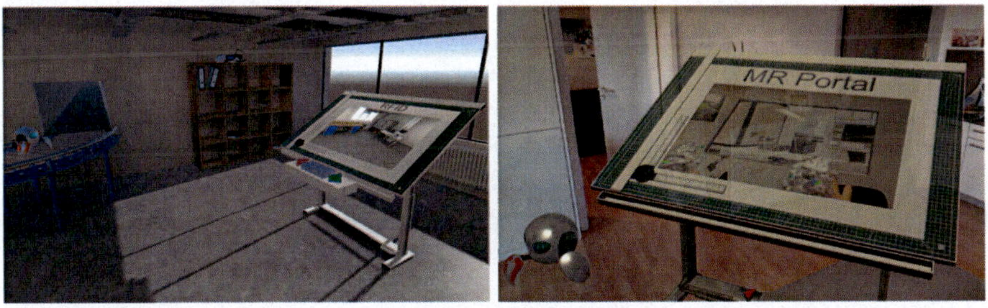

Fig. 1. Starting Rooms. The VR (left) and AR (right) starting rooms respectively.

M. E. Auer et al. (Eds.): REV 2022, LNNS 524, pp. 585–594, 2023.
https://doi.org/10.1007/978-3-031-17091-1_57

The comparatively novel Virtual (VR) and Augmented Reality (AR) applications have a similar profile to real existing labs: on the one hand users need to physically visit them, while on the other hand they are able to experience thematically different hands-on activities. For this reason, most existing AR & VR applications are only superficially integrated into learning platforms, oftentimes giving the user an impression of a "*nice addition*" without importance.

In earlier work pedagogical experts stated that a strong connection of VR-Applications with the course context both technically, as well as content wise should improve the motivation of students [7].

Therefore, our goal was to create a software architecture that allows to unify different AR & VR applications under a single launch application on the one hand and to create a closer connection to an existing online learning platform on the other. Ideally, teachers should be able to add VR or AR applications as modules to their courses on the learning platform. Resulting in the automatic distribution to the participants of the courses by the client software, based on the hardware used. After users have successfully completed the application, their learning platform should automatically display this in their account area. Ultimately this should lead to a more coherent feeling of the learning platform.

In addition to connection with a Learning Management System we decided to support OpenBadges as well as Learning Analytics since both topics became more important in recent years [4,8].

2 Related Work

There is many research on Learning Management Systems and their applications in general but very little focus on establishing connections to MR applications.

Mershad et al. [9], however, propose a framework for future learning management systems to enhance the support of IoT technology. In this context they developed the concept of Dynamic Virtual Reality which describes the combination of IoT sensors with VR content. They use their LMS to connect sensors and the NeuroVR [10] platform demonstrating the feasibility of such a system. Guimaraes et al. [6] focus more on the integration of web based AR content into a Learning Management System by utilizing the SCORM (Sharable Content Object Reference Model) [2] model. They first found a way of converting a web based AR app to the SCORM model which is supported by most Learning Management Systems.

Gradinaru et al. [5] proposed a architecture for connecting Moodle with MR applications with more focus on content creators. However, the system mostly consists of mockups and only parts of the architecture are implemented as prototype and tested.

There are also a few commercial systems like the Labster environment [1] which provides many virtual online labs and integrates well with several known Learning Management Systems. They, however, dropped the support for Head Mounted Displays since Google discontinued their Daydream platform. VRReddo [3] on the other hand focuses on Virtual Reality classes and provides its own content management system. An integration into existing Learning Management System is, however, not supported.

3 System Overview

Our architecture (as seen in Fig. 2) consists of a Mixed Reality (MR) Client software, a central Learning Management System (LMS) Server, as well as two external services (Open Badge Factory and a Learning Record Store). Additionally a plugin for content developers, as well as for the Moodle Learning Management System was developed (puzzle icon).

The heart of our system consists of the Moodle Server which contains the course content, user information as well as a data repository for all available Mixed Reality (MR) Applications. A user of our client software starts by entering his Moodle credentials independently of the hardware he uses. Afterwards the software communicates with the LMS to synchronize learner specific data as well as all available MR applications depending on the attended courses and the currently used hardware. The learner gets to choose the desired application which in turn is able to utilize our interfaces to Open Badge Factory (OBF) to automatically issue badges or the Learning Record Store (LRS) to store learning analytics (LA) data.

From the content developer perspective, we provide specific hardware related Unity prefabs and components to make existing applications compatible with our environment. Additionally a Unity plugin was developed which supports the user in creating AssetBundles of his system, as well as in the upload procedure.

4 Implementation Details

The software architecture was developed in a design-based research approach that ran alongside the development of specific virtual laboratory environments [7]. In this way, it was possible to ensure that realistic requirements were considered.

The software is based on both the Unity game engine and a plugin written for the Moodle Learning Management System. Consequently, we support AR & VR applications developed on this basis. As a starting point, we created an industrial-themed virtual launch room equipped with a control panel (AR only places the control panel in the world). Here the user has to log in with his Moodle credentials to receive information about his courses, as well as the available AR & VR applications (depending on the hardware used). On a technical level, the application communicates via a REST interface with the Moodle API and the web service of our plugin. After the application is aware of all available MR applications, they are downloaded from the web server and integrated into the client program. The user is now able to launch the application. From the software developer's point of view, we work internally with the Unity AssetBundle system. This means that the AR & VR software to be supported must first be integrated as subprojects in our launch application. Afterwards, a specifically developed Unity plugin can be used to create and export asset bundles, as well as to upload them to the server if required.

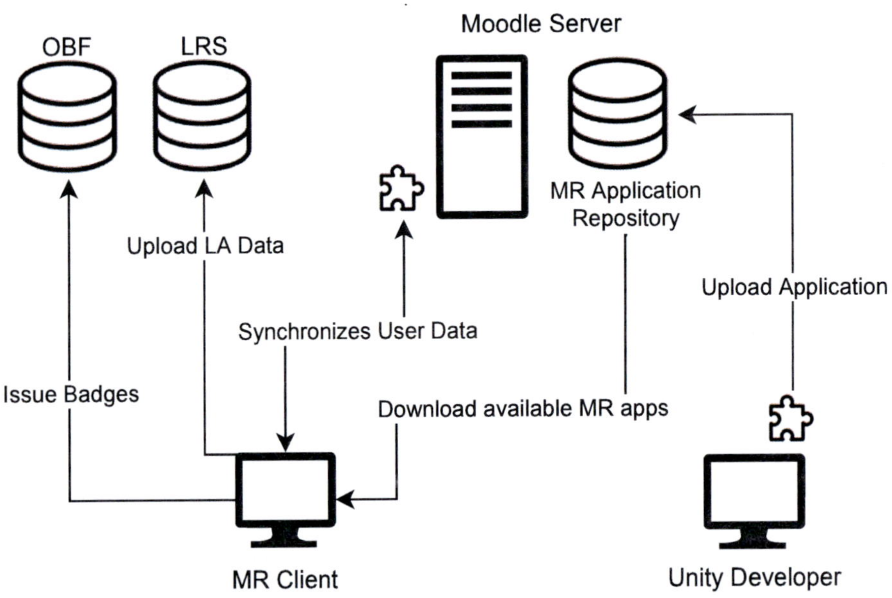

Fig. 2. System Overview. Visualization of the architecture from the point of view of the MR user and the Unity developer. The puzzle icon symbolizes our developed plugins.

4.1 Mixed Reality Client

Figure 3 shows the structure of the created MR Client. The Core of the application consists of three different scenes, namely the *Initialization Scene*, the *VR Scene* and the *AR Scene*. Additionally there are four main components which are available in every scene (*MoodleConnector*, *OBFConnector*, *LRSConnector* and the *Avatar System*). The actual MR applications are connected as *AssetBundles* to the corresponding (X)R Scene.

The **Initialization Scene** contains both the SteamVR plugin[1] for VR support and the MRTK plugin[2] for AR support as the runtime initialization of them

[1] https://valvesoftware.github.io/steamvr_unity_plugin. Accessed September 22, 2022.
[2] https://docs.microsoft.com/en-us/windows/mixed-reality/mrtk-unity/?view=mrtkunity-2021-05. Accessed September 22, 2022.

proves difficult for various reasons. Instead the first component to load inside the scene detects the connected hardware and removes the unnecessary game objects from the hierarchy before they are initialized. Afterwards the Connector Components as well as the Avatar system is created, followed by a hardware specific dialog for the user to login with his Moodle credentials. If the login is successful, the scene changes to the hardware-specific start environment.

The **VR and AR Scenes** as depicted in Fig. 1 act as starting environments for the learners. Both rooms consist of a control panel that can be operated with the currently used hardware as well as the Avatar that can be used throughout every application. The users are able to see their account information (username, image, enrolled courses etc.) on the control display. Additionally all available MR-Applications are shown with a small thumbnail preview. When the user wants to enter an MR application the AssetBundle is downloaded and cached locally. Depending on the scene a modified prefab of the SteamVR Player or the MRTK Playspace is used that contains a scene manager allowing the user to return to the start room at any time.

The **Avatar System** provides the content creator an optional animated robot avatar (see Fig. 1) with a event driven waypoint system. The content creator is able to add waypoints in his scene which the avatar automatically traverses based on predefined events. Furthermore the Avatar provides helper methods for easy integration of audio files for spoken dialog. Internally the system relies on the Unity NavMesh component

The **MoodleConnector** component deals with the communication between the MR Client and the Moodle LMS server. After successfully logging into the Moodle system the connector receives a user token from the server that is used for the communication with the Moodle webservice API. The component sends REST requests to the LMS to query information about the user, the enrolled courses and all available Moodle activities. The JSON based answers are serialized in a MoodleUser component that is available throughout all scenes, including the AssetBundles. Additionally the *MoodleConnector* offers interfaces to query or change activity states (e.g. mark as completed).

The **OBF- and LRSConnector** components use the same RESTful approach as the MoodleConnector to allow communication with external services. The OpenBadgeFactory[3] is a online platform that allows the management and issuing of open badges. Using the connector the MR applications are able to automatically issue badges based on defined criteria (e.g. successfully completing the MR application). The LRSConnector allows the connection to a Learning Locker[4] based Learning Record Store which is a data repository specialized in storing learning activity in form of xAPI statements. Both Connectors work internally with the UnityEvent system to permit easy integration.

[3] https://openbadgefactory.com/. Accessed September 22, 2022.
[4] https://docs.learninglocker.net/. Accessed September 22, 2022.

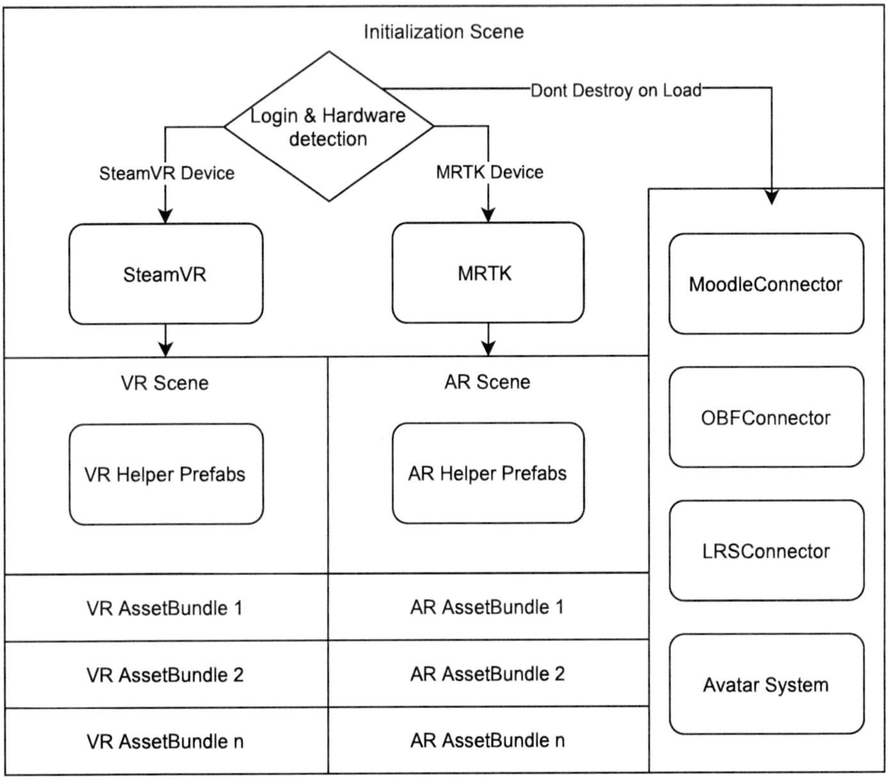

Fig. 3. Software Architecture. The architecture of the mixed reality learning management system.

4.2 Moodle Plugin

The Moodle plugin is a PHP plugin for the Moodle Learning management system which enables the connection between the MR Client and Moodle. From a technical standpoint the plugin is categorized as an *Activity Module* and offers special views for learners, teachers and the server admin. Figure 4 illustrates the Workflow between a teacher and a user of our system.

The Teacher is able to add the MR activity to his course identically to standard Moodle activities. Inside the settings dialog the plugin lists all available applications with a thumbnail and a short description for the teacher to choose from. After choosing the suitable MR application the activity integrates seamlessly into the course.

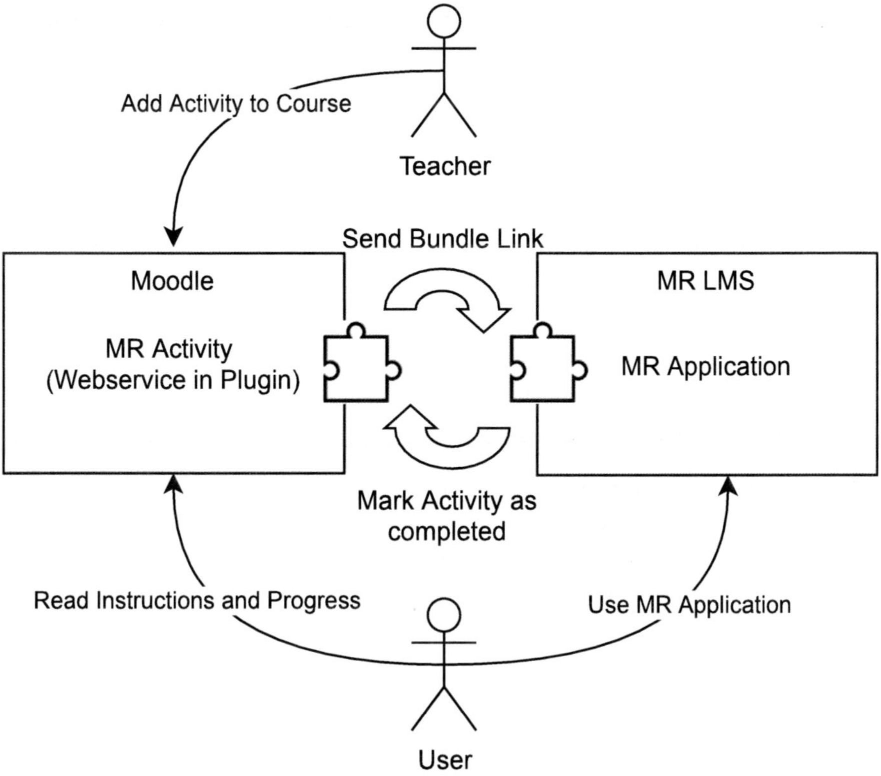

Fig. 4. Workflow. The workflow for users and teachers.

The Learner is now able to see the added activity in his course and is able to click it in order to read the description. After logging into the MR LMS the MoodleConnector (see Sect. 3.1) receives the Bundle Link and allows the user to enter the application. On the Plugin side this connection is realized by adding a Webservice function to the Moodle plugin enabling RESTful access from external sources. After finishing the application the activity is automatically marked as completed.

4.3 Content Creation

In Fig. 5 the workflow of the content creation has been extended from the perspective of the content creator and the admin.

The Admin of the Moodle server has access to the administration interface and is able to add new MR applications. To do this, he enters the link to the AssetBundle created by the Content Creator, adds a description and the Base64 variant of a thumbnail. Afterwards the Moodle plugin enters this data into its corresponding database making it available for teachers to use.

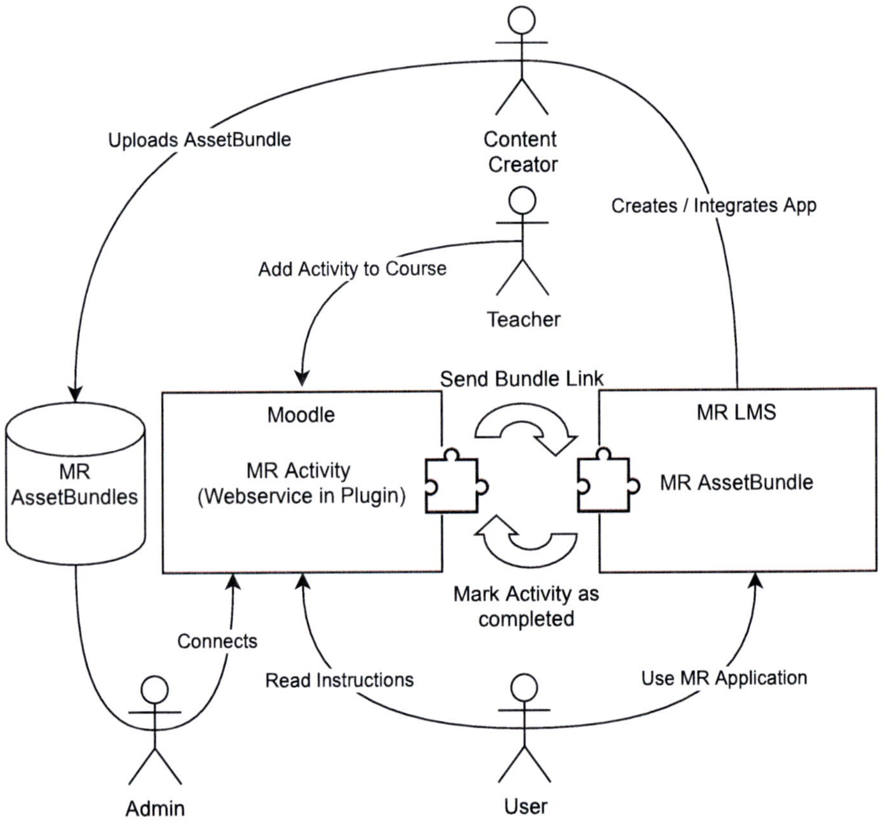

Fig. 5. Workflow. The workflow of the content creation process.

The Content Creator is responsible of creating the actual MR application and integrating it into the MR LMS. To support this process we created a Unity plugin that allows the automatical creation of AssetBundles as well as the upload to a specified ftp server.

Internally our system uses the Unity AssetBundle system[5]. An AssetBundle is an archive file that can contain non-code assets of a project that typically require large storage space (textures, models, materials etc.). During runtime, a Unity application can load AssetBundles and use the data inside. Since AssetBundles do not support the shipping of scripts, all code of the application to be integrated must be completely contained in the MR client.

[5] https://docs.unity3d.com/Manual/AssetBundlesIntro.html. Accessed September 22, 2022.

This demands that the external application has to be integrated into the project structure of the client. Figure 6 shows an example of the file structure that is recommended to enable a consistent layout of the resources. From a purely technical point of view, however, the complete project could also be copied into a single subfolder.

Therefore, all external scenes must be assigned to a new AssetBundle in the Unity Editor via an asset label. Afterwards, our Unity Editor plugin can be used to pack all scenes available in the client into AssetBundles. These scenes do not have to be present in the actual build of the client application.

In order to upload the bundle to the server the content creator can utilize our plugin which supports ftp upload to a specific server. To do this he or she has to specify the ip adress and credentials in the given dialog. Alternatively the user should be able to manually upload the bundle to a server.

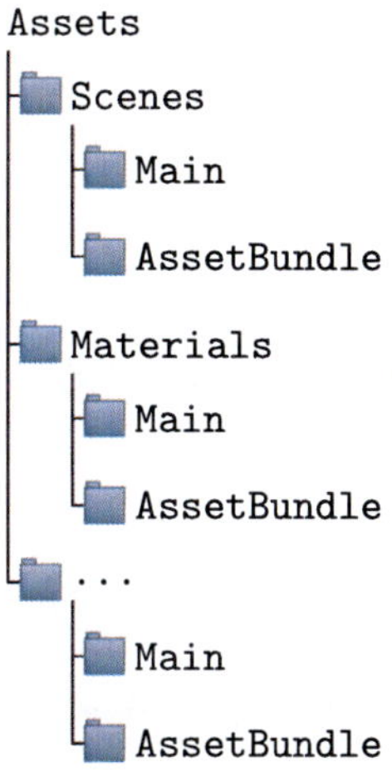

Fig. 6. Directory Structure. Example directory structure of the MR LMS application.

5 Lessons Learned

We were able to show that it is possible to successfully created a robust Mixed Reality application that supports different platforms and is able to unite thematically and technically different VR & AR applications in a single place. Specifically, we integrated one Virtual Reality laboratory [7] and one collaborative VR & AR application which utilizes different devices and several test scenes. Furthermore, we think it is plausible that the presented system gives users a more coherent feeling, increasing the motivational potential on the one hand, while enforcing the relevance of VR & AR applications in the context of the learning platform on the other. Both should theoretically lead to better learning outcomes for students. However, we suspect that it will take a long time for such effects to manifest themselves in an evaluation process as courses have to be followed over several periods.

6 Conclusion and Future Work

We strongly believe that new technology benefits immensely from being integrated in established connected environments early on. This refers not only to the content alone, but also includes good technical integration, so that users have a consistent overall experience. Therefore, the presented software architecture offers a standardized entry point for Unity Engine based AR & VR applications. We have put great emphasis on developing a generic system that can be used as a basis for similar environments.

Acknowledgement. Funded by the German Federal Ministry of Education and Research (BMBF), grants no. 16DHB2115 "DigiLab4U".

References

1. Labster. https://www.labster.com/. Accessed 13 Sept 2022
2. Shareable Content Object Reference Model. https://www.scorm.com/. Accessed 13 Sept 2022
3. Vreddo, Virtual Reality Education Solutions. https://vreddo.com.au/. Accessed 13 Sept 2022
4. Burghardt M, Ferdinand P, Pfeiffer A, Reverberi D, Romagnoli G (2021) Integration of new technologies and alternative methods in laboratory-based scenarios. In: Auer ME, May D (eds) REV 2020, vol 1231. AISC. Springer, Cham, pp 488–507. https://doi.org/10.1007/978-3-030-52575-0_40
5. Gradinaru A, Moldoveanu A, Moldoveanu F (2018) Designing a virtual reality learning management system. 2:11–16
6. Guimaraes M, Alves B, Martins V, Baglie L, Brega JR, Dias DR (2017) Embedding augmented reality applications into learning management systems, pp 585–594
7. Nils H (2020) Integrating virtual reality in a lab based learning environment. In: Patrick B, Victoria I, Regis K, Anne-Hélène O, Hideo S, Gabriel Z (eds) EuroVR 2020, vol 12499. LNCS. Springer, Cham, pp 99–114. https://doi.org/10.1007/978-3-030-62655-6_6
8. Hui Y, Kwok L (2019) A review on learning analytics. Int J Innov Learn 25:197
9. Mershad K, Wakim P (2018) A learning management system enhanced with internet of things applications. J Educ Learn 7:23
10. Riva G et al (2007) Neurovr: an open source virtual reality platform for clinical psychology and behavioral neurosciences. Stud Health Technol Inf 125:394–399

Autonomous Drone Path Scheduling and Management Strategy with Multi-agent Decision Support and Coordination

Laxshana Ganeshalingam and Ananda Maiti(✉)

University of Tasmania, Newnham, Tasmania 7248, Australia
anandamaiti@live.com

Abstract. In a harsh and isolated environment with limited communication and human interaction, e.g., Antarctica, multiple unmanned vehicles or multiple drone systems can use fuel/battery faster than normal conditions. This has led to many drone failures. Preservation of fuel level becomes an essential factor for the management of drones while performing the required set of tasks within an acceptable time. The uncertainty of the environment further challenges the autonomous task allocation and path planning for drones. The existing literature algorithm was not tested or used in these severe conditions, which creates the major gaps in the autonomous drone's management. This paper proposes a CBS-LSA algorithm, which is a combined algorithm of conflict-based search and linear sum assignment. This algorithm considers the drones' fuel consumption before allocating the tasks uniformly and proportionately to all the available agents whenever the new sets of tasks were generated randomly. It can proportionately allocate tasks to drones, even if the number of drones is lesser than the number of tasks while maintaining the optimal task allocation process. Moreover, the algorithm optimizes the fuel consumption of the drones by allocating them to the nearest available agents and providing the best possible path for the agents that avoid static obstacles such as mountains and terrains.

Keywords: Multi-unmanned vehicles · Dynamic task allocation · Path planning · Antarctic condition · Environmental monitoring

1 Introduction

In multi-agent task allocation and path planning, the tasks are allocated to the agents before the path are planned for the assigned destination. When it comes to environmental monitoring through autonomous drones, the distance should be considered before task allocation as it would reduce the time taken to perform a task and fuel consumption level. But most of the existing algorithms the tasks are assigned without consideration of the distance, i.e., the algorithm disregards the nearest tasks assignment [1, 3, 4]. These algorithms performance significantly reduces in the harsh colder regions, as the drones' battery consumption is higher [13]. This produces the potential risk of a higher drones'

M. E. Auer et al. (Eds.): REV 2022, LNNS 524, pp. 595–606, 2023.
https://doi.org/10.1007/978-3-031-17091-1_58

failure rate and reduces the significance of using the drones' technology in remote regions such as Antarctica.

A new combined algorithm is proposed that would allocate tasks uniformly to all the available drones based on the result from the path planning algorithm. Using this algorithm, the drones would be able to perform the nearest available tasks, which is calculated using the linear sum assignment to find the lowest cost for all the drones [11] [12]. The matrix for linear sum assignment is produced using the conflict-based search algorithm. The Jonker–Volgenant solver used in the linear sum assignment resolves the issues of needing an equal number of agents and tasks to assign the tasks to the drones [9, 11, 12]. Further, the algorithm could allocate the set of newly created tasks to the drones based on their location. This paper investigates and reviews the existing algorithm literature in Sect. 2. Section 3 discusses the proposed algorithm along with the simulation setup. Section 4 provides the result of the proposed algorithm, followed by a conclusion and future works.

2 Related Work

Multiagent pathfinding (MAPF), and task allocation are well-studied research areas in AI. This paper will investigate the relationship between pathfinding and task allocation to distribute tasks effectively and robustly among the drones and to reduce the failure of the drones in harsh cold conditions. Multiple works of literature were reviewed to obtain the problem statements and to determine if any of the existing pathfinding and task allocation algorithms could be implemented for drones in the harsh cold region. Firstly, this section will review the algorithm that was used in the Antarctic region and other colder regions. Secondly, it will review the literature on multiagent pathfinding task allocation. Lastly provide the problem statement, which is addressed in the research paper.

2.1 Drones Project in Antarctic Region

In the last decade, there was an increase in the use of unmanned vehicle projects in the Antarctic region. Some of these projects relied on static task allocation and path planning, i.e., the tasks and the path of the agents were configured manually into each drone before they were flown. In most cases, the semi-autonomous drones were used to perform tasks and they were controlled by humans from the ground station. In all these projects, the common problem was that the usage of the battery level is higher than in the polar region, which increased the failure rate of the drones [13]. So, the loss of drones in the Antarctic projects was higher than in the non-polar regions. In the recent research study on MAPF in Antarctic conditions, Path Optimization for Population Counting with Overhead Robotic Networks (POPCORN) was introduced to a path plan for drones to conduct an aerial survey to count the animals [13]. In this research study, the authors used static start points and endpoints to experiment.

2.2 Path Finding and Task Allocation Algorithms

A* is one of the most popular paths finding algorithms used for the MAPF. It is a heuristics-based algorithm that provides an optimal path for the multi-agents using the nodes-based search. The algorithm avoids the collision by extending the low-level nodes and it has the limit to extending the nodes [5]. This solution provides the optimal shortest path for the drones. But the limit in extending the nodes acts as the drawback for the algorithm because it allows the multiple drones to hold one or more paths when it reaches the maximum number of nodes extension [14]. This drawback further complicates by exhausting the memory and freezing while calculating the optimal path [14][14]. Similarly, the Dijkstra algorithm, another heuristic algorithm, extends the nodes until it reaches the target position from the start position. This algorithm computational time is higher as it moves back and forth between the nodes until it finds the best pathfinding solution. In the complex network, it does not calculate the optimal solution. Further Qadir et al. (2021) also provided evidence that the A* performance is better than the Dijkstra algorithm.

Bio-inspired algorithms are used in both pathfinding and task allocation. Ant Colony Optimization (ACO) algorithm is one of the well-known bioinspired MAPF algorithms, which mimics the pathfinding technique of the ants. Ants find the short path through continuous iteration to find the shortest and the best path. But this algorithm comes with multiple limitations such as the distribution change of the agents would affect the accuracy of the path and it does not provide better convergence for the agents [8]. Wu et al. 2018 modified the ant colony algorithm to perform task allocation in a dynamic environment for military purposes but this characteristic cannot solve the dynamic nature of the military tasks. So, the authors implemented reinforcement learning with the ant colony to address the dynamic tasks of the military. But the major drawback of reinforcement learning is that its robustness and flexibility decrease as the number of agents increases in the environment.

Another bio-inspired algorithm used for MAPF is the locust algorithm, Kurdi et al. (2018) proposed this algorithm that mimics the behaviour for locus. The locust stays away from each other when it finds other locusts in a grass patch, so this algorithm uses a similar technique to look for rescuers. So based on this method, the system is built known as Locust inspired approach for Multi-UAVs Task allocation. This algorithm uses an auction-based method to allocate tasks among the multi-agents. The drones assign themselves task when it discovers unassigned areas, the drones make an individual decision based on the information provided in the centralized system.

CBBA is an auction-based and consensus-based task allocation algorithm that provides a solution for conflicts [16]. It provides an accurate approximate solution for a single robot performing a single task in a group of heterogeneous agents [10]. It has two phases – the scoring function and the biding function. In the scoring function, the agents schedule the task and in the second phase, the agents communicate with each other to bids for the task to resolve the conflicts between each other [15]. However, the main drawback of this algorithm is that CBBA only provides an accurate solution if the scoring functions satisfy the time constraints of performing the single task by an agent [15].

3 Proposed Balanced Pathfinding

This section discusses the functionality of the distributed algorithm that was constructed using the Conflict based search and Linear sum algorithm to uniformly distribute tasks for the agents. The proposed algorithm combines the functionality of the Conflict-Based Search Algorithm (CBS) and Linear sum assignment (LSA) algorithm to allocate and to find the optimal path efficiently and robustly for the drones so that it would handle the harsh col conditions with the minimal failure rate. Under the Linear sum assignment, Jonker–Volgenant solver was considered as it can solve the non-squared matrix convergence, unlike Hungarian solver. This section will investigate the CBS-LSA algorithms separately and further propose a combined CBS-LSA algorithm.

3.1 Conflict-Based Search Algorithm (CBS)

A conflict-Based Search Algorithm is a pathfinding algorithm that uses a two-level algorithm to find the path for agents. The first level algorithm search focus on finding the shortest conflict-free path for agents, about the static obstacles and in the second level the algorithm focuses on resolving the conflicted path with other agents, also known as dynamic obstacles [14]. This algorithm resolves by extending the second level to provide the shortest conflict-free path for the multi-agents [14]. The major advantage of the algorithm is that it could extend its nodes to resolve the deadlocks between two or more agents, unlike the A* star algorithm [14]. This research paper considered this algorithm the predict the path for the agent in the simulation environment. This algorithm was considered over other pathfinding algorithms is because, in a simulation, there are high possibilities that two or more tasks appear near each other, which could lead to deadlocks between the agents especially when two or more agents appeared to be near the tasks' locations. Though A* is known to be a well-known optimal pathfinding algorithm, the CBS could provide a similar optimal result as the A* by avoiding the deadlock conditions.

CBS Algorithm

```
Low level search for the path
Insert path into route
If route not empty
    Check for the conflicts
        If no conflicts
            Return Path
        Else
            Extends the nodes until no conflicts
            Return Path
```

Algorithm 1: CBS Algorithm

3.2 Linear Sum Assignment (LSA)

Linear sum assignment is a linear programming method, which use the matrix to find the lowest cost between the row and column so that the sum of entries in the matrix is the lowest of all the other possible combinations [2]. This algorithm was considered over other task allocation as it allocates one task to one agent, based on the lowest possible cost.

Equation 1: Linear Sum Assignment

$$\min \sum_{i=0}^{m} \sum_{j=0}^{n} Cij$$

The ‘i’ in the algorithm represents the row in the matrix whereas the ‘j’ in the algorithm represents the column. For each row, it calculates each column to obtain the lowest cost. This iterative process is repeated until the lowest sum of the matrix is calculated, so that the optimal task assignment is produced for all the agents.

3.3 CBS-LSA Algorithm

Both the above-mentioned algorithm is combined to provide a distributed algorithm, which considers the fuel level of each agent before uniformly allocating tasks to all the available agents. The algorithm calculates the shortest path for each agent to each task, using the CBS algorithm method to create a matrix, which contains the distance value. The LSA algorithm uses the matrix to calculate the lowest possible distance between the agents and the tasks and allocate the tasks uniformly between them.

CBS-LSA algorithm

```
for each task in Task List
for each agent in Agent List
    If agent has enough fuel
        Calculate each task and each agent path using CBS
        Appends result in R
        Appends path in P
    Appends R in Cost Matrix
LSA to calculate the minimum cost
Assign agents to tasks
```

Algorithm 2: CBS-LSA algorithm

3.4 Simulation Setup

The simulation was built using the python language, 15x15 2-D grid was created, this grid represents the area of an Antarctic Map. The terrains and oceans in this represented Antarctic Maps were considered as the static obstacles in the simulation design. The performance of the simulated drones was measured in seconds. The simulation runs in the loops until it is manually stopped. The tasks for the drones were generated randomly at the interval of 10–15 s in the simulation. The algorithm is triggered when a set of new tasks are created in the simulation environment. In this research, the tasks were also generated on the obstacles as some of the scenarios collected from the literature included observing the animal behaviour near the ocean or measuring the glacier melting rate. The below table (see Table 1) shows the parameter used to build the simulation model.

Table 1. Simulation input parameter

Parameter	Values
Language	Python
Grid size	15 × 15
Data for Antarctic Map	U.S Antarctic Data Centre and Google Maps
Time	Seconds
Number of drones	10
Number of fixed ground stations(FGS)	10
Number of tasks generated	8–12
Task generation rate	10–15
Algorithms used	CBS-LSA

Simulation models are used as the substitute to create an artificial environment with different scenarios. The simulated drones were put to test under different scenarios, which are mentioned below. Then the numerical output which represents the behaviour of the drones were observed and analyzed using the descriptive statistical method.

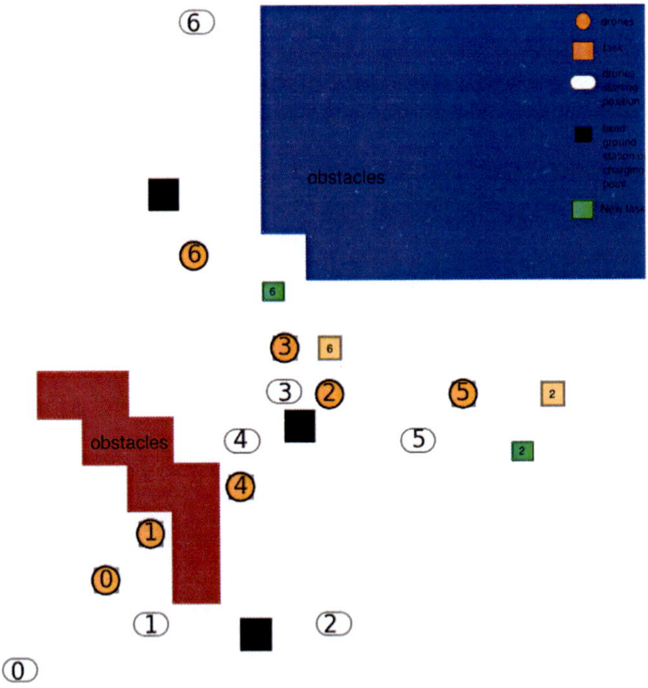

Fig. 1. Simulation environment

The above image (see Fig. 1) shows the computational simulation representing the rough map of geographic terrains and oceans, i.e., no-landing zones (blue) and obstacles or no-fly zones (red). The drones can move to a target location from their initial positions. The figure also shows the new in-coming tasks (green) are assigned to drones dynamically by calculating the drones' current locations.

3.4.1 Scenarios

Algorithm patterns were observed under different scenarios, to calculate the efficiency and the robustness of the algorithm.

Scenario 1: Task generated every 5 s, to calculate the effectiveness of the algorithm to allocate tasks to the agents based on their position.

Scenario 2: Fuel thresholds were set to 20, 15 and 10. When the fuel was set to these values, the agents were available for a longer period. So, this scenario was used to test if it has any impact on the drones' failure rate and the task waiting time.

Scenario 3: Task was created in the one corner of the simulation, to observe the algorithm tasks allocation and path planning.

4 Findings and Results

This distributed algorithm uses the Jonker–Volgenant solvers to allocate the tasks proportionately to all the available agents, using the matrix created by the CBS algorithm. The CBS algorithm considered static obstacles such as mountains and terrains to calculate the path. This algorithm uses Single Task – Single Agent allocate method to uniformly distribute the tasks, so it was assumed that in the open environment that the drones could fly over each other if there is a collision in the path. The CBS algorithm avoids the static obstacles by expanding the nodes, in the low-level search [6, 14]. Jonker – Volgenant solver in the LSA assignment assigns each agent to its closest task using the squared matrix. In this research, it is assumed that the distance (d) is directly proportionate to the fuel (f) [d μ f], under the condition that each step a drone takes to reach the task fuel is reduced by 1 for each step (f-1). So, it is considered that the f = d. The distance between the tasks and the agents were calculated using the CBS algorithm. Jonker – Volgenant uses the Hungarian algorithm and Dijkstra to calculate the optimal task allocation. Hungarian algorithm has known to provide an optimal task assignment but with high computational time whereas Dijkstra provides the optimal task by exploring all the nodes with less computational time [9, 11, 12]. Moreover, the CBS-LSA algorithm successfully planned the path to the tasks when it was created near the obstacles.

Moreover, the CBS-LSA algorithm allocated the nearest tasks for the agents. This conclusion was drawn based on the task waiting time in each agent's queue. From the image below (see Fig. 2), it could be seen that the task waiting time was less than two seconds until the iteration time of 50 s. The increase in the waiting time is due to the low fuel rate of the agents, as the tasks are not allocated to the agents when the fuel rate is below the threshold level. By allocating the nearest tasks to the agent, the fuel consumption of the drones could be reduced.

One of the advantages of the CBS-LSA algorithm is that it does not require a square matrix to assign the tasks to the agents as in most cases the LSA algorithm requires the

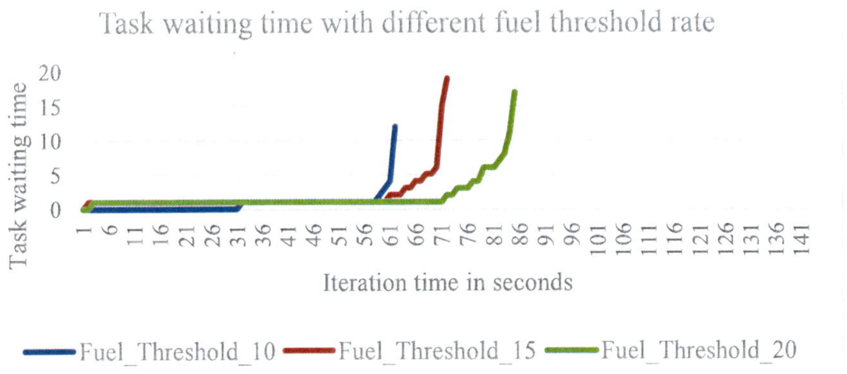

Fig. 2. Shows the task waiting for different threshold levels to estimate the waiting time of the tasks

square matrix to allocate tasks to the agents. But the Jonker–Volgenant solvers could eliminate the requirement of needing the square matrix for the LSA algorithm. Jonker-Volgenant solver as it can convert the non-square matrix into the square matrix by adding the zero value [11, 12]. So, this added dummy row and/or column allows the algorithm to perform the assignment when the number of agents (Na) < number of tasks (Nt). In most cases, the assignment algorithm fails when the Na < Nt or Na > Nt, but the Jonker – Volgenant solver in the Linear sum assignment uses the dual variable to provide the optimal assignment for the rectangular matrix. Crouse (2016) provided the dual variable modification for the Jonker – Volgenant solver to solve the rectangular matrix. This dual variable is also known as dual optimization, the use of the variable is to make dual and primal cost equal for the rectangular matrix as it is equal for the square matrix. The purpose of the dual and primal variable is to find the optimal task allocation for the rectangular matrix and solve the assignment problem [14]. Further, the author proposes that the Jonker – Volgenant solver uses these steps to solve the assignment issues and find the optimal assignment using the rectangle matrix.

Initialize: Converts rectangle matrix into the square matrix by adding dummy rows/columns. Calculate the lowest.

Cost: Find the lowest cost for each agent.

Update dual variables: Calculate the dual and primal cost, to solve the assignment problem. Augment: In this step, any overlapping assignment is fixed.

Optimal checking: Once, the augment is fixed then the assignment is checked for the optimal result if the optimal result is not found.

4.1 Comparison with the Multi-agent – Multi-task Assignment

CBBA algorithm uses auction method to bid on a few tasks per agent, this method is known as multitasking – multiagent assignment. The algorithm aims to provide a conflict-free solution for the multiagent. The algorithm uses two-step methods to provide this conflict-free solution, which is the bundle construction method, and the conflict resolution method. In the bundle construction method, each agent will add the tasks and the path to the tasks into its bundle without consensus, if the other agents are allocating tasks into their bundle [7]. In the second step, each agent will bid for the currently assigned bundle, the agent with the higher bidding wins the bundle. Through this method CBBA removes any conflicts found within the agents, this conflict resolution method predicts the trajectories, so the path provided by the CBBA algorithm shows that it does not collide with others at a given time.

The drawback of using the CBBA algorithm was that it did not consider the distance each agent needs to travel to perform the tasks. The tasks were bided by the agent without the consideration of the cost of the travel or the cost [15]. From the image (see Fig. 3), where the CBBA algorithm was analyzed using a 3-D graph, it could be seen that the Agent 3 (A3) is closer to task 0 (T0) and task 3 (T3) but whereas the CBBA algorithm allocated the tasks to agent 0 (A3). This finding was confirmed by another research paper, where it was proved that the CBBA algorithm task cost increases as the tasks and time increase.

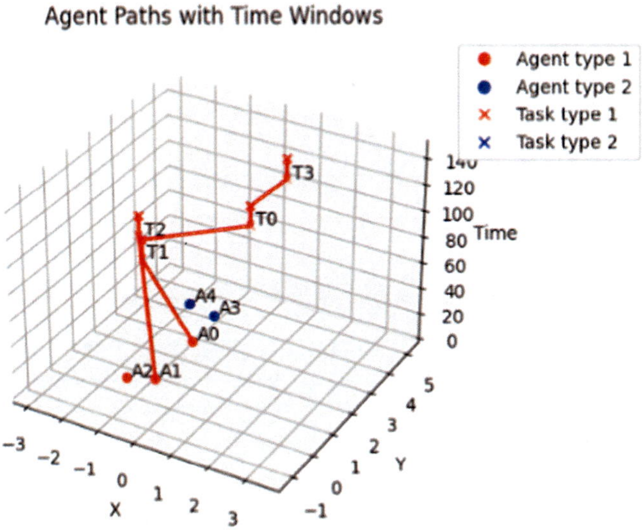

Fig. 3. Visualization of the CBBA task assignment for the drones

Further, the cost of the task allocation of CBBA was compared with the CBS-LSA algorithm. When the study was further conducted by using the CBBA algorithm for the prediction of the trajectories, the average cost assignment received was significantly higher when compared to the CBS-LSA algorithm. The result of the CBBA algorithm task allocation for 20 iterations was 212.9 whereas the cost of the task assignment of the CBS-LSA algorithm was 31.1. When the findings were further analyzed it was evident that the CBS-LSA algorithm allocated the tasks proportionately with all the available agents whereas the CBBA algorithm assigned the set of tasks to an agent through an auction method, which lead to a significant increase in the cost of the task allocation.

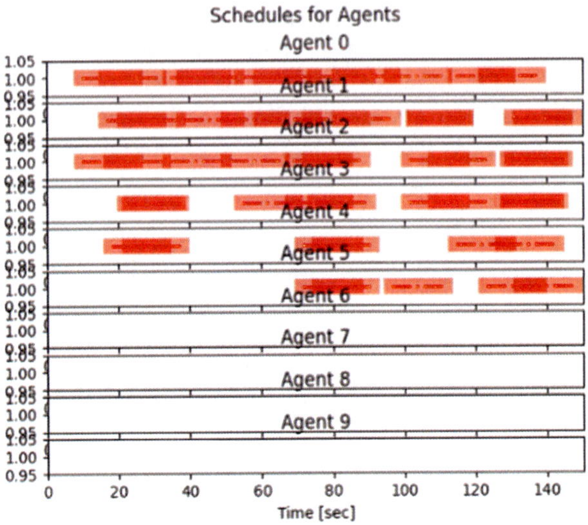

Fig. 4. Tasks distribution using CBBA algorithm for 10 agents

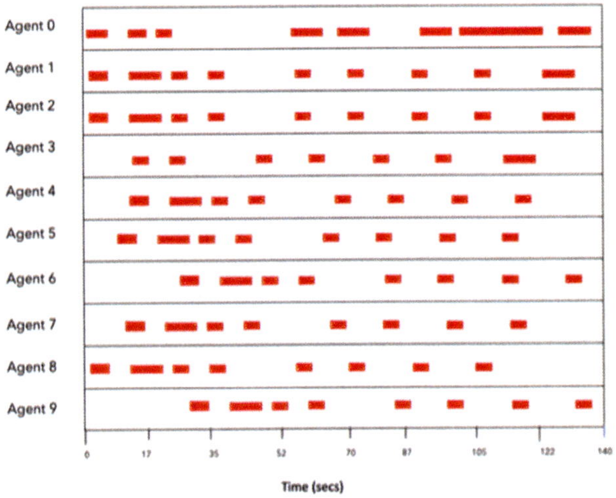

Fig. 5. Tasks distribution using CBS-LSA algorithm for 10 agents

The above image shows the different agents' task performance in each row (see Fig. 4 and Fig. 5). The CBBA task allocation, which uses the auction-based to allocate the tasks (see Fig. 4). The auction-based algorithm allocated the multiple tasks to agent 0 to agent 5 but did not allocate tasks to the other agents without considering the fuel consumption of the agents. In harsh weather conditions, the failure rate of the drones is higher as the fuel/battery level is consumed at a higher rate than in normal conditions. In comparison to CBBA, the CBS-LSA algorithm considers the fuel rate and distributes the task uniformly to all the available agents (see Fig. 5).

5 Conclusions

In the cold and harsh conditions, where the drones are known to utilize twice the amount of fuel, the fuel consumption level must be considered before allocating the tasks to the agents to reduce the failure rate of the drones. The CBS-LSA algorithm considers the fuel of each agent before it allocates the tasks proportionately and uniformly to them. When the tasks were dynamically created in the simulation, the CBS-LSA algorithm was successfully able to calculate the agents' current location and its assigned task location to allocate the nearest possible tasks to the agents.

Moreover, when the CBS-LSA algorithm was compared with the CBBA algorithm, the result produced by this algorithm was optimal. The total cost/distance of the agents for each task generation was higher for the CBBA algorithm as this auction algorithm is known to allocate multiple tasks, including the farthest task, to the single-agent when compared to the CBS-LSA algorithm as it can uniformly or proportionately allocate tasks to all the available agents. So, this algorithm is found to allocate the tasks dynamically to all the available autonomous agents considering the Antarctic conditions in the algorithms.

6 Future Work

In harsh weather pattern conditions, the failure of drones is unavoidable. So, in case of a failure, the nearest available autonomous agents should allocate the tasks of the failed drones to themselves. This will increase the reliability of the algorithm. Further, the agent would be able to calculate and estimate their nearest available tasks between themselves and adjust tasks allocation between themselves to improve the optimization of the task allocation.

References

1. Brunet L, Choi H-L, How JP (2008) Consensus-based auction approaches for decentralized task assignment. AIAA guidance, navigation, and control conference (GNC). Honolulu, Hawaii
2. Burkard R, Dell'Amico M, Martello S (2012) Assignment Problems: Revised Reprint, 1st edn. SIAM, Philadelphia, USA
3. Choi HL, Brunet L, How JP (2009) Consensus-based decentralized auctions for robust task allocation. IEEE Trans Rob 25(1):912–926
4. Crouse DF (2016) On implementing 2D rectangular assignment algorithms. IEEE Trans Aerosp Electron Syst 52(4):1679–1696
5. Fu Z, Yu J, Xie G, Chen Y, Mao Y (2018) A heuristic evolutionary algorithm of UAV path planning. Wirel Commun Mob Comput 2018(1):1–11
6. Hönig W, Kiesel S, Tinka A, Durham J, Ayanian N (2018) Conflict-based search with optimal task assignment. In: Dastani M, Sukthankar G, André E, Koenig S (eds) Proceedings of the international joint conference on autonomous agents and multiagent systems. AAMAS, Stockholm, Sweden (2018)
7. Hunt S, Meng Q, Hinde C, Huang T (2014) A consensus-based grouping algorithm for multi-agent cooperative task allocation with complex requirements. Cogn Comput 6(3):338–350

8. Qadir Z, Ullah F, Munawar HS, Al-Turjman F (2021) Addressing disasters in smart cities through UAVs path planning and 5G communications: a systematic review. Comput Commun 168(1):114–135
9. Rachmawati D, Gustin L (2020) Analysis of Dijkstra's algorithm and A* algorithm in shortest path problem. J Phys Conf Ser 1566:012061
10. Rossi F, Bandyopadhyay S, Wolf M, Pavone M (2018) Review of multi-agent algorithms for collective behavior: a structural taxonomy. IFAC-PapersOnLine 51(12):112–117
11. Serratosa F (2015) Computation of graph edit distance: reasoning about optimality and speed-up. Image Vis Comput 40(1):38–48
12. Serratosa F (2015) Speeding up fast bipartite graph matching through a new cost matrix. Int J Pat Recogn Artif Intell 29(2):1550010-1–1550010-17
13. Shah K, Ballard G, Schmidt A, Schwager M (2020) Multidrone aerial surveys of penguin colonies in Antarctica. Sci Robot 5(47):1–11
14. Sharon G, Stern R, Felner A, Sturtevant NR (2015) Conflict-based search for optimal multi-agent pathfinding. Artif Intell 219(40):40–66
15. Turner J, Meng Q, Schaefer G, Soltoggio A (2018) Fast consensus for fully distributed multi-agent task allocation. In Proceedings of the 33rd annual ACM symposium on applied computing, pp 832–839. ACM, France
16. Whitbrook A, Meng Q, Chung PWH (2018) Addressing robustness in time-critical, distributed, task allocation algorithms. Appl Intell 49(1):1–15. https://doi.org/10.1007/s10489-018-1169-3

Investigating the Impact of Variable Effects in Virtual Training on the Behavior of a Physical Autonomous Robot

Julian Rolf, Mario Wolf(✉), and Detlef Gerhard

Digital Engineering Chair, Ruhr-University Bochum, Bochum, Germany
{julian.rolf,mario.wolf,detlef.gerhard}@ruhr-uni-bochum.de

Abstract. Artificial Intelligence is a current megatrend in computer science and almost every aspect of digitization. A scenario for training robots in a virtual environment to fulfil tasks in the real world is created to offer engineering students relevant insights into the field of neural networks. The proposed system generates training data to train a convolutional neural network (CNN) to autonomously drive a mobile robot using computer vision. The physical aspect of the setup includes the mobile robot and two different race circuits to evaluate the driving characteristics. The digital aspect consists of a 3D environment and a digital representation of the physical robot, both of which are developed using the game engine Unity. Inside the 3D environment, an infinite, procedurally generated road is created for the digital robot to drive on. The generated images from the virtual camera of the virtual robots are the basis to train the CNN to maneuver the physical robot in the real-world experiment.

Keywords: Machine learning · Convolutional neural network · Autonomous robotics

1 Introduction

Artificial intelligence, or machine learning, is a term currently found in most publications regarding digitization. This is true for many aspects of innovation such as process optimization, data mining, speed recognition and analysis, computer vision and many more. Especially in the fields of autonomous robotics or autonomous driving, enabling the use of machine learning requires a giant amount of pre-labeled training data for a variety of scenarios to prepare the autonomous system for "every eventuality". Collecting such data is especially difficult in exceptional or emergency situations. Synthetically generating the needed training data can overcome this issue. This paper aims to evaluate the Sim to Real Domain Gap by training a small mobile robot to follow a winding road on an otherwise flat racing course without any junctions. The setup,

M. E. Auer et al. (Eds.): REV 2022, LNNS 524, pp. 607–614, 2023.
https://doi.org/10.1007/978-3-031-17091-1_59

metrics, image and video material of this use case act as a basis to offer undergraduate engineering students a motivational insight into both the complexity and the potential of machine learning in technical applications before starting their own, simpler machine vision project in the authors' lecture "IT in Mechanical Engineering". The next sections cover both the Sim to Real Domain Gap and the foundations for creating virtual training environments.

1.1 Generation of Synthetic Training Data Using 3D Environments

At GTC 2021, Nvidia introduced the latest version of its Nvidia Drive software package. This is a simulation tool to generate synthetic data to support deep neural network (DNN) training with its focus on autonomous driving [1]. In the robotics topic area, the latest version of the Nvidia Isaac software tool was also presented [2]. Both SDKs offer the possibility to digitally simulate real environments. Particular focus is placed on minimizing the Sim to Real Domain Gap, which describes the difference between a simulation and the reality that is to be reproduced, which is a difficult task to solve [5]. However, the software products presented are still quite at the beginning of their product life cycle in early 2022. The generation of synthetic data, as a way to improve the performance of machine learning algorithms, is a known challenge [3,4]. Synthetic data does not suffer from incorrect labeled data and can be generated in a short period of time, after building the 3D environment. Due to its reliable and instantly accessible nature, synthetically generated data can speed up the validation of new neural network architectures, e.g. evaluating performance benefits of machine learning solutions with added or improved sensor technology [1].

1.2 Robotics and Machine Learning in Unity

In contrast to the software solutions from Nvidia, Unity is mainly known as a game engine. It was released in the middle of 2005 and has since been regularly improved and expanded with updates. Due to the rapid development of machine learning and robotics in recent years, Unity has also been expanded with corresponding features. In May 2020, version 1.0 of the ML-Agents package was released, which can be used to train agents with reinforcement learning algorithms in 3D environments within Unity [6,7]. In robotics, Unity announced official support for Robot Operating System 2 in August 2021 [8]. Unity's physics engine has also been improved over time, narrowing the Sim to Real Domain Gap in robotics simulations [9]. In 2017 Enrico Sita et al. presented a framework to represent industrial processes executed by ROS-compatible devices in Unity without interfering with the real process [10].

2 Aims and Requirements

The goal of this paper is to develop a virtual 3D environment using Unity3D to generate synthetic training data. The 3D environment has to include a digital twin of an existing, physical robot, which is equipped with a camera for road detection. The generated data will be used to train a Convolutional Neural Network that will then be used to control the physical counterpart of the robot once trained properly. The performance of the network is to be evaluated using two physical race circuits. HTC Vive trackers will be used to capture the position of the robot for precise analysis of the movements. After the successful implementation of the desired setup, the learning outcomes for observing students are:

- Identification of the physical and virtual components required for a Sim2Real approach.
- Understanding of the relevant factors for the successful generation of synthetic training data.
- Insights into the challenges of transforming a virtual agent into a corresponding physical counterpart.

3 Concept

3.1 Physical Setup

The Robot used is based on the mBot2 from the manufacturer Makeblock. Two encoder motors enable the robot to move around. Due to the arrangement of the motors, rotations on the spot are also possible, so that the robot can maneuver even in tight spaces. The motors have a maximum RPM of 200 and have a torque of 1500 g *cm. The built-in lithium-ion battery allows the robot to operate for a period of three to six hours without an external power supply. The components are controlled and powered by the CyberPi, which is a programmable control unit based on the ESP32 microprocessor that can be programmed with Python. It also offers the possibility of wireless communication, thanks to the integrated Wi-Fi and Bluetooth module. To enable the processing of image files, the robot was extended with a Rasberry Pi camera module v2, connected to a Raspberry Pi Zero. A second lithium-ion battery is used to power the camera and the Pi Zero. To mount the camera and Pi Zero, suitable components were downloaded from the Thingiverse platform, then adapted, 3D-printed and mounted to the robot using additional components from Makeblock. The camera was placed so that it is orthogonal to the ground at a height of 300 mm.

The Test Tracks are based on the center-lines of two circuits from Amazon's AWS DeepRacer competition, but scaled to the robot's dimensions and with added non-uniform backgrounds. The two tracks were each printed on 1.5×2.3 m tarpaulins. The track itself has a dark gray background, which is bordered by two white, continuous lines to the left and right. Along the direction of travel, there is a light gray dashed center-line. The width of the lane is 30 cm. Identification/detection of the roadway is possible due to the contrast of the lines to the roadway. A light green was chosen as the background color of the surrounding area. The two race circuits differ significantly in some aspects. Course one has one long and several short straight track elements, while course two is made up exclusively of differently curved track elements. In addition, several textures have been placed near the track, which differ in shape and color.

(a) Track 1 (b) Track 2

Fig. 1. The two test tracks

The Physical Test Environment. Acts as the validation of the functionality of the wholistic system, after generating the training data and successfully training the CNN in the virtual environment. The CNN is set up on a local server, to which the Pi Zero and the robot are connected via Wifi. The captured images from the physical camera are then sent from the Pi Zero to the server for analysis. The evaluation result is then sent to the robot's control unit. The procedure is repeated ten times a second.

To be able to monitor the test drives, the robot is equipped with a Vive tracker. Additionally, two more trackers will be placed on the course to set up a fixed local coordinate system on the course.

3.2 Software Architecture

The virtual 3D environment mainly consists of two parts. The generated road and the digital twin of the robot. For road generation, the appearance of the road can be considered known. Accordingly, suitable virtual materials must be created to represent the appearance of the road in the virtual 3D environment. However, the physical course's layout is not transferred. The virtual environment must be able to generate a route at runtime, with which suitable synthetic training data can be generated.

For the successful creation of the digital twin, many properties of the physical counterpart often need to be considered and transferred. Due to the scale and complexity of capturing and transferring all physical characteristics, they are not been considered when creating the digital twin. Only the geometry of the robot is taken into account, together with the sensor size and focal length of the camera. To steer the digital twin along the generated roadway, the system determines the difference $\Delta\psi$ between the current forward vector and the "ideal" orientation ψ 15 cm ahead of the robot. In the real environment, a CNN is later used to estimate $\Delta\psi$. This approach has already been studied by Rodrigo F. Berriel et al. [11], who obtained training data through the Google Street View API and OpenStreetMap, and provided promising results.

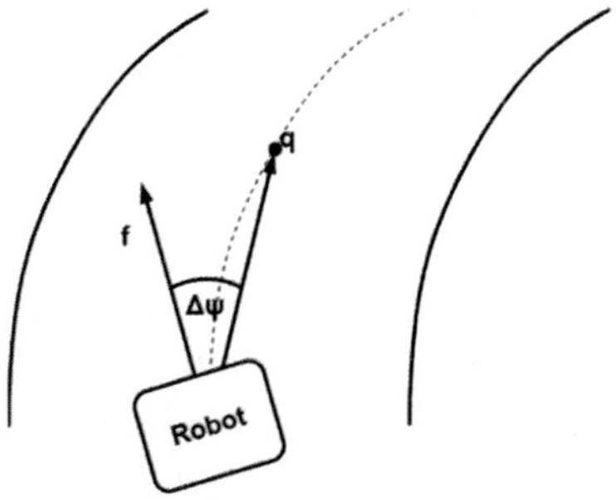

Fig. 2. Illustration of $\Delta\psi$

4 Implementation

4.1 Virtual 3D-Environment

The development of the 3D environment includes the development of a procedure for generating the training route and the creation of a digital twin of the robot described in the previous chapter. The route itself starts at a fixed starting point. Route elements are generated individually and attached one after the other. The number of track elements present in the virtual scene remains constant, meaning that after attaching a new element, the oldest one is removed. In this way, an arbitrarily large data set can be generated at one time, without memory or performance problems. In order to be able to adapt and describe the course of the roadway arbitrarily, cubic Bézer curves are used. The elements are placed in such a way that the omission of crossings and generally gapless transitions can be guaranteed. By the use of the De-Casteljau algorithm [12], points on the curves can be determined, which allows the generation of the mesh of the road. In order to visually represent the track as well, a material is used that also defines the reflective properties of the surface.

To map textures of random shape, color and rotation off the track, a shader based on the implementation of an author named Anas was used. This is to improve the generalization of the CNN in terms of the off-track textures on the physical courses.

4.2 Digital Twin

The digital twin is moved alongside its forward vector $\boldsymbol{f_t}$, multiplied by a constant S to adjust the speed of the twin.

$$\boldsymbol{p_{t+1}} = \boldsymbol{p_t} + \boldsymbol{f_t} * \Delta t * S \tag{1}$$

The orientation ϕ of the robot is determined based on the previously determined point q (Fig. 2).

$$\phi_{t+1} = \phi_t - A(\Delta\psi_t) * \Delta t * R \tag{2}$$

The constant R describes the rotation speed and the function $A(\Delta\psi_t)$ describes the steering behavior of the digital robot. Function $A(\Delta\psi_t)$ can be denoted as follows.

$$A(\Delta\psi_t) = \frac{2}{1 + \mathrm{e}^{\frac{-\Delta\psi_t}{10}}} - 1 \tag{3}$$

The orientation update occurs after a delay of 2.5 s to increase the span of the generated values for $\Delta\psi$.

Placing a light source above the digital twin was necessary in order to adjust the CNN's generalization for the lighting conditions in the test environment, so that the robot succeeds in following the road. The light source moves relative to the robot and serves to simulate the lighting conditions in the real environment. Since lighting conditions can vary in the real environment, the light source was randomized in its intensity and position, within a specified range.

4.3 Training of the CNN

Each sample of the dataset consists of a 128×96 grayscale image and the corresponding value of the angle $\Delta\psi$. In summary, 40.000 samples are generated. 20.000 with the random texture Shader and 20.000 without. Roughly 5% of the samples are filtered, because $\Delta\psi$ has to be greater than $-90°$ or less than $90°$ to be able to make a prediction for the given image. So only 36.980 samples are used for training.

The architecture of the CNN is a copy of the architecture used by Rodrigo F. Berriel er al., except for the dropout layers and a reduced number of parameters. The architecture is based on the AlexNet architecture. Mean Squared Error is used as the error function and Adam is used as the Optimizer, with an initial learning rate of 10^{-3}. 10% of the samples are used for validation. The learning rate is reduced by a factor of ten, after no improvement of the validation error for three epochs in a row. The training is stopped when the validation error could not be improved for ten epochs in a row. The model with the smallest validation error is used for testing.

5 Evaluation

By using the prediction of the CNN to estimate $\Delta\psi$ for the given observation from the PiCam, the physical Robot is able to adjust its heading direction accordingly. To evaluate the quality of the training, the robot runs for two minutes on each track, for five test runs in a row. On both test tracks, in all ten runs, the robot is able to keep the lane without crossing any of the boundary lines. Most of the time, the robot stays close to the center line, which shows that the system is able to detect the lane and adjusts its direction of travel accordingly. The same is true for the second, curvier course. On average, the robot is about two and a half centimeters away from the center-line. However, the data has to be considered under the aspect that the Vive Trackers were placed manually. Thus, deviations can occur when changing the test tracks. Nevertheless, the individual runs of a course can be compared without any problems, as the setup is static besides the robot. In addition, the tracker could not be mounted completely centered on the robot, resulting in a deviation of about 10 mm. The measured distances to the center-line can be seen in Table 1. Additionally, the first test drives on each track can be seen in Fig. 3, where the center-line is shown in black, while the driven line is shown in blue.

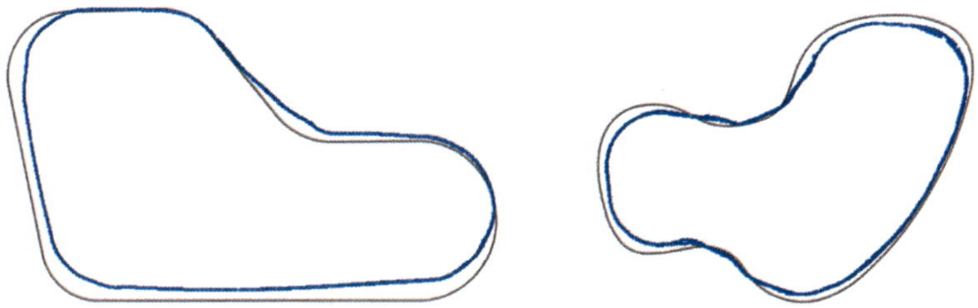

(a) Track 1 (b) Track 2

Fig. 3. First test run on each track. Center-line (black) vs. driven line (blue)

Table 1. Distances to the center-line on Track 1

	Track 1		Track 2	
Test run	Average	std. deviation	Average	std. deviation
1	37,70 mm	0,10 mm	25,33 mm	11,80 mm
2	35,35 mm	0,05 mm	17,31 mm	09,84 mm
3	17,70 mm	0,03 mm	21,46 mm	12,66 mm
4	26,40 mm	0,13 mm	14,98 mm	09,50 mm
5	22,18 mm	0,11 mm	25,87 mm	16,57 mm

6 Conclusion

The authors intent with the approach at hand was to show that virtual environments can be used to train robust CNN for real-world uses. The variable effects, in particular lighting and optical distractors, are often either overlooked or non-existent due to artificially tidy scenarios. The real-world race circuits, which are based on Amazon's Deep Racer competition tracks, added exactly what is missing in the original. The tracks offer a multitude of optical distractors besides the track, while the lighting is varying due to the room setup in the laboratory.

Those non-ideal circumstances challenged the CNN when trained with focus on the track alone. Adding abstract, randomized distractions to the virtual training scenario not only increased performance in the real world, but enabled the full navigation around the course in the first place. Future work will encompass the creation of a lecture about the basics of artificial intelligence, in particular computer vision, CNN and reinforcement learning, with engineering students as the target group. The approach at hand is the basis for several examples of use cases for AI, how to train AI in a virtual environment, how to implement AI in a physical robot and which problems might occur if one only trains for ideal circumstances.

References

1. Cragun M.: DRIVE Sim on Omniverse. https://reg.rainfocus.com/flow/nvidia/nvidiagtc/ap2/page/sessioncatalog/session/16301039199180013WUV
2. State G, Wawrzyniak L.: Isaac Gym and Omniverse: High Performance Reinforcement Learning Evolved. https://reg.rainfocus.com/flow/nvidia/nvidiagtc/ap2/page/sessioncatalog/session/1628832827500001qlRX
3. Adrien G, Qiao W, Yohann C, Eleonora V.: Virtual worlds as Proxy for Multi-Object Tracking Analysis. arxiv:1605.06457
4. Ros G, Sellart L, Materzynska J, Vazquez D, Lopez AM.: The SYNTHIA dataset: a large collection of synthetic images for semantic segmentation of Urban Scenes. https://doi.org/10.1109/CVPR.2016.352
5. Kar A.: Meta-Sim: Learning to generate synthetic datasets. https://doi.org/10.1109/ICCV.2019.00465
6. Unity Technologies, Unity ML-Agents Toolkit. https://github.com/Unity-Technologies/ml-agents
7. Wang S, Yang J, Hu R, Qingnian Z.: Research on unmanned ship simulation on the basis of Unity3d. https://doi.org/10.1145/3232651.3232652
8. Miller D, Gibson S, Navarro A.: Advance your robot autonomy with ROS 2 and Unity. https://blog.unity.com/manufacturing/advance-your-robot-autonomy-with-ros-2-and-unity
9. Yakovlev A, Navarro A.: Simulate robots with more realism: what's new in physics for Unity 2021.2 beta. https://blog.unity.com/technology/simulate-robots-with-more-realism-whats-new-in-physics-for-unity-20212-beta
10. Sita E, Horvath CM, Thomessen T, Korondi P, Pipe AG.: ROS-Unity3D based system for monitoring of an industrial robotic process. https://doi.org/10.1109/SII.2017.8279361
11. Berriel RF, Tabelini L, Cardoso VB, Guidolini,R.: Heading direction estimation using deep learning with automatic large-scale data acquisition. https://doi.org/10.1109/IJCNN.2018.8489154
12. Fischer K.: Piecewise linear approximation of Bézier curves. https://hcklbrrfnn.files.wordpress.com/2012/08/bez.pdf
13. Anas: URP material pack Vol 3 unity asset store. https://assetstore.unity.com/packages/vfx/shaders/urp-material-pack-vol-3-187839?aid=1101l4LJP&utm_campaign=unity_affiliate&utm_medium=affiliate&utm_source=partnerize-linkmaker

A Privacy Preserving Mixed Reality Approach for Industrial Remote Maintenance Applications

Jan Luca Siewert, Mario Wolf(✉), and Detlef Gerhard

Digital Engineering Chair, Ruhr-University Bochum, Bochum, Germany
{jan.siewert,mario.wolf,detlef.gerhard}@ruhr-uni-bochum.de

Abstract. Augmented Reality (AR) is used in various industrial contexts like on-site maintenance and can further provide the basis for remote maintenance and remote support applications. This work presents a novel coupled AR-VR approach that supports all common remote-service interactions without transmitting a live camera feed. Instead of that, the expert is using a VR headset and only sees the virtual 3D representation of the machinery. Both can interact with an avatar of each other in AR or VR respectively. A prototype has been implemented using the Microsoft HoloLens 2 AR HMD and the HP Reverb VR headset. Participants connect to a central server component that provides static assets, like the required 3D models, and a second server for real-time collaboration and synchronization of movements, voice, and annotations. By design, this application does not transfer camera data unless especially requested and allowed by the technician.

Keywords: Augmented reality · Mixed reality · Remote maintenance · Privacy · Data security

1 Introduction

The Covid-19 pandemic disrupted supply chains and travel across the globe. This accelerated the already existing trend towards remote maintenance. Instead of sending an expert to service or train workers to remote locations technology like Augmented and Virtual Reality (AR/VR) for service or training tasks can be utilized [2]. This, of cause, is not only useful during travel restrictions but also helps to cut costs and during shortages of skilled labor. Data security, privacy issues, and intellectual property protection are key barriers to the adoption of remote maintenance, particularly employing HMDs, as brought to the public eye by the planned and then redacted release of the Google Glass HMD [4].

2 Prior Works

2.1 Augmented and Mixed Reality

In his canonical definition of AR, Azuma [3] already describes maintenance and service tasks as a potential application for industrial uses. On the other side of

M. E. Auer et al. (Eds.): REV 2022, LNNS 524, pp. 615–621, 2023.
https://doi.org/10.1007/978-3-031-17091-1_60

the "virtuality continuum" [11] lies VR in which users are completely immersed into a virtual world. Industrial applications for VR focuses mostly on training [e.g., [1,9]].

AR has various industrial applications across the whole product development lifecycle [7]. Some examples include maintenance and assembly support, remote assistance, training, quality control, planning, and data visualization [14]. The main benefits of the use of AR are reduced execution time, higher process quality, and learnability [15].

Traditionally, remote maintenance and remote support revolve around a technician who is connected to a remote expert using a Head-Mounted-Display (HMD) or mobile device. The HMD or mobile device transmits a live camera-feed to the expert, who can give guidance using his voice or draw annotations in two or three dimensions [10].

2.2 Privacy Concerns

AR relies almost always on optical algorithms based on captured camera images to facility marker registration and tracking. The fact that the device's camera continually records, poses a threat both to privacy and data security both for private consumers [6,13] and the industry [12], where the protection of intellectual property and know-how require even more consideration when adopting AR technologies [5,8]. For industrial applications, data security needs to be ensured across the whole pipeline [17]. This becomes more complicated when the camera feed gets transferred to external an expert as it is common in remote service scenarios. These camera images can contain confidential information, e.g. details of manufactures products or other equipment across the factory that comes into the field of view while the worker moves around.

2.3 Possible Solutions

Varghese et al. [16] present a data security approach for AR applications with multiple stakeholders. The camera feed is partitioned into different regions of interest that only get presented to the stakeholders that have a justified interest in that particular region. To ensure that only the correct stakeholders get access to a region, each region gets encrypted with a key that can only be decrypted by authorized users.

The main drawback of such an approach is that regions of interest need to be predefined or computed on demand. Then, the effectiveness of the system to secure intellectual property is subject to the effectiveness and correctness of the algorithms used to extract those regions. Therefore, a better approach is to not transfer any data that might contain sensitive information at all.

3 Goals and Requirements

The main goal of the system is to combine the advantages of AR remote maintenance and service while securing confidential information that might be contained in the data transferred to the external expert. The system should work even without transferring a camera feed over the network. The presented approach relies on accurate, existing 3D models of the maintenance subject to facilitate such an application. Thus, the technician in the field can service the real machine using an AR headset while the expert can showcase problems in front of a virtual representation in VR.

The other requirements of the system are similar to requirements for other remote maintenance solutions. Both the technician and the expert must be connected in near real-time with minimal latency. To enable communication, a voice call is to be initiated so that participants can talk to each other. Additionally, both the expert and the technician must be able to use gestures and annotations to point to details around the machine to be serviced. Ideally, the system should reuse data generated in other engineering software as part of the product lifecycle like drawings, computer-aided design (CAD) models, or existing problem reports when one party explicitly shares this information with the other participants. The technician should have the option to consciously capture a still image or video and review it before it is shared with other participants. At the end of such a service call, the system should save 3D annotations as part of a service report and make those reports available to other engineering software or for future service tasks. Users may be able to choose between or create different avatars as a virtual representation of themselves that can be stored persistently. Of cause, all data must be stored and transmitted using state-of-the-art encryption standards.

4 Concept

The expert is using a VR headset and sees the virtual 3D representation of the machinery. Both can interact with an avatar of each other in AR or VR respectively (Fig. 1). The expert can add three-dimensional annotations, while the technician can capture and upload static images using the camera in the AR-HMD. With this approach, the users can interact and communicate over gestures that get captured using provided hand and gesture recognition provided by the HMDs or using controllers alternatively, and by voice. The gestures get mapped onto the virtual avatars that have been created by the users. The users can draw attention to specific parts of the machinery by placing free-form, three-dimensional annotation around important parts or by selecting individual components out of the provided CAD data.

Today, engineering software, like Enterprise Resource Planning (ERP), Product Data Management (PDM), and Product Lifecycle Management (PLM) software, creates, stores and manages a wide range of data that is relevant for workers and technicians during remote maintenance. However, this data usually needs to be processed before it can be used in MR contexts. Primarily, traditional 3D models are stored in CAD specific formats that need to be triangulated and optimized for MR hardware. To coordinate these pre-processing tasks as well as for an easier interface for the actual MR applications, a central persistent data server gets introduced (Fig. 2). The persistence server uses a Service Oriented

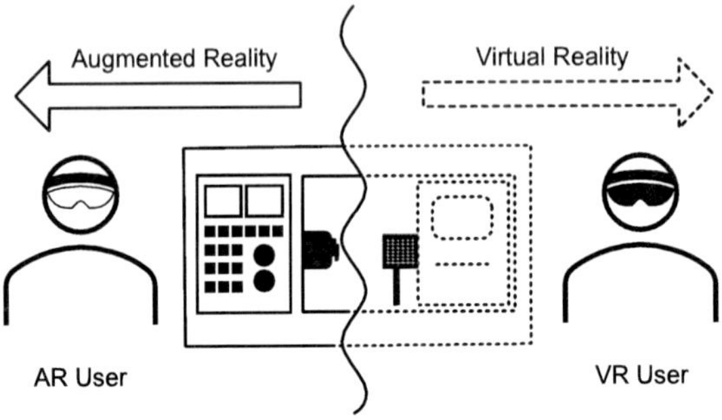

Fig. 1. Usage concept. A worker in front of the physical machine uses an AR HMD to connect to a remote expert collaborating using a VR HMD.

Architecture so that it can easily be expanded to cover more types of data as well as more native functionalities of different engineering software products. It also handles user management, because manufacturer and service providers rarely share engineering software systems. With this approach, third-party experts or workers can easily be given access to a limited subset of data relevant to the service task at hand. Here, users can import and convert 3D models. Problem reports and service calls can be created and scheduled using the persistent data server and all relevant information, users and models can be associated to a given task.

When the actual remote collaboration is about to start, a transient collaboration session gets started on a dedicated real-time collaboration server. This session is responsible for synchronizing the movement of the users that then gets mapped onto the virtual avatars. The positions are transferred relative to a shared origin on the machine so that the avatars can move around it and use their controllers or hands to point to particular parts of the machine. These avatars, as well as the pre-processed 3D models, and other relevant information are transferred to the users at the beginning of the session. During the session, the movements of the users, in particular the position of the headsets and the controllers or hands, are synchronized with low latency using the collaboration server. Additionally, the users can add virtual annotations, like texts or lines drawn into the room using a hand gesture or the controller, and can communicate using voice. These spatial and auditive information packages are synchronized by the collaboration server as well. An AR user stands in front of the real machine. The AR overlay shows added annotations as well as the avatar of the others users. The VR user sees the virtual 3D model of the machine as well as the avatars of the users and annotations. Therefore, the remote expert does not need access to a real machine to showcase problems or possible solutions but can use an existing model instead.

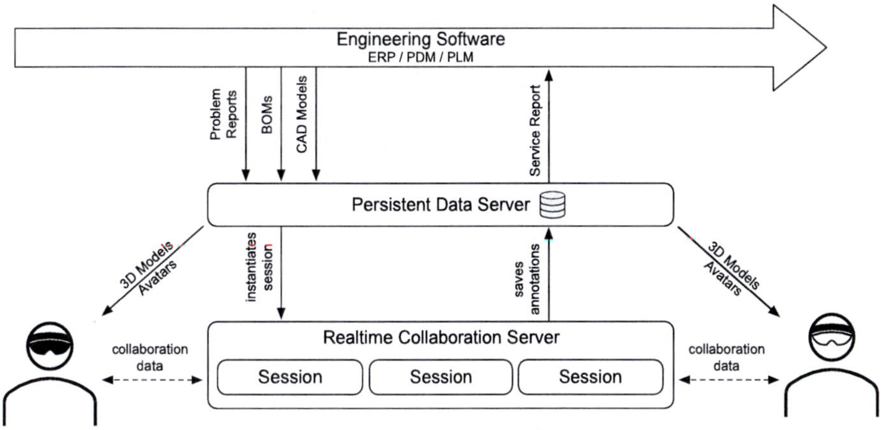

Fig. 2. System architecture. The system splits work between a persistent data server that imports and pre-processes engineering data, like BOMs and CAD models, and a real-time collaboration server that is used to share collaboration data between participants.

5 Implementation

A prototype has been implemented using the Microsoft HoloLens 2 AR HMD and HP Reverb G2 VR headset using the Microsoft Mixed Reality Toolkit (MRTK) in the Unity game engine. The main advantage of this hardware and software combination is that both the AR as well as the VR headset can be targeted by the same software development kit. Therefore, most of the logic between menus, presentation, and communication can be shared with only minimal adaptions between them. A sample scene from the point of view of both users is shown in Fig. 3.

For this prototype, only the real-time collaboration server has been fully implemented, with the persistence server only saving the reports of the finished sessions. Assets, like the avatars or 3D models, are made available in the Unity projects, so that they are part of the compiled applications, rather than being loaded from the persistent server. This also means that currently users initiate collaboration sessions manually.

The photon network engine is used for its reliable synchronization and deep integration into the Unity game engine in this prototype. The MR headsets are responsible for tracking the position of the headsets and the user's hands or controllers relative to the machine. For the VR scenario, the virtual model is presented and all orientations are transmitted relative to its origin. The real machine is equipped with an image marker as a point of reference. Of cause, the relative transformation between the model's origin and the position of the marker has to be known. The multiplayer engine transmits the user's heads and hands orientation, as well as voice and the 3D annotations that can be added to the scene using a gesture or the controller.

(a) AR user (b) VR user

Fig. 3. The system from the point of view of the AR user (left) and VR user (right).

At the end of the call, all shown annotations are saved as part of a service report on the persistent data server. These could later be overlayed on top of the virtual model as part of a 3D render or be replayed in AR by the technician or in VR by the expert.

The system is not limited to exactly one AR and one VR user in both concept and prototypical implementation. Instead, the current implementation could be deployed for virtual collaboration between multiple VR users or synchronous collaboration between one AR user and several VR users.

6 Conclusion

The prototype has been evaluated on an exemplary use-case in the author's laboratory. Here, an expert successfully explained a standard maintenance procedure using the VR-Interface to another person wearing the Hololens HMD. Even though the system is currently not implemented fully, the benefit of such a distributed system was clear to the users of the prototype, as all the collaborative interactions are functional. The presented hybrid approach uses the individual strength of the various Mixed Reality technologies and combines them to create a system with more manual control over the flow of possibly sensitive data. By design, this application does not transfer camera data unless especially requested and allowed by the technician or the technician's superior. This privacy-preserving approach therefore can be used in more sensitive environments and helps in building trust towards these innovative technologies. Only in finely tuning MR technologies to the desired use cases, can they show their true potential.

In the future, the authors plan to implement the complete systems and conduct a thorough user study on realistic, industrial use cases.

Acknowledgments. The authors thank ...[redacted]... for their prototyping and implementation work without which this contribution would not have been possible.

References

1. Daniel F, Abawi R, Haller M (2005) Efficient mixed reality: training applications development. In: Proceedings of TESI (training, education & simulation international), [Digest] // Institution of Electrical Engineers, p 289–294. IEE
2. Agrawal M, Eloot K, Mancini M, Patel A (2020) Industry 4.0: Reimagining manufacturing operations after covid-19
3. Azuma RT (1997) A survey of augmented reality. Presence Teleoper Virtual Environ 6(4):355–385
4. Berkemeier L, Zobel B, Werning S, Ickerott I, Thomas O (2019) Engineering of augmented reality-based information systems. Bus Inf Syst Eng 61(1):67–89
5. Rokka Chhetri S, Faezi S, Rashid N, Al Faruque MA (2017) Manufacturing supply chain and product lifecycle security in the era of industry 4.0. J Hardw Syst Secur 2(1):51–68
6. Denning T, Dehlawi Z, Kohno T (2014) In situ with bystanders of augmented reality glasses. In: Proceedings of the SIGCHI conference on human factors in computing systems
7. Fite-Georgel P (2011) Is there a reality in industrial augmented reality? In: International symposium on mixed and augmented reality, science and technology proceedings. IEEE
8. Langfinger M, Schneider M, Stricker D, Schotten HD (2017) Addressing security challenges in industrial augmented reality systems. In: 2017 IEEE 15th international conference on industrial informatics (INDIN)
9. Liagkou V, Salmas D (2018) Chrysostmos Stylios. Realizing virtual reality learning environment for industry 4.0. Procedia CIRP 79:712–717
10. Masoni R et al (2017) Supporting remote maintenance in industry 4.0 through augmented reality. Procedia Manuf 11:1296–1302
11. Milgram P, Kushino F (1994) A taxonomy of mixed reality visual displays. IEICE Trans Inf. Syst E77-D(12):1–15
12. Quandt M, Knoke B, Gorldt C, Freitag M, Thoben K-D (2018) General requirements for industrial augmented reality applications. Procedia CIRP 72:1130–1135
13. Roesner F, Franziska Roesner K, Kohno T, Molnar D (2014) Security and privacy for augmented reality systems. Commun ACM, 57(4):88–96
14. Röltgen D, Dumitrescu R (2020) Classification of industrial augmented reality use cases. Procedia CIRP 91:93–100
15. Fernando de Souza Cardoso L, Cristina Martins Queiroz Mariano F, Roberto Zorzal E (2019) A survey of industrial augmented reality. Comput Ind Eng 139:106159
16. Varghese A, et al (2015) Smart: secure mobile augmented reality for tele-assistance. In: 2015 Asia Pacific conference on multimedia and broadcasting
17. Vorraber W, Gasser J, Webb H, Neubacher D, Url P (2020) Assessing augmented reality in production: remote-assisted maintenance with hololens. Procedia CIRP 88:139–144

Biomedical Engineering

A Low Cost PC-Based System for Clinical Information Management in Infant Neurodevelopment Services of Santiago de Cuba Province

Sergio Daniel Cano-Ortiz(✉), Lienys Lombardía-Legra, Francisco Javier Deler-OFarril, and Liette Veranes-Bicet

Universidad de Oriente, Ave. Las Americas s/n, 90900 Santiago de Cuba, Cuba
scano@uo.edu.cu, scano1959@gmail.com

Abstract. Childhood Neurodevelopmental disorders (NDD) collectively represent a great challenge for health systems internationally, with a significant economic and social impact. Despite the existence of digital devices in Cuban hospitals, the clinical records of patients are still written in hard format and stored in unfavourable conditions. Moreover the international market for Electronic Hospital Records systems (EHR's) presents a wide variety of products but in addition to being expensive, they only prioritize clinical service. The main objective of this paper is to develop a low cost Web-based system oriented for research that automates the information management of the Services for Infant Neurodevelopment & Disability (SPNID) in Santiago de Cuba. The system was developed using open source tools and technologies, and the Extreme Programming as software development methodology. The ECR that responds to the services provided in the SPNID was developed in Visual Studio Code v1.39.2. The software has been implemented in the JavaScript programming language, using open source frameworks. Moreover a dedicated wireless network was designed for remote access of SPNID staff using mobile devices. The system was successfully evaluated in real conditions on the Intranet of the Southern Infant Hospital (HIS) by the SPNID specialists. Moreover a way of how to make the biomedical research about NDD closer to hospital was tackled by authors. The use of open source and the efficient use of low-cost solutions in the implemented system have proven to be viable alternatives for the improvement of health system in low-income countries like Cuba.

Keywords: Digital clinical history · Neuro-development · e-Health applications

1 Introduction

The health history or clinical records (CR) as it is normally known, is nothing more than the set of documents that emerged between the doctor and the patient, it also reflects the will of the patient and the valuable considerations, observations and judgments from the

M. E. Auer et al. (Eds.): REV 2022, LNNS 524, pp. 625–638, 2023.
https://doi.org/10.1007/978-3-031-17091-1_61

doctor. It is also considered as the record of the relationship established between user and the hospital in early care.

Several peculiarities of CR are ineludibly, such as:

- *Mandatory*: No hospital or office medical act should be carried out without its corresponding registration in the CR (except in some cases of extreme urgency and short time.
- Irreplaceable: the writing of the history cannot be replaced by the memory of the doctor, since it is logical that he cannot remember or know all the details of each patient.
- Private: It must be characterized by the confidentiality of its content [1, 2].

Cuba is among the countries of the third world and despite this, it has sufficient potential to improve the quality of medical care, which is why it has put its efforts into the digitization of the National Health System, as well as other spheres of the society. To do that, it has invested in a number of computer equipment for use in hospitals and polyclinics in order to improve the efficiency of the service.

Despite the existence of these digital tools, the clinical records of patients continue to be written in hard format and stored in unfavourable conditions that cause their loss and deterioration over time. In addition, the information that is reflected in the clinical history continues to be done manually, in some cases leading to duplication of information. This not only makes it difficult to read the CR content, but also to be able to use it as a data source for future research and longitudinal studies.

In Santiago de Cuba province both institutions, the Universidad de Oriente, together with the Southern Infant Hospital (where the Services of Infant Neurodevelopment and Disability (SPNID) is located) have joined forces in the development of software that allows digitization of all the processes and services are carried out within SPNID, having as a core a digital CR. The purpose of this clinical service is to investigate new-borns who have finished their postnatal life with some type of sequelae, both from the Neurological point of view as well as Motor or Functional Disability [3].

Therefore, the main objective of this paper is to develop a PC-based system with a digital medical history as core that automates the information management of SPNID in Santiago de Cuba. Initially the main features of SPNID are explained, and the information flow of the SPNID, which was obtained in the initial stage of the investigation, was properly described. The tools and technologies used as well as the benefits obtained with this work are too discussed.

2 Methods and Materials

It is valid to highlight below some references that have been found on the meaning of electronic (digital) clinical record (ECR). The ECR, also known as the electronic medical record, is a computerized (digital) record of a patient's social, preventive, and medical data. It must integrate all the multimedia information that is used in clinical practice [4]. It is the collection of a patient's medical information in digital format that is saved on a computer (diagnoses, medications, tests, treatments). This information can be seen and used by the physicians in charge of the patient, thus let physicians help them and make recommendations about patient care [5].

To carry out this work, different methods and data collection techniques were applied to obtain the information, such as participant observation, review of the documents used in the clinical information processing, interviews with medical personnel from different specialties, interviews with parents and relatives of patients and the analysis of bibliographic sources involved within the SPNID dynamics.

2.1 Design of the System

The SPNID in Santiago de Cuba is located at the Infant Southern (HIS). It is responsible for detecting deviations in neurodevelopment of children between 0–6 years of life, as well as ensuring the monitoring and study of these deviations, for a dislocated child population in the 9 municipalities of Santiago de Cuba province. It covers the following services:

- *Attention to children with risk factors to debut with a disability or deviation in their neurodevelopment.*
- *Early diagnosis of deviations in neonatal neurodevelopment.*
- *Rehabilitation of children with visual and hearing limitations.*
- *Facilitate the infant education within a family environment.*
- *The qualification of the personnel in charge of infant caring.*
- *Promote the integration of institutions aimed at raising the quality and effectiveness of the care and monitoring process.*

All the nine municipalities of the province are covered by the SPNID, in which the following specialties intervene: Neonatology, Paediatrics, Ophthalmology, Physiatry, Genetics, Neurology, Neurophysiology, Psychology, Audiology, Speech Therapy and Phoniatry. About specialties named above, only Neonatology and Paediatrics are those that govern the admission and follow-up process. From them are derived the specialties through which the patient's attention must pass for its diagnosis and treatment. In this consultation, the patients are studied and followed up on their pathology and make the appropriate assessment. For a better understanding, the following Fig. 1 is shown.

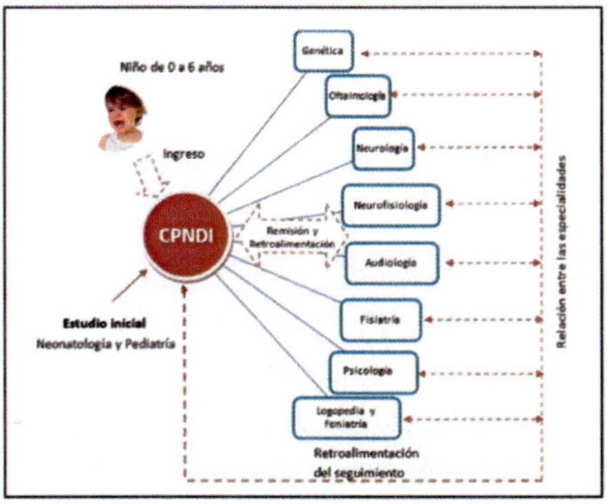

Fig. 1. The SPNID information flux

From the above, it is assumed that the Neonatal and Paediatric consultation keep track of the child's evolution process until he is discharged. That is why these two consultations constitute the pivot point and around them all clinical information flows. Digitizing that stream represents an important goal within work.

Methodology, Tools and Technologies Used in the Development of a Digital Record. The digital CR (ECR) that responds to the services provided in the SPNID was developed in Visual Studio Code v1.39.2; it is a free and open source code editor. The Extreme Programming (XP) was used as software development methodology, which is based on continuous feedback between the client and the development team, fluid communication between all participants and simplicity in the implemented solutions. [6]. Node.js was used because it is a runtime environment that is designed to generate highly optimized web applications, as well as being lightweight and scalable [7].

The software has been implemented in the JavaScript programming language, using open source frame-works, such as AngularJS for the front of the system, which provides a well-integrated set of tools that allow building flexible applications of the On the client side [8]. SailsJS was used for the backend, it makes development easier, it has several layers of abstraction and it is very effective for developing real time applications [9].

In order to store all the information in the digital medical record, MongoDB was used as database manager. Its data storage method provides greater speed and ease of data integration, considering it to be one of the managers of greater use and popularity [10].

An ECR Architecture for the SPNID. The system is a web application with a client-server architecture, governed by a Model-View-Controller (MVC) architectural pattern, as shown below (Fig. 2).

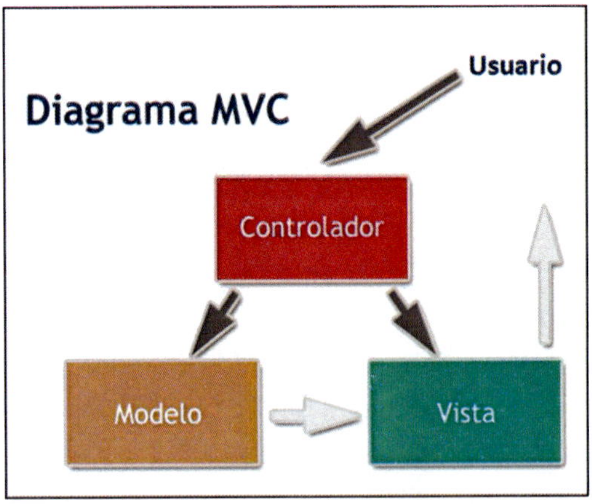

Fig. 2. A Model-View-Controller (MVC) architecture

In this architecture, the client makes the requests to the server and obtains the responses, which makes it possible to use them easily in low-performance computers, also allowing an easier communication between different systems.

Software Development. The designed system allows the management of the patient's data in an ECR. It starts from the filling in of forms where the user identified as Doctor of the Central Consultation (Neonatology and Paediatrics specialties) is the only one authorized to create new patient and add his login information to the Database. The information that is recorded for each patient is in correspondence with the general data, Antenatal History, Perinatal History, Postnatal History, as well as the information of the family member who accompanies the patient to the first consultation. The consultations are registered through forms that collect the corresponding data, allow assigning medical examinations, monitor the patient's evolution by viewing their consultation history, refer to other specialists, and assign medical treatment. The system also has a Chat option, which allows collaborative exchange among the specialists of SPNID just in real time.

User Roles and Functionalities. For more details, the functionalities are broken down below according to the different user roles:

- *Medical-Consultation-Driver*: They are medical specialties in Neonatology and Paediatrics. They have the following privileges:
  - Insert, modify and unsubscribe a patient.
  - Create query. Within the consultation view you can: indicate medical examination, write results of pending exams, plan the date of next consultation, refer patient.
  - View the medical history of a patient.
  - See list of patients in the consultation.
  - See patients referred to other specialties.
  - Export the ECR to PDF file format.
  - Access the chat service in real time.
  - See consultation calendar for planning.
  - Search patient by ECR number or by ID.
  - See statistical reports.

- *Physician-Consultation-Passive:* They are the doctors of the other specialties that make up the Neuro-development Consultation; they have access to all the privileges of the previous role except inserting, modifying or withdrawing a patient:
- *Executive:* They are the users who have administrative positions and require statistical reports for monitoring and control of the general state of the Neuro-development Consultation. The statistics provided by the software supports decision-making that can be carried out later.
- *Administrator:* It is the hospital computer scientist who has the privilege of administering the users of the system according to the doctors who intervene in the consultation. In addition, it manages the established database nomenclatures.
- *Cry Corpus collection:* development of an interface to acquire biomedical signals (in a first approach, patient cries are recorded in the SPNID that make up a database for the development of biomedical research). It will be linked to R&D research currently under development by the CENPIS[1] in the SPNID.

[1] Research Centre for Neuroscience, Image and Signal Processing.

A glossary of terms and a user manual were pre-pared and can be accessed once they have been registered in the system. Here are some views of the system (Fig. 3):

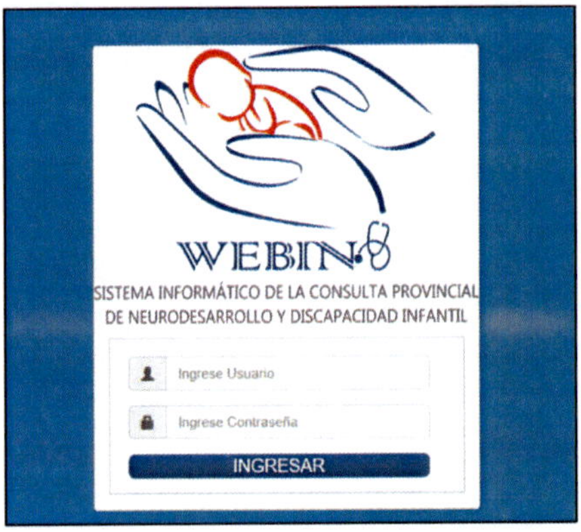

Fig. 3. Authentication screen

The figure above shows the view that allows users to authenticate themselves to the system, for this they must enter their username and password. In this way, each of them has sole and exclusive access to the privileges according to the role they play. It responds to the security requirement with which the software complies to protect the information. To avoid vulnerabilities, the password encryption was used through the Bcrypt library, it is written in JavaScript and is responsible for generating much more secure passwords against brute force attacks [11].

In Fig. 4, the main view for the Doctor-Consultation-Central role is shown, it consists of a main menu with a set of functionalities that can be accessed, such as the Insert Patient option in Fig. 5.

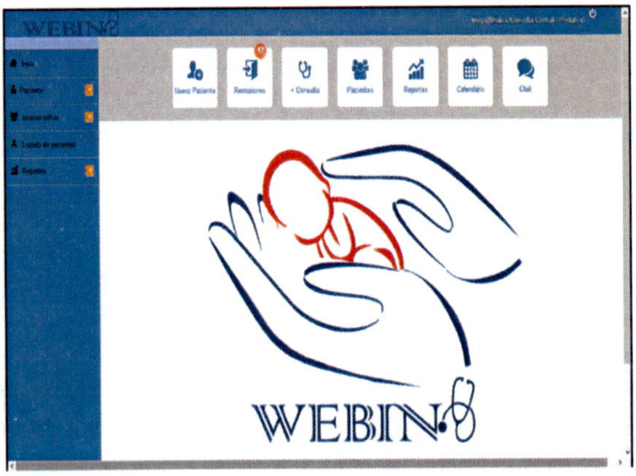

Fig. 4. Main screen for the Doctor-Consultation-Central role

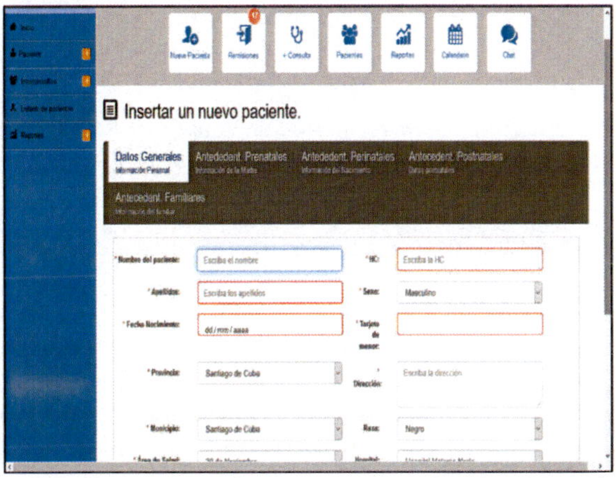

Fig. 5. Screen for patient insertion

A patient's medical record contains abundant information; this begins with general data that includes personal information, prenatal, perinatal, postnatal antecedents and family data, up to the history of each consultation through which he has passed.

Below is a view where you can read patient data (see Fig. 6).

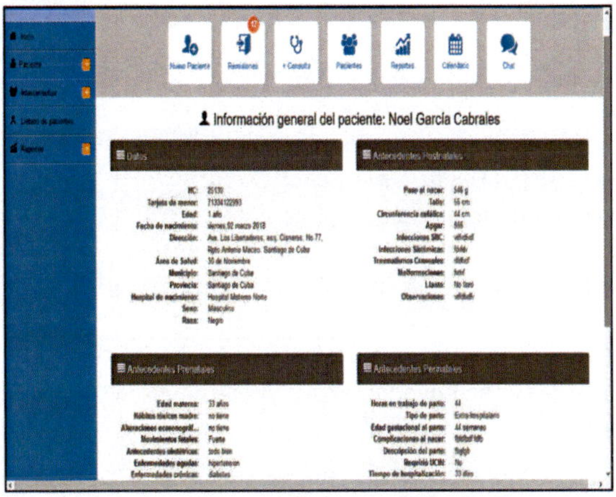

Fig. 6. Screen for general data of patient

3 Results

After concluding the implementation and testing stage, several trainings were carried out within the Southern Infant Hospital (HIS), where the SPNID currently resides and outside it, at the Research Centre for Neuroscience, Image and Signal Processing (CENPIS) at the Universidad de Oriente, as shown in Fig. 7.

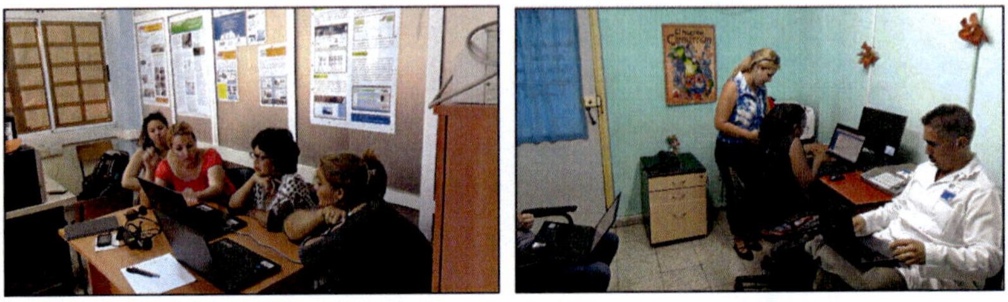

Fig. 7. Testing the system at UO-CENPIS and HIS

In all the tests done with the medical personnel, evaluations were carried out to verify the correct use and operation of the options implemented in the software. Moreover, numerous benefits were found with the use of the system:

- Accessing to the appropriate information in the shortest possible time.
- According to the design of the Database, the system is capable of being deployed to the rest of the country's provinces.
- Facilitates the exchange of knowledge between doctors during consultation hours without the need to leave their job.
- Remote access from mobile devices by medical personnel in the consultation.
- •A new and low-cost ECR capable of providing different neurodevelopment services can be counted on in the Santiago de Cuba province for first time.

3.1 A Wi-Fi Network for Supporting Mobile Communications Within the SPNID

A Wi-Fi network was designed to guarantee remote access to the system for members of the multidisciplinary SPNID team. Several steps were developed: characterization of the SPNID that included defining the structural characteristics of the building as well as the needs of specialists who work in it, characterization of the hospital network (technical characteristics to be used in the design of the Wi-Fi network for the management of digital medical records).

Despite the fact that the structures of networks based on Ethernet are the most widespread, in the case of the hospital environment wireless technology acquires great importance, since medical personnel are constantly on the move and require facilities for continuous access to patient medical records (it could be achieved by internet searches). Factors such as the characteristics of the work environment, the type of application to be implemented and the regulations for the implementation of wireless networks in hospitals define the type of configuration to be used. For these reasons, special attention was devoted when choosing the network standard to be implemented, since not everyone can provide the same facilities when using a certain application [12].

The SPNID Location. The SPNID is located in a four-floors building located at the bottom of the HIS. For its characterization, a study was made from an observation process, as well as a consultation of structural plans (see Fig. 8).

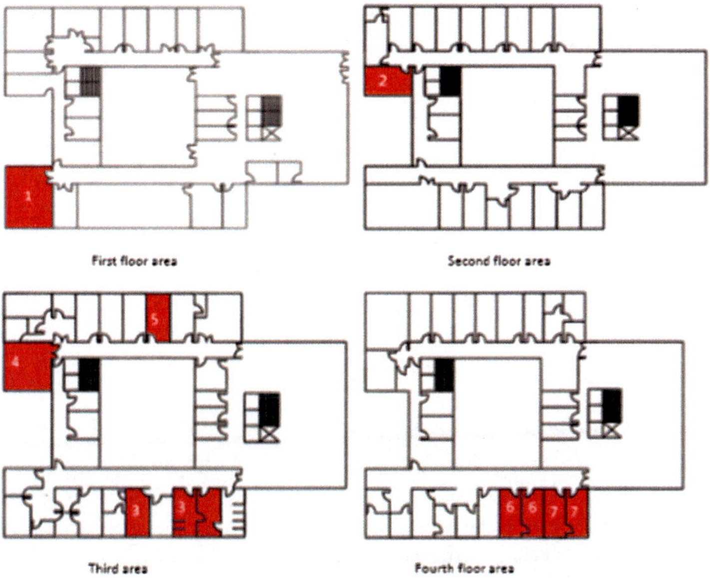

Fig. 8. Location of the SPNID sites. Where 1-Physiatry, 2-Audiology, 3-Ophthalmology, 4-Neurodevelopment consultation (Neonatology, Paediatrics), 5-Head of SPNID, 6- Speech Therapy and Phoniatry, 7-Othorhyno

Network Design. The 802.11n standard was used for the design since it is compatible with any wireless device that can be used by SPNID specialists. In the 5 GHz band, it allows working with 20 MHz channels, which enables compliance with the recommendations of the Cuban Ministry of Communications and Informatics. It also uses strong modulation techniques, which provides robustness against interference. Firstly, the sizing of the system was made where the number of users on each floor and the simultaneity factor were defined. The radio planning was carried out where the channels to be used were defined too, and then equipment distribution was also made. Finally it was set what security system will be used in order to prevent intruders from accessing the network content.

The deployment of access points should cover the entire surface of the specialists' work area. Each specialist will have a minimum signal level of −85 dBM in their work area, which will always provide a real bandwidth of 100 Mbps. Due to the number of access points required for deployment, frequency reuse is necessary. The frequencies to be used in the deployment are those corresponding to channels 1, 6 and 11 for the 2.4 GHz band and channels 149, 153 and 157 in the 5 GHz band. These frequencies will be assigned, so that the interferences between access points that use the same frequency, be minimal. The frequency assignment was not made arbitrarily since the recommendations of the Cuban Ministry of Communications were taken into account. The width of the channels will be 20 MHz for the two bands.

In addition, the roaming network is provided, for which it must be taken into account when placing the access points that the coverage area between adjacent access points must overlap.

Equipment to Use. Cuba for more than 50 years has been the victim of an economic blockade by the United States of America, which has limited its commercial relations with the main providers of network equipment in the world, limiting access to different equipment. TP Link has been chosen as the main supplier for this project, considering that it has commercial relations with Cuba, in addition to its equipment having good benefits.

The equipments to be used are the following:

1. **Auranet EAP 220 access points:** They will be located in the consultation areas. These access points were chosen because they work with Dual Band technology, they have 4 internal antennas of 4 dBi of gain each, a transmission power of up to 20 dBm and they support personal and business WPA/WPA2 with support on RADIUS server.
2. **Auranet AC 500 wireless controller**: It will be located in the Main Node. Its main characteristics are that it can handle up to 500 access points, it has 5 Gigabit Ethernet ports and it allows rack mounting.
3. **TL-SG2424P switch**: It will replace the switch that is currently in the query building, since the current one has insufficient ports to ensure the scalability of the network. Its main characteristics are that it has 24 ports, supports POE 802.3af/at and has four SFP slots.
4. **QNAP TS-453 server**. It will be located in the Main Node. It was chosen because it has capacity for 4 hard disks, with expansion units for 4 more disks, Sata connections of up to 6 Gbps, and can be used as FTP, RADIUS and file servers.

Location of Access Points and Coverage Analysis. For the location of each access point, an iterative process was carried out, through which a location was selected and it was verified if all the requirements were met, if not, a change of location, a power readjustment in the transmitter or an offset of the position of the antennas until the result is satisfactory. In this way it was possible to cover the total work areas of the specialists, guaranteeing an acceptable coverage for them. Figure 9 shows the case corresponding to the first floor. The minimum level values, at the manufacturer's suggestion, were set at −60 dBm (red) for 150 Mbps and −80 dBm (green, yellow and orange) for 65 Mbps.

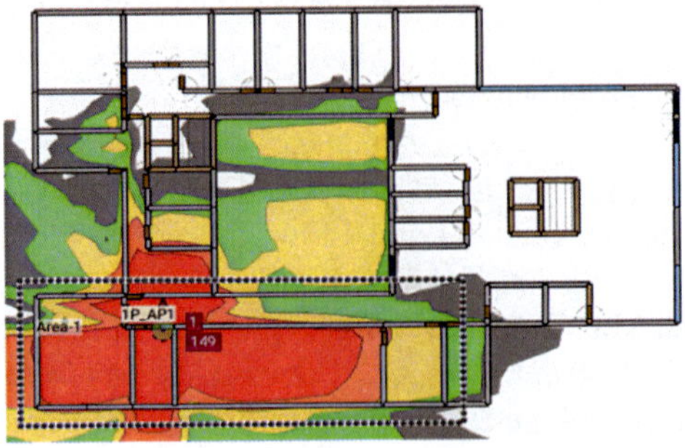

Fig. 9. 1st floor locations

Network Security. In view of the fundamental importance of network security for a hospital environment, a protocol is needed that has the best and most robust security methods for the data transmitted, in addition to allowing access to the network to be restricted to only relevant users. For these reasons, it was decided to implement the WPA2 protocol with 802.1x through the implementation of a RADIUS (Remote Authentication Dial-In User Service) server. In this way, there is strong user authentication, in addition to reducing the complexity of access for users, since after the first access, their wireless equipment will store all the information necessary to connect with the access points. The enterprise WPA2 protocol was also activated with each of the access points with 802.1x, which ensured that the entire authentication process was done through the RADIUS server. The hospital network administrator will be in charge of managing the RADIUS server databases. This will ensure that patients' medical records are only accessed by authorized personnel.

Final Scheme of the Network. All access points will be connected to the building switch. The wireless controller will be located in the central node of the hospital network, as well as the servers and the RADIUS server, guaranteeing the security of the equipment. The network will be designated as the Neurodevelopment SSID (Fig. 10).

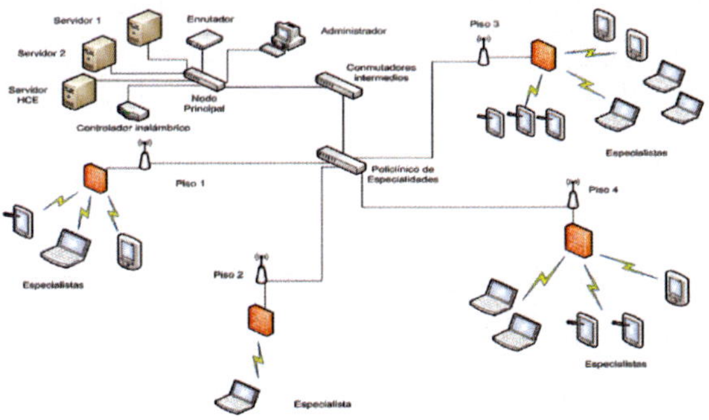

Fig. 10. Final schema of network

Validation of the Proposed Design. The proposed design was validated through the simulation of the different areas of the SPNID using the *Ekahau Site Survey 9.0.2* software, professional tool from the Aruba's Networks company, a subsidiary of Hewlett Packard. Thus the following results were obtained:

Wireless Coverage. The wireless coverage of each floor was analysed, both for the 2.4 GHz and 5 GHz bands, with which it was verified that the proposed design meets the needs and requirements of the SPNID specialists. For example see in Figs. 11 and 12 coverages corresponding to 1st floor. The design carried out managed to endow with a power of −85 dBm or better 98.3% of the consultations in the 2.4 GHz band (Fig. 11) and 98.6% in the 5 GHz band (Fig. 12), which gave response to the requirements of Physiology specialists for this specific case. In addition, the signal range outside the building was reduced to the minimum possible.

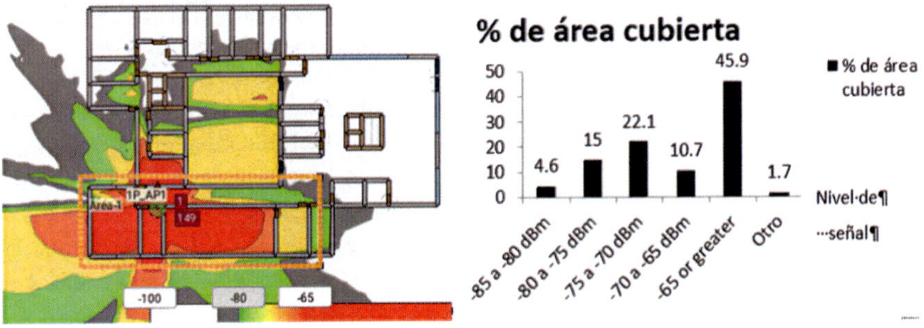

Fig. 11. Wireless coverage for 2.4 GHz in 1st floor

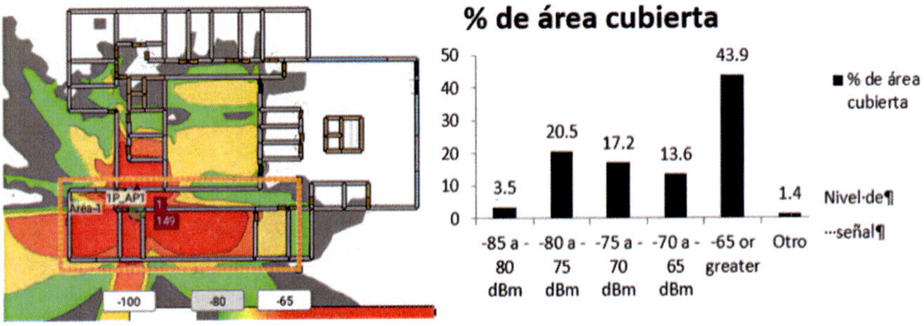

Fig. 12. Wireless coverage for 5 GHz in 1st floor

For the rest of the floors the result was as follows:

For the second floor, 97.3% of queries in the 2.4 GHz band and 97.7% in the 5 GHz band were endowed with a power of −85 dBm or better, thus responding to the requirements of Audiology specialists for this specific case. In addition, the signal range outside the building was reduced to the minimum possible. For the third floor, 97.3% of the consultations in the 2.4 GHz band and 97.7% in the 5 GHz band were provided with a power of −85 dBm or better, thus responding to the requirements of the specialists of Ophthalmology and the Neurodevelopment office itself, managing to reduce the range of the signal outside the building to the minimum possible. Finally, on the fourth floor, a power of −85 dBm or better is reached for 99.1% of the queries in the 2.4 GHz band and 98.2% in the 5 GHz band, thus responding to the requirements of the Audis and Othorrino specialists. Here it is also possible to reduce the signal range outside the building to the minimum possible.

Bandwidth per Zone. Here it was possible to obtain a map of the bandwidth in each of the areas of the building. For the first floor, it was found that more than 96.4% of the areas where queries are carried out on the first floor (Fig. 13) have a signal strength strong enough to provide it with a bandwidth greater than 100 Mbps.

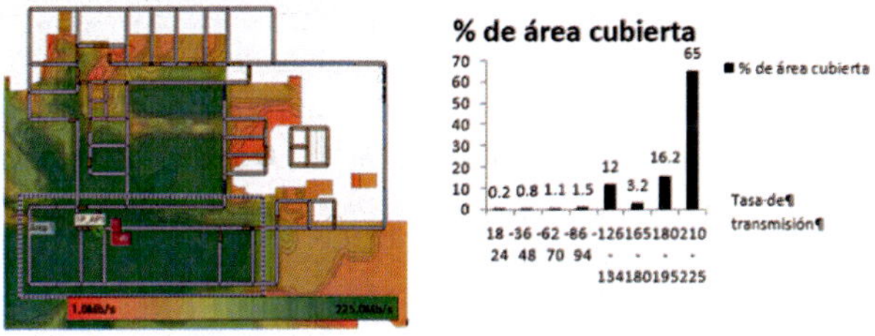

Fig. 13. A bandwidth map for 1st floor

For the rest of the floors it was found that 93% (for floor 2), 93.7% (for floor 3) and 100% (for floor 4) of the respective areas where the consultations take place have a signal intensity strong enough to provide them with a bandwidth greater than 100 Mbps.

Channel Coverage. It was verified that there is overlap between the signals of the different access points, both in the 2.4 and 5 GHz bands, so that roaming will be possible, providing users with the ability to move through the different areas of SPNID without feeling that the connection has been lost. Figure 14 shows the distribution on the fourth floor, area where there is presence of more than one access point.

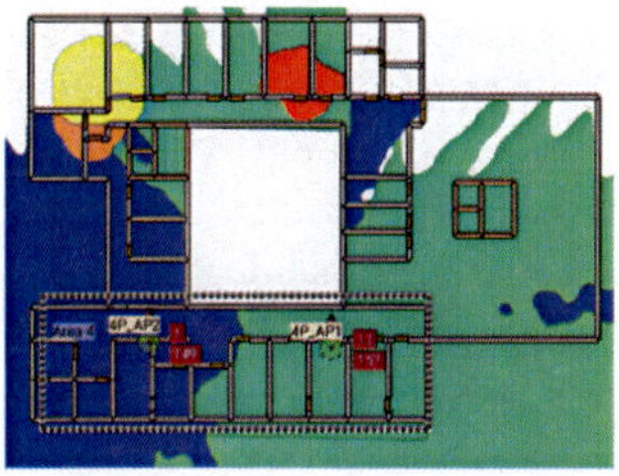

Fig. 14. Coverage of every access point in 4th floor.

Economic Valuation. The amount of the investment to be made for the WiFi network reaches 2066.6 euros. This cost is below similar standards in the international market, meaning savings in financial resources for the country.

4 Conclusions

A PC-based system and its core (the electronic clinical record) were developed in order to respond to problems posed by the SPNID in Santiago de Cuba province. It includes several facilities to SPNID specialists as: an easy-and-fast management of patient information, queries, search performing, statistical analysis for decision making and scaled-feature filtering. A real-time exchange of ideas and criteria among doctors during consultation hours through chat service was also included. Moreover the upgrading and cyclic storage of clinical information of patient let physicians hold faster decision making during clinical process. An added value to the developed system is the possibility of recording audio signals, thus allowing the acquisition of voice and child cry signals to feed databases related to the biomedical research carried out by the CENPIS at the SPNID. Moreover a WiFi network was also designed and tested just to support a full access by SPNID staff to all clinical information managed by the system, using their own mobile devices. The use of open source and the efficient use of low-cost solutions in the implemented system are viable alternatives for the improvement of health system in low-income countries like Cuba.

Acknowledgments. This work was partially supported by the National Basic Sciences Program PNCB-Cuba under project (PAP-10513) as well as by the DAAD of Germany (in close collaboration with the Applied Sciences University of Dusseldorf). The authors thank the Dept. of Telecommunications Systems and the manager and medical staff of the Southern Infant Hospital in Santiago de Cuba because of providing their valuable time and knowledge during the investigation process, in gathering information, as well as in testing the system within the hospital.

References

1. Alcaraz M, Nápoles Y, Chaveco I, Martínez M, Coello (2010) La historia clínica: un documento básico para el personal médico, Medisan, vol 14, no 7, Santiago de Cuba
2. Ramos JM, Cuchí M, Sánchez MA (2009) Archivo de historias clínicas Digitalizado, una solución previa a la Historia Clínica Electrónica. Papeles Médicos 18(2):4–10
3. Cano-Ortiz SD, Martinez-Cañete Y, Lombardía-Legrá L, Langmann R, Jacques H (2019) A support system for information management oriented for the infant neurodevelopment study. In: Auer ME, Langmann R (eds) REV 2018, vol 47. LNNS. Springer, Cham, pp 528–535. https://doi.org/10.1007/978-3-319-95678-7_59
4. Prego A (2019) Historia Clínica Digital o Electrónica. Definición y ventajas. https://gestionmedica.org/historia-clinica-digital/. Accessed 2 Nov 2019
5. Prego A (2019) National Cancer Institute. Diccionario de Cancer. https://www.cancer.gov/espanol/publicaciones/diccionario/def/historia-clinica-electronica. Accessed 2 Nov 2019
6. Gómez A, Duarte A, Guevara CD (2014) Desarrollo ágil de software aplicando programación extrema. Rev Ingen Univ Francisco de Paula Santander Ocaña 5(1):24–29
7. Madsen MS, Tip F, Lhoták O (2015) Static analysis of event-driven Node. JS JavaScript applications. In: ACM SIGPLAN notices. ACM, pp 505–519
8. Jain N, Bhansali A, Mehta D, Angular JS (2015) A modern MVC framework in JavaScript. J Global Res Comput Sci 5(12):17–23
9. Mcneil RM, Nathan I (2017) Sails.js in Action. Manning Publications Company
10. Chodorow K (2013) MongoDB: the definitive guide: powerful and scalable data storage. O'Reilly Media, Inc.
11. Duuna K (2015) Secure your node. Js web application: keep attackers out and users happy. Pragmatic Bookshelf
12. Krupinski A, Bernard J (2014) Standards and guidelines in telemedicine and telehealth. Healthcare 2:74–93

Assessment of Psychological Impact Due to the Pandemic in Society and Education in Colombia with Machine Learning

Hernando Vélez Sánchez(✉)

Universidad Distrital Francisco José de Caldas, Bogotá, Colombia
hvelezs@udistrital.edu.co

Abstract. After 20 months of the pandemic, the health situation in the world has evolved favorably with vaccination in most of the countries that have had access to it. However, the impact on society from the point of view of people's behavior and their relationship with physical health is being evaluated by stakeholders. It helps states and institutions to formulate public policies to achieve social adaptation to new conditions, after pandemic. Sociodemographic conditions interact with the environment produced by the pandemic and can cause physical and psychological affectation that is reflected in the behavior of the person, such as the reaction to a new threatening situation (stress). This response is used to evaluate the degree of mentioned affectation of people. Models based on machine learning are proposed to evaluate and predict this level as well as the consequence in the educational process.

Keywords: Pandemic · Psychological impact · Education process

1 Introduction

The pandemic has affected the performance of human activities, including educational processes. The objective of the study is to evaluate and relate the psychosocial impact of the pandemic on people's health and in particular their performance and school permanence with sociodemographic characteristics in Colombia using machine learning, so that it serves as a basis for the design of public policies that enable their reincorporation to productive life. Situations such as the pandemic influence people's behavior and their ability to perform in their social environment and individually can be impacted positively or negatively. Specialists in this area have developed a methodology to evaluate this level of affectation from the point of view of epidemiological surveillance. For this, they define some dimensions in which psychosocial affectation can be measured and associated with some indicators that can be identified through an interview or survey with the people involved. These dimensions can be summarized in concepts such as somatization, fear, anxiety, intolerance, aggressiveness, depression, stress, hopelessness, loneliness, family maladjustment. It is about estimating the level of psychosocial involvement for each dimension and for all of these (global).

M. E. Auer et al. (Eds.): REV 2022, LNNS 524, pp. 639–650, 2023.
https://doi.org/10.1007/978-3-031-17091-1_62

Valdivieso Miranda et al. [1] conducted a study on the impact of confinement on academic performance in higher-level institutions. They found an impact on physical and emotional health, in particular on mood, and consequently academic performance decreased by approximately 20% compared to before of the pandemic. This caused spaces for reflection to increase their level of resilienc.

There are some studies on the application of artificial intelligence on the psychosocial impact of the pandemic, but not on educational activity.

Bottan [2] in a study for the IDB/Cornell analyzed the impacts of the pandemic on food security, the labor market, social distancing, and awareness of the pandemic. They studied socioeconomic patterns between regions and countries too, using weighting factors to correct for biases when comparing countries of different sizes.

Sánchez et al. [3], used percentiles to delimit the levels of affectation in the dimensions considered; finding hopelessness with higher value in a study on mental health in Peru.

Zabala et al. [4], found that confinement and virtual methodology produced an increase in the educational gap, school dropouts, stress, and learning difficulties [5, 6].

Sandoval et al. [7] carry out a work on the evaluation of social maladjustment in grade 4 students and in an institution in Chiclayo, Peru. He applied a survey of 80 questions corresponding to 9 dimensions of mismatch. The dimensions varied with the level of schooling.

2 Approach

The study carried out is field, explanatory, and correlational. Two main topics were addressed; the first is the relationship between the sociodemographic variables and the level of psychosocial affectation due to the pandemic in Colombia, and the second is the relationship between school dropout and the pandemic on the one hand, in addition to the impact of virtual education on students and educational processes.

For the first case, a database consisting of 1900 interviews with people from different cities, regions, socioeconomic strata in Colombia has been used. Within the chain of affectation, the causes, and situations of the variable environment that lead to the manifestation of physical and psychological symptoms are identified. This serve as a basis for predicting the level of affectation, as well as the behaviors of the individual that can manifest themselves in substance use as a mechanism to cope with stress.

The procedure defined in the scales established for this type of study was followed. The data obtained include sociodemographic information including those related to education processes (sociodemographic variables) from questions related to eight different dimensions of behavior impairment (indicators of unadjusted psychosocial behavior) on physical and mental health, these are; anxiety, depression, aggressiveness, somatization, hopelessness, intolerance, family maladjustment and fears. The questions are directly related to these dimensions and are unique to each of them.

There are three response values for each. The dimensions related to mood were assessed according to the EVEA standard that includes four general moods; anxiety, anger-hostility sadness-depression and joy. [Sánchez Carlessi]. Database structure is shown in Table 1.

Table 1. Data structure

Dimension	Indicators-questions	Level
Anxiety	It has taken me a lot to release the tension. I felt my hands shaking. I felt my heart beat despite not having made any physical effort. I have felt restless. I felt my heart beat despite not having made any physical effort. How satisfied are you with the quality of your sleep during quarantine?	H, M, L
Tolerance	I did not tolerate anything that would not allow me to continue what I was doing. I have tended to get angry easily	H, M, L
Fear	I overreacted in certain situations. I was worried about situations in which he might panic or make a fool of himself. I felt like I was on the verge of panic. I was scared for no reason	H, M, L
Depression	I couldn't feel any positive feelings. It was difficult for me to take the initiative to do things. I have felt restless. I felt sad and depressed. I couldn't get excited about anything	H, M, L
Agressivity	I have tended to get angry easily. I overreacted in certain situations	H, M, L
Hopeleness	I felt that life had no meaning. I felt that I was worth very little as a person. I couldn't feel any positive feelings	H, M, L
Family mismatch	Physical assaults at home verbal assaults at home economic aggressions Psychological aggression Sexual aggression	H, M, L
Somatization	How satisfied are you with the quality of your sleep during quarantine? I felt my heart beat despite not having made any physical effort. I noticed that my mouth was dry. It was hard for me to breathe. I felt my hands shaking	H, M, L

The eight defined dimensions can be summarized in a global mismatch indicator, which in this case is described by the low, medium and high levels, which have been defined considering the 50th and 75th percentiles of the series of values as limits. The free software Weka 3.8 was used to debug and process the database.

The data analysis was carried out establishing between the sociodemographic variables and the index of affectation on the other. To establish descriptive relationships between the sociodemographic condition and the level of affectation, unsupervised algorithms such as clustering k mean were applied. On the other hand, two classification models were built to make predictions about the global variable of affectation, having as explanatory sociodemographic variables and the dimensions of affectation separately. Supervised classification models as neural networks were constructed. One data set was used for training and another for testing. Known metrics were used to select the model with the best fit, such as confusion matrix, precision, RMSE, relative error, Kappa index, margin an ROC curves.

In the second case, a database was used from a survey conducted by the national consulting center with 7,600 interviews with students, parents, and teachers from technical and technological, undergraduate, and graduate education institutions. The percentage of people who withdraw or change institutions, the cause of it and the inconveniences found with virtual classes is analyzed. For this, a decision tree algorithm was used. Table 2 shows the variables related to school dropout and their values. The value 0 represents answer no and 1 answer yes.

Table 2. Variables related to student dropout

Variable	Values
Health problem including risk of acquiring Covid 19	0, 1
Prefer no contact with other students or teachers	0, 1
You perceive that there is a greater cost to the benefit that is received	0, 1
Financial problem	0, 1
Doesn't like remote classes	0, 1
Family problem	0, 1

3 Outcomes

3.1 Results

The results and discussion are presented for the two cases considered; Sociodemographic relationship-degree of affectation and impact of COVID 19 on education. An analysis of the results found for each of the cases is presented below.

Case 1: Socio-demographic Relationship Level of Affectation
It is about establishing relationships between the level of affectation and the sociodemographic characteristics of the population. For this, it is first analyzed through an unsupervised association model, such as clustering.

Clustering
To establish associations between the dimensions of affectation and the global index, the unsupervised k-mean algorithm was configured with four clusters, it was applied to the dimensions of affectation. The algorithm configuration is as follows:

weka.clusterers.SimpleKMeans -init 0 -max-candidates 100 -periodic-pruning 10000 -min-density 2.0 -t1 -1.25 -t2 -1.0 -N 4 -A "weka.core.EuclideanDistance -R first-last" -I 500 -num-slots 1 -S 1.

Figure 1 shows the distribution of the clusters. A greater number of records is observed for cluster 2 in green color. Once the descriptive and predictive models were adjusted, a close relationship was found between some sociodemographic attributes related to education and work aspects and the psychological affectation.

The first cluster, with a weight of 45%, is characterized by Bogotá department, region, a low level of affectation for 95.54% of its members, single marital status, age 38 years, female biological gender, single marital status. Postgraduate educational level, socioeconomic level 3, family of 3, works remotely. Has not had rotating shifts, has not been fired or gone on unpaid leave, has left very little, 47.6 average days of confinement, level of information on COVID-19 high, always follow the protection recommendations.

Cluster 2 presents a participation of 23% of the total, by department Bogotá, region, a high level of affectation for 52.7% of its members, single marital status, age 29.88 years, female biological sex, state single civilian. Professional educational level, socioeconomic level 3, family of 4, works and studies remotely, has not had rotating shifts, has not been fired or taken unpaid leave, has left very little, 60.5 average days of confinement. Level of information about COVID-19 high, follow the protection recommendations almost always.

The participation of cluster 3 is 7% of the total population. The level of affectation is high, by department Chocó, Arauca and Amazonas Pacific region, a high level of affectation for 54.55% of its members, single marital status or free union, age 25.09 years, female biological sex, single marital status, technical education level. Socioeconomic level 2, family of 6, studies remotely and have quit the study, has been fired or taken unpaid leave, has left very little, 50.04 average days of confinement, high level of information about COVID-19, it almost always keeps the protection recommendations.

Cluster 4 participates with 25% of the total and has a low rate of affectation in 20% of its individuals.

Figure 2 shows the level of affectation for different sociodemographic variables.

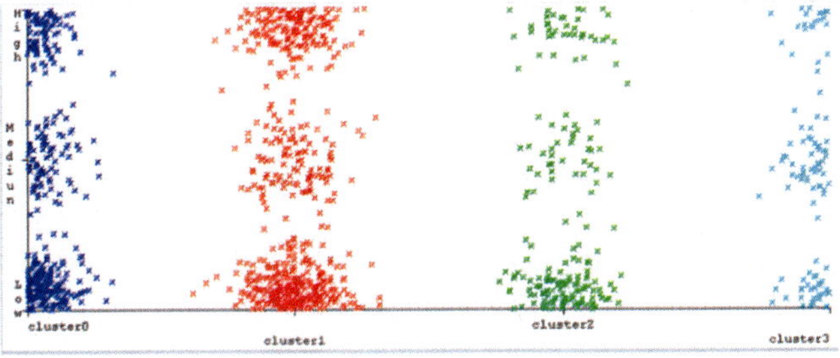

Fig. 1. Clusters for level of psychosocial affectation

Classification

To explore the correlation between the sociodemographic variables and the level of global affectation, several classification algorithms were used; Naive Bayes, Artificial neural network (ANN), Support Vector Machine (SVM). The one with the best performance in terms of metrics: (SMO) is presented.

To perform the classification of the response variable; level of psychosocial affectation with respect to the sociodemographic variables, an SMO algorithm configured in the following way was used: weka. Classifiers. Functions. SMO -C 1.0 -L 0.001 -P 1.0E-12 -N 0 -V -1 -W 1 -K "weka. Classifiers. Functions. Support Vector. PolyKernel -E 1.0 -C 250007" -calibrator "weka.classifiers.functions.Logistic -R 1.0E-8 -M -1 -num-decimal-places 4".

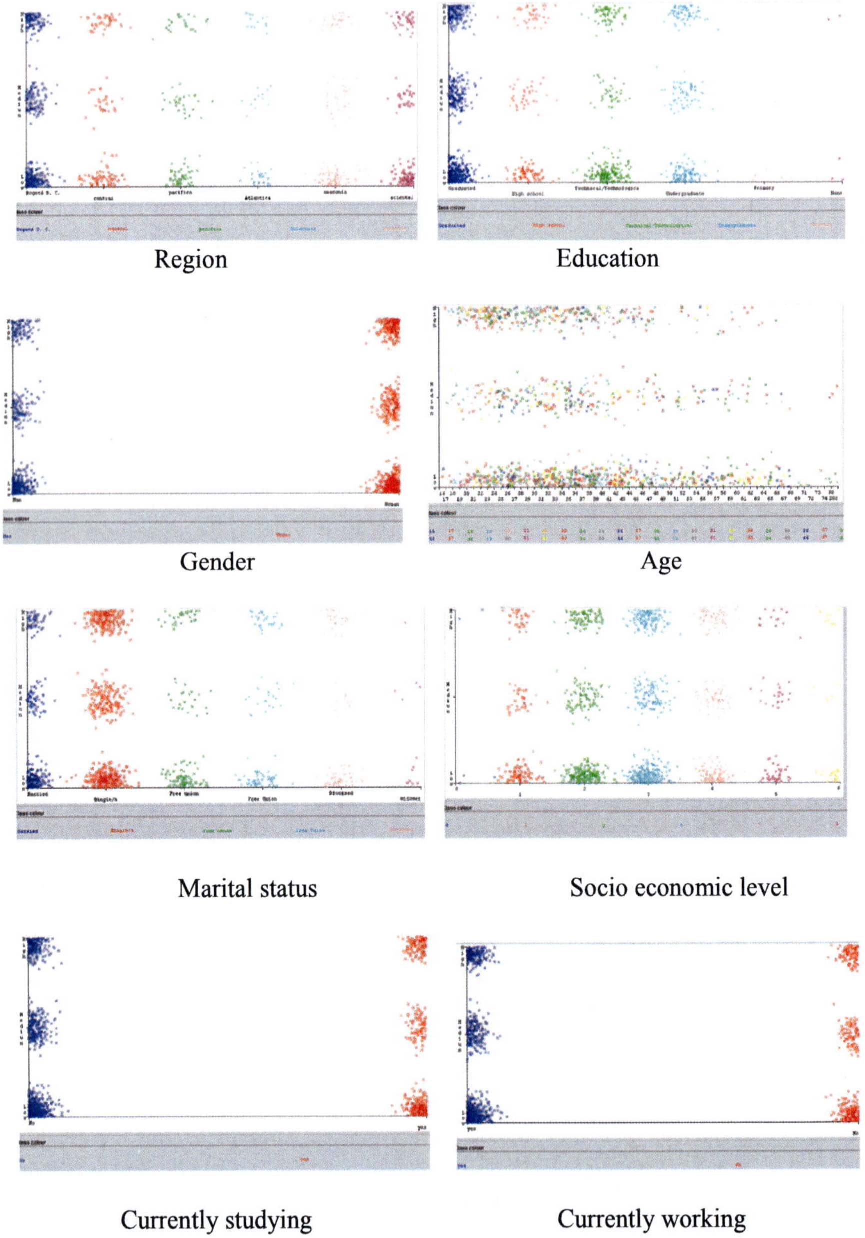

Fig. 2. Level of psychosocial involvement for sociodemographic variables

Figure 3 shows the graphs of different criteria used to evaluate the classifier SMO for level of psychosocial affectation. For the ROC curve, an area of 0.814 was obtained, while precision recall, which indicates a good precision of the classifier algorithm. Margin curve.

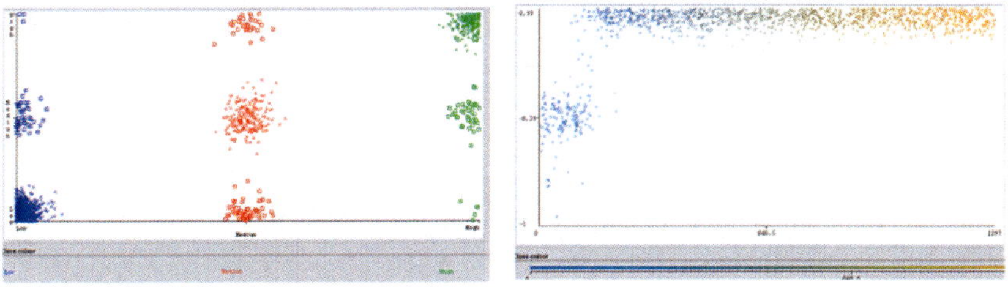

Classifier errors Predicted vs true

b margin curve – instance number

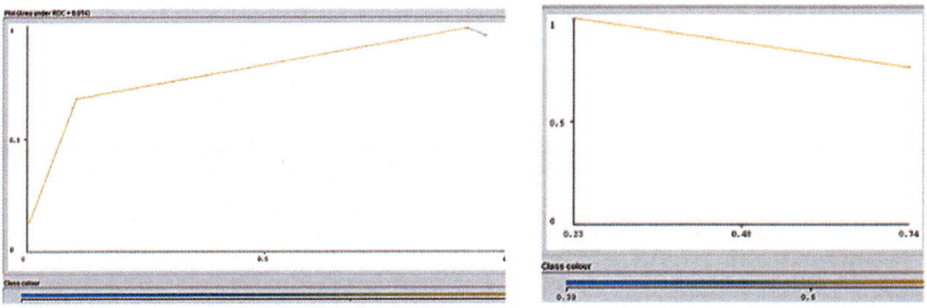

ROC TruePositive-FalsePositive

Presicion Recall

Fig. 3. SMO algorithm metric curves

Table 3 shows the metrics used to assess the performance of SMO algorithm. Kappa index is close to 1 which indicates a good fit, the same for RMSE and precision.

Table 3. Metrics for SMO algorithm

Metric	Score
Kappa statistic	0.8142
Mean absolute error	0.2483
Root mean squared error	0.3159
Relative absolute error	60.4829%
Root relative squared error	69.7283%
Total Number of Instances	1289
Precision	88.5958%

When developing the classification algorithms for this case, it can be observed that the relationship between the sociodemographic variables and the level of psychosocial involvement is not so clear. However, when questions about what happens to people in their family and personal environment during the pandemic are used as predictor variables, a relationship can be established. This indicates that the mere fact of belonging to a social conglomerate does not explain the affectation. This seems to be more related to the individual psychological characteristics of people than to their social environment. Here, characteristics such as character, temperament, resilience and in general all his individual and cultural experience are put in place.

Case 2: Covid 19 Impact on Education

For people who continued their studies, a decrease in their academic performance of close to 20% is observed due to the difficulty in accessing personalized tutorials, the formation of working groups, lack of feedback on evaluation results and face-to-face practice. This produced a feeling of frustration in about 30% of the university population and 55% in technical and technological. In rural areas, desertion was higher than in cities, with 25%, while in urban areas it was higher for strata 4 and below. Family coexistence improves performance for technical and technological students and did not have a significant impact on higher education students.

Tables 4 and 5 show the behavior of the variables associated with dropout and those with the difficulties of the virtual education, the correlation between dropout and its cause and between the opinion about virtuality and its drawbacks.

Table 4. Variables associated with desertion

Variable	Technique-technology	Undergraduate	Graduated
Health problem including risk of acquiring covid 19	52	53	66
Prefer no contact with other students or teachers	32	23	20
You perceive that there is a greater cost to the benefit that is received	0	10	7
Financial problem	8	8	3
Doesn't like remote classes	8	6	4
Family problem	0	1	0

Table 5. Problem with virtuality

Problem	Technical/Technological	Undergraduate	Graduated
Internet connection	52	56	56
Concentration problem on the part of students	38	40	44
Digital ownership	26	25	30
Not having a computer tablet or cell phone	28	22	24

(continued)

Table 5. (*continued*)

Problem	Technical/Technological	Undergraduate	Graduated
Device use capacity	20	15	18
Difficulty following instructions in class	21	19	20
Technical quality or class preparation	18	18	13
Inattention teacher time	14	12	11
Academic agenda	16	12	11
None	14	12	7

Classification Algorithms

To analyze the relationships between dropout and its possible causes, as well as those related to virtual education, the tree algorithm J48 which is a variant of C4.5 has been used.

Variable Desertion Cause Relationship

The algorithm setting is weka. Classifiers.trees.J48 -C 0.25 -M 2 size 17. In Fig. 4 it is identified that the most important attribute for dropping out is they did not like remote classes followed by COVID-19, higher cost than benefit, financial problems, and they did not want contact with teachers and students. In the branch, they observe the values 0 (no) and 1 (yes), while the leaves show the number of records associated with the node and the total number of records.

The root node is disliked remote classes and decision nodes: financial problem and prefers to avoid contact with students and teachers.

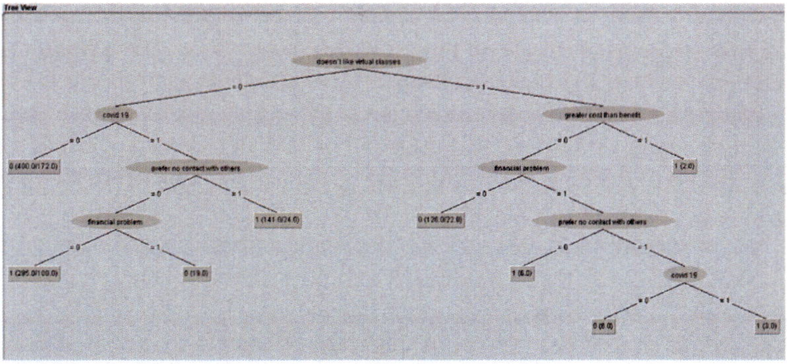

Fig. 4. J48 algorithm application for Weka 3.8 withdrawal class classification

Relationship of Variable Virtual Education Problems

A J48 algorithm is used with the following configuration: weka.classifiers.trees. J48 -C 0.25 -M 2 with 31 sheets and size 55. In Fig. 5 we can observe than the root node is a problem of concentration in the classes and the decision nodes internet connection difficulty following instructions and class quality.

Table 6 presents the performance of algorithm J48 for the relationship between dropout and it cause, and between the difficulties of virtual classes and with-drawl.

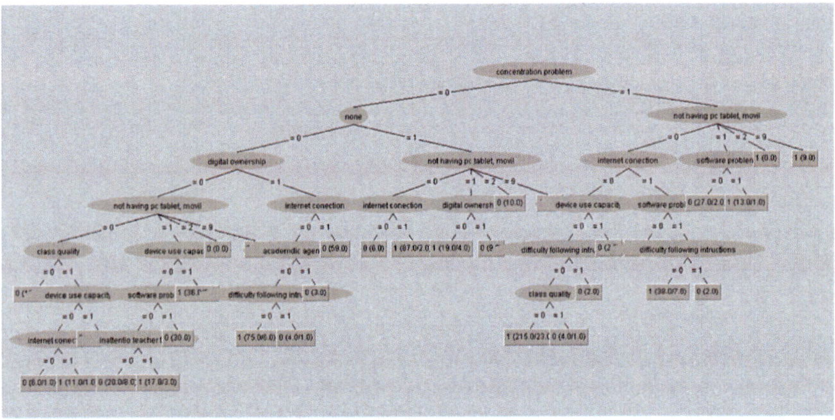

Fig. 5. Decision tree J48 for relationship cause withdrawal

When analyzing the cases of possible withdrawal of people from the institutions and relating them to their causes, it is observed that, contrary to what might be thought, financial issues are not the main cause of desertion. This has focused more on the fear of the pandemic and its consequences, as well as in a way the use of the virtual class system. From the analysis of the causes associated with virtual classes, it is observed that a significant percentage of students and parents consider that the quality of education provided in this way is deficient. In this topic, when the causes for which people do not find this type of relationship and instruction convenient are analyzed, it is observed. That is in part it due to the organization of classes and evaluations and on the other hand to the equipment and facilities related to the virtual media (computers, tablet, internet access, etc.). In the first case, the institutions must ensure that there are the necessary resources and logistics to take advantage of the potentiality of virtuality.

The use of industry 4.0 tools can improve students performance. Among others; augmented reality, virtual reality, online evaluations with immediate feedback such as grader, the use of digital twins as well as the deployment of resources in the cloud that can be accessible at any time. Some of them can be integrated with platforms such as moodle They are tools that can complement the treacherous face-to-face methods and interaction between students and between them and teachers. Another aspect that may be relevant is that due to the lack of face-to-face tutoring, students' autonomous work time has increased, a situation conducive to stimulating creativity, dynamism, and self-management.

Table 6. J48 Classification algorithm performance

Metrics	Cause withdrawl relationship	Virtual class withdrawl relationship
Kappa statistic	0.358	0.747
MAE	0.4094	0.1857
RMSE	0.4543	0.3206
RAE	81.98	37.1876
RRSE	90.92	64.13
Precision	67.76	87.38

From the pedagogical point of view, it can change the balance of time dedicated to learning, particularly between autonomous work activities and face-to-face work. Additionally, the savings in travel means more time for other activities.

In Colombia, in particular, there is a difficulty that internet coverage is limited, especially in rural areas that represent 20% of the school-age population. Digital appropriation is close to 83%, but at an advanced level only 6% use this medium. The indicator of technical appropriation is 0.2 while in developing country is 0.4. The intensity, speed, and frequency of the internet signal are not adequate for these uses. The digital Gini index in Colombia at 0.53. Due to these inequalities in the digital subject, the gap in the learning process has increased with virtuality.

4 Conclusions/Recommendations/Summary

Association and classification algorithms were used to determine the relationship between the sociodemographic variables and the level of psychosocial affectation due to the COVID-19 pandemic, in inhabitants of different regions of Colombia.

The K-means association algorithm divided the population into four clusters and a high level of involvement is observed in 2 and 3.

The classification algorithm used to predict the level of affectation depending on the sociodemographic variables, obtained high precision scores, indicating a good fit of the SMO model.

The use of machine learning allows the design of public policies aimed at the care of people with high psychosocial affectation in work, school, and family settings.

The observed impact is greater in the female population between 23 and 40 years old belonging to middle-income socioeconomic strata, the dimensions most sensitive to the affectation are anxiety and depression.

About 19% of people have dropped out of the study, due to difficulties like job loss, lack of access to practical and internet sessions and lack of conformity with the quality of education.

The development of evaluation programs and virtual practices as a complement to face-to-face systems can contribute to improving performance.

Prediction algorithms make it possible to establish relationships between sociodemographic variables, the rate of impairment and dropout, as well as academic performance.

References

1. Valdivieso MM, Burbano Pantoja VM, Burbano Valdivieso AM (2020) Percepción de estudiantes universitarios colombianos sobre el efecto el confinamiento por el coronavirus y u rendimiento académico. Rev Espacio 41
2. Mundial B (S.f.) Personas que usan Internet con base en Unión Internacional de Telecomunicaciones. In: Informe sobre el Desarrollo mundial de las telecomunicaciones/TIC y base de datos. https://datos.bancomundial.org/indicador/IT.NET.USER.ZS
3. Hector Hugo SC, Katia Betty MS. Investigaciones en salud mental en condiciones de pandemia por el covid 19 In memoria el primer encuentro virtual univeria ricaro palma
4. Castells M (2001) Internet y la sociedad red. La factoría 14(15):1–12
5. García-Espinosa P, Ortiz-Jiménezb X, Botello-Hernándeza E, Aguayo-Samaniegoa R, Leija-Herreraa J, Góngora-Riverac F (2021) Psychosocial impact on health-related and non-health related university students during the COVID-19 pandemic. Results of an electronic survey. Rev Colombian Psiquiat 50(3):214–224
6. Huerta Fredy Z, Taylor Gisselee R, Maria Paola G, Yuli B. Diagnostico de la situación de los estudiantes universitarios en Bogotá por impacto de coronavirus (Covid 19). https://repository.ean.edu.co/handle/10882/10707. Accessed July 2021
7. https://tesis.usat.edu.pe/bitstream/20.500.12423/1826/1/TL_SandovalValenzuelaCarolain.pdf. Accessed Aug 2021

A LabVIEW Application Implemented for Simulating the Working Principle of the Brain-Computer Interface

Oana Andreea Rusanu(✉) and Ileana Constanta Rosca

Product Design, Mechatronics and Environment Department, Transilvania University of Brasov, Brasov, Romania

oana.rusanu@unitbv.ro

Abstract. Electroencephalography offers the possibility of exploring the electrical activity of the brain. A brain-computer interface system helps the neuromotor disabled persons, establishes an alternative communication pathway, and offers a non-muscular control channel without involving the peripheral nerves. The biopotentials triggered within neurons are analyzed and translated into control signals for a computer application, versatile assistive mechatronic systems, and other devices like neuroprosthesis. The EEG technique is recommended for recording the signals used by a brain-computer interface. This paper's primary purpose is to develop a software application providing a virtual simulation ("BCIRobot_VirtualSIM") using NI LabVIEW graphical programming environment for EEG signals based on controlling some small 3D robots. This virtual instrument represents the original contribution of this paper, whose aim is to demonstrate the brain-computer interface (BCI) based operating principle. The LabVIEW application accomplished the following stages: the design of two 3D humanoid robots composed of cylindrical and rectangular geometrical shapes, the simulation of the EEG signals acquisition by inserting the raw numerical values gathered from an online source, the manual and the automatic control of the virtual humanoid robots by selecting the Knob and the Radio buttons.

Keywords: Brain-computer interface · Electroencephalographic signal · LabVIEW simulation

1 Introduction

The brain-computer interface (BCI) is a technological system that develops biomedical applications and improves nonclinical activities. A BCI system is vital due to its strong impact on people with neuromotor disabilities and communication incapacities. These persons need permanent supervision and assistance from a relative during their daily activities. The BCI has the potential to allow people previously diagnosed with severe diseases (lateral amyotrophic sclerosis, cerebral stroke, spinal cord injuries, tetraplegia) to get recovered. The BCI users could benefit from replacing the supervisor with

M. E. Auer et al. (Eds.): REV 2022, LNNS 524, pp. 651–663, 2023.
https://doi.org/10.1007/978-3-031-17091-1_63

a mechatronics system (motorized wheelchair [1], robotic arm, neuroprosthesis, multi-functional robot [2]) that can be mentally controlled by the signals acquired from the brain's electrical activity.

The role of a BCI system is to provide an alternative communication and control channel, offering the advantage of not involving the peripheral nerves and the muscles. Therefore, a BCI analyzes the biopotentials generated by the cerebral neurons and translates them into control signals of computer and mechatronics systems. The BCI associates the patient's mental intentions or thoughts with a command transmitted to a device from the outside environment. Like the previous technologies (eye movement tracking, touch screen, hand gesture, and voice recognition), the BCI offers a more natural and attractive interaction between the human and the computer. Therefore, nonclinical and commercial applications involving virtual reality, artificial intelligence, cognitive states, human vigilance, and entertainment were created.

This paper presents an original LabVIEW virtual instrument to illustrate the concept of a BCI based on the EEG technique and the motor imagery mental task. The developed software application allows the understanding and the simulation of the working principle underlying a BCI prototype by highlighting its innovator potential in extending the applications of the biomedical engineering research field. The LabVIEW application was implemented by simulating the acquisition of the EEG signals using the default values determined after real-life experiments conducted to create online databases [3]. The values transmitted by the EEG signals allow the control of simple 3D humanoid robots included by the LabVIEW application. Considering the situation when the complete EEG equipment is available, it is possible to implement the EEG signals' advanced processing techniques. The neuromotor disabled patients could mentally control the 3D robots to gather the cognitive abilities necessary to express their mental intentions related to commands aimed for an assistive mechatronic system.

The 3D robots designed in the LabVIEW simulation presented in this paper are classic and straightforward, targeting learning and teaching purposes. The LabVIEW software application does not intend to show the rigorous robots' construction but to demonstrate the BCI system's working principle. The initial construction and manual animation of the simple 3D robots were accomplished considering a YouTube video tutorial [4]. The EEG data were simulated based on the previously mentioned online source and the LabVIEW methods indicated by a video tutorial [5]. The design and development of the automatic control of the virtual 3D robots in correlation with the numerical values transmitted by the simulated EEG signals are graphically displayed to represent the original contribution of this paper. This programming stage is fundamental because it proves the BCI system's working principle.

It is difficult to identify clear evidence [6] in the current scientific literature showing other research work simulating a brain-computer interface [7]. The virtual demonstration of the sequential stages corresponding to using a BCI system [8] could be helpful in the beginning level of developing BCI projects and necessary for students and novice researchers that need an inexpensive, attractive, and efficient virtual instrument.

The paper is structured as it follows: Sect. 2 highlights the previous research work on simulated BCI systems, Sect. 3 presents the design of the front panel (the graphical user interface) and the implementation of the block diagram (the source code) of the original

LabVIEW application aimed for the BCI simulation, Sect. 4 is related to the results and discussions, and Sect. 5 highlights the conclusions and the future research directions.

2 Previous Research Work on BCI Based Virtual Simulations

Most frequently, researchers reported their results [9, 10] underlying research of combining the virtual reality systems (3D smart home) with the P300 or SSVEPs (Steady-State Visual Evoked Potentials) EEG signals based BCI systems. Regarding the previous work of developing virtual simulations underlying the brain-computer interface, the paper [11] presents the integration between a cVEP-BCIs approach and a 3D kitchen scenario. There were also published papers [12, 13] presenting immersive desktop-based virtual environments for training on achieving BCI-based control of the surrounding objects used in daily activities. The previously mentioned research articles implied the development of complex software applications, employing expensive EEG acquisition systems, and further experimentation involving several subjects. Otherwise, the paper [14] focused on validating and testing the proposed wheelchair simulator game for BCI controlling tasks using an online database comprising the related EEG signals.

3 BCIRobot_VirtualSIM – A LabVIEW Software Application Enabling the Simulation and the Fitness Training Aimed for the BCI Underlying the Control of Simple 3D Humanoid Robots

3.1 The Simulation of a Brain-Computer Interface – The Basic Mode

The Front Panel is the graphical user interface (Fig. 1) of the LabVIEW software application, which comprises the following features:

- The 3D Scene – includes a simple 3D humanoid robot composed of cylindrical geometrical shapes;
- A graphical representation – called Waveform Graph in LabVIEW – for showing the numerical content of the EEG signal;
- A panel with Knob buttons;
- An area with Radio buttons of LEDs shapes;
- A virtual chronometer built of a display system with seven segments

The role of the chronometer is to indicate the time interval a patient spent on the mental task by imagining the movement of an arm or a leg. The patient should spend a specific time interval learning how to accurately execute the motor imagery of an arm or a leg to achieve the mental command of the movement of an element from the 3D structure underlying the simple 3D humanoid robot. Thus, the virtual chronometer assigns the additional functionality of training to the LabVIEW software application.

The following paragraphs present a general overview of the LabVIEW software application – Basic Mode to highlight the role of this virtual instrument.

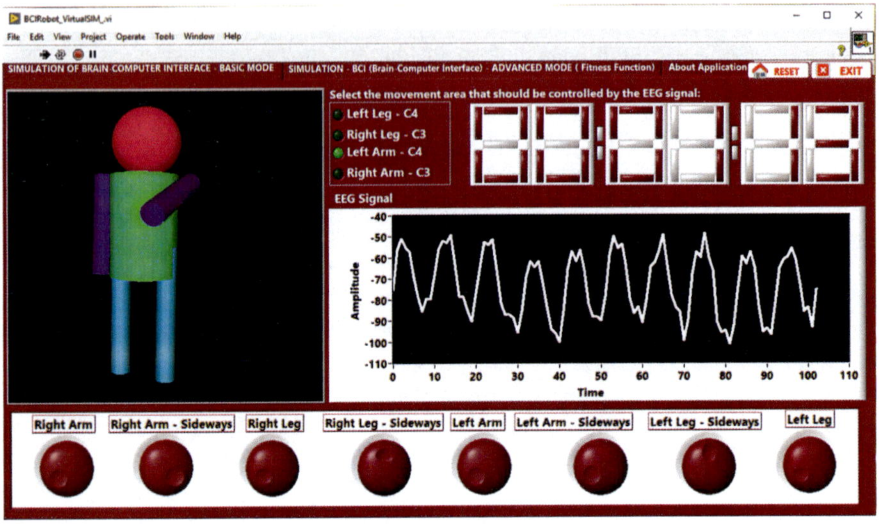

Fig. 1. The Front Panel (the graphical user interface) of the LabVIEW application aimed to simulate the brain-computer interface – the basic mode.

The graphical representation displays the EEG signal acquired by monitoring the cerebral electrical activity when the patient imagined the movement of his/her arm or leg. At the moment of visualizing the EEG waveform graphical representation, it is considered that it is monitored the neuronal activity of the patient, who imagined the movement of the superior or inferior limbs by the direction forward-backward. The simulation of the EEG signal was preferred considering the situation when it is impossible to enable the real-time EEG signal acquisition by using the proper equipment composed of a headset with electrodes, and there are no available subjects who may accept to be involved in such an experiment.

The LabVIEW software application simulates the mental task of imagery by selecting a specific Radio button. Identifying the limb, whose movement the patient imagined, is possible by choosing the corresponding Radio button called in an intuitively way: “Left Leg – C4”, “Left Arm – C4”, “Right Leg – C3”, and “Right Arm – C3”. These names include the central areas C4 and C3, which are the locations where EEG signal variations occur, according to the imagined movement of an arm or a leg.

As a result of the mental task associated with the motor imagery of the superior or inferior limbs or the intention to execute the natural motion of these limbs (arms or legs), the simple 3D robot is animated. In parallel with graphical displaying of the EEG signal, it can be is observed how the robot moves a specific limb (arm or leg). The input values are introduced either by the raw values of the EEG signal graphically displayed or by adjusting the Knob buttons. The EEG data occurs after selecting the corresponding Radio buttons when the movement of the limbs is executed in an automatic, repetitive, and continuous way. The EEG variations are not graphically displayed by adjusting the Knob buttons when the movement of the limbs is executed in a saccadic and manual way.

The simple virtual 3D robot replaces a real robot or a specific application of the brain-computer interface based on the electroencephalographic technique and the motor imagery. A robotic arm can substitute the virtual 3D Scene (robot) that the patient can mentally control. The 3D figure can be further linked to a mechatronic system that should execute a command considering the patient's intention. An example of a command is related to catching and lifting a cup with the help of the robotic arm so that the patient benefits from support for drinking water. Likewise, the necessary mental command is represented by the same task of motor imagery of a specific limb (the left or right arm, as well as the left or right leg) that was considered in the case related to the control of the simple robot designed in the simulation implemented in the LabVIEW graphical programming environment. The output value, which proves the achievement of the mental task associated with the motor imagery of the superior or inferior limbs, is determined by the view that shows how the robot moves the corresponding limbs.

3.2 The Simulation of a Brain-Computer Interface – The Advanced Mode Featured with the Fitness Training Functionality

The Front Panel or the graphical user interface (Fig. 2) of the LabVIEW software application comprises the following features:

- A 3D scene composed of a bit more versatile virtual humanoid robot (made of rectangular geometrical shapes)
- A graphical display – called Waveform Graph in LabVIEW – for showing the numerical features of the EEG signal;
- Panels that include Knob buttons;
- A combo box input control represented by a list with several options;
- A series of buttons of Boolean type (that are located in the Controls Palette – Silver/Boolean/OK Button);
- A picture showing a map of the cerebral areas that indicate the locations where the EEG signal is acquired;
- Input controls of LED type for selecting the area (C4, C3, Cz, P4, P3) are associated with the user's intention to display the EEG waveform.

The additional functions, intuitively defined as "Fitness" that simulate simple sports activities or some exercises for moving the arm and forearms, are helpful during a medical rehabilitation schedule aimed at old age people. These persons would benefit from the assistance of the virtual robotic instructor that should determine them to perform specific movements towards the superior and inferior arms limbs.

The "Fitness" functionality added in the Advanced Mode of the LabVIEW application highlights the additional feature of providing physical training. Unlike the Basic Mode that allows the patients to have mental training for gaining the ability of motor imagery towards a specific limb, the Advanced Mode enables the patients to have physical training for the real-life movement of the left or right leg or arm.

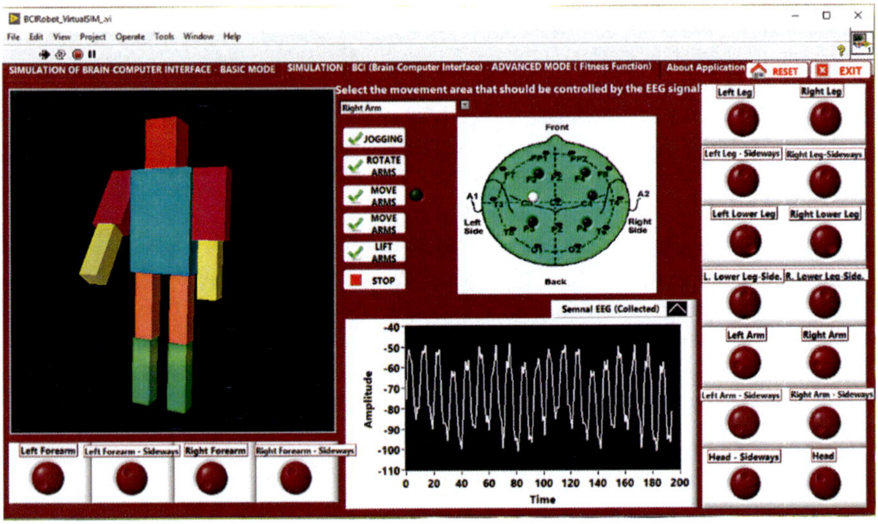

Fig. 2. The Front Panel (the graphical user interface) of the LabVIEW application aimed to simulate the brain-computer interface – the advanced mode.

3.3 The Simulation of a Brain-Computer Interface – Tab: About Software

The Front Panel includes the third tab window called "About Software," which comprises a short description of the working principle underlying the proposed LabVIEW application to simulate a brain-computer interface. The presented information is structured concerning the basic and advanced BCI simulation modes.

3.4 The Block Diagram of the LabVIEW Application for BCI Simulation

The block-diagram (source code) of the virtual instrument allows the below features:

A). Design of the geometrical figures from the construction of the 3D virtual simple humanoid robots. This stage involves:

a). Using the specific functions of the 3D Picture Palette from LabVIEW: Create Object, Create Box, Create Sphere, Create Cylinder;
b). Using the property nodes: Set Drawable, Add Object.

B). Manual and automatic control of the 3D humanoid robots. This stage includes:

a). A While Loop (repetitive programming structure with a stop condition) comprises a Flat Sequence structure with two frames. The first frame contains:
b). Certain functions for enabling the animation or the controlling of the 3D robots: Clear Transformation, Rotate Object, Translate Object, and Scale Object;
c). Several While Loops that enable the graphical displaying of the EEG signal by using the functions: Simulate Arbitrary Signal, Collector and the filter Butterworth;

d). Several Case Structures whose sequences are accessed in the Front Panel (graphical user interface) by selecting the Radio buttons (in the Basic Mode of the BCI simulation) and the Combo Box with the scrolling list (in the Advanced Mode);
e). A While Loop containing particular functions for implementing the algorithm underlying the chronometer shown as a graphical display system with seven segments;
f). Programming sequences composed of the Case Structure, the While Loop, and the Flat Sequence Structure for running the additional "Fitness" functionality that simulates some activities such as jogging and different exercises for physical training towards arms and forearms.

C). Display the result – the 3D Scene for animating or controlling the simple 3D humanoid robots, which comprises the programming structure and functions added into the third window of the Flat Sequence Structure.

The programming structures, functions, and property nodes (Fig. 3) added in the block diagram are presented in the paragraphs below.

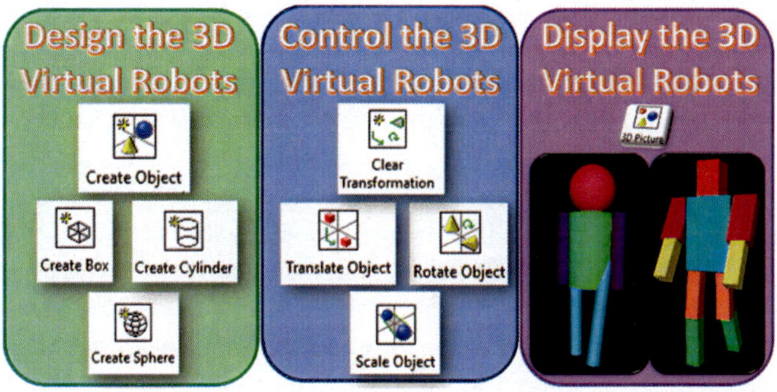

Fig. 3. The structure of the software implementation underlying the proposed BCI simulations.

Designing the elements (head, body, inferior and superior limbs, legs, and forearms) composing the structure of the simple 3D humanoid robots
The function Create Object (Fig. 4) was used to build a 3D element displayed in the Front Panel (graphical user interface). This function was necessary to build each segment from the two virtual humanoid robots. The elements can be specified at the input string parameter called Name featured by Create Object function.

By opening the contextual menu (right-click) of the function Create Object, the following methods are selected: Create/Method for Scene Object Class/Set Drawable. The Drawable attribute (Fig. 4) sets the cylindrical or rectangular shape of the elements from the structure of the 3D humanoid robots included by the basic and advanced modes of the LabVIEW application aimed for the brain-computer interface-based simulation.

The function Create Cylinder was called to set the cylindrical shape, and the corresponding values were introduced for sizes of details, height, and radius. To set the rectangular shape, the function Create Box was used, and the necessary values were applied to obtain the lengths by the directions of the axis X, Y, and Z (height, length, and the base width). The function Create Sphere was added to set the spherical shape, and the values were defined for input dimensions of details and radius. The color input parameter was specified for each function: Create Cylinder, Create Box, and Create Sphere (Fig. 4).

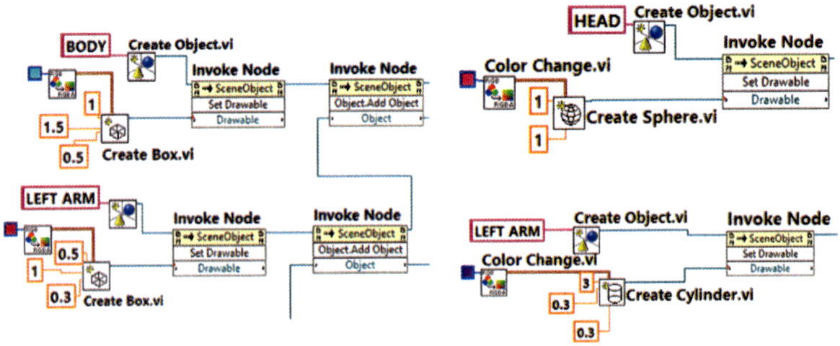

Fig. 4. A sequence of the LabVIEW block diagram showing the Create Object, Create Box, Create Sphere, Create Cylinder functions and the Invoke Nodes – Set Drawable and Add Object.

The corresponding values were established based on different trials and after successive experiments to obtain the humanoid appearance of the 3D robots. Moreover, the target was to balance and get symmetry between the dimensions of the elements composing the 3D humanoid robots.

The construction of the simple 3D robots, included by the basic and advanced modes of the LabVIEW application for BCI simulation, involved using another property node, called Add Object. The Add Object property node can connect many elements to the main component. This property node was introduced in the Block Diagram by accessing the contextual menu (right-click) and selecting the following sequence of methods: Create/Method for SceneObject/Class/Object/Add Object.

By considering the example of building the 3D robot included by the BCI simulation's primary mode, a series of elements were connected to the "body" fundamental component: head, right arm, left arm, left leg, left leg. A similar approach was considered for creating the 3D robot included by the advanced mode of the BCI simulation, but in this case, an additional condition had to be accomplished: the arms were linked to the forearms, and the inferior limbs were connected to the legs. The output parameter of the property node Add Object was connected to the input terminal of the 3D Picture function aimed to display the 3D Scene of the simple humanoid robot shown in the Front Panel.

The manual control (by using the numerical values transmitted by selecting the corresponding Knob buttons) and the automatic control (by using the numerical values sent by the graphical representation of the EEG signals) of the 3D robots
It should be emphasized that the EEG signal is displayed on the graphical element added to the Front Panel only after selecting the Radio button corresponding to controlling the movement of the desired segment: "Left leg – C4", "Left-arm – C4", "Right leg – C3"

and "Right arm – C3". No signal variations will be displayed after selecting a specific Knob button. The Knob buttons allow both the movement by the forward-backward direction and the sideways direction of a particular segment from the structure of the humanoid 3D robots.

The EEG signals are displayed only after selecting the Radio buttons that determine a single movement by forward-backward direction of a particular segment included by the construction of the 3D robots. The role of the Knob buttons is revealed in the programming stage of the LabVIEW-based BCI simulation. At that stage, adjustments and calibration operations were applied for the automatic movement of a particular segment included by the 3D structure of the robot after transmitting the numerical values of the EEG signal.

Likewise, the Knob buttons are necessary for making a comparative analysis between the automatic control of the 3D robot as it was triggered by the EEG signals embedding the features corresponding to executing a mental task of motor imagery with the manual control of the 3D robot as it would be executed in a usual way by involving the peripheral nerves and the muscles. Thus, there can be observed differences and similarities between the imagined movement (the automatic movement of the virtual limbs obtained by selecting the Radio buttons that send the values corresponding to the EEG signals) with the natural movement (the manual movement of the virtual 3D structures resulting by selecting the Knob buttons that do not trigger any signal variation or graphical displaying).

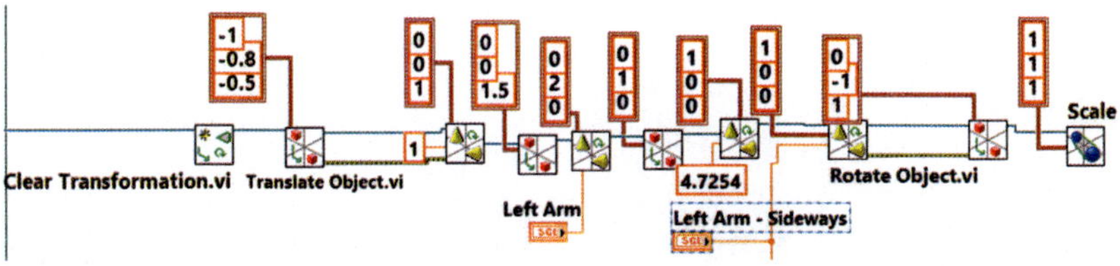

Fig. 5. The block diagram (the source code) of the LabVIEW-based BCI simulation – the sequence of functions aimed for controlling the 3D robots by using the following functions: Clear Transformation, Translate Object, Rotate Object, and Scale Object.

Orienting the movement of the inferior and superior limbs included by the construction of the 3D humanoid robots to reproduce a human character's actions was achieved with the help of the function Rotate Object (Fig. 5). The function Rotate Object plays an essential role in controlling the 3D humanoid robots. The numerical constant identifies the rotational axis and the numerical input variable used to send the values necessary to move a specific segment (leg, arm, or head). The input parameters of this function get updated to specify the corresponding values for the rotation axis (X, Y, Z) and the rotational angle. These input values can be changed or renewed during the running of the BCI simulation by accessing a Knob button which is the input parameter of the function Rotate Object, as it was previously stated. The input values can be introduced not only by using this numerical control but also by using the value transmitted by the EEG signal. In this last case, the control attached to the function Rotate Object will be replaced by another control used for transmitting the numerical values resulting after

EEG signal displaying embedding the features of a particular mental task of imagining the movement of the inferior or superior limbs from the 3D robots.

It should be highlighted that the numerical values, which define these specific features of the EEG signal, were introduced by accessing the functions palette Express/Input/Simulate Arbitrary Signal. It should be emphasized that the numerical values were acquired from the online source [3], and they correspond precisely to the cerebral variations that are triggered after imagining the movement of an inferior or a superior limb. This way, it is considered that the EEG signals, which were graphically displayed in the Front Panel, were previously undertaken to the stages of amplification, analog-digital conversion, processing, filtering, features extraction, and classification. The mentioned stages were not considered or implemented in the block diagram of the BCI simulation, except for the filtering stage that involved the use of a digital filter called Butterworth Filter/Highpass necessary for the EEG signal corresponding to the imagined movement of the left leg and the right leg. Nevertheless, this filter was added for a teaching or experimental purpose without involving the experimental results.

The waveform that resulted after the filtering stage was graphically displayed. The function Signal Collector provides the numerical values necessary in the graphical representation of the EEG signal at each moment. Both the function Signal Collector and the function Simulate Arbitrary Signal are included in the content of the While Loop (Fig. 6), which allows the repetitive representation of the EEG signal variations during the running of the LabVIEW-based BCI simulation.

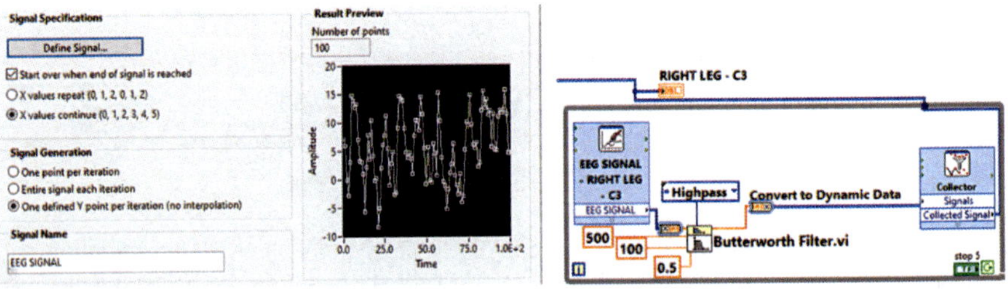

Fig. 6. The sequence of the block diagram that includes the functions necessary to display the EEG signal: Simulated Arbitrary Signal, Collector, and Butterworth Filter.

The additional functions called Fitness, which are related to the simulation of activities such as jogging or several physical exercises for arms and forearms were implemented in the advanced mode of the BCI simulation by providing parallel control of a series of segments from the 3D robot. The simple 3D robot alternatively moves its superior and inferior limbs regarding the jogging activity. For achieving the parallel control of a minimum of two segments from the construction of the 3D robot, several programming sequences were introduced, including a case structure, a while loop, and a stacked sequence. A vital aspect consisted of setting default values for the numerical Knob input control (for example, "Left leg - C4", "Right arm – sideways," attached to the angle parameter of the function Rotate Object. These default values should allow the synchronous movement of the controlled limbs determining the jogging activity.

Displaying the result - the 3D Scene for animating or controlling the simple 3D humanoid robots
By using the indicator 3D picture, found in the second window of the Flat Sequence structure from the block diagram, it is displayed the result from applying the functions of translating and rotating (by using the numerical values transmitted by the EEG signal) towards the constructed model of the 3D simple robots.

4 Results and Discussions

The YouTube Unlisted link [15] includes a video demonstration with the running mode of the LabVIEW application aimed for the simulation of the brain-computer interface simulations. This way, there can be viewed the results obtained by designing the Front Panel (the graphical user interface) and the Block Diagram (the source code) of the presented LabVIEW virtual instrument allowing the manual and automatic control of the 3D humanoid robots to demonstrate the working principle underlying the brain-computer interface. It should be noticed that the proposed solution of BCI simulation, discussed in this paper, does not include the advanced techniques of processing the electroencephalographic signals, the methods of features extraction, and the artificial intelligence-based classification of the neuronal biopotentials.

This paper presented a convenient software application based on defining the EEG signal by using the raw numerical values provided by a beneficial online database obtained after performing real-life experiments of motor imagery. Likewise, the built 3D robots represent simple models composed of general geometrical shapes aimed at teaching and learning. Several challenges need to be overcome in a real-life LabVIEW control of a complex mechatronic system, such as a mobile robot, a robotic arm, or a robotic hand. Moreover, further improvements will be required to the current software framework to integrate the EEG data acquisition systems with the hardware components, including the microcontroller or FPGA-based development boards.

5 Conclusions

This paper presented a LabVIEW application necessary for simulating and demonstrating an understanding of the brain-computer interface system's working principle. The LabVIEW virtual instrument is composed of two fundamental and advanced running modes. The primary stage includes a simple 3D humanoid robot composed of cylindrical geometrical shapes that can be manually controlled in two movement directions by using some Knob buttons. The advanced stage comprises a little bit complex 3D humanoid robot composed of rectangular geometrical elements that can also be automatically controlled on a single movement direction by using some Radio buttons that transmit the raw numerical values corresponding to the electroencephalographic signals acquired executing actual experiments of motor imagery. The LabVIEW application constitutes the original contribution of this paper, and it was implemented by using both general programming structures (while loop, flat sequence, case structures) and express palettes of functions (3D Picture Control, Advanced Signal Processing).

Future research directions are related to developing a natural brain-computer interface system composed of all the necessary stages requiring software and hardware components to enable the EEG signals acquisition, filtering, processing, features extraction, and classification to control a mechatronic device. Moreover, the aim is to integrate the LabVIEW virtual simulation with a natural BCI assistive system to enable an additional feature related to the mental training of the user who needs to achieve a specific cognitive ability.

References

1. Isah RO, Olaniyi OM, et al. (2020) A smart omnidirectional controlled wheelchair. J Eng Sci 4:88–102. https://doi.org/10.5281/zenodo.4288291
2. Izzuddin TA, Safri NM, Othman MA (2021) Mental imagery classification using one-dimensional convolutional neural network for target selection in single-channel BCI-controlled mobile robot. Neural Comput Appl 33(11):6233–6246. https://doi.org/10.1007/s00521-020-05393-6
3. ElectroencephaloGram (EEG) – Laboratory Simulation – College of Engineering, Dune (An Autonomous Institute of Government of Maharashtra). https://bmsp-coep.vlabs.ac.in/Electroencephalogram/Electroencephalogram.html. Accessed 12 Jan 21
4. Objets 3D dans LabVIEW – VBTech YouTube Tutorial. https://www.youtube.com/watch?v=CCztaBkEFMw. Accessed 12 Jan 21
5. Simulated ACI – TheHelloiii YouTube Tutorial. https://www.youtube.com/watch?v=Phw9Sosyz2M. Accessed 12 Jan 21
6. Lindgren J, Merlini A, et al (2018) simBCI-A framework for studying BCI methods by simulated EEG. In: IEEE transactions on neural systems and rehabilitation engineering, institute of electrical and electronics engineers, vol **26**, no 11, pp 2096–2105. https://doi.org/10.1109/TNSRE.2018.2873061
7. Huang X, Xue X, Yuan Z (2020) Simulation platform for the brain-computer interface (BCI) based smart wheelchair. In: Sun X, Wang J, Bertino E (eds) Artificial intelligence and security. ICAIS 2020. Lecture Notes in Computer Science, vol 12239. Springer, Cham (2020). https://doi.org/10.1007/978-3-030-57884-8_23
8. Alchalabi B, Faubert J (2019) A comparison between BCI simulation and neurofeedback for forward/backward navigation in virtual reality. Comput. Intell. Neuroscience. Hindawi J. https://doi.org/10.1155/2019/2503431
9. Zhao W, Zhang X, Qu J, Xiao J, Xiao Y, Huang A (2019) Virtual smart home based on EEG control. In: 2019 IEEE 9th international conference on electronics information and emergency communication (ICEIEC), pp 85–89. https://doi.org/10.1109/ICEIEC.2019.8784485
10. Stawicki P, et al (2018) SSVEP-based BCI in virtual reality - control of a vacuum cleaner robot. In 2018 IEEE international conference on systems, man, and cybernetics (SMC), pp 534–537. https://doi.org/10.1109/SMC.2018.00749
11. Riechmann H, Finke A, Ritter H (2016) Using a cVEP-based brain-computer interface to control a virtual agent. In: IEEE Trans Neural Syst Rehab Eng, vol **24**, no 6, pp 692–699, June 2016 https://doi.org/10.1109/TNSRE.2015.2490621
12. Schwarz A, et al (2020) Decoding hand movements from human EEG to control a robotic arm in a simulation environment. IOP Sci J Neural Eng 17:036010. https://doi.org/10.1088/1741-2552/ab882e
13. Yang B, Zhang T, Duan K (2015) Development of a BCI simulated application system based on Unity3D. In: 2015 international conference on virtual reality and visualization (ICVRV), pp 277–280. https://doi.org/10.1109/ICVRV.2015.30

14. Pinheiro OR, Alves LRG, Romero MFM, de Souza JR (2016) Wheelchair simulator game for training people with severe disabilities. In: The 1st international conference on technology and innovation in sports, health and wellbeing (TISHW), pp 1–8 (2016). https://doi.org/10.1109/TISHW.2016.7847792
15. The YouTube video with the original LabVIEW application's running mode – BCI_VirtualSIM – presented in this paper. https://youtu.be/b4POGPlYPOA. Accessed 12 Jan 21

Food Carbon Footprint: Challenges and Opportunities

Nourhan Magdy Abdelrhim[1], Frédéric Andrès[2(✉)], and Andreas Pester[1]

[1] The British University in Egypt, P.O. Box 43, El Sherouk City, Cairo 11837, Egypt
[2] National Institute of Informatics, 2–1-2 Hitotsubashi, Chiyoda-ku, Tokyo 101-8430, Japan
andres@nii.ac.jp

Abstract. "Carbon Footprint" (CF) term has become popular over the last decade due to the high rate of climate change. Food related GHG emissions is a main cause of climate change, which needs to be reduced. Following the raising awareness of the food impact on carbon footprint of consumers, a number of recent studies have analysed how the citizen contribute to the reduction of emission of carbon 'footprint'. This paper identifies the challenges and opportunities in being able to evaluate carbon 'footprint' (named CF) of food choices and we reviewed some current approaches of CF calculator tools at dish ingredient level, or recipe level, and finally, applications that help to reduce Carbon Footprint of foods.

Keywords: Food carbon footprint · GHG emissions · CF optimisation

1 Introduction

Climate change is happening at the fastest rate we have ever witnessed before. This phenomenon causes evolutionary troubles and dangers, and nature cannot always adapt fast enough to overcome those issues. GHG can have many forms, the carbon dioxide form being the most recognized one. Those emissions play a key role in accelerating the rate at which our climate changes [5]. Numerous reports [4,6,7] emphasized the relationship between food consumption and climate change patterns. Food production has also been recognized to greatly impact climate change [9]. Also, the way we prepare recipes and how we choose ingredients to cook them has a sharp effect on the environment, given that it contributes to the total carbon footprint citizens leave on earth. Hence, becoming more conscious on how we influence climate change with our food choices is a priority. Thinking that the ingredients available today will remain available forever obviously endangers their sustainability. Food Carbon Footprint is being increasingly noticed and studied around the globe. However, a broader knowledge on this issue would lead to a more prosperous and safer future if correctly exposed to the general public. This serves to conclude that consumers and professionals need to raise their awareness on it. Broad literature and programs on the evaluation of food's carbon footprint [3,5,10] are accessible nowadays. There

M. E. Auer et al. (Eds.): REV 2022, LNNS 524, pp. 664–670, 2023.
https://doi.org/10.1007/978-3-031-17091-1_64

are two characteristics that are important in carbon footprint calculators, which are the granularity of CF evaluation for ingredient level of recipe and the diversity of ingredients. Most calculators concern individual evaluation of ingredients. In addition, they do not seem to offer a very complete diversity of ingredients as options in their database. Recipes as a whole are sometimes but more rarely assessed [11]. In these recipe evaluators, measurements are often made with logos, describing the impact on the environment. The usefulness of these kinds of tools is undeniable, but a more suitable measure as the GHG emissions of foods would be of the utmost interest. These limitations bring Challenges such **recipe scalability**, **ingredient scalability** and **user's localisation scalability**. It is important to provide flexibility in the selection of recipes, the choice of ingredients and the localisation of the user as parameter of the evaluation of the Carbon footprint. Carbon footprint Calculation also brings opportunities of providing better way to calculate the Carbon footprint as applying **optimisation** or **machine learning**. This paper is structured as follows. Section 2 reviews the state of art in carbon footprint calculation including CROPPER, our Carbon Footprint recipe optimizer. Section 3 evaluates the CROPPER tool. Finally, Sect. 4 concludes and introduces the future works.

2 State of the Art

In this section, we aim to go through multiple domains concerned with our study and review their current advancements. We will firstly analyse the carbon footprint of foods. Finally we will be interested in the existing computing tools for such carbon footprints.

2.1 Introducing the Carbon Footprint of Foods

The term "Carbon Footprint" emerged around the year 2006. However the environmental influence of food products has been assessed from before, as part of their Life Cycle Assessment (LCA)([10, 15]). Various articles concerning the specific carbon footprint of some ingredients ([10, 11]), Danish diets [1] as well as

Table 1. Review of the consumer activities and CF

	Where ?	What ?	When ?	Why ?	Result
Finland (Kaup-pinen et al., 2010) [8]	Houses in the city and isolated places in Finland.	Household food-stuff management	2010	To identify the main activities	Each house = 170 kg of CO2-equivalent per person 50% derived from food preservation 27% from food transportation 23% from food preparation
Denmark (Bruno et al., 2019) [1]	Denmark	1) Estimation of the consumed amount of food by an average person. 2) Transformation of food supply values into real food intakes	2019	Unhealthy danish diet	Comparison of the CF of the 4 diets and evaluation of the main consumption in LCA cycle Vegan diet: 0.89t CO2eq pers-1 y-1 Vegetarian diet: 1.37t CO2eq pers-1 y-1 Carnivore diets: 1.83t and 1.59t
Mahayana buddhists practicing vegetarian diets (Tseng, 2020) [2]	Buddhists in seven regions (China, Japan, Vietnam, South Korea, Taiwan, Hong Kong and Singapore)	To get the accurate greenhouse gas emission (GHGe) of vegetarian diets	2017	to evaluate the GHGe of different diets by Mahayana Buddhists by Mahayana Buddhists	Correlation between the GHGe of the vegan diet and other diets

LCAs of food products in Finish households can be found nowadays in [8]. The overall review is given in Table 1.

Relevant researches that link dietary patterns with the emitted Carbon Footprint in foods are also available in [13]. In those findings Italian people are said to have heavier diets in terms of carbon footprint emissions than English people. Other specifications on individual ingredients can also be read and mostly follow the same pattern: vegetables have the lowest carbon footprint while dairies and meat seat at the top in terms of carbon footprint.

Other findings [14] point out a new method to investigate the composition of "current" or "contemporary" recipes: the meta-analysis of internet recipes to create a statistically representative recipe.

2.2 Computing Carbon Footprint and Online Calculators

Computing Carbon Footprints is a difficult task. Online Carbon Footprint calculators have been spotlighted recently and most of them only evaluate the carbon footprint of ingredients individually [3,5]. Other, more advanced technologies [11] try to offer a richer approach by changing the ingredients of a recipe and optimizing its carbon footprint. The problem with those technologies is that they lack diversity in their options of ingredients and, most of the times, are not adequate for household cooking. In addition, language options are limited.

E. Röös [10] states that computing the carbon footprint of ingredients has a certain attached uncertainty to it and, nowadays, the passing of research insights in this area to the general public is not very efficient. Our study aims to break this barrier of misunderstanding or confusion that lies in between scientific information and the consumer/citizen's understanding.

Other researchers as H. Ritchie [16] cite that the specific foods of a diet have more importance in reducing environmental impact than their actual origins. Specifically, it suggests that practicing a plant-based diet entails much less GHG emissions than a diet including dairy and meat.

The Mossy Earth website [17] presents a broader vision of the calculation of the carbon footprint of an individual by including other origins than food consumption. When discussing about the specifics of the food GHG emissions during an interview, it is pointed that foods can be evaluated carbon footprint wise by determining their LCA and that it is important to take into consideration land use change in the final result. For example, deforestation can be viewed as a form of land use change since forests are removed so that new crop fields can be established.

Drewnowski et al. [18] pointed out that, generally speaking, high nutrient content foods, measured in kcal per 100 g, have a high GHG emission rate. As an example, high nutrient content foods are meats or dairies, whilst low nutrient content foods are sweets and vegetables.

The architecture and features of Carbon Footprint calculators are described in Table 2.

Table 2. CF calculator architecture and features

	Creation Date	Online Offline	Open Close	Point of interest	Calculation metric
BBC	2018	online	closed	UK	Calculates the CF of ingredients based on how often a person eats an ingredient each week. The output is the CF value and equivalent such as driving car And the consumption of water
Food emission	-	Online	Closed	point based on number of miles (2800 max)	Ingredient CF calculation
EATS	2018	Offline	Closed	UK	calculate values of CF and WF For a school meal
Cool climate network	-	Online	Closed	USA	Provides CO2 emissions per year for a household of specific number of people an specific income and how it affects City, Energy, Food, Population and Planet
Cropper	2020	Offline	Open	Japan	Recipe CF calculation
Food miles calculator	-	Online	Closed	worldwide	Calculation of CF of an ingredient taking into consideration the miles of transportation by Aeroplan, Car, or train

3 Calibration of the CROPPER Service

In this section, we aim to evaluate the calibration of the CROPPER carbon footprint calculator as a mono-dish system. For this purpose, for a given recipe, we compare the values of carbon footprint obtained with the CROPPER with those obtained using other calculators.

3.1 CROPPER's Computation of Carbon Footprint

The CROPPER service [12] computes a recipe's carbon footprint by summing each of the individual carbon footprints of its ingredients, as explained in the following formula:

$$CF_i = \sum_{j=1}^{n} I_{i,j} x_j, i \in [1\,; p]$$

where $CF_i, I_{i,j}, x_j$ and n are the carbon footprint of the recipe i, the jth ingredient of the ith recipe, the carbon footprint of the jth ingredient and the number of ingredients for this recipe, respectively.

3.2 Comparison of CROPPER with Other Online Calculators

Tests were made with the CROPPER algorithm to evaluate its reliability. To accomplish this goal, we firstly performed manual web scraping on various websites [22–25] in order to build our dummy data basis. Secondly, we performed the computation of the carbon footprint due to the ingredient's transportation. Indeed, the carbon footprint of transport is calculated as a function of the import quantities of a given product to the country we adopted as a reference point of view (geographical position of the cook). In our case, the studies have been conducted for a user located in Japan.

Lastly, the reliability tests were conducted by comparing the carbon footprint estimation of CROPPER with the estimation of other selected calculators accessible online that, as we have pointed out in the state of the art (see Sect. 2), presented a narrow variety of ingredients. We thus had to craft recipes that would conciliate the different calculators according to the available ingredients in their database. The Fig. 1 illustrates this comparison with three major tools available online: The Vegan Society [3], The BBC calculator [19] and Food Emissions [20].

	BBC	Food Emissions	The Vegan Society	CROPPER
Oatmeal	0.09	0.03	0.03	0.05
Milk	0.24	0.12	0.09	0.08
Black Chocolat	0.74	0.66	0.15	1.08
Blueberries	0.3	0.11	0.14	0.16
Total for the recipe	1.37 kgCO2	0.92 kgCO2	0.41 kgCO2	1.37 kgCO2

Fig. 1. Carbon footprint comparative array between CROPPER and multiple online tools for the recipe "Oatmeal bowl with almond milk, black chocolate and blueberries". All values are given in kg of CO_2.

The comparative array presented in Fig. 1 is color coded according to the level of each ingredient's carbon footprint, expressed in kgCO2 (per kG of food) and attributed as we explain next. No color is used when CO2 is less than 100 g CO2. We use yellow color when the value varies from 100 to 250 g CO2; orange color when the value ranges from 250 to 500 g CO2; red color when the values ranges from 500 g to 2 kg CO2; and dark red color when the values are greater than 2 kg CO2. In this color code, since higher values of CO2 correspond to higher levels of carbon footprint, the closer the color is to dark red the more harmful is the carbon footprint for the environment.

We observe that the final results (in total) are in the same order of magnitude (around 10^0). The total value of carbon footprint found using CROPPER even matches identically the value obtained using the BBC calculator. These results are encouraging by themselves but they also show that the scrapping method employed (manual retrieval from websites) is an appropriate path to take.

In Fig. 2, we present a comparison between the results obtained using CROPPER and those given by the WWF (World Wildlife Fund) report: "Food in a Warming World Report" [21], corresponding to the recipe "Welsh Cawl".

Welsh Cawl Recipe		
Ingredient \ Tool	FWWR	CROPPER
Butter	0.20	0.01
Cheese	0.83	2.60
Lamb	4.51	5.40
Wheat flour	0.10	0.04
Others	0.29	0.11
Total for the recipe	5.9	9.36

Fig. 2. Carbon footprint comparison between CROPPER evaluation and results from the report of WWF for the "Welsh Cawl" recipe. All values are expressed in kg of CO2.

We can observe in Fig. 2 that the CROPPER tool provides once again results in the same order of magnitude for the value of the total carbon footprint. Furthermore, the results indicate the same tendencies regarding the costs of carbon footprint for each ingredient, i.e., the lamb is the ingredient that shows the biggest carbon footprint in both cases, and cheese is second.

4 Conclusion and Future Work

Consumers as well as professional cookers are becoming increasingly aware of how they produce carbon footprint when using food products or cooking certain ingredients. After reviewing the state of the art in the field of carbon food print, we compared the CROPPER tool to other existing carbon footprint calculators. In next steps of this research, we will extend the current single dish evaluation of Carbon Footprint in the CROPPER system to evaluate the Carbon Footprint of multiple dishes.

References

1. Bruno M, Thomsen M, Pulselli FM, Patrizi N, Marini M, Caro D (2019) The carbon footprint of Danish diets. Climatic Change 156(4):489-507. Available at: https://ideas.repec.org/a/spr/climat/v156y2019i4d10.1007_s10584-019-02508-4.html. Accessed 22 Nov 2021
2. Tseng AA (2020) Equivalent reduction in greenhouse gas emissions by Mahayana Buddhists practicing vegetarian diets. J Relig Health 59(1):598. https://ixtheo.de/Record/1690960132. Accessed 22 Nov 2021
3. Carbon Food Calculator-The Vegan Society. https://bit.ly/2Xwo5U2. Accessed 29 Nov
4. Clark M, et al. (2019) Multiple health and environmental impacts of foods In: Proceedings of the national academy of sciences, vol 116
5. Climate Change Indicators: Greenhouse Gases. https://www.epa.gov/climate-indicators/greenhouse-gases. Accessed 29 May 20
6. Folk and Knife (2019) Food & CO2, What are the carbon emissions of different foods? https://folkandknife.com/wp-content/uploads/2019/05/Food-CO2-2.pdf

7. Guide to PAS 2050 (2008) How to assess the carbon footprint of goods and services, British Standard PAS 2050, pp58, ISBN 978-0-580-64636-2
8. Kauppinen T, Pesonen I, Katajajuuri JM, Kurppa S (2010) Carbon footprint of food-related activities in Finnish households. Prog Ind Ecol Int J 7(3):257. https://doi.org/10.1504/PIE.2010.037779. Accessed 22 Nov 2021
9. Poore J, Nemecek T (2018) Reducing food's environmental impacts through producers and consumers. Science 360(6392):987–992
10. Röös, E (2013) Analysing the carbon footprint of food. Acta Universitatis Agriculturae Sueciae (1652-6880) 2013(56):0–96
11. Speck M, et al (2020) Creating sustainable meals supported by the NAHGAST online tool - approach and effects on GHG emissions and use of natural resources. Sustainability (12)(1136)
12. de Toledo DA, d'Orazio L, Andres F, Leite MCA (2020) Cooking related carbon footprint evaluation and optimisation. TPDL/ADBIS -2020, vol 1260. CCIS. Springer, Cham, pp 122–128. https://doi.org/10.1007/978-3-030-55814-7_10
13. Christian R, et al (2019) A pilot method linking greenhouse gas emission databases to the FoodEx2 classification https://doi.org/10.13140/RG.2.2.15990.34889
14. Reynolds C, Reynolds C (2017) Can statistics tell us what is a representative recipe? The case of Yorkshire pudding, H-Nutrition. Kristen Ann Ehrenberger. SSRN: https://ssrn.com/abstract=2989646
15. Murphy-Bokern D (2010) Understanding the carbon footprint of our food. Complete Nutr 10:61–63
16. Ritchie H (2020) https://ourworldindata.org; You want to reduce the carbon footprint of your food? Focus on what you eat, not whether your food is local
17. Carbon Footprint Calculator. https://mossy.earth/methodologies/carbon-footprint-calculator. Last Visited 17 October 2021
18. Drewnowski A, Rehm, CD, Martin, A, Verger, EO, Voinnesson, M, Imbert, P (2015) Energy and nutrient density of foods in relation to their carbon footprint. Am J Clin Nutr 101(1):184–191. https://doi.org/10.3945/ajcn.114.092486
19. BBC carbon footprint calculator. https://www.bbc.com/news/science-environment-46459714. Accessed 18 Oct 2021
20. Food Emissions's carbon footprint calculator. http://www.foodemissions.com/Calculator. Accessed 18 Oct 2021
21. Food in a warming world (2018) https://www.wwf.org.uk/sites/default/files/2018-03/Food_in_a_warming_world_report.PDF. Last Visited 26 September 2021
22. https://healabel.com/. Last Visited 7 November 2021
23. www.walmart.com. Last Visited 7 November 2021
24. www.carrefour.fr. Last Visited 7 November 2021
25. www.exportgenius.in. Last Visited 7 November 2021

Hardware Design of a Smart Wristband Used for Epileptic Seizure Detection

Hasan Smajić and Toni Duspara(✉)

Faculty of Vehicle Systems and Production, TH Köln, Betzdorfer Str. 2, 50679 Köln, Germany
{hasan.smajic,toni.duspara}@th-koeln.de

Abstract. It is a commonly known fact that one out of 26 people worldwide suffers from epilepsy or will develop epilepsy at some point in their life. Epilepsy is a common condition that affects the brain and causes frequent seizures. Seizures are bursts of electrical activity in the brain that temporarily affect how it works. Most of the people that suffer from epilepsy cannot predict when the seizure is going to occur. If seizure occurs for example, when someone is driving, or when the patient is climbing the stairs, outcome could be fatal for the patient. Also, patients do not always know that they experienced a seizure, due to memory loss during the seizure itself. Commonly used method for detecting seizures is video electroencephalography. This method is utilized with clinical monitoring, but seizures can be detected only while patients are admitted and while testing occurs. Since clinical monitoring is not an everyday activity, a great number of seizures goes unnoticed. Advancements in biosensors and machine learning principles allow us the opportunity to design an affordable mobile device, which would be used daily to monitor and register epileptic seizures.

Keywords: Epilepsy · Seizure · Biosensors · Machine learning

1 Approach

Idea of this project is to develop a device which would act as a mobile epilepsy detector in a form of a wearable wristband. Electroencelography (EEG) as a conventional method of recognition, pinpoints electrical activity of the surface layer of the brain. studies have shown that instead of eeg, combination of various other biosensors could output the same result as conventionally used method. These biosensors include: electrodermal activity sensor, body temperature sensor, body volume pulse sensor (BVP) and accelerometry sensor [1]. These components, including a microcontroller and power source would be incorporated inside a wearable device. Device would be designed in a way that its aesthetically pleasing. Patient would wear this device at all times, and the measurement data would be transferred directly to the application on patient's smartphone. The application would do the processing of the data and create notifications based on the results of measurement. Additional benefit of this study is the low-cost factor. devices similar to the one described can already be found on the market. An existing wristband with similar

M. E. Auer et al. (Eds.): REV 2022, LNNS 524, pp. 671–678, 2023.
https://doi.org/10.1007/978-3-031-17091-1_65

functions called "EMpatica E4" could be purchased for 1,690.00€ [2]. When designing medical equipment and accessories, affordability is an important factor. medical equipment that could help someone to save their wellbeing and health should be equally affordable to every single person. Due to the advancements in technology, prices for the materials and electronics have significantly dropped over the years. To summarize, idea is to build an affordable wearable medical device, that could help the humanity preserve their health and wellbeing [3].

2 Hardware Requirements

Every mechatronic system used for data acquisition must consist of two main groups. Those are; sensors, that acquire data from the outside, and control which computes the acquired data and outputs the measurement result.

2.1 Data Acquisition Through Biosensors

Earlier, four main biosensors used for this project were mentioned. Important note, is that these sensors, due to the massive manufacture of electronics can be cheaply purchased and enveloped into a wristband.

Electrodermal Activity Sensor. EDA is the property of the human body that causes continuous variation in the electrical characteristics of the skin. The traditional theory of eda holds that skin resistance varies with the state of sweat glands in the skin. Sweating is controlled by a sympathetic nervous system (Fig. 1), and skin conductance is an indication of psychological or physiological arousal. If the sympathetic branch of nervous system is highly aroused, then sweat gland activity also increases, which increases sweat conductance. In this way, skin conductance can be a measure of emotional and sympathetic responses [4].

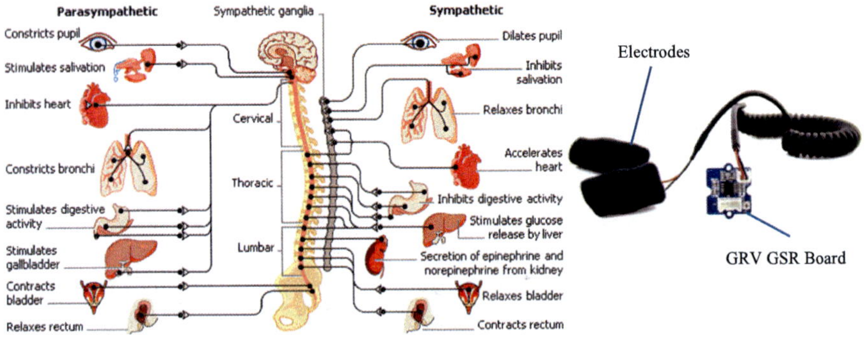

Fig. 1. Left – Sympathetic and parasympathetic nervous system. Right – EDA sensor (GRV GSR SENS) with development board and two electrodes

When something important happens, or is about to happen, the sympathetic arousal is heightened. Also, when the body is exposed to stress, whether it is cognitive, physical or emotional, same increase is noticeable. In the case of epilepsy, this increase relates to the activation of certain parts of the brain. Skin, as the largest organ in the human body is instigated through the sympathetic nervous system. Increase of sympathetic activation can be observed by registering electrical disruptions on the skin surface. In the Fig. 1 an EDA senor (also known as Galvanic skin response sensor) is shown. This sensor would be incorporated inside the wristband in order to measure the galvanic response of the skin. Sensor consists of controller board and electrodes. Controller board used for this project is the GRV GSR SENS. Board is linked on one side to the control system (microcontroller) and to the electrodes on the other side. Electrodes are placed firmly onto the skin through the silicone wristband.

Body Volume Pulse Sensor (BVP). Studies have shown that in 93% of cases, heart rate is increased at least 10 beats per minute around seizure onset. Therefore, in order to have an accurate seizure recognition, a bvp sensor should be incorporated into the wristband. In combination with eda (GSR) sensor, addition of other sensors, such as BVP, would decrease the probability of a misread diagnosis. BVP sensor works on the principles of photoplethysmography. It emits infrared light which bounces of the skin and then it measures the quantity of reflected light. This quantity varies according to the amount of blood registered in the skin. Sensor chosen for this project is the ky-039 development board (Fig. 2). This is an analog sensor which sends the analog read data to the microcontroller. This data is later on calculated through a code, and outputted to the user's smartphone in a form of a number and a graph.

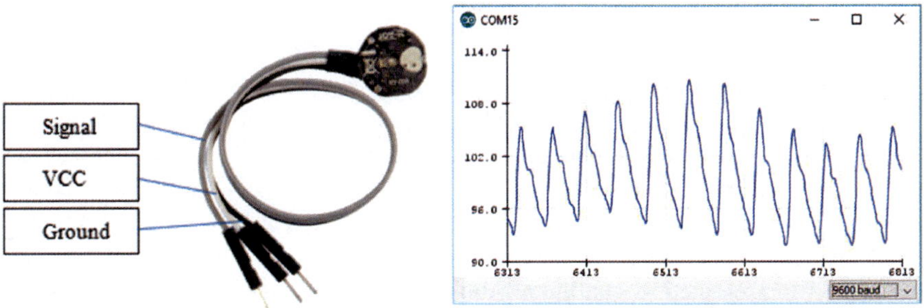

Fig. 2. Body volume pulse sensor (KY-039) and example graph readings from the sensor

Body Temperature Sensor. Tonic-clonic seizure is known to increase the body temperature.according to the latest studies, in order to describe the course of body temperature elevation due to a generalized seizure, the temperature curves of 93 hospitalized patients were reviewed. More than forty percent of the patients had an elevation of body temperature above 37.8 °C.[1] According to these studies, registering body temperature plays an enormous role when it comes to epileptic seizure detection. For this purpose, infrared body temperature sensors are utilized. These sensors use a lens to focus a ray of infrared

[1] Wachtel TJ, Steele GH, Day JA. Natural history of fever following seizure. Arch Intern Med. 1987 Jun;147(6):1153–5. PMID: 3592881.

light emitting from the surface of the skin onto a detector called thermopile. Thermopile represents a range of thermocouples connected parallel or in series. Infrared radiation fallen on the thermopile surface is absorbed and later on converted into heat. Sensor used to measure body temperature is MLX90614 single zone infrared thermometer (Fig. 3 - Left). This sensor consists of four pins, which are through various resistors and capacitors (Fig. 3 – Right) linked to analog pins of the microcontroller. sensor has compact dimensions and is easily placed into the wristband. Top part of the sensor, where the lens is located is in direct contact with skin through the wristband, so body temperature can be accurately measured.

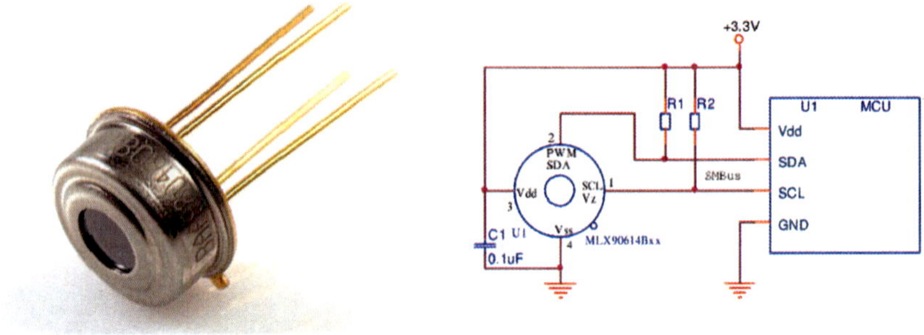

Fig. 3. Infrared body temperature sensor (MLX90614) and wiring schematic

Accelerometry Sensor. In the case of a tonic-clonic seizure, recognition can be made with a use of accelerometry sensors. the detection of rhythmic movement components, such as the clonic part of tonic–clonic seizures, is essential in all seizure detection based on movement sensors.[2] Movement detection is also a commonly known method used in clinical praxis for seizure detection. Motoric components of a seizure are different one from another due to frequency of epileptic impulses that brain sends to the muscles. Impulses with low frequency are represented as rhythmical jerks. high frequency impulses are represented as continuous muscular contractions. In order to envelop an accelerometry sensor to detect epileptic seizures a preliminary machine learning model must be trained. Model training requires use of accelerometry sensor while the seizure occurs in order to create a database, which would be used as a reference point to classify a certain movement as a possible epileptic seizure. [5] An accelerometer used for this project is LIS3LV02DL. This is a micro three axis accelerometer that includes a sensing interface and an ic communication protocol. It is compact, therefor allows the installation inside the wristband. four digital pins on the sensor are used for connection with microcontroller. Those are pins 2, 3, 5 and 6 (Fig. 4).

[2] Arends, JBAM. Movement-based seizure detection. *Epilepsia.* 2018; 59(S1): 30– 35.

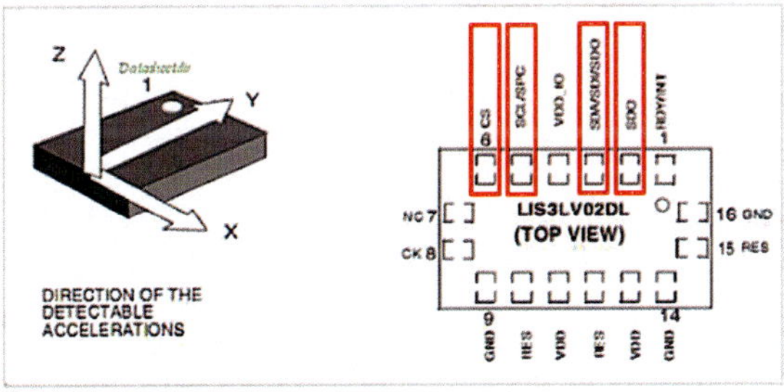

Fig. 4. Micro accelerometry sensor (LI3SLV0DL) axis and pin layout

2.2 Communication with Smart-Phone

To communicate with smartphone and transmit measurement data, a Bluetooth communication protocol is used. Microcontroller must either include an integrated Bluetooth module, or must be connected with a separate Bluetooth module. In order to contain the low-cost factor, and preserve the compactness of the wearable device, a separate Bluetooth module is purchased and linked with microcontroller. Chosen module is the HC-05 BLE 5.0 development board. This board connects to the microcontroller through serial communication using RX and TX pins (which on microcontroller can be digital inputs).

2.3 Microcontroller and Power Source

To read the measurement data from the sensors incorporated into a wristband, an adequate microcontroller must be chosen. While choosing the microcontroller, one must keep in mind the necessary number of inputs and outputs. Sensor outputs with numbers of required connections (pins) are shown in Table 1.

Table 1. List of electrical components and sensors with required analog and digital inputs for the microcontroller.

Sensor	Required analog inputs	Required digital inputs
Electrodermal activity sensor	1	0
Body temperature sensor	2	0
Body volume pulse sensor	1	0
Accelerometry sensor	0	4
Bluetooth module	0	2
Altogether	**4**	**6**

According to the table above, main characteristics that microcontroller needs to posses are:

- At least 4 analog inputs,
- At least 6 digital inputs and
- Compact size

According to said characteristics microcontroller chosen for this project is the Seeeduino Xiao microcontroller. Due to its compact design and pinout, this controller was found adequate for development of wristband.

Power Source (Battery). To supply the wristband with enough power required by sensors, controllers and communication modules an adequate power source in a form of battery must be chosen. for mobile devices, usual choice are lithium-ion batteries. Due to its compactness and required capacity, a 3.7 V battery with 400 mAh capacity is chosen.

Housing and Design. Enclosure for the wristband is made using combination of plastic and silicone. Top part of the wristband is made out of plastic. For development and prototyping, a 3D printed enclosure will be made and later on, when designing a final product, moulds would be created to increase the design aesthetics. The band itself would be designed out of silicone, with grooves and holes for the other components, that need to be in direct contact with the skin.

3 Wiring Schematics and Principles

Wiring schematic for the wristband is shown on Fig. 5. All components are connected to the same power source and controlled through Seeeduino Xiao Microcontroller. Figure shows only plain wiring, whereas later on, wires are going to be replaced with custom flex PCB board which would run through the wristband.

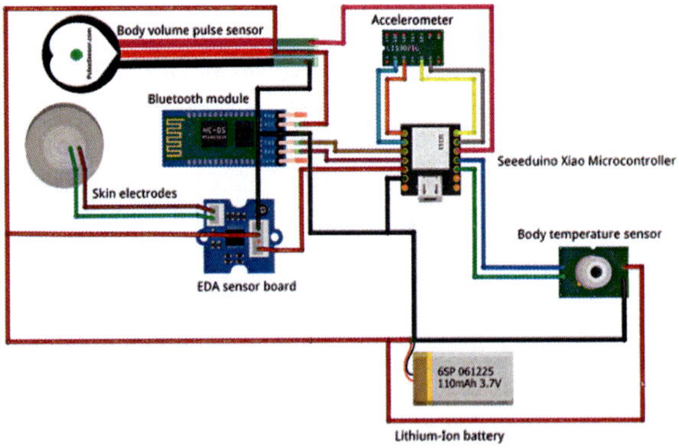

Fig. 5. Wiring schematic with all electrical components

4 Material Costs for the Wristband

As earlier mentioned, great benefit of this project is the low-cost factor. Possible solutions for the same device could be found on the market, but for a greatly larger price. Medical devices, which help the humanity improve and preserve their wellbeing, should be equally available and affordable for everyone. On the Table 2, material costs for the wristband are shown. Table does not include pricing of the material for wristband enclosure.

Table 2. A figure caption is always placed below the illustration. Short captions are centered

Part	Description	Reseller	Price [€]
Seeeduino Xiao	Microcontroller	OKDO	10,84
LIS3LV02DL	Accelerometer	Reichelt Elektronik	7,50
Blue Sensor VL	Skin electrodes (2x)	Amazon	2,32
400 mAh, 3,7V	Battery	Amazon	12,99
GRV GSR Sens	EDA Sensor	Reichelt Elektronik	9,23
DEBO SENS Heart	BVP Sensor	Reichelt Elektronik	5,35
HC-05 BLE5.0	Bluetooth module	Amazon	7,99
MLX90614	IR Thermometer	Amazon	14,39
Altogether			**70,91**

According to the table above, overall price for the parts of wristband components is just a little above 70,00 €. With this in mind, low-cost factor of the project is achieved.

5 Conclusion

In the end of the research, a device should be created which successfully after recognition of epileptic activity, notifies the patient that epileptic seizure is about to occur or notifies the caretaker that epileptic seizure has already occurred. wristband registers epileptic seizures using machine learning principles [6]. By the use of biosensors, conventional electroencelography method is replaced. In this paper, hardware requirements and main design of the wristband are explained, including selection of components, basic work principle and wiring schematic. after the physical creation of wristband, a smartphone app based on machine learning principles is going to be developed. To fully develop the wristband, it is necessary to train the machine learning model. this model requires a database of sensor readings that indicate epileptic seizure. idea behind the paper is to develop a functioning prototype of a wristband, and later on to develop an algorithm used for seizure recognition based on sensor readings [7].

References

1. Regalia G et al (2019) Multimodal wrist-worn devices for seizure detection and advancing research: focus on the Empatica wrist-bands. Epilepsy Res 153:79–82
2. Rukasha T et al (2020) Evaluation of wearable electronics for epilepsy: a systematic review. Electronics 9(6):968
3. Vossler DG (2021) Forecasting seizure storms using epilepsy wrist-band sensors. Epilepsy Curr. 21(2):99–101
4. Nagai Y, Jones CI, Sen A (2019) Galvanic skin response (GSR)/electrodermal/skin conductance biofeedback on epilepsy: a systematic review and meta-analysis. Front Neurol 10:377
5. Jagtap, P.T., Bhosale, N.P.: IOT based epilepsy monitoring using accelerometer sensor. In: 2018 International Conference on Information, Communication, Engineering and Technology (ICICET). IEEE (2018)
6. Tang J et al (2021) Seizure detection using wearable sensors and machine learning: setting a benchmark. Epilepsia 62(8):1807–1819
7. Meisel C et al (2020) Machine learning from wristband sensor data for wearable, noninvasive seizure forecasting. Epilepsia 61(12):2653–2666

Author Index

M. E. Auer et al. (Eds.): REV 2022, LNNS 524, pp. 679–681, 2023.
https://doi.org/10.1007/978-3-031-17091-1